Convex Optimization for Signal Processing and Communications

From Fundamentals to Applications

Convex Optimization for Signal Processing and Communications

From Fundamentals to Applications

Chong-Yung-Chi · Wei-Chiang Li · Chia-Hsiang Lin

CRC Press
Taylor & Francis Group
Boca Raton London New York

CRC Press is an imprint of the
Taylor & Francis Group, an **informa** business

CRC Press
Taylor & Francis Group
6000 Broken Sound Parkway NW, Suite 300
Boca Raton, FL 33487-2742

First issued in paperback 2020

© 2017 by Taylor & Francis Group, LLC
CRC Press is an imprint of Taylor & Francis Group, an Informa business

No claim to original U.S. Government works

ISBN 13: 978-0-367-57392-8 (pbk)
ISBN 13: 978-1-4987-7645-5 (hbk)

Visit the Taylor & Francis Web site at
http://www.taylorandfrancis.com

and the CRC Press Web site at
http://www.crcpress.com

Contents

Preface

Convex Optimization has been recognized as a powerful tool for solving many science and engineering problems. Over the last two decades, convex optimization has been successfully and extensively applied to various problems in signal processing such as blind source separation (BSS) for biomedical and hyperspectral image analysis, and in multiple-input multiple-output (MIMO) wireless communications and networking, such as coherent/noncoherent detection, transmit/robust/distributed beamforming, and physical-layer secret communications. Particularly, fourth generation (4G) wireless communication systems have been in operation, and various researches for fifth generation (5G) systems, e.g., massive MIMO, millimeter wave wireless communications, full-duplex MIMO, energy harvesting, and multicell coordinated beamforming, have been intensively studied and reported in the open literature, where the convex optimization tool is extensively wielded, validating its central role to the development of 5G systems and to many interdisciplinary science and engineering applications.

Next, let us address the motivation, organization of the book, suggestions for instructors, and acknowledgment of writing the book *"Convex Optimization for Signal processing and Communication: From Fundamentals to Applications,"* respectively.

Motivation:

Since Spring 2008, I have been teaching the graduate-level course *"Optimization for Communications"* at National Tsing Hua University (NTHU), Hsinchu, Taiwan. As in teaching any other course, I prepared my own lecture notes for this course. My lecture notes are primarily based on the seminal textbook, *Convex Optimization* (by Stephen Boyd and Lieven Vandenberghe), Cambridge University Press, 2004; some research results published in the open literature; and some materials offered by my former colleague, Prof. Wing-Kin Ma (Chinese University of Hong Kong), who taught this course at NTHU from August 2005 through July 2007.

From my teaching experience, many engineering students are often at a loss in abstract mathematics due to lack of tangible linkage between mathematical theory and applications. Consequently they would gradually lose the motivation of learning powerful mathematical theory and tools, thereby leading to losing momentum to solve research problems by using the mathematics they were trying to learn. In order to help students to be fully equipped with this powerful tool, my lecture notes on convex optimization are molded into a bridge from the fundamental mathematical theory to practical applications. I have assembled my lecture notes in this book, and sincerely hope that the readers, especially the student community, will benefit from the materials presented here.

Over the last decade, my lecture notes have been successfully used 12 times for my intensive 2-week short course *"Convex Optimization for Signal Processing and Communications"* at major universities in China, including Shandong University (January 2010), Tsinghua University (August 2010 and August 2012), Tianjin University (August 2011), Beijing Jiaotong University (July 2013 and July 2015), University of Electronic Science and Technology of China (November 2013, September 2014, and September 2015), Xiamen University (December 2013), Sun Yat-Sen University (August 2015), and Beijing University of Posts and Telecommunications (July 2016). These short courses differed from traditional short courses in conferences, workshops, and symposia (usually using a set of synoptic slides without enough details due to limited time). In each short course I offered in China, I spent around 32 lecture hours over two consecutive weeks, going through almost all the theories, proofs, illustrative examples, algorithm design and implementation, and some state-of-the-art research applications in detail, like a guided journey/exploration from fundamental mathematics to cutting-edge researches and applications rather than pure mathematics. Finally, a post term project was offered for the attendees to get hands-on experience of solving some advisable problems afterwards. I have received many positive feedbacks from the short-course attendees, and now many of them are good at using convex optimization in solving research problems, leading to many research breakthroughs and successful applications.

Organization of the book:

With a balance between mathematical theory and applications, this book provides an introduction to convex optimization from fundamentals to applications. It is suitable for the first-year graduate course *"Convex Optimization"* or *"Nonlinear Optimization"* for engineering students who need to solve optimization problems, and meanwhile wish to clearly see the link between mathematics and applications in hands. Some mathematical prerequisites such as linear algebra, matrix theory, and calculus are surely much help in reading this book.

The book contains 10 chapters and an appendix, basically written in a causally sequential fashion; namely, to have in-depth learning in each chapter, one needs to absorb the materials introduced in early chapters. Chapter 1 provides some mathematical background materials that will be used in the ensuing chapters. Chapter 2 introduces convex sets and Chapter 3 introduces convex functions that are essential to the subsequent introduction of convex problems and problem reformulations in Chapter 4, along with many examples in each of the these chapters.

Some widely known convex optimization problems (or simply termed as convex problems) are introduced next, including geometric programming (GP) that is introduced in Chapter 5 (where a geometric program, nonconvex at first glance, can be easily reformulated into a convex problem); linear programming (LP), quadratic programming (QP) and quadratically constrained quadratic programming (QCQP) that are introduced in Chapter 6; second-order cone programming (SOCP) that is introduced in Chapter 7; and semidefinite programming (SDP) that is introduced in Chapter 8. Each of these chapters presents how the essential materials (introduced in Chapter 2 to 4) are advisably and effectively applied to practical problems in communications and/or signal processing. However, we only present key ideas, philosophies, and major reformulations for solving the problem under consideration. Some simulation results, also real data experiments (in biomedical and hyperspectral image analysis), are presented for the readers to visually see the solution accuracy and efficiency of the designed algorithms. Readers can refer to the associated research papers for full details to ascertain whether he/she can understand/apply the convex optimization theory comprehensively. Because SDP has been extensively used in wireless communications and networking, we especially introduce more challenging applications in Chapter 8, where various intricate optimization problems involving SDP have been prevalent in the evolution towards 5G.

In Chapter 9, we introduce *"duality"* which is of paramount importance and a perfect complement to Chapters 2 through 4, because some convex problems can be solved more efficiently by using Karush–Kuhn–Tucker (KKT) conditions introduced in Chapter 9, comparing using the optimality conditions introduced in the early chapters, and vice versa. In our experience, analytical performance evaluation and complexity analysis of the designed algorithm for solving an optimization problem is crucial not only to the algorithm design in a perspective and insightful manner, but also to the future direction/clues for further research breakthroughs. These analyses can justify and interpret the simulation and experimental results qualitatively and quantitatively, thereby providing a concrete foundation for the applicability of the designed algorithm. However, these analyses heavily rely on the delicate duality theory. On the other hand, once an optimization problem is reformulated into a convex problem, it can be readily solved by using off-the-shelf convex solvers, e.g., CVX and SeDuMi, which are briefly introduced in the Appendix. This may be adequate during the research stage, but not necessarily suitable for practical applications, where real-time processing or on-line processing is highly desired or required. Chapter 10 introduces the interior-point method that actually tries to numerically solve the KKT conditions introduced in Chapter 9, which has been widely used for the realization of obtaining a solution of a specific convex problem in a more computationally efficient manner.

Suggestions for instructors:

For instructors who consider teaching this subject with this book for a one-semester course, I have a few suggestions based on my years of teaching experience. First of all, Chapter 1 through 4 can be covered followed by a midterm examination. Next, some selected applications in Chapter 5 to 8 can be covered, and then a term project of studying a research paper can be announced. The purpose of the term project is for students (1 to 2 students as a group) to experience how a practical problem can be solved by using what they have learned to verify all the theory, analysis and simulation/experimental results of the assigned paper. Then the instructor can continue to teach Chapter 9 and 10. Finally, students would take the final examination, followed by an oral presentation from each term project group. After several implementations myself, I found this practice quite inspirational and beneficial to students.

Acknowledgment:

Over the last eight years' accumulation of my lecture notes, this book was accomplished through tremendous voluntary efforts from many of my former students, including my former PhD students, Dr. ArulMurugan Ambikapathi, Dr. Kun-Yu Wang, Dr. Wei-Chiang Li, and Dr. Chia-Hsiang Lin and my former Master students, Yi-Lin Chiou, Yu-Shiuan Shen, Tung-Chi Ye and Yu-Ping Chang who helped draw many figures in the book. I would also like to thank my former colleague, Prof. Wing-Kin Ma, my former PhD students, Dr. Tsung-Hui Chang and Dr. Tsung-Han Chan, and former visiting scholar, Dr. Fei Ma, and former visiting PhD students, Dr. Xiang Chen, Dr. Chao Shen, Dr. Haohao Qin, Dr. Fei He, Gui-Xian Xu, Kai Zhang, Yang Lu, Christian Weiss, and visiting Master students Lei Li and Ze-Liang Ou, and my PhD student Yao-Rong Syu and Master student Amin Jalili, and all of my graduate students who have offered voluntary assistance, either directly or indirectly.

I would like particularly to express my deep appreciation to those participants of my short courses offered in the above-mentioned major universities in Mainland China over the last seven years, for their numerous questions, interactions, and comments that have been taken into account during the writing of this book, thereby significantly improving the readability of the book, especially to engineering students and professionals.

This book is also supported by my university over the last two years (2015-2016). Finally, I would like to thank my wife, Yi-Teh, for her patience and understanding during the preparation of the book over the last eight years.

Chong-Yung Chi

National Tsing Hua University, Hsinchu, Taiwan   December 2016

1 Mathematical Background

Convex (CVX) optimization is an important class of optimization techniques that includes least squares and linear programs as special cases, and has been extensively used in various science and engineering areas. If one can formulate a practical problem as a convex optimization problem, then actually he (she) has solved the original problem (for an optimal solution either analytically or numerically), like least squares (LS) or linear program, (almost) technology. This chapter provides some essential mathematical basics of vector spaces, norms, sets, functions, matrices, and linear algebra, etc., in order to smoothly introduce the CVX optimization theory from fundamentals to applications in each of the following chapters. It is expected that the CVX optimization theory will be more straightforward and readily understood and learned.

1.1 Mathematical prerequisites

In this section, let us introduce all the notations and abbreviations and some mathematical preliminaries that will be used in the remainder of the book. Our notations and abbreviations are standard, following those widely used in convex optimization for signal processing and communications, that are defined, respectively, as follows:

Notations:

$\mathbb{R}$, $\mathbb{R}^n$, $\mathbb{R}^{m \times n}$	Set of real numbers, n-vectors, $m \times n$ matrices
$\mathbb{C}$, $\mathbb{C}^n$, $\mathbb{C}^{m \times n}$	Set of complex numbers, n-vectors, $m \times n$ matrices
$\mathbb{R}_+$, $\mathbb{R}_+^n$, $\mathbb{R}_+^{m \times n}$	Set of nonnegative real numbers, n-vectors, $m \times n$ matrices
$\mathbb{R}_{++}$, $\mathbb{R}_{++}^n$, $\mathbb{R}_{++}^{m \times n}$	Set of positive real numbers, n-vectors, $m \times n$ matrices
$\mathbb{Z}$, $\mathbb{Z}_+$, $\mathbb{Z}_{++}$	Set of integers, nonnegative integers, positive integers

$\mathbb{S}^n$, $\mathbb{S}^n_+$, $\mathbb{S}^n_{++}$	Set of $n \times n$ real symmetric matrices, positive semidefinite matrices, positive definite matrices
$\mathbb{H}^n$, $\mathbb{H}^n_+$, $\mathbb{H}^n_{++}$	Set of $n \times n$ Hermitian matrices, positive semidefinite matrices, positive definite matrices
$\{x_i\}_{i=1}^N$	The set $\{x_1, \ldots, x_N\}$
$\mathbf{x} = [x_1, \ldots, x_n]^T$ $= (x_1, \ldots, x_n)$	n-dimensional column vector $\mathbf{x}$
$[\mathbf{x}]_i$	ith component of a vector $\mathbf{x}$
$[\mathbf{x}]_{i:j}$	A column vector constituted by partial elements of the vector $\mathbf{x}$, containing $[\mathbf{x}]_i, [\mathbf{x}]_{i+1}, \ldots, [\mathbf{x}]_j$
$\mathrm{card}(\mathbf{x})$	Cardinality (number of nonzero elements) of a vector $\mathbf{x}$
$\mathbf{Diag}(\mathbf{x})$	Diagonal (square) matrix whose ith diagonal element is the ith element of a vector $\mathbf{x}$
$\mathbf{X} = \{x_{ij}\}_{M \times N}$ $= \{[\mathbf{X}]_{ij}\}_{M \times N}$	$M \times N$ matrix $\mathbf{X}$ with the (i,j)th component $[\mathbf{X}]_{ij} = x_{ij}$
$\mathbf{X}^*$	Complex conjugate of a matrix $\mathbf{X}$
$\mathbf{X}^T$	Transpose of a matrix $\mathbf{X}$
$\mathbf{X}^H = (\mathbf{X}^*)^T$	Hermitian (i.e., conjugate transpose) of a matrix $\mathbf{X}$
$\mathrm{Re}\{\cdot\}$	Real part of the argument
$\mathrm{Im}\{\cdot\}$	Imaginary part of the argument
$\mathbf{X}^\dagger$	Pseudo-inverse of a matrix $\mathbf{X}$
$\mathrm{Tr}(\mathbf{X})$	Trace of a square matrix $\mathbf{X}$
$\mathbf{vec}(\mathbf{X})$	Column vector formed by sequentially stacking all the columns of a square matrix $\mathbf{X}$
$\mathbf{vecdiag}(\mathbf{X})$	Column vector whose elements are the diagonal elements of a square matrix $\mathbf{X}$
$\mathbf{DIAG}(\mathbf{X}_1, \ldots, \mathbf{X}_n)$	Block-diagonal matrix (not necessarily a square matrix), with $\mathbf{X}_1, \ldots, \mathbf{X}_n$ as its diagonal blocks, where $\mathbf{X}_1, \ldots, \mathbf{X}_n$ may not be square matrices
$\mathrm{rank}(\mathbf{X})$	Rank of a matrix $\mathbf{X}$
$\det(\mathbf{X})$	Determinant of a square matrix $\mathbf{X}$
$\lambda_i(\mathbf{X})$	The ith eigenvalue (or ith principal eigenvalue if specified) of a real symmetric (or *Hermitian*) matrix $\mathbf{X}$
$\mathcal{R}(\mathbf{X})$	Range space of a matrix $\mathbf{X}$
$\mathcal{N}(\mathbf{X})$	Null space of a matrix $\mathbf{X}$
$\dim(V)$	Dimension of a subspace V
$\|\cdot\|$	Norm
$\mathrm{span}[\mathbf{v}_1, \ldots, \mathbf{v}_n]$	Subspace spanned by vectors $\mathbf{v}_1, \ldots, \mathbf{v}_n$
$\mathbf{1}_n$	All-one column vector of dimension n

$\mathbf{0}_m$	All-zero column vector of dimension m
$\mathbf{0}_{m \times n}$	All-zero matrix of dimension $m \times n$
$\mathbf{I}_n$	$n \times n$ identity matrix
$\mathbf{e}_i$	Unit column vector of proper dimension with the ith entry equal to 1
$\boldsymbol{f}$	Function defined from $\mathbb{R}^n \to \mathbb{R}^m$
f	Function defined from $\mathbb{R}^n \to \mathbb{R}$ or $\mathbb{R} \to \mathbb{R}$
$\mathbf{dom}\ f$	Domain of a function f
$\mathbf{epi}\ f$	Epigraph of a function f
$|C|$	Size of a finite set C (i.e., total number of elements in C)
$\sup C$	Supremum of a set C
$\inf C$	Infimum of a set C
$\mathbf{int}\ C$	Interior of a set C
$\mathbf{cl}\ C$	Closure of a set C
$\mathbf{bd}\ C$	Boundary of a set C
$\mathbf{relint}\ C$	Relative interior of a set C
$\mathbf{relbd}\ C$	Relative boundary of a set C
$\mathbf{aff}\ C$	Affine hull of a set C
$\mathbf{conv}\ C$	Convex hull of a set C
$\mathbf{conic}\ C$	Conic hull of a set C
$\mathbf{affdim}(C)$	Affine dimension of a set C
K	Proper cone
K^*	Dual cone associated with the proper cone K
$\succeq_K$	Generalized inequality defined on the proper cone K
$\succeq$	Componentwise inequality for vector comparison (i.e., the proper cone $K = \mathbb{R}^n_+$); generalized inequality for symmetric matrix comparison (i.e., the proper cone $K = \mathbb{S}^n_+$)
$\mathbb{E}\{\cdot\}$	Expectation operator
$\mathrm{Prob}\{\cdot\}$	Probability function
$\mathcal{N}(\boldsymbol{\mu}, \boldsymbol{\Sigma})$	Real Gaussian distribution with mean $\boldsymbol{\mu}$ and covariance matrix $\boldsymbol{\Sigma}$
$\mathcal{CN}(\boldsymbol{\mu}, \boldsymbol{\Sigma})$	Complex Gaussian distribution with mean $\boldsymbol{\mu}$ and covariance matrix $\boldsymbol{\Sigma}$
$\Leftrightarrow$	If and only if
$\Rightarrow$	Implies
$\nRightarrow$	Does not imply
$\triangleq$	Is defined as or denoted by
$:=$	Is updated by

$\equiv$	Equivalence of two optimization problems with identical solutions but different objective functions
$\log x$	Natural log function of x (i.e., $\ln x$)
$\operatorname{sgn} x$	Sign function of x
$[x]^{+}$	Maximum of x and 0
$\lceil x \rceil$	Smallest integer greater than or equal to x

Abbreviations:

ADMM	Alternating direction method of multipliers
AWGN	Additive white Gaussian noise
BQP	Boolean quadratic program
BS	Base station
BSS	Blind source separation
BSUM	Block successive upper bound minimization
CDI	Channel distribution information
CSI	Channel state information
DoF	Degree of freedom
DR	Dimension reduced
EVD	Eigenvalue decomposition
FBS	Femtocell base station
FCLS	Fully constrained least squares
FUE	Femtocell user equipment
GP	Geometric program
HU	Hyperspectral unmixing
HyperCSI	Hyperplane-based Craig simplex identification
IFC	Interference channel
IPM	Interior-point method
LFSDR	Linear fractional SDR
LMI	Linear matrix inequality
LMMSE	Linear minimum mean-squared estimator
LP	Linear program
LS	Least squares
MBS	Macrocell base station
MCBF	Multicell beamforming
MIMO	Multiple-input multiple-output
MISO	Multiple-input single-output
ML	Maximum-likelihood

MMF	Max-min fairness
MMSE	Minimum mean-squared error
MSE	Mean-squared error
MUE	Macrocell user equipment
MVES	Minimum-volume enclosing simplex
nBSS	Nonnegative blind source separation
OFDM	Othogonal frequency division multiplexing
OSTBC	Orthogonal space-time block code
PCA	Principal component analysis
PD	Positive definite
PSD	Positive semidefinite
QCQP	Quadratic constrained quadratic program
QP	Quadratic program
ROI	Regions of interest
SCA	Successive convex approximation
SDP	Semidefinite program
SDR	Semidefinite relaxation
SIMO	Single-input multiple-output
SINR	Signal-to-interference-plus-noise ratio
SISO	Single-input single-output
SNR	Signal-to-noise ratio
SOCP	Second-order cone program
SPA	Successive projection algorithm
SVD	Singular value decomposition
TDMA	Time division multiple access
WSR	Weighted sum rate
w.r.t.	With respect to

Remark 1.1 The notations and abbreviations listed above are used throughout the entire book. We would like to mention that different notations stand for different variables, though some of them may look quite similar (e.g., $\mathbf{x}$ and x, $\mathbf{X}$ and X), and that $(\mathbf{x}, \mathbf{y})$ and $[\mathbf{x}^T, \mathbf{y}^T]^T$ represent the same column vector. $\square$

1.1.1   Vector norm

In linear algebra, functional analysis and related areas of mathematics, *norm* is a function that assigns a strictly positive length or size to all vectors (other than the zero vector) in a vector space. A vector space with a norm is called a *normed vector space*. A simple example is the 2-dimensional Euclidean space "$\mathbb{R}^2$" equipped with the Euclidean norm or 2-norm. Elements in this vector space are usually drawn as arrows in a 2-dimensional Cartesian coordinate system starting at the origin $\mathbf{0}_2$. The Euclidean norm assigns to each vector the length from the origin to the vector end. Because of this, the Euclidean norm is often known as the magnitude of the vector.

Given a vector space V over a subfield F of real (or complex) numbers, norm of a vector in V is a function $\|\cdot\| : V \to \mathbb{R}_+$ with the following axioms: For all a in F and all $\mathbf{u}$ and $\mathbf{v} \in V$,

- $\|a\mathbf{v}\| = |a| \cdot \|\mathbf{v}\|$ (positive homogeneity or positive scalability).
- $\|\mathbf{u} + \mathbf{v}\| \le \|\mathbf{u}\| + \|\mathbf{v}\|$ (triangle inequality or subadditivity).
- $\|\mathbf{v}\| = 0$ if and only if $\mathbf{v}$ is the zero vector (positive definiteness).

A simple consequence of the first two axioms, positive homogeneity and the triangle inequality, is $\|\mathbf{0}\| = 0$ and thus $\|\mathbf{v}\| \ge 0$ (positivity).

The ℓ_p-*norm* (or *p-norm*) of a vector $\mathbf{v}$ is usually denoted as $\|\mathbf{v}\|_p$ and is defined as:

$$\|\mathbf{v}\|_p = \left(\sum_{i=1}^{n} |v_i|^p \right)^{1/p}, \tag{1.1}$$

where $p \ge 1$. The above formula for $0 < p < 1$ is a well-defined function of $\mathbf{v}$, but it is not a norm of $\mathbf{v}$, because it violates the triangle inequality. For $p = 1$ and $p = 2$,

$$\|\mathbf{v}\|_1 = \sum_{i=1}^{n} |v_i|, \tag{1.2}$$

$$\|\mathbf{v}\|_2 = \left(\sum_{i=1}^{n} |v_i|^2 \right)^{1/2}. \tag{1.3}$$

When $p = \infty$, the norm is called maximum norm or infinity norm or uniform norm or supremum norm and can be expressed as

$$\|\mathbf{v}\|_\infty = \max\{|v_1|, |v_2|, \dots, |v_n|\}. \tag{1.4}$$

Note that 1-norm, 2-norm (which is also called the Euclidean norm), and ∞-norm have been widely used in various science and engineering problems, while for other values of p (e.g., $p = 3, 4, 5, \dots$) p-norm remain theoretical but not yet practical.

Remark 1.2 Every norm is a convex function (which will be introduced in Chapter 3). As a result, finding a global optimum of a norm-based objective function is often tractable. □

1.1.2   Matrix norm

In mathematics, a matrix norm is a natural extension of the notion of a vector norm to matrices. Some useful matrix norms needed throughout the book are introduced next.

The *Frobenius norm* of an $m \times n$ matrix $\mathbf{A}$ is defined as

$$\|\mathbf{A}\|_{\mathrm{F}} = \left(\sum_{i=1}^{m} \sum_{j=1}^{n} |[\mathbf{A}]_{ij}|^2 \right)^{1/2} = \sqrt{\mathrm{Tr}(\mathbf{A}^T \mathbf{A})}, \tag{1.5}$$

where

$$\mathrm{Tr}(\mathbf{X}) = \sum_{i=1}^{n} [\mathbf{X}]_{ii} \tag{1.6}$$

denotes the trace of a square matrix $\mathbf{X} \in \mathbb{R}^{n \times n}$. As $n = 1$, $\mathbf{A}$ reduces to a column vector of dimension m and its Frobenius norm also reduces to the 2-norm of the vector.

The other class of norm is known as the *induced norm* or *operator norm*. Suppose that $\|\cdot\|_a$ and $\|\cdot\|_b$ are norms on $\mathbb{R}^m$ and $\mathbb{R}^n$, respectively. Then the operator/induced norm of $\mathbf{A} \in \mathbb{R}^{m \times n}$, induced by the norms $\|\cdot\|_a$ and $\|\cdot\|_b$, is defined as

$$\|\mathbf{A}\|_{a,b} = \sup \left\{ \|\mathbf{A}\mathbf{u}\|_a \mid \|\mathbf{u}\|_b \leq 1 \right\}, \tag{1.7}$$

where $\sup(C)$ denotes the least upper bound of the set C. As $a = b$, we simply denote $\|\mathbf{A}\|_{a,b}$ by $\|\mathbf{A}\|_a$.

Commonly used induced norms of an $m \times n$ matrix

$$\mathbf{A} = \{a_{ij}\}_{m \times n} = [\mathbf{a}_1, \ldots, \mathbf{a}_n]$$

are as follows:

$$\|\mathbf{A}\|_1 = \max_{\|\mathbf{u}\|_1 \leq 1} \left\| \sum_{j=1}^{n} u_j \mathbf{a}_j \right\|_1, \quad (a = b = 1)$$

$$\leq \max_{\|\mathbf{u}\|_1 \leq 1} \sum_{j=1}^{n} |u_j| \cdot \|\mathbf{a}_j\|_1 \quad \text{(by triangle inequality)}$$

$$= \max_{1 \leq j \leq n} \|\mathbf{a}_j\|_1 = \max_{1 \leq j \leq n} \sum_{i=1}^{m} |a_{ij}| \tag{1.8}$$

(with the inequality to hold with equality for $\mathbf{u} = \mathbf{e}_l$ where $l = \arg\max_{1 \leq j \leq n} \|\mathbf{a}_j\|_1$) which is simply the maximum absolute column sum of the matrix.

$$\|\mathbf{A}\|_\infty = \max_{\|\mathbf{u}\|_\infty \le 1} \left\{ \max_{1 \le i \le m} \left| \sum_{j=1}^n a_{ij} u_j \right| \right\}, \quad (a = b = \infty)$$

$$= \max_{1 \le i \le m} \left\{ \max_{\|\mathbf{u}\|_\infty \le 1} \left| \sum_{j=1}^n a_{ij} u_j \right| \right\}$$

$$= \max_{1 \le i \le m} \sum_{j=1}^n |a_{ij}|, \quad (\text{i.e., } u_j = \text{sgn}\{a_{ij}\} \ \forall j) \tag{1.9}$$

which is simply the maximum absolute row sum of the matrix.

In the special case of $a = b = 2$, the induced norm is called the *spectral norm* or ℓ_2 *norm*. The spectral norm of a matrix $\mathbf{A}$ is the largest singular value of $\mathbf{A}$ or the square root of the largest eigenvalue of the positive semidefinite matrix $\mathbf{A}^T\mathbf{A}$, i.e.,

$$\|\mathbf{A}\|_2 = \sup\{\|\mathbf{A}\mathbf{u}\|_2 \mid \|\mathbf{u}\|_2 \le 1\} = \sigma_{\max}(\mathbf{A}) = \sqrt{\lambda_{\max}(\mathbf{A}^T\mathbf{A})}. \tag{1.10}$$

The singular values of a matrix $\mathbf{A}$ will be defined in (1.109) and their relation to the corresponding eigenvalues of $\mathbf{A}^T\mathbf{A}$ (or $\mathbf{A}\mathbf{A}^T$) is given by (1.116) in Subsection 1.2.6 later.

1.1.3   Inner product

The inner product of two real vectors $\mathbf{x} \in \mathbb{R}^n$ and $\mathbf{y} \in \mathbb{R}^n$ is a real scalar and is defined as

$$\langle \mathbf{x}, \mathbf{y} \rangle = \mathbf{y}^T \mathbf{x} = \sum_{i=1}^n x_i y_i. \tag{1.11}$$

If $\mathbf{x}$ and $\mathbf{y}$ are complex vectors, then the transpose in the above equation will be replaced by Hermitian. Note that the square root of the inner product of a vector $\mathbf{x}$ with itself gives the Euclidean norm of that vector.

Cauchy Schwartz inequality: For any two vectors $\mathbf{x}$ and $\mathbf{y}$ in $\mathbb{R}^n$, the Cauchy–Schwartz inequality

$$|\langle \mathbf{x}, \mathbf{y} \rangle| \le \|\mathbf{x}\|_2 \cdot \|\mathbf{y}\|_2 \tag{1.12}$$

holds. Furthermore, the equality holds if and only if $\mathbf{x} = \alpha \mathbf{y}$ for some $\alpha \in \mathbb{R}$.

Pythagorean theorem: If two vectors $\mathbf{x}$ and $\mathbf{y}$ in $\mathbb{R}^n$ are orthogonal, i.e., $\langle \mathbf{x}, \mathbf{y} \rangle = 0$, then

$$\|\mathbf{x} + \mathbf{y}\|_2^2 = (\mathbf{x} + \mathbf{y})^T(\mathbf{x} + \mathbf{y}) = \|\mathbf{x}\|_2^2 + 2\langle \mathbf{x}, \mathbf{y} \rangle + \|\mathbf{y}\|_2^2 = \|\mathbf{x}\|_2^2 + \|\mathbf{y}\|_2^2. \tag{1.13}$$

Similarly, the inner product of two real matrices $\mathbf{X} = \{x_{ij}\}_{m \times n} \in \mathbb{R}^{m \times n}$ and $\mathbf{Y} = \{y_{ij}\}_{m \times n} \in \mathbb{R}^{m \times n}$ can be defined as:

$$\langle \mathbf{X}, \mathbf{Y} \rangle = \sum_{i=1}^{m} \sum_{j=1}^{n} x_{ij} y_{ij} = \text{Tr}(\mathbf{X}^T \mathbf{Y}) = \text{Tr}(\mathbf{Y}^T \mathbf{X}). \qquad (1.14)$$

Note that $\langle \mathbf{X}, \mathbf{Y} \rangle = \langle \mathbf{vec}(\mathbf{X}), \mathbf{vec}(\mathbf{Y}) \rangle$, i.e., inner product of the associated two column vectors $\mathbf{vec}(\mathbf{X})$ and $\mathbf{vec}(\mathbf{Y})$ of dimension mn (column-wise stacking of columns of $\mathbf{X}$ and $\mathbf{Y}$, respectively).

Remark 1.3 An inner product space is a vector space with the additional structure called the inner product which associates each pair of vectors in the space with a scalar quantity known as the inner product of the two vectors. □

Remark 1.4 An inequality similar to Cauchy–Schwartz inequality involving the spectral norm of $\mathbf{A}$ (see (1.10)) is given by

$$\|\mathbf{A}\mathbf{x}\|_2 = \left\| \mathbf{A} \frac{\mathbf{x}}{\|\mathbf{x}\|_2} \right\|_2 \cdot \|\mathbf{x}\|_2 \leq \|\mathbf{A}\|_2 \cdot \|\mathbf{x}\|_2 = \sigma_{\max}(\mathbf{A}) \cdot \|\mathbf{x}\|_2, \qquad (1.15)$$

for which the inequality holds with equality when $\mathbf{x} = \alpha \mathbf{v}$ where $\mathbf{v}$ is the right singular vector associated with the maximum singular value of $\mathbf{A}$. □

1.1.4   Norm ball

The norm ball of a point $\mathbf{x} \in \mathbb{R}^n$ is defined as the following set[1]

$$B(\mathbf{x}, r) = \{ \mathbf{y} \in \mathbb{R}^n \mid \|\mathbf{y} - \mathbf{x}\| \leq r \}, \qquad (1.16)$$

where r is the radius and $\mathbf{x}$ is the center of the norm ball. It is also called the *neighborhood* of the point $\mathbf{x}$. For the case of $n = 2$, $\mathbf{x} = \mathbf{0}_2$, and $r = 1$, the 2-norm ball is $B(\mathbf{x}, r) = \{ \mathbf{y} \mid y_1^2 + y_2^2 \leq 1 \}$ (a circular disk of radius equal to 1), the 1-norm ball is $B(\mathbf{x}, r) = \{ \mathbf{y} \mid |y_1| + |y_2| \leq 1 \}$ (a 2-dimensional cross-polytope of area equal to 2), and the ∞-norm ball is $B(\mathbf{x}, r) = \{ \mathbf{y} \mid |y_1| \leq 1, |y_2| \leq 1 \}$ (a square of area equal to 4) (see Figure 1.1). Note that the norm ball is symmetric with respect to (w.r.t.) the origin, convex, closed, bounded and has nonempty interior. Moreover, the 1-norm ball is a subset of the 2-norm ball which is a subset of ∞-norm ball, due to the following inequality:

$$\|\mathbf{v}\|_p \leq \|\mathbf{v}\|_q, \qquad (1.17)$$

where $\mathbf{v} \in \mathbb{R}^n$, p and q are real and $p > q \geq 1$, and the equality holds when $\mathbf{v} = r\mathbf{e}_i$, i.e., all the p-norm balls of constant radius r have intersections at $r\mathbf{e}_i$, $i = 1, \ldots, n$. For instance, in Figure 1.1, $\|\mathbf{x}_1\|_p = 1$ for all $p \geq 1$, and $\|\mathbf{x}_2\|_\infty = 1 < \|\mathbf{x}_2\|_2 = \sqrt{2} < \|\mathbf{x}_2\|_1 = 2$. The inequality (1.17) is proven as follows.

---

[1]The norm ball $B(\mathbf{x}, r)$ will be used frequently hereafter without explicitly mentioning the associated norm in (1.16) for simplicity, meaning that the norm in (1.16) is 2-norm.

Proof of (1.17): Assume that $\mathbf{v} \neq \mathbf{0}_n$. Let $\beta = \|\mathbf{v}\|_\infty = \max\{|v_1|, \ldots, |v_n|\} > 0$. Then $|v_i|/\beta \leq 1$ for all i. It can be easily inferred that, for $p > q \geq 1$,

$$1 \leq \sum_{i=1}^{n} |v_i/\beta|^p \leq \sum_{i=1}^{n} |v_i/\beta|^q, \tag{1.18}$$

which straightforwardly leads to

$$\left\{ \sum_{i=1}^{n} |v_i/\beta|^p \right\}^{1/p} \leq \left\{ \sum_{i=1}^{n} |v_i/\beta|^q \right\}^{1/q}. \tag{1.19}$$

Canceling the common term β on both sides of (1.19) gives rise to the inequality (1.17). ∎

Remark 1.5 The inequality of (1.17) can be explained as the the height or the weight of a person (corresponding to the vector $\mathbf{v}$) by a measuring device $\|\cdot\|_p$ with a different unit (e.g., in meters or kilograms (larger p) or in feet or pounds (smaller p)). So $\|\mathbf{v}\|_p$ is smaller for larger p. One can observe, from Figure 1.1, that $\|\mathbf{U}\mathbf{v}\|_p = \|\mathbf{v}\|_p$ for any orthogonal matrix $\mathbf{U}$ (i.e., $\mathbf{U}\mathbf{U}^T = \mathbf{U}^T\mathbf{U} = \mathbf{I}$, cf. (1.87)) for $p = 2$, implying that $\|\mathbf{v}\|_p$ is orientation-invariant for the case of $p = 2$. This property is not applicable for the case of $p \neq 2$. □

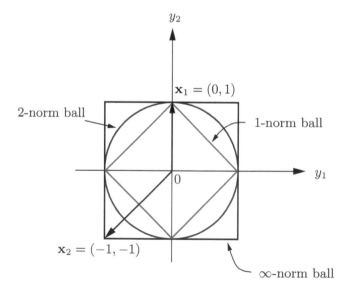

Figure 1.1 Illustration for 1-norm ball, 2-norm ball, and ∞-norm ball, for the case of $n = 2$, $\mathbf{x} = \mathbf{0}_2$ and radius $r = 1$ (conceptually corresponding to a different unit for each ball, e.g., centimeter, decimeter, and meter). It can be seen that $\|\mathbf{x}_1\|_p = 1$ for all $p \geq 1$, and $\|\mathbf{x}_2\|_\infty = 1 < \|\mathbf{x}_2\|_2 = \sqrt{2} < \|\mathbf{x}_2\|_1 = 2$.

Remark 1.6 The induced norm defined in (1.7) can be illustrated from a geometric perspective. Consider the linear function

$$f(\mathbf{x}) = \mathbf{A}\mathbf{x}, \ \mathbf{A} = [\mathbf{a}_1 \ \mathbf{a}_2] = \begin{bmatrix} 5 & 11 \\ 7 & 3 \end{bmatrix}.$$

An illustration for $\|\mathbf{A}\|_p$ is given in Figure 1.2 for $p = 1, 2, \infty$, where the function f maps the ∞-norm ball, 2-norm ball, and 1-norm ball (all with identical radius of unity) shown in Figure 1.1 to the parallelogram containing $\mathbf{y}_1 = \mathbf{a}_1 + \mathbf{a}_2$, the ellipsoid containing $\mathbf{y}_2$ with $\|\mathbf{y}_2\|_2 = \max\{\|f(\mathbf{u})\|_2 \mid \|\mathbf{u}\|_2 \le 1)\}$ (a subset of the former), and the parallelogram containing $\mathbf{a}_1$ and $\mathbf{a}_2$ (a subset of the ellipsoid), respectively. It can been seen by (1.7) that

$$\|\mathbf{A}\|_1 = \|\mathbf{a}_2\|_1 = 14 \quad (\text{cf. } (1.8))$$
$$\|\mathbf{A}\|_2 = \sigma_{\max}(\mathbf{A}) = \|\mathbf{y}_2\|_2 \approx 13.5275 \quad (\text{cf. } (1.10))$$
$$\|\mathbf{A}\|_\infty = \|\mathbf{y}_1\|_\infty = 16 \quad (\text{cf. } (1.9)).$$

Note that an ellipsoid in $\mathbb{R}^n$ can be characterized by lengths of n semiaxes (identical to singular values of $\mathbf{A}$), and will be introduced in more detail in Subsection 2.2.2 later. □

1.1.5   Interior point

A point $\mathbf{x}$ in a set $C \subseteq \mathbb{R}^n$ is an interior point of the set C if there exists an $\epsilon > 0$ for which $B(\mathbf{x}, \epsilon) \subseteq C$ (see Figure 1.3). In other words, a point $\mathbf{x} \in C$ is said to be an *interior point* of the set C if the set C contains some neighborhood of $\mathbf{x}$, that is, if all points within some neighborhood of $\mathbf{x}$ are also in C.

Remark 1.7 The set of all the interior points of C is called the *interior* of C and is represented as **int** C, which can also be expressed as

$$\mathbf{int} \ C = \{\mathbf{x} \in C \mid B(\mathbf{x}, r) \subseteq C, \text{ for some } r > 0\}, \tag{1.20}$$

which will be frequently used in many proofs directly or indirectly in the ensuing chapters. □

1.1.6   Complement, scaled sets, and sum of sets

The *complement* of a set $C \subset \mathbb{R}^n$ is defined as follows (see Figure 1.3):

$$\mathbb{R}^n \setminus C = \{\mathbf{x} \in \mathbb{R}^n \mid \mathbf{x} \notin C\}, \tag{1.21}$$

where "\" denotes the *set difference*, i.e., $A \setminus B = \{\mathbf{x} \in A \mid \mathbf{x} \notin B\}$. The set $C \subset \mathbb{R}^n$ scaled by a real number α is a set defined as

$$\alpha \cdot C \triangleq \{\alpha\mathbf{x} \mid \mathbf{x} \in C\}. \tag{1.22}$$

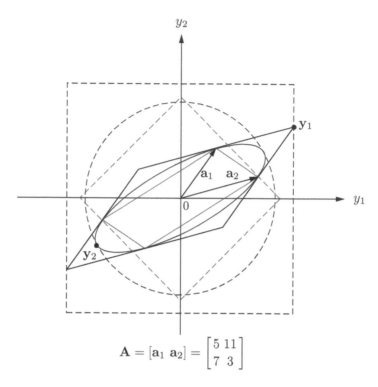

$$\mathbf{A} = [\mathbf{a}_1 \ \mathbf{a}_2] = \begin{bmatrix} 5 & 11 \\ 7 & 3 \end{bmatrix}$$

Figure 1.2 An illustration of the induced norm of $\mathbf{A} \in \mathbb{R}^{2\times 2}$, $\|\mathbf{A}\|_1 = \|\mathbf{a}_2\|_1$, $\|\mathbf{A}\|_2 = \|\mathbf{y}_2\|_2$ (the largest semiaxis of the ellipsoid), and $\|\mathbf{A}\|_\infty = \|\mathbf{y}_1\|_\infty$, where the parallelograms and the ellipsoid (denoted by solid lines) are the images of the associated unit-radius 1-norm, 2-norm, and ∞-norm balls shown in Figure 1.1, via the linear transformation $\mathbf{A}\mathbf{x}$, together with the associated norm balls (denoted by dashed lines) with radii being $\|\mathbf{A}\|_1$, $\|\mathbf{A}\|_2$, and $\|\mathbf{A}\|_\infty$, respectively.

The *sum* of sets $C_1 \subset \mathbb{R}^n$ and $C_2 \subset \mathbb{R}^n$ is a set defined as

$$C_1 + C_2 \triangleq \{\mathbf{x} = \mathbf{x}_1 + \mathbf{x}_2 \mid \mathbf{x}_1 \in C_1, \ \mathbf{x}_2 \in C_2\}. \tag{1.23}$$

1.1.7   Closure and boundary

The *closure* of a set C is defined as the set of all limit points of convergent sequences in the set C, which can also be expressed as

$$\mathbf{cl} \ C = \mathbb{R}^n \setminus \mathbf{int} \ (\mathbb{R}^n \setminus C). \tag{1.24}$$

A point $\mathbf{x}$ is in the closure of C, if for every $\epsilon > 0$, there exists a $\mathbf{y} \in C$ and $\mathbf{y} \neq \mathbf{x}$ such that $\|\mathbf{x} - \mathbf{y}\| \leq \epsilon$.

A point $\mathbf{x}$ is said to be a *boundary point* of the set C if every neighborhood of $\mathbf{x}$ contains a point in C and a point not in C (cf. Figure 1.3). Note that a boundary point of C may or may not be an element of C. The set of all boundary

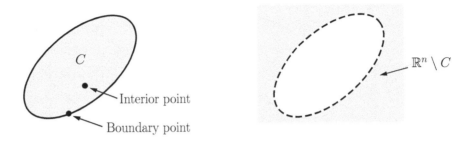

Figure 1.3 Interior point (left plot) and complement of a set C (right plot).

points of C is called the *boundary* of C. The boundary of a set C can be also expressed as

$$\mathbf{bd}\ C = \mathbf{cl}\ C \setminus \mathbf{int}\ C. \qquad (1.25)$$

The concepts of closure and boundary of a set are illustrated in Figure 1.4.

A set C is said to be *open* if it contains a neighborhood of each of its points, that is, if every point in C is an interior point, or equivalently, if C contains no boundary points. In other words, C is open if either $\mathbf{int}\ C = C$ or $C \cap \mathbf{bd}\ C = \emptyset$ (empty set) is true. For instance, $C = (a, b) = \{x \in \mathbb{R} \mid a < x < b\}$ is open and $\mathbf{int}\ C = C$; $C = (a, b] = \{x \in \mathbb{R} \mid a < x \leq b\}$ is not open; $C = [a, b] = \{x \in \mathbb{R} \mid a \leq x \leq b\}$ is not open; and $C = \{a, b, c\}$ is not open.

A set C is said to be *closed* if it contains its boundary. It can be shown that a set is closed if and only if its complement $\mathbb{R}^n \setminus C$ is open. A set C is closed if and only if it contains the limit point of every convergent sequence in it; and it is closed if $\mathbf{bd}\ C \subseteq C$. For instance, $C = [a, b]$ is closed; $C = (a, b]$ is not open and not closed; the empty set $\emptyset$ and $\mathbb{R}^n$ are both open and closed. For two closed sets C_1 and C_2, $C_1 \cup C_2$ is also a closed set even if their intersection is empty, and so $\{1\}$ and $\mathbb{Z}_+$ are also closed sets.

A set that is contained in a ball of finite radius is said to be *bounded*. A set is *compact* if it is both closed and bounded. For instance, the norm ball $B(\mathbf{x}, r)$ is compact, $\mathbb{R}_+$ is not bounded; $C = (a, b]$ is bounded but not compact.

1.1.8 Supremum and infimum

In mathematics, given a subset S of a partially ordered set T, the supremum (sup) of S, if it exists, is the least element of T that is greater than or equal to each element of S. Consequently, the supremum is also referred to as the least upper bound, *lub* or *LUB*. If the supremum exists, it may or may not belong to S. On the other hand, the infimum (inf) of S is the greatest element in T, not necessarily in S, that is less than or equal to all elements of S. Consequently the

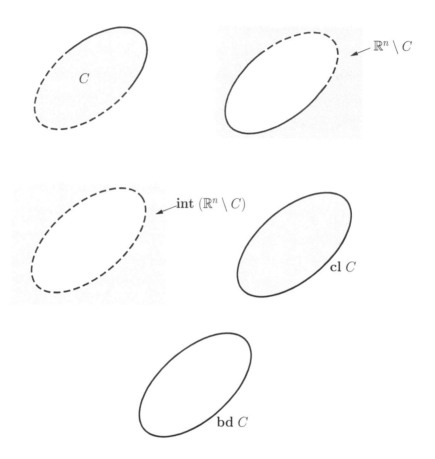

Figure 1.4 Closure and boundary of a set C.

term greatest lower bound (also abbreviated as *glb* or *GLB*) is also commonly used. Consider a set $C \subseteq \mathbb{R}$.

- A number a is an upper bound (lower bound) on C if for each $x \in C, x \leq a$ $(x \geq a)$.
- A number b is the least upper bound (greatest lower bound) or the supremum (infimum) of C if
  (i) b is an upper bound (lower bound) on C, and
  (ii) $b \leq a$ $(b \geq a)$ for every upper bound (lower bound) a on C.

Remark 1.8 An infimum is in a precise sense dual to the concept of a supremum and vice versa. For instance, $\sup C = \infty$ if C is unbounded above and $\inf C = -\infty$ if C is unbounded below.   □

Remark 1.9 If $C = \emptyset$, then $\sup C = -\infty$ (where "$-\infty$" can be "thought of as" the starting point of $\sup C$ ascending towards $+\infty$ for any nonempty set C) and $\inf C = +\infty$ (where "$+\infty$" can be "thought of as" the starting point of $\inf C$ descending towards $-\infty$ for any nonempty set C).   □

1.1.9   Function

The notation $f : X \to Y$ indicates that f is a function with domain X and codomain Y. For instance, $f(x) = \log x$, where the domain of f is $X = \mathbf{dom}\ f = \mathbb{R}_{++}$ and $Y = \mathbb{R}$.

1.1.10   Continuity

A function $\boldsymbol{f} : \mathbb{R}^n \to \mathbb{R}^m$ is continuous at $\mathbf{x} \in \mathbf{dom}\ \boldsymbol{f}$, if for any $\epsilon > 0$ there exists a $\delta > 0$ such that

$$\mathbf{y} \in \mathbf{dom}\ \boldsymbol{f} \cap\ B(\mathbf{x}, \delta) \Rightarrow \|\boldsymbol{f}(\mathbf{y}) - \boldsymbol{f}(\mathbf{x})\|_2 \le \epsilon. \tag{1.26}$$

A function $\boldsymbol{f}$ is continuous if it is continuous at every point in its domain. Suppose that $\boldsymbol{f}$ is continuous. Whenever the sequence $\mathbf{x}_1, \mathbf{x}_2, \ldots$, in $\mathbf{dom}\ \boldsymbol{f}$ converges to a point $\mathbf{x} \in \mathbf{dom}\ \boldsymbol{f}$, the sequence $\boldsymbol{f}(\mathbf{x}_1), \boldsymbol{f}(\mathbf{x}_2), \ldots$, converges to $\boldsymbol{f}(\mathbf{x})$, i.e.,

$$\lim_{i \to \infty} \boldsymbol{f}(\mathbf{x}_i) = \boldsymbol{f}(\lim_{i \to \infty} \mathbf{x}_i) = \boldsymbol{f}(\mathbf{x}). \tag{1.27}$$

One example of a continuous function and one example of a discontinuous function are shown in Figure 1.5 where

$$\mathrm{sgn}(x) \triangleq \begin{cases} 1, & \text{if } x > 0 \\ 0, & \text{if } x = 0 \\ -1, & \text{if } x < 0. \end{cases} \tag{1.28}$$

For ease of later use, some more definitions or forms about continuity of a function $f : \mathbb{R}^n \to \mathbb{R}$, that may not be a continuous function, are as follows. f is *Lipschitz continuous* (which is also continuous) [Ber09] if there exists a finite positive constant L such that

$$\|f(\mathbf{x}_1) - f(\mathbf{x}_2)\|_2 \le L\|\mathbf{x}_1 - \mathbf{x}_2\|_2\ \forall \mathbf{x}_1, \mathbf{x}_2 \in \mathbf{dom}\ f. \tag{1.29}$$

For instance, any function with bounded first derivative is Lipschitz continuous. Also, there exist two weaker forms of continuity: upper semi-continuity and lower semi-continuity. f is *upper semi-continuous* [Ber09] if

$$\limsup_{\mathbf{x} \to \mathbf{x}_0} f(\mathbf{x}) = \lim_{\delta \to 0}\ \sup\ \{f(\mathbf{x}) \mid \mathbf{x} \in B(\mathbf{x}_0, \delta) \setminus \{\mathbf{x}_0\}\} \le f(\mathbf{x}_0),\ \forall \mathbf{x}_0, \tag{1.30}$$

and *lower semi-continuous* if

$$\liminf_{\mathbf{x} \to \mathbf{x}_0} f(\mathbf{x}) = \lim_{\delta \to 0}\ \inf\ \{f(\mathbf{x}) \mid \mathbf{x} \in B(\mathbf{x}_0, \delta) \setminus \{\mathbf{x}_0\}\} \ge f(\mathbf{x}_0),\ \forall \mathbf{x}_0. \tag{1.31}$$

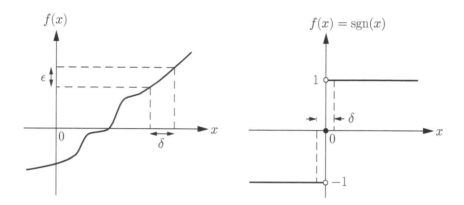

Figure 1.5 An example of a continuous function (left plot) and an example of a noncontinuous function (right plot).

It can be inferred that f is continuous if and only if it is upper semi-continuous and lower semi-continuous. For instance, supposing that $u(x) = 1$ for $x \geq 0$ and $u(x) = 0$ for $x < 0$, $u(x)$ is upper semi-continuous while $-u(x)$ is lower semi-continuous; $\text{sgn}(x)$ is neither upper nor lower semi-continuous.

1.1.11   Derivative and gradient

Since vector limits are computed by taking the limit of each coordinate function, we can write the function $\boldsymbol{f} : \mathbb{R}^n \to \mathbb{R}^m$ for a point $\mathbf{x} \in \mathbb{R}^n$ as follows:

$$\boldsymbol{f}(\mathbf{x}) = \begin{bmatrix} f_1(\mathbf{x}) \\ f_2(\mathbf{x}) \\ \vdots \\ f_m(\mathbf{x}) \end{bmatrix} = (f_1(\mathbf{x}), f_2(\mathbf{x}), \ldots, f_m(\mathbf{x})) \qquad (1.32)$$

where each $f_i(\mathbf{x})$ is a function from $\mathbb{R}^n$ to $\mathbb{R}$. Now, $\dfrac{\partial \boldsymbol{f}(\mathbf{x})}{\partial x_j}$ can be defined as

$$\frac{\partial \boldsymbol{f}(\mathbf{x})}{\partial x_j} = \begin{bmatrix} \dfrac{\partial f_1(\mathbf{x})}{\partial x_j} \\[2mm] \dfrac{\partial f_2(\mathbf{x})}{\partial x_j} \\[2mm] \vdots \\[2mm] \dfrac{\partial f_m(\mathbf{x})}{\partial x_j} \end{bmatrix} = \left(\frac{\partial f_1(\mathbf{x})}{\partial x_j}, \frac{\partial f_2(\mathbf{x})}{\partial x_j}, \ldots, \frac{\partial f_m(\mathbf{x})}{\partial x_j} \right). \qquad (1.33)$$

The above vector is a tangent vector at the point $\mathbf{x}$ of the curve $\boldsymbol{f}$ obtained by varying only x_j (the jth coordinate of $\mathbf{x}$) with x_i fixed for all $i \neq j$.

The derivative of a differentiable function $\boldsymbol{f} : \mathbb{R}^n \to \mathbb{R}^m$ can be represented by $D\boldsymbol{f}(\mathbf{x})$—an $m \times n$ matrix defined as

$$D\boldsymbol{f}(\mathbf{x}) = \begin{bmatrix} \dfrac{\partial \boldsymbol{f}(\mathbf{x})}{\partial x_1} & \dfrac{\partial \boldsymbol{f}(\mathbf{x})}{\partial x_2} & \cdots & \dfrac{\partial \boldsymbol{f}(\mathbf{x})}{\partial x_n} \end{bmatrix} = \begin{bmatrix} \nabla f_1(\mathbf{x})^T \\ \nabla f_2(\mathbf{x})^T \\ \vdots \\ \nabla f_m(\mathbf{x})^T \end{bmatrix}$$

$$= \begin{bmatrix} \dfrac{\partial f_1(\mathbf{x})}{\partial x_1} & \cdots & \dfrac{\partial f_1(\mathbf{x})}{\partial x_n} \\ \vdots & \vdots & \vdots \\ \dfrac{\partial f_m(\mathbf{x})}{\partial x_1} & \cdots & \dfrac{\partial f_m(\mathbf{x})}{\partial x_n} \end{bmatrix} \in \mathbb{R}^{m \times n} \qquad (1.34)$$

where $\nabla f_i(\mathbf{x})$ will be defined below. The above matrix $D\boldsymbol{f}(\mathbf{x})$ is called the *Jacobian matrix* or *derivative matrix* of $\boldsymbol{f}$ at the point $\mathbf{x}$.

If the function $f : \mathbb{R}^n \to \mathbb{R}$ is differentiable, then its gradient $\nabla f(\mathbf{x})$ at a point $\mathbf{x}$ can be defined as

$$\nabla f(\mathbf{x}) = Df(\mathbf{x})^T = \begin{bmatrix} \dfrac{\partial f(\mathbf{x})}{\partial x_1} \\ \dfrac{\partial f(\mathbf{x})}{\partial x_2} \\ \vdots \\ \dfrac{\partial f(\mathbf{x})}{\partial x_n} \end{bmatrix} \in \mathbb{R}^n. \qquad (1.35)$$

Note that $\nabla f(\mathbf{x})$ and $\mathbf{x}$ have the same dimension (i.e., both are column vectors of dimension n). Moreover, if the function $f : \mathbb{R}^{n \times m} \to \mathbb{R}$ is differentiable, then its gradient $\nabla f(\mathbf{X})$ at a point $\mathbf{X}$ can be defined as

$$\nabla f(\mathbf{X}) = Df(\mathbf{X})^T = \begin{bmatrix} \dfrac{\partial f(\mathbf{X})}{\partial x_{1,1}} & \cdots & \dfrac{\partial f(\mathbf{X})}{\partial x_{1,m}} \\ \vdots & \vdots & \vdots \\ \dfrac{\partial f(\mathbf{X})}{\partial x_{n,1}} & \cdots & \dfrac{\partial f(\mathbf{X})}{\partial x_{n,m}} \end{bmatrix} \in \mathbb{R}^{n \times m}, \qquad (1.36)$$

which also has the same dimension with $\mathbf{X} \in \mathbf{dom}\, f$.

Remark 1.10 Consider a differentiable function $f : \mathbb{R}^n \to \mathbb{R}$ and $\mathbf{x} = (\mathbf{x}_1, \ldots, \mathbf{x}_m)$ where $\mathbf{x}_i \in \mathbb{R}^{n_i}$ and $n = n_1 + \cdots + n_m$. One may express its gradient and derivative explicitly w.r.t. $\mathbf{x}$ as

$$\nabla_{\mathbf{x}} f(\mathbf{x}) = \begin{bmatrix} \nabla_{\mathbf{x}_1} f(\mathbf{x})^T, \ldots, \nabla_{\mathbf{x}_m} f(\mathbf{x})^T \end{bmatrix}^T$$
$$= [D_{\mathbf{x}_1} f(\mathbf{x}), \ldots, D_{\mathbf{x}_m} f(\mathbf{x})]^T = D_{\mathbf{x}} f(\mathbf{x})^T. \qquad (1.37)$$

Similarly, as for a differentiable function $f : \mathbb{R}^{n \times m} \to \mathbb{R}$, its gradient and derivative w.r.t. $\mathbf{X}$ can also be expressed as

$$\nabla_{\mathbf{X}} f(\mathbf{X}) = [\nabla_{\mathbf{x}_1} f(\mathbf{X}), \ldots, \nabla_{\mathbf{x}_m} f(\mathbf{X})]$$
$$= \left[D_{\mathbf{x}_1} f(\mathbf{X})^T, \ldots, D_{\mathbf{x}_m} f(\mathbf{X})^T \right] = D_{\mathbf{X}} f(\mathbf{X})^T, \qquad (1.38)$$

where $\mathbf{x}_i \in \mathbb{R}^n$ is the ith column of $\mathbf{X} \in \mathbb{R}^{n \times m}$. $\qquad\square$

Remark 1.11 The gradient matrix defined by (1.36) or the derivative of a function defined by (1.34) is under the premise that all of the entries in $\mathbf{x}$ and $\mathbf{X}$ are independent variables. If some of them are dependent (e.g., $\mathbf{X} \in \mathbb{R}^{n \times n}$ being symmetric ($\mathbf{X} = \mathbf{X}^T$), or antisymmetric ($\mathbf{X} = -\mathbf{X}^T$), or Toeplitz ($[\mathbf{X}]_{i,j} = [\mathbf{X}]_{i+1,j+1}$ for all i, j)), the total number of independent variables is actually reduced, thus violating the premise. To meet the premise, $\mathbf{x}$ and $\mathbf{X}$ need be redefined such that they only contain independent variables. Instead, we still maintain (1.36) and (1.34) under the above premise, while the corresponding gradient matrix will be considered as a special case of the result of the one defined by (1.36) or (1.34) with the entries of $\mathbf{x}$ and $\mathbf{X}$ taking any values in any dependent manner.

For a given function $f : \mathbb{R}^n \to \mathbb{R}$ (or $f : \mathbb{R}^{n \times m} \to \mathbb{R}$), an indirect approach for obtaining its gradient (rather than by (1.36) or (1.34)) is to find its first-order Taylor series approximation from which to find the corresponding term $\nabla f(\mathbf{x})$ (or $\nabla f(\mathbf{X})$). This approach will be proven useful in the ensuing chapters. $\qquad\square$

For the case that f is a complex-differentiable function of complex variable

$$\mathbf{x} = \mathbf{u} + j\mathbf{v} = \text{Re}\{\mathbf{x}\} + j\text{Im}\{\mathbf{x}\},$$

where $\text{Re}\{\cdot\}$ and $\text{Im}\{\cdot\}$ denote the real part and imaginary part of the argument, respectively, and $\mathbf{u}$ and $\mathbf{v}$ are real vectors. Then Cauchy–Riemann equations must be satisfied as follows [Sch10]:

$$\begin{cases} \nabla_{\mathbf{u}} \text{Re}\{f(\mathbf{x})\} = \nabla_{\mathbf{v}} \text{Im}\{f(\mathbf{x})\} \\ \nabla_{\mathbf{v}} \text{Re}\{f(\mathbf{x})\} = -\nabla_{\mathbf{u}} \text{Im}\{f(\mathbf{x})\} \end{cases} \text{ or } \nabla_{\mathbf{u}} f(\mathbf{x}) = -j \nabla_{\mathbf{v}} f(\mathbf{x}) = \nabla_{\mathbf{x}} f(\mathbf{x}). \quad (1.39)$$

The gradient of f w.r.t. $\mathbf{x}$ and $\mathbf{x}^*$ can also be defined as

$$\nabla_{\mathbf{x}} f(\mathbf{x}) \triangleq \frac{1}{2} \left(\nabla_{\mathbf{u}} f(\mathbf{x}) - j \nabla_{\mathbf{v}} f(\mathbf{x}) \right) \qquad (1.40)$$

$$\nabla_{\mathbf{x}^*} f(\mathbf{x}) \triangleq \frac{1}{2} \left(\nabla_{\mathbf{u}} f(\mathbf{x}) + j \nabla_{\mathbf{v}} f(\mathbf{x}) \right) \qquad (1.41)$$

where $\mathbf{x}$ and its complex conjugate $\mathbf{x}^*$ are viewed as independent variables. Note that when $f(\mathbf{x})$ is analytic (i.e., (1.39) is satisfied), $\nabla_{\mathbf{x}} f(\mathbf{x})$ equals the gradient of f as if $\mathbf{x}$ were real (cf. (1.35)), and $\nabla_{\mathbf{x}^*} f(\mathbf{x}) = \mathbf{0}$; when $f(\mathbf{x})$ is not analytic,

$\nabla_{\mathbf{x}} f(\mathbf{x}) = \mathbf{0}$. For instance, for $\mathbf{x}, \mathbf{a} \in \mathbb{C}^n$ and $\mathbf{X}, \mathbf{C} \in \mathbb{C}^{m \times n}$,

$$\nabla_{\mathbf{x}} \, \mathbf{a}^T \mathbf{x} = \mathbf{a}, \quad \nabla_{\mathbf{X}} \, \mathrm{Tr}(\mathbf{C}^T \mathbf{X}) = \mathbf{C}$$
$$\nabla_{\mathbf{x}} \, \mathbf{a}^T \mathbf{x}^* = \mathbf{0}, \quad \nabla_{\mathbf{X}^*} \, \mathrm{Tr}(\mathbf{C} \mathbf{X}^H) = \mathbf{C}. \tag{1.42}$$

For the case that $f : \mathbb{C}^n \to \mathbb{R}$ (which must be a function of both $\mathbf{x}$ and $\mathbf{x}^*$, and not analytic in $\mathbf{x}$, but it is analytic in $\mathbf{x}$ when $\mathbf{x}^*$ is treated as an independent variable and vice versa [Bra83], [Hay96]), the gradient of f is defined as

$$\nabla_{\mathbf{x}} f(\mathbf{x}) \triangleq \nabla_{\mathbf{u}} f(\mathbf{x}) + j \nabla_{\mathbf{v}} f(\mathbf{x}) = 2 \nabla_{\mathbf{x}^*} f(\mathbf{x}), \tag{1.43}$$

which is actually solved from (1.40) and (1.41). For instance, for the complex vector variable $\mathbf{x} \in \mathbb{C}^n$ and a constant vector $\mathbf{a} \in \mathbb{C}^n$ and a constant matrix $\mathbf{A} \in \mathbb{H}^n$, it can be easily shown that

$$\nabla_{\mathbf{x}} \, \mathbf{x}^H \mathbf{A} \mathbf{x} = 2 \mathbf{A} \mathbf{x}, \quad \nabla_{\mathbf{x}} \left(\mathbf{x}^H \mathbf{a} + \mathbf{a}^H \mathbf{x} \right) = 2 \mathbf{a}, \tag{1.44}$$

and for complex matrix variables $\mathbf{X} \in \mathbb{C}^{m \times n}$ and $\mathbf{Y} \in \mathbb{H}^n$,

$$\nabla_{\mathbf{X}} \, \mathrm{Tr}(\mathbf{X} \mathbf{A} \mathbf{X}^H) = 2 \mathbf{X} \mathbf{A}, \quad \nabla_{\mathbf{Y}} \left(\mathrm{Tr}(\mathbf{A} \mathbf{Y}) + \mathrm{Tr}(\mathbf{A}^* \mathbf{Y}^*) \right) = 2 \mathbf{A}. \tag{1.45}$$

1.1.12 Hessian

Suppose that $f : \mathbb{R}^n \to \mathbb{R}$ is twice differentiable, and all its second partial derivatives exist and are continuous over the domain of f. The *Hessian* $\nabla^2 f(\mathbf{x})$ of f is defined as follows:

$$\nabla^2 f(\mathbf{x}) = D(\nabla f(\mathbf{x})) = \left\{ \frac{\partial^2 f(\mathbf{x})}{\partial x_i \partial x_j} \right\}_{n \times n}$$

$$= \begin{bmatrix} \dfrac{\partial^2 f(\mathbf{x})}{\partial x_1^2} & \dfrac{\partial^2 f(\mathbf{x})}{\partial x_1 \partial x_2} & \cdots & \dfrac{\partial^2 f(\mathbf{x})}{\partial x_1 \partial x_n} \\[2mm] \dfrac{\partial^2 f(\mathbf{x})}{\partial x_2 \partial x_1} & \dfrac{\partial^2 f(\mathbf{x})}{\partial x_2^2} & \cdots & \dfrac{\partial^2 f(\mathbf{x})}{\partial x_2 \partial x_n} \\[2mm] \vdots & \vdots & & \vdots \\[2mm] \dfrac{\partial^2 f(\mathbf{x})}{\partial x_n \partial x_1} & \dfrac{\partial^2 f(\mathbf{x})}{\partial x_n \partial x_2} & \cdots & \dfrac{\partial^2 f(\mathbf{x})}{\partial x_n^2} \end{bmatrix} \in \mathbb{S}^n. \tag{1.46}$$

The Hessian of a function can be used for verifying the convexity of a twice differentiable function, so its calculation is needed quite often. For instance, assuming that

$$f(\mathbf{x}) = \mathbf{x}^T \mathbf{P} \mathbf{x} + \mathbf{x}^T \mathbf{q} + c,$$

where $\mathbf{P} \in \mathbb{R}^{n \times n}$, $\mathbf{q} \in \mathbb{R}^n$, and $c \in \mathbb{R}$, one can easily obtain

$$\nabla f(\mathbf{x}) = (\mathbf{P} + \mathbf{P}^T)\mathbf{x} + \mathbf{q}, \quad \nabla^2 f(\mathbf{x}) = D(\nabla f(\mathbf{x})) = \mathbf{P} + \mathbf{P}^T.$$

For the case of $\mathbf{P} = \mathbf{P}^T \in \mathbb{S}^n$,

$$\nabla f(\mathbf{x}) = 2\mathbf{P}\mathbf{x} + \mathbf{q}, \quad \nabla^2 f(\mathbf{x}) = D(\nabla f(\mathbf{x})) = 2\mathbf{P}.$$

Consider another example as follows:

$$g(\mathbf{y}) = \|\mathbf{A}\mathbf{x} - \mathbf{z}\|_2^2, \ \mathbf{y} = (\mathbf{x}, \mathbf{z}) \in \mathbb{R}^{n+m}, \ \mathbf{A} \in \mathbb{R}^{m \times n}$$

$$\implies \nabla g(\mathbf{y}) = \begin{bmatrix} \nabla_{\mathbf{x}} g(\mathbf{y}) \\ \nabla_{\mathbf{z}} g(\mathbf{y}) \end{bmatrix} = \begin{bmatrix} 2\mathbf{A}^T \mathbf{A}\mathbf{x} - 2\mathbf{A}^T \mathbf{z} \\ 2\mathbf{z} - 2\mathbf{A}\mathbf{x} \end{bmatrix} \in \mathbb{R}^{n+m} \tag{1.47}$$

$$\implies \nabla^2 g(\mathbf{y}) = \begin{bmatrix} D(\nabla_{\mathbf{x}} g(\mathbf{y})) \\ D(\nabla_{\mathbf{z}} g(\mathbf{y})) \end{bmatrix} = \begin{bmatrix} \nabla_{\mathbf{x}}^2 g(\mathbf{y}) & D_{\mathbf{z}}(\nabla_{\mathbf{x}} g(\mathbf{y})) \\ D_{\mathbf{x}}(\nabla_{\mathbf{z}} g(\mathbf{y})) & \nabla_{\mathbf{z}}^2 g(\mathbf{y}) \end{bmatrix}$$

$$= \begin{bmatrix} 2\mathbf{A}^T \mathbf{A} & -2\mathbf{A}^T \\ -2\mathbf{A} & 2\mathbf{I}_m \end{bmatrix} \in \mathbb{S}^{n+m}. \tag{1.48}$$

What is the gradient of $f(\mathbf{X}) = \log \det(\mathbf{X})$ for $\mathbf{X} \in \mathbb{S}_{++}^n$ (the set of positive definite matrices)? The answer will be given in Chapter 3 (cf. Remark 3.20).

1.1.13   Taylor series

Assume that a function $f : \mathbb{R} \to \mathbb{R}$ is m times continuously differentiable. Then

$$f(x + h) = f(x) + \frac{h}{1!} f^{(1)}(x) + \frac{h^2}{2!} f^{(2)}(x) + \cdots$$

$$+ \frac{h^{m-1}}{(m-1)!} f^{(m-1)}(x) + R_m \tag{1.49}$$

is called the *Taylor series expansion*, where $f^{(i)}$ is the ith derivative of f, and

$$R_m = \frac{h^m}{m!} f^{(m)}(x + \theta h) \tag{1.50}$$

is the residual where $\theta \in [0, 1]$. If $x = 0$, then the series is called *Maclaurin series*.

On the other hand, if a function is defined as $f : \mathbb{R}^n \to \mathbb{R}$ and if f is m times continuously differentiable, then the Taylor series expansion is given by

$$f(\mathbf{x} + \mathbf{h}) = f(\mathbf{x}) + \frac{df(\mathbf{x})}{1!} + \frac{1}{2!} d^2 f(\mathbf{x}) + \cdots$$

$$+ \frac{1}{(m-1)!} d^{(m-1)} f(\mathbf{x}) + R_m, \tag{1.51}$$

where

$$d^r f(\mathbf{x}) = \underbrace{\sum_{i=1}^{n} \sum_{j=1}^{n} \cdots \sum_{k=1}^{n}}_{r \text{ terms}} h_i h_j \cdots h_k \underbrace{\frac{\partial^r f(\mathbf{x})}{\partial x_i \partial x_j \cdots \partial x_k}}_{r \text{ terms}}$$

(h_i and x_i, respectively, denoting the ith element of $\mathbf{h}$ and $\mathbf{x}$) and

$$R_m = \frac{1}{m!} d^m f(\mathbf{x} + \theta \mathbf{h}) \tag{1.52}$$

for some $\theta \in [0, 1]$.

Remark 1.12 The first-order and second-order Taylor series expansions of a function $f : \mathbb{R}^n \to \mathbb{R}$ are given by

$$f(\mathbf{x} + \mathbf{h}) = f(\mathbf{x}) + \nabla f(\mathbf{x} + \theta_1 \mathbf{h})^T \mathbf{h} = f(\mathbf{x}) + Df(\mathbf{x} + \theta_1 \mathbf{h})\mathbf{h} \qquad (1.53)$$

$$= f(\mathbf{x}) + \nabla f(\mathbf{x})^T \mathbf{h} + \frac{1}{2}\mathbf{h}^T \nabla^2 f(\mathbf{x} + \theta_2 \mathbf{h})\mathbf{h} \qquad (1.54)$$

for some $\theta_1, \theta_2 \in [0, 1]$. When $f : \mathbb{R}^{n \times m} \to \mathbb{R}$, let $\mathbf{X} = \{x_{ij}\}_{n \times m} = [\mathbf{x}_1, \ldots, \mathbf{x}_m]$ and $\mathbf{H} = [\mathbf{h}_1, \ldots, \mathbf{h}_m] \in \mathbb{R}^{n \times m}$. Then the first-order and second-order Taylor series expansions of f are given by

$$
\begin{aligned}
f(\mathbf{X} + \mathbf{H}) &= f(\mathbf{X}) + \mathrm{Tr}\big(\nabla f(\mathbf{X} + \theta_1 \mathbf{H})^T \mathbf{H}\big) \\
&= f(\mathbf{X}) + \mathrm{Tr}\big(Df(\mathbf{X} + \theta_1 \mathbf{H})\mathbf{H}\big) \qquad (1.55)
\end{aligned}
$$

$$= f(\mathbf{X}) + \mathrm{Tr}\big(\nabla f(\mathbf{X})^T \mathbf{H}\big) + \sum_{j=1}^{m}\sum_{l=1}^{m} \mathbf{h}_j^T D_{\mathbf{x}_l}\left(\nabla_{\mathbf{x}_j} f(\mathbf{X} + \theta_2 \mathbf{H})\right)\mathbf{h}_l$$

$$
\begin{aligned}
&= f(\mathbf{X}) + \mathrm{Tr}\big(\nabla f(\mathbf{X})^T \mathbf{H}\big) \\
&\quad + \sum_{j=1}^{m}\sum_{l=1}^{m} \mathbf{h}_j^T \left\{\frac{\partial^2 f(\mathbf{X} + \theta_2 \mathbf{H})}{\partial x_{ij}\partial x_{kl}}\right\}_{n \times n} \mathbf{h}_l \qquad (1.56)
\end{aligned}
$$

for some $\theta_1, \theta_2 \in [0, 1]$. Moreover, (1.53) and (1.55) are also the corresponding first-order Taylor series approximations, and (1.54) and (1.56) are the corresponding second-order Taylor series approximations, if θ_1 and θ_2 are set to zero.

For a differentiable function $\boldsymbol{f} : \mathbb{R}^n \to \mathbb{R}^m$, the first-order Taylor series expansion is given by

$$\boldsymbol{f}(\mathbf{x} + \mathbf{h}) = \boldsymbol{f}(\mathbf{x}) + \big(D\boldsymbol{f}(\mathbf{x} + \theta \mathbf{h})\big)\mathbf{h}, \qquad (1.57)$$

for some $\theta \in [0, 1]$, which is also the corresponding first-order Taylor series approximation of $\boldsymbol{f}(\mathbf{x})$ if θ is set to zero. $\qquad\square$

Remark 1.13 The Taylor series of a real-valued function of a complex variable, $f : \mathbb{C}^n \to \mathbb{R}$, is quite complicated in terms of complex variable $\mathbf{x} = \mathbf{u} + j\mathbf{v} \in \mathbb{C}^n$ in general, except for the first-order Taylor series approximation which is given by

$$f(\mathbf{x} + \mathbf{h}) \simeq f(\mathbf{x}) + \mathrm{Re}\left\{\nabla f(\mathbf{x})^H \mathbf{h}\right\}. \qquad (1.58)$$

Similarly, the one for the case of complex matrix variable is given by

$$f(\mathbf{X} + \mathbf{H}) \simeq f(\mathbf{X}) + \mathrm{Tr}\left(\mathrm{Re}\left\{\nabla f(\mathbf{X})^H \mathbf{H}\right\}\right). \qquad (1.59)$$

However, the Hessian matrix of f becomes quite involved in general such that the second-order Taylor series approximation is limited in applications. An alternative is to re-express f as a real-valued function of variable $\boldsymbol{x} = (\mathbf{u}, \mathbf{v}) \in \mathbb{R}^{2n}$ to which the preceding Taylor series can be easily applied. $\qquad\square$

1.2   Linear algebra revisited

1.2.1   Vector subspace

A set of vectors $\{\mathbf{a}_1, \ldots, \mathbf{a}_k\}$ is said to be *linearly independent* if the following equality holds only when $\alpha_1 = \alpha_2 = \cdots = \alpha_k = 0$,

$$\alpha_1 \mathbf{a}_1 + \alpha_2 \mathbf{a}_2 + \cdots + \alpha_k \mathbf{a}_k = \mathbf{0}. \tag{1.60}$$

A set of vectors $\{\mathbf{a}_1, \ldots, \mathbf{a}_k\}$ is said to be *linearly dependent* if any one of the vectors from the set is a linear combination of the remaining vectors or if one of the vectors is a zero vector. The vector set $\{\mathbf{a}_1, \ldots, \mathbf{a}_k\}$ is linearly dependent if it is not linearly independent and vice versa.

A subset V of $\mathbb{R}^n$ is called a *subspace* of $\mathbb{R}^n$ if V is closed under the operations of vector addition and scalar multiplication (i.e., $\alpha \mathbf{v}_1 + \beta \mathbf{v}_2 \in V$ for all $\alpha, \beta \in \mathbb{R}$ and $\mathbf{v}_1, \mathbf{v}_2 \in V$). Note that every subspace must contain the zero vector.

Let $\mathbf{a}_1, \mathbf{a}_2, \ldots, \mathbf{a}_k$ be arbitrary vectors in $\mathbb{R}^n$. The set of *all* their linear combinations is called the span of $\mathbf{a}_1, \mathbf{a}_2, \ldots, \mathbf{a}_k$ and is denoted as

$$\text{span}[\mathbf{a}_1, \ldots, \mathbf{a}_k] = \left\{ \sum_{i=1}^{k} \alpha_i \mathbf{a}_i \mid \alpha_1, \alpha_2, \ldots, \alpha_k \in \mathbb{R} \right\}. \tag{1.61}$$

Note that the span of any set of vectors is a subspace.

Given a subspace V, any set of linearly independent vectors $\{\mathbf{a}_1, \ldots, \mathbf{a}_k\} \subset V$ such that $V = \text{span}[\mathbf{a}_1, \ldots, \mathbf{a}_k]$ is referred to as a *basis* of the subspace V. All bases of a subspace V contain the same number of vectors and this number is called the *dimension* of V and is denoted as $\dim(V)$. Any vector in V can be represented uniquely by a linear combination of the vectors of any basis of V.

1.2.2   Range space, null space, and orthogonal projection

Let $\mathbf{A} = [\mathbf{a}_1, \ldots, \mathbf{a}_n] \in \mathbb{R}^{m \times n}$. The *range space* or *image* (also a subspace) of the matrix $\mathbf{A}$ is defined as

$$\mathcal{R}(\mathbf{A}) = \{ \mathbf{y} \in \mathbb{R}^m \mid \mathbf{y} = \mathbf{A}\mathbf{x}, \mathbf{x} \in \mathbb{R}^n \} = \text{span}[\mathbf{a}_1, \ldots, \mathbf{a}_n], \tag{1.62}$$

and the *rank* of $\mathbf{A}$, denoted as $\text{rank}(\mathbf{A})$, is the maximum number of independent columns (or independent rows) of $\mathbf{A}$. In fact, $\dim(\mathcal{R}(\mathbf{A})) = \text{rank}(\mathbf{A})$. Some facts about matrix rank are as follows:

- If $\mathbf{A} \in \mathbb{R}^{m \times k}$, $\mathbf{B} \in \mathbb{R}^{k \times n}$, then

$$\text{rank}(\mathbf{A}) + \text{rank}(\mathbf{B}) - k \leq \text{rank}(\mathbf{AB}) \leq \min\{\text{rank}(\mathbf{A}), \text{rank}(\mathbf{B})\}. \tag{1.63}$$

- If $\mathbf{A} \in \mathbb{R}^{m \times m}$, $\mathbf{C} \in \mathbb{R}^{n \times n}$ are both nonsingular, and $\mathbf{B} \in \mathbb{R}^{m \times n}$, then

$$\text{rank}(\mathbf{B}) = \text{rank}(\mathbf{AB}) = \text{rank}(\mathbf{BC}) = \text{rank}(\mathbf{ABC}). \tag{1.64}$$

The *null space* or *kernel* (also a subspace) of $\mathbf{A} \in \mathbb{R}^{m \times n}$ is defined as

$$\mathcal{N}(\mathbf{A}) = \left\{ \mathbf{x} \in \mathbb{R}^n \mid \mathbf{A}\mathbf{x} = \mathbf{0}_m \right\}. \tag{1.65}$$

The dimension of $\mathcal{N}(\mathbf{A})$ is called the *nullity* of $\mathbf{A}$.

A linear transformation $\mathbf{P} \in \mathbb{R}^{n \times n}$ is called an *orthogonal projector* (or projection matrix) onto a subspace V if for all $\mathbf{x} \in \mathbb{R}^n$, $\mathbf{v}_1 = \mathbf{P}\mathbf{x} \in V$ (i.e., $V = \mathcal{R}(\mathbf{P})$ is the range space of $\mathbf{P}$) and $(\mathbf{I}_n - \mathbf{P})\mathbf{x} = \mathbf{x} - \mathbf{v}_1 = \mathbf{v}_2 \in V^{\perp}$ where

$$V^{\perp} = \left\{ \mathbf{x} \in \mathbb{R}^n \mid \mathbf{z}^T \mathbf{x} = 0, \ \forall \mathbf{z} \in V \right\} \tag{1.66}$$

represents the subspace orthogonal to the subspace V. The sum of the projections $\mathbf{v}_1$ and $\mathbf{v}_2$ uniquely represents the vector $\mathbf{x}$. In other words, the *sum* of the set V and the set $V^{\perp}$ is given by

$$V + V^{\perp} = \{\mathbf{x} + \mathbf{y} \mid \mathbf{x} \in V, \ \mathbf{y} \in V^{\perp}\} = \mathbb{R}^n. \tag{1.67}$$

Remark 1.14 For a given matrix $\mathbf{A} \in \mathbb{R}^{m \times n}$,

$$\mathcal{R}(\mathbf{A})^{\perp} = \mathcal{N}(\mathbf{A}^T) \tag{1.68}$$

$$\mathcal{N}(\mathbf{A})^{\perp} = \mathcal{R}(\mathbf{A}^T) \tag{1.69}$$

$$\dim(\mathcal{R}(\mathbf{A})) = \dim(\mathcal{R}(\mathbf{A}^T)) = \operatorname{rank}(\mathbf{A}) \tag{1.70}$$

$$\dim(\mathcal{N}(\mathbf{A})) = n - \operatorname{rank}(\mathbf{A}) \tag{1.71}$$

$$\dim(\mathcal{N}(\mathbf{A}^T)) = m - \operatorname{rank}(\mathbf{A}) \tag{1.72}$$

are useful properties with regard to the rank of $\mathbf{A}$. □

Remark 1.15 A matrix $\mathbf{P}$ is said to be an orthogonal projector onto the subspace $V = \mathcal{R}(\mathbf{P})$ if and only if $\mathbf{P}^2 = \mathbf{P} = \mathbf{P}^T$. For the case that $\mathbf{A} \in \mathbb{R}^{m \times n}$, the projection matrix for $V = \mathcal{R}(\mathbf{A})$ is given by

$$\mathbf{P_A} = \mathbf{A}\mathbf{A}^{\dagger} = \begin{cases} \mathbf{A}(\mathbf{A}^T \mathbf{A})^{-1}\mathbf{A}^T, & \text{if } \operatorname{rank}(\mathbf{A}) = n \ \ (\text{cf. } (1.125)) \\ \mathbf{I}_m, & \text{if } \operatorname{rank}(\mathbf{A}) = m \ \ (\text{cf. } (1.126)) \end{cases} \tag{1.73}$$

where $\mathbf{A}^{\dagger}$ denotes the pseudo-inverse of $\mathbf{A}$ (cf. (1.114)). Note that $\mathbf{P_A}\boldsymbol{x} \in \mathcal{R}(\mathbf{A})$ for all $\boldsymbol{x} \in \mathbb{R}^m$ (implying $\|\mathbf{P_A}\boldsymbol{x}\|_2 \leq \|\boldsymbol{x}\|_2$), and $\mathbf{P_A}\boldsymbol{x} = \boldsymbol{x}$ for all $\boldsymbol{x} \in \mathcal{R}(\mathbf{A})$. The corresponding *orthogonal complement projector* of $\mathbf{A} \in \mathbb{R}^{m \times n}$ is known to be

$$\mathbf{P_A}^{\perp} = \mathbf{I}_m - \mathbf{P_A} \tag{1.74}$$

Note that $\mathcal{R}(\mathbf{P_A}^{\perp}) = \mathcal{R}(\mathbf{A})^{\perp}$. Moreover, when $\mathbf{A}$ is *semi-unitary*, i.e.,

$$\mathbf{A}^T \mathbf{A} = \mathbf{I}_n, \text{ or } \mathbf{A}\mathbf{A}^T = \mathbf{I}_m, \tag{1.75}$$

it can be seen, from (1.73) and (1.74), that $\mathbf{P_A} = \mathbf{A}\mathbf{A}^T$ and $\mathbf{P_A}^{\perp} = \mathbf{I}_m - \mathbf{A}\mathbf{A}^T$ for the former, and $\mathbf{P_A} = \mathbf{I}_m$ and $\mathbf{P_A}^{\perp} = \mathbf{0}_{m \times m}$ for the latter. □

1.2.3   Matrix determinant and inverse

Let $\mathbf{A} = \{a_{i,j}\}_{n \times n} \in \mathbb{R}^{n \times n}$ and $\boldsymbol{\mathcal{A}}_{ij} \in \mathbb{R}^{(n-1) \times (n-1)}$ be the submatrix of $\mathbf{A}$ by deleting the ith row and jth column of $\mathbf{A}$. Then the *determinant* of $\mathbf{A}$ is defined as

$$\det(\mathbf{A}) = \begin{cases} \sum_{j=1}^{n} a_{ij} \cdot (-1)^{i+j} \det(\boldsymbol{\mathcal{A}}_{ij}), & \forall i \in \{1, \dots, n\} \\ \sum_{i=1}^{n} a_{ij} \cdot (-1)^{i+j} \det(\boldsymbol{\mathcal{A}}_{ij}), & \forall j \in \{1, \dots, n\} \end{cases} \tag{1.76}$$

which is called the cofactor expansion since the term in each summation $(-1)^{i+j} \det(\boldsymbol{\mathcal{A}}_{ij})$ is the (i,j)th cofactor of $\mathbf{A}$.

The inverse of $\mathbf{A}$ is defined as

$$\mathbf{A}^{-1} = \frac{1}{\det(\mathbf{A})} \cdot \mathrm{adj}(\mathbf{A}) \tag{1.77}$$

where $\mathrm{adj}(\mathbf{A}) \in \mathbb{R}^{n \times n}$ denotes the adjoint matrix of $\mathbf{A}$ with the (j,i)th element given by

$$\{\mathrm{adj}(\mathbf{A})\}_{ji} = (-1)^{i+j} \det(\boldsymbol{\mathcal{A}}_{ij}). \tag{1.78}$$

A useful matrix inverse identity, called the *Woodbury identity*, is given by

$$(\mathbf{A} + \mathbf{UBV})^{-1} = \mathbf{A}^{-1} - \mathbf{A}^{-1}\mathbf{U}(\mathbf{B}^{-1} + \mathbf{VA}^{-1}\mathbf{U})^{-1}\mathbf{VA}^{-1}. \tag{1.79}$$

Some other useful matrix inverse identities and matrix determinants are given as follows:

$$(\mathbf{AB})^{-1} = \mathbf{B}^{-1}\mathbf{A}^{-1} \tag{1.80}$$
$$(\mathbf{A}^T)^{-1} = (\mathbf{A}^{-1})^T \tag{1.81}$$
$$\det(\mathbf{A}^T) = \det(\mathbf{A}) \tag{1.82}$$
$$\det(\mathbf{A}^{-1}) = 1/\det(\mathbf{A}) \tag{1.83}$$
$$\det(\mathbf{AB}) = \det(\mathbf{A}) \cdot \det(\mathbf{B}) \tag{1.84}$$
$$\det(\mathbf{I}_n + \mathbf{uv}^T) = 1 + \mathbf{u}^T\mathbf{v}, \quad \mathbf{u}, \mathbf{v} \in \mathbb{R}^n. \tag{1.85}$$

Note that $\mathbf{uv}^T$ in (1.85) is a rank-1 asymmetric $n \times n$ matrix with one nonzero eigenvalue equal to $\mathbf{u}^T\mathbf{v}$ and a corresponding eigenvector $\mathbf{u}$ (to be introduced in Subsection 1.2.5).

1.2.4   Positive definiteness and semidefiniteness

An $n \times n$ real symmetric matrix $\mathbf{M}$ is *positive definite* (PD) (i.e., $\mathbf{M} \in \mathbb{S}_{++}^n$) if $\mathbf{z}^T\mathbf{Mz} > 0$ for any nonzero vector $\mathbf{z} \in \mathbb{R}^n$, where $\mathbf{z}^T$ denotes the transpose of $\mathbf{z}$. $\mathbf{M} \succ \mathbf{0}$ is also used to denote that $\mathbf{M}$ is a PD matrix. For complex matrices, this definition becomes: a Hermitian matrix $\mathbf{M} = \mathbf{M}^H = (\mathbf{M}^*)^T \in \mathbb{H}_{++}^n$ is positive definite if $\mathbf{z}^H\mathbf{Mz} > 0$ for any nonzero complex vector $\mathbf{z} \in \mathbb{C}^n$, where $\mathbf{z}^H$ denotes the conjugate transpose of $\mathbf{z}$.

Remark 1.16 An $n \times n$ real symmetric matrix $\mathbf{M}$ is said to be *positive semidefinite* (PSD) (i.e., $\mathbf{M} \in \mathbb{S}_+^n$) and *negative definite* if $\mathbf{z}^T \mathbf{M} \mathbf{z} \geq 0$ and $\mathbf{z}^T \mathbf{M} \mathbf{z} < 0$, respectively, for any nonzero vector $\mathbf{z} \in \mathbb{R}^n$. An $n \times n$ Hermitian PSD matrix can be defined similarly, and $\mathbf{M} \succeq \mathbf{0}$ is also used to denote that $\mathbf{M}$ is a PSD matrix. A real symmetric $n \times n$ matrix $\mathbf{X}$ is called *indefinite* if there exist $\mathbf{z}_1, \mathbf{z}_2 \in \mathbb{R}^n$ such that $\mathbf{z}_1^T \mathbf{X} \mathbf{z}_1 > 0$ and $\mathbf{z}_2^T \mathbf{X} \mathbf{z}_2 < 0$; so is the case of indefinite $n \times n$ Hermitian matrix $\mathbf{X}$ for which $\mathbf{z}_1^H \mathbf{X} \mathbf{z}_1 > 0$ and $\mathbf{z}_2^H \mathbf{X} \mathbf{z}_2 < 0$ where $\mathbf{z}_1, \mathbf{z}_2 \in \mathbb{C}^n$.   □

Remark 1.17 The mathematical definitions of PD and PSD matrices do not require the matrix to be symmetric or Hermitian. However, we only concentrate on the real symmetric matrices or complex Hermitian matrices because this is the case in most practical applications by our experience on one hand, and a lot of available mathematical results on symmetric or Hermitian matrices can be utilized in the development and analysis of convex optimization algorithms on the other hand.   □

1.2.5   Eigenvalue decomposition

Eigenvalue decomposition (EVD) or sometimes spectral decomposition is the factorization of a matrix into a canonical form (standard form), whereby the matrix can be represented in terms of its eigenvalues and eigenvectors.

Let $\mathbf{A} \in \mathbb{S}^n$, i.e., an $n \times n$ symmetric matrix. Then it can be decomposed as

$$\mathbf{A} = \mathbf{Q} \mathbf{\Lambda} \mathbf{Q}^T = \sum_{i=1}^{n} \lambda_i \mathbf{q}_i \mathbf{q}_i^T \Leftrightarrow \mathbf{A} \mathbf{q}_i = \lambda_i \mathbf{q}_i \; \forall i \tag{1.86}$$

where $\mathbf{Q} = [\mathbf{q}_1, \mathbf{q}_2, \ldots, \mathbf{q}_n] \in \mathbb{R}^{n \times n}$ constituted by the n *orthonormal* eigenvectors $\mathbf{q}_i$ of $\mathbf{A}$ (i.e., $\mathbf{q}_i^T \mathbf{q}_j = 0$ for all $i \neq j$ and $\|\mathbf{q}_i\|_2 = 1$) is *orthogonal*, i.e.,

$$\mathbf{Q}\mathbf{Q}^T = \mathbf{Q}^T\mathbf{Q} = \mathbf{I}_n, \tag{1.87}$$

and

$$\mathbf{\Lambda} = \mathbf{Diag}(\lambda_1, \ldots, \lambda_n) \tag{1.88}$$

is a diagonal matrix with the n real eigenvalues $\lambda_1, \ldots, \lambda_n$ of $\mathbf{A}$ as its diagonal entries. Note that all the eigenvalues of a symmetric matrix (or a Hermitian matrix) are the roots of the following nth-order polynomial of λ

$$\det(\lambda \mathbf{I}_n - \mathbf{A}) = (\lambda - \lambda_1) \cdots (\lambda - \lambda_n) = 0 \tag{1.89}$$

and they are all real. If the symmetric matrix is also positive definite (positive semidefinite), then all its eigenvalues are all positive (nonnegative), while indefinite symmetric matrices must have both positive and negative eigenvalues. It can be inferred that for $\mathbf{A} \in \mathbb{S}_+^n$ with eigenvalue-eigenvector pairs

$(\lambda_1, \mathbf{q}_i), \ldots, (\lambda_n, \mathbf{q}_n)$, where $\lambda_i > 0$ for $i \leq r$ and $\lambda_i = 0$ for $i > r$,

$$\begin{cases} \mathcal{R}(\mathbf{A}) = \text{span}[\mathbf{q}_1, \ldots, \mathbf{q}_r], \\ \mathcal{N}(\mathbf{A}) = \text{span}[\mathbf{q}_{r+1}, \ldots, \mathbf{q}_n] = \mathcal{R}(\mathbf{A})^\perp. \end{cases} \tag{1.90}$$

Either of maximum eigenvalue and minimum eigenvalue of $\mathbf{A} \in \mathbb{S}^n$ can also be redefined as an optimization problem

$$\lambda_{\max}(\mathbf{A}) = \sup\left\{\mathbf{q}^T\mathbf{A}\mathbf{q} \mid \|\mathbf{q}\|_2 = 1\right\}$$
$$\geq \inf\left\{\mathbf{q}^T\mathbf{A}\mathbf{q} \mid \|\mathbf{q}\|_2 = 1\right\} = \lambda_{\min}(\mathbf{A}) = -\lambda_{\max}(-\mathbf{A}). \tag{1.91}$$

The *spectral radius* of $\mathbf{A} \in \mathbb{S}^n$ is defined as

$$\rho(\mathbf{A}) = \max\left\{|\lambda_1|, \ldots, |\lambda_n|\right\}, \tag{1.92}$$

and the *condition number* of $\mathbf{A} \in \mathbb{S}^n$ is defined as

$$\kappa(\mathbf{A}) = \frac{|\lambda_{\max}(\mathbf{A})|}{|\lambda_{\min}(\mathbf{A})|}. \tag{1.93}$$

Suppose that $\mathbf{A} \in \mathbb{S}^n$ with the eigenvalue and eigenvector pairs $(\lambda_1, \mathbf{q}_1), \ldots, (\lambda_n, \mathbf{q}_n)$ and λ_i in nonincreasing order. A useful result that will be needed in the ensuing chapters is

$$\sup\left\{\text{Tr}(\mathbf{Q}_\ell^T\mathbf{A}\mathbf{Q}_\ell) \mid \mathbf{Q}_\ell \in \mathbb{R}^{n\times\ell}, \ \mathbf{Q}_\ell^T\mathbf{Q}_\ell = \mathbf{I}_\ell\right\} = \sum_{i=1}^{\ell}\lambda_i, \quad \ell \leq n \tag{1.94}$$

which can be easily proven by EVD of $\mathbf{A}$ given by (1.86), and $\mathbf{Q}_\ell = [\mathbf{q}_1, \ldots, \mathbf{q}_\ell]$ is an optimal solution to problem (1.94).

The following are some important properties associated with EVD of a symmetric matrix and can be proved conveniently from the respective definitions.

Property 1.1 The determinant of $\mathbf{A} \in \mathbb{S}^n$ is the product of all its eigenvalues λ_i, i.e.,

$$\det(\mathbf{A}) = \prod_{i=1}^{n}\lambda_i. \tag{1.95}$$

Property 1.2 The trace of $\mathbf{A} \in \mathbb{S}^n$ is the sum of all its eigenvalues λ_i, i.e.,

$$\text{Tr}(\mathbf{A}) = \sum_{i=1}^{n}\lambda_i. \tag{1.96}$$

Property 1.3 The squared Frobenius norm of $\mathbf{A} \in \mathbb{S}^n$ is the sum of squares of all its eigenvalues λ_i, i.e.,

$$\|\mathbf{A}\|_F^2 = \text{Tr}(\mathbf{A}^T\mathbf{A}) = \sum_{i=1}^{n}\lambda_i^2. \tag{1.97}$$

Suppose that $\mathbf{A} \in \mathbb{S}^n$ and $\lambda(\mathbf{A})$ (or λ_i) denotes an eigenvalue of $\mathbf{A}$. Some useful properties regarding matrix eigenvalues are given below:

$$\text{rank}(\mathbf{A}) = r \Leftrightarrow r \text{ nonzero } \lambda_i \tag{1.98}$$

$$\{\lambda(\mathbf{I}_n + c\mathbf{A})\} = \{1 + c\lambda_i, \ i = 1, \ldots, n\}, \ c \in \mathbb{R} \tag{1.99}$$

$$\{\lambda(\mathbf{A}^{-1})\} = \{\lambda_i^{-1}, \ i = 1, \ldots, n\} \tag{1.100}$$

$$\text{Tr}(\mathbf{A}^p) = \sum_{i=1}^{n} \lambda_i^p, \ p \in \mathbb{Z}_{++}. \tag{1.101}$$

Remark 1.18 Let $(\lambda_i, \mathbf{q}_i)$ be a pair of eigenvalue and eigenvector of $\mathbf{A} \in \mathbb{S}^n$. Then $(1 + c\lambda_i, \mathbf{q}_i)$, $(\lambda_i^{-1}, \mathbf{q}_i)$, and $(\lambda_i^p, \mathbf{q}_i)$ are the corresponding ones of $\mathbf{I}_n + c\mathbf{A}$, $\mathbf{A}^{-1}$, and $\mathbf{A}^p$, respectively. □

Remark 1.19 [BV04] Suppose that $\mathbf{A}, \mathbf{B} \in \mathbb{S}^n$. Then for every $\mathbf{X} \in \mathbb{S}_+^n$, there exists an $\mathbf{x} \in \mathbb{R}^n$ such that $\text{Tr}(\mathbf{AX}) = \mathbf{x}^T \mathbf{A} \mathbf{x}$ and $\text{Tr}(\mathbf{BX}) = \mathbf{x}^T \mathbf{B} \mathbf{x}$. □

Remark 1.20 Suppose that $\mathbf{A} = \mathbf{A}_R + j\mathbf{A}_I \in \mathbb{H}^n$, where $\mathbf{A}_R = \text{Re}\{\mathbf{A}\}$ and $\mathbf{A}_I = \text{Im}\{\mathbf{A}\}$. Then $\mathbf{A}_R = \mathbf{A}_R^T \in \mathbb{S}^n$ and $\mathbf{A}_I = -\mathbf{A}_I^T \in \mathbb{R}^{n \times n}$. Let $\mathbf{q} = \mathbf{q}_R + j\mathbf{q}_I \in \mathbb{C}^n$ be an eigenvector of $\mathbf{A}$ with the associated eigenvalue $\lambda \in \mathbb{R}$, where $\mathbf{q}_R, \mathbf{q}_I \in \mathbb{R}^n$. Let

$$\mathcal{A} = \begin{bmatrix} \mathbf{A}_R & -\mathbf{A}_I \\ \mathbf{A}_I & \mathbf{A}_R \end{bmatrix} = \begin{bmatrix} \mathbf{A}_R & \mathbf{A}_I^T \\ \mathbf{A}_I & \mathbf{A}_R \end{bmatrix} \in \mathbb{S}^{2n} \tag{1.102}$$

$$\boldsymbol{q}_1 = \begin{bmatrix} \mathbf{q}_R \\ \mathbf{q}_I \end{bmatrix} \in \mathbb{R}^{2n}, \quad \boldsymbol{q}_2 = \begin{bmatrix} -\mathbf{q}_I \\ \mathbf{q}_R \end{bmatrix} \in \mathbb{R}^{2n}. \tag{1.103}$$

Then $\lambda \in \mathbb{R}$ is an eigenvalue of $\mathcal{A}$ with multiplicity of two, and $\boldsymbol{q}_1$ and $\boldsymbol{q}_2$ are the associated eigenvectors. Furthermore, it can be shown that if and only if $\mathbf{A}, \mathbf{B} \in \mathbb{H}^n$ ($\mathbb{H}_+^n$) and the corresponding $\mathcal{A}$ and $\mathcal{B}$ are defined as above (cf. (1.102)), then $\mathcal{A}, \mathcal{B} \in \mathbb{S}^{2n}$ ($\mathbb{S}_+^{2n}$) and $\text{Tr}(\mathcal{A}\mathcal{B}) = 2\text{Tr}(\mathbf{AB}) \in \mathbb{R}$ ($\mathbb{R}_+$). □

1.2.6   Square root factorization of PSD matrices

A PSD square matrix $\mathbf{A} \in \mathbb{S}_+^n$, i.e., $\mathbf{A} \succeq \mathbf{0}$ with all its eigenvalues $\lambda_i \geq 0$, $i = 1, \ldots, n$, can be decomposed into the form of

$$\mathbf{A} = \mathbf{B}^T \mathbf{B} \tag{1.104}$$

where $\mathbf{B} \in \mathbb{R}^{n \times n}$ is not unique. Let $\mathbf{U} \in \mathbb{R}^{n \times n}$ be an orthogonal matrix (i.e., $\mathbf{U}^T \mathbf{U} = \mathbf{U} \mathbf{U}^T = \mathbf{I}_n$). An asymmetric choice (1.105) and a symmetric choice (1.106) for $\mathbf{B}$ via the use of EVD of $\mathbf{A}$ are given as follows.

$$\mathbf{A} = \mathbf{Q} \boldsymbol{\Lambda} \mathbf{Q}^T = \mathbf{Q} \boldsymbol{\Lambda}^{1/2} \mathbf{U}^T \mathbf{U} \boldsymbol{\Lambda}^{1/2} \mathbf{Q}^T, \quad (\boldsymbol{\Lambda}^{1/2} = \mathbf{Diag}(\lambda_1^{1/2}, \ldots, \lambda_n^{1/2}))$$

$$\Rightarrow \mathbf{B}_1 = \boldsymbol{\Lambda}^{1/2} \mathbf{Q}^T \neq \mathbf{B}_1^T = \mathbf{Q} \boldsymbol{\Lambda}^{1/2} \Rightarrow \mathbf{A} = \mathbf{B}_1^T \mathbf{B}_1 \ (\text{i.e., } \mathbf{U} = \mathbf{I}_n) \tag{1.105}$$

$$\Rightarrow \mathbf{B}_2 = \mathbf{Q} \boldsymbol{\Lambda}^{1/2} \mathbf{Q}^T = \mathbf{B}_2^T \succeq \mathbf{0} \Rightarrow \mathbf{A} = \mathbf{B}_2^T \mathbf{B}_2 \ (\text{i.e., } \mathbf{U} = \mathbf{Q}). \tag{1.106}$$

It is quite often to express the decomposition $\mathbf{A} = (\mathbf{A}^{1/2})^T \mathbf{A}^{1/2}$, and both the asymmetric $\mathbf{B}_1$ and the PSD $\mathbf{B}_2$ above can serve as $\mathbf{A}^{1/2}$.

1.2.7   Singular value decomposition

The *singular value decomposition* (SVD) is an important factorization of a rectangular real or complex matrix, with many applications in signal processing and communications. Applications which employ the SVD include computation of pseudo-inverse, least-squares fitting of data, matrix approximation, and determination of rank, range space, and null space of a matrix and so on and so forth.

Let $\mathbf{A} \in \mathbb{R}^{m \times n}$ with rank$(\mathbf{A}) = r$. The SVD of $\mathbf{A}$ is expressed in the form

$$\mathbf{A} = \mathbf{U}\boldsymbol{\Sigma}\mathbf{V}^T. \tag{1.107}$$

In the full SVD (1.107) for $\mathbf{A}$, $\mathbf{U} = [\mathbf{U}_r, \mathbf{U}'] \in \mathbb{R}^{m \times m}$ and $\mathbf{V} = [\mathbf{V}_r, \mathbf{V}'] \in \mathbb{R}^{n \times n}$ are orthogonal matrices in which $\mathbf{U}_r = [\mathbf{u}_1, \ldots, \mathbf{u}_r]$ (consisting of r left singular vectors) is an $m \times r$ semi-unitary matrix (cf. (1.75)) due to

$$\mathbf{U}_r^T \mathbf{U}_r = \mathbf{I}_r \tag{1.108}$$

and $\mathbf{V}_r = [\mathbf{v}_1, \ldots, \mathbf{v}_r]$ (consisting of r right singular vectors) is an $n \times r$ semi-unitary matrix (i.e., $\mathbf{V}_r^T \mathbf{V}_r = \mathbf{I}_r$), and

$$\boldsymbol{\Sigma} = \left[\begin{array}{c|c} \mathbf{Diag}(\sigma_1, \ldots, \sigma_r) & \mathbf{0}_{r \times (n-r)} \\ \hline \mathbf{0}_{(m-r) \times r} & \mathbf{0}_{(m-r) \times (n-r)} \end{array} \right] \in \mathbb{R}^{m \times n} \tag{1.109}$$

is a rectangular matrix with r positive *singular values* (supposedly arranged in nonincreasing order), denoted as σ_i, as the first r diagonal elements and zeros elsewhere. Moreover, the range space of $\mathbf{A}$ and that of $\mathbf{U}_r$ are identical, i.e.,

$$\mathcal{R}(\mathbf{A}) = \mathcal{R}(\mathbf{U}_r). \tag{1.110}$$

The *thin SVD* of an $m \times n$ matrix $\mathbf{A}$ with rank r is given by

$$\mathbf{A} = \mathbf{U}_r \boldsymbol{\Sigma}_r \mathbf{V}_r^T = \sum_{i=1}^{r} \sigma_i \mathbf{u}_i \mathbf{v}_i^T \tag{1.111}$$

(i.e., sum of r rank-1 matrices $\mathbf{u}_i \mathbf{v}_i^T$ weighted by the associated singular value σ_i) where

$$\boldsymbol{\Sigma}_r = \mathbf{Diag}(\sigma_1, \ldots, \sigma_r) \tag{1.112}$$

is a diagonal matrix whose diagonal terms σ_i are the r positive singular values of $\mathbf{A}$. It is noticeable that

$$\begin{aligned} \sigma_i &= \mathbf{e}_i^T \boldsymbol{\Sigma}_r \mathbf{e}_i = \mathbf{e}_i^T \mathbf{U}_r^T \mathbf{A} \mathbf{V}_r \mathbf{e}_i \quad \text{(by (1.111))} \\ &= (\mathbf{U}_r \mathbf{e}_i)^T \mathbf{A} (\mathbf{V}_r \mathbf{e}_i) = \mathbf{u}_i^T \mathbf{A} \mathbf{v}_i. \end{aligned} \tag{1.113}$$

The thin SVD above is computationally more economical than the full SVD. For instance, the *pseudo-inverse* of a matrix $\mathbf{A} \in \mathbb{R}^{m \times n}$ with rank r can be calculated as

$$\mathbf{A}^\dagger = \mathbf{V}_r \mathbf{\Sigma}_r^{-1} \mathbf{U}_r^T \in \mathbb{R}^{n \times m}, \tag{1.114}$$

and so

$$\begin{cases} \mathbf{A}\mathbf{A}^\dagger = \mathbf{I}_m, \text{ if } \operatorname{rank}(\mathbf{A}) = m \\ \mathbf{A}^\dagger\mathbf{A} = \mathbf{I}_n, \text{ if } \operatorname{rank}(\mathbf{A}) = n. \end{cases} \tag{1.115}$$

It can be easily seen that singular values of $\mathbf{A}$ are related to eigenvalues of $\mathbf{A}^T\mathbf{A}$ or $\mathbf{A}\mathbf{A}^T$ by

$$\sigma_i(\mathbf{A}) = \sqrt{\lambda_i(\mathbf{A}^T\mathbf{A})} = \sqrt{\lambda_i(\mathbf{A}\mathbf{A}^T)} \tag{1.116}$$

provided that both the singular values σ_i and the eigenvalues λ_i are in nonincreasing order. Meanwhile, (1.116) also implies that for $\sigma_i(\mathbf{A}) > 0$, the ith right singular vector of $\mathbf{A}$ and the ith eigenvector of $\mathbf{A}^T\mathbf{A}$ are also identical, and so are the ith left singular vector of $\mathbf{A}$ and the ith eigenvector of $\mathbf{A}\mathbf{A}^T$. It can be seen from (1.116) and (1.96) that

$$\|\mathbf{A}\|_F^2 = \operatorname{Tr}(\mathbf{A}\mathbf{A}^T) = \sum_{i=1}^{\operatorname{rank}(\mathbf{A})} \sigma_i^2(\mathbf{A}). \tag{1.117}$$

Moreover, when $\mathbf{A} \in \mathbb{S}_+^n$, its SVD and EVD are identical and so $\lambda_i(\mathbf{A}) = \sigma_i(\mathbf{A})$.

The SVD has been widely used to solve a set of linear equations characterized by the matrix $\mathbf{A} \in \mathbb{R}^{m \times n}$,

$$\mathbf{A}\mathbf{x} = \mathbf{b}. \tag{1.118}$$

The solution of (1.118), denoted as $\widehat{\mathbf{x}}$, exists only when $\mathbf{b} \in \mathcal{R}(\mathbf{A})$ and it is given by

$$\widehat{\mathbf{x}} = \mathbf{A}^\dagger\mathbf{b} + \mathbf{v}, \quad \mathbf{v} \in \mathcal{N}(\mathbf{A}) \tag{1.119}$$

which is unique only when $\mathbf{A}$ is of full column rank. However, if $\mathbf{b} \notin \mathcal{R}(\mathbf{A})$, the solution of (1.118) does not exist, since

$$\mathbf{A}\widehat{\mathbf{x}} = \mathbf{A}\mathbf{A}^\dagger\mathbf{b} = \mathbf{P_A}\mathbf{b} \neq \mathbf{b} \text{ for } \mathbf{b} \notin \mathcal{R}(\mathbf{A}) \text{ (cf. (1.73)).} \tag{1.120}$$

Suppose that $\mathbf{A} \in \mathbb{R}^{m \times n}$ with $\operatorname{rank}(\mathbf{A}) = r$ and the thin SVD of $\mathbf{A}$ is given by (1.111). Let $\mathbf{X}_\ell \in \mathbb{R}^{m \times n}$ with $\operatorname{rank}(\mathbf{X}) \leq \ell$ denote the optimal low rank approximation to the matrix $\mathbf{A} \in \mathbb{R}^{m \times n}$ by minimizing $\|\mathbf{X} - \mathbf{A}\|_F^2$, which has been widely known as

$$\mathbf{X}_\ell = \arg\min_{\operatorname{rank}(\mathbf{X}) \leq \ell} \|\mathbf{X} - \mathbf{A}\|_F^2 = \sum_{i=1}^{\ell} \sigma_i \mathbf{u}_i \mathbf{v}_i^T \tag{1.121}$$

(which will also be proven via the use of a convex optimization condition and EVD in Chapter 4) and the associated approximation error is given by

$$\rho_\ell = \left\| \mathbf{X}_\ell - \mathbf{A} \right\|_{\mathrm{F}}^2 = \sum_{i=\ell+1}^{r} \sigma_i^{\,2}, \tag{1.122}$$

which will be zero as $\ell \geq r$. This is also an example illustrating LS approximation via SVD, which has been popularly used in various applications in science and engineering areas. Some more introduction to LS approximation is given in the next subsection.

1.2.8   Least-squares approximation

The method of least squares is extensively used to approximately solve for the unknown variables of a linear system with the given set of noisy measurements. Least squares can be interpreted as a method of data fitting. The best fit, between modeled and observed data, in the LS sense is that the sum of squared residuals reaches the least value, where a residual is the difference between an observed value and the value computed from the model.

Consider a system characterized by a set of linear equations,

$$\mathbf{b} = \mathbf{A}\mathbf{x} + \boldsymbol{\epsilon}, \tag{1.123}$$

where $\mathbf{A} \in \mathbb{R}^{m \times n}$ is the given system matrix, $\mathbf{b}$ is the given data vector, and $\boldsymbol{\epsilon} \in \mathbb{R}^m$ is the measurement noise vector. The LS problem is to find an optimal $\mathbf{x} \in \mathbb{R}^n$ by minimizing $\|\mathbf{A}\mathbf{x} - \mathbf{b}\|_2^2$. The LS solution, denoted as $\mathbf{x}_{\mathrm{LS}}$, that minimizes $\|\mathbf{A}\mathbf{x} - \mathbf{b}\|_2^2$, is known as

$$\mathbf{x}_{\mathrm{LS}} \triangleq \arg \min_{\mathbf{x} \in \mathbb{R}^n} \left\{ \left\| \mathbf{A}\mathbf{x} - \mathbf{b} \right\|_2^2 \right\} = \mathbf{A}^\dagger \mathbf{b} + \mathbf{v}, \ \ \mathbf{v} \in \mathcal{N}(\mathbf{A}) \tag{1.124}$$

which is actually an unconstrained optimization problem and the solution (with the same form as the solution (1.119) to the linear equations (1.118)) may not be unique.

As $m \geq n$, the system (1.123) is an over-determined system (i.e., more equations than unknowns), otherwise an under-determined system (i.e., more unknowns than equations). Suppose that $\mathbf{A}$ is of full column rank for the over-determined case $(m \geq n)$, and thus

$$\mathbf{A}^\dagger = (\mathbf{A}^T \mathbf{A})^{-1} \mathbf{A}^T, \ \ \mathbf{A}^\dagger \mathbf{A} = \mathbf{I}_n. \tag{1.125}$$

Then the optimal $\mathbf{x}_{\mathrm{LS}} = \mathbf{A}^\dagger \mathbf{b}$ is unique with the approximation error

$$\left\| \mathbf{A}\mathbf{x}_{\mathrm{LS}} - \mathbf{b} \right\|_2^2 = \left\| \mathbf{P}_{\mathbf{A}}^\perp \mathbf{b} \right\|_2^2 > 0, \ \text{if } \mathbf{b} \notin \mathcal{R}(\mathbf{A}) \ \ (\text{by (1.125) and (1.74)}).$$

In other words, $\mathbf{A}\mathbf{x}_{\mathrm{LS}}$ is the image vector of $\mathbf{b}$ projected on the range space $\mathcal{R}(\mathbf{A})$, and so the approximation error is equal to zero only if $\mathbf{b} \in \mathcal{R}(\mathbf{A})$.

For the under-determined case $(m \leq n)$, suppose that $\mathbf{A}$ is of full row rank, and thus

$$\mathbf{A}^{\dagger} = \mathbf{A}^T \left(\mathbf{A}\mathbf{A}^T\right)^{-1}, \quad \mathbf{A}\mathbf{A}^{\dagger} = \mathbf{I}_m. \tag{1.126}$$

Then the optimal $\mathbf{x}_{\mathrm{LS}}$ given by (1.124) is not unique, but $\mathbf{x}_{\mathrm{LS}} = \mathbf{A}^{\dagger}\mathbf{b}$ is also the minimum-norm solution due to $(\mathbf{A}^{\dagger}\mathbf{b})^T \mathbf{v} = 0$ and $\mathbf{A}\mathbf{x}_{\mathrm{LS}} = \mathbf{b}$. The minimum-norm solution $\mathbf{x}_{\mathrm{LS}} = \mathbf{A}^{\dagger}\mathbf{b}$ is also exactly the optimal solution (9.123) (which can be easily obtained by the associated KKT conditions) to a convex optimization problem defined in (9.8). However, if the system model is not linear, closed-form LS solutions usually do not exist.

1.3   Summary and discussion

In this chapter, we have revisited some mathematical basics of sets, functions, matrices, and vector spaces that will be very useful to understand the remaining chapters and we also introduced the notations that will be used throughout this book. The mathematical preliminaries reviewed in this chapter are by no means complete. For further details, the readers can refer to [Apo07] and [WZ97] for Section 1.1, and [HJ85] and [MS00] for Section 1.2, and other related textbooks.

Suppose that we are given an optimization problem in the following form:

$$\begin{aligned} \text{minimize} \quad & f(\boldsymbol{x}) \\ \text{subject to} \quad & \boldsymbol{x} \in \mathcal{C} \end{aligned} \tag{1.127}$$

where $f(\boldsymbol{x})$ is the objective function to be minimized and $\mathcal{C}$ is the feasible set from which we try to find an optimal solution. Convex optimization itself is a powerful mathematical tool for optimally solving a well-defined convex optimization problem (i.e., $f(\boldsymbol{x})$ is a convex function and $\mathcal{C}$ is a convex set in problem (1.127)), or for handling a nonconvex optimization problem (that can be approximated as a convex one). However, the problem (1.127) under investigation may often appear to be a nonconvex optimization problem (with various camouflages) or a nonconvex and nondeterministic polynomial-time hard (NP-hard) problem that forces us to find an approximate solution with some performance or computational efficiency merits and characteristics instead. Furthermore, reformulation of the considered optimization problem into a convex optimization problem can be quite challenging. Fortunately, there are many problem reformulation approaches (e.g., function transformation, change of variables, and equivalent representations) to conversion of a nonconvex problem into a convex problem (i.e., unveiling of all the camouflages of the original problem).

The bridge between the pure mathematical convex optimization theory and how to use it in practical applications is the key for a successful researcher or professional who can efficiently exert his (her) efforts on solving a challenging scientific and engineering problem to which he (she) is dedicated. For a given opti-

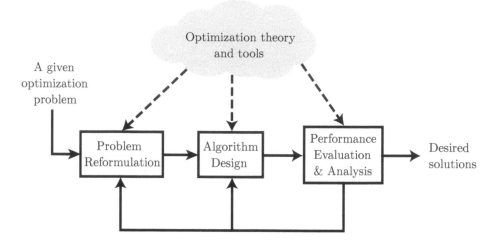

Figure 1.6 Optimization theory and tools aided algorithm design for a given optimization problem.

mization problem, we aim to design an algorithm (e.g., transmit beamforming algorithm and resource allocation algorithm in communications and networking, nonnegative blind source separation algorithm for the analysis of biomedical and hyperspectral images) to efficiently and reliably yield a desired solution (that may just be an approximate solution rather than an optimal solution), as shown in Figure 1.6, where the block "Problem Reformulation," the block "Algorithm Design," and the block "Performance Evaluation and Analysis" are essential design steps before an algorithm that meets our goal is obtained. These design steps rely on *smart use of advisable optimization theory and tools* that remain in the cloud, like a military commander who needs not only ammunition and weapons but also an intelligent fighting strategy. It is quite helpful to build a bridge so that one can readily use any suitable mathematical theory (e.g., convex sets and functions, optimality conditions, duality, KKT conditions, Schur complement, S-procedure, etc.) and convex solvers (e.g., CVX and SeDuMi) to accomplish these design steps.

The ensuing chapters will introduce fundamental elements of the convex optimization theory in the cloud on one hand and illustrate how these elements were collectively applied in some successful cutting edge researches in communications and signal processing through the design procedure shown in Figure 1.6 on the other hand, provided that the solid bridges between the cloud and all the design blocks have been constructed.

References for Chapter 1

[Apo07]   T. M. Apostol, *Mathematical Analysis*, 2nd ed. Pearson Edu. Taiwan Ltd., 2007.

[Ber09]   D. P. Bertsekas, *Convex Optimization Theory*. Belmont, MA, USA: Athena Scientific, 2009.

[Bra83]   D. Brandwood, "A complex gradient operator and its application in adaptive array theory," *IEE Proc. F, Commun., Radar Signal Process.*, vol. 130, no. 1, pp. 11–16, Feb. 1983.

[BV04]   S. Boyd and L. Vandenberghe, *Convex Optimization*. Cambridge, UK: Cambridge University Press, 2004.

[Hay96]   M. H. Hayes, *Statistical Digital Signal Processing and Modeling*. New York, USA: John Wiley & Sons, Inc., 1996.

[HJ85]   R. A. Horn and C. R. Johnson, *Matrix Analysis*. New York, USA: Cambridge University Press, 1985.

[MS00]   T. K. Moon and W. C. Stirling, *Mathematical Methods and Application for Signal Processing*. Upper Saddle River, NJ, USA: Prentice Hall, 2000.

[Sch10]   P. J. Schreier, *Statistical Signal Processing of Complex-Valued Data*. Cambridge, UK: Cambridge University Press, 2010.

[WZ97]   R. L. Wheeden and A. Zygmund, *Measure and Integral: An Introduction to Real Analysis*. New York, USA: Marcel Dekker, 1997.

2 Convex Sets

In this chapter we introduce convex sets and their representations, properties, illustrative examples, convexity preserving operations, and geometry of convex sets which have proven very useful in signal processing applications such as hyperspectral and biomedical image analysis. Then we introduce proper cones (convex cones), dual norms and dual cones, generalized inequalities, and separating and supporting hyperplanes. All the materials on convex sets introduced in this chapter are essential to convex functions, convex problems, and duality to be introduced in the ensuing chapters. From this chapter on, for simplicity, we may use $\mathbf{x}$ to denote a vector in $\mathbb{R}^n$ and $x_1, \ldots, x_n$ for its components without explicitly mentioning $\mathbf{x} \in \mathbb{R}^n$.

2.1 Affine and convex sets

2.1.1 Lines and line segments

Mathematically, a line $\mathcal{L}(\mathbf{x}_1, \mathbf{x}_2)$ passing through two points $\mathbf{x}_1$ and $\mathbf{x}_2$ in $\mathbb{R}^n$ is the set defined as

$$\mathcal{L}(\mathbf{x}_1, \mathbf{x}_2) = \{\theta\mathbf{x}_1 + (1 - \theta)\mathbf{x}_2, \ \theta \in \mathbb{R}\}, \ \mathbf{x}_1, \mathbf{x}_2 \in \mathbb{R}^n. \tag{2.1}$$

If $0 \le \theta \le 1$, then it is a line segment connecting $\mathbf{x}_1$ and $\mathbf{x}_2$. Note that the linear combination $\theta\mathbf{x}_1 + (1 - \theta)\mathbf{x}_2$ of two points $\mathbf{x}_1$ and $\mathbf{x}_2$ with the coefficient sum equal to unity as in (2.1) plays an essential role in defining affine sets and convex sets, and hence the one with $\theta \in \mathbb{R}$ is referred to as the *affine combination* and the one with $\theta \in [0, 1]$ is referred to as the *convex combination*. Affine combination and convex combination can be extended to the case of more than two points in the same fashion.

2.1.2 Affine sets and affine hulls

A set C is said to be an *affine set* if for any $\mathbf{x}_1, \mathbf{x}_2 \in C$ and for any $\theta_1, \theta_2 \in \mathbb{R}$ such that $\theta_1 + \theta_2 = 1$, the point $\theta_1\mathbf{x}_1 + \theta_2\mathbf{x}_2$ also belongs to the set C. For instance, the line defined in (2.1) is an affine set. This concept can be extended to more than two points, as illustrated in the following example.

Example 2.1 If a set C is affine with $\mathbf{x}_1, \mathbf{x}_2, \ldots, \mathbf{x}_k \in C$, then $\sum_{i=1}^{k} \theta_i \mathbf{x}_i \in C$ for every $\boldsymbol{\theta} = [\theta_1, \ldots, \theta_k]^T \in \mathbb{R}^k$ satisfying $\sum_{i=1}^{k} \theta_i = 1$.

Proof: Assume that $\mathbf{x}_1$, $\mathbf{x}_2$, and $\mathbf{x}_3 \in C$. Then, it is true that

$$x_2 = \theta_2 \mathbf{x}_2 + (1 - \theta_2)\mathbf{x}_3 \in C, \ \theta_2 \in \mathbb{R}, \tag{2.2}$$

and $\mathbf{x} = \theta_1 \mathbf{x}_1 + (1 - \theta_1)x_2 \in C, \ \theta_1 \in \mathbb{R}$. Hence

$$\begin{aligned} \mathbf{x} &= \theta_1 \mathbf{x}_1 + (1 - \theta_1)\theta_2 \mathbf{x}_2 + (1 - \theta_1)(1 - \theta_2)\mathbf{x}_3 \in C \\ &= \alpha_1 \mathbf{x}_1 + \alpha_2 \mathbf{x}_2 + \alpha_3 \mathbf{x}_3 \end{aligned} \tag{2.3}$$

where $\alpha_1 = \theta_1$, $\alpha_2 = (1 - \theta_1)\theta_2$, and $\alpha_3 = (1 - \theta_1)(1 - \theta_2) \in \mathbb{R}$, and $\alpha_1 + \alpha_2 + \alpha_3 = 1$. This can be extended to any k points $\in C$ by induction. Thus we have completed the proof.   ∎ □

That is to say that the affine set contains all the affine combinations (linear combinations of points with sum of the *real* coefficients equal to one) of the points in it. It is worthwhile to mention that the affine set need not contain the origin, whereas a subspace must contain the origin. In fact, it is very straightforward to show that the following set is a subspace:

$$V = C - \{\mathbf{x}_0\} = \{\mathbf{x} - \mathbf{x}_0 \mid \mathbf{x} \in C\} \ \ (\text{cf. (1.22) and (1.23)}), \tag{2.4}$$

where C is an affine set and $\mathbf{x}_0 \in C$. Also note that V and C have the same dimension (as illustrated in Figure 2.1). Hence the dimension of the affine set C, denoted as affdim(C), is defined as the dimension of the associated subspace V given by (2.4), i.e.,

$$\text{affdim}(C) \triangleq \dim(V). \tag{2.5}$$

Given a set of vectors $\{\mathbf{s}_1, \ldots, \mathbf{s}_n\} \subset \mathbb{R}^\ell$, the *affine hull* of this set is defined as

$$\mathbf{aff}\{\mathbf{s}_1, \ldots, \mathbf{s}_n\} = \left\{ \mathbf{x} = \sum_{i=1}^{n} \theta_i \mathbf{s}_i \mid (\theta_1, \ldots, \theta_n) \in \mathbb{R}^n, \ \sum_{i=1}^{n} \theta_i = 1 \right\}. \tag{2.6}$$

Two typical affine sets are illustrated in Figure 2.2, where $\mathbf{aff}\{\mathbf{s}_1, \mathbf{s}_2\}$ is a line passing through $\mathbf{s}_1$ and $\mathbf{s}_2$, and $\mathbf{aff}\{\mathbf{s}_1, \mathbf{s}_2, \mathbf{s}_3\}$ is a 2-dimensional hyperplane passing $\mathbf{s}_1$, $\mathbf{s}_2$, and $\mathbf{s}_3$.

The affine hull of $\{\mathbf{s}_1, \ldots, \mathbf{s}_n\} \subset \mathbb{R}^\ell$ is also the smallest affine set containing the vectors $\mathbf{s}_1, \ldots, \mathbf{s}_n$, and can be alternatively expressed as

$$\mathbf{aff}\{\mathbf{s}_1, \ldots, \mathbf{s}_n\} = \{\mathbf{x} = \mathbf{C}\boldsymbol{\alpha} + \mathbf{d} \mid \boldsymbol{\alpha} \in \mathbb{R}^p\} \subseteq \mathbb{R}^\ell \quad (\text{by (2.4)}) \tag{2.7}$$

for an arbitrary $\mathbf{d} \in \mathbf{aff}\{\mathbf{s}_1, \ldots, \mathbf{s}_n\}$ (nonunique) and some full column rank $\mathbf{C} \in \mathbb{R}^{\ell \times p}$ (also nonunique), and for some $p \geq 0$. Note that the representation of the affine hull given by (2.7) is not unique, namely, the affine set parameters $(\mathbf{C}, \mathbf{d})$

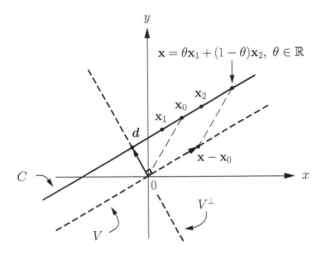

Figure 2.1 An affine set C and the associated subspace V in $\mathbb{R}^2$, where $V = C - \{\mathbf{x}_0\}$ for any $\mathbf{x}_0 \in C$, and $\boldsymbol{d}$ is the orthogonal projection of any $\mathbf{x} \in C$ on $V^\perp$, i.e., $C \cap V^\perp = \{\boldsymbol{d}\}$.

are not unique, and that

$$\mathbf{P}_\mathbf{C}^\perp \mathbf{x} = \mathbf{P}_\mathbf{C}^\perp \mathbf{d} \text{ (a constant vector)}, \quad \forall \mathbf{x} \in \mathbf{aff}\{\mathbf{s}_1, \dots, \mathbf{s}_n\} \qquad (2.8)$$

is invariant for all $(\mathbf{C}, \mathbf{d})$ associated with the same affine hull. The affine set representation for $\{\mathbf{s}_1, \dots, \mathbf{s}_n\}$ given by (2.7) can also be illustrated as in Figure 2.1, with $\mathbf{aff}\{\mathbf{s}_1, \dots, \mathbf{s}_n\}$ corresponding to the affine set C, $\mathcal{R}(\mathbf{C})$ corresponding to the subspace V, and $\mathbf{P}_\mathbf{C}^\perp \mathbf{x}$ corresponding to the vector $\boldsymbol{d}$ (a constant vector). Note that

$$\mathrm{affdim}(\mathbf{aff}\{\mathbf{s}_1, \dots, \mathbf{s}_n\}) = \mathrm{rank}(\mathbf{C}) = p \le \min\{n - 1, \ell\}, \qquad (2.9)$$

which characterizes the effective dimension of the affine hull. Note that as $p = \ell$, $\mathbf{aff}\{\mathbf{s}_1, \dots, \mathbf{s}_n\} = \mathbb{R}^\ell$.

The affine hull given by (2.6) can be rewritten in the same form as (2.7), with

$$\mathbf{d} = \mathbf{s}_n, \quad \mathbf{C} = [\, \mathbf{s}_1 - \mathbf{s}_n, \mathbf{s}_2 - \mathbf{s}_n, \dots, \mathbf{s}_{n-1} - \mathbf{s}_n \,] \in \mathbb{R}^{\ell \times (n-1)},$$
$$p = n - 1, \quad \boldsymbol{\alpha} = (\alpha_1, \dots, \alpha_{n-1}) = (\theta_1, \dots, \theta_{n-1}). \qquad (2.10)$$

However, the matrix $\mathbf{C}$ may not be of full column rank and thus $p = n - 1$ here may not represent the affine dimension of the affine hull, meanwhile implying that the affine dimension p of an affine hull constituted by n different vectors must be less than or equal to $n - 1$. A sufficient condition for the affine hull to have maximum affine dimension is stated in the following property.

Property 2.1 If a finite set $S \triangleq \{\mathbf{s}_1, \dots, \mathbf{s}_n\}$ is *affinely independent*, which means that $S_j \triangleq \{\mathbf{s} - \mathbf{s}_j \mid \mathbf{s} \in S, \mathbf{s} \ne \mathbf{s}_j\}$ is linearly independent for any j, then

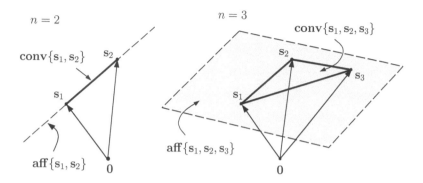

Figure 2.2 Affine hulls and convex hulls in $\mathbb{R}^2$ (left plot) and $\mathbb{R}^3$ (right plot).

its affine dimension (cf. (2.14)) is $n-1$; any subset $\mathcal{S} \subset S$ is also affinely independent with affine dimension $|\mathcal{S}| - 1$.

For any given vectors $\mathbf{x}_1, \mathbf{x}_2 \in \mathbb{R}^n$ and a linearly independent set $\{\mathbf{s}_1, \ldots, \mathbf{s}_n\} \subseteq \mathbb{R}^\ell$, it is widely known that $[\mathbf{s}_1, \ldots, \mathbf{s}_n]\mathbf{x}_1 = [\mathbf{s}_1, \ldots, \mathbf{s}_n]\mathbf{x}_2$ holds true if and only if $\mathbf{x}_1 = \mathbf{x}_2$. A similar property also holds for any affinely independent set as follows.

Property 2.2 Given $\mathbf{x}_1, \mathbf{x}_2 \in \mathbb{R}^n$ that satisfy $\mathbf{1}_n^T\mathbf{x}_1 = \mathbf{1}_n^T\mathbf{x}_2$. If $\{\mathbf{s}_1, \ldots, \mathbf{s}_n\} \subseteq \mathbb{R}^\ell$ is an affinely independent set, then the following implication holds true:

$$[\mathbf{s}_1, \ldots, \mathbf{s}_n]\mathbf{x}_1 = [\mathbf{s}_1, \ldots, \mathbf{s}_n]\mathbf{x}_2 \iff \mathbf{x}_1 = \mathbf{x}_2. \tag{2.11}$$

Proof: As the proof of the sufficiency is trivial, we only need to prove the necessity. As $\mathbf{1}_n^T\mathbf{x}_1 = \mathbf{1}_n^T\mathbf{x}_2$, the condition $[\mathbf{s}_1, \ldots, \mathbf{s}_n]\mathbf{x}_1 = [\mathbf{s}_1, \ldots, \mathbf{s}_n]\mathbf{x}_2$ can be alternatively expressed as

$$[\mathbf{s}_1 - \mathbf{s}_n, \ldots, \mathbf{s}_{n-1} - \mathbf{s}_n][\mathbf{x}_1]_{1:n-1} = [\mathbf{s}_1 - \mathbf{s}_n, \ldots, \mathbf{s}_{n-1} - \mathbf{s}_n][\mathbf{x}_2]_{1:n-1}, \tag{2.12}$$

where

$$[\mathbf{x}]_{i:j} \triangleq [x_i, \ldots, x_j]^T \in \mathbb{R}^{j-i+1}$$

for any given vector $\mathbf{x} = [x_1, \ldots, x_n]^T \in \mathbb{R}^n$. Then, owing to the *affine independence* of the set $\{\mathbf{s}_1, \ldots, \mathbf{s}_n\} \subseteq \mathbb{R}^\ell$ (cf. Property 2.1), the matrix $[\mathbf{s}_1 - \mathbf{s}_n, \ldots, \mathbf{s}_{n-1} - \mathbf{s}_n]$ is of full column rank, which together with (2.12) implies $[\mathbf{x}_1]_{1:n-1} = [\mathbf{x}_2]_{1:n-1}$. Hence, under the premise of $\mathbf{1}_n^T\mathbf{x}_1 = \mathbf{1}_n^T\mathbf{x}_2$, we have $\mathbf{x}_1 = \mathbf{x}_2$. Thus the proof is completed. ∎

The affine hull of an arbitrary set $C \subset \mathbb{R}^n$ (either continuous or discrete), denoted as **aff** C, is defined as the smallest affine set containing C (implying **aff** $C = C$ if C is an affine set), which is exactly the set of all affine combinations of elements of C and can be expressed as

$$\textbf{aff }C = \left\{ \sum_{i=1}^{k} \theta_i \mathbf{x}_i \mid \{\mathbf{x}_i\}_{i=1}^{k} \subset C, \ \{\theta_i\}_{i=1}^{k} \subset \mathbb{R}, \ \sum_{i=1}^{k} \theta_i = 1, k \in \mathbb{Z}_{++} \right\}. \quad (2.13)$$

The affine dimension of C (which may not be an affine set) is defined as the affine dimension of its affine hull, i.e.,

$$\text{affdim } C \triangleq \text{affdim}(\textbf{aff }C) = \dim\big((\textbf{aff }C) - \{\mathbf{x}_0\}\big) \ \text{(by (2.5))}, \quad (2.14)$$

where $\mathbf{x}_0 \in \textbf{aff }C$. For instance, $\textbf{aff }C = \mathbb{R}$ and affdim $C = 1$, for $C = (0,1]$ or $C = \{0,1\}$; $\textbf{aff }C = \{(x,y,z) \in \mathbb{R}^3 \mid x+y+z=1\}$ and affdim $C = 2$ for $C = \{(1,0,0),(0,1,0),(0,0,1)\}$ or $C = \{(x,y,z) \in \mathbb{R}^3 \mid x+y+z = 1, 0 < x < 1, 0 < y < 1, 0 < z < 1\}$; and $\textbf{aff }C = \mathbb{R}^3$ and affdim $C = 3$ for $C = \{(0,0,0),(1,0,0),(0,1,0),(0,0,1)\}$.

Suppose that affdim $C = p$. Then $\textbf{aff }C$ can be alternatively represented by

$$\textbf{aff }C = \left\{ \sum_{i=1}^{p+1} \theta_i \mathbf{x}_i \mid \{\mathbf{x}_i\}_{i=1}^{p+1} \subset C, \ \{\theta_i\}_{i=1}^{p+1} \subset \mathbb{R}, \ \sum_{i=1}^{p+1} \theta_i = 1 \right\}$$
$$= \big\{ \mathbf{x} = \mathbf{C}\boldsymbol{\alpha} + \mathbf{d} \mid \boldsymbol{\alpha} \in \mathbb{R}^p \big\} \quad \text{(by (2.7))} \quad (2.15)$$

for some $(\mathbf{C}, \mathbf{d})$ where $\mathbf{d} \in \textbf{aff }C$, $\text{rank}(\mathbf{C} = [\mathbf{c}_1, \ldots, \mathbf{c}_p]) = p$, and $\mathbf{c}_i + \mathbf{d} \in \textbf{aff }C$ for all i.

The two affine hull representations given by (2.13) and (2.15) can be proven to be the same by Property 2.1 and Property 2.2. According to Property 2.1, there exists an affinely independent set $\{\mathbf{y}_1, \ldots, \mathbf{y}_{p+1}\} \subseteq C$ such that $\textbf{aff }\{\mathbf{y}_1, \ldots, \mathbf{y}_{p+1}\} = \textbf{aff }C$ due to affdim $C = p$. Furthermore, by Property 2.2, every vector $\mathbf{x}$ in the affine hull given by (2.13) can be uniquely represented as an affine combination of $\mathbf{y}_1, \ldots, \mathbf{y}_{p+1}$, implying that $\mathbf{x}$ belongs to the affine hull given by (2.15). On the other hand, it is trivial to see that every component of the latter also belongs to the former. Hence they represent the same affine hull $\textbf{aff }C$.

2.1.3   Relative interior and relative boundary

Affine hull defined in (2.13) and affine dimension of a set defined in (2.14) play an essential role in convex geometric analysis, and have been applied to dimension reduction in many signal processing applications such as blind separation (or unmixing) of biomedical and hyperspectral image signals (to be introduced in Chapter 6). To further illustrate their characteristics, it would be useful to address the interior and the boundary of a set w.r.t. its affine hull, which are, respectively, termed as *relative interior* and *relative boundary*, and are defined below.

The relative interior of $C \subseteq \mathbb{R}^n$ is defined as

$$\text{relint } C = \{\mathbf{x} \in C \mid B(\mathbf{x}, r) \cap \text{aff } C \subseteq C, \text{ for some } r > 0\} \qquad (2.16)$$
$$= \text{int } C \text{ if aff } C = \mathbb{R}^n \quad \text{(cf. (1.20))},$$

where $B(\mathbf{x}, r)$ is a 2-norm ball with center at $\mathbf{x}$ and radius r. It can be inferred from (2.16) that

$$\text{int } C = \begin{cases} \text{relint } C, & \text{if affdim } C = n \\ \emptyset, & \text{otherwise.} \end{cases} \qquad (2.17)$$

The relative boundary of a set C is defined as

$$\text{relbd } C = \text{cl } C \setminus \text{relint } C$$
$$= \text{bd } C, \text{ if int } C \neq \emptyset \quad \text{(by (2.17))} \qquad (2.18)$$

For instance, for $C = \{\mathbf{x} \in \mathbb{R}^n \mid \|\mathbf{x}\|_\infty \leq 1\}$ (an infinity-norm ball), its interior and relative interior are identical, so are its boundary and relative boundary; for $C = \{\mathbf{x}_0\} \subset \mathbb{R}^n$ (a singleton set), $\text{int } C = \emptyset$ and $\text{bd } C = C$, but $\text{relbd } C = \emptyset$. Note that $\text{affdim}(C) = n$ for the former but $\text{affdim}(C) = 0 \neq n$ for the latter, thereby providing the information of differentiating the interior (boundary) and the relative interior (relative boundary) of a set. Some more examples about the relative interior (relative boundary) of C and the interior (boundary) of C, are illustrated in the following examples.

Example 2.2 Let $C = \{\mathbf{x} \in \mathbb{R}^3 \mid x_1^2 + x_3^2 \leq 1, x_2 = 0\} = \text{cl } C$. Then $\text{relint } C = \{\mathbf{x} \in \mathbb{R}^3 \mid x_1^2 + x_3^2 < 1, x_2 = 0\}$ and $\text{relbd } C = \{\mathbf{x} \in \mathbb{R}^3 \mid x_1^2 + x_3^2 = 1, x_2 = 0\}$ as shown in Figure 2.3. Note that $\text{int } C = \emptyset$ since $\text{affdim}(C) = 2 < 3$, while $\text{bd } C = \text{cl } C \setminus \text{int } C = C$. $\qquad \square$

Example 2.3 Let $C_1 = \{\mathbf{x} \in \mathbb{R}^3 \mid \|\mathbf{x}\|_2 \leq 1\}$ and $C_2 = \{\mathbf{x} \in \mathbb{R}^3 \mid \|\mathbf{x}\|_2 = 1\}$. Then $\text{int } C_1 = \{\mathbf{x} \in \mathbb{R}^3 \mid \|\mathbf{x}\|_2 < 1\} = \text{relint } C_1$ and $\text{int } C_2 = \text{relint } C_2 = \emptyset$ due to $\text{affdim}(C_1) = \text{affdim}(C_2) = 3$. $\qquad \square$

From now on, for the conceptual conciseness and clarity in the following introduction to convex sets, sometimes we address the pair $(\text{int } C, \text{bd } C)$ in the context without explicitly mentioning that a convex set C has nonempty interior. However, when $\text{int } C = \emptyset$ in the context, one can interpret the pair $(\text{int } C, \text{bd } C)$ as the pair $(\text{relint } C, \text{relbd } C)$.

2.1.4   Convex sets and convex hulls

A set C is said to be a *convex set* if for any $\mathbf{x}_1, \mathbf{x}_2 \in C$ and for any $\theta_1, \theta_2 \in \mathbb{R}_+$ such that $\theta_1 + \theta_2 = 1$, the point $\theta_1 \mathbf{x}_1 + \theta_2 \mathbf{x}_2$ also belongs to the set C; the set C is called *strictly convex* if $\theta \mathbf{x}_1 + (1 - \theta) \mathbf{x}_2 \in \text{int } C$ for all $\mathbf{x}_1 \neq \mathbf{x}_2 \in C$ and

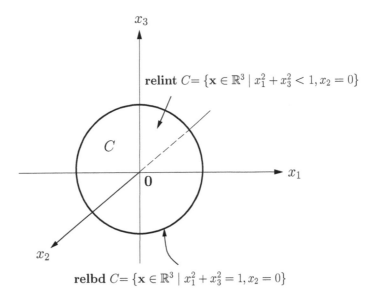

relint $C = \{\mathbf{x} \in \mathbb{R}^3 \mid x_1^2 + x_3^2 < 1, x_2 = 0\}$

relbd $C = \{\mathbf{x} \in \mathbb{R}^3 \mid x_1^2 + x_3^2 = 1, x_2 = 0\}$

Figure 2.3 The relative boundary and relative interior of the set $C = \{\mathbf{x} \in \mathbb{R}^3 \mid x_1^2 + x_3^2 \le 1, x_2 = 0\}$. Note that **int** $C = \emptyset$ and **bd** $C = \textbf{cl}\ C = C$.

$0 < \theta < 1$, implying that

$$\textbf{int}\ C = C_{+\text{int}}$$
$$\triangleq \{\theta\mathbf{x}_1 + (1-\theta)\mathbf{x}_2 \mid \mathbf{x}_1 \ne \mathbf{x}_2, \ \mathbf{x}_1, \mathbf{x}_2 \in C, \ 0 < \theta < 1\} \qquad (2.19)$$

for a strictly convex set C.

Simply we can say that a set is convex if every point in the set can be seen by every other point in the set, along an unobstructed straight path between them, where "unobstructed" means lying in the set. A convex compact set $C \subseteq \mathbb{R}^n$ with nonempty interior is strictly convex if its boundary does not contain any line segments. For instance, a line segment in $\mathbb{R}$ is strictly convex; 2-norm ball is strictly convex in $\mathbb{R}^n$, while 1-norm ball and ∞-norm ball in $\mathbb{R}^n$ for $n = 1$ are strictly convex (a line segment), and convex but not strictly convex for $n \ge 2$; a hemisphere in $\mathbb{R}^3$ is convex but not strictly convex.

Every affine set can be easily shown to be convex. Further, there is a very useful property as stated below.

Property 2.3 Let

$$\mathcal{L}(\mathbf{x}_0, \mathbf{v}) = \{\mathbf{x}_0 + \alpha\mathbf{v} \mid \alpha \in \mathbb{R}\}. \qquad (2.20)$$

If $C \cap \mathcal{L}(\mathbf{x}_0, \mathbf{v})$ is convex, for all $\mathbf{x}_0$, $\mathbf{v} \neq \mathbf{0}$, then C is convex. Conversely, if C is convex, then the set

$$G = \{\alpha \in \mathbb{R} \mid \mathbf{x}_0 + \alpha\mathbf{v} \in C\} \tag{2.21}$$

is convex for all $\mathbf{x}_0, \mathbf{v} \neq \mathbf{0}$, i.e., $C \cap \mathcal{L}(\mathbf{x}_0, \mathbf{v})$ is convex for all $\mathbf{x}_0$ and $\mathbf{v} \neq \mathbf{0}$. In other words, a set is convex if and only if the intersection of this set and any line (or line segment or ray) crossing it is convex.

Proof: We prove sufficiency followed by necessity.

- Sufficiency: Suppose C is nonconvex. Then there exist $\mathbf{x}_1, \mathbf{x}_2 \in C, \theta \in (0,1)$ such that $\mathbf{y} = \theta\mathbf{x}_1 + (1-\theta)\mathbf{x}_2 = \mathbf{x}_2 + \theta(\mathbf{x}_1 - \mathbf{x}_2) \notin C$. Let $\mathbf{x}_0 = \mathbf{x}_2$ and $\mathbf{v} = \mathbf{x}_1 - \mathbf{x}_2$, implying that $\mathbf{x}_1, \mathbf{x}_2 \in \mathcal{L}(\mathbf{x}_0, \mathbf{v})$ and therefore $\mathbf{x}_1, \mathbf{x}_2 \in C \cap \mathcal{L}(\mathbf{x}_0, \mathbf{v})$. Since $\mathbf{x}_1, \mathbf{x}_2 \in C \cap \mathcal{L}(\mathbf{x}_0, \mathbf{v})$ and $C \cap \mathcal{L}(\mathbf{x}_0, \mathbf{v})$ is convex, $\mathbf{y} = \theta\mathbf{x}_1 + (1-\theta)\mathbf{x}_2 \in C \cap \mathcal{L}(\mathbf{x}_0, \mathbf{v})$ leading to $\mathbf{y} \in C$ (contradiction with $\mathbf{y} \notin C$).

- Necessity: Assume G is not convex. Then there exist α_1 and α_2 in G such that $\theta\alpha_1 + (1-\theta)\alpha_2 \notin G$, for some $\theta \in (0,1)$. Hence,

$$\mathbf{x}_0 + (\theta\alpha_1 + (1-\theta)\alpha_2)\mathbf{v} \notin C$$
$$\Rightarrow \theta\mathbf{x}_0 + (1-\theta)\mathbf{x}_0 + \theta\alpha_1\mathbf{v} + (1-\theta)\alpha_2\mathbf{v} \notin C$$
$$\Rightarrow \theta(\mathbf{x}_0 + \alpha_1\mathbf{v}) + (1-\theta)(\mathbf{x}_0 + \alpha_2\mathbf{v}) \notin C.$$

So C is a nonconvex set (as $\mathbf{x}_0 + \alpha_1\mathbf{v} \in C$ due to $\alpha_1 \in G$ and $\mathbf{x}_0 + \alpha_2\mathbf{v} \in C$ due to $\alpha_2 \in G$) which contradicts with the assumption that C is convex. Hence G is convex. ■

Actually, $\mathcal{L}(\mathbf{x}_0, \mathbf{v})$ in (2.20) can be easily shown to be convex, and so the intersection of the convex C and convex $\mathcal{L}(\mathbf{x}_0, \mathbf{v})$ is surely convex (which is a convexity preserving operation of convex sets to be introduced in Section 2.3).

Given a set of vectors $\{\mathbf{s}_1, \ldots, \mathbf{s}_n\} \subset \mathbb{R}^\ell$, the smallest convex set containing these vectors is called its *convex hull*, which can be defined as

$$\mathbf{conv}\{\mathbf{s}_1, \ldots, \mathbf{s}_n\} = \left\{ \mathbf{x} = \sum_{i=1}^n \theta_i \mathbf{s}_i \mid (\theta_1, \ldots, \theta_n) \in \mathbb{R}_+^n, \ \sum_{i=1}^n \theta_i = 1 \right\}. \tag{2.22}$$

Note that $\mathbf{conv}\{\mathbf{s}_1, \ldots, \mathbf{s}_n\}$ is a convex, compact, but not strictly convex set. A convex hull would be a line segment for $n = 2$, a triangle and its interior for $n = 3$ provided that $\{\mathbf{s}_1, \ldots, \mathbf{s}_n\}$ is affinely independent. An example illustrating the concept of affine hull and convex hull is shown in Figure 2.2.

Convex hull of an arbitrary set (either discrete or continuous) $C \subset \mathbb{R}^n$, denoted as $\mathbf{conv}\, C$, is defined as the smallest convex set containing C (implying $\mathbf{conv}\, C = C$ if C is a convex set), which is exactly the set of all convex combinations of elements of C and can be expressed as

$$\mathbf{conv}\ C = \left\{ \sum_{i=1}^{k} \theta_i \mathbf{x}_i \ \Big|\ \{\mathbf{x}_i\}_{i=1}^{k} \subset C,\ \{\theta_i\}_{i=1}^{k} \subset \mathbb{R}_+,\ \sum_{i=1}^{k} \theta_i = 1, k \in \mathbb{Z}_{++} \right\}. \tag{2.23}$$

A point $\mathbf{x}$ in a convex set C is said to be an *extreme point* of C if there are no two distinct points $\mathbf{x}_1, \mathbf{x}_2$ such that $\mathbf{x} = \theta \mathbf{x}_1 + (1 - \theta) \mathbf{x}_2$ for some $0 < \theta < 1$. In other words, every extreme point is essential and unique to the convex set C since it cannot be reconstructed from any other points in C via convex combinations, implying that extreme points of a closed convex set must be on the boundary of the set, but not in any open line segments on the boundary. Supposing that $C_{\text{extr}} \subseteq \mathbf{bd}\,C$ denotes the set of all the extreme points of a closed convex set C, it can be expressed as

$$C_{\text{extr}} = C \setminus C_{+\text{int}} \quad (\text{cf. } (2.19)) \tag{2.24}$$

$$= \mathbf{bd}\,C \quad \text{if } C \text{ is strictly convex.} \tag{2.25}$$

Furthermore, provided that C is a compact convex set, it can be shown that $\mathbf{conv}\,C_{\text{extr}} = C$, implying that C_{extr} is also the smallest subset of C for reconstructing C via convex combinations.

Two examples about convex sets and their extreme points are illustrated in Figure 2.4, where both sets C and B are nonconvex sets, and $\mathbf{conv}\,C$ and $\mathbf{conv}\,B$ are convex but not strictly convex sets, and $C \subset \mathbf{conv}\,C$, $B \subset \mathbf{conv}\,B$. In Figure 2.4, the extreme points of $\mathbf{conv}\,C$ are the corner points $\mathbf{a}_1, \ldots, \mathbf{a}_5$, while the extreme points of $\mathbf{conv}\,B$ consist of all its boundary points excluding the open line segment connecting the two extreme points $\mathbf{y}_1$ and $\mathbf{y}_2$ on the boundary. One can easily see that both $\mathbf{conv}\,C$ and $\mathbf{conv}\,B$ are exactly the convex hulls of their respective extreme points.

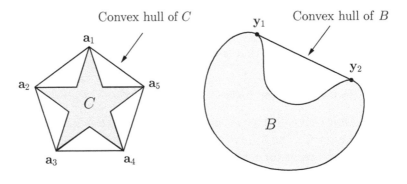

Figure 2.4 Convex hull of nonconvex sets C (left plot) and B (right plot).

Extreme points of $\mathbf{conv}\{\mathbf{s}_1,\ldots,\mathbf{s}_n\}$ defined by (2.22) are also its *vertices*, or "corner points." Equivalently, a point $\mathbf{x} \in \mathbf{conv}\{\mathbf{s}_1,\ldots,\mathbf{s}_n\}$ is an extreme point if $\mathbf{x}$ can never be a convex combination of $\mathbf{s}_1,\ldots,\mathbf{s}_n$ in a non-trivial manner, i.e.,

$$\mathbf{x} \neq \sum_{i=1}^{n} \theta_i \mathbf{s}_i$$

for all $\boldsymbol{\theta} \in \mathbb{R}_+^n$, $\sum_{i=1}^{n} \theta_i = 1$, and $\boldsymbol{\theta} \neq \mathbf{e}_i$ for any i. For instance, in Figure 2.2, $\{\mathbf{s}_1,\mathbf{s}_2\}$ are the extreme points of $\mathbf{conv}\{\mathbf{s}_1,\mathbf{s}_2\}$ for $n=2$, and $\{\mathbf{s}_1,\mathbf{s}_2,\mathbf{s}_3\}$ are those of $\mathbf{conv}\{\mathbf{s}_1,\mathbf{s}_2,\mathbf{s}_3\}$ for $n = 3$. An interesting property about extreme points is as follows:

Property 2.4 Let $S = \{\mathbf{s}_1,\ldots,\mathbf{s}_n\}$. The set of extreme points of $\mathbf{conv}\ S$ must be the full set of S when S is affinely independent, otherwise a subset of S.

Finding all the extreme points of the convex hull of a finite set C is useful in problem size reduction of an optimization problem with a feasible set being a convex hull of C, especially when $|C|$ (the number of elements in the set C) is large but the total number of the extreme points of $\mathbf{conv}\ C$ is much smaller than $|C|$. Let us conclude this subsection with the following property of convex sets.

Property 2.5 If C is a convex set, then $\mathbf{cl}\ C$ and $\mathbf{int}\ C$ are also convex sets.

Proof: First of all, let us prove that $\mathbf{cl}\ C$ is convex. Let $\mathbf{x},\mathbf{y} \in \mathbf{cl}\ C$ and $\theta \in [0,1]$. Then there exist two sequences $\{\mathbf{x}_i\} \subseteq C$ and $\{\mathbf{y}_i\} \subseteq C$, and they converge to $\mathbf{x}$ and $\mathbf{y}$, respectively [Apo07]. Because the sequence $\{\theta \cdot \mathbf{x}_i + (1-\theta) \cdot \mathbf{y}_i\}$ is a subset of C (since C is convex) and converges to $\theta \cdot \mathbf{x} + (1-\theta) \cdot \mathbf{y}$, we have $\theta \cdot \mathbf{x} + (1-\theta) \cdot \mathbf{y} \in \mathbf{cl}\ C$. Therefore, $\mathbf{cl}\ C$ is convex.

Next, we show that $\mathbf{int}\ C$ is convex. Let $\mathbf{x},\mathbf{y} \in \mathbf{int}\ C$ and $\theta \in [0,1]$. Then there exists an $r > 0$ such that the 2-norm balls (one with center at $\mathbf{x}$ and the other with center at $\mathbf{y}$ and both with radius r) $B(\mathbf{x},r) \subseteq C$ and $B(\mathbf{y},r) \subseteq C$. Then this together with the convexity of C gives rise to $B(\theta \cdot \mathbf{x} + (1-\theta) \cdot \mathbf{y}, r) \subseteq \mathbf{conv}\{B(\mathbf{x},r), B(\mathbf{y},r)\} \subseteq C$, i.e., $\theta \cdot \mathbf{x} + (1-\theta) \cdot \mathbf{y} \in \mathbf{int}\ C$. Thus we have completed the proof that $\mathbf{int}\ C$ is convex. ∎

2.1.5   Cones and conic hulls

A set C is called a *cone* if for any $\mathbf{x} \in C$ and $\theta \in \mathbb{R}_+$, the point $\theta\mathbf{x}$ also belongs to the set C. The form $\theta_1\mathbf{x}_1 + \theta_2\mathbf{x}_2 + \cdots + \theta_k\mathbf{x}_k$ for all θ_i's in $\mathbb{R}_+$ is called a *conic combination* (or a nonnegative linear combination) of $\mathbf{x}_1, \mathbf{x}_2, \ldots, \mathbf{x}_k$. A set C is a *convex cone* if and only if it contains all conic combinations of its elements. A convex cone formed by two distinct points (vectors) is illustrated in the left plot of Figure 2.5.

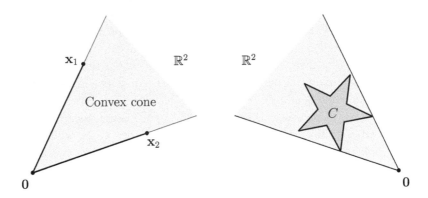

Figure 2.5 Left plot: convex cone formed by two distinct vectors (points) $\mathbf{x}_1$ and $\mathbf{x}_2$ via conic combinations, i.e., conic hull of the finite set $C = \{\mathbf{x}_1, \mathbf{x}_2\}$; right plot: conic hull of another nonconvex set C (star).

Given a set of vectors $\{\mathbf{s}_1, \ldots, \mathbf{s}_n\} \subset \mathbb{R}^\ell$, the *conic hull* of $\{\mathbf{s}_1, \ldots, \mathbf{s}_n\}$ is defined as

$$\mathbf{conic}\,\{\mathbf{s}_1, \ldots, \mathbf{s}_n\} = \left\{ \mathbf{x} = \sum_{i=1}^{n} \theta_i \mathbf{s}_i \mid \theta_i \in \mathbb{R}_+, i = 1, \ldots, n \right\}. \qquad (2.26)$$

The conic hull of an arbitrary set $C \subset \mathbb{R}^n$ is the smallest convex cone that contains C, which is exactly the set of all conic combinations of points in C, i.e.,

$$\mathbf{conic}\, C = \left\{ \sum_{i=1}^{k} \theta_i \mathbf{x}_i \mid \mathbf{x}_i \in C,\ \theta_i \in \mathbb{R}_+,\ i = 1, \ldots, k,\ k \in \mathbb{Z}_{++} \right\}. \qquad (2.27)$$

A conic hull of a nonconvex set C (star) is also illustrated in the right plot of Figure 2.5.

It can be readily inferred, from the definitions of convex set, conic hull, and affine hull, that

$$\mathbf{conic}\, C = \{\theta \mathbf{x} \mid \mathbf{x} \in \mathbf{conv}\, C,\ \theta \in \mathbb{R}_+\} \quad \text{(cf. (2.27))}$$

$$= \mathbf{conic}\,(\mathbf{conv}\, C), \qquad (2.28)$$

which is also a set of all the rays, each passing a point of $\mathbf{conv}\, C$ from the common base point at the origin, and that

$$\mathbf{conv}\, C \subseteq \mathbf{conic}\, C \cap \mathbf{aff}\, C$$

$$= \mathbf{conv}\, C \quad \text{if}\ \mathbf{0} \notin \mathbf{aff}\, C. \qquad (2.29)$$

As shown in Figure 2.5, $\mathbf{0} \notin \mathbf{aff}\, C$ (a line passing $\mathbf{x}_1$ and $\mathbf{x}_2$) and the equality in (2.29) holds true for the left plot, but not true for the right plot where $\mathbf{aff}\, C = \mathbb{R}^2$ contains the origin. Some characteristics about cones and conic hulls are discussed in the following remark.

Remark 2.1 A conic hull of a set is also a convex cone. Suppose that $C \subset \mathbb{R}^n$ and $C \neq \emptyset$. For $n > 1$, if the set C is closed (open and $\mathbf{0}_n \notin \mathbf{conv}\, C$), then $\mathbf{conic}\, C$ is a closed (neither closed nor open) and convex (strictly convex) cone; for $n = 1$, $\mathbf{conic}\, C$ is always a closed convex cone. A convex set does not necessarily contain the origin, but a cone must contain the origin which is also the only extreme point if the cone is convex. For instance, the set $\mathbb{R}^n_+ + \{\mathbf{1}_n\}$ is not a cone, though it has the same shape as the cone $\mathbb{R}^n_+$. A cone is not necessarily a convex cone. For instance, $\mathbb{R}^2 \setminus \mathbb{R}^2_{++}$ is a cone but not a convex cone, while $\mathbb{R}^2_+$ is a convex cone and surely a cone. □

2.2   Examples of convex sets

Some very useful and important examples of convex sets will be explained in this section. To begin with, we consider some simple examples:

- The empty set $\emptyset$, any single point (i.e., singleton) $\{\mathbf{x}_0\}$, and the whole space $\mathbb{R}^n$ are affine (hence, convex) subsets of $\mathbb{R}^n$. An abstract interpretation for the convexity of $\emptyset$ is that $\theta \times$ "nothing" $+ (1 - \theta) \times$ "nothing" $=$ "nothing" $\in \emptyset$. Mathematically, the "necessary" condition of the convex combination of points required by any convex set need not be checked due to no points in $\emptyset$, so $\emptyset$ is a convex set.

- Any line is affine. If it passes through the origin, it is a subspace, hence also a convex cone.

- A line segment is convex, but not affine (unless it reduces to a point).

- A ray, which has the form $\{\mathbf{x}_0 + \theta\mathbf{v} \mid \theta \geq 0\}$ where $\mathbf{v} \neq \mathbf{0}$, is convex but not affine. It is a convex cone if its base $\mathbf{x}_0$ is the origin.

- Any subspace is affine, and a convex cone (hence convex).

- Any norm ball of radius r and center $\mathbf{x}_c$, i.e., the set $\{\mathbf{x} \mid \|\mathbf{x} - \mathbf{x}_c\| \leq r\}$ (cf. (1.16)) is convex.

Now, we will discuss other convex sets, which will be useful in solving convex optimization problems.

2.2.1   Hyperplanes and halfspaces

A *hyperplane* is an affine set (hence a convex set) and is of the form

$$H = \{\mathbf{x} \mid \mathbf{a}^T\mathbf{x} = b\} \subset \mathbb{R}^n, \tag{2.30}$$

where $\mathbf{a} \in \mathbb{R}^n \setminus \{\mathbf{0}_n\}$ is a normal vector of the hyperplane, and $b \in \mathbb{R}$. Analytically it is the solution set of a linear equation of the components of $\mathbf{x}$. In geometrical sense, a hyperplane can be interpreted as the set of points having a constant inner product (b) with the normal vector ($\mathbf{a}$). Since affdim(H) $= n - 1$, the hyperplane (2.30) can also be expressed as

$$H = \mathbf{aff}\,\{\mathbf{s}_1,\dots,\mathbf{s}_n\} \subset \mathbb{R}^n, \tag{2.31}$$

where $\{\mathbf{s}_1,\dots,\mathbf{s}_n\} \subset H$ is any affinely independent set. Then it can be seen that

$$\begin{cases} \mathbf{B} \triangleq [\mathbf{s}_2 - \mathbf{s}_1,\dots,\mathbf{s}_n - \mathbf{s}_1] \in \mathbb{R}^{n \times (n-1)} \text{ and } \dim(\mathcal{R}(\mathbf{B})) = n - 1 \\ \mathbf{B}^T \mathbf{a} = \mathbf{0}_{n-1}, \text{ i.e., } \mathcal{R}(\mathbf{B})^{\perp} = \mathcal{R}(\mathbf{a}) \end{cases} \tag{2.32}$$

implying that the normal vector $\mathbf{a}$ can be determined from $\{\mathbf{s}_1,\dots,\mathbf{s}_n\}$ up to a scale factor.

The hyperplane H defined in (2.30) divides $\mathbb{R}^n$ into two closed *halfspaces* as follows:

$$\begin{aligned} H_- &= \{\mathbf{x} \mid \mathbf{a}^T\mathbf{x} \leq b\} \\ H_+ &= \{\mathbf{x} \mid \mathbf{a}^T\mathbf{x} \geq b\} \end{aligned} \tag{2.33}$$

and so each of them is the solution set of one (non-trivial) linear inequality. Note that $\mathbf{a} = \nabla(\mathbf{a}^T\mathbf{x})$ denotes the maximally increasing direction of the linear function $\mathbf{a}^T\mathbf{x}$. The above representations for both H_- and H_+ for a given $\mathbf{a} \neq \mathbf{0}$, are not unique, while they are unique if $\mathbf{a}$ is normalized such that $\|\mathbf{a}\|_2 = 1$. Moreover, $H_- \cap H_+ = H$.

An *open* halfspace is a set of the form

$$H_{--} = \{\mathbf{x} \mid \mathbf{a}^T\mathbf{x} < b\} \text{ or } H_{++} = \{\mathbf{x} \mid \mathbf{a}^T\mathbf{x} > b\}, \tag{2.34}$$

where $\mathbf{a} \in \mathbb{R}^n$, $\mathbf{a} \neq \mathbf{0}$, and $b \in \mathbb{R}$.

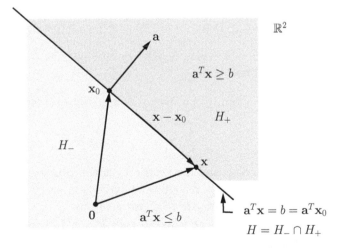

Figure 2.6 A hyperplane H in $\mathbb{R}^2$, with normal vector $\mathbf{a}$ and a point $\mathbf{x}_0$ on the hyperplane, is illustrated together with the associated two closed halfspaces, H_+ extending in the direction $\mathbf{a}$ and H_- extending in the direction $-\mathbf{a}$. For any point $\mathbf{x}$ on the hyperplane, $\mathbf{x} - \mathbf{x}_0$ is orthogonal to $\mathbf{a}$.

Remark 2.2 Halfspaces are convex but not affine. The boundary of a halfspace is a hyperplane. Hyperplanes and closed halfspaces are also cones if the origin is contained in their boundaries. An illustration of a hyperplane and associated closed halfspaces in $\mathbb{R}^2$ is shown in Figure 2.6.   □

2.2.2   Euclidean balls and ellipsoids

A *Euclidean ball* (or, simply, ball) in $\mathbb{R}^n$ has the following form:

$$B(\mathbf{x}_c, r) = \{\mathbf{x} \mid \|\mathbf{x} - \mathbf{x}_c\|_2 \leq r\} = \{\mathbf{x} \mid (\mathbf{x} - \mathbf{x}_c)^T(\mathbf{x} - \mathbf{x}_c) \leq r^2\}, \qquad (2.35)$$

where $r > 0$. The vector $\mathbf{x}_c$ is the center of the ball and the positive scalar r is its radius (see Figure 2.7). The Euclidean ball is also a 2-norm ball, and, for simplicity, a ball without explicitly mentioning the associated norm, means the Euclidean ball hereafter.

Another common representation for the Euclidean ball is

$$B(\mathbf{x}_c, r) = \{\mathbf{x}_c + r\mathbf{u} \mid \|\mathbf{u}\|_2 \leq 1\}. \qquad (2.36)$$

It can be easily proved that the Euclidean ball is a convex set.

Proof of convexity: Let $\mathbf{x}_1$ and $\mathbf{x}_2 \in B(\mathbf{x}_c, r)$, i.e., $\|\mathbf{x}_1 - \mathbf{x}_c\|_2 \leq r$ and $\|\mathbf{x}_2 - \mathbf{x}_c\|_2 \leq r$. Then,

$$\begin{aligned}
\|\theta\mathbf{x}_1 + (1 - \theta)\mathbf{x}_2 - \mathbf{x}_c\|_2 &= \|\theta\mathbf{x}_1 + (1 - \theta)\mathbf{x}_2 - [\theta\mathbf{x}_c + (1 - \theta)\mathbf{x}_c]\|_2 \\
&= \|\theta(\mathbf{x}_1 - \mathbf{x}_c) + (1 - \theta)(\mathbf{x}_2 - \mathbf{x}_c)\|_2 \\
&\leq \|\theta(\mathbf{x}_1 - \mathbf{x}_c)\|_2 + \|(1 - \theta)(\mathbf{x}_2 - \mathbf{x}_c)\|_2 \\
&\leq \theta r + (1 - \theta)r \\
&= r, \text{ for all } 0 \leq \theta \leq 1.
\end{aligned}$$

Hence, $\theta\mathbf{x}_1 + (1 - \theta)\mathbf{x}_2 \in B(\mathbf{x}_c, r)$ for all $\theta \in [0, 1]$, and thus we have proven that $B(\mathbf{x}_c, r)$ is convex.   ∎

A related family of convex sets are *ellipsoids* (see Figure 2.7), which have the form

$$\mathcal{E} = \left\{\mathbf{x} \mid (\mathbf{x} - \mathbf{x}_c)^T \mathbf{P}^{-1}(\mathbf{x} - \mathbf{x}_c) \leq 1\right\}, \qquad (2.37)$$

where $\mathbf{P} \in \mathbb{S}_{++}^n$ and the vector $\mathbf{x}_c$ is the center of the ellipsoid. The matrix $\mathbf{P}$ determines how far the ellipsoid extends in every direction from the center; the lengths of the semiaxes of $\mathcal{E}$ are given by $\sqrt{\lambda_i}$, where λ_i are eigenvalues of $\mathbf{P}$. Note that a ball is an ellipsoid with $\mathbf{P} = r^2\mathbf{I}_n$ where $r > 0$.

Another common representation of an ellipsoid is

$$\mathcal{E} = \{\mathbf{x}_c + \mathbf{A}\mathbf{u} \mid \|\mathbf{u}\|_2 \leq 1\}, \qquad (2.38)$$

where $\mathbf{A}$ is a square matrix and nonsingular. The ellipsoid $\mathcal{E}$ expressed by (2.38) is actually the image of the 2-norm ball $B(\mathbf{0}, 1) = \{\mathbf{u} \in \mathbb{R}^n \mid \|\mathbf{u}\|_2 \leq 1\}$ via an

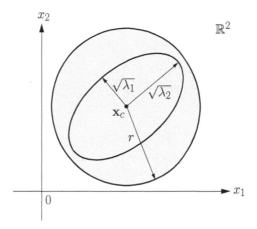

Figure 2.7 An ellipsoid centered at $\mathbf{x}_c$ with semiaxes $\sqrt{\lambda}_1$, $\sqrt{\lambda}_2$, and an Euclidean ball centered at $\mathbf{x}_c$ with radius r in $\mathbb{R}^2$.

affine mapping $\mathbf{x}_c + \mathbf{A}\mathbf{u}$ (cf. (2.58)), where $\mathbf{A} = (\mathbf{P}^{1/2})^T$, and the proof will be presented below. With the expression (2.38) for an ellipsoid, the singular values of $\mathbf{A}$, $\sigma_i(\mathbf{A}) = \sqrt{\lambda}_i$ (lengths of semiaxes) characterize the structure of the ellipsoid in a more straightforward fashion than the expression (2.37).

Proof of the ellipsoid representation (2.38) and convexity: Let

$$\mathbf{P} = \mathbf{Q}\mathbf{\Lambda}\mathbf{Q}^T = (\mathbf{P}^{1/2})^T\mathbf{P}^{1/2}$$

(EVD of $\mathbf{P} \succ \mathbf{0}$), where $\mathbf{\Lambda} = \mathbf{Diag}(\lambda_1, \lambda_2, \ldots, \lambda_n)$ and

$$\mathbf{P}^{1/2} = \mathbf{\Lambda}^{1/2}\mathbf{Q}^T,$$

in which $\mathbf{\Lambda}^{1/2} = \mathbf{Diag}(\sqrt{\lambda_1}, \sqrt{\lambda_2}, \ldots, \sqrt{\lambda_n})$. Then

$$\mathbf{P}^{-1} = \mathbf{Q}\mathbf{\Lambda}^{-1}\mathbf{Q}^T = \mathbf{P}^{-1/2}(\mathbf{P}^{-1/2})^T, \quad \mathbf{P}^{-1/2} = \mathbf{Q}\mathbf{\Lambda}^{-1/2}.$$

From the definition of ellipsoid, we have

$$\begin{aligned}
\mathcal{E} &= \left\{\mathbf{x} \mid (\mathbf{x} - \mathbf{x}_c)^T\mathbf{P}^{-1}(\mathbf{x} - \mathbf{x}_c) \leq 1\right\} \\
&= \left\{\mathbf{x} \mid (\mathbf{x} - \mathbf{x}_c)^T\mathbf{Q}\mathbf{\Lambda}^{-1}\mathbf{Q}^T(\mathbf{x} - \mathbf{x}_c) \leq 1\right\}.
\end{aligned} \tag{2.39}$$

Let $\mathbf{z} = \mathbf{x} - \mathbf{x}_c$. Then

$$\mathcal{E} = \left\{\mathbf{x}_c + \mathbf{z} \mid \mathbf{z}^T\mathbf{Q}\mathbf{\Lambda}^{-1}\mathbf{Q}^T\mathbf{z} \leq 1\right\}. \tag{2.40}$$

Now, by letting $\mathbf{u} = \mathbf{\Lambda}^{-1/2}\mathbf{Q}^T\mathbf{z} = (\mathbf{P}^{-1/2})^T\mathbf{z}$, we then obtain

$$\mathcal{E} = \left\{\mathbf{x}_c + (\mathbf{P}^{1/2})^T\mathbf{u} \mid \|\mathbf{u}\|_2 \leq 1\right\}$$

which is exactly (2.38) with $(\mathbf{P}^{1/2})^T$ replaced by $\mathbf{A}$.

Assume that $\mathbf{x}_1 = \mathbf{x}_c + \mathbf{A}\mathbf{u}_1$, $\mathbf{x}_2 = \mathbf{x}_c + \mathbf{A}\mathbf{u}_2 \in \mathcal{E}$. Thus $\|\mathbf{u}_1\|_2 \le 1$ and $\|\mathbf{u}_2\|_2 \le 1$. Then for any $0 \le \theta \le 1$, we have

$$\theta\mathbf{x}_1 + (1-\theta)\mathbf{x}_2 = \mathbf{x}_c + \mathbf{A}(\theta\mathbf{u}_1 + (1-\theta)\mathbf{u}_2),$$

where $\|\theta\mathbf{u}_1 + (1-\theta)\mathbf{u}_2\|_2 \le \theta\|\mathbf{u}_1\|_2 + (1-\theta)\|\mathbf{u}_2\|_2 \le 1$. So $\theta\mathbf{x}_1 + (1-\theta)\mathbf{x}_2 \in \mathcal{E}$. Hence we have completed the proof that $\mathcal{E}$ is convex. ∎

2.2.3   Polyhedra

A *polyhedron* is a nonempty convex set and is defined as the solution set of a finite number of linear equalities and inequalities:

$$\mathcal{P} = \left\{ \mathbf{x} \mid \mathbf{a}_i^T\mathbf{x} \le b_i, i = 1, 2, \ldots, m,\ \mathbf{c}_j^T\mathbf{x} = d_j, j = 1, 2, \ldots, p \right\}$$
$$= \left\{ \mathbf{x} \mid \mathbf{A}\mathbf{x} \preceq \mathbf{b} = (b_1, \ldots, b_m), \mathbf{C}\mathbf{x} = \mathbf{d} = (d_1, \ldots, d_p) \right\}, \qquad (2.41)$$

where "$\preceq$" stands for componentwise inequality, $\mathbf{A} \in \mathbb{R}^{m \times n}$ and $\mathbf{C} \in \mathbb{R}^{p \times n}$ are matrices whose rows are $\mathbf{a}_j^T$s and $\mathbf{c}_j^T$s, respectively, and $\mathbf{A} \ne \mathbf{0}$ or $\mathbf{C} \ne \mathbf{0}$ must be true. Note that either $m = 0$ or $p = 0$ is allowed as long as the other parameter is finite and nonzero.

A polyhedron is just the intersection of some halfspaces and hyperplanes (see Figure 2.8). A polyhedron can be unbounded, while a bounded polyhedron is called a *polytope*, e.g., any 1-norm ball and ∞-norm ball of finite radius are polytopes.

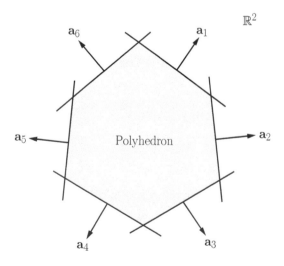

Figure 2.8 The polyhedron (shown shaded) is the intersection of six halfspaces in $\mathbb{R}^2$, with outward-pointing normal vectors $\mathbf{a}_1, \ldots, \mathbf{a}_6$.

Remark 2.3 It can be seen that the ℓ-dimensional space $\mathbb{R}^\ell$ (also an affine set) cannot be expressed in the standard form (2.41) with either nonzero $\mathbf{A}$ or nonzero $\mathbf{C}$, and so it is not a polyhedron for any $\ell \in \mathbb{Z}_{++}$. However, because any affine set in $\mathbb{R}^\ell$ can be expressed as (2.7) and each $\mathbf{x}$ in the set satisfies (2.8), this implies that the affine set must be a polyhedron if its affine dimension is strictly less than ℓ. For instance, any subspaces with dimension less than n in $\mathbb{R}^n$, and any hyperplane that is defined by a normal vector $\mathbf{a} \neq \mathbf{0}$ and a point $\mathbf{x}_0$ on the hyperplane, i.e.,

$$\mathcal{H}(\mathbf{a}, \mathbf{x}_0) = \{\mathbf{x} \mid \mathbf{a}^T(\mathbf{x} - \mathbf{x}_0) = 0\} = \{\mathbf{x}_0\} + \mathcal{R}(\mathbf{a})^\perp \qquad (2.42)$$

in $\mathbb{R}^n$, and rays, line segments, and halfspaces are all polyhedra. $\qquad \square$

Remark 2.4 The nonnegative orthant:

$$\mathbb{R}_+^n = \{\mathbf{x} \in \mathbb{R}^n \mid x_i \geq 0, \ \forall i = 1, 2, \ldots, n\} \qquad (2.43)$$

is both a polyhedron and a convex cone; hence it is called a *polyhedral cone* (see Figure 2.9). $\qquad \square$

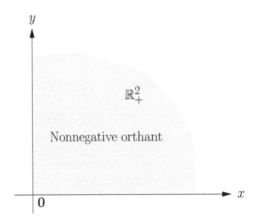

Figure 2.9 Nonnegative orthant in $\mathbb{R}^2$.

2.2.4   Simplexes

A special, but representative, case of polyhedra is *simplex*. For a finite set $\mathcal{S} = \{\mathbf{s}_1, \ldots, \mathbf{s}_n\} \subset \mathbb{R}^\ell$, the convex hull of an affinely independent subset $S \subseteq \mathcal{S}$ such that $\mathbf{conv}\, S = \mathbf{conv}\, \mathcal{S}$ is called a simplex. Hence, the simplex $\mathbf{conv}\, \mathcal{S}$ is compact and convex with $|S|$ extreme points (i.e., $\mathbf{conv}\, \mathcal{S}$ has $|S|$ vertices) where $|S| \leq n$ (cf. Property 2.4), and it is also a polytope. Meanwhile, if $\mathbf{conv}\, \mathcal{S}$ is a simplex, then S is unique and $\mathrm{affdim}(\mathcal{S}) = |S| - 1$ (cf. Property 2.1). However, if such an affinely independent subset S does not exist, then $\mathbf{conv}\, \mathcal{S}$ is not a simplex.

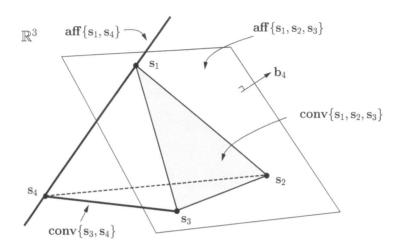

Figure 2.10 A graphical illustration in $\mathbb{R}^3$ for some convex geometry concepts, where $\mathbf{conv}\{\mathbf{s}_3, \mathbf{s}_4\}$ is a simplex defined by the line segment connecting $\mathbf{s}_3$ and $\mathbf{s}_4$, $\mathbf{conv}\{\mathbf{s}_1, \mathbf{s}_2, \mathbf{s}_3\}$ is a simplex defined by the shaded triangle, and $\mathbf{conv}\{\mathbf{s}_1, \mathbf{s}_2, \mathbf{s}_3, \mathbf{s}_4\}$ is a simplex (and also a simplest simplex) defined by the tetrahedron with the four extreme points $\{\mathbf{s}_1, \mathbf{s}_2, \mathbf{s}_3, \mathbf{s}_4\}$. Moreover, $\mathbf{aff}\{\mathbf{s}_1, \mathbf{s}_4\}$ is the straight line passing $\mathbf{s}_1$ and $\mathbf{s}_4$, $\mathbf{aff}\{\mathbf{s}_1, \mathbf{s}_2, \mathbf{s}_3\}$ is the plane passing the three points $\{\mathbf{s}_1, \mathbf{s}_2, \mathbf{s}_3\}$, and $\mathbf{aff}\{\mathbf{s}_1, \mathbf{s}_2, \mathbf{s}_3, \mathbf{s}_4\}$ is the whole $\mathbb{R}^3$ space.

An n-vertex simplex $\mathbf{conv}\{\mathbf{s}_1, \ldots, \mathbf{s}_n\} \subset \mathbb{R}^\ell$ is called the *simplest simplex* if $n = \ell + 1$ (due to minimum $\ell = n - 1$ for an n-vertex simplex by (2.9)) and $\text{affdim}(\{\mathbf{s}_1, \ldots, \mathbf{s}_n\}) = n - 1 = \ell$ (i.e., $\mathbf{aff}\{\mathbf{s}_1, \ldots, \mathbf{s}_n\} = \mathbb{R}^\ell$ for the simplest simplex of $n = \ell + 1$ vertices), with the property about its extreme points as follows:

Property 2.6 The set of extreme points of a simplest simplex $\mathbf{conv}\{\mathbf{s}_1, \ldots, \mathbf{s}_n\} \subset \mathbb{R}^{n-1}$ is $\{\mathbf{s}_1, \ldots, \mathbf{s}_n\}$.

A graphical illustration of a simplest simplex (a tetrahedron) in $\mathbb{R}^3$ together with some previously introduced convex geometry concepts is given in Figure 2.10. It can also be seen, from this figure, that a simplest simplex in $\mathbb{R}$ is a line segment, while a simplest simplex in $\mathbb{R}^2$ is a triangle and its interior. Of particular interest is that a simplest simplex with n extreme points can be equivalently reconstructed by the n hyperplanes that tightly enclose it as stated in the following property:

Property 2.7 If $\{\mathbf{s}_1, \ldots, \mathbf{s}_n\} \subseteq \mathbb{R}^{n-1}$ is affinely independent, then the simplest simplex $\mathcal{T} = \mathbf{conv}\{\mathbf{s}_1, \ldots, \mathbf{s}_n\} \subseteq \mathbb{R}^{n-1}$ can be reconstructed from the n hyperplanes $\{\mathcal{H}_1, \ldots, \mathcal{H}_n\}$ and vice versa, where

$$\mathcal{H}_i \triangleq \mathbf{aff}(\{\mathbf{s}_1, \ldots, \mathbf{s}_n\} \setminus \{\mathbf{s}_i\}) \quad (\text{cf. (2.31)}). \tag{2.44}$$

Proof: By (2.30), the hyperplane $\mathcal{H}_i$, with affine dimension $n - 2$, can be parameterized by an "outward-pointing normal vector" $\mathbf{b}_i \in \mathbb{R}^{n-1}$ such that

$\mathcal{T} \subset \mathcal{H}_{i-}$ (cf. (2.46)) and an inner product constant $h_i \in \mathbb{R}$ as follows:

$$\mathcal{H}_i = \left\{ \mathbf{x} \in \mathbb{R}^{n-1} \mid \mathbf{b}_i^T \mathbf{x} = h_i \right\} \tag{2.45}$$

and

$$\mathbf{s}_i \in \mathcal{H}_{i-} = \left\{ \mathbf{x} \in \mathbb{R}^{n-1} \mid \mathbf{b}_i^T \mathbf{x} \le h_i \right\}. \tag{2.46}$$

We will show that

$$\mathcal{T} = \mathbf{conv}\{\mathbf{s}_1, \dots, \mathbf{s}_n\} = \bigcap_{i=1}^{n} \mathcal{H}_{i-}. \tag{2.47}$$

The reconstruction formula (2.47) implies that all the extreme points of $\mathcal{T}$ can be obtained from $\mathcal{H}_i, i = 1, \dots, n$, and vice versa. In other words, we need to derive the expressions of $\{\mathbf{s}_1, \dots, \mathbf{s}_n\}$ in terms of $(\mathbf{b}_i, h_i), i = 1, \dots, n$, and vice versa.

By the fact of $\mathbf{s}_i \in \mathbf{aff}(\{\mathbf{s}_1, \dots, \mathbf{s}_n\} \setminus \{\mathbf{s}_j\}) = \mathcal{H}_j$ for all $j \ne i$, we have from (2.45) that $\mathbf{b}_j^T \mathbf{s}_i = h_j$ for all $j \ne i$, or equivalently

$$\mathbf{B}_{-i}\mathbf{s}_i = \mathbf{h}_{-i}, \tag{2.48}$$

where

$$\begin{aligned} \mathbf{B}_{-i} &\triangleq [\mathbf{b}_1, \dots, \mathbf{b}_{i-1}, \mathbf{b}_{i+1}, \dots, \mathbf{b}_n]^T \in \mathbb{R}^{(n-1) \times (n-1)}, \\ \mathbf{h}_{-i} &\triangleq [h_1, \dots, h_{i-1}, h_{i+1}, \dots, h_n]^T \in \mathbb{R}^{n-1}. \end{aligned} \tag{2.49}$$

As $\mathcal{T}_{\text{extr}} = \{\mathbf{s}_1, \dots, \mathbf{s}_n\}$ is a unique set for the simplest simplex $\mathcal{T} \subset \mathbb{R}^{n-1}$ (deduced by Property 2.6), $\mathbf{B}_{-i}$ must be of full rank and hence invertible (otherwise, $\mathbf{s}_i$ that satisfies (2.48) is nonunique). So the unique solution of (2.48) is given by

$$\mathbf{s}_i = \mathbf{B}_{-i}^{-1} \mathbf{h}_{-i}, \ \forall i. \tag{2.50}$$

Alternatively, each extreme point $\mathbf{s}_i$ of the simplest simplex $\mathcal{T}$ can also be expressed as

$$\{\mathbf{s}_i\} = \bigcap_{j=1, j \ne i}^{n} \mathcal{H}_j, \ \forall i. \tag{2.51}$$

Next, we drive the expressions of $(\mathbf{b}_i, h_i), i = 1, \dots, n$ in terms of $\{\mathbf{s}_1, \dots, \mathbf{s}_n\}$. The normal vector $\mathbf{b}_i$ of $\mathcal{H}_i$ can be obtained by projecting the vector $\mathbf{s}_j - \mathbf{s}_i$ (for any $j \ne i$) onto the one-dimensional subspace of $\mathbb{R}^{n-1}$, that is orthogonal to the $(n-2)$-dimensional hyperplane $\mathcal{H}_i$; precisely,

$$\begin{aligned} \mathbf{b}_i &\triangleq \boldsymbol{v}_i(\mathbf{s}_1, \dots, \mathbf{s}_n) \\ &= \mathbf{P}_{\mathbf{A}}^{\perp}(\mathbf{s}_j - \mathbf{s}_i) \text{ for any } j \ne i \ \ (\text{cf. (2.32)}) \\ &= \left(\mathbf{I}_{n-1} - \mathbf{A}(\mathbf{A}^T\mathbf{A})^{-1}\mathbf{A}^T\right) \cdot (\mathbf{s}_j - \mathbf{s}_i) \ \ (\text{cf. (1.74) and (1.73)}), \end{aligned} \tag{2.52}$$

where $\boldsymbol{v}_i(\mathbf{s}_1, \dots, \mathbf{s}_n)$ simply indicates that $\mathbf{b}_i$ is a function of $\mathcal{T}_{\text{extr}}$, $\mathbf{A}$ is a matrix formed by the $n-2$ vectors $\{\mathbf{s}_k - \mathbf{s}_j \mid k \in \{1, \dots, n\} \setminus \{i, j\}\}$ (which is also of full column rank by Property 2.1), namely,

$$\mathbf{A} \triangleq \mathbf{Q}_{ij} - \mathbf{s}_j \cdot \mathbf{1}_{n-2}^T \in \mathbb{R}^{(n-1) \times (n-2)}, \tag{2.53}$$

and $\mathbf{Q}_{ij} \in \mathbb{R}^{(n-1) \times (n-2)}$ is the matrix $[\mathbf{s}_1, \ldots, \mathbf{s}_n] \in \mathbb{R}^{(n-1) \times n}$ with its ith and jth columns removed. In other words, the $(n-2)$-dimensional subspace $\mathcal{R}(\mathbf{A})$, spanned by the $n-2$ linearly independent vectors, $\mathbf{s}_k - \mathbf{s}_j, \forall k \neq j, i$, is also a hyperplane passing through the origin and in parallel with $\mathcal{H}_i$, i.e.,

$$\mathcal{R}(\mathbf{A}) = \mathcal{H}_i - \{\mathbf{s}_j\}, \ \forall j \neq i.$$

Moreover, the outward-pointing normal vector $\mathbf{b}_i \in \mathcal{R}(\mathbf{A})^{\perp} \setminus \{\mathbf{0}_{n-1}\}$ (cf. (2.52)) is also a vector that satisfies

$$\mathbf{b}_i^T(\mathbf{s}_j - \mathbf{s}_i) > 0 \text{ for any } j \neq i.$$

On the other hand, as $\mathbf{s}_j \in \mathcal{H}_i$ (for any $j \neq i$), the inner product constant h_i can be calculated by (cf. (2.45) and (2.52))

$$h_i = \mathbf{b}_i^T \mathbf{s}_j = \boldsymbol{v}_i(\mathbf{s}_1, \ldots, \mathbf{s}_n)^T \mathbf{s}_j, \ j \neq i. \tag{2.54}$$

Consequently, by (2.52) and (2.54) we have proven that $(\mathbf{b}_i, h_i), i = 1, \ldots, n$ can be expressed in terms of $\{\mathbf{s}_1, \ldots, \mathbf{s}_n\}$. Thus we have completed the proof. ■

Some cutting-edge researches in hyperspectral image analysis by applying the preceding neat simplex geometry (e.g., (2.50), (2.52) and (2.54)) will be introduced in Subsection 6.3.3.

2.2.5   Norm cones

Let $\| \cdot \|$ represent a norm on $\mathbb{R}^n$. A lot of commonly known convex sets involve norm, such as norm balls (cf. (1.16)). Norm cones (convex sets to be defined next) are particularly important in many applications.

The *norm cone* associated with the norm $\| \cdot \|$ is the convex set

$$C = \{(\mathbf{x}, t) \in \mathbb{R}^{n+1} \mid \|\mathbf{x}\| \leq t\} \subseteq \mathbb{R}^{n+1}. \tag{2.55}$$

Its convexity can be easily proven by the definition of convex set. Note that it is not strictly convex because the boundary of this cone is a set of rays

$$\mathbf{bd} \ C = \bigcup_{\|\mathbf{u}\|=1} \{t(\mathbf{u}, 1) \in \mathbb{R}^{n+1}, \ t \geq 0\} \subseteq \mathbb{R}^{n+1} \tag{2.56}$$

(containing line segments).

Remark 2.5 In the special case when the norm in (2.55) is Euclidean norm (i.e., 2-norm), the norm cone is called Lorentz cone or quadratic cone, second-order cone, or ice-cream cone (see Figure 2.11). The second-order cone has been prevalent in many signal processing and communication problems that will be presented in the subsequent chapters. □

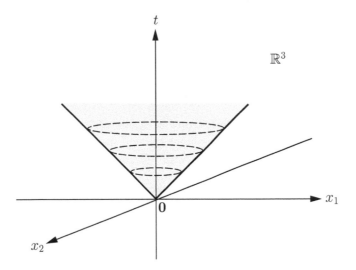

Figure 2.11 Second-order cone in $\mathbb{R}^3$.

2.2.6   Positive semidefinite cones

The set of positive semidefinite matrices,

$$\mathbb{S}_+^n = \left\{ \mathbf{A} \in \mathbb{S}^n \mid \mathbf{x}^T \mathbf{A} \mathbf{x} \geq 0, \ \forall \mathbf{x} \in \mathbb{R}^n \right\}, \tag{2.57}$$

is a convex cone, that is to say, for $\theta_1, \theta_2 \geq 0$ and for any $\mathbf{A}, \mathbf{B} \in \mathbb{S}_+^n$, we have $\theta_1 \mathbf{A} + \theta_2 \mathbf{B} \in \mathbb{S}_+^n$ (i.e., the PSD cone $\mathbb{S}_+^n$ is closed under conic combination). This can be proved directly from the definition of positive semidefiniteness. For any $\mathbf{x} \in \mathbb{R}^n$, it is straightforward to see

$$\mathbf{x}^T (\theta_1 \mathbf{A} + \theta_2 \mathbf{B}) \mathbf{x} = \theta_1 (\mathbf{x}^T \mathbf{A} \mathbf{x}) + \theta_2 (\mathbf{x}^T \mathbf{B} \mathbf{x}) \geq 0,$$

if $\mathbf{A}, \mathbf{B} \in \mathbb{S}_+^n$ and $\theta_1, \theta_2 \geq 0$. In other words, $\theta_1 \mathbf{A} + \theta_2 \mathbf{B} \in \mathbb{S}_+^n$ for all $\theta_1, \theta_2 \geq 0$, i.e., $\mathbb{S}_+^n$ is a convex cone, but not strictly convex for $n \geq 2$ because all the points on a ray $\{t\mathbf{u}\mathbf{u}^T \mid \mathbf{u} \in \mathbb{R}^n, t \geq 0\}$ are rank-1 PSD matrices on the boundary of $\mathbb{S}_+^n$. Actually, **bd** $\mathbb{S}_+^n$ will be shown to contain those PSD matrices with at least one eigenvalue equal to zero in Remark 3.28 later. As the second-order cone, the PSD cone $\mathbb{S}_+^n$ has been prevalent in multiple-input multiple-output (MIMO) wireless communications.

2.3   Convexity preserving operations

Some operations that preserve convexity of sets help us construct convex sets from other convex sets and such operators are discussed in this section.

2.3.1   Intersection

If S_1 and S_2 are convex sets, then $S_1 \cap S_2$ is also convex. This property extends
to the intersection of an infinite number of convex sets, i.e., if S_α is convex for
every $\alpha \in \mathcal{A}$, then $\cap_{\alpha \in \mathcal{A}} S_\alpha$ is convex. Let us illuminate the usefulness of this
convexity preserving operation with the following remarks and examples.

Remark 2.6 A polyhedron can be considered as intersection of a finite number
of halfspaces and hyperplanes (which are convex) and hence the polyhedron is
convex.   □

Remark 2.7 Subspaces are closed under arbitrary intersections; so are affine
sets and convex cones. So they all are convex sets.   □

Remark 2.8 A closed convex set S is the intersection of all (possibly an infinite
number of) closed halfspaces that contain S. This can be proven by the separating
hyperplane theory (to be introduced in Subsection 2.6.1).   □

Example 2.4 The PSD cone $\mathbb{S}_+^n$ is known to be convex. The proof of its convexity by the intersection property is given as follows. It is easy to see that $\mathbb{S}_+^n$
can be expressed as

$$\mathbb{S}_+^n = \{\mathbf{X} \in \mathbb{S}^n \mid \mathbf{z}^T \mathbf{X} \mathbf{z} \geq 0, \ \forall \mathbf{z} \in \mathbb{R}^n\} = \bigcap_{\mathbf{z} \in \mathbb{R}^n} S_{\mathbf{z}}$$

where

$$S_{\mathbf{z}} = \{\mathbf{X} \in \mathbb{S}^n \mid \mathbf{z}^T \mathbf{X} \mathbf{z} \geq 0\} = \{\mathbf{X} \in \mathbb{S}^n \mid \mathrm{Tr}(\mathbf{z}^T \mathbf{X} \mathbf{z}) \geq 0\}$$
$$= \{\mathbf{X} \in \mathbb{S}^n \mid \mathrm{Tr}(\mathbf{X} \mathbf{z} \mathbf{z}^T) \geq 0\} = \{\mathbf{X} \in \mathbb{S}^n \mid \mathrm{Tr}(\mathbf{X} \mathbf{Z}) \geq 0\},$$

in which $\mathbf{Z} = \mathbf{z}\mathbf{z}^T$, implying that $S_{\mathbf{z}}$ is a halfspace if $\mathbf{z} \neq \mathbf{0}_n$. As the intersection
of halfspaces is also convex, $\mathbb{S}_+^n$ (intersection of infinite number of halfspaces) is
a convex set. It is even easier to prove the convexity of $S_{\mathbf{z}}$ by the definition of
convex sets.   □

Example 2.5 Consider

$$P(\mathbf{x}, \omega) = \sum_{i=1}^{n} x_i \cos(i\omega)$$

and a set

$$C = \{\mathbf{x} \in \mathbb{R}^n \mid l(\omega) \leq P(\mathbf{x}, \omega) \leq u(\omega) \ \forall \omega \in \mathbf{\Omega}\}$$
$$= \bigcap_{\omega \in \mathbf{\Omega}} \{\mathbf{x} \in \mathbb{R}^n \mid l(\omega) \leq \sum_{i=1}^{n} x_i \cos(i\omega) \leq u(\omega)\}.$$

Let

$$\mathbf{a}(\omega) = [\cos(\omega), \cos(2\omega), \dots, \cos(n\omega)]^T.$$

Then we have

$$C = \bigcap_{\omega \in \Omega} \{ \mathbf{x} \in \mathbb{R}^n \mid \mathbf{a}^T(\omega)\mathbf{x} \geq l(\omega), \mathbf{a}^T(\omega)\mathbf{x} \leq u(\omega) \} \text{ (intersection of halfspaces)},$$

which implies that C is convex. Note that the set C is a polyhedron only when the set size $|\Omega|$ is finite. $\qquad\qquad\square$

2.3.2   Affine function

A function $\boldsymbol{f} : \mathbb{R}^n \to \mathbb{R}^m$ is *affine* if it takes the form

$$\boldsymbol{f}(\mathbf{x}) = \mathbf{A}\mathbf{x} + \mathbf{b}, \qquad\qquad (2.58)$$

where $\mathbf{A} \in \mathbb{R}^{m \times n}$ and $\mathbf{b} \in \mathbb{R}^m$. The affine function, for which $\boldsymbol{f}(\mathbf{dom}\ \boldsymbol{f})$ is an affine set if $\mathbf{dom}\ \boldsymbol{f}$ is an affine set, also called the affine transformation or the affine mapping, has been implicitly used in defining the affine hull given by (2.7) in the preceding Subsection 2.1.2. It preserves points, straight lines, and planes, but not necessarily preserves angles between lines or distances between points. The affine mapping plays an important role in a variety of convex sets and convex functions, problem reformulations to be introduced in the subsequent chapters.

Suppose $S \subseteq \mathbb{R}^n$ is convex and $\boldsymbol{f} : \mathbb{R}^n \to \mathbb{R}^m$ is an affine function (see Figure 2.12). Then the image of S under $\boldsymbol{f}$,

$$\boldsymbol{f}(S) = \{\ \boldsymbol{f}(\mathbf{x}) \mid \mathbf{x} \in S\ \}, \qquad\qquad (2.59)$$

is convex. The converse is also true, i.e., the inverse image of the convex set C

$$\boldsymbol{f}^{-1}(C) = \{\ \mathbf{x} \mid \boldsymbol{f}(\mathbf{x}) \in C\ \} \qquad\qquad (2.60)$$

is convex. The proof is given below.

Proof: Let $\mathbf{y}_1$ and $\mathbf{y}_2 \in C$. Then there exist $\mathbf{x}_1$ and $\mathbf{x}_2 \in \boldsymbol{f}^{-1}(C)$ such that $\mathbf{y}_1 = \mathbf{A}\mathbf{x}_1 + \mathbf{b}$ and $\mathbf{y}_2 = \mathbf{A}\mathbf{x}_2 + \mathbf{b}$. Our aim is to show that the set $\boldsymbol{f}^{-1}(C)$, which is the inverse image of $\boldsymbol{f}$, is convex. For $\theta \in [0, 1]$,

$$\theta\mathbf{y}_1 + (1 - \theta)\mathbf{y}_2 = \theta(\mathbf{A}\mathbf{x}_1 + \mathbf{b}) + (1 - \theta)(\mathbf{A}\mathbf{x}_2 + \mathbf{b})$$
$$= \mathbf{A}(\theta\mathbf{x}_1 + (1 - \theta)\mathbf{x}_2) + \mathbf{b} \in C,$$

which implies that $\theta\mathbf{x}_1 + (1 - \theta)\mathbf{x}_2 \in \boldsymbol{f}^{-1}(C)$, and that the convex combination of $\mathbf{x}_1$ and $\mathbf{x}_2$ is in $\boldsymbol{f}^{-1}(C)$, and hence $\boldsymbol{f}^{-1}(C)$ is convex. $\qquad\blacksquare$

Remark 2.9 If $S_1 \subset \mathbb{R}^n$ and $S_2 \subset \mathbb{R}^n$ are convex and $\alpha_1, \alpha_2 \in \mathbb{R}$, then the set $S = \{(\mathbf{x}, \mathbf{y}) \mid \mathbf{x} \in S_1, \mathbf{y} \in S_2\}$ is convex. Furthermore, the set

$$\alpha_1 S_1 + \alpha_2 S_2 = \{\mathbf{z} = \alpha_1 \mathbf{x} + \alpha_2 \mathbf{y} \mid \mathbf{x} \in S_1, \mathbf{y} \in S_2\} \quad \text{(cf. (1.22) and (1.23))}$$
$$(2.61)$$

is also convex (since this set can be thought of as the image of the convex set S through the affine mapping given by (2.58) from S to $\alpha_1 S_1 + \alpha_2 S_2$ with $\mathbf{A} = [\alpha_1 \mathbf{I}_n \ \alpha_2 \mathbf{I}_n]$ and $\mathbf{b} = \mathbf{0}$). $\qquad\qquad\square$

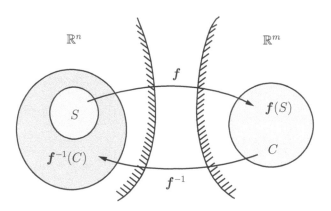

Figure 2.12 For an affine function $\boldsymbol{f} : \mathbb{R}^n \to \mathbb{R}^m$, the set $S \subset \boldsymbol{f}^{-1}(C) \subseteq \mathbb{R}^n$ (or the inverse image $\boldsymbol{f}^{-1}(C)$) and its image $\boldsymbol{f}(S) \subset \mathbb{R}^m$ (or $C \subset \mathbb{R}^m$) are both convex sets if any one of them is convex, where $\boldsymbol{f}(S) = C$ is assumed.

Remark 2.10 Let us present an alternative proof for the inverse image $\boldsymbol{f}^{-1}(C)$ given by (2.60) to be convex if C is convex and $\boldsymbol{f}$ is the affine mapping given by (2.58). Note that $\boldsymbol{f}^{-1}(C)$ can be expressed as

$$\boldsymbol{f}^{-1}(C) = \left\{ \mathbf{A}^{\dagger}(\mathbf{y} - \mathbf{b}) \mid \mathbf{y} \in C \right\} + \mathcal{N}(\mathbf{A}), \qquad (2.62)$$

where the first set $\left\{ \mathbf{A}^{\dagger}(\mathbf{y} - \mathbf{b}) \mid \mathbf{y} \in C \right\}$ is convex due to the convexity preservation property of the affine mapping, and the second set is a subspace which is convex. So $\boldsymbol{f}^{-1}(C)$ is convex by (2.61). Moreover, it is noticeable that when $\mathbf{A} \in \mathbb{R}^{m \times n}$ is of full column rank, $\boldsymbol{f}(S) = C$ and $\boldsymbol{f}^{-1}(C) = S$ (due to $\mathcal{N}(\mathbf{A}) = \{\mathbf{0}\}$), and furthermore, $\boldsymbol{f}$ is distance preserving (i.e., $\|\boldsymbol{f}(\mathbf{x}_1) - \boldsymbol{f}(\mathbf{x}_2)\|_2 = \|\mathbf{x}_1 - \mathbf{x}_2\|_2$ for all $\mathbf{x}_1, \mathbf{x}_2 \in S$) (due to $\mathbf{A}^T \mathbf{A} = \mathbf{I}_n$) if $\mathbf{A}$ is also semi-unitary (cf. (1.75)). $\square$

Remark 2.11 Suppose that C is a convex set in $\mathbb{R}^{n+m}$. Then

$$C_1 = \left\{ \mathbf{x} \in \mathbb{R}^n \mid (\mathbf{x}, \mathbf{y}) \in C \subset \mathbb{R}^{n+m} \right\} \qquad (2.63)$$

is convex through the linear mapping

$$\mathbf{x} = [\mathbf{I}_n \ \mathbf{0}_{n \times m}][\mathbf{x}^T \ \mathbf{y}^T]^T$$

from C to C_1. In other words, the projection of a convex set onto some of its coordinates forms a convex set, as illustrated by the next example. $\square$

Example 2.6 Consider a convex set $C_1 = \{(x_1, x_2, x_3) \mid x_1^2 + x_2^2 + x_3^2 \leq 1\}$, a Euclidean ball with radius equal to unity. Then, the projection of C_1 on the

x_1-x_2 plane is

$$C_2 = \{(x_1, x_2) \mid (x_1, x_2, x_3) \in C_1 \text{ for some } x_3\}$$
$$= \bigcup_{x_3} \{(x_1, x_2) \mid (x_1, x_2, x_3) \in C_1\}$$
$$= \{(x_1, x_2) \mid x_1^2 + x_2^2 \le 1\}.$$

Note that C_2 is also a convex set which can surely be shown by the definition of the convex set. This can also be simply proved by the affine mapping f from C_1 to C_2 given by (2.58) where $\mathbf{A} = [\mathbf{I}_2, \mathbf{0}_2]$ and $\mathbf{b} = \mathbf{0}_2$. Next, let us check if $f^{-1}(C_2)$ is convex. It can be easily shown that

$$\mathbf{A}^\dagger = \mathbf{A}^T(\mathbf{A}\mathbf{A}^T)^{-1} = \mathbf{A}^T \text{ (by (1.126))},$$
$$\mathcal{N}(\mathbf{A}) = \{(0, 0, x_3) \mid x_3 \in \mathbb{R}\}.$$

Then by (2.62), we have

$$f^{-1}(C_2) = \{(x_1, x_2, x_3) \in \mathbb{R}^3 \mid x_1^2 + x_2^2 \le 1\}$$

which is a cylinder (a convex set but not the set C_1). $\quad\square$

Example 2.7 Hyperbolic cone defined as:

$$C = \{\mathbf{x} \in \mathbb{R}^n \mid \mathbf{x}^T\mathbf{P}\mathbf{x} \le (\mathbf{c}^T\mathbf{x})^2, \; \mathbf{c}^T\mathbf{x} \ge 0\} \tag{2.64}$$

is convex, where $\mathbf{P} \in \mathbb{S}_+^n$ and $\mathbf{c} \in \mathbb{R}^n$.

Proof: The convexity of C can be proven by the convex combination of any two points in C belonging to C. Let $\mathbf{x}_1, \mathbf{x}_2 \in C$ such that

$$\mathbf{x}_i^T\mathbf{P}\mathbf{x}_i \le (\mathbf{c}^T\mathbf{x}_i)^2 \;\Rightarrow\; \|\mathbf{P}^{1/2}\mathbf{x}_i\|_2 \le \mathbf{c}^T\mathbf{x}_i, \; i = 1, 2$$

where $\mathbf{P}^{1/2} \in \mathbb{S}_+^n$. Then for any $\theta \in [0, 1]$ we have

$$\begin{aligned}
&(\theta\mathbf{x}_1 + (1-\theta)\mathbf{x}_2)^T\mathbf{P}(\theta\mathbf{x}_1 + (1-\theta)\mathbf{x}_2) \\
&= \theta^2\mathbf{x}_1^T\mathbf{P}\mathbf{x}_1 + (1-\theta)^2\mathbf{x}_2^T\mathbf{P}\mathbf{x}_2 + 2\theta(1-\theta)\mathbf{x}_1^T\mathbf{P}\mathbf{x}_2 \\
&\le \theta^2(\mathbf{c}^T\mathbf{x}_1)^2 + (1-\theta)^2(\mathbf{c}^T\mathbf{x}_2)^2 + 2\theta(1-\theta)(\mathbf{P}^{1/2}\mathbf{x}_1)^T(\mathbf{P}^{1/2}\mathbf{x}_2) \\
&\le \theta^2(\mathbf{c}^T\mathbf{x}_1)^2 + (1-\theta)^2(\mathbf{c}^T\mathbf{x}_2)^2 + 2\theta(1-\theta)(\mathbf{c}^T\mathbf{x}_1) \cdot (\mathbf{c}^T\mathbf{x}_2) \\
&= (\mathbf{c}^T(\theta\mathbf{x}_1 + (1-\theta)\mathbf{x}_2))^2,
\end{aligned}$$

implying $\theta\mathbf{x}_1 + (1-\theta)\mathbf{x}_2 \in C$. So C is convex.

An alternative proof via the affine set mapping is given next. Recall the convex second-order cone has been defined as (cf. (2.55))

$$D = \{(\mathbf{y}, t) \in \mathbb{R}^{n+1} \mid \|\mathbf{y}\|_2 \le t\} = \{(\mathbf{y}, t) \in \mathbb{R}^{n+1} \mid \|\mathbf{y}\|_2^2 \le t^2, t \ge 0\}.$$

Now, we define an affine mapping from $\mathbb{R}^n$ to $\mathbb{R}^{n+1}$ as

$$(\mathbf{z}, r) = f(\mathbf{x}) = (\mathbf{P}^{1/2}\mathbf{x}, \mathbf{c}^T\mathbf{x}) = \mathbf{A}\mathbf{x}, \; \mathbf{x} \in C \subset \mathbb{R}^n \tag{2.65}$$

where $\mathbf{A} = [\mathbf{P}^{1/2}\ \mathbf{c}]^T \in \mathbb{R}^{(n+1)\times n}$ and $(\mathbf{z}, r) \in D \subset \mathbb{R}^{n+1}$ by the above defined sets C and D. Then it can be inferred, from Remark 2.10, that

$$\boldsymbol{f}(C) = D \cap \mathcal{R}(\mathbf{A}) \triangleq \mathcal{D} \subset \mathbb{R}^{n+1} \tag{2.66}$$

$$C = \boldsymbol{f}^{-1}(\mathcal{D}) \subset \mathbb{R}^n \tag{2.67}$$

provided that $\mathbf{A}$ is of full column rank. Note that $\mathcal{D}$ is convex since both D and $\mathcal{R}(\mathbf{A})$ are convex. In other words, the inverse image of the convex $\mathcal{D}$ under the affine transformation $\boldsymbol{f}$ is the hyperbolic cone which must be convex as well. An illustration is given in Figure 2.13, where

$$\mathbf{P} = \begin{bmatrix} 3 & -1 \\ -1 & 3 \end{bmatrix}, \quad \mathbf{c} = \begin{bmatrix} 0 \\ 2 \end{bmatrix} \tag{2.68}$$

for which (2.66) and (2.67) are valid.   ■ □

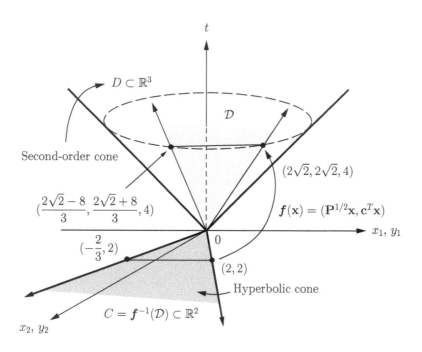

Figure 2.13 A hyperbolic cone C in $\mathbb{R}^2$ and its image $\mathcal{D} \subset D$ (second-order cone) in $\mathbb{R}^3$ through the affine mapping $\boldsymbol{f}(\mathbf{x})$ given by (2.65), with $\mathbf{P}$ and $\mathbf{c}$ given by (2.68).

Example 2.8 Let $\mathbf{A}, \mathbf{B} \in \mathbb{S}^n$ and $\boldsymbol{f} : \mathbb{S}^n \to \mathbb{R}^2$ defined by

$$\boldsymbol{f}(\mathbf{X}) \triangleq (\mathrm{Tr}(\mathbf{AX}), \mathrm{Tr}(\mathbf{BX})). \tag{2.69}$$

Then the set

$$W(\mathbf{A}, \mathbf{B}) = \boldsymbol{f}(\mathbb{S}^n_+) \tag{2.70}$$

is a convex cone due to the fact that $W(\mathbf{A}, \mathbf{B})$ is the image of the convex cone $\mathbb{S}^n_+$ through the linear transformation $\boldsymbol{f}$ defined in (2.69).

Furthermore, it can be seen, by Remark 1.19, that the set $W(\mathbf{A}, \mathbf{B})$ can also be expressed as

$$W(\mathbf{A}, \mathbf{B}) = \{(\mathbf{x}^T \mathbf{A} \mathbf{x}, \mathbf{x}^T \mathbf{B} \mathbf{x}) \mid \mathbf{x} \in \mathbb{R}^n\}, \tag{2.71}$$

implying that it is also the image of the set of rank-one PSD matrices (i.e., the set $\{\mathbf{X} = \mathbf{x}\mathbf{x}^T \mid \mathbf{x} \in \mathbb{R}^n\}$) under the linear mapping $\boldsymbol{f}$. □

2.3.3   Perspective function and linear-fractional function

Linear-fractional functions are functions which are more general than affine but still preserve convexity. The perspective function scales or normalizes vectors so that the last component is one, and then drops the last component.

The *perspective function* $\boldsymbol{p}: \mathbb{R}^{n+1} \to \mathbb{R}^n$, with $\mathbf{dom}\, \boldsymbol{p} = \mathbb{R}^n \times \mathbb{R}_{++}$, is defined as

$$\boldsymbol{p}(\mathbf{z}, t) = \frac{\mathbf{z}}{t}. \tag{2.72}$$

The perspective function $\boldsymbol{p}$ preserves the convexity of the convex set.

Proof. Consider two points $(\mathbf{z}_1, t_1)$ and $(\mathbf{z}_2, t_2)$ in a convex set C and so $\mathbf{z}_1/t_1$ and $\mathbf{z}_2/t_2 \in \boldsymbol{p}(C)$. Then

$$\theta(\mathbf{z}_1, t_1) + (1 - \theta)(\mathbf{z}_2, t_2) = (\theta \mathbf{z}_1 + (1 - \theta)\mathbf{z}_2, \; \theta t_1 + (1 - \theta)t_2) \in C, \tag{2.73}$$

for any $\theta \in [0, 1]$ implying

$$\frac{\theta \mathbf{z}_1 + (1 - \theta)\mathbf{z}_2}{\theta t_1 + (1 - \theta)t_2} \in \boldsymbol{p}(C).$$

Now, by defining

$$\mu = \frac{\theta t_1}{\theta t_1 + (1 - \theta)t_2} \in [0, 1],$$

we get

$$\frac{\theta \mathbf{z}_1 + (1 - \theta)\mathbf{z}_2}{\theta t_1 + (1 - \theta)t_2} = \mu \frac{\mathbf{z}_1}{t_1} + (1 - \mu)\frac{\mathbf{z}_2}{t_2} \in \boldsymbol{p}(C), \tag{2.74}$$

which implies $\boldsymbol{p}(C)$ is convex. ∎

A *linear-fractional function* is formed by composing the perspective function with an affine function. Suppose $\boldsymbol{g}: \mathbb{R}^n \to \mathbb{R}^{m+1}$ is affine, i.e.,

$$\boldsymbol{g}(\mathbf{x}) = \begin{bmatrix} \mathbf{A} \\ \mathbf{c}^T \end{bmatrix} \mathbf{x} + \begin{bmatrix} \mathbf{b} \\ d \end{bmatrix}, \tag{2.75}$$

where $\mathbf{A} \in \mathbb{R}^{m \times n}, \mathbf{b} \in \mathbb{R}^m, \mathbf{c} \in \mathbb{R}^n$, and $d \in \mathbb{R}$. The function $\boldsymbol{f} : \ \mathbb{R}^n \to \mathbb{R}^m$ is given by $\boldsymbol{f} = \boldsymbol{p} \circ \boldsymbol{g}$, i.e.,

$$\boldsymbol{f}(\mathbf{x}) = \boldsymbol{p}(\boldsymbol{g}(\mathbf{x})) = \frac{\mathbf{A}\mathbf{x} + \mathbf{b}}{\mathbf{c}^T\mathbf{x} + d}, \quad \mathbf{dom} \ \boldsymbol{f} = \{\mathbf{x} \mid \mathbf{c}^T\mathbf{x} + d > 0\}, \tag{2.76}$$

is called a linear-fractional (or projective) function. Hence, linear-fractional functions preserve the convexity.

Remark 2.12 A perspective function can be viewed as the action of a pinhole camera [BV04]. Consider the second-order cone

$$C = \left\{ (\mathbf{x}, t) \in \mathbb{R}^3 \mid \|\mathbf{x}\|_2 \le t \right\}.$$

It can be readily seen that

$$\boldsymbol{p}(C \setminus \mathbf{0}_3) = \{\mathbf{x} \in \mathbb{R}^2 \mid \|\mathbf{x}\|_2 \le 1\}$$

is a convex set since $C \setminus \mathbf{0}_3$ is a convex set. As illustrated in Figure 2.14, $\boldsymbol{p}(C \setminus \mathbf{0}_3)$ corresponds to the unit disk in the hyperplane $\mathcal{H} = \{(\mathbf{x}, t = -1) \in \mathbb{R}^3\}$. The image of a tetrahedron (a convex object) in C through the perspective mapping is a triangle (a convex image) in $\mathcal{H}$. □

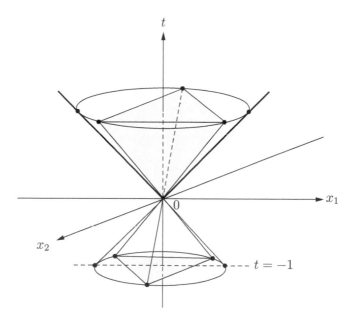

Figure 2.14 Pinhole camera interpretation of the perspective function, where the pinhole is the origin in $\mathbb{R}^3$ and the image of the second-order cone excluding the origin is a unit disk and the image of a tetrahedron excluding the origin is a triangle in the unit disk in the hyperplane $\mathcal{H} = \{(\mathbf{x}, t = -1) \in \mathbb{R}^3\}$.

Remark 2.13 The inverse image of a convex set under the perspective function is also convex, i.e., if $C \subseteq \mathbb{R}^n$ is convex, then

$$p^{-1}(C) = \{(\mathbf{x}, t) \in \mathbb{R}^{n+1} \mid \mathbf{x}/t \in C, \, t > 0\} \qquad (2.77)$$

is also convex. $\square$

2.4   Generalized inequalities

2.4.1   Proper cones and generalized inequalities

A cone $K \subseteq \mathbb{R}^n$ is called a *proper cone* if it satisfies the following conditions:

- K is convex.

- K is closed.

- K is solid, which means it has nonempty interior.

- K is pointed, which means that it contains no line (or equivalently, $\mathbf{x} \in K, -\mathbf{x} \in K \Rightarrow \mathbf{x} = \mathbf{0}$).

A proper cone K can be used to define a *"generalized inequality,"* which is a partial ordering on $\mathbb{R}^n$ (meaning that not all pairs of vectors in $\mathbb{R}^n$ are comparable) that has many of the properties of ordering on $\mathbb{R}$. The generalized inequality associated with proper cone K can be defined by

$$\textit{Non-strict generalized inequality: } \mathbf{x} \preceq_K \mathbf{y} \iff \mathbf{y} - \mathbf{x} \in K. \qquad (2.78a)$$
$$\textit{Strict generalized inequality: } \mathbf{x} \prec_K \mathbf{y} \iff \mathbf{y} - \mathbf{x} \in \mathbf{int} \; K. \qquad (2.78b)$$

Remark 2.14 The nonnegative orthant $\mathbb{R}^n_+$, which is a polyhedral cone, is also a proper cone (see Figure 2.9). The second-order cone $C = \{(\mathbf{x}, t) \in \mathbb{R}^{n+1} \mid \|\mathbf{x}\|_2 \leq t\} \subseteq \mathbb{R}^{n+1}$ is also a proper cone (see Figure 2.11). $\square$

Example 2.9 The positive semidefinite cone $K = \mathbb{S}^n_+$ is a proper cone in $\mathbb{S}^n$. Therefore it can be seen that for any $\mathbf{A}, \mathbf{B} \in \mathbb{S}^n$, $\mathbf{A} \preceq_K \mathbf{B}$ (or simply denoted by $\mathbf{A} \preceq \mathbf{B}$ only when $K = \mathbb{S}^n_+$) if and only if $\mathbf{B} - \mathbf{A} \in K = \mathbb{S}^n_+$ by the definition of $\preceq_K$, i.e., $\lambda_i(\mathbf{B} - \mathbf{A}) \geq 0$ for all i, where $\lambda_i(\cdot)$ denotes the ith eigenvalue of a matrix. It can also be shown that for any $\mathbf{A}, \mathbf{B} \in \mathbb{S}^n_{++}$,

$$\mathbf{A} \preceq \mathbf{B} \Leftrightarrow \mathcal{E}_{\mathbf{A}} = \{\mathbf{u} \mid \mathbf{u}^T \mathbf{A}^{-1} \mathbf{u} \leq 1\} \subseteq \mathcal{E}_{\mathbf{B}}, \qquad (2.79)$$

where both $\mathcal{E}_{\mathbf{A}}$ and $\mathcal{E}_{\mathbf{B}}$ are ellipsoids centered at the origin (cf. (2.37)). The proof of (2.79) is straightforward by (2.80) which is proven in Remark 2.15 below.

For instance, supposing that $\mathbf{B} = \mathbf{I}_2 \in \mathbb{S}^2_{++}$, and $\mathbf{A} \in \mathbb{S}^2_{++}$ with $\lambda_i(\mathbf{A}) \leq 1$, $i = 1, 2$, it can be seen that $\mathbf{A} \preceq \mathbf{B}$; $\mathcal{E}_{\mathbf{B}}$ is a Euclidean ball with center at the origin and radius equal to unity, and $\mathcal{E}_{\mathbf{A}}$, an ellipsoid with center at the origin and the maximum semiaxis length less than or equal to unity, is contained in $\mathcal{E}_{\mathbf{B}}$. $\square$

Remark 2.15 For $\mathbf{A}, \mathbf{B} \in \mathbb{S}_{++}^n$,

$$\mathbf{A} \preceq \mathbf{B} \;\Leftrightarrow\; \mathbf{A}^{-1} \succeq \mathbf{B}^{-1}. \tag{2.80}$$

Proof: Let us prove the sufficiency first. Because $\mathbf{A} \preceq \mathbf{B}$, we have $\mathbf{B} - \mathbf{A} \succeq \mathbf{0}$ by the definition of the generalized inequality defined on the proper cone $\mathbb{S}_+^n$. Thus, by EVD of $\mathbf{B} - \mathbf{A}$, we have

$$\mathbf{B} = \mathbf{A} + \mathbf{Q}\boldsymbol{\Lambda}\mathbf{Q}^T \tag{2.81}$$

where $\mathbf{Q} = [\mathbf{q}_1, \dots, \mathbf{q}_m] \in \mathbb{R}^{n \times m}$ is a semi-unitary matrix and $\boldsymbol{\Lambda} = \mathbf{Diag}(\lambda_1, \dots, \lambda_m) \succ \mathbf{0}$, and $m \leq n$. Note that when $m = n$, $\mathbf{B} - \mathbf{A} \succ \mathbf{0}$ and thus is nonsingular; otherwise it is singular. Applying Woodbury identity (1.79) to (2.81) yields

$$\mathbf{A}^{-1} - \mathbf{B}^{-1} = \mathbf{A}^{-1}\mathbf{Q}(\boldsymbol{\Lambda}^{-1} + \mathbf{Q}^T\mathbf{A}^{-1}\mathbf{Q})^{-1}\mathbf{Q}^T\mathbf{A}^{-1} \succeq \mathbf{0}, \tag{2.82}$$

implying $\mathbf{A}^{-1} \succeq \mathbf{B}^{-1}$.

The proof of necessity is straightforward by replacing $\mathbf{A}$ with $\mathbf{B}^{-1}$ and $\mathbf{B}$ with $\mathbf{A}^{-1}$ in the above sufficiency proof. Hence $\mathbf{A} \preceq \mathbf{B}$ if and only if $\mathbf{A}^{-1} \succeq \mathbf{B}^{-1}$. ∎ □

2.4.2   Properties of generalized inequalities

The generalized inequality has the following properties:

- It is preserved under addition: if $\mathbf{x} \preceq_K \mathbf{y}$ and $\mathbf{u} \preceq_K \mathbf{v}$, then $\mathbf{x} + \mathbf{u} \preceq_K \mathbf{y} + \mathbf{v}$.
- It is transitive: if $\mathbf{x} \preceq_K \mathbf{y}$ and $\mathbf{y} \preceq_K \mathbf{z}$, then $\mathbf{x} \preceq_K \mathbf{z}$.
- It is preserved under nonnegative scaling: if $\mathbf{x} \preceq_K \mathbf{y}$ and $\alpha \geq 0$, then $\alpha\mathbf{x} \preceq_K \alpha\mathbf{y}$.
- It is reflexive: $\mathbf{x} \preceq_K \mathbf{x}$.
- It is antisymmetric: if $\mathbf{x} \preceq_K \mathbf{y}$ and $\mathbf{y} \preceq_K \mathbf{x}$, then $\mathbf{x} = \mathbf{y}$.
- It is preserved under limits: if $\mathbf{x}_i \preceq_K \mathbf{y}_i$ for $i = 1, 2, \dots$, and $\mathbf{x}_i \to \mathbf{x}$ and $\mathbf{y}_i \to \mathbf{y}$ as $i \to \infty$, then $\mathbf{x} \preceq_K \mathbf{y}$.

Remark 2.16 Consider $K = \mathbb{R}_+^n$. Then,

$$\mathbf{x} \preceq_K \mathbf{y} \text{ if and only if } x_i \leq y_i, i = 1, \dots, n,$$
$$\mathbf{x} \prec_K \mathbf{y} \text{ if and only if } x_i < y_i, i = 1, \dots, n.$$

Also note that for any $x, y \in \mathbb{R}$, $x \not\leq y$ implies $x > y$ but for any $\mathbf{x}, \mathbf{y} \in \mathbb{R}^n$, $\mathbf{x} \not\preceq_K \mathbf{y}$ does not imply $\mathbf{x} \succ_K \mathbf{y}$. This is because $\mathbf{x} \not\preceq_K \mathbf{y}$ implies $x_i > y_i$, only for some i but not for all i. □

2.4.3   Minimum and minimal elements

With regard to the generalized inequality defined by a proper cone K, $\mathbf{x} \in S$ is a *minimum element* (*maximum element*) of S if $\mathbf{x} \preceq_K \mathbf{y}$ (or $\mathbf{x} \succeq_K \mathbf{y}$), for all

$y \in S$. Then it can be shown that

$$\mathbf{x} \in S \text{ is a minimum element of } S \text{ if and only if } S \subseteq \{\mathbf{x}\} + K; \qquad (2.83)$$

$$\mathbf{y} \in S \text{ is a maximum element of } S \text{ if and only if } S \subseteq \{\mathbf{y}\} - K. \qquad (2.84)$$

In other words, all the points in S are comparable to $\mathbf{x}$ and are greater than or equal to $\mathbf{x}$ (according to $\preceq_K$), and all the points in S are comparable to $\mathbf{y}$ and are smaller than or equal to $\mathbf{y}$. It can be inferred that a minimum element of a set S may not exist, and that it is unique if it does exist because the origin is the unique minimum element of K, and the set $\{\mathbf{x}\} + K$ is just a translation of the proper cone K with the orientation unchanged. Similarly, the maximum point of a set S is also unique if it exists (cf. Figure 2.15, where a unique minimum element (left plot) and a unique maximum element (right plot) of a set S in $\mathbb{R}^2$ are illustrated).

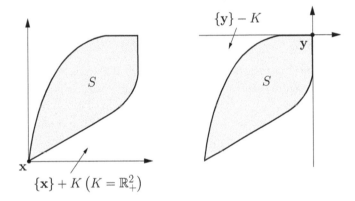

Figure 2.15 Minimum element ($\mathbf{x}$) (left plot) and maximum element ($\mathbf{y}$) (right plot) of a set $S \subset \mathbb{R}^2$, where $K = \mathbb{R}^2_+$.

Proof of (2.83): By definition of minimum element, we have $\mathbf{x} \preceq_K \mathbf{y}$, $\forall \mathbf{y} \in S$. This means

$$\mathbf{y} - \mathbf{x} \in K, \ \forall \mathbf{y} \in S \quad \text{(by definition of } \preceq_K)$$
$$\Leftrightarrow \{\mathbf{y} - \mathbf{x} \mid \mathbf{y} \in S\} \subseteq K$$
$$\Leftrightarrow S - \{\mathbf{x}\} \subseteq K$$
$$\Leftrightarrow S \subseteq \{\mathbf{x}\} + K.$$

Thus we have completed the proof that $\mathbf{x}$ is the minimum element of S if and only if $S \subseteq \{\mathbf{x}\} + K$. The proof for the maximum point $\mathbf{y}$ of S is similar to that for the minimum element of S and so omitted here. ∎

An element $\mathbf{x}$ in a set S is said to be a *minimal element* of S if $\mathbf{y} \in S$, $\mathbf{y} \preceq_K \mathbf{x} \Rightarrow \mathbf{y} = \mathbf{x}$. In other words, $\mathbf{x}$ is a minimal element if

$$\{\mathbf{y} \mid \mathbf{y} \preceq_K \mathbf{x} \; \forall \mathbf{y} \in S\} = \{\mathbf{x}\}$$
$$\Leftrightarrow \{\mathbf{y} \mid \mathbf{y} - \mathbf{x} = \mathbf{k} \in -K\} \cap S = \{\mathbf{x}\}$$
$$\Leftrightarrow (\{\mathbf{x}\} - K) \cap S = \{\mathbf{x}\} \tag{2.85}$$

(cf. Figure 2.16, where infinitely many minimal elements of a set S are illustrated). An alternative condition for $\mathbf{x}$ to be a minimal point of S is

$$\mathbf{y} \not\preceq \mathbf{x} \; \forall \mathbf{y} \in S, \; \mathbf{y} \neq \mathbf{x}. \tag{2.86}$$

It can be seen that when the set of all the minimal elements of S only contains a unique element, that must be the minimum element of S.

In contrast to minimal elements of S, an element $\mathbf{x}$ of S is said to be a *maximal element* of S if $\mathbf{y} \in S$, $\mathbf{y} \succeq_K \mathbf{x} \Rightarrow \mathbf{y} = \mathbf{x}$. Similarly, it can be shown that $\mathbf{x} \in S$ is a maximal element if

$$(\{\mathbf{x}\} + K) \cap S = \{\mathbf{x}\}. \tag{2.87}$$

The maximum element is also unique if it exists, while maximal elements are always nonunique as illustrated in Figure 2.16. The condition for maximum element (2.84) and that for maximal elements (2.87) actually corresponds to that for minimum element (2.83) and that for minimal elements (2.85), respectively, with the proper cone K replaced by $-K$.

A set $S \subseteq \mathbb{R}_+^N$ is called a polyblock [Tuy00] if it is the union of a finite number of boxes. A polyblock $S \in \mathbb{R}_+^2$ containing two minimal elements, $\mathbf{x}_1$, $\mathbf{x}_2$, and a unique maximum element (which is also a maximal point), $\mathbf{y}$, is depicted in Figure 2.17.

2.5   Dual norms and dual cones

Dual of a norm is also a norm. Dual of a cone K is also a cone denoted by K^*. When K and K^* are proper cones, each of them can define a generalized inequality. The one defined by K involved in a convex optimization problem will induce the other involved in the associated dual optimization problem, and vice versa. Because cones and dual cones may often involve norms and the associated dual norms, let us introduce the concepts of dual norms and those of dual cones in the following two subsections, respectively.

2.5.1   Dual norms

Let $\|\cdot\|$ be a norm on $\mathbb{R}^n$. The associated *dual norm* $\|\cdot\|_*$ is defined as

$$\|\mathbf{u}\|_* = \sup \{\mathbf{u}^T \mathbf{x} \mid \|\mathbf{x}\| \leq 1\}, \tag{2.88}$$

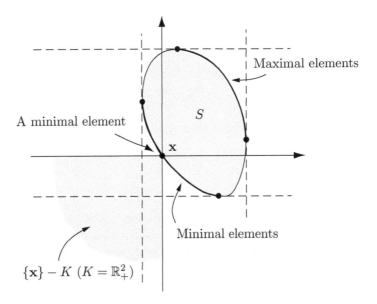

Figure 2.16 Minimal elements and maximal elements of a set $S \subset \mathbb{R}^2$.

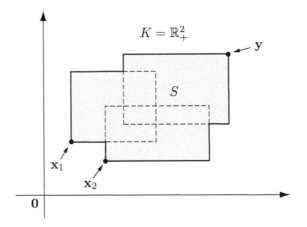

Figure 2.17 Minimal elements, $\mathbf{x}_1$, $\mathbf{x}_2$, and maximum element, $\mathbf{y}$, of a polyblock $S \subset \mathbb{R}_+^2$.

that is, the maximum inner product of $\mathbf{u}$ with $\mathbf{x}$ over all $\mathbf{x}$ in the norm ball of unit radius. It can be shown that the dual of the p-norm is the q-norm by Hölder's inequality (to be introduced in Chapter 3), where q satisfies $1/p + 1/q = 1$. In other words, for $\| \cdot \| = \| \cdot \|_p$,

$$\|\mathbf{u}\|_* = \sup \left\{ \mathbf{u}^T\mathbf{x} \mid \|\mathbf{x}\|_p \le 1 \right\}$$
$$\le \sup_{\|\mathbf{x}\|_p \le 1} \|\mathbf{x}\|_p \|\mathbf{u}\|_q = \|\mathbf{u}\|_q \quad \text{(by Hölder's inequality),}$$

where the upper bound on the right-hand side of the inequality is achievable. Therefore, $\| \cdot \|_* = \| \cdot \|_q$ for $\| \cdot \| = \| \cdot \|_p$ and vice versa where $1/p + 1/q = 1$ and $p > 1$ (e.g., $\| \cdot \|_* = \| \cdot \|_1$ for $\| \cdot \| = \| \cdot \|_\infty$ and vice versa).

Let us illustrate $\|\mathbf{u}\|_*$ for $p = \infty$ in Figure 2.18, where $\mathbf{u}$ is in the first quadrant of $\mathbb{R}^2$,

$$\mathbf{x}^\star = \arg\sup_{\|\mathbf{x}\|_\infty \le 1} \mathbf{u}^T\mathbf{x} = \big(\mathrm{sgn}(u_1), \mathrm{sgn}(u_2) \big)$$
$$\|\mathbf{u}\|_* = \mathbf{u}^T\mathbf{x}^\star = \|\mathbf{u}\|_1 \tag{2.89}$$

where $\mathrm{sgn}(x)$ is the sign function of x, and $\mathbf{x}^\star$ is the column vector with maximum projection (rather than in absolute value) on the vector $\mathbf{u}$. From the above illustration, one can infer that $\|\mathbf{x}^\star\|_\infty = 1$ for any nonzero $\mathbf{u}$, and $\| \cdot \|_* = \| \cdot \|_1$ is the dual of $\| \cdot \|_\infty$. Some more interesting characteristics about the norm and dual norm are given in the following remark.

Remark 2.17 Suppose that $\| \cdot \|$ is a norm and $\mathcal{B} \triangleq \{\mathbf{x} \mid \|\mathbf{x}\| \le 1\}$ is the associated unit-radius norm ball centered at the origin. The dual of $\| \cdot \|$ defined in (2.88) is actually a convex optimization problem (to be introduced in Chapter 4) to maximize the linear objective function $\mathbf{u}^T\mathbf{x}$ subject to $\mathbf{x} \in \mathcal{B}$, and the dual norm $\|\mathbf{u}\|_*$ is the optimal value, namely, there exists an optimal $\mathbf{x}^\star \in \mathbf{bd}\,\mathcal{B}$ (and thus $\|\mathbf{x}^\star\| = 1$) such that $\mathbf{u}^T\mathbf{x} \le \|\mathbf{u}\|_* = \mathbf{u}^T\mathbf{x}^\star$ for all $\mathbf{x} \in \mathcal{B}$. Due to the fact that $\mathcal{B}$ is compact and convex, the dual norm of $\| \cdot \|$ can be further characterized as follows:

$$\|\mathbf{u}\|_* = \mathbf{u}^T\mathbf{x}^\star, \quad \exists \mathbf{x}^\star \in \mathcal{B}_{\mathrm{extr}} \subseteq \mathbf{bd}\,\mathcal{B}$$
$$\mathcal{B} = \mathbf{conv}\,\mathcal{B}_{\mathrm{extr}} \subset \mathcal{H}_-(\mathbf{u}, \mathbf{x}^\star) \triangleq \{\mathbf{x} \mid \mathbf{u}^T(\mathbf{x} - \mathbf{x}^\star) \le 0\} \tag{2.90}$$

where the halfspace $\mathcal{H}_-(\mathbf{u}, \mathbf{x}^\star)$ has defined a supporting hyperplane $\mathcal{H}(\mathbf{u}, \mathbf{x}^\star)$ (the boundary hyperplane of $\mathcal{H}_-(\mathbf{u}, \mathbf{x}^\star)$ such that $\mathbf{x}^\star \in \mathbf{bd}\,\mathcal{B} \cap \mathcal{H}(\mathbf{u}, \mathbf{x}^\star)$) of the compact convex set $\mathcal{B}$ (cf. (2.137)) to be introduced in Subsection 2.6.2 as illustrated in Figure 2.18. □

Consider a matrix $\mathbf{X} \in \mathbb{R}^{M \times N}$ with the SVD given by

$$\mathbf{X} = \mathbf{U}_r \boldsymbol{\Sigma}_r \mathbf{V}_r^T = \sum_{i=1}^{r} \sigma_i \mathbf{u}_i \mathbf{v}_i^T \quad \text{(by (1.111)),} \tag{2.91}$$

where $\boldsymbol{\Sigma}_r = \mathbf{Diag}(\sigma_1, \ldots, \sigma_r)$ contains all the positive singular values of $\mathbf{X}$, $\mathbf{U}_r = [\mathbf{u}_1, \ldots, \mathbf{u}_r] \in \mathbb{R}^{M \times r}$ contains the associated left singular vectors, and $\mathbf{V}_r = [\mathbf{v}_1, \ldots, \mathbf{v}_r] \in \mathbb{R}^{N \times r}$ contains the associated right singular vectors. Note that $\mathbf{U}_r^T\mathbf{U}_r = \mathbf{V}_r^T\mathbf{V}_r = \mathbf{I}_r$. The ℓ_2-norm (spectral norm) of $\mathbf{X}$ (cf. (1.10)) has

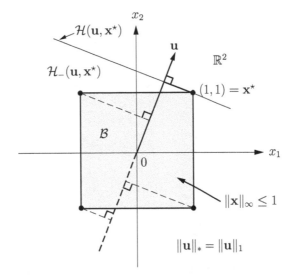

Figure 2.18 Associated with the ∞-norm of a vector $\mathbf{u} \in \mathbb{R}^2$, the dual norm is $\|\mathbf{u}\|_* = \mathbf{u}^T \mathbf{x}^\star = \|\mathbf{u}\|_1$, where $\mathbf{x}^\star$ is an extreme point of the norm ball $\mathcal{B}$ (unit-radius ∞-norm ball centered at the origin for this case); $(\mathbf{u}, \mathbf{x}^\star)$ also defines a supporting hyperplane $\mathcal{H}$ of the compact convex set $\mathcal{B}$ such that $\mathcal{B}$ is contained in the associated halfspace $\mathcal{H}_-$ (cf. (2.90)).

been shown to be the maximum singular value of $\mathbf{X}$, i.e.,

$$\|\mathbf{X}\|_2 = \sigma_{\max}(\mathbf{X}) = \max\{\sigma_1, \ldots, \sigma_r\} \quad \text{(by (1.10))}. \tag{2.92}$$

Next, let us find the dual norm of $\|\mathbf{X}\|_2$, which is defined as

$$\|\mathbf{X}\|_* = \sup \ \{\text{Tr}(\mathbf{X}^T \mathbf{Y}) \mid \|\mathbf{Y}\|_2 \leq 1\}. \tag{2.93}$$

Let

$$\mathbf{Y} = \boldsymbol{U}_{r'} \boldsymbol{\Sigma}_{r'} \boldsymbol{V}_{r'}^T = \sum_{i=1}^{r'} \sigma_i' \boldsymbol{u}_i \boldsymbol{v}_i^T \tag{2.94}$$

where $(\sigma_i', \boldsymbol{u}_i, \boldsymbol{v}_i), i = 1, \ldots, r'$, are the 3-tuples of the positive singular values, left singular vectors and right singular vectors of $\mathbf{Y}$, and

$$\boldsymbol{\Sigma}_{r'} = \mathbf{Diag}(\sigma_1', \ldots, \sigma_{r'}'),$$
$$\boldsymbol{U}_{r'} = [\boldsymbol{u}_1, \ldots, \boldsymbol{u}_{r'}] \in \mathbb{R}^{M \times r'},$$
$$\boldsymbol{V}_{r'} = [\boldsymbol{v}_1, \ldots, \boldsymbol{v}_{r'}] \in \mathbb{R}^{N \times r'}.$$

Then

$$\text{Tr}(\mathbf{X}^T\mathbf{Y}) = \sum_{i=1}^{r}\sum_{j=1}^{r'}\sigma_i\sigma_j'\text{Tr}(\mathbf{v}_i\mathbf{u}_i^T\mathbf{u}_j\mathbf{v}_j^T) \quad (\text{by (2.91) and (2.94)})$$

$$= \sum_{i=1}^{r}\sigma_i\sum_{j=1}^{r'}\left(\sqrt{\sigma_j'}\mathbf{u}_i^T\mathbf{u}_j\right)\left(\sqrt{\sigma_j'}\mathbf{v}_i^T\mathbf{v}_j\right)$$

$$\leq \sum_{i=1}^{r}\sigma_i\left(\sum_{j=1}^{r'}\sigma_j'(\mathbf{u}_i^T\mathbf{u}_j)^2\right)^{1/2}\left(\sum_{j=1}^{r'}\sigma_j'(\mathbf{v}_i^T\mathbf{v}_j)^2\right)^{1/2}$$

$$\leq \sum_{i=1}^{r}\sigma_i\left(\mathbf{u}_i^T(\boldsymbol{U}_{r'}\boldsymbol{U}_{r'}^T)\mathbf{u}_i\right)^{1/2}\left(\mathbf{v}_i^T(\boldsymbol{V}_{r'}\boldsymbol{V}_{r'}^T)\mathbf{v}_i\right)^{1/2}$$

$$\leq \sum_{i=1}^{r}\sigma_i \tag{2.95}$$

where the first inequality is due to Cauchy–Schwartz inequality, the second inequality is due to $0 < \sigma_j' \leq \sigma_{\max}(\mathbf{Y}) = \|\mathbf{Y}\|_2 \leq 1$ for all j, and the last inequality is due to the fact that $\boldsymbol{U}_{r'}$ and $\boldsymbol{V}_{r'}$ are semi-unitary matrices with identical rank $r' \leq \min\{M, N\}$, and thus $\boldsymbol{U}_{r'}\boldsymbol{U}_{r'}^T$ and $\boldsymbol{V}_{r'}\boldsymbol{V}_{r'}^T$ are $M \times M$ and $N \times N$ projection matrices with the same rank r' (cf. Remark 1.15). Hence, all the inequalities in (2.95) hold with equality if (i) $\mathbf{u}_i^T\mathbf{u}_j = \alpha\mathbf{v}_i^T\mathbf{v}_j$ (where $\alpha > 0$ is a constant) for all $j = 1, \ldots, r'$, (ii) $\sigma_i' = 1$ for all $i = 1, \ldots, r'$, and (iii) $\mathbf{u}_i \in \mathcal{R}(\boldsymbol{U}_{r'})$, and $\mathbf{v}_i \in \mathcal{R}(\boldsymbol{V}_{r'})$ for all $i = 1, \ldots, r$, accordingly requiring $r' \geq r$. For instance, when $r' = r$, $\mathbf{u}_i = \mathbf{u}_i$ and $\mathbf{v}_i = \mathbf{v}_i$, for all i, and thus $\alpha = 1$, an optimal $\mathbf{Y}^\star = \mathbf{U}_r\mathbf{V}_r^T$ to (2.93) yields

$$\|\mathbf{X}\|_* = \text{Tr}(\mathbf{X}^T\mathbf{Y}^\star) = \sum_{i=1}^{r}\sigma_i, \tag{2.96}$$

also called the *nuclear norm*, which is the sum of all the singular values of $\mathbf{X}$. Actually, the dual norm of $\|\mathbf{X}\|_2$ defined by (2.93) itself is also a convex optimization problem (to be introduced in Chapter 4). This dual norm is also a special case of the atomic norm to be introduced next.

The *atomic norm*, denoted as $\|\mathbf{x}\|_{\mathcal{A}}$, is defined as

$$\|\mathbf{x}\|_{\mathcal{A}} = \inf\{t > 0 \mid \mathbf{x} \in t \cdot \mathbf{conv}\,\mathcal{A}\} \quad (\text{cf. (1.22)}) \tag{2.97}$$

where $\mathcal{A}$ (called atomic set) is required to be a centrally symmetric set (i.e., symmetric w.r.t. the origin) that contains all the extreme points of $\mathbf{conv}\,\mathcal{A}$. Note that the atomic set $\mathcal{A}$ can be a finite set but it does not need to be finite, and the set

$$\mathbf{conv}\,\mathcal{A} = \{\mathbf{x} \mid \|\mathbf{x}\|_{\mathcal{A}} \leq 1\} \tag{2.98}$$

denotes an atomic-norm ball of radius equal to unity. $\|\mathbf{x}\|_{\mathcal{A}}$ is also the optimal value $t^\star \geq 0$ of the minimization problem (2.97) such that $\mathbf{x} \in \mathbf{bd}\,(t^\star\mathbf{conv}\,\mathcal{A})$.

The atomic norm is actually a more general vector or matrix norm as illustrated in the following cases.

- $\mathcal{A}$ constituted by sparse vectors:

$$\mathcal{A} = \{\pm \mathbf{e}_i \in \mathbb{R}^N\}_{i=1}^N \tag{2.99}$$

for which **conv** $\mathcal{A}$ is a 1-norm ball of radius equal to one in $\mathbb{R}^N$, also a cross-polytope. Then $\|\mathbf{x}\|_{\mathcal{A}} = \|\mathbf{x}\|_1$ (cf. Figure 1.1).

- $\mathcal{A}$ constituted by binary vectors:

$$\mathcal{A} = \{(a_1, \ldots, a_N) \in \{\pm 1\}^N\} \tag{2.100}$$

for which **conv** $\mathcal{A}$ is an ∞-norm ball of radius equal to one in $\mathbb{R}^N$, also a hypercube. Then $\|\mathbf{x}\|_{\mathcal{A}} = \|\mathbf{x}\|_\infty$ (cf. Figure 1.1).

- $\mathcal{A}$ constituted by the boundary of a 2-norm ball with center at the origin and radius equal to unity:

$$\mathcal{A} = \{\mathbf{a} \in \mathbb{R}^N \mid \|\mathbf{a}\|_2 = 1\}. \tag{2.101}$$

Then $\|\mathbf{x}\|_{\mathcal{A}} = \|\mathbf{x}\|_2$.

- $\mathcal{A}$ constituted by normalized rank-one matrices:

$$\begin{aligned} \mathcal{A} &= \{\mathbf{A} \in \mathbb{R}^{M \times N} \mid \mathrm{rank}(\mathbf{A}) = 1, \|\mathbf{A}\|_{\mathrm{F}} = 1\} \\ &= \{\mathbf{u}\mathbf{v}^T \mid \|\mathbf{u}\|_2 = \|\mathbf{v}\|_2 = 1, \mathbf{u} \in \mathbb{R}^M, \mathbf{v} \in \mathbb{R}^N\} \end{aligned} \tag{2.102}$$

for which **conv** $\mathcal{A}$ is a nuclear-norm ball of radius equal to one. Let $\mathbf{X} \in \mathbb{R}^{M \times N}$, with $\mathrm{rank}(\mathbf{X}) = r$ and positive singular values $\sigma_i, i = 1, \ldots, r$. Then

$$\|\mathbf{X}\|_{\mathcal{A}} = \sum_{i=1}^r \sigma_i \quad \text{(nuclear norm of } \mathbf{X}), \tag{2.103}$$

i.e., the dual norm of the ℓ_2-norm of $\mathbf{X}$.

Proof of (2.103): The SVD of $\mathbf{X} \in \mathbb{R}^{M \times N}$ is given by

$$\mathbf{X} = \sum_{i=1}^r \sigma_i \mathbf{u}_i \mathbf{v}_i^T,$$

where $\mathbf{u}_i$ and $\mathbf{v}_i$, respectively, are the left and right singular vectors associated with σ_i. Let

$$t^\star \triangleq \sum_{i=1}^r \sigma_i = \|\mathbf{X}\|_* \quad \text{(cf. (2.96))}.$$

Then it can be seen that

$$\sum_{i=1}^r \frac{\sigma_i}{t} \begin{cases} \leq 1, & \text{if } t \geq t^\star, \\ > 1, & \text{if } 0 < t < t^\star. \end{cases} \tag{2.104}$$

Let $\mathbf{A}_i = \boldsymbol{u}_i \boldsymbol{v}_i^T \in \mathcal{A}$. Then it can be easily shown that $\mathbf{0}_{M \times N} \in \mathbf{conv}\,\mathcal{A}$ (since $\mathcal{A}$ is symmetric w.r.t. the origin, i.e., the zero matrix), and further it can be inferred from (2.104) that

$$\mathbf{A}(t) \triangleq \sum_{i=1}^{r} \frac{\sigma_i}{t} \mathbf{A}_i \begin{cases} \in \mathbf{conv}\,\mathcal{A}, \text{ if } t \geq t^\star, \\ \notin \mathbf{conv}\,\mathcal{A}, \text{ if } 0 < t < t^\star \end{cases}$$

$$\Rightarrow \ \mathbf{A}(t^\star) \in \mathbf{bd}\,(\mathbf{conv}\,\mathcal{A}).$$

Therefore,

$$\mathbf{X} = t^\star \mathbf{A}(t^\star) \in t \cdot \mathbf{conv}\,\mathcal{A}$$

only when $t \geq t^\star$, and so we have proven $\|\mathbf{X}\|_{\mathcal{A}} = t^\star = \|\mathbf{X}\|_*$ by (2.97). ■

Remark 2.18 *(Convex hull of normalized rank-one matrices)* Similar to (2.98) for the vector case, $\mathbf{conv}\,\mathcal{A}$ represents the unit norm ball defined by the atomic set $\mathcal{A}$ composed of normalized rank-one matrices (cf. (2.102)). In this remark, we show that

$$\mathbf{conv}\,\mathcal{A} = \left\{ \mathbf{X} \in \mathbb{R}^{M \times N} \mid \|\mathbf{X}\|_{\mathcal{A}} \leq 1 \right\}$$

$$= \left\{ \mathbf{X} \in \mathbb{R}^{M \times N} \mid \sum_{i=1}^{r} \sigma_i \leq 1 \right\} \quad \text{(by (2.103))}, \tag{2.105}$$

where $\sigma_1, \ldots, \sigma_r$ denote the singular values of $\mathbf{X}$.

Proof of (2.105): Let

$$\mathbf{X} = \sum_{i=1}^{k} \theta_i \mathbf{u}_i \mathbf{v}_i^T = \sum_{i=1}^{r} \sigma_i \boldsymbol{u}_i \boldsymbol{v}_i^T \in \mathbf{conv}\,\mathcal{A}, \ k \in \mathbb{Z}_{++} \tag{2.106}$$

where $\|\mathbf{u}_i\|_2 = \|\mathbf{v}_i\|_2 = 1$ and $\theta_i \geq 0$ for all i, and $\sum_i \theta_i = 1$; $\boldsymbol{u}_i$ and $\boldsymbol{v}_i$ are the left singular vector and right singular vector of $\mathbf{X}$ associated with σ_i, respectively. Then we have

$$\sum_{i=1}^{r} \sigma_i = \sum_{i=1}^{r} \boldsymbol{u}_i^T \mathbf{X} \boldsymbol{v}_i \quad \text{(cf. (1.113))}$$

$$= \sum_{i=1}^{r} \boldsymbol{u}_i^T \left(\sum_{j=1}^{k} \theta_j \mathbf{u}_j \mathbf{v}_j^T \right) \boldsymbol{v}_i = \sum_{j=1}^{k} \theta_j \sum_{i=1}^{r} (\boldsymbol{u}_i^T \mathbf{u}_j)(\boldsymbol{v}_i^T \mathbf{v}_j)$$

$$\leq \sum_{j=1}^{k} \theta_j \left\{ \sum_{i=1}^{r} (\boldsymbol{u}_i^T \mathbf{u}_j)^2 \right\}^{1/2} \left\{ \sum_{i=1}^{r} (\boldsymbol{v}_i^T \mathbf{v}_j)^2 \right\}^{1/2} \leq \sum_{j=1}^{k} \theta_j = 1$$

where the first inequality is due to Cauchy–Schwartz inequality, and the second inequality is because $\{\boldsymbol{u}_1, \ldots, \boldsymbol{u}_r\}$ and $\{\boldsymbol{v}_1, \ldots, \boldsymbol{v}_r\}$ are an orthonormal basis of a subspace of $\mathbb{R}^M$ and $\mathbb{R}^N$, respectively, leading to the terms inside the braces less than or equal to $\|\mathbf{u}_j\|_2^2 = \|\mathbf{v}_j\|_2^2 = 1$. Thus $\mathbf{conv}\,\mathcal{A}$ is a subset on the right-hand side of (2.105).

On the other hand, it can be easily seen, from (2.106), that any $M \times N$ matrix $\mathbf{X}$ with the sum of singular values σ_i no larger than unity belongs to $\mathbf{conv}\,\mathcal{A}$ due to the fact that $\mathbf{0}_{M \times N} \in \mathbf{conv}\,\mathcal{A}$ and all σ_i are strictly positive. Thus, the set on the right-hand side of (2.105) is also a subset of $\mathbf{conv}\,\mathcal{A}$. Therefore, the proof has been completed. ■ □

Remark 2.19 *(Dual and bidual of atomic norm)* The dual of the atomic norm of a vector $\mathbf{u} \in \mathbb{R}^n$ defined by $\mathcal{A} \subset \mathbb{R}^n$, denoted as $\|\mathbf{u}\|_{\mathcal{A}}^*$, is given by

$$\|\mathbf{u}\|_{\mathcal{A}}^* = \sup\{\mathbf{u}^T\mathbf{x} \mid \|\mathbf{x}\|_{\mathcal{A}} \leq 1\} = \sup\{\mathbf{u}^T\mathbf{x} \mid \mathbf{x} \in \mathcal{A}\} \quad \text{(cf. (2.90))}$$
$$\Rightarrow \mathbf{u}^T\mathbf{x} \leq \|\mathbf{u}\|_{\mathcal{A}}^* \cdot \|\mathbf{x}\|_{\mathcal{A}}, \ \forall \mathbf{u}, \mathbf{x} \in \mathbb{R}^n \tag{2.107}$$

where the inequality is tight with the Hölder's inequality (cf. Example 3.2) as a special case. Similarly, for a matrix $\mathbf{U} \in \mathbb{R}^{m \times n}$ along with an atomic set $\mathcal{A} \subset \mathbb{R}^{m \times n}$, its atomic norm denoted as $\|\mathbf{U}\|_{\mathcal{A}}^*$ is given by

$$\|\mathbf{U}\|_{\mathcal{A}}^* = \sup\{\mathrm{Tr}(\mathbf{U}^T\mathbf{X}) \mid \|\mathbf{X}\|_{\mathcal{A}} \leq 1\} = \sup\{\mathrm{Tr}(\mathbf{U}^T\mathbf{X}) \mid \mathbf{X} \in \mathcal{A}\}$$
$$\Rightarrow \mathrm{Tr}(\mathbf{U}^T\mathbf{X}) \leq \|\mathbf{U}\|_{\mathcal{A}}^* \cdot \|\mathbf{X}\|_{\mathcal{A}}, \ \forall \mathbf{U}, \mathbf{X} \in \mathbb{R}^{m \times n}. \tag{2.108}$$

By (2.107), one can infer that the following implications hold

$$\mathbf{u}^T\mathbf{x} \leq \|\mathbf{u}\|_{\mathcal{A}}^* \cdot \|\mathbf{x}\|_{\mathcal{A}} \ \forall \mathbf{u}, \mathbf{x} \in \mathbb{R}^n \Rightarrow \sup\{\mathbf{u}^T\mathbf{x} \mid \mathbf{x} \in \mathbb{A}\} = \|\mathbf{u}\|_{\mathcal{A}}^* \ \forall \mathbf{u} \in \mathbb{R}^n$$
$$\mathbf{u}^T\mathbf{x} \leq \|\mathbf{u}\|_{\mathcal{A}}^* \cdot \|\mathbf{x}\|_{\mathcal{A}}^{**} \ \forall \mathbf{u}, \mathbf{x} \in \mathbb{R}^n \Rightarrow \sup\{\mathbf{u}^T\mathbf{x} \mid \mathbf{x} \in \mathbb{B}\} = \|\mathbf{u}\|_{\mathcal{A}}^* \ \forall \mathbf{u} \in \mathbb{R}^n$$

where $\mathbb{A} = \{\mathbf{x} \in \mathbb{R}^n \mid \|\mathbf{x}\|_{\mathcal{A}} \leq 1\}$ and $\mathbb{B} = \{\mathbf{x} \in \mathbb{R}^n \mid \|\mathbf{x}\|_{\mathcal{A}}^{**} \leq 1\}$. Note that both $\mathbb{A}$ and $\mathbb{B}$ are compact convex sets. It can be further inferred, by (2.90), that for each nonzero $\mathbf{u} \in \mathbb{R}^n$, there must exist an extreme point $\boldsymbol{\alpha} \in \mathbb{A}_{\mathrm{extr}}$ and an extreme point $\boldsymbol{\beta} \in \mathbb{B}_{\mathrm{extr}}$ such that $\mathbf{u}^T\boldsymbol{\alpha} = \mathbf{u}^T\boldsymbol{\beta} = \|\mathbf{u}\|_{\mathcal{A}}^*$, $\boldsymbol{\alpha}, \boldsymbol{\beta} \in \mathcal{H}(\mathbf{u}) = \{\boldsymbol{x} \mid \mathbf{u}^T\boldsymbol{x} = \|\mathbf{u}\|_{\mathcal{A}}^*\}$ and meanwhile $\mathbb{A} \subset \mathcal{H}_-(\mathbf{u}) = \{\boldsymbol{x} \mid \mathbf{u}^T\boldsymbol{x} \leq \|\mathbf{u}\|_{\mathcal{A}}^*\}$ and $\mathbb{B} \subset \mathcal{H}_-(\mathbf{u})$. Hence,

$$\mathbb{A} = \bigcap_{\mathbf{u} \in \mathbb{R}^n \setminus \{\mathbf{0}\}} \mathcal{H}_-(\mathbf{u}) = \mathbb{B} \quad \text{(by (2.137))}$$
$$\Rightarrow \|\mathbf{x}\|_{\mathcal{A}}^{**} = \|\mathbf{x}\|_{\mathcal{A}} \ \forall \mathbf{x} \in \mathbb{R}^n, \tag{2.109}$$

namely, the bidual of any atomic norm is itself. For example, for $\mathcal{A}$ being the set of low-rank matrices given by (2.102), $\|\mathbf{X}\|_{\mathcal{A}}$ (cf. (2.103)) is the nuclear norm $\|\mathbf{X}\|_*$ (cf. (2.96)), which is also the dual of $\|\mathbf{X}\|_2$. Hence, $\|\mathbf{X}\|_{\mathcal{A}}^* = \|\mathbf{X}\|_2$ (spectral norm of $\mathbf{X}$). □

2.5.2 Dual cones

The *dual cone* of a cone K is defined as

$$K^* \triangleq \{\mathbf{y} \mid \mathbf{x}^T\mathbf{y} \geq 0 \ \forall \mathbf{x} \in K\}. \tag{2.110}$$

For instance, the set $\{\mathbf{x} = (x_1, x_2) \in \mathbb{R}_+^2 \mid x_1 \geq x_2\}$ is a cone of angle 45 degree, and its dual is the set $\{\mathbf{x} = (x_1, x_2) \in \mathbb{R}^2 \mid x_1 \geq 0, \ x_1 + x_2 \geq 0\}$ which is a cone of angle 135 degree. Moreover, $\mathbb{R}^2 \setminus \mathbb{R}^2_{++}$ is a cone, and its dual is the origin $\mathbf{0}_2$ whose dual is $\mathbb{R}^2$.

The dual cone K^* of any cone K is unique, and K^* is always convex and closed, even if the original cone K is not. For instance, assume that $\mathbf{a}$ and $\mathbf{b}$ are two distinct nonzero vectors in $\mathbb{R}^2$, and let

$$K_1 = \mathbf{conic}\{\mathbf{a}\} = \{\mathbf{x} = \theta\mathbf{a} \mid \theta \geq 0\}, \tag{2.111}$$

$$K_2 = \mathbf{conic}\{\mathbf{b}\} = \{\mathbf{x} = \theta\mathbf{b} \mid \theta \geq 0\}, \tag{2.112}$$

$$K = K_1 \cup K_2 = \{\mathbf{x} \mid \mathbf{x} = \theta_1\mathbf{a} + \theta_2\mathbf{b}, \ (\theta_1, \theta_2) \in \mathbb{R}^2_+, \ \theta_1\theta_2 = 0\}, \tag{2.113}$$

$$\mathcal{K} = \mathbf{conv} \ K = \mathbf{conic}\{\mathbf{a}, \mathbf{b}\}, \tag{2.114}$$

$$\mathcal{K}_1 = (\mathbf{int} \ \mathcal{K}) \cup \{\mathbf{0}_2\}. \tag{2.115}$$

Note that K_1, K_2, and $\mathcal{K}$ are closed convex cones, K is a closed nonconvex cone, and $\mathcal{K}_1$ a convex cone but not closed. Their duals can be easily seen by (2.110) to be

$$K_1^* = H_+(\mathbf{a}) \triangleq \{\mathbf{y} \mid \mathbf{a}^T\mathbf{y} \geq 0\} \tag{2.116}$$

$$K_2^* = H_+(\mathbf{b}) \triangleq \{\mathbf{y} \mid \mathbf{b}^T\mathbf{y} \geq 0\} \tag{2.117}$$

$$K^* = \{\mathbf{y} \mid \mathbf{a}^T\mathbf{y} \geq 0, \ \mathbf{b}^T\mathbf{y} \geq 0\} = \mathcal{K}^* = \mathcal{K}_1^* \tag{2.118}$$

which are all closed convex cones as illustrated in Figure 2.19. Next, by the following examples, we illustrate some widely used cones and three of them are *self-dual* (i.e., a cone and its dual are identical).

Example 2.10 The dual cone of a subspace $V \subseteq \mathbb{R}^n$ (which is a convex cone) is its orthogonal complement $V^\perp = \{\mathbf{y} \mid \mathbf{y}^T\mathbf{v} = 0, \ \forall\mathbf{v} \in V\}$. $\quad\square$

Example 2.11 *(Nonnegative orthant)* The dual cone of the nonnegative orthant cone $\mathbb{R}^n_+$ is also the nonnegative orthant cone $\mathbb{R}^n_+$. That is to say, the cone $\mathbb{R}^n_+$ is self-dual. The result can be easily proved by the definition of dual cone. $\quad\square$

Example 2.12 $K = \mathbb{S}^n_+$, $K^* = K$ (self-dual), where

$$K^* = \{\mathbf{Y} \mid \mathrm{Tr}(\mathbf{XY}) \geq 0, \ \forall\mathbf{X} \succeq \mathbf{0}\}$$

is the dual of K.

Proof: First of all, let us prove $K^* \subseteq \mathbb{S}^n_+$. Let $\mathbf{Y} \in K^*$. We need to prove $\mathbf{Y} \in \mathbb{S}^n_+$. Suppose that $\mathbf{Y} \notin \mathbb{S}^n_+$. Then there must exist a $\mathbf{q} \in \mathbb{R}^n$ such that

$$\mathbf{q}^T\mathbf{Y}\mathbf{q} = \mathrm{Tr}\left(\mathbf{q}\mathbf{q}^T\mathbf{Y}\right) < 0.$$

Let $\mathbf{X} = \mathbf{q}\mathbf{q}^T \in \mathbb{S}^n_+$. Then $\mathrm{Tr}(\mathbf{XY}) = \mathrm{Tr}(\mathbf{q}\mathbf{q}^T\mathbf{Y}) < 0 \Rightarrow \mathbf{Y} \notin K^*$. Therefore, it contradicts with $\mathbf{Y} \in K^*$. Therefore, $\mathbf{Y} \in \mathbb{S}^n_+$, and thus we have completed the proof of $K^* \subseteq \mathbb{S}^n_+$.

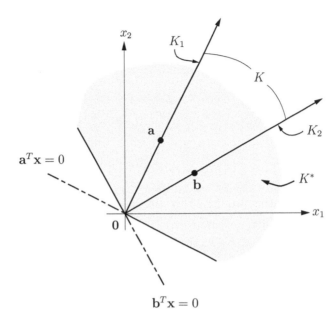

Figure 2.19 The dual cone K^* of a nonconvex cone $K = (K_1 \cup K_2) \subseteq \mathbb{R}^2$ is a closed convex cone which is also the dual of both $\mathcal{K} = \mathbf{conv}\, K$ and $\mathcal{K}_1 = (\mathbf{int}\, \mathcal{K}) \cup \{\mathbf{0}_2\}$.

Next, let us prove $\mathbb{S}_+^n \subseteq K^*$. Suppose that $\mathbf{Y} \in \mathbb{S}_+^n$. Then, by expressing $\mathbf{X} = \sum_{i=1}^{n} \lambda_i \mathbf{q}_i \mathbf{q}_i^T \in K$ with $\lambda_i \geq 0$ (EVD of $\mathbf{X}$), we have

$$\mathrm{Tr}(\mathbf{X}\mathbf{Y}) = \mathrm{Tr}\left(\sum_{i=1}^{n} \lambda_i \mathbf{q}_i \mathbf{q}_i^T \mathbf{Y} \right) = \sum_{i=1}^{n} (\mathbf{q}_i^T \mathbf{Y} \mathbf{q}_i) \lambda_i \geq 0, \ \forall \mathbf{X} \in K,$$

which implies $\mathbf{Y} \in K^*$. Hence, we have completed the proof that $\mathbb{S}_+^n \subseteq K^*$. Thus $K^* = K$ has been proved. ∎ □

Example 2.13 *(Dual of a norm cone)* Consider a cone defined by an atomic norm as follows

$$K = \{(\mathbf{x}, t) \in \mathbb{R}^{n+1} \mid \|\mathbf{x}\|_{\mathcal{A}} \leq t\} \tag{2.119}$$

where $\mathcal{A} \subset \mathbb{R}^n$ is the associated atomic set. Then its dual is given by

$$K^* = \{(\mathbf{u}, v) \in \mathbb{R}^{n+1} \mid \|\mathbf{u}\|_{\mathcal{A}}^* \leq v\}. \tag{2.120}$$

Proof of (2.120): By the definition of the dual cone (cf. (2.110)), we have

$$
\begin{aligned}
K^* &= \left\{ (\mathbf{u}, v) \in \mathbb{R}^{n+1} \mid \mathbf{x}^T \mathbf{u} + t v \geq 0 \ \forall (\mathbf{x}, t) \in K \right\} \\
&= \left\{ (\mathbf{u}, v) \in \mathbb{R}^{n+1} \mid -\mathbf{x}^T \mathbf{u} + t v \geq 0 \ \forall (\mathbf{x}, t) \in K \right\} \\
&= \left\{ (\mathbf{u}, v) \in \mathbb{R}^{n+1} \mid \boldsymbol{x}^T \mathbf{u} \leq v \ \forall \boldsymbol{x} \in \mathbf{conv}\, \mathcal{A} \right\} \quad (\text{where } \boldsymbol{x} = \mathbf{x}/t) \\
&= \left\{ (\mathbf{u}, v) \in \mathbb{R}^{n+1} \mid \|\mathbf{u}\|_{\mathcal{A}}^* \leq v \right\} \quad (\text{by definition of dual norm})
\end{aligned}
$$

where the second equality is due to the fact of $(\mathbf{x}, t) \in K \Leftrightarrow (-\mathbf{x}, t) \in K$, and in the third equality, we have applied (2.98), and meanwhile $t \neq 0$ is required with no harm in the ensuing derivations (since $\mathbf{x} = \mathbf{0}_n$ when $t = 0$, for which $\mathbf{x}^T \mathbf{u} + tv = 0$ is admissible for any $(\mathbf{u}, v)$ in the first two equalities).   ■ □

Example 2.14 *(Dual of a Lorentz cone)* Let $\mathbf{x} \in \mathbb{R}^n$, and $t \in \mathbb{R}$. The dual of a Lorentz cone $K = \{(\mathbf{x}, t) \in \mathbb{R}^{n+1} \mid \|\mathbf{x}\|_2 \leq t\}$ is

$$K^* = \{(\mathbf{u}, v) \in \mathbb{R}^{n+1} \mid \|\mathbf{u}\|_* \leq v\}, \quad \text{(cf. (2.120))}$$

where $\|\cdot\|_* = \|\cdot\|_2$, implying $K^* = K$ (self-dual). This example is actually a special case of $\|\cdot\|_{\mathcal{A}} = \|\cdot\|_2 = \|\cdot\|_{\mathcal{A}}^*$ in Example 2.13. Nevertheless, we provide an alternative proof for this case in this example.

Proof: First, let us prove $K^* \subseteq K$. Assume that $(\mathbf{y}, t') \in K^*$. Then $\mathbf{y}^T \mathbf{x} + t't \geq 0 \; \forall (\mathbf{x}, t) \in K$ by the definition of dual cone, and we have the following implications

$$(\mathbf{y}, t') \in K^* \Rightarrow \min_{\|\mathbf{x}\|_2 \leq t} \mathbf{y}^T \mathbf{x} + t't = \mathbf{y}^T \left(t \frac{-\mathbf{y}}{\|\mathbf{y}\|_2} \right) + t't$$
$$= -\|\mathbf{y}\|_2 t + t't$$
$$= (t' - \|\mathbf{y}\|_2)t \geq 0, \; \forall t \geq 0 \; \Rightarrow (\mathbf{y}, t') \in K$$

which implies $K^* \subseteq K$. On the other hand, if $(\mathbf{y}, t') \in K$, i.e., $\|\mathbf{y}\|_2 \leq t'$ and $t' \geq 0$, it can be seen that the reverse implications are also true, implying that $K \subseteq K^*$. Therefore, we have completed the proof that $K^* = K$ for the case of Lorentz cone.   ■ □

Dual cones satisfy the following properties:

(d1) K^* is closed and convex.

As illustrated in Figure 2.19, where K^* given by (2.118) is a closed convex cone, which is also the common dual of the closed nonconvex cone K given by (2.113), the closed convex cone $\mathcal{K}$ given by (2.114), and the nonclosed convex cone $\mathcal{K}_1$ given by (2.115).

(d2) $K_1 \subseteq K_2$ implies $K_2^* \subseteq K_1^*$.

For instance, for the convex cones $K = K_1 \cup K_2$ given in (2.113) and K_1 given by (2.111), $K^* \subset K_1^* = H_+(\mathbf{a})$ (cf. (2.115)).

(d3) If K has nonempty interior, then K^* is pointed.

For instance, $K = \mathbb{R}_+^n = K^*$ has nonempty interior and it is pointed. If K has empty interior, then K^* may not be pointed. An example is that $K = \mathbf{0}_2$ with **int** $K = \emptyset$, and $K^* = \mathbb{R}^2$ is not pointed. Another example is that for K_1 given by (2.111), **int** $K_1 = \emptyset$, and $K_1^* = H_+(\mathbf{a})$ (cf. (2.116)) is a closed halfspace but not pointed.

(d4) If the closure of K is pointed, then K^* has nonempty interior.

For instance, the closure of the cone K given in (2.113) is pointed, and its dual K^* given in (2.118) has nonempty interior. It is noticeable that if the closure of K is not pointed, then K^* may have empty interior. For example, for the cone $\widetilde{K} = \{\mathbf{x} \in \mathbb{R}^2 \mid \boldsymbol{a}^T\mathbf{x} > 0\} \cup \{\mathbf{0}_2\}$, $\mathbf{cl}\ \widetilde{K} = H_+(\boldsymbol{a})$ is a closed halfspace but not pointed, and its dual $\widetilde{K}^* = \mathbf{conic}\{\boldsymbol{a}\}$ is a cone with empty interior, where $\boldsymbol{a}$ is a nonzero vector in $\mathbb{R}^2$.

($d5$) K^{**} is the closure of the convex hull of K. (Hence, if K is convex and closed, $K^{**} = K$.)

For instance, for the closed nonconvex cone K given by (2.113) and the nonclosed convex cone $\mathcal{K}_1$ given by (2.115), they have identical closed and convex bidual $K^{**} = \mathcal{K}_1^{**} = \mathcal{K} = \mathbf{cl}\ \mathcal{K}_1$ (cf. (2.118), (2.114), and (2.115)).

These properties show that if K is a proper cone, then K^* is also a proper cone and $K^{**} = K$. For this case, two important properties about the relationship between a generalized inequality (defined by K) and an ordinary inequality (involving K^*) are as follows:

($P1$) $\mathbf{x} \preceq_K \mathbf{y}$ if and only if $\boldsymbol{\lambda}^T\mathbf{x} \leq \boldsymbol{\lambda}^T\mathbf{y}, \forall \boldsymbol{\lambda} \succeq_{K^*} \mathbf{0}$.

Alternatively, $\mathbf{z} \in K$ if and only if $\boldsymbol{\lambda}^T\mathbf{z} \geq 0, \forall \boldsymbol{\lambda} \in K^*$. $\qquad$ (2.121a)

($P2$) $\mathbf{x} \prec_K \mathbf{y}$ if and only if $\boldsymbol{\lambda}^T\mathbf{x} < \boldsymbol{\lambda}^T\mathbf{y}, \forall \boldsymbol{\lambda} \succeq_{K^*} \mathbf{0}, \boldsymbol{\lambda} \neq \mathbf{0}$.

Alternatively, $\mathbf{z} \in \mathbf{int}\ K$ if and only if $\boldsymbol{\lambda}^T\mathbf{z} > 0, \forall \boldsymbol{\lambda} \in K^* \setminus \{\mathbf{0}\}$. (2.121b)

These two properties can be shown by the definitions of $\prec_K$, $\preceq_K$, and $\succeq_{K^*}$ (cf. (2.78)), and they are still true if K and K^* are interchanged due to $K^{**} = (K^*)^* = K$. We only prove the second one.

Proof of (P2): (Necessity) Let us prove by contradiction. Assume that there exists a $\boldsymbol{\lambda} \in K^* \setminus \{\mathbf{0}\}$ such that $\boldsymbol{\lambda}^T(\mathbf{y} - \mathbf{x}) \leq 0$. Since $\mathbf{y} - \mathbf{x} \in \mathbf{int}\ K$, there must exist a Euclidean ball $B(\mathbf{y} - \mathbf{x}, \varepsilon) \subset K$ with center at $\mathbf{y} - \mathbf{x}$ and radius $\varepsilon > 0$. Thus we have

$$(\mathbf{y} - \mathbf{x}) + (-\varepsilon\boldsymbol{\lambda}/\|\boldsymbol{\lambda}\|_2) \in K$$
$$\Rightarrow \boldsymbol{\lambda}^T[(\mathbf{y} - \mathbf{x}) + (-\varepsilon\boldsymbol{\lambda}/\|\boldsymbol{\lambda}\|_2)] \geq 0 \quad (\text{since } \boldsymbol{\lambda} \in K^* \setminus \{\mathbf{0}\})$$
$$\Rightarrow \boldsymbol{\lambda}^T(\mathbf{y} - \mathbf{x}) \geq \varepsilon\|\boldsymbol{\lambda}\|_2 > 0,$$

which is a contradiction with the premise $\boldsymbol{\lambda}^T(\mathbf{y} - \mathbf{x}) \leq 0$.

(Sufficiency) Again, let us prove by contradiction. Assume that $\mathbf{x} \not\prec_K \mathbf{y}$ where $K = K^{**}$, i.e., $\mathbf{y} - \mathbf{x} \notin \mathbf{int}\ K = \mathbf{int}\ K^{**}$. Then $\mathbf{y} - \mathbf{x} \notin K^{**}$ or $\mathbf{y} - \mathbf{x} \in \mathbf{bd}\ K^{**}$. For the case of $\mathbf{y} - \mathbf{x} \notin K^{**}$, there must exist a $\boldsymbol{\lambda} \in K^*$ such that $\boldsymbol{\lambda}^T(\mathbf{y} - \mathbf{x}) < 0$, thus violating the assumption of $\boldsymbol{\lambda}^T\mathbf{x} < \boldsymbol{\lambda}^T\mathbf{y}$. For the case of $\mathbf{y} - \mathbf{x} \in \mathbf{bd}\ K^{**}$, there must exist a $\boldsymbol{\lambda} \in K^* \setminus \{\mathbf{0}\}$ such that $\boldsymbol{\lambda}^T(\mathbf{y} - \mathbf{x}) = 0$ since both K^* and K^{**} are closed, again violating the assumption of $\boldsymbol{\lambda}^T\mathbf{x} < \boldsymbol{\lambda}^T\mathbf{y}$. Thus the proof is completed. $\blacksquare$

Remark 2.20 The above two properties *(P1)* and *(P2)* in (2.121) about the generalized inequality (defined on a proper cone K) based comparison of two vectors also imply that it can be equivalently converted into the ordinary comparison of the inner products of the two vectors and every element in the proper cone K^*. □

Remark 2.21 It has been commonly presumed that, if the proper cone K is not explicitly mentioned for the associated generalized inequality used, the "$\mathbf{x} \succeq \mathbf{y}$" denotes the componentwise inequality defined on the proper cone $K = \mathbb{R}^n_+$, and "$\mathbf{X} \succeq \mathbf{0}$" means $\mathbf{X} \in \mathbb{S}^n_+$ (i.e., $K = \mathbb{S}^n_+$). □

2.6   Separating and supporting hyperplanes

2.6.1   Separating hyperplane theorem

Suppose that C and D are convex sets in $\mathbb{R}^n$ and $C \cap D = \emptyset$. Then there exists a hyperplane $H(\mathbf{a}, b) = \{\mathbf{x} \mid \mathbf{a}^T \mathbf{x} = b\}$ where $\mathbf{a} \in \mathbb{R}^n$ is a nonzero vector and $b \in \mathbb{R}$ such that

$$C \subseteq H_-(\mathbf{a}, b), \quad \text{i.e.,} \quad \mathbf{a}^T \mathbf{x} \leq b, \ \forall \mathbf{x} \in C \tag{2.122}$$

and

$$D \subseteq H_+(\mathbf{a}, b), \quad \text{i.e.,} \quad \mathbf{a}^T \mathbf{x} \geq b, \ \forall \mathbf{x} \in D. \tag{2.123}$$

Only the proof of (2.123) will be given below since the proof of (2.122) can be proven similarly. In the proof, we implicitly assume that the two convex sets C and D are closed without loss of generality. The reasons are that **cl** C, **int** C, **cl** D, and **int** D are also convex by Property 2.5 in Subsection 2.1.4, and thus the same hyperplane that separates **cl** C (or **int** C) and **cl** D (or **int** D) can also separate C and D.

Proof: Let

$$\mathbf{dist}(C, D) = \inf \ \{\|\mathbf{u} - \mathbf{v}\|_2 \mid \mathbf{v} \in C, \ \mathbf{u} \in D\}. \tag{2.124}$$

Assume that $\mathbf{dist}(C, D) > 0$, and that there exists a point $\mathbf{c} \in C$ and a point $\mathbf{d} \in D$ such that

$$\|\mathbf{c} - \mathbf{d}\|_2 = \mathbf{dist}(C, D) \tag{2.125}$$

(as illustrated in Figure 2.20). These assumptions will be satisfied if C and D are closed, and one of C and D is bounded. Note that it is possible that if both C and D are not bounded, such $\mathbf{c} \in C$ and $\mathbf{d} \in D$ may not exist. For instance, $C = \{(x, y) \in \mathbb{R}^2 \mid y \geq e^{-x} + 1, x \geq 0\}$ and $D = \{(x, y) \in \mathbb{R}^2 \mid y \leq -e^{-x}, x \geq 0\}$ are convex, closed, and unbounded with $\mathbf{dist}(C, D) = 1$, but $\mathbf{c} \in C$ and $\mathbf{d} \in D$ satisfying (2.125) do not exist.

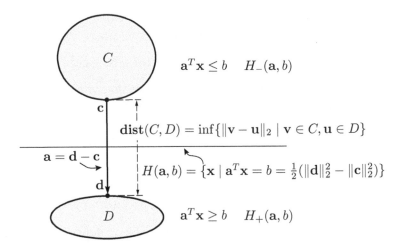

Figure 2.20 Separating hyperplane for two closed convex sets C and D.

Under the preceding premise that the two points $\mathbf{c} \in C$ and $\mathbf{d} \in D$ exist such that (2.125) is true (cf. Figure 2.20), let

$$\begin{cases} \mathbf{a} = \mathbf{d} - \mathbf{c} \\ b = \frac{1}{2}(\|\mathbf{d}\|_2^2 - \|\mathbf{c}\|_2^2) \end{cases} \tag{2.126}$$

Then, $H(\mathbf{a}, b)$ will be a hyperplane separating C and D. For ease of the proof below, let

$$\begin{aligned} f(\mathbf{x}) &= \mathbf{a}^T\mathbf{x} - b \\ &= (\mathbf{d} \quad \mathbf{c})^T\mathbf{x} - \frac{\|\mathbf{d}\|_2^2}{2} + \frac{\|\mathbf{c}\|_2^2}{2} \\ &= (\mathbf{d} - \mathbf{c})^T\left(\mathbf{x} - \frac{\mathbf{d} + \mathbf{c}}{2}\right). \end{aligned} \tag{2.127}$$

Next, let us prove (2.123), i.e., $f(\mathbf{x}) \geq 0$ for all $\mathbf{x} \in D$. Suppose that there exists a $\mathbf{u} \in D$ such that $f(\mathbf{u}) < 0$. Then,

$$\begin{aligned} 0 &> f(\mathbf{u}) \\ &= (\mathbf{d} - \mathbf{c})^T\left(\mathbf{u} - \frac{\mathbf{d} + \mathbf{c}}{2}\right) \\ &= (\mathbf{d} - \mathbf{c})^T\left(\mathbf{u} - \mathbf{d} + \frac{\mathbf{d} - \mathbf{c}}{2}\right) \\ &= (\mathbf{d} - \mathbf{c})^T(\mathbf{u} - \mathbf{d}) + \frac{1}{2}\|\mathbf{d} - \mathbf{c}\|_2^2 \end{aligned} \tag{2.128}$$

implying

$$(\mathbf{d} - \mathbf{c})^T(\mathbf{u} - \mathbf{d}) < 0 \tag{2.129}$$

as we have assumed that $\mathbf{dist}(C, D) > 0$.

Consider $\mathbf{z}(t) = (1 - t)\mathbf{d} + t\mathbf{u} \in D$, where $t \in [0, 1]$. Then

$$\|\mathbf{z}(t) - \mathbf{c}\|_2^2 = \|(1 - t)\mathbf{d} + t\mathbf{u} - \mathbf{c}\|_2^2 \geq \mathbf{dist}^2(C, D), \qquad (2.130)$$

which is a differentiable quadratic function of t with the minimum value being $\mathbf{dist}^2(C, D)$ when $t = 0$. Then we have

$$
\begin{aligned}
\frac{d}{dt}\|\mathbf{z}(t) - \mathbf{c}\|_2^2 &= \frac{d}{dt}\|(1 - t)\mathbf{d} + t\mathbf{u} - \mathbf{c}\|_2^2 \\
&= \frac{d}{dt}\|\mathbf{d} - \mathbf{c} + t(\mathbf{u} - \mathbf{d})\|_2^2 \\
&= \frac{d}{dt}\left\{ \|\mathbf{d} - \mathbf{c}\|_2^2 + 2t(\mathbf{d} - \mathbf{c})^T(\mathbf{u} - \mathbf{d}) + t^2\|\mathbf{u} - \mathbf{d}\|_2^2 \right\} \\
&= 2(\mathbf{d} - \mathbf{c})^T(\mathbf{u} - \mathbf{d}) + 2t\|\mathbf{u} - \mathbf{d}\|_2^2.
\end{aligned}
\qquad (2.131)
$$

Note that $t \to 0$, $\mathbf{z}(t) \to \mathbf{d}$, and

$$\left.\frac{d}{dt}\|\mathbf{z}(t) - \mathbf{c}\|_2^2\right|_{t=0} = 2(\mathbf{d} - \mathbf{c})^T(\mathbf{u} - \mathbf{d}) < 0 \quad \text{(by (2.129))}. \qquad (2.132)$$

This means that for some sufficiently small t,

$$\|\mathbf{z}(t) - \mathbf{c}\|_2^2 < \|\mathbf{d} - \mathbf{c}\|_2^2 = \mathbf{dist}^2(C, D), \qquad (2.133)$$

which contradicts with (2.130). Thus we have completed the proof of (2.123). Similarly, one can prove (2.122). ∎

Note that the above proof is under the premise that C and D are disjoint closed convex sets with $\mathbf{dist}(C, D) > 0$. Let us continue the proof for the case that $\mathbf{dist}(C, D) = 0$, thereby requiring either C or D to be open, using the preceding result of separating hyperplane theorem for $\mathbf{dist}(C, D) > 0$ with C and D assumed to be closed.

Proof (continued): Let $\{C_i\}$ and $\{D_i\}$ be two sequences of convex compact sets which satisfy:

$$\lim_{i \to \infty} C_i = \mathbf{int}\, C \text{ and } C_i \subseteq C_{i+1}, \ \forall i \in \mathbb{Z}_{++},$$

$$\lim_{i \to \infty} D_i = \mathbf{int}\, D \text{ and } D_i \subseteq D_{i+1}, \ \forall i \in \mathbb{Z}_{++}.$$

Since C_i and D_i are disjoint compact sets, we have

$$\mathbf{dist}(C_i, D_i) = \|\mathbf{c}_i - \mathbf{d}_i\|_2 > 0$$

for some vectors $\mathbf{c}_i \in C_i$ and $\mathbf{d}_i \in D_i$ and $\{\mathbf{c}_i\}$ and $\{\mathbf{d}_i\}$ are bounded sequences. By the previous proof and by (2.126), there must exist a hyperplane $H(\mathbf{a}_i, b_i)$

that can separate C_i and D_i, where

$$\mathbf{a}_i \triangleq \frac{\mathbf{d}_i - \mathbf{c}_i}{\|\mathbf{d}_i - \mathbf{c}_i\|_2},$$

$$b_i \triangleq \frac{\|\mathbf{d}_i\|_2^2 - \|\mathbf{c}_i\|_2^2}{2 \cdot \|\mathbf{d}_i - \mathbf{c}_i\|_2} = \frac{1}{2}(\mathbf{d}_i + \mathbf{c}_i)^T \mathbf{a}_i. \qquad \text{(cf. (2.126))}$$

According to the Bolzano–Weierstrass theorem [Apo07, Ber09], every bounded sequence in $\mathbb{R}^n$ has a convergent subsequence. Since $\{(\mathbf{a}_j, b_j)\}$ is a bounded sequence, there must exist a subsequence $\{(\mathbf{a}_j, b_j) \mid j \in Z \subset \mathbb{Z}_{++}\}$ such that

$$\mathbf{a} \triangleq \lim_{j \to \infty, j \in Z} \mathbf{a}_j \neq \mathbf{0} \quad (\text{since } \|\mathbf{a}_j\|_2 = 1 \ \forall j)$$

$$b \triangleq \lim_{j \to \infty, j \in Z} b_j.$$

This together with

$$\lim_{j \to \infty, j \in Z} C_j = \lim_{j \to \infty} C_j = \text{int } C,$$

$$\lim_{j \to \infty, j \in Z} D_j = \lim_{j \to \infty} D_j = \text{int } D,$$

and $\lim_{j \to \infty} \mathbf{dist}(C_j, D_j) = 0$ leads to the existence of a hyperplane

$$H(\mathbf{a}, b) \triangleq \lim_{j \to \infty, j \in Z} H(\mathbf{a}_j, b_j)$$

which separates **int** C and **int** D. The hyperplane $H(\mathbf{a}, b)$ also separates C and D by Remark 2.23 below. ∎

Remark 2.22 Any two convex sets C and D, at least one of which is open, are disjoint if there exists a separating hyperplane between the two sets. □

Remark 2.23 Assume that C (not necessarily closed) is a convex set with nonempty interior. If $H = \{\mathbf{x} \mid \mathbf{a}^T \mathbf{x} = b\}$ is a hyperplane and **int** $C \subseteq H_- = \{\mathbf{x} \mid \mathbf{a}^T \mathbf{x} \leq b\}$, then $C \subseteq H_-$ and **bd** $C \subset H_-$.

Proof: Assume that $\mathbf{x} \in \text{int } C$ and $\mathbf{y} \in \text{bd } C$. Then $\mathbf{z} = \theta \mathbf{x} + (1 - \theta)\mathbf{y} \in \text{int } C$ for all $0 < \theta < 1$. Hence, $\mathbf{a}^T \mathbf{x} \leq b$ and $\mathbf{a}^T \mathbf{z} \leq b$, and then

$$\mathbf{a}^T \mathbf{z} = \theta \mathbf{a}^T \mathbf{x} + (1 - \theta)\mathbf{a}^T \mathbf{y} \leq b = \theta b + (1 - \theta)b$$

$$\Rightarrow \mathbf{a}^T \mathbf{y} \leq b + \frac{\theta}{1 - \theta}(b - \mathbf{a}^T \mathbf{x}),$$

for all $0 < \theta < 1$, which implies $\mathbf{a}^T \mathbf{y} \leq b$, and thus $C \subseteq H_-$ and **bd** $C \subset H_-$ are true. Therefore, the proof is completed. ∎ □

2.6.2 Supporting hyperplanes

For any nonempty convex set C and for any $\mathbf{x}_0 \in \text{bd } C$, there exists an $\mathbf{a} \neq \mathbf{0}$, such that $\mathbf{a}^T \mathbf{x} \leq \mathbf{a}^T \mathbf{x}_0$, for all $\mathbf{x} \in C$; namely, the convex set C is supported by the hyperplane $H = \{\mathbf{x} \mid \mathbf{a}^T\mathbf{x} = \mathbf{a}^T\mathbf{x}_0\}$ such that $C \subseteq H_- = \{\mathbf{x} \mid \mathbf{a}^T\mathbf{x} \leq \mathbf{a}^T\mathbf{x}_0\}$ (see Figure 2.21).

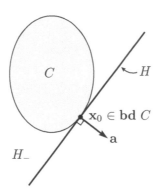

Figure 2.21 A hyperplane $H = \{\mathbf{x} \mid \mathbf{a}^T\mathbf{x} = \mathbf{a}^T\mathbf{x}_0\}$ that supports the set C at $\mathbf{x}_0$, and thus $C \subseteq H_- = \{\mathbf{x} \mid \mathbf{a}^T\mathbf{x} \leq \mathbf{a}^T\mathbf{x}_0\}$.

Proof: Assume that C is a convex set, $A = \textbf{int } C$ (which is open and convex), and $\mathbf{x}_0 \in \textbf{bd } C$. Let $B = \{\mathbf{x}_0\}$ (which is convex). Then $A \cap B = \emptyset$. By the separating hyperplane theorem, there exists a separating hyperplane $H = \{\mathbf{x} \mid \mathbf{a}^T\mathbf{x} = \mathbf{a}^T\mathbf{x}_0\}$ (since the distance between the set A and the set B is equal to zero), where $\mathbf{a} \neq \mathbf{0}$, between A and B, such that $\mathbf{a}^T(\mathbf{x} - \mathbf{x}_0) \leq 0$ for all $\mathbf{x} \in C$ (i.e., $C \subseteq H_-$ by Remark 2.23). Therefore, the hyperplane H is a supporting hyperplane of the convex set C which passes $\mathbf{x}_0 \in \textbf{bd } C$. ∎

It is now easy to prove, by the supporting hyperplane theorem, that a closed convex set S with $\textbf{int } S \neq \emptyset$ is the intersection of all (possibly an infinite number of) closed halfspaces that contain it (cf. Remark 2.8). Let

$$\mathcal{H}(\mathbf{x}_0) \triangleq \left\{\mathbf{x} \mid \mathbf{a}^T(\mathbf{x} - \mathbf{x}_0) = 0\right\} \tag{2.134}$$

be a supporting hyperplane of S passing $\mathbf{x}_0 \in \textbf{bd } S$. This implies, by the hyperplane supporting theorem, that the associated closed halfspace $\mathcal{H}_-(\mathbf{x}_0)$, which contains the closed convex set S, is given by

$$\mathcal{H}_-(\mathbf{x}_0) \triangleq \left\{\mathbf{x} \mid \mathbf{a}^T(\mathbf{x} - \mathbf{x}_0) \leq 0\right\}, \quad \mathbf{x}_0 \in \textbf{bd } S. \tag{2.135}$$

Thus it must be true that

$$S = \bigcap_{\mathbf{x}_0 \in \textbf{bd } S} \mathcal{H}_-(\mathbf{x}_0), \tag{2.136}$$

implying that a closed convex set S can be defined by all of its supporting hyperplanes $\mathcal{H}(\mathbf{x}_0)$, though the expression (2.136) may not be unique, thereby justifying Remark 2.8. When the number of supporting halfspaces containing the closed convex set S is finite, S is a polyhedron. When S is compact and convex, the supporting hyperplane representation (2.136) can also be expressed as

$$S = \bigcap_{\mathbf{x}_0 \in S_{\mathrm{extr}}} \mathcal{H}_-(\mathbf{x}_0) \quad (\text{cf. (2.24)}). \qquad (2.137)$$

where the intersection also contains those halfspaces whose boundaries may contain multiple extreme points of S. Let us conclude this section with the following three remarks.

Remark 2.24 If a set S can be expressed in the form of (2.136), then it must be a closed convex set because a closed nonconvex set must contain a boundary point that never belongs to any supporting hyperplane. In other words, the boundary of S and that of **conv** S must be different if S is closed and nonconvex.   □

Remark 2.25 If the supporting hyperplane defined in (2.134) is unique (i.e., $\mathbf{a}/\|\mathbf{a}\|_2$ is unique) for any $\mathbf{x}_0 \in \mathbf{bd}\ S \neq \emptyset$, then **bd** S is said to be *smooth*, otherwise *nonsmooth*. Closed convex sets with nonempty interior and smooth boundary include

- Strictly convex sets: Any strictly convex set can be shown to have a smooth boundary. For instance, 2-norm balls with nonzero radii and ellipsoids with nonzero lengths of semiaxes in $\mathbb{R}^n$ for $n \geq 2$ have smooth boundaries. As an illustrative example, for a 2-norm ball with center at the origin and radius equal to unity

$$C_1 = \{\mathbf{x} \in \mathbb{R}^n \mid \|\mathbf{x}\|_2 \leq 1\},$$

the corresponding halfspace $\mathcal{H}_-(\mathbf{x}_0)$ defined in (2.135) that contains C_1 is uniquely given by

$$\mathcal{H}_-(\mathbf{x}_0) \triangleq \{\mathbf{x} \mid \mathbf{x}_0^T(\mathbf{x} - \mathbf{x}_0) \leq 0\}, \quad \|\mathbf{x}_0\|_2 = 1, \qquad (2.138)$$

and the intersection of the supporting hyperplane $\mathcal{H}(\mathbf{x}_0)$ and C_1 is exactly the point $\mathbf{x}_0 \in \mathbf{bd}\ C_1$. So **bd** C_1 is smooth.

- Some nonstrictly convex sets: For instance, **conv** B in Figure 2.4 and closed halfspaces have smooth boundaries.

On the other hand, some nonstrictly convex sets, e.g., simplexes, 1-norm balls, and ∞-norm balls with nonzero radii, and the second-order cone in $\mathbb{R}^n$ for $n \geq 2$ have nonsmooth boundaries. However, their boundaries are not smooth at finite boundary points.   □

Remark 2.26 For a closed convex set $S \subset \mathbb{R}^n$ with **int** $S = \emptyset$, any hyperplane (whose affine dimension is equal to $n - 1$) in $\mathbb{R}^n$ containing **aff** S is a supporting hyperplane of S due to affdim $S \leq n - 1$; (2.136) with **bd** S replaced by **relbd** S still holds true if the intersection also contains those halfspaces whose boundaries may contain multiple relative boundary points of S; (2.137) still holds true if S is also compact.   □

2.7   Summary and discussion

In this chapter, we have introduced convex sets and their properties (mostly geometric properties). Various convexity preserving operations were introduced together with many examples. In addition, the concepts of proper cones on which the generalized equality is defined, dual norms, and dual cones were introduced in detail. Finally, we presented the separating hyperplane theorem, which corroborates the existence of a hyperplane separating two disjoint convex sets, and the existence of the supporting hyperplane of any nonempty convex set. These fundamentals on convex sets along with convex functions to be introduced in the next chapter will be highly instrumental in understanding the concepts of convex optimization. The convex geometry properties introduced in this chapter have been applied to blind hyperspectral unmixing for material identification in remote sensing. Some will be introduced in Chapter 6.

References for Chapter 2

[Apo07]   T. M. Apostol, *Mathematical Analysis*, 2nd ed. Pearson Edu. Taiwan Ltd., 2007.

[Ber09]   D. P. Bertsekas, *Convex Optimization Theory*. Belmont, MA, USA: Athena Scientific, 2009.

[BV04]   S. Boyd and L. Vandenberghe, *Convex Optimization*. Cambridge, UK: Cambridge University Press, 2004.

[Tuy00]   H. Tuy, "Monotonic optimization: Problems and solution approaches," *SIAM J. Optimization*, vol. 11, no. 2, pp. 464–49, 2000.

3   Convex Functions

Together with the convex set introduced in Chapter 2, convex function is required to define a convex optimization problem to be introduced in Chapter 4. This chapter introduces the basics of convex functions, and quasiconvex functions including definitions, properties, representations and convexity preserving operations, and various conditions for proving or disproving if a function is convex or quasiconvex. These concepts are also extended to K-convex functions defined on a proper cone K. Many examples are provided to illustrate how to prove the convexity and quasiconvexity of functions.

3.1   Basic properties and examples of convex functions

Prior to introducing the definition, properties and various conditions of convex functions together with illustrative examples, we need to clarify the role of $+\infty$ and $-\infty$ for a function $f : \mathbb{R}^n \to \mathbb{R}$. In spite of $+\infty, -\infty \notin \mathbb{R}$, $f(\mathbf{x})$ is allowed to take a value of $+\infty$ or $-\infty$ for some $\mathbf{x} \in \mathbf{dom}\ f$, hereafter. For instance, the following functions

$$f_1(\mathbf{x}) = \begin{cases} \|\mathbf{x}\|_2^2, \|\mathbf{x}\|_2 \leq 1 \\ +\infty, \ 1 < \|\mathbf{x}\|_2 \leq 2 \end{cases}, \quad \mathbf{dom}\ f_1 = \{\mathbf{x} \in \mathbb{R}^n \mid \|\mathbf{x}\|_2 \leq 2\} \qquad (3.1)$$

$$f_2(x) = \begin{cases} -\infty, \ x = 0 \\ \log x, \ x > 0 \end{cases}, \quad \mathbf{dom}\ f_2 = \mathbb{R}_+ \qquad (3.2)$$

are well-defined functions, and f_1 is a convex function and f_2 is a concave function. The convexity of functions will be presented next in detail.

3.1.1   Definition and fundamental properties

A function $f : \mathbb{R}^n \to \mathbb{R}$ is said to be *convex* if the following conditions are satisfied

- $\mathbf{dom}\ f$ is convex.
- For all $\mathbf{x}, \mathbf{y} \in \mathbf{dom}\ f, \theta \in [0, 1]$,

$$f(\theta \mathbf{x} + (1 - \theta)\mathbf{y}) \leq \theta f(\mathbf{x}) + (1 - \theta)f(\mathbf{y}). \qquad (3.3)$$

85

A convex function basically looks like a faceup bowl as illustrated in Figure 3.1, and it may be differentiable, or continuous but nonsmooth or a nondifferentiable function (e.g., with some discontinuities or with $f(\mathbf{x}) = +\infty$ for some $\mathbf{x}$). Note that for a given $\theta \in [0, 1]$, $\mathbf{z} \triangleq \theta\mathbf{x} + (1 - \theta)\mathbf{y}$ is a point on the line segment from $\mathbf{x}$ to $\mathbf{y}$ with

$$\frac{\|\mathbf{z} - \mathbf{y}\|_2}{\|\mathbf{y} - \mathbf{x}\|_2} = \theta, \text{ and } \frac{\|\mathbf{z} - \mathbf{x}\|_2}{\|\mathbf{y} - \mathbf{x}\|_2} = 1 - \theta, \tag{3.4}$$

and $f(\mathbf{z})$ is upper bounded by the sum of $100 \times \theta\%$ of $f(\mathbf{x})$ and $100 \times (1 - \theta)\%$ of $f(\mathbf{y})$ (i.e., the closer (further) the $\mathbf{z}$ to $\mathbf{x}$, the larger (smaller) the contribution of $f(\mathbf{x})$ to the upper bound of $f(\mathbf{z})$, and this also applies to the contribution of $f(\mathbf{y})$ as shown in Figure 3.1). Note that when $\mathbf{z}$ is given instead of θ, the value of θ in the upper bound of $f(\mathbf{z})$ can also be determined by (3.4). Various convex function examples will be provided in Subsection 3.1.4.

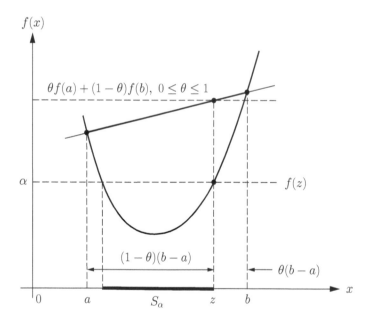

Figure 3.1 A convex function (also a strictly convex function) and a sublevel set.

Remark 3.1 f is said to be *strictly convex* if $\mathbf{dom}\ f$ is convex and for all distinct $\mathbf{x}, \mathbf{y} \in \mathbf{dom}\ f$ with $f(\mathbf{x}) < \infty$, $f(\mathbf{y}) < \infty$, the inequality (3.3) holds strictly, i.e.,

$$f(\theta\mathbf{x} + (1 - \theta)\mathbf{y}) < \theta f(\mathbf{x}) + (1 - \theta)f(\mathbf{y}), \quad \theta \in (0, 1). \tag{3.5}$$

The inequality defined by (3.3) holds with equality only when f is an affine function over the line segment between $\mathbf{x}$ and $\mathbf{y}$. Hence, a strictly convex function f must not contain any affine function over any subset of $\mathbf{dom}\ f$. $\qquad\square$

Remark 3.2 f is said to be concave if $-f$ is convex. In other words, the statement that f is concave means that $\mathbf{dom} f$ is convex and

$$f(\theta\mathbf{x} + (1-\theta)\mathbf{y}) \geq \theta f(\mathbf{x}) + (1-\theta)f(\mathbf{y}), \tag{3.6}$$

for all $\mathbf{x}, \mathbf{y} \in \mathbf{dom} f$, and for all $\theta \in [0,1]$. $\square$

Remark 3.3 When $f : \mathbb{C}^n \to \mathbb{R}$ is a real-valued function with a convex $\mathbf{dom} f \subseteq \mathbb{C}^n$ (i.e., convex combination of any $\mathbf{x}, \mathbf{y} \in \mathbf{dom} f$ still belongs to $\mathbf{dom} f$ though $\mathbf{dom} f$ is a set of complex vectors) and (3.3) holds true for all $\mathbf{x}, \mathbf{y} \in \mathbf{dom} f$, f is also a convex function because it can be converted into a convex function with the corresponding domain being a convex subset of $\mathbb{R}^{2n}$. $\square$

Remark 3.4 If $f(\mathbf{x}, \mathbf{y})$ is convex in $(\mathbf{x}, \mathbf{y})$, then it is convex in $\mathbf{x}$ and convex in $\mathbf{y}$. This can be easily shown by the definition of convex functions, but the converse is not necessarily true. Let us prove this by a counterexample. Consider

$$g(x, y) = x^2 y, \ x \in \mathbb{R}, \ y \in \mathbb{R}_+.$$

It can be easily shown that $g(x, y)$ is convex in x and convex in y because for $\theta \in [0,1]$,

$$\theta g(x_1, y) + (1-\theta)g(x_2, y) - g(\theta x_1 + (1-\theta)x_2, y)$$
$$= y\theta(1-\theta)(x_1 - x_2)^2 \geq 0 \ \forall x_1, x_2 \in \mathbb{R}, \ y \in \mathbb{R}_+,$$
$$g(x, \theta y_1 + (1-\theta)y_2) = \theta g(x, y_1) + (1-\theta)g(x, y_2) \ \forall x \in \mathbb{R}, \ y_1, y_2 \in \mathbb{R}_+,$$

namely, the associated inequality (3.3) is satisfied for each of them. However, for $\mathbf{z}_1 = (x_1, y_1) = (0, 2)$, $\mathbf{z}_2 = (x_2, y_2) = (2, 0)$, and $\theta = 1/2$, the inequality (3.3) is violated due to

$$g(0.5\mathbf{z}_1 + 0.5\mathbf{z}_2) = 1 \nleq 0.5g(\mathbf{z}_1) + 0.5g(\mathbf{z}_2) = 0.$$

So g is not convex in (x, y). $\square$

Remark 3.5 If f is convex, then the sublevel set

$$S_\alpha = \{\mathbf{x} \mid \mathbf{x} \in \mathbf{dom} \ f, f(\mathbf{x}) \leq \alpha\} \subseteq \mathbf{dom} \ f$$

(cf. (3.104)) is convex for all α (see Figure 3.1) which can be readily proven by the definition of convex sets. However, a function may not be convex even if all of its sublevel-sets are convex; such a function is called a quasiconvex function (which will be addressed later in Section 3.3 in detail). $\square$

Remark 3.6 *(Effective domain of convex functions)* Suppose that f is convex and

$$\mathcal{A}_\infty = \{\mathbf{x} \in \mathbf{dom} \ f \mid f(\mathbf{x}) = \infty\} \subseteq \mathbf{dom} \ f,$$

which may not be a convex set. Because the sublevel set S_α of f must be convex for all $\alpha < +\infty$ by Remark 3.5, it can be seen that $S_\alpha \subseteq S_{\alpha'} \subseteq \mathbf{dom} \ f \setminus \mathcal{A}_\infty$ for all $\alpha \leq \alpha' < +\infty$, and

$$S_\alpha \to \mathbf{dom}\, f \setminus \mathcal{A}_\infty \text{ as } \alpha \to \infty,$$

implying that the *effective domain*, defined as [Ber09]

$$\text{Eff-}\mathbf{dom}\, f \triangleq \mathbf{dom}\, f \setminus \mathcal{A}_\infty = \{\mathbf{x} \in \mathbf{dom}\, f \mid f(\mathbf{x}) < \infty\}, \qquad (3.7)$$

must be a convex set on which f is also convex. For instance, consider the convex function f_1 given by (3.1) that is not continuous on $\mathbf{bd}\, B(\mathbf{0}, 1)$ and $f_1(\mathbf{x}) = \infty$ for all $\mathbf{x} \in \{\mathbf{x} \mid 1 < \|\mathbf{x}\|_2 \leq 2\}$. Its effective domain

$$\text{Eff-}\mathbf{dom}\, f_1 = B(\mathbf{0}, 1) \text{ (Eucledian ball)}$$

is a convex set on which f_1 is a convex function.

Moreover, for a convex function f, if $f(\mathbf{x}) = -\infty$ for some $\mathbf{x} \in \mathbf{dom}\, f$, then $f(\mathbf{x}) = -\infty$ for all $\mathbf{x} \in \mathbf{dom}\, f$ by (3.3). This special case is of little interest in applications. A convex function f is said to be *proper convex* if $f(\mathbf{x}) < +\infty$ for at least one $\mathbf{x}$ and $f(\mathbf{x}) > -\infty$ for every $\mathbf{x}$, i.e., the effective domain of a proper convex function must be nonempty. $\qquad \square$

Remark 3.7 Suppose that $f : \mathbb{R}^n \to \mathbb{R}$ is a convex function with $\mathbf{int}(\mathbf{dom}\, f) \neq \emptyset$, and $f(\mathbf{x}) < \infty$ for all $\mathbf{x} \in \mathbf{int}(\mathbf{dom}\, f)$. Then f must be continuous over $\mathbf{int}(\mathbf{dom}\, f)$ [Ber09].

Proof: Let $\mathbf{x}_0 \in \mathbf{int}(\mathbf{dom}\, f)$. Then there exists an $\varepsilon > 0$ such that the 2-norm ball (with center $\mathbf{x}_0$ and radius ε) $B(\mathbf{x}_0, \varepsilon) \subseteq \mathbf{int}(\mathbf{dom}\, f)$. Let $\{\mathbf{x}_k\}$ be any sequence that converges to $\mathbf{x}_0$. It suffices to show that $\lim_{k \to \infty} f(\mathbf{x}_k) = f(\mathbf{x}_0)$.

Without loss of generality, assume $\{\mathbf{x}_k\} \subseteq B(\mathbf{x}_0, \varepsilon)$. Define the following two sequences $\{\mathbf{y}_k\}$ and $\{\mathbf{z}_k\}$ on the boundary of the 2-norm ball $B(\mathbf{x}_0, \varepsilon)$:

$$\mathbf{y}_k \triangleq \mathbf{x}_0 + \varepsilon \cdot \frac{\mathbf{x}_k - \mathbf{x}_0}{\|\mathbf{x}_k - \mathbf{x}_0\|_2} \quad \text{and} \quad \mathbf{z}_k \triangleq \mathbf{x}_0 - \varepsilon \cdot \frac{\mathbf{x}_k - \mathbf{x}_0}{\|\mathbf{x}_k - \mathbf{x}_0\|_2}. \qquad (3.8)$$

Note that $\mathbf{y}_k$, $\mathbf{x}_k$, $\mathbf{x}_0$, and $\mathbf{z}_k$ lie in order on the line segment from $\mathbf{y}_k$ to $\mathbf{z}_k$. The orientation of this line segment can change for different k, as illustrated in Figure 3.2 for the case of $n = 2$. Then we have from (3.3), (3.4), and (3.8) that

$$f(\mathbf{x}_k) \leq \frac{\|\mathbf{x}_k - \mathbf{x}_0\|_2}{\varepsilon} \cdot f(\mathbf{y}_k) + \frac{\varepsilon - \|\mathbf{x}_k - \mathbf{x}_0\|_2}{\varepsilon} \cdot f(\mathbf{x}_0); \qquad (3.9)$$

$$f(\mathbf{x}_0) \leq \frac{\|\mathbf{x}_k - \mathbf{x}_0\|_2}{\varepsilon + \|\mathbf{x}_k - \mathbf{x}_0\|_2} \cdot f(\mathbf{z}_k) + \frac{\varepsilon}{\varepsilon + \|\mathbf{x}_k - \mathbf{x}_0\|_2} \cdot f(\mathbf{x}_k). \qquad (3.10)$$

By taking limit superior (or called supremum limit) and limit inferior (or called infimum limit) on both sides of inequalities (3.9) and (3.10), respectively, together with the fact that $\lim_{k \to \infty} \|\mathbf{x}_k - \mathbf{x}_0\|_2 = 0$, we come up with

$$\limsup_{k \to \infty} f(\mathbf{x}_k) \triangleq \lim_{k \to \infty} \sup_{n \geq k} f(\mathbf{x}_n) \leq 0 + f(\mathbf{x}_0) = f(\mathbf{x}_0); \qquad (3.11)$$

$$f(\mathbf{x}_0) \leq 0 + \liminf_{k \to \infty} f(\mathbf{x}_k) = \liminf_{k \to \infty} f(\mathbf{x}_k). \qquad (3.12)$$

By the fact that $\liminf_{k\to\infty} f(\mathbf{x}_k) \le \limsup_{k\to\infty} f(\mathbf{x}_k)$, together with (3.11) and (3.12), it can be concluded that

$$\liminf_{k\to\infty} f(\mathbf{x}_k) = \limsup_{k\to\infty} f(\mathbf{x}_k) = f(\mathbf{x}_0), \tag{3.13}$$

namely, $\lim_{k\to\infty} f(\mathbf{x}_k) = f(\mathbf{x}_0)$, for each $\mathbf{x}_0 \in \mathbf{int}(\mathbf{dom}\ f)$. Thus we have completed the proof. ∎ □

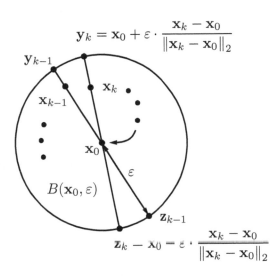

Figure 3.2 The converging sequence $\{\mathbf{x}_k\}$, its limit $\mathbf{x}_0$, and the two sequences $\{\mathbf{y}_k\} \subset \mathbf{bd}\ B(\mathbf{x}_0, \varepsilon)$ and $\{\mathbf{z}_k\} \subset \mathbf{bd}\ B(\mathbf{x}_0, \varepsilon)$ in $\mathbb{R}^2$.

Remark 3.8 The continuity of a convex function f with $f(\mathbf{x}) < \infty$ over $\mathbf{int}(\mathbf{dom}\ f) \neq \emptyset$ presented in Remark 3.7 can be extended to the case that $f(\mathbf{x}) < \infty$ for all $\mathbf{x} \in \mathbf{relint}(\mathbf{dom}\ f) \neq \emptyset$ due to the isomorphism (i.e., there exists a one-to-one, onto, and distance-preserving mapping) between $\mathbf{aff}(\mathbf{dom}\ f)$ with affine dimension k and the k-dimensional Euclidean space (cf. (2.15) as $\mathbf{C}$ in the affine mapping is a semi-unitary matrix). Therefore, it can be inferred that any convex function g with $\mathbf{relint}(\text{Eff-}\mathbf{dom}\ g) \neq \emptyset$ is continuous over $\mathbf{relint}(\text{Eff-}\mathbf{dom}\ g)$. □

Remark 3.9 Let

$$C = \{\mathbf{x} \in \mathbb{R}^n \mid f(\mathbf{x}) \le 0\} \tag{3.14}$$
$$H = \{\mathbf{x} \in \mathbb{R}^n \mid h(\mathbf{x}) = 0\}. \tag{3.15}$$

It can be easily shown, by the definition of convex sets, that C is a convex set if f is a convex function and that H is a convex set if h is an affine function. These two sets will be used to define an optimization problem in Chapter 4 later (cf. (4.1)), where an objective function $f(\mathbf{x})$ is minimized subject to $\mathbf{x} \in C$ with the constraint set C of the problem expressed in terms of these two sets. □

Fact 3.1 A function $f : \mathbb{R}^n \to \mathbb{R}$ is convex (strictly convex) if and only if for all $\mathbf{x} \in \mathbf{dom}\ f$ and for all $\mathbf{v} \neq \mathbf{0}$, the function

$$g(t) = f(\mathbf{x} + t\mathbf{v})$$

defined from $\mathbb{R} \to \mathbb{R}$ is convex (strictly convex) on $\mathbf{dom}\ g = \{t \mid \mathbf{x} + t\mathbf{v} \in \mathbf{dom}\ f\} \neq \emptyset$.

Proof: Let us prove the necessity followed by sufficiency.

- Necessity: Assume $g(t)$ is nonconvex for some $\mathbf{x}$ and $\mathbf{v}$. There exist $t_1, t_2 \in \mathbf{dom}\ g$ (and $\mathbf{x} + t_1\mathbf{v},\ \mathbf{x} + t_2\mathbf{v} \in \mathbf{dom}\ f$) such that

$$g(\theta t_1 + (1-\theta)t_2) > \theta g(t_1) + (1-\theta)g(t_2),\ 0 \leq \theta \leq 1,$$

  implying

$$f(\theta(\mathbf{x} + t_1\mathbf{v}) + (1-\theta)(\mathbf{x} + t_2\mathbf{v})) > \theta f(\mathbf{x} + t_1\mathbf{v}) + (1-\theta)f(\mathbf{x} + t_2\mathbf{v}).$$

  Therefore, f is nonconvex (contradiction with "f is convex"). Hence g is convex.

- Sufficiency: Suppose that $f(\mathbf{x})$ is nonconvex. Then, there exist $\mathbf{x}_1, \mathbf{x}_2 \in \mathbf{dom}\ f$ and some $0 < \theta < 1$ such that

$$f(\theta\mathbf{x}_1 + (1-\theta)\mathbf{x}_2) > \theta f(\mathbf{x}_1) + (1-\theta)f(\mathbf{x}_2).$$

  Now $g(t) = f(\mathbf{x} + t\mathbf{v})$ must be a convex function. Let $\mathbf{x} = \mathbf{x}_1$ and $\mathbf{v} = \mathbf{x}_2 - \mathbf{x}_1$. Thus $0, 1 \in \mathbf{dom}\ g$ and also $[0, 1] \subset \mathbf{dom}\ g$ since $\mathbf{dom}\ g$ is a convex set. Then we have

$$\begin{aligned} g(1-\theta) &= f(\theta\mathbf{x}_1 + (1-\theta)\mathbf{x}_2) \\ &> \theta f(\mathbf{x}_1) + (1-\theta)f(\mathbf{x}_2) \\ &= \theta g(0) + (1-\theta)g(1). \end{aligned}$$

  Therefore, $g(t)$ is nonconvex (contradiction with the premise of $g(t)$ is convex). Thus $f(\mathbf{x})$ must be convex. ∎

The above fact is very useful, since it allows us to prove or check whether a function is convex or not, by restricting it to a line in its domain, thereby significantly simplifying the convexity verification of a high-dimensional function into that of a one-dimensional function.

An important question that arises when discussing convex function is how to check whether a function is convex or not. Besides the verification by the definition of convex function introduced above, in the ensuing subsections, let us introduce some useful conditions that are very powerful to check or verify the convexity of a given function.

3.1.2 First-order condition

Suppose that f is differentiable. Then f is convex if and only if $\mathbf{dom}\, f$ is convex and

$$f(\mathbf{y}) \geq f(\mathbf{x}) + \nabla f(\mathbf{x})^T (\mathbf{y} - \mathbf{x}) \quad \forall \mathbf{x}, \mathbf{y} \in \mathbf{dom}\, f. \tag{3.16}$$

This is called the first-order condition, which means that the first-order Taylor series approximation of $f(\mathbf{y})$ w.r.t. $\mathbf{y} = \mathbf{x}$ is always below the original function (see Figure 3.3 for the one-dimensional case), i.e., the first-order condition (3.16) provides a tight lower bound (which is an affine function in $\mathbf{y}$) over the entire domain for a differentiable convex function. Moreover, it can be seen from (3.16) that

$$f(\mathbf{y}) = \max_{\mathbf{x} \in \mathbf{dom}\, f} f(\mathbf{x}) + \nabla f(\mathbf{x})^T (\mathbf{y} - \mathbf{x}) \quad \forall \mathbf{y} \in \mathbf{dom}\, f. \tag{3.17}$$

For instance, as illustrated in Figure 3.3, $f(b) \geq f(a) + f'(a)(b - a)$ for any a and the equality holds only when $a = b$. Next, let us prove the first-order condition.

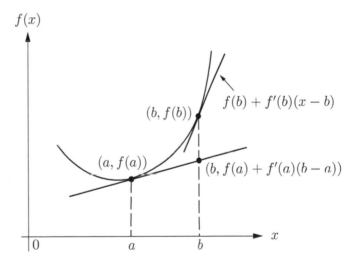

Figure 3.3 First-order condition for a convex function f for the one-dimensional case: $f(b) \geq f(a) + f'(a)(b - a)$ and $(f'(b) - f'(a))(b - a) \geq 0$ for all $a, b \in \mathbf{dom}\, f$.

Proof of (3.16): Let us prove the sufficiency followed by necessity.

- Sufficiency: (i.e., if (3.16) holds, then f is convex) From (3.16), we have, for all $\mathbf{x}, \mathbf{y}, \mathbf{z} \in \mathbf{dom}\, f$ which is convex and $0 \leq \lambda \leq 1$,

$$f(\mathbf{y}) \geq f(\mathbf{x}) + \nabla f(\mathbf{x})^T (\mathbf{y} - \mathbf{x}),$$
$$f(\mathbf{z}) \geq f(\mathbf{x}) + \nabla f(\mathbf{x})^T (\mathbf{z} - \mathbf{x}),$$
$$\Rightarrow \lambda f(\mathbf{y}) + (1 - \lambda)f(\mathbf{z}) \geq f(\mathbf{x}) + \nabla f(\mathbf{x})^T (\lambda \mathbf{y} + (1 - \lambda)\mathbf{z} - \mathbf{x}).$$

By setting $\mathbf{x} = \lambda \mathbf{y} + (1 - \lambda)\mathbf{z} \in \mathbf{dom}\ f$ in the above inequality, we obtain $\lambda f(\mathbf{y}) + (1 - \lambda)f(\mathbf{z}) \geq f(\lambda \mathbf{y} + (1 - \lambda)\mathbf{z})$. So f is convex.

- Necessity: (i.e., if f is convex, then (3.16) holds) For $\mathbf{x}, \mathbf{y} \in \mathbf{dom}\ f$ and $0 \leq \lambda \leq 1$,

$$f((1 - \lambda)\mathbf{x} + \lambda \mathbf{y}) = f(\mathbf{x} + \lambda(\mathbf{y} - \mathbf{x}))$$
$$= f(\mathbf{x}) + \lambda \nabla f(\mathbf{x} + \theta \lambda(\mathbf{y} - \mathbf{x}))^T(\mathbf{y} - \mathbf{x}), \qquad (3.18)$$

for some $\theta \in [0, 1]$ (from the first-order expansion of Taylor series (1.53)). Since f is convex, we have

$$f((1 - \lambda)\mathbf{x} + \lambda \mathbf{y}) \leq (1 - \lambda)f(\mathbf{x}) + \lambda f(\mathbf{y}).$$

Substituting (3.18) on the left-hand side of this inequality yields

$$\lambda f(\mathbf{y}) \geq \lambda f(\mathbf{x}) + \lambda \nabla f(\mathbf{x} + \theta \lambda(\mathbf{y} - \mathbf{x}))^T(\mathbf{y} - \mathbf{x}).$$

For $\lambda > 0$, we get (after dividing by λ),

$$f(\mathbf{y}) \geq f(\mathbf{x}) + \nabla f(\mathbf{x} + \theta \lambda(\mathbf{y} - \mathbf{x}))^T(\mathbf{y} - \mathbf{x})$$
$$= f(\mathbf{x}) + \nabla f(\mathbf{x})^T(\mathbf{y} - \mathbf{x}) \quad (\text{as } \lambda \to 0^+)$$

because ∇f is continuous due to the fact that f is differentiable and convex (cf. Remark 3.13 below). Hence (3.16) has been proved. ∎

Alternative proof for Necessity: Consider a convex function $f : \mathbb{R} \to \mathbb{R}$ and let $x, y \in \mathbf{dom}\ f$ (which is a convex set). Then for all $0 < t \leq 1$, $x + t(y - x) \in \mathbf{dom}\ f$,

$$f(x + t(y - x)) \leq tf(y) + (1 - t)f(x)$$
$$\Rightarrow\ f(y) \geq f(x) + \frac{f(x + t(y - x)) - f(x)}{t(y - x)}(y - x)$$
$$\Rightarrow\ f(y) \geq f(x) + f'(x)(y - x) \text{ as } t \to 0. \qquad (3.19)$$

By using the above idea, let us prove the necessity of the first-order condition for a general case, with $f : \mathbb{R}^n \to \mathbb{R}$. Let $\mathbf{x}, \mathbf{y} \in \mathbf{dom}\ f \subset \mathbb{R}^n$ and consider f restricted to any line passing through $\mathbf{dom}\ f$, i.e., consider the function defined by

$$g(t) = f(t\mathbf{y} + (1 - t)\mathbf{x}),$$

with $\mathbf{dom}\ g = \{t \mid t\mathbf{y} + (1 - t)\mathbf{x} \in \mathbf{dom}\ f\} \neq \emptyset$ due to $[0, 1] \subset \mathbf{dom}\ g$. The derivative of the function is

$$g'(t) = \nabla f(t\mathbf{y} + (1 - t)\mathbf{x})^T(\mathbf{y} - \mathbf{x}).$$

Now since f is convex, $g(t)$ is also convex (by Fact 3.1). So

$$g(1) \geq g(0) + g'(0) \quad (\text{by (3.19)})$$

which leads to $f(\mathbf{y}) \geq f(\mathbf{x}) + \nabla f(\mathbf{x})^T(\mathbf{y} - \mathbf{x})$. ∎

Remark 3.10 If the first-order condition given by (3.16) holds true with the strict inequality for any distinct points $\mathbf{x}, \mathbf{y} \in \mathbf{dom}\ f$, the function f is strictly convex. $\qquad\square$

Remark 3.11 Suppose that $f : \mathbb{R}^{m \times n} \to \mathbb{R}$ is a differentiable function of matrix $\mathbf{X} \in \mathbb{R}^{m \times n}$. Then the corresponding first-order condition for f to be convex is given by

$$f(\mathbf{Y}) \geq f(\mathbf{X}) + \text{Tr}\big(\nabla f(\mathbf{X})^T (\mathbf{Y} - \mathbf{X})\big), \ \forall \mathbf{X}, \mathbf{Y} \in \mathbf{dom}\ f. \tag{3.20}$$

Similarly, if (3.20) holds true with the strict inequality for all distinct $\mathbf{X}, \mathbf{Y} \in \mathbf{dom}\ f$, the function f is strictly convex. $\qquad\square$

Remark 3.12 For a convex nonsmooth function f, the corresponding first-order condition is given by

$$f(\mathbf{y}) \geq f(\mathbf{x}) + \bar{\nabla} f(\mathbf{x})^T (\mathbf{y} - \mathbf{x}), \quad \forall \mathbf{x}, \mathbf{y} \in \mathbf{dom}\ f, \tag{3.21}$$

where $\bar{\nabla} f(\mathbf{x})$ is a subgradient of f at a point $\mathbf{x}$. A vector is called a *subgradient* of f at a point $\mathbf{x}$ if it satisfies (3.21) for all $\mathbf{y} \in \mathbf{dom}\ f$. Let

$$\mathcal{G}_f(\mathbf{x}) = \{\bar{\nabla} f(\mathbf{x})\}$$

denote the set of all the subgradients of $f(\mathbf{x})$ at $\mathbf{x}$, also known as the subdifferential of f at $\mathbf{x}$. Then $\mathcal{G}_f(\mathbf{x}) = \{\nabla f(\mathbf{x})\}$ when $f(\mathbf{x})$ is differentiable at $\mathbf{x}$ (i.e., the subgradient is exactly the gradient of $f(\mathbf{x})$); otherwise the subgradients of $f(\mathbf{x})$ at $\mathbf{x}$ are nonunique.

Notice that the subdifferential $\mathcal{G}_f(\mathbf{x}) = \{\bar{\nabla} f(\mathbf{x})\}$ is a convex set. This can be illustrated with a simple convex function $f(x) = |x|$, which is not differentiable at $x = 0$. For this case, $\mathcal{G}_f(0) = [-1, 1]$ is a convex set, and so the subgradients of $f(x)$ at $x = 0$ are nonunique. $\qquad\square$

Remark 3.13 For a convex differentiable function $f : \mathbb{R} \to \mathbb{R}$, it can be easily inferred by the first-order condition given by (3.16) that

$$\frac{f(x) - f(x_0)}{x - x_0} \leq f'(x_0) \leq \frac{f(y) - f(x_0)}{y - x_0} \ \forall\ x < x_0 < y$$

$$\Rightarrow \begin{cases} f'(x) \leq \dfrac{f(y) - f(x)}{y - x}, & \text{as } x_0 \to x \\[2mm] \dfrac{f(y) - f(x)}{y - x} \leq f'(y), & \text{as } x_0 \to y \\[2mm] f'(x_0^-) = f'(x_0) = f'(x_0^+), & \text{as } x \to x_0\ \&\ y \to x_0 \end{cases}$$

where the two equalities are due to the fact that $f(x)$ is differentiable at $x = x_0$, implying that $f'(x)$ is a nondecreasing continuous function, and thus the second derivative $f''(x) \geq 0$ if f is twice differentiable. Similarly, it can be inferred that if $f(x)$ is strictly convex, $f'(x)$ is also a strictly increasing continuous function and vice versa. When $f : \mathbb{R} \to \mathbb{R}$ is a nonsmooth convex function, $f''(x) \geq 0$ is still true over its domain excluding the points where f is not differentiable. For instance, $f(x) = |x|$ is a nonsmooth convex function, and $f'(x) = \text{sgn}(x)$ for $x \neq 0$ is nondecreasing continuous for all $x \neq 0$.

Furthermore, for a differentiable convex function $f : \mathbb{R}^n \to \mathbb{R}$, by Fact 3.1, $g(t) = f(\mathbf{x}_1 + t\mathbf{v})$ is convex and differentiable in $t \in \textbf{dom } g = \{t \mid \mathbf{x}_1 + t\mathbf{v} \in \textbf{dom } f\}$. Let $\mathbf{v} = \mathbf{x}_2 - \mathbf{x}_1$. Then $[0,1] \subset \textbf{dom } g$ and we have $g'(1) - g'(0) \geq 0$ since $g'(t)$ must be a nondecreasing differentiable function as proved above, leading to the following inequality:

$$(\nabla f(\mathbf{x}_2) - \nabla f(\mathbf{x}_1))^T (\mathbf{x}_2 - \mathbf{x}_1) \geq 0 \; \forall \mathbf{x}_1, \mathbf{x}_2 \in \textbf{dom } f. \tag{3.22}$$

The inequality given by (3.22) implies that the correlation of any change of a point $\mathbf{x} \in \textbf{dom } f$ and the corresponding change of $\nabla f(\mathbf{x})$ is nonnegative, as illustrated in Figure 3.3.

Next, we prove that ∇f is continuous over the interior of its domain for a convex differentiable function $f : \mathbb{R}^n \to \mathbb{R}$. Assume that $\{\mathbf{x}_k\} \subset \textbf{int}(\textbf{dom } f)$ is a sequence that converges to $\mathbf{x} \in \textbf{int}(\textbf{dom } f)$. Let us prove that the first component of ∇f, i.e., $\partial f(\mathbf{x})/\partial x_1$ is continuous as follows:

$$
\begin{aligned}
\lim_{k \to \infty} \frac{\partial f(\mathbf{x}_k)}{\partial x_1} &= \lim_{k \to \infty} \lim_{\varepsilon \to 0} \frac{f(\mathbf{x}_k + \varepsilon \mathbf{e}_1) - f(\mathbf{x}_k)}{\varepsilon} \\
&= \lim_{\varepsilon \to 0} \lim_{k \to \infty} \frac{f(\mathbf{x}_k + \varepsilon \mathbf{e}_1) - f(\mathbf{x}_k)}{\varepsilon} \\
&= \lim_{\varepsilon \to 0} \frac{f(\mathbf{x} + \varepsilon \mathbf{e}_1) - f(\mathbf{x})}{\varepsilon} = \frac{\partial f(\mathbf{x})}{\partial x_1},
\end{aligned}
$$

where $\mathbf{e}_1 \in \mathbb{R}^n$ is a unit column vector with the first element equal to unity, and the second and third equalities are due to the fact that f is continuous over $\textbf{int}(\textbf{dom } f)$ for a convex function. Hence $\partial f(\mathbf{x})/\partial x_1$ is continuous. Similarly, all the other $n - 1$ components of ∇f can be shown to be continuous, and hence ∇f is continuous over the interior of $\textbf{dom } f$. $\qquad\square$

Remark 3.14 Suppose that $f(\mathbf{x})$ is convex, nonsmooth, or differentiable. Then the set $C = \{\mathbf{x} \in \mathbb{R}^n \mid f(\mathbf{x}) \leq 0\}$ (3.14) is convex and closed (cf. Remark 3.9). Provided that $\textbf{int } C \neq \emptyset$, by the first-order condition of convex functions and the supporting hyperplane theory, it can be easily shown that for any $\mathbf{x}_0 \in \textbf{bd } C$ (i.e., $f(\mathbf{x}_0) = 0$),

$$C \subset \mathcal{H}_-(\mathbf{x}_0) = \left\{ \mathbf{x} \mid \bar{\nabla} f(\mathbf{x}_0)^T (\mathbf{x} - \mathbf{x}_0) \leq 0 \right\}, \tag{3.23}$$

and so

$$C = \bigcap_{\mathbf{x}_0 \in \textbf{bd } C} \mathcal{H}_-(\mathbf{x}_0). \tag{3.24}$$

When $f(\mathbf{x})$ is convex and differentiable, the set C has a smooth boundary since the halfspace $\mathcal{H}_-(\mathbf{x}_0)$ in (3.23) is unique due to $\bar{\nabla} f(\mathbf{x}_0) = \nabla f(\mathbf{x}_0)$. $\qquad\square$

Remark 3.15 For the case that $f : \mathbb{C}^n \to \mathbb{R}$, the first-order inequalities corresponding to (3.16) and (3.22) are given by

$$
\begin{aligned}
f(\mathbf{y}) &\geq f(\mathbf{x}) + \text{Re}\left\{\nabla f(\mathbf{x})^H(\mathbf{y} - \mathbf{x})\right\}, \quad \forall \mathbf{x}, \mathbf{y} \in \textbf{dom } f \\
\text{Re}&\left\{(\nabla f(\mathbf{x}_2) - \nabla f(\mathbf{x}_1))^H(\mathbf{x}_2 - \mathbf{x}_1)\right\} \geq 0, \quad \forall \mathbf{x}_1, \mathbf{x}_2 \in \textbf{dom } f.
\end{aligned}
\tag{3.25}
$$

For the case that $f : \mathbb{C}^{m \times n} \to \mathbb{R}$, the first-order inequality corresponding to (3.20) is given by

$$
f(\mathbf{Y}) \geq f(\mathbf{X}) + \text{Tr}\big(\text{Re}\left\{\nabla f(\mathbf{X})^H(\mathbf{Y} - \mathbf{X})\right\}\big), \quad \forall \mathbf{X}, \mathbf{Y} \in \textbf{dom } f,
\tag{3.26}
$$

which is in a form quite similar to (3.20) for the real-variable case.   □

3.1.3 Second-order condition

Suppose that f is twice differentiable. Then f is convex if and only if $\textbf{dom } f$ is convex and the Hessian of f is PSD for all $\mathbf{x} \in \textbf{dom } f$, that is,

$$
\nabla^2 f(\mathbf{x}) \succeq \mathbf{0}, \quad \forall \mathbf{x} \in \textbf{dom } f.
\tag{3.27}
$$

Proof: Let us prove the sufficiency followed by necessity.

- Sufficiency: (i.e., if $\nabla^2 f(\mathbf{x}) \succeq \mathbf{0}, \forall \mathbf{x} \in \textbf{dom } f$, then f is convex) From the second-order expansion of Taylor series of $f(\mathbf{x})$ (cf. (1.54)), we have

$$
\begin{aligned}
f(\mathbf{x} + \mathbf{v}) &= f(\mathbf{x}) + \nabla f(\mathbf{x})^T \mathbf{v} + \frac{1}{2}\mathbf{v}^T \nabla^2 f(\mathbf{x} + \theta\mathbf{v})\mathbf{v} \\
&\geq f(\mathbf{x}) + \nabla f(\mathbf{x})^T \mathbf{v} \quad \text{(by (3.27))}
\end{aligned}
\tag{3.28}
$$

for some $\theta \in [0, 1]$. Let $\mathbf{y} = \mathbf{x} + \mathbf{v}$, i.e., $\mathbf{v} = \mathbf{y} - \mathbf{x}$. Then we have

$$
f(\mathbf{y}) \geq f(\mathbf{x}) + \nabla f(\mathbf{x})^T(\mathbf{y} - \mathbf{x})
$$

which is the exactly first-order condition for the convexity of $f(\mathbf{x})$, implying that f is convex.

- Necessity: Since $f(\mathbf{x})$ is convex, from the first-order condition we have

$$
f(\mathbf{x} + \mathbf{v}) \geq f(\mathbf{x}) + \nabla f(\mathbf{x})^T \mathbf{v}
$$

which together with the second-order expansion of Taylor series of $f(\mathbf{x})$ given by (3.28) implies

$$
\mathbf{v}^T \nabla^2 f(\mathbf{x} + \theta\mathbf{v})\mathbf{v} \geq 0.
$$

By letting $\|\mathbf{v}\|_2 \to 0$, it can be inferred that $\nabla^2 f(\mathbf{x}) \succeq \mathbf{0}$ because $\nabla^2 f(\mathbf{x})$ is continuous for a convex twice differentiable function $f(\mathbf{x})$.   ■

Remark 3.16 If the second-order condition given by (3.27) holds true with the strict inequality for all $\mathbf{x} \in \textbf{dom } f$, the function f is strictly convex; moreover, under the second-order condition given by (3.27) for the case that $f : \mathbb{R} \to \mathbb{R}$, the first derivative f' must be continuous and nondecreasing if f is convex, and continuous and strictly increasing if f is strictly convex. $\qquad\square$

Remark 3.17 *(Strong convexity)* A convex function f is *strongly convex* on a set C if there exists an $m > 0$ such that either $\nabla^2 f(\mathbf{x}) \succeq m\mathbf{I}$ for all $\mathbf{x} \in C$, or equivalently the following second-order condition holds true:

$$f(\mathbf{y}) \geq f(\mathbf{x}) + \nabla f(\mathbf{x})^T(\mathbf{y} - \mathbf{x}) + \frac{m}{2}\|\mathbf{y} - \mathbf{x}\|_2^2 \quad \forall \mathbf{x}, \mathbf{y} \in C \qquad (3.29)$$

which is directly implied from (3.28). So if f is strongly convex, it must be strictly convex, but the reverse is not necessarily true. $\qquad\square$

3.1.4 Examples

Some examples of convex and concave functions on $\mathbb{R}$ are as follows:

- $f(x) = e^{ax}$ is strictly convex on $\mathbb{R}$ for any nonzero $a \in \mathbb{R}$ by second-order condition; $f'(x) = ae^{ax}$ is continuous and strictly increasing for any $a \neq 0$. However, for a convex function $g : \mathbb{R}_+ \to \mathbb{R}$ defined as

$$g(x) = \begin{cases} 2, & x = 0 \\ e^{ax}, & x > 0 \end{cases}$$

it can be seen that $g(x)$ is continuous and $g'(x)$ is strictly increasing for $a \neq 0$ and $x > 0$ (i.e., the interior of **dom** g). Because $g(x)$ itself is not continuous (so not differentiable), let us prove its convexity by the inequality (3.3) as follows. Let $x_1 = 0$ and $x_2 > 0$ and $\theta \in [0, 1)$. Then

$$\theta g(x_1) + (1-\theta)g(x_2) = \theta \cdot 2 + (1-\theta)e^{ax_2}$$
$$\geq 2^\theta \cdot e^{(1-\theta)ax_2} \quad \text{(by (3.67))}$$
$$\geq e^{a(1-\theta)x_2} = g(\theta x_1 + (1-\theta)x_2).$$

For the case $x_1 \neq 0$, $x_2 \neq 0$, the inequality (3.3) will be satisfied due to the fact that e^{ax} is convex on $\mathbb{R}_{++}$. So g is convex.

- $f(x) = \log x$ is strictly concave on $\mathbb{R}_{++}$ by second-order condition; $f'(x) = 1/x$ is continuous and strictly decreasing over $\mathbb{R}_{++}$.
- $f(x) = x\log x$ is strictly convex on $\mathbb{R}_+$ (where $f(0) = 0$) by second-order condition; $f'(x) = 1 + \log x$ is continuous and strictly increasing over $\mathbb{R}_{++} =$ int $\mathbb{R}_+$.
- $\log Q(x)$ is strictly concave on $\mathbb{R}$, where

$$Q(x) = \frac{1}{\sqrt{2\pi}}\int_x^\infty e^{-t^2/2}dt.$$

Its first derivative

$$\frac{d\log Q(x)}{dx} = -\frac{e^{-x^2/2}}{\sqrt{2\pi}Q(x)}$$

is strictly decreasing. Its second derivative can be shown to be

$$\frac{d^2 \log Q(x)}{dx^2} = \frac{xe^{-x^2/2} \int_x^\infty e^{-t^2/2} dt - e^{-x^2}}{\left(\int_x^\infty e^{-t^2/2} dt\right)^2},$$

and the numerator is less than zero since

$$xe^{-x^2/2} \int_x^\infty e^{-t^2/2} dt - e^{-x^2} < e^{-x^2/2} \int_x^\infty te^{-t^2/2} dt - e^{-x^2} = 0.$$

By second-order condition, $\log Q(x)$ is strictly concave.

Some more examples of convex functions on $\mathbb{R}^n$ are as follows:

- An affine function $f(\mathbf{x}) = \mathbf{a}^T \mathbf{x} + b$ is both convex and concave. This can be easily proved either by the first-order condition

$$f(\mathbf{y}) = \mathbf{a}^T \mathbf{y} + b = \mathbf{a}^T \mathbf{x} + b + \mathbf{a}^T(\mathbf{y} - \mathbf{x}), \tag{3.30}$$

or by the definition of convex functions. However, f is not strictly convex.

- $f(\mathbf{x}) = \mathbf{x}^T \mathbf{P} \mathbf{x} + 2\mathbf{q}^T \mathbf{x} + r$, where $\mathbf{P} \in \mathbb{S}^n$, $\mathbf{q} \in \mathbb{R}^n$ and $r \in \mathbb{R}$ (quadratic function). Then

$$\nabla f(\mathbf{x}) = 2\mathbf{P}\mathbf{x} + 2\mathbf{q}, \tag{3.31}$$
$$\nabla^2 f(\mathbf{x}) = 2\mathbf{P}. \tag{3.32}$$

So $f(\mathbf{x})$ is convex if and only if $\mathbf{P} \succeq \mathbf{0}$ by second-order condition.

- Every norm $\|\cdot\|$ on $\mathbb{R}^n$ is convex since for $0 \le \theta \le 1$,

$$\|\theta\mathbf{x} + (1-\theta)\mathbf{y}\| \le \|\theta\mathbf{x}\| + \|(1-\theta)\mathbf{y}\| = \theta\|\mathbf{x}\| + (1-\theta)\|\mathbf{y}\|, \tag{3.33}$$

where the inequality is due to the triangle inequality and the equality is due to the homogeneity of a norm operator.

- $f(\mathbf{x}) = \max\{x_1, x_2, \ldots, x_n\}$ is convex since

$$f(\theta\mathbf{x} + (1-\theta)\mathbf{y}) = \max_i\{\theta x_i + (1-\theta)y_i\}$$
$$\le \max_i\{\theta x_i\} + \max_i\{(1-\theta)y_i\}$$
$$= \theta f(\mathbf{x}) + (1-\theta)f(\mathbf{y}). \tag{3.34}$$

- Geometric mean:

$$f(\mathbf{x}) = \left(\prod_{i=1}^n x_i\right)^{1/n} \tag{3.35}$$

is concave on $\mathbb{R}_{++}^n$ (which can be proven by second-order condition [BV04]).

- log-sum-exponential:

$$f(\mathbf{x}) = \log\left(\sum_{i=1}^n e^{x_i}\right) \tag{3.36}$$

is convex on $\mathbb{R}^n$ (which can be proven by second-order condition [BV04]).

Remark 3.18 The log-sum-exponential function $f(\mathbf{x}) = \log(\sum_{i=1}^n e^{x_i})$ is an approximation of the max function as follows:

$$\max\{x_1, x_2, \ldots, x_n\} = \log\left(\max_i \{e^{x_i}\}\right) \leq \log\left(\sum_{i=1}^n e^{x_i}\right)$$

$$\leq \log\left(n e^{\max_i\{x_i\}}\right)$$

$$= \log n + \max\{x_1, x_2, \ldots, x_n\}. \tag{3.37}$$

With the approximation error no larger than $\log n$, the differentiable log-sum-exponential approximation to the nondifferentiable $\max\{x_1, \ldots, x_n\}$ has been applied to noncoherent decoding [YCM+12] in communications.

An approximation to $\min\{a_1, \ldots, a_N\}$ using the log-sum-exponential function can be obtained by (3.37) as follows. For any positive real γ,

$$\min_{n \in \{1,\ldots,N\}} a_n = -\frac{1}{\gamma} \max_{n \in \{1,\ldots,N\}} \{-\gamma a_n\} \geq -\frac{1}{\gamma} \log_2\left(\sum_{n=1}^N 2^{-\gamma a_n}\right)$$

$$\geq -\frac{1}{\gamma} \max_{n \in \{1,\ldots,N\}} \{-\gamma a_n\} - \frac{1}{\gamma}\log_2 N$$

$$= \min_{n \in \{1,\ldots,N\}} a_n - \frac{1}{\gamma}\log_2 N. \tag{3.38}$$

The inequalities in (3.38) show that $-\frac{1}{\gamma}\log_2\left(\sum_{n=1}^N 2^{-\gamma a_n}\right)$ can be used as an approximation of $\min\{a_1, \ldots, a_N\}$, and the approximation error is no larger than $\frac{1}{\gamma}\log_2 N$. The log-sum-exponential function in the first line in (3.38) has been used as an approximation to the weighted minimum rate in a K-user coordinated beamforming design [LCC15] as follows:

$$\min_{i \in \{1,\ldots,K\}} \frac{R_i}{\alpha_i} \approx -\frac{1}{\gamma}\log_2\left(\sum_{i=1}^K 2^{-\gamma R_i / \alpha_i}\right), \tag{3.39}$$

where R_i denotes the transmission rate for the ith user, $\alpha_i \in [0, 1]$ represents the user priority, and $\sum_{i=1}^K \alpha_i = 1$. Maximizing the nondifferentiable weighted minimum rate (though a concave function) for obtaining the max-min-fairness (MMF) rate is hard to handle. Instead, maximizing its approximation on the right-hand side of (3.39) (a differentiable concave function) has been practically applied in the coordinated beamforming design. □

Some examples of convex functions of matrices are as follows:

- Linear function $f(\mathbf{X}) = \text{Tr}(\mathbf{AX})$ is both convex and concave on $\mathbb{R}^{m \times n}$, where $\mathbf{A} \in \mathbb{R}^{n \times m}$. It can be seen that the first-order condition

$$f(\mathbf{Y}) = \text{Tr}(\mathbf{AY}) = \text{Tr}(\mathbf{AX}) + \text{Tr}(\mathbf{A}(\mathbf{Y} - \mathbf{X}))$$
$$= \text{Tr}(\mathbf{AX}) + \text{Tr}(\nabla f(\mathbf{X})^T (\mathbf{Y} - \mathbf{X}))$$

  holds with equality, where

$$\nabla f(\mathbf{X}) = \mathbf{A}^T \quad \text{(by (3.20))}. \tag{3.40}$$

  So $f(\mathbf{X})$ is convex.

- Quadratic function $f(\mathbf{X}) = \text{Tr}(\mathbf{X}^2)$ is strictly convex on $\mathbb{S}^n$. This can be easily shown by the first-order condition given by (3.20). Because $\text{Tr}((\mathbf{Y} - \mathbf{X})^2) > 0$ for $\mathbf{X} \neq \mathbf{Y}$ and $\nabla f(\mathbf{X}) = 2\mathbf{X}$, the first-order condition

$$f(\mathbf{Y}) = \text{Tr}(\mathbf{Y}^2) > \text{Tr}(2\mathbf{YX}) - \text{Tr}(\mathbf{X}^2)$$
$$= f(\mathbf{X}) + \text{Tr}(\nabla f(\mathbf{X})^T (\mathbf{Y} - \mathbf{X})) \quad \forall \mathbf{X} \neq \mathbf{Y}$$

  holds with the strictly inequality.

- $f(\mathbf{X}) = -\log \det(\mathbf{X})$ is strictly convex on $\mathbb{S}^n_{++}$.

  Proof. We will use Fact 3.1 to prove the convexity of $f(\mathbf{X})$. Let

$$g(t) = -\log \det(\mathbf{X} + t\mathbf{V}), \tag{3.41}$$

  where $\mathbf{X} \succ \mathbf{0}$, $\mathbf{V} \neq \mathbf{0}$, and $\mathbf{V} \in \mathbb{S}^n$, $\textbf{dom}\, g = \{t \mid \mathbf{X} + t\mathbf{V} \succ \mathbf{0}\} \neq \emptyset$. Since $\mathbf{X} \succ \mathbf{0}$, we can factorize $\mathbf{X}$ as $\mathbf{X}^{1/2}\mathbf{X}^{1/2}$, where $\mathbf{X}^{1/2} \succ \mathbf{0}$ is invertible.

  Now, let

$$\mathbf{Z} = \mathbf{X}^{-1/2}\mathbf{V}\mathbf{X}^{-1/2} \in \mathbb{S}^n.$$

  It can be easily verified that for $\mathbf{X} \succ \mathbf{0}$ and $\mathbf{V} \in \mathbb{S}^n$,

$$\mathbf{X} + t\mathbf{V} \succ \mathbf{0} \Leftrightarrow \mathbf{I}_n + t\mathbf{Z} \succ \mathbf{0} \tag{3.42}$$

  (see the proof for (3.42) in Remark 3.19 below). Then,

$$g(t) = -\log \det(\mathbf{X}^{1/2}(\mathbf{I}_n + t\mathbf{Z})\mathbf{X}^{1/2})$$
$$= -\log \left[\det(\mathbf{X}^{1/2}) \cdot \det(\mathbf{I}_n + t\mathbf{Z}) \cdot \det(\mathbf{X}^{1/2})\right]$$
$$= -\log \det(\mathbf{X}^{1/2}) - \log \det(\mathbf{I}_n + t\mathbf{Z}) - \log \det(\mathbf{X}^{1/2})$$
$$= -\log \det(\mathbf{X}) - \log \det(\mathbf{I}_n + t\mathbf{Z}).$$

  Since $\mathbf{Z} \in \mathbb{S}^n$, we can perform EVD on $\mathbf{Z}$ as

$$\mathbf{Z} = \mathbf{Q}\mathbf{\Lambda}\mathbf{Q}^T,$$

  where $\mathbf{\Lambda} = \textbf{Diag}\,(\lambda_1, \lambda_2, \ldots, \lambda_n)$ in which all the λ_is are real, and $\mathbf{Q}\mathbf{Q}^T = \mathbf{Q}^T\mathbf{Q} = \mathbf{I}_n$. Moreover, $\mathbf{I}_n + t\mathbf{Z} \succ \mathbf{0} \Leftrightarrow 1 + t\lambda_i > 0, \forall i$ by (1.99). In other words,

$$\mathbf{dom}\ g = \{t \mid \mathbf{X} + t\mathbf{V} \succ \mathbf{0}\} = \{t \mid 1 + t\lambda_i > 0,\ i = 1, \ldots, n\}. \qquad (3.43)$$

Then $g(t)$ can be further simplified as follows:

$$\begin{aligned}
g(t) &= -\log\ \det(\mathbf{X}) - \log\ \det(\mathbf{I}_n + t\mathbf{Q}\boldsymbol{\Lambda}\mathbf{Q}^T) \\
&= -\log\ \det(\mathbf{X}) - \log\ \det(\mathbf{Q}(\mathbf{I}_n + t\boldsymbol{\Lambda})\mathbf{Q}^T) \\
&= -\log\ \det(\mathbf{X}) - \log\ \det(\mathbf{I}_n + t\boldsymbol{\Lambda})\ \ (\text{since } \det(\mathbf{Q}) = 1) \\
&= -\log\ \det(\mathbf{X}) - \log \prod_{i=1}^{n}(1 + t\lambda_i) \\
&= -\log\ \det(\mathbf{X}) - \sum_{i=1}^{n}\log(1 + t\lambda_i), \qquad (3.44)
\end{aligned}$$

and

$$\frac{dg(t)}{dt} = -\sum_{i=1}^{n}\frac{\lambda_i}{1 + t\lambda_i},$$

$$\frac{d^2 g(t)}{dt^2} = \sum_{i=1}^{n}\frac{\lambda_i^2}{(1 + t\lambda_i)^2} > 0.$$

Since $\mathbf{V} \neq \mathbf{0}$ and $\mathbf{X} \in \mathbb{S}_{++}^n$, we have $\mathbf{Z} \neq \mathbf{0}$. Then $\mathrm{Tr}(\mathbf{Z}^T\mathbf{Z}) = \sum_{i=1}^{n}\lambda_i^2 > 0$ by (1.97), implying that at least one λ_i is not equal to zero. Therefore, we have proved that $g(t)$ is strictly convex by second-order condition. Thus we have completed the proof that $f(\mathbf{X})$ is strictly convex on $\mathbb{S}_{++}^n$ by Fact 3.1. ∎

Remark 3.19 By observing the following mutual implications,

$$\begin{aligned}
\mathbf{X} + t\mathbf{V} \succ \mathbf{0} &\Leftrightarrow\ \mathbf{a}^T(\mathbf{X} + t\mathbf{V})\mathbf{a} > 0\ \forall \mathbf{a} \neq \mathbf{0} \in \mathbb{R}^n \\
&\Leftrightarrow\ \mathbf{a}^T\mathbf{X}^{1/2}(\mathbf{I}_n + t\mathbf{Z})\mathbf{X}^{1/2}\mathbf{a} > 0\ \forall \mathbf{a} \neq \mathbf{0} \in \mathbb{R}^n \\
&\Leftrightarrow\ \mathbf{b}^T(\mathbf{I}_n + t\mathbf{Z})\mathbf{b} > 0\ \forall \mathbf{b} = \mathbf{X}^{1/2}\mathbf{a} \neq \mathbf{0}\ \ (\text{since } \mathbf{X}^{1/2} \succ \mathbf{0}) \\
&\Leftrightarrow\ \mathbf{I}_n + t\mathbf{Z} \succ \mathbf{0},
\end{aligned}$$

one can easily prove (3.42). □

Remark 3.20 In the above proof of the convexity of $-\log\ \det(\mathbf{X})$ on $\mathbb{S}_{++}^n$, replacing $t\mathbf{V}$ by $\Delta\mathbf{X}$ in (3.41) (i.e., $t\mathbf{Z} = \mathbf{X}^{-1/2}(\Delta\mathbf{X})\mathbf{X}^{-1/2}$), and considering the following first-order Taylor series approximation to (3.44) for small $t\lambda_i$s (eigenvalues of $t\mathbf{Z}$), we have

$$-g(t) = \log \det(\mathbf{X} + \Delta\mathbf{X})$$

$$\approx \log \det(\mathbf{X}) + \sum_{i=1}^{n} t\lambda_i \quad \text{(first-order Taylor series approximation)}$$

$$= \log \det(\mathbf{X}) + \text{Tr}(\mathbf{X}^{-1/2}(\Delta\mathbf{X})\mathbf{X}^{-1/2})$$

$$= \log \det(\mathbf{X}) + \text{Tr}(\mathbf{X}^{-1}\Delta\mathbf{X}), \tag{3.45}$$

which implies that

$$\nabla_{\mathbf{X}} \log \det(\mathbf{X}) = (\mathbf{X}^{-1})^T = \mathbf{X}^{-1}, \quad \mathbf{X} \in \mathbb{S}^n_{++}. \tag{3.46}$$

Then by (3.45), the first-order Taylor series approximation of $\det(\mathbf{X}) = \exp\{\log \det(\mathbf{X})\}$ further yields

$$\nabla_{\mathbf{X}} \det(\mathbf{X}) = \det(\mathbf{X}) \cdot \mathbf{X}^{-1}, \quad \mathbf{X} \in \mathbb{S}^n_{++}. \tag{3.47}$$

Alternatively, it can also be shown, by the first-order condition (3.20) together with (3.46), that $-\log \det(\mathbf{X})$ is strictly convex on $\mathbb{S}^n_{++}$ as presented in the next remark. $\qquad\square$

Remark 3.21 By Fact 3.1, we have shown above that $\log \det(\mathbf{X})$ (or equivalently $-\log \det(\mathbf{X})$) is strictly concave (convex) on $\mathbb{S}^n_{++}$. By the first-order condition (cf. (3.20)) and the use of (3.46), a more concise proof is to prove

$$\log \det(\mathbf{Y}) < \log \det(\mathbf{X}) + \text{Tr}(\mathbf{X}^{-1}(\mathbf{Y} - \mathbf{X}))$$

$$\Longleftrightarrow \log \det(\mathbf{X}^{-1}\mathbf{Y}) < \text{Tr}(\mathbf{X}^{-1}\mathbf{Y}) - n, \ \forall \mathbf{X} \neq \mathbf{Y} \quad \text{(by (1.83))}.$$

Let $\mathbf{X}^{1/2} \succ \mathbf{0}$ be a square root of $\mathbf{X}$ and $\mathbf{Z} \triangleq \mathbf{X}^{-1/2}\mathbf{Y}\mathbf{X}^{-1/2} \neq \mathbf{I}_n$ (due to $\mathbf{X} \neq \mathbf{Y}$). Then the above inequality can be further simplified into

$$\log \det(\mathbf{Z}) < \text{Tr}(\mathbf{Z}) - n \quad \text{(by (1.84))}$$

$$\Longleftrightarrow \sum_{i=1}^{n} \log \lambda_i < \sum_{i=1}^{n} \lambda_i - n \quad \text{(by (1.95) and (1.96))} \tag{3.48}$$

where $\lambda_i > 0, i = 1, \ldots, n$ are the eigenvalues of $\mathbf{Z}$. Note that at least one $\lambda_i \neq 1$ since $\mathbf{Z} \neq \mathbf{I}_n$. On the other hand, $\log x$ is a strictly concave function which must satisfy the first-order inequality w.r.t. $x = 1$, i.e.,

$$\log x < x - 1, \quad \forall x \neq 1, \ x > 0$$

from which one can infer that (3.48) must be true for $\mathbf{Z} \neq \mathbf{I}_n$. Thus we have completed the proof that $\log \det(\mathbf{X})$ is strictly concave on $\mathbb{S}^n_{++}$. $\qquad\square$

3.1.5   Epigraph

The *epigraph* of a function $f : \mathbb{R}^n \to \mathbb{R}$ is defined as

$$\textbf{epi } f = \{(\mathbf{x}, t) \mid \mathbf{x} \in \textbf{dom } f, \ f(\mathbf{x}) \leq t\} \subseteq \mathbb{R}^{n+1}. \tag{3.49}$$

"Epi" means "above", so epigraph means "above the graph." The epigraph of a function is a set defined by the associated function, which looks like a faceup bowl full of fluid as illustrated in Figure 3.4.

The convexity of the set **epi** f has a strong relation to the convexity of the associated function f as stated in the following fact.

Fact 3.2 f is convex if and only if **epi** f is convex.

Proof: Let us prove the sufficiency followed by necessity.

- Sufficiency: Assume that f is nonconvex. Then there exist $\mathbf{x}_1, \mathbf{x}_2 \in$ **dom** f and $\theta \in [0,1]$, such that

$$f(\theta\mathbf{x}_1 + (1-\theta)\mathbf{x}_2) > \theta f(\mathbf{x}_1) + (1-\theta)f(\mathbf{x}_2). \qquad (3.50)$$

  Let $(\mathbf{x}_1, t_1), (\mathbf{x}_2, t_2) \in$ **epi** f. Then we have $f(\mathbf{x}_1) \leq t_1$, $f(\mathbf{x}_2) \leq t_2$, and

$$\theta(\mathbf{x}_1, t_1) + (1-\theta)(\mathbf{x}_2, t_2) = (\theta\mathbf{x}_1 + (1-\theta)\mathbf{x}_2, \theta t_1 + (1-\theta)t_2) \in \textbf{epi } f$$
  (since **epi** f is convex)
$$\Rightarrow f(\theta\mathbf{x}_1 + (1-\theta)\mathbf{x}_2) \leq \theta t_1 + (1-\theta)t_2.$$

  Combining the above equation and (3.50), we get

$$\theta f(\mathbf{x}_1) + (1-\theta)f(\mathbf{x}_2) < \theta t_1 + (1-\theta)t_2.$$

  Letting $t_1 = f(\mathbf{x}_1)$ and $t_2 = f(\mathbf{x}_2)$ in this inequality yields

$$\theta f(\mathbf{x}_1) + (1-\theta)f(\mathbf{x}_2) < \theta f(\mathbf{x}_1) + (1-\theta)f(\mathbf{x}_2),$$

  which is an impossibility and hence our assumption is not valid. Therefore, f must be convex.

- Necessity: Since f is convex, we have

$$f(\theta\mathbf{x}_1 + (1-\theta)\mathbf{x}_2) \leq \theta f(\mathbf{x}_1) + (1-\theta)f(\mathbf{x}_2), \ \forall \mathbf{x}_1, \mathbf{x}_2 \in \textbf{dom } f, \ \theta \in [0,1].$$

  Let $(\mathbf{x}_1, t_1), (\mathbf{x}_2, t_2) \in$ **epi** f. Then $f(\mathbf{x}_1) \leq t_1, f(\mathbf{x}_2) \leq t_2$, and

$$f(\theta\mathbf{x}_1 + (1-\theta)\mathbf{x}_2) \leq \theta f(\mathbf{x}_1) + (1-\theta)f(\mathbf{x}_2) \leq \theta t_1 + (1-\theta)t_2.$$

  By the definition of epigraph, we have

$$(\theta\mathbf{x}_1 + (1-\theta)\mathbf{x}_2, \theta t_1 + (1-\theta)t_2) \in \textbf{epi } f$$
$$\Rightarrow \theta(\mathbf{x}_1, t_1) + (1-\theta)(\mathbf{x}_2, t_2) \in \textbf{epi } f$$
$$\Rightarrow \textbf{epi } f \text{ is convex.}$$

Thus we have completed the proof of Fact 3.2.   ∎

Remark 3.22 It can be shown that f is strictly convex if and only if **epi** f is a strictly convex set.   □

Remark 3.23 A function $f : \mathbb{R}^n \to \mathbb{R}$ is said to be closed if its epigraph $\mathbf{epi}\, f \subseteq \mathbb{R}^{n+1}$ is a closed set. That is, for every sequence $\{(\mathbf{x}_k, \alpha_k)\} \subset \mathbf{epi}\, f$ converging to some $(\hat{\mathbf{x}}, \hat{\alpha})$, we have $(\hat{\mathbf{x}}, \hat{\alpha}) \in \mathbf{epi}\, f$. Moreover, all continuous functions with closed domain are closed, but the converse may not be true. For instance, for the lower semi-continuous function $f_1(x)$ and the upper semi-continuous function $f_2(x)$ defined as

$$f_1(x) = \begin{cases} 1, & x < 0 \\ -1, & x \geq 0 \end{cases}, \quad f_2(x) = \begin{cases} 1, & x \leq 0 \\ -1, & x > 0 \end{cases},$$

it can be seen that $\mathbf{epi}\, f_1 = \{(x,t) \mid t \geq 1, x < 0\} \cup \{(x,t) \mid t \geq -1, x \geq 0\}$ is a closed set but f_1 is not continuous, and that $\mathbf{epi}\, f_2 = \{(x,t) \mid t \geq 1, x \leq 0\} \cup \{(x,t) \mid t \geq -1, x > 0\}$ is not a closed set and f_2 is not continuous either. □

Remark 3.24 For a differentiable convex function f with closed domain, the $\mathbf{epi}\, f$ defined in (3.49) is a closed convex set (as illustrated in Figure 3.4 for a one-dimensional function) since f is continuous by Remark 3.23. By the first-order condition given by (3.16) w.r.t. any $\mathbf{x}_0 \in \mathbf{dom}\, f$, it can be easily seen that

$$t \geq f(\mathbf{x}) \geq f(\mathbf{x}_0) + \nabla f(\mathbf{x}_0)^T (\mathbf{x} - \mathbf{x}_0)$$

$$\Leftrightarrow h(\mathbf{y}_0) \triangleq [\nabla f(\mathbf{x}_0)^T, \ -1] \left\{ \begin{bmatrix} \mathbf{x} \\ t \end{bmatrix} - \begin{bmatrix} \mathbf{x}_0 \\ f(\mathbf{x}_0) \end{bmatrix} \right\} \leq 0 \ \ \forall (\mathbf{x}, t) \in \mathbf{epi}\, f \quad (3.51)$$

which actually defines a supporting hyperplane

$$\mathcal{H}(\mathbf{y}_0) = \{\mathbf{y} = (\mathbf{x}, t) \in \mathbb{R}^{n+1} \mid h(\mathbf{y}_0) = 0\}$$

of the set $\mathbf{epi}\, f$ at any boundary point $\mathbf{y}_0 = (\mathbf{x}_0, f(\mathbf{x}_0))$ of $\mathbf{epi}\, f$ (cf. Figure 3.4), and thus $\mathbf{epi}\, f$ can be expressed in the intersection form of halfspaces as follows (cf. (2.136)):

$$\mathbf{epi}\, f = \bigcap_{\mathbf{y}_0 \in \mathbf{bd}\, \mathbf{epi}\, f} \mathcal{H}_-(\mathbf{y}_0), \quad (3.52)$$

where

$$\mathcal{H}_-(\mathbf{y}_0) = \{\mathbf{y} \in \mathbb{R}^{n+1} \mid \mathbf{a}(\mathbf{y}_0)^T(\mathbf{y} - \mathbf{y}_0) \leq 0\} \quad \text{(by (3.51))} \quad (3.53)$$

in which $\mathbf{a}(\mathbf{y}_0) = [\nabla f(\mathbf{x}_0)^T, -1]^T$. Thus $\mathbf{epi}\, f$ is convex if the first-order condition holds true, and vice versa. Moreover, each boundary hyperplane of $\mathcal{H}_-(\mathbf{y}_0)$ is uniquely defined by $\mathbf{y}_0$, and thus the boundary of $\mathbf{epi}\, f$ is smooth (cf. Remark 2.25). □

Remark 3.25 *(Convex envelope of a nonconvex function)* The convex envelope of a function $f : \mathbb{R}^n \to \mathbb{R}$ is defined as

$$g_f(\mathbf{x}) = \inf\, \{t \mid (\mathbf{x}, t) \in \mathbf{conv}\, \mathbf{epi}\, f\}, \quad \mathbf{dom}\, g_f = \mathbf{dom}\, f, \quad (3.54)$$

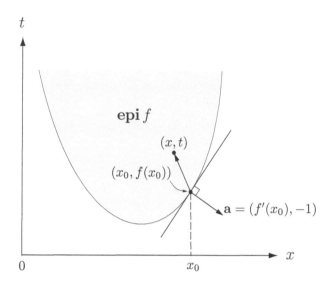

Figure 3.4 The epigraph of a convex function $f : \mathbb{R} \to \mathbb{R}$ and the hyperplane with the normal vector $\mathbf{a} = (f'(x_0), -1)$ supports $\mathbf{epi}\, f$ at boundary point $(x_0, f(x_0))$.

provided that $\mathbf{dom}\, f$ is convex. It can be inferred from (3.54) that $\mathbf{epi}\, g_f = \mathbf{conv\ epi}\, f$ and $g_f(\mathbf{x}) \le f(\mathbf{x})$ for all $\mathbf{x}$.

Two illustrative examples are given in Figure 3.5, where the two nonconvex functions are given by

$$f_1(x) = \begin{cases} 2\sqrt{x}, & 0 \le x \le 1 \\ 2(x-1)^2 + 2, & x \ge 1 \end{cases} \qquad \mathbf{dom}\, f = \mathbb{R}_+ \qquad (3.55)$$

$$f_2(x) = \begin{cases} 0, & x = 0 \\ 1, & -1 \le x < 0, \text{ or } 0 < x \le 1 \end{cases} \qquad \mathbf{dom}\, f = [-1, 1]. \qquad (3.56)$$

It can be readily proven from (3.54) that

$$g_{f_1}(x) = \begin{cases} 4(\sqrt{2} - 1)x, & 0 \le x \le \sqrt{2} \\ 2(x-1)^2 + 2, & x \ge \sqrt{2} \end{cases} \qquad (3.57)$$

$$g_{f_2}(x) = |x|, \quad -1 \le x \le 1. \qquad (3.58)$$

Note that $f_2(x)$ is not continuous and not differentiable, but it is actually a 1-dimensional special case of $f(\mathbf{x}) = \|\mathbf{x}\|_0$ (defined as the number of nonzero elements in $\mathbf{x}$), $\mathbf{dom}\, f = \mathcal{B} \triangleq \{\mathbf{x} \in \mathbb{R}^n \mid \|\mathbf{x}\|_1 \le 1\}$. Because the nonconvex function $f(\mathbf{x}) = \|\mathbf{x}\|_0$ is lower semi-continuous and nonconvex, its convex envelope can also be obtained by computing its bi-conjugate (i.e., the conjugate of the conjugate of f) (cf. Remark 9.1) besides applying the definition of the convex envelope (3.54). As a result,

$$g_f(\mathbf{x}) = \|\mathbf{x}\|_1, \quad \mathbf{x} \in \mathcal{B} \qquad (3.59)$$

(and the detailed proof will be presented in Subsection 9.1.2; cf. Remark 9.2 and (9.29)), which has been used in modeling sparsity (sparseness) of a vector.

Another interesting lower semi-continuous nonconvex function is $f(\mathbf{X}) = \mathrm{rank}(\mathbf{X})$, $\mathbf{dom}\ f = \{\mathbf{X} \in \mathbb{R}^{M \times N} \mid \|\mathbf{X}\|_* = \|\mathbf{X}\|_{\mathcal{A}} \le 1\} = \mathbf{conv}\ \mathcal{A}$ (cf. (2.105)), where $\mathcal{A}$ is the set of rank-1 matrices defined in (2.102). The convex envelope of f can be shown to be

$$g_f(\mathbf{X}) = \|\mathbf{X}\|_* = \sum_{i=1}^{\mathrm{rank}(\mathbf{X})} \sigma_i(\mathbf{X}), \ \mathbf{X} \in \mathbf{conv}\ \mathcal{A} \ \ (\text{cf. } (2.96)) \tag{3.60}$$

which has been used in low-rank approximations of high-dimensional data matrices in compressive sensing. □

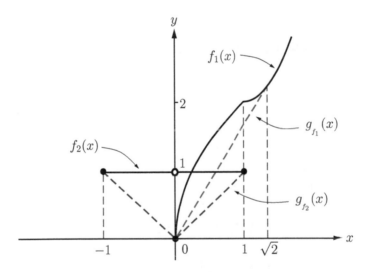

Figure 3.5 Convex envelopes $g_{f_1}(x)$ and $g_{f_2}(x)$ (cf. (3.57) and (3.58)) of two nonconvex functions $f_1(x)$ and $f_2(x)$ defined in (3.55) and (3.56), respectively.

As we will see later, the concept of epigraph, and the epigraph representation of optimization problems to be introduced in the next chapter, will be very useful in solving optimization problems.

3.1.6   Jensen's inequality

For a convex function $f : \mathbb{R}^n \to \mathbb{R}$, the inequality

$$f(\theta\mathbf{x} + (1 - \theta)\mathbf{y}) \le \theta f(\mathbf{x}) + (1 - \theta)f(\mathbf{y}), \ \ \theta \in [0, 1] \tag{3.61}$$

is sometimes called *Jensen's inequality*. Substituting $\mathbf{y} = \alpha\mathbf{z} + (1 - \alpha)\mathbf{w} \in \mathbf{dom}\ f$ with $\mathbf{z}$, $\mathbf{w} \in \mathbf{dom}\ f$ and $\alpha \in [0, 1]$ into (3.61) yields

$$f(\theta\mathbf{x} + (1-\theta)\mathbf{y}) = f(\theta\mathbf{x} + (1-\theta)\alpha\mathbf{z} + (1-\theta)(1-\alpha)\mathbf{w})$$
$$\leq \theta f(\mathbf{x}) + (1-\theta)f(\alpha\mathbf{z} + (1-\alpha)\mathbf{w})$$
$$\leq \theta f(\mathbf{x}) + (1-\theta)(\alpha f(\mathbf{z}) + (1-\alpha)f(\mathbf{w}))$$
$$= \theta f(\mathbf{x}) + (1-\theta)\alpha f(\mathbf{z}) + (1-\theta)(1-\alpha)f(\mathbf{w})$$
$$= \theta_1 f(\mathbf{x}) + \theta_2 f(\mathbf{z}) + \theta_3 f(\mathbf{w})$$

where $\theta_1 = \theta \in [0,1]$, $\theta_2 = (1-\theta)\alpha \in [0,1]$, $\theta_3 = (1-\theta)(1-\alpha) \in [0,1]$, and $\theta_1 + \theta_2 + \theta_3 = 1$. For convex f one can infer by induction that, for any positive integer $k \geq 2$,

$$f\left(\sum_{i=1}^{k} \theta_i \mathbf{x}_i\right) \leq \sum_{i=1}^{k} \theta_i f(\mathbf{x}_i), \ \forall \theta_i \geq 0, \ \sum_{i=1}^{k} \theta_i = 1. \tag{3.62}$$

Example 3.1 For all $\mathbf{x} \in \mathbb{R}_{++}^n$,

$$\left(\prod_{i=1}^{n} x_i\right)^{1/n} \leq \frac{1}{n}\left(\sum_{i=1}^{n} x_i\right), \tag{3.63}$$

which is the commonly known arithmetic-geometric mean inequality, and its general form is given in (3.64) below.

Proof of (3.63): Recall that $\log x$ is concave on $\mathbb{R}_{++}$. Hence, for $\theta_i \geq 0$ and $\sum_{i=1}^{n} \theta_i = 1$, we have

$$\log\left(\sum_{i=1}^{n} \theta_i x_i\right) \geq \sum_{i=1}^{n} \theta_i \log x_i = \sum_{i=1}^{n} \log x_i^{\theta_i} = \log\left(\prod_{i=1}^{n} x_i^{\theta_i}\right)$$
$$\Rightarrow \sum_{i=1}^{n} \theta_i x_i \geq \prod_{i=1}^{n} x_i^{\theta_i}. \tag{3.64}$$

Substituting $\theta_i = \frac{1}{n}$ for all i into (3.64) yields (3.63). ∎

Note that (3.64) holds true for all $x_i > 0$, $\theta_i \geq 0$, and $\sum_{i=1}^{n} \theta_i = 1$. An alternative representation by replacing x_i with $x_i/\theta_i > 0$ in (3.64) is given by

$$\sum_{i=1}^{n} x_i \geq \prod_{i=1}^{n}\left(\frac{x_i}{\theta_i}\right)^{\theta_i}, \tag{3.65}$$

where $\theta_i \neq 0$ for all i. The inequalities of (3.64) and (3.65) have been widely used in various optimization problems. Some examples will be given in Chapters 6 and 8 later.   □

Example 3.2 *(Hölder's inequality)*

$$\mathbf{y}^T\mathbf{x} \le \|\mathbf{x}\|_p \cdot \|\mathbf{y}\|_q, \tag{3.66}$$

where $\frac{1}{p} + \frac{1}{q} = 1$, $p \ge 1$, and $q \ge 1$. This inequality is actually a special case of (2.107) since $\|\cdot\|_p$ is the dual of $\|\cdot\|_q$, and vice versa. Nevertheless, this example provides an alternative proof using Jensen's inequality, followed by a discussion when the equality in (3.66) holds true.

Proof of (3.66): By (3.64), we have

$$\theta a + (1-\theta)b \ge a^\theta b^{1-\theta} \tag{3.67}$$

for $a, b > 0$ and $0 \le \theta \le 1$. Let

$$a_i = \frac{|x_i|^p}{\sum_{j=1}^n |x_j|^p}, \quad b_i = \frac{|y_i|^q}{\sum_{j=1}^n |y_j|^q}, \quad \theta = \frac{1}{p}.$$

Then we have

$$
\begin{aligned}
\sum_{i=1}^n \theta a_i + (1-\theta)b_i &= \sum_{i=1}^n \left(\frac{|x_i|^p/p}{\sum_{j=1}^n |x_j|^p} + \frac{|y_i|^q/q}{\sum_{j=1}^n |y_j|^q} \right) = \frac{1}{p} + \frac{1}{q} = 1 \\
&\ge \sum_{i=1}^n \left(\frac{|x_i|^p}{\sum_{j=1}^n |x_j|^p} \right)^{1/p} \left(\frac{|y_i|^q}{\sum_{j=1}^n |y_i|^q} \right)^{1/q} \\
&= \frac{\sum_{i=1}^n |x_i| \cdot |y_i|}{(\sum_{j=1}^n |x_j|^p)^{1/p}(\sum_{j=1}^n |y_j|^q)^{1/q}} \\
&\ge \frac{\mathbf{y}^T\mathbf{x}}{\|\mathbf{x}\|_p \cdot \|\mathbf{y}\|_q},
\end{aligned}
$$

which yields Hölder's inequality given by (3.66). ∎

A special case is the Cauchy–Schwartz inequality $(p = q = 2)$ as follows

$$\mathbf{y}^T\mathbf{x} \le \|\mathbf{x}\|_2 \cdot \|\mathbf{y}\|_2.$$

The equality holds as $\mathbf{y} = \alpha\mathbf{x}$ for all $\alpha \ge 0$. Another case of interest $(p = \infty, q = 1)$ is as follows

$$\mathbf{y}^T\mathbf{x} \le \|\mathbf{x}\|_\infty \cdot \|\mathbf{y}\|_1. \tag{3.68}$$

Some instances where the equality in (3.68) holds are (i) when $\mathbf{x} = \alpha\mathbf{1}_n$, $\alpha > 0$, $\mathbf{y} \in \mathbb{R}_+^n$, and (ii) when $\mathbf{y} = \alpha\mathbf{e}_i$, $\alpha > 0$, $\mathbf{x} \in \mathbb{R}_+^n$ and $x_i \ge x_j$, for all $j \ne i$. □

Remark 3.26 Let $p(\mathbf{x}) \ge 0$ for all $\mathbf{x} \in S \subseteq \mathbf{dom}\, f$. Supposing that $\mathbf{x}$ is a random vector with probability density function $p(\mathbf{x}) \ge 0$ over S, i.e., $\int_S p(\mathbf{x})d\mathbf{x} = 1$, then for a convex f, we have

$$f(\mathbb{E}\{\mathbf{x}\}) = f\left(\int_S \mathbf{x}p(\mathbf{x})d\mathbf{x} \right) \le \int_S f(\mathbf{x})p(\mathbf{x})d\mathbf{x} = \mathbb{E}\{f(\mathbf{x})\}, \tag{3.69}$$

where the inequality directly follows from the Jensen's inequality. □

3.2 Convexity preserving operations

In this section we describe some operations that preserve convexity or concavity of functions, or operations that will allow us to construct new convex and concave functions.

3.2.1 Nonnegative weighted sum

Let $f_1, \ldots, f_m$ be convex functions and $w_1, \ldots, w_m \geq 0$. Then $\sum_{i=1}^{m} w_i f_i$ is convex.

Proof: $\mathbf{dom}(\sum_{i=1}^{m} w_i f_i) = \bigcap_{i=1}^{m} \mathbf{dom}\ f_i$ is convex because $\mathbf{dom}\ f_i$ is convex for all i. For $0 \leq \theta \leq 1$, and $\mathbf{x}, \mathbf{y} \in \mathbf{dom}(\sum_{i=1}^{m} w_i f_i)$, we have

$$\sum_{i=1}^{m} w_i f_i(\theta \mathbf{x} + (1-\theta)\mathbf{y}) \leq \sum_{i=1}^{m} w_i(\theta f_i(\mathbf{x}) + (1-\theta)f_i(\mathbf{y}))$$

$$= \theta \sum_{i=1}^{m} w_i f_i(\mathbf{x}) + (1-\theta) \sum_{i=1}^{m} w_i f_i(\mathbf{y}).$$

Hence proved. ∎

Remark 3.27 $f(\mathbf{x}, \mathbf{y})$ is convex in $\mathbf{x}$ for each $\mathbf{y} \in \mathcal{A}$ and $w(\mathbf{y}) \geq 0$. Then,

$$g(\mathbf{x}) = \int_{\mathcal{A}} w(\mathbf{y}) f(\mathbf{x}, \mathbf{y}) d\mathbf{y} \tag{3.70}$$

is convex on $\bigcap_{\mathbf{y} \in \mathcal{A}} \mathbf{dom}\ f$. □

3.2.2 Composition with affine mapping

If $f : \mathbb{R}^n \to \mathbb{R}$ is a convex function, then for $\mathbf{A} \in \mathbb{R}^{n \times m}$ and $\mathbf{b} \in \mathbb{R}^n$, the function $g : \mathbb{R}^m \to \mathbb{R}$, defined as

$$g(\mathbf{x}) = f(\mathbf{A}\mathbf{x} + \mathbf{b}), \tag{3.71}$$

is also convex and its domain can be expressed as

$$\mathbf{dom}\ g = \{\mathbf{x} \in \mathbb{R}^m \mid \mathbf{A}\mathbf{x} + \mathbf{b} \in \mathbf{dom}\ f\}$$
$$= \{\mathbf{A}^\dagger(\mathbf{y} - \mathbf{b}) \mid \mathbf{y} \in \mathbf{dom}\ f\} + \mathcal{N}(\mathbf{A}) \quad \text{(cf. (2.62))} \tag{3.72}$$

which is also a convex set by Remark 2.9.

Proof (using epigraph): Since $g(\mathbf{x}) = f(\mathbf{A}\mathbf{x} + \mathbf{b})$ and $\mathbf{epi}\ f = \{(\mathbf{y}, t) \mid f(\mathbf{y}) \leq t\}$, we have

$$\mathbf{epi}\ g = \{(\mathbf{x}, t) \in \mathbb{R}^{m+1} \mid f(\mathbf{A}\mathbf{x} + \mathbf{b}) \leq t\}$$
$$= \{(\mathbf{x}, t) \in \mathbb{R}^{m+1} \mid (\mathbf{A}\mathbf{x} + \mathbf{b}, t) \in \mathbf{epi}\ f\}.$$

Now, define

$$S = \{(\mathbf{x}, \mathbf{y}, t) \in \mathbb{R}^{m+n+1} \mid \mathbf{y} = \mathbf{A}\mathbf{x} + \mathbf{b}, f(\mathbf{y}) \leq t\},$$

so that

$$\mathbf{epi}\ g = \left\{ \begin{bmatrix} \mathbf{I}_m & \mathbf{0}_{m \times n} & \mathbf{0}_m \\ \mathbf{0}_m^T & \mathbf{0}_n^T & 1 \end{bmatrix} (\mathbf{x}, \mathbf{y}, t) \mid (\mathbf{x}, \mathbf{y}, t) \in S \right\}$$

which is nothing but the image of S via an affine mapping. It can be easily shown, by the definition of convex sets, that S is convex if f is convex. Therefore **epi** g is convex (due to affine mapping from the convex set S) implying that g is convex (by Fact 3.2). ∎

Alternative proof: For $0 \leq \theta \leq 1$, we have

$$\begin{aligned}
g(\theta \mathbf{x}_1 + (1-\theta)\mathbf{x}_2) &= f(\mathbf{A}(\theta \mathbf{x}_1 + (1-\theta)\mathbf{x}_2) + \mathbf{b}) \\
&= f(\theta(\mathbf{A}\mathbf{x}_1 + \mathbf{b}) + (1-\theta)(\mathbf{A}\mathbf{x}_2 + \mathbf{b})) \\
&\leq \theta f(\mathbf{A}\mathbf{x}_1 + \mathbf{b}) + (1-\theta)f(\mathbf{A}\mathbf{x}_2 + \mathbf{b}) \\
&= \theta g(\mathbf{x}_1) + (1-\theta)g(\mathbf{x}_2).
\end{aligned}$$

Moreover, **dom** g (cf. (3.72)) is also a convex set, and so we conclude that $f(\mathbf{A}\mathbf{x} + \mathbf{b})$ is a convex function. ∎

3.2.3   Composition (scalar)

Suppose that $h : \mathbf{dom}\ h \to \mathbb{R}$ is a convex (concave) function and $\mathbf{dom}\ h \subset \mathbb{R}^n$. The extended-value extension of h, denoted as $\tilde{h}$, with $\mathbf{dom}\ \tilde{h} = \mathbb{R}^n$ aids in simple representation as its domain is the entire $\mathbb{R}^n$, which need not be explicitly mentioned. The extended-valued function $\tilde{h}$ is a function taking the same value of $h(\mathbf{x})$ for $\mathbf{x} \in \mathbf{dom}\ h$, otherwise taking the value of $+\infty$ ($-\infty$). Specifically, if h is convex,

$$\tilde{h}(\mathbf{x}) = \begin{cases} h(\mathbf{x}), & \mathbf{x} \in \mathbf{dom}\ h \\ +\infty, & \mathbf{x} \notin \mathbf{dom}\ h, \end{cases} \tag{3.73}$$

and if h is concave,

$$\tilde{h}(\mathbf{x}) = \begin{cases} h(\mathbf{x}), & \mathbf{x} \in \mathbf{dom}\ h \\ -\infty, & \mathbf{x} \notin \mathbf{dom}\ h. \end{cases} \tag{3.74}$$

Then the extended-valued function $\tilde{h}$ does not affect the convexity (or concavity) of the original function h and Eff-**dom** $\tilde{h}$ = Eff-**dom** h.

Some examples for illustrating properties of an extended-value extension of a function are as follows.

- $h(x) = \log x$, **dom** $h = \mathbb{R}_{++}$. Then $h(x)$ is concave and $\tilde{h}(x)$ is concave and nondecreasing.

- $h(x) = x^{1/2}$, **dom** $h = \mathbb{R}_+$. Then $h(x)$ is concave and $\tilde{h}(x)$ is concave and nondecreasing.

- In the function

$$h(x) = x^2, \; x \geq 0, \tag{3.75}$$

i.e., **dom** $h = \mathbb{R}_+$, $h(x)$ is convex and $\tilde{h}(x)$ is convex but neither nondecreasing nor nonincreasing.

Let $f(\mathbf{x}) = h(g(\mathbf{x}))$, where $h : \mathbb{R} \to \mathbb{R}$ and $g : \mathbb{R}^n \to \mathbb{R}$. Then we have the following four composition rules about the convexity or concavity of f.

(a) f is convex if h is convex, $\tilde{h}$ nondecreasing, and g convex.   (3.76a)

(b) f is convex if h is convex, $\tilde{h}$ nonincreasing, and g concave.   (3.76b)

(c) f is concave if h is concave, $\tilde{h}$ nondecreasing, and g concave.   (3.76c)

(d) f is concave if h is concave, $\tilde{h}$ nonincreasing, and g convex.   (3.76d)

Consider the case that g and h are twice differentiable and $\tilde{h}(x) = h(x)$. Then,

$$\nabla f(\mathbf{x}) = h'(g(\mathbf{x}))\nabla g(\mathbf{x}) \tag{3.77}$$

and

$$\begin{aligned}
\nabla^2 f(\mathbf{x}) &= D(\nabla f(\mathbf{x})) = D\big(h'(g(\mathbf{x})) \cdot \nabla g(\mathbf{x})\big) \quad \text{(by (1.46))} \\
&= \nabla g(\mathbf{x})D\big(h'(g(\mathbf{x}))\big) + h'(g(\mathbf{x})) \cdot D(\nabla g(\mathbf{x})) \\
&= h''(g(\mathbf{x}))\nabla g(\mathbf{x})\nabla g(\mathbf{x})^T + h'(g(\mathbf{x}))\nabla^2 g(\mathbf{x}).
\end{aligned} \tag{3.78}$$

The composition rules (a) (cf. (3.76a)) and (b) (cf. (3.76b)) can be proven for convexity of f by checking if $\nabla^2 f(\mathbf{x}) \succeq \mathbf{0}$, and the composition rules (c) (cf. (3.76c)) and (d) (cf. (3.76d)) for concavity of f by checking if $\nabla^2 f(\mathbf{x}) \preceq \mathbf{0}$. Let us conclude this subsection with a simple example.

Example 3.3 Let $g(\mathbf{x}) = \|\mathbf{x}\|_2$ (convex) and

$$h(x) = \begin{cases} x^2, \, x \geq 0 \\ 0, \quad \text{otherwise} \end{cases}$$

which is convex. So $\tilde{h}(x) = h(x)$ is nondecreasing. Then, $f(\mathbf{x}) = h(g(\mathbf{x})) = \|\mathbf{x}\|_2^2 = \mathbf{x}^T\mathbf{x}$ is convex by (3.76a), or by the second-order condition $\nabla^2 f(\mathbf{x}) = 2\mathbf{I}_n \succ \mathbf{0}$, f is indeed convex. ☐

3.2.4   Pointwise maximum and supremum

If f_1 and f_2 are convex, then $f(\mathbf{x}) = \max\{f_1(\mathbf{x}), f_2(\mathbf{x})\}$ with **dom** $f = $ **dom** $f_1 \cap$ **dom** f_2, which is not differentiable in general, is convex. This can be easily proved by showing that **epi** f is convex.

Example 3.4 Piecewise-linear function (see Figure 3.6)

$$f(\mathbf{x}) = \max_{i=1,\ldots,L} \{\mathbf{a}_i^T \mathbf{x} + b_i\} \tag{3.79}$$

is convex since affine functions are known to be convex. Note that this function is not differentiable and so its convexity cannot be proved by neither the first-order condition (3.16) nor the second-order condition (3.27). For a piecewise-linear function as given in this example, **epi** f (which is a polyhedron) is a convex but not a strictly convex set, so $f(\mathbf{x})$ is convex but not a strictly convex function either. □

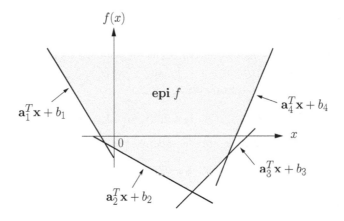

Figure 3.6 Epigraph of a piecewise linear function $f(x)$ given by (3.79) for $L = 4$ in $\mathbb{R}^2$.

If $f(\mathbf{x}, \mathbf{y})$ is convex in $\mathbf{x}$ with **dom** $f_{\mathbf{y}}$ for each $\mathbf{y} \in \mathcal{A}$, then

$$g(\mathbf{x}) = \sup_{\mathbf{y} \in \mathcal{A}} f(\mathbf{x}, \mathbf{y}) \tag{3.80}$$

is convex on **dom** $g = \cap_{\mathbf{y} \in \mathcal{A}} \mathbf{dom}\, f_{\mathbf{y}}$. Similarly, if $f(\mathbf{x}, \mathbf{y})$ is concave in $\mathbf{x}$ for each $\mathbf{y} \in \mathcal{A}$, then

$$\tilde{g}(\mathbf{x}) = \inf_{\mathbf{y} \in \mathcal{A}} f(\mathbf{x}, \mathbf{y}) \tag{3.81}$$

is also concave on **dom** $\tilde{g} = \mathbf{dom}\, g$. Let us prove (3.80) next but skip the proof of (3.81).

Proof of (3.80): Let $S_{\mathbf{y}} = \{(\mathbf{x}, t) \mid f(\mathbf{x}, \mathbf{y}) \leq t\}$ be the epigraph of f (treating $\mathbf{y}$ as a parameter), which is convex since $f(\mathbf{x}, \mathbf{y})$ is convex in $\mathbf{x}$.

$$\mathbf{epi}\, g = \{(\mathbf{x}, t) \mid \sup_{\mathbf{y} \in \mathcal{A}} f(\mathbf{x}, \mathbf{y}) \leq t\} = \{(\mathbf{x}, t) \mid f(\mathbf{x}, \mathbf{y}) \leq t\ \forall \mathbf{y} \in \mathcal{A}\}$$

$$= \bigcap_{\mathbf{y} \in \mathcal{A}} \{(\mathbf{x}, t) \mid f(\mathbf{x}, \mathbf{y}) \leq t\} = \bigcap_{\mathbf{y} \in \mathcal{A}} S_{\mathbf{y}}.$$

Because $S_{\mathbf{y}}$ is convex, **epi** g is convex.   ■

Example 3.5 *(Support function of a set)* Let $C \subseteq \mathbb{R}^n$ and $C \neq \emptyset$. The support function of C is defined as

$$S_C(\mathbf{x}) = \sup_{\mathbf{y} \in C} \mathbf{x}^T \mathbf{y} \qquad (3.82)$$

with **dom** $S_C = \{\mathbf{x} \mid \sup_{\mathbf{y} \in C} \mathbf{x}^T \mathbf{y} < \infty\}$. It can be seen that $S_C(\mathbf{x})$ is a convex function by the pointwise maximum property because the linear function $\mathbf{x}^T \mathbf{y}$ is convex for every $\mathbf{y} \in C$. Consider the case that $C = \{\mathbf{y} \in \mathbb{R}^n \mid \|\mathbf{y}\| \leq 1\}$ is a norm ball with unity radius. Then $S_C(\mathbf{x}) = \|\mathbf{x}\|_*$ (cf. (2.88)), the dual of the norm $\| \cdot \|$ on $\mathbb{R}^n$.   □

Example 3.6 $f(\mathbf{x}) = \sup_{\mathbf{y} \in A} \|\mathbf{x} - \mathbf{y}\|$ is convex, where $\| \cdot \|$ is a norm. This can be proved by using the convexity of pointwise supremum and the convexity of norm operator (that is to say, since $\|\mathbf{x}\|$ is convex, $\|\mathbf{x} - \mathbf{y}\|$ must be convex in $\mathbf{x}$ for a given $\mathbf{y}$, see Subsection 3.2.2).   □

Example 3.7 $\lambda_{\max}(\mathbf{X})$ is convex and $\lambda_{\min}(\mathbf{X})$ is concave on $\mathbb{S}^n$.

Proof: The maximum eigenvalue of $\mathbf{X}$ can be expressed as

$$\lambda_{\max}(\mathbf{X}) = \sup\{\mathbf{y}^T \mathbf{X} \mathbf{y} \mid \|\mathbf{y}\|_2 = 1\} = \sup_{\|\mathbf{y}\|_2 = 1} \mathrm{Tr}(\mathbf{X}\mathbf{y}\mathbf{y}^T) \quad \text{(cf. (1.91)),}$$

where $\mathrm{Tr}(\mathbf{X}\mathbf{y}\mathbf{y}^T)$ is linear in $\mathbf{X}$ for each $\mathbf{y}$. Therefore $\lambda_{\max}(\mathbf{X})$ is convex. So $\lambda_{\min}(\mathbf{X}) = -\lambda_{\max}(-\mathbf{X})$ is concave.   ■ □

By the above example, it can be further shown that

$$\lambda_{\max}(\mathbf{X} + \mathbf{Y}) \leq \lambda_{\max}(\mathbf{X}) + \lambda_{\max}(\mathbf{Y}) \quad \forall \mathbf{X}, \mathbf{Y} \in \mathbb{S}^n \qquad (3.83)$$

$$\lambda_{\min}(\mathbf{X} + \mathbf{Y}) \geq \lambda_{\min}(\mathbf{X}) + \lambda_{\min}(\mathbf{Y}) \quad \forall \mathbf{X}, \mathbf{Y} \in \mathbb{S}^n. \qquad (3.84)$$

Let us conclude this subsection with the following remark.

Remark 3.28 It can be shown, by (3.84), that the interior of the set of $n \times n$ symmetric PSD matrices is given by

$$\mathbf{int}\ \mathbb{S}_+^n = \mathbb{S}_{++}^n = \{\mathbf{X} \in \mathbb{S}^n \mid \lambda_{\min}(\mathbf{X}) > 0\}. \qquad (3.85)$$

Thus, its boundary is given by

$$\mathbf{bd}\ \mathbb{S}_+^n = \mathbb{S}_+^n \setminus \mathbb{S}_{++}^n = \{\mathbf{X} \in \mathbb{S}^n \mid \lambda_{\min}(\mathbf{X}) = 0\}. \qquad (3.86)$$

Proof of (3.85): Suppose that $\mathbf{X} \in \mathbf{int}\ \mathbb{S}_+^n$. Let $\lambda_{\min}(\mathbf{X})$ and $\boldsymbol{\nu}_{\min}(\mathbf{X})$ denote the minimal eigenvalue of $\mathbf{X}$ and the associated eigenvector with $\|\boldsymbol{\nu}_{\min}(\mathbf{X})\|_2 = 1$. Consider a norm ball with center $\mathbf{X}$ and radius r, defined as

$$\mathcal{B}(\mathbf{X}, r) \triangleq \{\mathbf{Y} = \mathbf{X} + r\mathbf{U} \mid \|\mathbf{U}\|_F \leq 1,\ \mathbf{U} \in \mathbb{S}^n\}.$$

Then

$$\min_{\mathbf{Y} \in \mathcal{B}(\mathbf{X}, r)} \lambda_{\min}(\mathbf{Y}) = \min_{\|\mathbf{U}\|_F \leq 1, \mathbf{U} \in \mathbb{S}^n} \lambda_{\min}(\mathbf{X} + r\mathbf{U})$$

$$\geq \lambda_{\min}(\mathbf{X}) + r \min_{\|\mathbf{U}\|_F \leq 1, \mathbf{U} \in \mathbb{S}^n} \lambda_{\min}(\mathbf{U}) \quad \text{(by (3.84))}$$

$$= \lambda_{\min}(\mathbf{X}) - r, \quad \text{(by (1.97))}$$

where the inequality holds with the equality as $\mathbf{U} = -\boldsymbol{\nu}_{\min}(\mathbf{X})\boldsymbol{\nu}_{\min}(\mathbf{X})^T$. It can be seen that $\mathcal{B}(\mathbf{X}, r) \subset \mathbb{S}_+^n$ if and only if $\lambda_{\min}(\mathbf{X}) \geq r > 0$, implying that $\mathbf{X} \in \textbf{int } \mathbb{S}_+^n$ if and only if $\lambda_{\min}(\mathbf{X}) > 0$. ■ □

3.2.5 Pointwise minimum and infimum

If $f(\mathbf{x}, \mathbf{y})$ is convex in $(\mathbf{x}, \mathbf{y}) \in \mathbb{R}^m \times \mathbb{R}^n$ and $C \subset \mathbb{R}^n$ is convex and nonempty, then

$$g(\mathbf{x}) = \inf_{\mathbf{y} \in C} f(\mathbf{x}, \mathbf{y}) \tag{3.87}$$

is convex, provided that $g(\mathbf{x}) > -\infty$ for some $\mathbf{x}$. Similarly, if $f(\mathbf{x}, \mathbf{y})$ is concave in $(\mathbf{x}, \mathbf{y}) \in \mathbb{R}^m \times \mathbb{R}^n$, then

$$\tilde{g}(\mathbf{x}) = \sup_{\mathbf{y} \in C} f(\mathbf{x}, \mathbf{y}) \tag{3.88}$$

is concave provided that $C \subset \mathbb{R}^n$ is convex and nonempty and $\tilde{g}(\mathbf{x}) < \infty$ for some $\mathbf{x}$. Next, we present the proof for the former.

Proof of (3.87): Since f is continuous over $\textbf{int}(\textbf{dom } f)$ (cf. Remark 3.7), for any $\epsilon > 0$ and $\mathbf{x}_1, \mathbf{x}_2 \in \textbf{dom } g$, there exist $\mathbf{y}_1, \mathbf{y}_2 \in C$ (depending on ϵ) such that

$$f(\mathbf{x}_i, \mathbf{y}_i) \leq g(\mathbf{x}_i) + \epsilon, \quad i = 1, 2. \tag{3.89}$$

Let $(\mathbf{x}_1, t_1), (\mathbf{x}_2, t_2) \in \textbf{epi } g$. Then $g(\mathbf{x}_i) = \inf_{\mathbf{y} \in C} f(\mathbf{x}_i, \mathbf{y}) \leq t_i$, $i = 1, 2$. Then for any $\theta \in [0, 1]$, we have

$$g(\theta \mathbf{x}_1 + (1 - \theta)\mathbf{x}_2) = \inf_{\mathbf{y} \in C} f(\theta \mathbf{x}_1 + (1 - \theta)\mathbf{x}_2, \mathbf{y})$$

$$\leq f(\theta \mathbf{x}_1 + (1 - \theta)\mathbf{x}_2, \theta \mathbf{y}_1 + (1 - \theta)\mathbf{y}_2)$$

$$\leq \theta f(\mathbf{x}_1, \mathbf{y}_1) + (1 - \theta)f(\mathbf{x}_2, \mathbf{y}_2) \quad \text{(since } f \text{ is convex)}$$

$$\leq \theta g(\mathbf{x}_1) + (1 - \theta)g(\mathbf{x}_2) + \epsilon \quad \text{(by (3.89))} \tag{3.90}$$

$$\leq \theta t_1 + (1 - \theta)t_2 + \epsilon.$$

It can be seen that as $\epsilon \to 0$, $g(\theta \mathbf{x}_1 + (1 - \theta)\mathbf{x}_2) \leq \theta t_1 + (1 - \theta)t_2$, implying $(\theta \mathbf{x}_1 + (1 - \theta)\mathbf{x}_2, \theta t_1 + (1 - \theta)t_2) \in \textbf{epi } g$. Hence $\textbf{epi } g$ is a convex set, and thus $g(\mathbf{x})$ is a convex function by Fact 3.2. ■

Alternative proof of (3.87): Because $\textbf{dom } g = \{\mathbf{x} \mid (\mathbf{x}, \mathbf{y}) \in \textbf{dom } f, \ \mathbf{y} \in C\}$ is the projection of the convex set $\{(\mathbf{x}, \mathbf{y}) \mid (\mathbf{x}, \mathbf{y}) \in \textbf{dom } f, \ \mathbf{y} \in C\}$ on the $\mathbf{x}$-coordinate, it must be a convex set (cf. Remark 2.11).

Let $\theta \in [0, 1]$. Under the same premise about $\epsilon > 0$, $\mathbf{x}_1, \mathbf{x}_2 \in \mathbf{dom}\ g$, and $\mathbf{y}_1, \mathbf{y}_2 \in C$ such that (3.89) is true, it can be easily inferred, by letting $\epsilon \to 0$, that (3.90) yields $g(\theta \mathbf{x}_1 + (1 - \theta)\mathbf{x}_2) \leq \theta g(\mathbf{x}_1) + (1 - \theta)g(\mathbf{x}_2)$. Hence, we have completed the proof that $g(\mathbf{x})$ is convex.   ∎

Let us present some examples to illustrate the usefulness of the convexity of the pointwise minimum or infimum function defined in (3.87).

- The minimum distance between a point $\mathbf{x} \in \mathbb{R}^n$ and a convex set $C \subset \mathbb{R}^n$ can be expressed as

$$\mathbf{dist}_C(\mathbf{x}) = \inf_{\mathbf{y} \in C} \|\mathbf{x} - \mathbf{y}\|_2 \quad (\text{cf. } (2.124)), \tag{3.91}$$

which, by the pointwise infimum property, is a convex function (where $\|\mathbf{x} - \mathbf{y}\|_2$ can be shown to be convex in $(\mathbf{x}, \mathbf{y})$ by definition, or by the composition of the affine mapping $\mathbf{x} - \mathbf{y}$ and the convex function of $\|(\mathbf{x}, \mathbf{y})\|_2$).

- Schur complement: Suppose that $\mathbf{C} \in \mathbb{S}_{++}^m$ and $\mathbf{A} \in \mathbb{S}^n$. Then

$$\mathbf{S} \triangleq \begin{bmatrix} \mathbf{A} & \mathbf{B} \\ \mathbf{B}^T & \mathbf{C} \end{bmatrix} \succeq \mathbf{0} \text{ if and only if } \mathbf{S}_{\mathbf{C}} \triangleq \mathbf{A} - \mathbf{B}\mathbf{C}^{-1}\mathbf{B}^T \succeq \mathbf{0}, \tag{3.92}$$

where $\mathbf{S}_{\mathbf{C}}$ is called the Schur complement of $\mathbf{C}$ in $\mathbf{S}$.

Proof: The necessity of Schur complement is proved through the pointwise infimum property. Let

$$f(\mathbf{x}, \mathbf{y}) = [\mathbf{x}^T\ \mathbf{y}^T]\ \mathbf{S} \begin{bmatrix} \mathbf{x} \\ \mathbf{y} \end{bmatrix}$$

$$= [\mathbf{x}^T\ \mathbf{y}^T] \begin{bmatrix} \mathbf{A} & \mathbf{B} \\ \mathbf{B}^T & \mathbf{C} \end{bmatrix} \begin{bmatrix} \mathbf{x} \\ \mathbf{y} \end{bmatrix} \geq 0\ \forall (\mathbf{x}, \mathbf{y}) \in \mathbb{R}^{n+m}, \tag{3.93}$$

which is convex in $(\mathbf{x}, \mathbf{y})$ since $\mathbf{S} \succeq \mathbf{0}$. Consider

$$g(\mathbf{x}) = \inf_{\mathbf{y} \in \mathbb{R}^m} f(\mathbf{x}, \mathbf{y}) \geq 0, \tag{3.94}$$

which is convex in $\mathbf{x}$ since f is convex in $(\mathbf{x}, \mathbf{y})$ and $\mathbb{R}^m$ is a nonempty convex set. Moreover, the computation of $g(\mathbf{x})$ itself is a minimization problem with $f(\mathbf{x}, \mathbf{y})$ treated as the objective function of $\mathbf{y}$ for any fixed $\mathbf{x}$. Furthermore, $f(\mathbf{x}, \mathbf{y})$ is also convex in $\mathbf{y}$ since $\mathbf{C} \in \mathbb{S}_{++}^m$, and

$$f(\mathbf{x}, \mathbf{y}) = \mathbf{x}^T\mathbf{A}\mathbf{x} + 2\mathbf{x}^T\mathbf{B}\mathbf{y} + \mathbf{y}^T\mathbf{C}\mathbf{y} \geq g(\mathbf{x}) \geq 0.$$

By the first-order condition for finding the optimal solution (cf. (4.28)), denoted as $\mathbf{y}^\star$, of the unconstrained convex problem (3.94) (to be introduced in Chapter 4), we have

$$\nabla_{\mathbf{y}} f(\mathbf{x}, \mathbf{y}) = 2\mathbf{B}^T\mathbf{x} + 2\mathbf{C}\mathbf{y} = \mathbf{0} \Rightarrow \mathbf{y}^\star = -\mathbf{C}^{-1}\mathbf{B}^T\mathbf{x},$$

and

$$g(\mathbf{x}) = f(\mathbf{x}, \mathbf{y}^\star)$$
$$= \mathbf{x}^T \mathbf{A}\mathbf{x} - 2\mathbf{x}^T \mathbf{B}\mathbf{C}^{-1}\mathbf{B}^T\mathbf{x} + \mathbf{x}^T \mathbf{B}\mathbf{C}^{-1}\mathbf{B}^T\mathbf{x}$$
$$= \mathbf{x}^T (\mathbf{A} - \mathbf{B}\mathbf{C}^{-1}\mathbf{B}^T)\mathbf{x} = \mathbf{x}^T \mathbf{S_C}\mathbf{x} \geq 0 \; \forall \mathbf{x} \in \mathbb{R}^n, \qquad (3.95)$$

implying that Schur complement $\mathbf{S_C}$ is a PSD matrix.

As for the proof of sufficiency, it can be readily seen that if $\mathbf{S_C} \succeq \mathbf{0}$, the reverse implications from (3.95) to (3.93) are also true, so $\mathbf{S} \succeq \mathbf{0}$ holds true. Thus, we have completed the proof that $\mathbf{S} \succeq \mathbf{0}$ if and only if $\mathbf{S_C} \succeq \mathbf{0}$. ∎

Similarly, suppose that $\mathbf{A} \in \mathbb{S}^n_{++}$ and $\mathbf{C} \in \mathbb{S}^m$. Then

$$\mathbf{S} \triangleq \begin{bmatrix} \mathbf{A} & \mathbf{B} \\ \mathbf{B}^T & \mathbf{C} \end{bmatrix} \succeq \mathbf{0} \; \text{ if and only if } \; \mathbf{S_A} \triangleq \mathbf{C} - \mathbf{B}^T\mathbf{A}^{-1}\mathbf{B} \succeq \mathbf{0}. \qquad (3.96)$$

Remark 3.29 It can be shown that when $\mathbf{C} \in \mathbb{S}^m_+$, $\mathcal{R}(\mathbf{B}^T) \subset \mathcal{R}(\mathbf{C})$ and $\mathbf{A} \in \mathbb{S}^n$, (3.92) is still true except that $\mathbf{S_C} = \mathbf{A} - \mathbf{B}\mathbf{C}^\dagger\mathbf{B}^T$; when $\mathbf{A} \in \mathbb{S}^n_+$, $\mathcal{R}(\mathbf{B}) \subset \mathcal{R}(\mathbf{A})$, and $\mathbf{C} \in \mathbb{S}^m$, (3.96) is still true except that $\mathbf{S_A} = \mathbf{C} - \mathbf{B}^T\mathbf{A}^\dagger\mathbf{B}$. □

3.2.6   Perspective of a function

The *perspective* of a function $f : \mathbb{R}^n \to \mathbb{R}$ is defined as

$$g(\mathbf{x}, t) = tf(\mathbf{x}/t) \qquad (3.97)$$

with

$$\mathbf{dom}\, g = \{(\mathbf{x}, t) \mid \mathbf{x}/t \in \mathbf{dom}\, f, \; t > 0\}. \qquad (3.98)$$

If $f(\mathbf{x})$ is convex, then its perspective $g(\mathbf{x}, t)$ is convex.

Proof: The $\mathbf{dom}\, g$ is the inverse image of $\mathbf{dom}\, f$ under the perspective mapping $p(\mathbf{x}, t) = \mathbf{x}/t$ for $t > 0$ (cf. (2.72) in Subsection 2.3.3), so it is convex. For any two points $(\mathbf{x}, t), (\mathbf{y}, s) \in \mathbf{dom}\, g$ (i.e., $s, t > 0$, $\mathbf{x}/t$ and $\mathbf{y}/s \in \mathbf{dom}\, f$), and $0 \leq \theta \leq 1$, we have

$$g(\theta\mathbf{x} + (1-\theta)\mathbf{y}, \theta t + (1-\theta)s) = (\theta t + (1-\theta)s)f\left(\frac{\theta\mathbf{x} + (1-\theta)\mathbf{y}}{\theta t + (1-\theta)s}\right)$$
$$= (\theta t + (1-\theta)s)\, f\left(\frac{\theta t(\mathbf{x}/t) + (1-\theta)s(\mathbf{y}/s)}{\theta t + (1-\theta)s}\right)$$
$$\leq \theta t f(\mathbf{x}/t) + (1-\theta)s f(\mathbf{y}/s) \quad (\text{since } f \text{ is convex})$$
$$= \theta g(\mathbf{x}, t) + (1-\theta)g(\mathbf{y}, s).$$

Thus, we have completed the proof that $g(\mathbf{x}, t)$ is convex. ∎

Alternative proof: Another way of proving the convexity is by showing that the epigraph of the function is convex as follows:

$$(\mathbf{x}, t, s) \in \mathbf{epi}\ g \Leftrightarrow tf(\mathbf{x}/t) \le s$$
$$\Leftrightarrow f(\mathbf{x}/t) \le s/t$$
$$\Leftrightarrow (\mathbf{x}/t, s/t) \in \mathbf{epi}\ f.$$

Therefore, **epi** g is the inverse image of **epi** f under the perspective mapping. So **epi** g is convex since **epi** f is convex. ∎

Remark 3.30 If $g(\mathbf{x}, t)$ defined by (3.97) is convex (concave), then $f(\mathbf{x})$ is also convex (concave), and vice versa. Note that the convex set **epi** g does not contain the origin $\mathbf{0}_{n+2}$, and so it is not a cone. However, **epi** $g \cup \{\mathbf{0}_{n+2}\}$ can be shown to be a convex cone and thus the closure of **epi** g is a closed convex cone. Moreover, for a ray $C = \{(t\mathbf{a}, t), t > 0\} \subset \mathbf{dom}\ g$ where $\mathbf{a} \ne \mathbf{0}_n$, its image $g(C) = \{tf(\mathbf{a}), t > 0\}$ together with C forms a ray in $\mathbb{R}^{n+2}$ for every nonzero vector $\mathbf{a} \in \mathbb{R}^n$. □

Example 3.8 Consider

$$f(\mathbf{x}) = \|\mathbf{x}\|_2^2 = \mathbf{x}^T\mathbf{x}\ \text{(convex)}. \tag{3.99}$$

Then the perspective of $f(\mathbf{x})$ given by

$$g(\mathbf{x}, t) = t \cdot \frac{\|\mathbf{x}\|_2^2}{t^2} = \frac{\|\mathbf{x}\|_2^2}{t},\ t > 0 \tag{3.100}$$

is also convex. Figure 3.7 shows the perspective $g(x, t) = x^2/t$ for $f(x) = x^2$, and the convex cone of **epi** $g \cup \{\mathbf{0}_3\}$ which is not closed. Note that **epi** $g \cup \{(0, 0, s) \mid s \ge 0\}$ (exactly the closure of **epi** g) is a closed convex cone. □

Example 3.9 Consider

$$h(\mathbf{x}) = \frac{\|\mathbf{A}\mathbf{x} + \mathbf{b}\|_2^2}{\mathbf{c}^T\mathbf{x} + d} \tag{3.101}$$

with **dom** $h = \{\mathbf{x} \mid \mathbf{c}^T\mathbf{x} + d > 0\}$. Because $g(\mathbf{x}, t)$ given by (3.100) is convex in $(\mathbf{x}, t)$, it can be inferred that $h(\mathbf{x})$ is also convex due to the affine mapping $(\mathbf{y}, t) = (\mathbf{A}\mathbf{x} + \mathbf{b}, \mathbf{c}^T\mathbf{x} + d)$. □

Example 3.10 *(Relative entropy or Kullback–Leibler divergence)* It has been proven that $f(x) = -\log x$ is a convex function with **dom** $f = \mathbb{R}_{++}$. Then the perspective of f

$$g(x, t) = -t\log(x/t) = t\log(t/x) \tag{3.102}$$

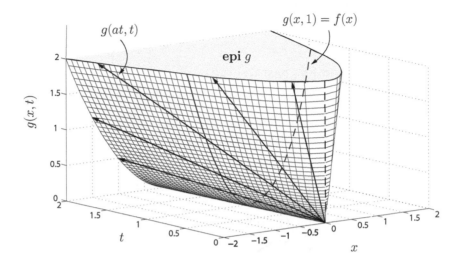

Figure 3.7 Epigraph of $g(x,t) = x^2/t$, the perspective of $f(x) = x^2$, where the red curve denotes $g(x,1) = f(x)$, and each ray in blue color is associated with $g(at,t) = a^2 t$ for a different value of a.

is convex on $\mathbb{R}_{++}^2$. Furthermore, for any two vectors $\boldsymbol{u}, \boldsymbol{v} \in \mathbb{R}_{++}^n$ with $\boldsymbol{u}^T \mathbf{1}_n = \boldsymbol{v}^T \mathbf{1}_n = 1$, their relative entropy or Kullback–Leibler divergence, defined as

$$D_{\mathrm{KL}}(\boldsymbol{u}, \boldsymbol{v}) = \sum_{i=1}^{n} u_i \log(u_i/v_i) \geq 0, \qquad (3.103)$$

is also convex in $(\boldsymbol{u}, \boldsymbol{v})$ since $g(x,t)$ given by (3.102) is convex. It has been used as a measure of deviation between two probability distributions $\boldsymbol{u}$ and $\boldsymbol{v}$. $\square$

3.3   Quasiconvex functions

3.3.1   Definition and examples

A function $f : \mathbb{R}^n \to \mathbb{R}$ is said to be *quasiconvex* if its domain and all its α-sublevel sets defined as

$$S_\alpha = \{\mathbf{x} \mid \mathbf{x} \in \mathbf{dom}\ f,\ f(\mathbf{x}) \leq \alpha\} \qquad (3.104)$$

(see Figure 3.8) are convex for every α. Moreover,

- f is quasiconvex if f is convex since every sublevel set of convex functions is a convex set (cf. Remark 3.5), but the converse is not necessarily true.

- f is *quasiconcave* if $-f$ is quasiconvex. It is also true that f is quasiconcave if its domain and all the α-superlevel sets defined as

$$S_\alpha = \{\mathbf{x} \mid \mathbf{x} \in \mathbf{dom}\ f,\ f(\mathbf{x}) \geq \alpha\} \tag{3.105}$$

are convex for every α.

- f is *quasilinear* if f is both quasiconvex and quasiconcave.

The relationships among convex functions, quasiconvex functions, concave functions, and quasiconcave functions are illustrated in Figure 3.9.

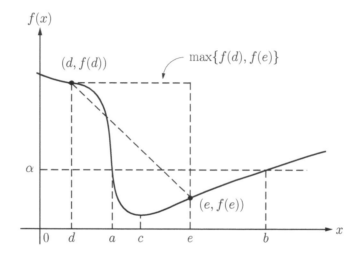

Figure 3.8 Illustration of a quasiconvex function $f : \mathbb{R} \to \mathbb{R}$ and the modified Jensen's inequality given by (3.106), where the sublevel set $S_\alpha = [a, b]$ is convex, $x = d$ is a saddle point, and $x = c$ is a globally minimum point.

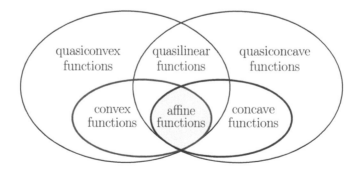

Figure 3.9 Relationships among the set of convex functions, the set of quasiconvex functions, the set of concave functions, and the set of quasiconcave functions.

Remark 3.31 As Fact 3.1 for convex functions, a function f is quasiconvex, if and only if, for all $\mathbf{x} \in \mathbf{dom}\ f$ and for all $\mathbf{v}$, the function $g(t) = f(\mathbf{x} + t\mathbf{v})$ is quasiconvex on its domain $\{t \mid \mathbf{x} + t\mathbf{v} \in \mathbf{dom}\ f\}$. $\qquad\square$

Fact 3.3 A continuous function $f : \mathbb{R} \to \mathbb{R}$ is quasiconvex if and only if at least one of the following conditions holds:

- f is nondecreasing;
- f is nonincreasing;
- there is a point $c \in \mathbf{dom}\ f$ such that for $x \leq c$ (and $x \in \mathbf{dom}\ f$), f is nonincreasing, and for $x \geq c$ (and $x \in \mathbf{dom}\ f$), f is nondecreasing (see Figure 3.8). This implies that $x = c$ is a global minimum.

Proof: The sufficiency of the conditions in Fact 3.3 can be straightforwardly proven by the definition of quasiconvex functions. The necessity of the conditions can be proved as follows. Assume that none of the conditions in Fact 3.3 is true. Then for the continuous function f, there exist two points a and b with $f(a) = f(b)$ and $(a + \epsilon, b - \epsilon) \in \mathbf{dom}\ f$ for some $\epsilon > 0$, such that $f(a + \epsilon) > f(a)$ and $f(b - \epsilon) > f(b)$. Then, it can be easily seen that its sublevel set $S_\alpha = \{x \mid f(x) \leq \alpha\}$, where $f(a) < \alpha < \min\{f(a + \epsilon), f(b - \epsilon)\}$ is disjoint and hence a nonconvex set, implying that f is not a quasiconvex function. $\qquad\blacksquare$

Remark 3.32 Some properties about a quasiconvex function $f : \mathbb{R} \to \mathbb{R}$ can be inferred from Fact 3.3 and illustrated in Figure 3.8. It is clear to see that f is quasilinear if f is either nondecreasing or nonincreasing. If f is twice differentiable with $f'(t) = 0$, then $f''(t) \geq 0$. Specifically, t must be a saddle point, or a locally minimal point, or a locally maximal point, or a globally minimal/maximal point as $f''(t) = 0$, and must be a unique strictly and globally minimal point as $f''(t) > 0$. $\qquad\square$

Some examples of quasiconvex functions are given next, where some of them are easy and some of them are non-trivial to prove their quasi-convexity.

- $f_1(x) = e^{-x}$ is quasilinear on $\mathbb{R}$, and $f_2(x) = \log x$ is quasilinear on $\mathbb{R}_{++}$. This can be easily verified either by the definition of the quasiconvex functions or by Fact 3.3.
- Ceiling function:

$$\lceil x \rceil = \inf\{z \in \mathbb{Z} \mid z \geq x\}$$

is lower semi-continuous, nonconvex but quasiconvex (and quasiconcave and thus quasilinear), since its sublevel set

$$S_\alpha = \{x \mid \lceil x \rceil \leq \alpha\}$$

is convex (see Figure 3.10). This example also shows that a quasiconvex function may not be continuous over the interior of its domain, although a convex function must be continuous over the interior of its effective domain (cf. (3.7)).

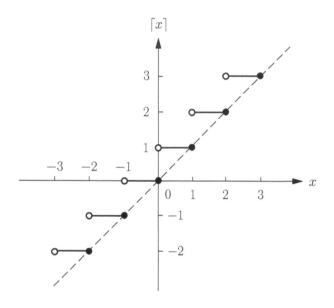

Figure 3.10 A ceiling function.

- Linear-fractional function (cf. (2.76))

$$f(\mathbf{x}) = \frac{\mathbf{a}^T\mathbf{x} + b}{\mathbf{c}^T\mathbf{x} + d}$$

is quasiconvex on $\{\mathbf{x} \mid \mathbf{c}^T\mathbf{x} + d > 0\}$.

Proof: Consider the sublevel set of the function

$$S_\alpha = \left\{\mathbf{x} \ \middle| \ \frac{\mathbf{a}^T\mathbf{x} + b}{\mathbf{c}^T\mathbf{x} + d} \leq \alpha, \mathbf{c}^T\mathbf{x} + d > 0\right\}$$
$$= \{\mathbf{x} \mid \mathbf{a}^T\mathbf{x} + b \leq \alpha\mathbf{c}^T\mathbf{x} + \alpha d, \mathbf{c}^T\mathbf{x} + d > 0\}$$
$$= \{\mathbf{x} \mid (\mathbf{a} - \alpha\mathbf{c})^T\mathbf{x} + (b - \alpha d) \leq 0, \mathbf{c}^T\mathbf{x} + d > 0\} .$$

Note that S_α is a polyhedron and is thus convex. Hence $f(\mathbf{x})$ is quasiconvex by definition. ∎

It can be shown that a linear-fractional function is also quasiconcave, and so is quasilinear.

- $f(\mathbf{X}) = \text{rank}(\mathbf{X})$ is quasiconcave on $\mathbb{S}_+^n$.

  Proof: The superlevel set of $\text{rank}(\mathbf{X})$ is

  $$\mathcal{S}_\alpha = \{\mathbf{X} \mid \text{rank}(\mathbf{X}) \geq \alpha, \mathbf{X} \in \mathbb{S}_+^n\}.$$

  Let $\mathbf{z} \in \mathbb{R}^n$, $\mathbf{X}_1, \mathbf{X}_2 \in \mathcal{S}_\alpha$ with $\text{rank}(\mathbf{X}_1) \geq \alpha$, $\text{rank}(\mathbf{X}_2) \geq \alpha$. For any $0 < \theta < 1$, it can be seen that

  $$\mathbf{z}^T(\theta\mathbf{X}_1 + (1-\theta)\mathbf{X}_2)\mathbf{z} = 0 \text{ if and only if } \mathbf{z}^T\mathbf{X}_1\mathbf{z} = 0 \text{ and } \mathbf{z}^T\mathbf{X}_2\mathbf{z} = 0,$$

  which implies

  $$\mathcal{R}(\mathbf{U})^\perp = \mathcal{R}(\mathbf{U}_1)^\perp \cap \mathcal{R}(\mathbf{U}_2)^\perp,$$

  or equivalently

  $$\mathcal{R}(\mathbf{U}) = \mathcal{R}([\mathbf{U}_1, \mathbf{U}_2]),$$

  where $\mathbf{U}$, $\mathbf{U}_1$, $\mathbf{U}_2$ are semi-unitary matrices formed by the eigenvectors associated with all the positive eigenvalues of $\theta\mathbf{X}_1 + (1-\theta)\mathbf{X}_2$, $\mathbf{X}_1$, $\mathbf{X}_2$, respectively. Therefore,

  $$\text{rank}(\mathbf{U}) \geq \max\{\text{rank}(\mathbf{U}_1), \text{rank}(\mathbf{U}_2)\} \geq \alpha$$
  $$\Rightarrow \text{rank}(\theta\mathbf{X}_1 + (1-\theta)\mathbf{X}_2) \geq \max\{\text{rank}(\mathbf{X}_1), \text{rank}(\mathbf{X}_2)\} \geq \alpha, \text{ for } 0 < \theta < 1.$$

  Also,

  $$\text{rank}(\theta\mathbf{X}_1 + (1-\theta)\mathbf{X}_2) = \text{rank}(\mathbf{X}_2) \geq \alpha, \text{ for } \theta = 0,$$
  $$\text{rank}(\theta\mathbf{X}_1 + (1-\theta)\mathbf{X}_2) = \text{rank}(\mathbf{X}_1) \geq \alpha, \text{ for } \theta = 1.$$

  Hence,

  $$\text{rank}(\theta\mathbf{X}_1 + (1-\theta)\mathbf{X}_2) \geq \alpha \Rightarrow \theta\mathbf{X}_1 + (1-\theta)\mathbf{X}_2 \in \mathcal{S}_\alpha, \forall\theta \in [0,1],$$

  implying that the superlevel set $\mathcal{S}_\alpha$ is convex for every α. Thus, $f(\mathbf{X}) = \text{rank}(\mathbf{X})$ is quasiconcave on $\mathbb{S}_+^n$. ∎

- $\text{card}(\mathbf{x})$ denotes the cardinality, i.e., the number of nonzero elements in a vector $\mathbf{x} \in \mathbb{R}^n$. The function $\text{card}(\mathbf{x})$ is quasiconcave on $\mathbb{R}_+^n$.

  Proof: The superlevel set of $\text{card}(\mathbf{x})$ is

  $$\mathcal{S}_\alpha = \{\mathbf{x} \mid \text{card}(\mathbf{x}) \geq \alpha, \mathbf{x} \in \mathbb{R}_+^n\}.$$

  Let $\mathbf{x}_1, \mathbf{x}_2 \in \mathcal{S}_\alpha$, implying that $\text{card}(\mathbf{x}_1) \geq \alpha$ and $\text{card}(\mathbf{x}_2) \geq \alpha$. Then for any $0 \leq \theta \leq 1$, $\theta\mathbf{x}_1 + (1-\theta)\mathbf{x}_2 \in \mathbb{R}_+^n$ and

  $$\text{card}(\theta\mathbf{x}_1 + (1-\theta)\mathbf{x}_2) \geq \min\{\text{card}(\mathbf{x}_1), \text{card}(\mathbf{x}_2)\} \geq \alpha,$$

  implying that $\theta\mathbf{x}_1 + (1-\theta)\mathbf{x}_2 \in \mathcal{S}_\alpha$. Therefore, $\mathcal{S}_\alpha$ is a convex set which in turn implies that $\text{card}(\mathbf{x})$ is quasiconcave on $\mathbb{R}_+^n$. ∎

3.3.2   Modified Jensen's inequality

A function $f : \mathbb{R}^n \to \mathbb{R}$ is quasiconvex if and only if

$$f(\theta\mathbf{x} + (1-\theta)\mathbf{y}) \leq \max\{f(\mathbf{x}), f(\mathbf{y})\}, \qquad (3.106)$$

for all $\mathbf{x}, \mathbf{y} \in \mathbf{dom}\ f$, and $0 \leq \theta \leq 1$ (see Figure 3.8).

Proof: Let us prove the necessity followed by sufficiency.

- Necessity: Let $\mathbf{x}, \mathbf{y} \in \mathbf{dom}\ f$. Choose $\alpha = \max\{f(\mathbf{x}), f(\mathbf{y})\}$. Then $\mathbf{x}, \mathbf{y} \in S_\alpha$. Since f is quasiconvex by assumption, S_α is convex, that is, for $\theta \in [0,1]$,

$$\theta\mathbf{x} + (1-\theta)\mathbf{y} \in S_\alpha$$
$$\Rightarrow f(\theta\mathbf{x} + (1-\theta)\mathbf{y}) \leq \alpha = \max\{f(\mathbf{x}), f(\mathbf{y})\}.$$

- Sufficiency: For every α, pick two points $\mathbf{x}, \mathbf{y} \in S_\alpha \Rightarrow f(\mathbf{x}) \leq \alpha,\ f(\mathbf{y}) \leq \alpha$. Since for $0 \leq \theta \leq 1$, $f(\theta\mathbf{x} + (1-\theta)\mathbf{y}) \leq \max\{f(\mathbf{x}), f(\mathbf{y})\} \leq \alpha$ (by (3.106)), we have

$$\theta\mathbf{x} + (1-\theta)\mathbf{y} \in S_\alpha.$$

Therefore, S_α is convex and thus the function f is quasiconvex.   ∎

Remark 3.33 f is quasiconcave if and only if

$$f(\theta\mathbf{x} + (1-\theta)\mathbf{y}) \geq \min\{f(\mathbf{x}), f(\mathbf{y})\}, \qquad (3.107)$$

for all $\mathbf{x}, \mathbf{y} \in \mathbf{dom}\ f$, and $0 \leq \theta \leq 1$. This is also the modified Jensen's inequality for quasiconcave functions. If the inequality (3.106) holds strictly for $0 < \theta < 1$, then f is strictly quasiconvex. Similarly, if the inequality (3.107) holds strictly for $0 < \theta < 1$, then f is strictly quasiconcave.   □

Remark 3.34   Since the rank of a PSD matrix is quasiconcave,

$$\mathrm{rank}(\mathbf{X} + \mathbf{Y}) \geq \min\{\mathrm{rank}(\mathbf{X}), \mathrm{rank}(\mathbf{Y})\},\ \mathbf{X}, \mathbf{Y} \in \mathbb{S}_+^n, \qquad (3.108)$$

holds true. This can be proved by (3.107), by which we get

$$\mathrm{rank}(\theta\mathbf{X} + (1-\theta)\mathbf{Y}) \geq \min\{\mathrm{rank}(\mathbf{X}), \mathrm{rank}(\mathbf{Y})\}, \qquad (3.109)$$

for all $\mathbf{X} \in \mathbb{S}_+^n, \mathbf{Y} \in \mathbb{S}_+^n$, and $0 \leq \theta \leq 1$. Then replacing $\mathbf{X}$ by $\mathbf{X}/\theta$ and $\mathbf{Y}$ by $\mathbf{Y}/(1-\theta)$ where $\theta \neq 0$ and $\theta \neq 1$ gives rise to (3.108).   □

Remark 3.35 $\mathrm{card}(\mathbf{x} + \mathbf{y}) \geq \min\{\mathrm{card}(\mathbf{x}), \mathrm{card}(\mathbf{y})\}$, $\mathbf{x}, \mathbf{y} \in \mathbb{R}_+^n$. Similar to the proof of (3.108) in Remark 3.34, this inequality can be shown to be true by using (3.107) again, since $\mathrm{card}(\mathbf{x})$ is quasiconcave.   □

3.3.3   First-order condition

Suppose that f is differentiable. Then f is quasiconvex if and only if $\textbf{dom } f$ is convex and for all $\mathbf{x}, \mathbf{y} \in \textbf{dom } f$

$$f(\mathbf{y}) \leq f(\mathbf{x}) \Rightarrow \nabla f(\mathbf{x})^T(\mathbf{y} - \mathbf{x}) \leq 0, \tag{3.110}$$

that is, $\nabla f(\mathbf{x})$ defines a supporting hyperplane to the sublevel set

$$S_{\alpha = f(\mathbf{x})} = \{\mathbf{y} \mid f(\mathbf{y}) \leq \alpha = f(\mathbf{x})\} \tag{3.111}$$

at the point $\mathbf{x}$ (see Figure 3.11). Moreover, the first-order condition given by (3.110) means that the first-order term in the Taylor series of $f(\mathbf{y})$ at the point $\mathbf{x}$ is no greater than zero whenever $f(\mathbf{y}) \leq f(\mathbf{x})$.

Proof: Let us prove the necessity followed by the sufficiency.

- Necessity: Suppose $f(\mathbf{x}) \geq f(\mathbf{y})$. Then, by modified Jensen's inequality, we have

$$f(t\mathbf{y} + (1 - t)\mathbf{x}) \leq f(\mathbf{x}) \text{ for all } 0 \leq t \leq 1,$$

Therefore,

$$\lim_{t \to 0^+} \frac{f(\mathbf{x} + t(\mathbf{y} - \mathbf{x})) - f(\mathbf{x})}{t} = \lim_{t \to 0^+} \frac{1}{t} \left(f(\mathbf{x}) + t\nabla f(\mathbf{x})^T(\mathbf{y} - \mathbf{x}) - f(\mathbf{x}) \right)$$
$$= \nabla f(\mathbf{x})^T(\mathbf{y} - \mathbf{x}) \leq 0,$$

where we have used the first-order Taylor series approximation in the first equality.

- Sufficiency: Suppose that $f(\mathbf{x})$ is not quasiconvex. Then there exists a non-convex sublevel set of f,

$$S_\alpha = \{\mathbf{x} \mid f(\mathbf{x}) \leq \alpha\},$$

and two distinct points $\mathbf{x}_1, \mathbf{x}_2 \in S_\alpha$ such that $\theta\mathbf{x}_1 + (1 - \theta)\mathbf{x}_2 \notin S_\alpha$, for some $0 < \theta < 1$, i.e.,

$$f(\theta\mathbf{x}_1 + (1 - \theta)\mathbf{x}_2) > \alpha \text{ for some } 0 < \theta < 1. \tag{3.112}$$

Since f is differentiable, hence continuous, (3.112) implies that, as illustrated in Figure 3.12, there exist distinct $\theta_1, \theta_2 \in (0, 1)$ such that

$$f(\theta\mathbf{x}_1 + (1 - \theta)\mathbf{x}_2) > \alpha \text{ for all } \theta_1 < \theta < \theta_2,$$
$$f(\theta_1\mathbf{x}_1 + (1 - \theta_1)\mathbf{x}_2) = f(\theta_2\mathbf{x}_1 + (1 - \theta_2)\mathbf{x}_2) = \alpha.$$

Let $\mathbf{x} = \theta_1\mathbf{x}_1 + (1 - \theta_1)\mathbf{x}_2$ and $\mathbf{y} = \theta_2\mathbf{x}_1 + (1 - \theta_2)\mathbf{x}_2$, and so

$$f(\mathbf{x}) = f(\mathbf{y}) = \alpha, \tag{3.113}$$

and

$$g(t) = f(t\mathbf{y} + (1 - t)\mathbf{x}) > \alpha \text{ for all } 0 < t < 1 \tag{3.114}$$

is a differentiable function of t and $\partial g(t)/\partial t > 0$ for $t \in [0, \varepsilon)$ where $0 < \varepsilon \ll 1$, as illustrated in Figure 3.12. Then, it can be inferred that

$$(1-t)\frac{\partial g(t)}{\partial t} = \nabla f(\mathbf{x} + t(\mathbf{y} - \mathbf{x}))^T [(1-t)(\mathbf{y} - \mathbf{x})] > 0 \quad \text{for all } t \in [0, \varepsilon)$$

$$= \nabla f(\boldsymbol{x})^T (\mathbf{y} - \boldsymbol{x}) > 0 \quad \text{for all } t \in [0, \varepsilon),$$

where

$$\boldsymbol{x} = \mathbf{x} + t(\mathbf{y} - \mathbf{x}), \quad t \in [0, \varepsilon]$$

$$\Rightarrow \quad g(t) = f(\boldsymbol{x}) \geq f(\mathbf{y}) = \alpha \quad \text{(by (3.113) and (3.114))}.$$

Therefore, if f is not quasiconvex, there exist $\boldsymbol{x}, \mathbf{y}$ such that $f(\mathbf{y}) \leq f(\boldsymbol{x})$ and $\nabla f(\boldsymbol{x})^T (\mathbf{y} - \boldsymbol{x}) > 0$, which contradicts with the implication (3.110). Thus we have completed the proof of sufficiency. ∎

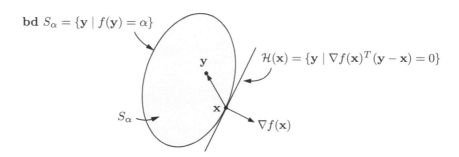

Figure 3.11 Illustration of the first-order condition (3.110) for a quasiconvex function f, where $\mathcal{H}(\mathbf{x})$ is a supporting hyperplane of the sublevel set S_α for $\alpha = f(\mathbf{x})$.

Remark 3.36 Let S_α be the sublevel set of the differentiable $f(\mathbf{y})$. Suppose that $\mathbf{dom}\, f = \mathbb{R}^n$ for simplicity. Then S_α is a closed set and $f(\mathbf{y}) = \alpha$ for all $\mathbf{y} \in \mathbf{bd}\, S_\alpha$. Then an alternative proof for sufficiency of the quasiconvexity condition (3.110) of f is much simpler than the previous proof and given below.

Let $\mathbf{x}_0$ be any arbitrary point in $\mathbf{bd}\, S_\alpha$. When $S_\alpha = \{\mathbf{x}_0\}$ (a singleton set), it is a convex set and the first-order condition (3.110) is also true for $\mathbf{x} = \mathbf{y} = \mathbf{x}_0$. Next, let us prove the sufficiency of (3.110) for the case of $\mathbf{int}\, S_\alpha \neq \emptyset$.

By the first-order condition (3.110), $f(\mathbf{y}) \leq f(\mathbf{x}_0) = \alpha$ for all $\mathbf{y} \in S_\alpha$, it can be inferred that S_α must be contained in the following halfspace:

$$\mathcal{H}_-(\mathbf{x}_0) = \left\{ \mathbf{y} \mid \nabla f(\mathbf{x}_0)^T (\mathbf{y} - \mathbf{x}_0) \leq 0 \right\}$$

(cf. Figure 3.11 where $\mathbf{x}$ corresponds to $\mathbf{x}_0$ here) and

$$\mathcal{H}(\mathbf{x}_0) = \left\{ \mathbf{y} \mid \nabla f(\mathbf{x}_0)^T (\mathbf{y} - \mathbf{x}_0) = 0 \right\}$$

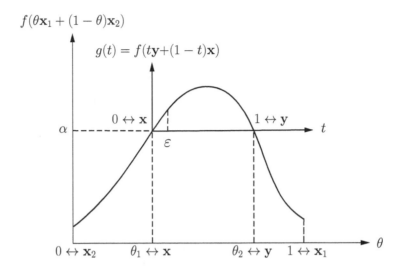

Figure 3.12 Illustration of $f(\theta \mathbf{x}_1 + (1-\theta)\mathbf{x}_2)$ and $g(t) = f(t\mathbf{y} + (1-t)\mathbf{x})$ where $\mathbf{x} = \theta_1 \mathbf{x}_1 + (1-\theta_1)\mathbf{x}_2$ and $\mathbf{y} = \theta_2 \mathbf{x}_1 + (1-\theta_2)\mathbf{x}_2$.

is the associated supporting hyperplane of S_α passing $\mathbf{x}_0$. Hence, we have

$$S_\alpha \subseteq \tilde{S}_\alpha \triangleq \bigcap_{\mathbf{x}_0 \in \mathbf{bd}\ S_\alpha} \mathcal{H}_-(\mathbf{x}_0), \tag{3.115}$$

where $\tilde{S}_\alpha$ must be a closed convex set. However, (3.115) holds only when $\tilde{S}_\alpha = S_\alpha$ is convex by Remark 2.24. Thus we have finished the proof that S_α is a closed convex set for every α. Hence f is quasiconvex. $\qquad \square$

3.3.4   Second-order condition

Suppose that f is twice differentiable. Unlike the first-order condition given by (3.110), the second-order condition for quaiconvex functions is not a necessary and sufficient condition, thus downgrading its usefulness. Next, the necessary one (cf. (3.116)) and the sufficient one (cf. (3.120)) (though look similar but different) are introduced, respectively.

- If f is quasiconvex, then for all $\mathbf{x} \in \mathbf{dom}\ f$, and for all $\mathbf{y} \in \mathbb{R}^n$,

$$\mathbf{y}^T \nabla f(\mathbf{x}) = 0 \ \Rightarrow \ \mathbf{y}^T \nabla^2 f(\mathbf{x})\mathbf{y} \geq 0, \forall \mathbf{y} \neq \mathbf{0}. \tag{3.116}$$

The physical meanings for (3.116) are described in the following two cases.

Case 1: $\nabla f(\mathbf{x}) = \mathbf{0}$. Then,

$$\mathbf{y}^T \nabla^2 f(\mathbf{x})\mathbf{y} \geq 0, \forall \mathbf{y} \in \mathbb{R}^n$$
$$\Rightarrow \nabla^2 f(\mathbf{x}) \succeq \mathbf{0},$$

which implies that $\mathbf{x}$ is a saddle point or a locally minimal point, a locally maximal point, or a globally minimal/maximal point.

Case 2: $\nabla f(\mathbf{x}) \neq \mathbf{0}$, $\mathbf{y} \in \{\nabla f(\mathbf{x})\}^{\perp} = \{\mathbf{y} \mid \mathbf{y}^T \nabla f(\mathbf{x}) = 0\}$. It means that $\nabla^2 f(\mathbf{x})$ is positive semidefinite on the $(n-1)$-dimensional subspace $\{\nabla f(\mathbf{x})\}^{\perp}$ (cf. (1.67)), implying that $\nabla^2 f(\mathbf{x})$ must have $n-1$ nonnegative eigenvalues and the associated $n-1$ eigenvectors also span the $(n-1)$-dimensional subspace $\{\nabla f(\mathbf{x})\}^{\perp}$. Thus the nth eigenvector of $\nabla^2 f(\mathbf{x})$ must be a nonzero vector (including $\nabla f(\mathbf{x})$ itself) in the one-dimensional subspace spanned by $\{\nabla f(\mathbf{x})\}$ simply because all the n eigenvectors of $\nabla^2 f(\mathbf{x}) \in \mathbb{S}^n$ are orthognal to each other (cf. (1.87)), but its eigenvalue can be any real number. In other words, $\nabla^2 f(\mathbf{x})$ can have at most one negative eigenvalue.

Proof of (3.116): If f is quasiconvex, then

$$g(t) = f(\mathbf{x} + t\mathbf{y}) \tag{3.117}$$

is quasiconvex over $\mathbf{dom}\ g = \{t \mid \mathbf{x} + t\mathbf{y} \in \mathbf{dom}\ f\}$ (by Remark 3.31, Subsection 3.3.1). Then

$$g'(t) = \mathbf{y}^T \nabla f(\mathbf{x} + t\mathbf{y}), \tag{3.118}$$
$$g''(t) = \mathbf{y}^T \nabla^2 f(\mathbf{x} + t\mathbf{y})\mathbf{y}. \tag{3.119}$$

It must be true that $g'(0) = \mathbf{y}^T \nabla f(\mathbf{x}) = 0$ implies $g''(0) = \mathbf{y}^T \nabla^2 f(\mathbf{x})\mathbf{y} \geq 0$ by Remark 3.32 in Subsection 3.3.1. Therefore,

$$\mathbf{y}^T \nabla f(\mathbf{x}) = 0 \;\Rightarrow\; \mathbf{y}^T \nabla^2 f(\mathbf{x})\mathbf{y} \geq 0, \; \forall \mathbf{y} \neq \mathbf{0},$$

namely, (3.116) is true.   ∎

- If f satisfies

$$\mathbf{y}^T \nabla f(\mathbf{x}) = 0 \;\Rightarrow\; \mathbf{y}^T \nabla^2 f(\mathbf{x})\mathbf{y} > 0, \tag{3.120}$$

for all $\mathbf{x} \in \mathbf{dom}\ f$ and all $\mathbf{y} \in \mathbb{R}^n$, $\mathbf{y} \neq \mathbf{0}$, then f is quasiconvex.

Proof of (3.120): All the functions $g(t)$, $g'(t)$, and $g''(t)$ (cf. (3.117), (3.118), and (3.119)) used in the proof of (3.116) are still used here. Suppose that $g'(t_1) = \mathbf{y}^T \nabla f(\mathbf{x} + t_1\mathbf{y}) = 0$. Then

$$g''(t_1) = \mathbf{y}^T \nabla^2 f(\mathbf{x} + t_1\mathbf{y})\mathbf{y} > 0 \;\; (\text{by } (3.120)),$$

implying that $g(t)$ has a strictly local minimum at $t = t_1$. It can be shown that $g(t)$ can have only one strictly local minimum. Let us prove this by contradiction.

Suppose that $g(t)$ has another strictly local minimum at $t = t_2 \neq t_1$. Then there must exist a locally maximal point t_3 between t_1 and t_2 such that

$$g'(t_3) = \mathbf{y}^T \nabla f(\mathbf{x} + t_3 \mathbf{y}) = 0, \quad g''(t_3) = \mathbf{y}^T \nabla^2 f(\mathbf{x} + t_3 \mathbf{y})\mathbf{y} \leq 0.$$

Let $\boldsymbol{y} = \mathbf{y}$ and $\boldsymbol{x} = \mathbf{x} + t_3\mathbf{y}$. We have

$$g'(t_3) = \boldsymbol{y}^T \nabla f(\boldsymbol{x}) = 0, \quad g''(t_3) = \boldsymbol{y}^T \nabla^2 f(\boldsymbol{x})\boldsymbol{y} \leq 0$$

which contradict with (3.120). Therefore, $g'(t) < 0$ for $t < t_1$ and $g'(t) > 0$ for $t > t_1$, implying that $t = t_1$ is a unique global minimum of $g(t)$. For the case that $g(t)$ does not have any local minimum (i.e., $g'(0) \neq 0$), $g(t)$ must be strictly decreasing or strictly increasing. Hence $g(t)$ is quasiconvex by Fact 3.3. Thus we have completed the proof that f is quasiconvex (by Remark 3.31, Subsection 3.3.1). ∎

3.4 Monotonicity on generalized inequalities

A function $f : \mathbb{R}^n \to \mathbb{R}$ is K-*nondecreasing* if

$$\mathbf{x} \preceq_K \mathbf{y} \text{ (i.e., } \mathbf{y} - \mathbf{x} \in K) \Rightarrow f(\mathbf{x}) \leq f(\mathbf{y})$$

and K-*increasing* if

$$\mathbf{x} \preceq_K \mathbf{y}, \ \mathbf{x} \neq \mathbf{y} \text{ (i.e., } \mathbf{y} - \mathbf{x} \in K \setminus \{\mathbf{0}\}) \Rightarrow f(\mathbf{x}) < f(\mathbf{y}),$$

where $K \subseteq \mathbb{R}^n$ is a proper cone; f is K-*nonincreasing* if

$$\mathbf{x} \preceq_K \mathbf{y} \Rightarrow f(\mathbf{x}) \geq f(\mathbf{y})$$

and K-*decreasing* if

$$\mathbf{x} \preceq_K \mathbf{y}, \ \mathbf{x} \neq \mathbf{y} \Rightarrow f(\mathbf{x}) > f(\mathbf{y}).$$

Some examples are given next to illustrate the monotonicity properties of the generalized inequality.

- $K = \mathbb{S}^n_+$, $f(\mathbf{X}) = \text{Tr}(\mathbf{WX})$ is K-nondecreasing on $\mathbb{S}^n$ if $\mathbf{W} \succeq \mathbf{0}$.

  Proof: Suppose that $\mathbf{W} = \mathbf{W}^{1/2}\mathbf{W}^{1/2} \succeq \mathbf{0}$ where $\mathbf{W}^{1/2} \succeq \mathbf{0}$, and

  $$\mathbf{X} \preceq_K \mathbf{Y} \Rightarrow (\mathbf{Y} - \mathbf{X}) \in K = \mathbb{S}^n_+.$$

  Then, $\mathbf{W}^{1/2}(\mathbf{Y} - \mathbf{X})\mathbf{W}^{1/2} \succeq \mathbf{0}$ and

  $$f(\mathbf{Y}) - f(\mathbf{X}) = \text{Tr}(\mathbf{W}(\mathbf{Y} - \mathbf{X})) = \text{Tr}(\mathbf{W}^{1/2}[\mathbf{W}^{1/2}(\mathbf{Y} - \mathbf{X})])$$
  $$= \text{Tr}(\mathbf{W}^{1/2}(\mathbf{Y} - \mathbf{X})\mathbf{W}^{1/2}) \geq 0 \quad (\text{since } \text{Tr}(\mathbf{AB}) = \text{Tr}(\mathbf{BA})).$$

  Therefore, $f(\mathbf{X})$ is K-nondecreasing if $\mathbf{W} \succeq \mathbf{0}$. ∎

- $K = \mathbb{S}^n_+$, $\text{Tr}(\mathbf{X}^{-1})$ is K-decreasing on $\mathbb{S}^n_{++}$.

  Proof: Suppose that $\mathbf{Y} \succeq \mathbf{X} \succ \mathbf{0}$. By (2.80), we have $\mathbf{X}^{-1} \succeq \mathbf{Y}^{-1} \succ \mathbf{0}$. Let $\mathbf{X}^{-1} = \mathbf{Y}^{-1} + \mathbf{Z}$, where $\mathbf{Z} \succeq \mathbf{0}$ and $\mathbf{Z} \neq \mathbf{0}$. Then $\text{Tr}(\mathbf{X}^{-1}) = \text{Tr}(\mathbf{Y}^{-1} + \mathbf{Z}) > \text{Tr}(\mathbf{Y}^{-1})$ due to $\text{Tr}(\mathbf{Z}) > 0$. Hence, $\text{Tr}(\mathbf{X}^{-1})$ is K-decreasing. ∎

- $K = \mathbb{S}^n_+$, $\det(\mathbf{X})$ is K-increasing on $\mathbb{S}^n_{++}$.

  Proof: Let $\mathbf{X} \succ \mathbf{0}$ and $\mathbf{Y} = \mathbf{X} + \mathbf{Z} \succ \mathbf{0}$, where $\mathbf{Z} \succeq \mathbf{0}$ and $\mathbf{Z} \neq \mathbf{0}$. Then it can be inferred that

  $$
  \begin{aligned}
  \det(\mathbf{Y}) &= \det(\mathbf{X} + \mathbf{Z}) \\
  &= \det\left(\mathbf{X}^{1/2}(\mathbf{I} + \mathbf{X}^{-1/2}\mathbf{Z}\mathbf{X}^{-1/2})\mathbf{X}^{1/2} \right) \\
  &= \det(\mathbf{X}) \cdot \det(\mathbf{I} + \mathbf{X}^{-1/2}\mathbf{Z}\mathbf{X}^{-1/2}) \\
  &= \det(\mathbf{X}) \cdot \prod_{i=1}^{n}(1 + \lambda_i(\mathbf{X}^{-1/2}\mathbf{Z}\mathbf{X}^{-1/2})) > \det(\mathbf{X}) \quad \text{(cf. (1.99))},
  \end{aligned}
  $$

  where the inequality is due to $\mathbf{X}^{-1/2}\mathbf{Z}\mathbf{X}^{-1/2} \neq \mathbf{0}$ and $\mathbf{X}^{-1/2}\mathbf{Z}\mathbf{X}^{-1/2} \succeq \mathbf{0}$. Hence we have completed the proof that $\det(\mathbf{X})$ is K-increasing. ∎

Remark 3.37 The preceding proofs that $\text{Tr}(\mathbf{WX})$ is K-nondecreasing on $\mathbb{S}^n$ for $\mathbf{W} \succeq \mathbf{0}$, that $\det(\mathbf{X})$ is K-increasing, and that $\text{Tr}(\mathbf{X}^{-1})$ is K-decreasing on $\mathbb{S}^n_{++}$ can be alternatively done by Remark 3.38 due to $\nabla\text{Tr}(\mathbf{WX}) = \mathbf{W} \succeq \mathbf{0}$ by (3.40), $\nabla\det(\mathbf{X}) = \det(\mathbf{X}) \cdot \mathbf{X}^{-1} \succ \mathbf{0}$ for $\mathbf{X} \succ \mathbf{0}$ by (3.47), and the gradient [PP08]

$$
\nabla\text{Tr}(\mathbf{X}^{-1}) = -\mathbf{X}^{-2} \prec \mathbf{0} \quad \text{for } \mathbf{X} \succ \mathbf{0}, \tag{3.121}
$$

respectively. In other words, Remark 3.38 provides an efficient alternative means for verifying if a function is K-nondecreasing, K-increasing, K-nonincreasing, or K-decreasing. □

Remark 3.38 (i) A differentiable function f, with convex domain, is K-*nondecreasing* if and only if $\nabla f(\mathbf{x}) \succeq_{K^*} \mathbf{0}$ for all $\mathbf{x} \in \textbf{dom } f$. (ii) For the strict case, if $\nabla f(\mathbf{x}) \succ_{K^*} \mathbf{0}$ for all $\mathbf{x} \in \textbf{dom } f$, then f is K-*increasing*, but the converse is not true. □

Proof of (i): (Necessity) We prove by contradiction. Assume that $\nabla f(\mathbf{x}) \notin K^*$ for some $\mathbf{x} \in \textbf{dom } f$. Then there exists a $\mathbf{z} \in K, \mathbf{z} \neq \mathbf{0}$ such that

$$
\nabla f(\mathbf{x})^T \mathbf{z} < 0. \tag{3.122}
$$

Let $g : \mathbb{R} \to \mathbb{R}$ with $\textbf{dom } g \triangleq \{t \in \mathbb{R} \mid \mathbf{x} + t\mathbf{z} \in \textbf{dom } f\}$ where

$$
g(t) \triangleq f(\mathbf{x} + t\mathbf{z}). \tag{3.123}
$$

Then by (3.122) and (3.123), we have

$$
g'(0) = \nabla f(\mathbf{x})^T \mathbf{z} < 0.
$$

On the other hand, since $\mathbf{x} \preceq \mathbf{x} + t\mathbf{z}$ for $t > 0$ and $\mathbf{x} + t\mathbf{z} \preceq \mathbf{x}$ for $t < 0$ owing to $\mathbf{z} \in K$, $g(t)$ must be a nondecreasing function for all t by the assumption that $f(\mathbf{x})$ is K-*nondecreasing*, thereby leading to $g'(0) \geq 0$, a contradiction with the above inferred result. Hence, we have proved that $\nabla f(\mathbf{x}) \succeq_{K^*} \mathbf{0}$.

(Sufficiency) Let us only prove for the case that $\mathbf{x} \preceq_K \mathbf{y}$, $\mathbf{y} \neq \mathbf{x}$, and $\mathbf{x}, \mathbf{y} \in$ **dom** f since the case of $\mathbf{y} = \mathbf{x}$ is trivial. Let

$$\mathbf{z} \triangleq \frac{\mathbf{y} - \mathbf{x}}{\|\mathbf{y} - \mathbf{x}\|_2} \in K. \tag{3.124}$$

Because $\mathbf{x} + t\mathbf{z} \in$ **dom** f and thus $g'(t) = \nabla f(\mathbf{x} + t\mathbf{z})^T \mathbf{z}$ exists, we have

$$f(\mathbf{y}) - f(\mathbf{x}) = g(\|\mathbf{y} - \mathbf{x}\|_2) - g(0) = \int_0^{\|\mathbf{y}-\mathbf{x}\|_2} g'(t) \, dt. \tag{3.125}$$

Since $\nabla f(\mathbf{x} + t\mathbf{z}) \succeq_{K^*} \mathbf{0}$ for all $t \in [0, \|\mathbf{y} - \mathbf{x}\|] \subset$ **dom** g and $\mathbf{z} \in K$, we have

$$g'(t) = \nabla f(\mathbf{x} + t\mathbf{z})^T \mathbf{z} \geq 0, \; \forall t \in [0, \|\mathbf{y} - \mathbf{x}\|_2], \tag{3.126}$$

which together with (3.125) leads to $f(\mathbf{x}) \leq f(\mathbf{y})$. ∎

Proof of (ii): Again, consider that $\mathbf{x} \preceq_K \mathbf{y}$ and $\mathbf{x} \neq \mathbf{y}$, and $\mathbf{x}, \mathbf{y} \in$ **dom** f. Let

$$\mathbf{z} \triangleq \frac{\mathbf{y} - \mathbf{x}}{\|\mathbf{y} - \mathbf{x}\|_2} \in K \setminus \{\mathbf{0}\}. \tag{3.127}$$

Since $\nabla f(\boldsymbol{x}) \in$ **int** K^* for all $\boldsymbol{x} \in$ **dom** f, and by (P2) (cf. (2.121b)), we have

$$\nabla f(\boldsymbol{x})^T \mathbf{z} > 0, \; \forall \boldsymbol{x} \in \text{ } \mathbf{dom} \; f. \tag{3.128}$$

Through the same procedure as we proved (3.126) above, we can also show that

$$g'(t) = \nabla f(\mathbf{x} + t\mathbf{z})^T \mathbf{z} > 0, \; \forall t \in [0, \|\mathbf{y} - \mathbf{x}\|_2], \tag{3.129}$$

which together with (3.125) leads to $f(\mathbf{x}) < f(\mathbf{y})$.

Finally, let us prove that the converse statement is not true using a counterexample. Consider $f(\mathbf{x}) = \|\mathbf{x}\|_2^2$ with **dom** $f = \mathbb{R}_+^n = K = K^*$. It can be easily shown that f is K-*increasing* but $\nabla f = 2\mathbf{x} \succ_{K^*} \mathbf{0}$ for all $\mathbf{x} \in$ **dom** $f = \mathbb{R}_+^n$ is not true. ∎

3.5 Convexity on generalized inequalities

A function $\boldsymbol{f} : \mathbb{R}^n \to \mathbb{R}^m$ is K-*convex* (where $K \subseteq \mathbb{R}^m$ is a proper cone) if **dom** $\boldsymbol{f}$ is a convex set and

$$\boldsymbol{f}(\theta\mathbf{x} + (1-\theta)\mathbf{y}) \preceq_K \theta\boldsymbol{f}(\mathbf{x}) + (1-\theta)\boldsymbol{f}(\mathbf{y}) \tag{3.130}$$

for all $\mathbf{x}, \mathbf{y} \in$ **dom** $\boldsymbol{f}$ and $\theta \in [0,1]$; $\boldsymbol{f}$ is *strictly K-convex* if

$$\boldsymbol{f}(\theta\mathbf{x} + (1-\theta)\mathbf{y}) \prec_K \theta\boldsymbol{f}(\mathbf{x}) + (1-\theta)\boldsymbol{f}(\mathbf{y}) \tag{3.131}$$

for all $\mathbf{x} \neq \mathbf{y} \in \mathbf{dom}\ \boldsymbol{f}$ and $\theta \in (0,1)$. As the concave function defined previously for the ordinary case, a function $\boldsymbol{f}$ is K-concave (*strictly K-concave*) if $-\boldsymbol{f}$ is K-convex (strictly K-convex). It can be shown that $\boldsymbol{f}(\mathbf{x})$ is K-convex if, and only if, the associated epigraph defined as

$$\mathbf{epi}_K\ \boldsymbol{f} = \{(\mathbf{x},\boldsymbol{t}) \in \mathbb{R}^{n+m} \mid \boldsymbol{f}(\mathbf{x}) \preceq_K \boldsymbol{t}\} \tag{3.132}$$

is a convex set.

Note that the K-convex function $\boldsymbol{f}(\mathbf{x}) \in \mathbb{R}^m$ which satisfies the inequality (3.130) may not be very conceivable for the case of $m > 1$ as for the case of $m = 1$; it is conceptually illustrated in Figure 3.13, where $\boldsymbol{f} : \mathbb{R}^n \to \mathbb{R}^2$ is K-convex and the proper cone $K = \mathbb{R}^2_+$, and $\boldsymbol{f}$ looks like a bowl facing northeast in $\mathbb{R}^2$; it will face north for $m = 1$ if the horizontal axis is replaced by the domain of $\boldsymbol{f}$ (i.e., $\mathbb{R}^n$), thus reducing to the ordinary case (cf. Figure 3.1).

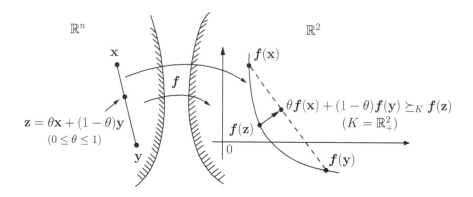

Figure 3.13 An illustration of a K-convex function $\boldsymbol{f} : \mathbb{R}^n \to \mathbb{R}^2$ where $K = \mathbb{R}^2_+$.

Example 3.11 Let $K = \mathbb{R}^m_+$, and

$$\boldsymbol{f}(\mathbf{x}) = \begin{bmatrix} f_1(\mathbf{x}) \\ f_2(\mathbf{x}) \\ \vdots \\ f_m(\mathbf{x}) \end{bmatrix},$$

where f_i is convex, i.e., $f_i(\theta \mathbf{x} + (1-\theta)\mathbf{y}) \leq \theta f_i(\mathbf{x}) + (1-\theta)f_i(\mathbf{y})$, for all $i = 1, \ldots, m$. Then, $\boldsymbol{f}(\mathbf{x})$ is K-convex. $\qquad\square$

Example 3.12 Let $K = \mathbb{S}^n_+$. Then, $\boldsymbol{f}(\mathbf{X}) = \mathbf{X}\mathbf{X}^T$ is K-convex on $\mathbb{R}^{n \times m}$, where $\boldsymbol{f} : \mathbb{R}^{n \times m} \to \mathbb{R}^{n \times n}$.

Proof: Consider for $\mathbf{z} \in \mathbb{R}^n$,

$$\mathbf{z}^T \boldsymbol{f}(\mathbf{X})\mathbf{z} = \mathbf{z}^T \mathbf{X}\mathbf{X}^T \mathbf{z} = \left\| \mathbf{X}^T \mathbf{z} \right\|^2_2.$$

Since $\|\cdot\|_2^2$ is convex, for $\theta \in [0,1]$, we have

$$
\begin{aligned}
\mathbf{z}^T \boldsymbol{f}(\theta\mathbf{X} + (1-\theta)\mathbf{Y})\mathbf{z} &= \left\|\theta\mathbf{X}^T\mathbf{z} + (1-\theta)\mathbf{Y}^T\mathbf{z}\right\|_2^2 \\
&\leq \theta\left\|\mathbf{X}^T\mathbf{z}\right\|_2^2 + (1-\theta)\left\|\mathbf{Y}^T\mathbf{z}\right\|_2^2 \quad (\text{since } \|\cdot\|_2^2 \text{ is convex}) \\
&= \theta\mathbf{z}^T\boldsymbol{f}(\mathbf{X})\mathbf{z} + (1-\theta)\mathbf{z}^T\boldsymbol{f}(\mathbf{Y})\mathbf{z} \\
&= \mathbf{z}^T\left(\theta\boldsymbol{f}(\mathbf{X}) + (1-\theta)\boldsymbol{f}(\mathbf{Y})\right)\mathbf{z}, \quad\quad (3.133)
\end{aligned}
$$

which holds for every $\mathbf{z} \in \mathbb{R}^n$, i.e., $\boldsymbol{f}(\theta\mathbf{X} + (1-\theta)\mathbf{Y}) \preceq_K \theta\boldsymbol{f}(\mathbf{X}) + (1-\theta)\boldsymbol{f}(\mathbf{Y})$. Hence $f(\mathbf{X}) = \mathbf{X}\mathbf{X}^T$ is K-convex. ∎ □

Fact 3.4 Let $K = \mathbb{S}_+^n$. Then $\boldsymbol{f}$ is K-convex if and only if $\mathbf{z}^T\boldsymbol{f}(\mathbf{X})\mathbf{z}$ is convex for every $\mathbf{z} \in \mathbb{R}^n$ by (3.133).

Example 3.13 Let $K = \mathbb{S}_+^n$. Then $\boldsymbol{f}(\mathbf{X}) = \mathbf{X}^{-1}$ is K-convex on $\mathbb{S}_{++}^n$. Prior to the proof of K-convexity of $\boldsymbol{f}$, it is interesting to see the degenerate case of $n = 1$, i.e., K-convexity is exactly the convexity. For this case, $f(x) = 1/x$ for $x > 0$. Then

$$
f'(x) = -\frac{1}{x^2}; \quad f''(x) = \frac{2}{x^3} > 0, \; \forall x > 0,
$$

implying that $f(x)$ is actually strictly convex by the second-order condition. Next, let us prove the K-convexity of $\boldsymbol{f}$ for $n \geq 2$.

Proof: Let us prove that $\mathbf{z}^T\boldsymbol{f}(\mathbf{X})\mathbf{z}$ is convex for every $\mathbf{z} \in \mathbb{R}^n$, by using Fact 3.4. Let $\mathbf{X} \succ \mathbf{0}$, $\mathbf{V} \neq \mathbf{0}$, $\mathbf{V} \in \mathbb{S}^n$ and let $\mathbf{z} \in \mathbb{R}^n$ and

$$
g(t) = \mathbf{z}^T(\mathbf{X} + t\mathbf{V})^{-1}\mathbf{z}
$$

for $t \in \mathbf{dom}\, g = \{t \mid \mathbf{X} + t\mathbf{V} \succ \mathbf{0}\}$. Next, let us prove that g is convex on $\mathbf{dom}\, g$. Let

$$
\mathbf{X}^{-1/2}\mathbf{V}\mathbf{X}^{-1/2} = \mathbf{Q}\boldsymbol{\Lambda}\mathbf{Q}^T \quad (\text{EVD of } \mathbf{X}^{-1/2}\mathbf{V}\mathbf{X}^{-1/2}),
$$

where $\boldsymbol{\Lambda} = \mathbf{Diag}(\lambda_1, \ldots, \lambda_n)$ and $\mathbf{Q}$ is an orthogonal matrix. Then

$$
\begin{aligned}
g(t) &= \mathbf{z}^T\mathbf{X}^{-1/2}(\mathbf{I}_n + t\mathbf{X}^{-1/2}\mathbf{V}\mathbf{X}^{-1/2})^{-1}\mathbf{X}^{-1/2}\mathbf{z} \\
&= \mathbf{z}^T\mathbf{X}^{-1/2}[\mathbf{Q}(\mathbf{I}_n + t\boldsymbol{\Lambda})\mathbf{Q}^T]^{-1}\mathbf{X}^{-1/2}\mathbf{z} \\
&= \mathbf{z}^T\mathbf{X}^{-1/2}\mathbf{Q}(\mathbf{I}_n + t\boldsymbol{\Lambda})^{-1}\mathbf{Q}^T\mathbf{X}^{-1/2}\mathbf{z} \quad (\text{since } \mathbf{Q}^{-1} = \mathbf{Q}^T) \\
&= \mathbf{y}^T(\mathbf{I}_n + t\boldsymbol{\Lambda})^{-1}\mathbf{y} \\
&= \sum_{i=1}^n y_i^2 \frac{1}{1 + t\lambda_i},
\end{aligned}
$$

where $\mathbf{y} = (y_1, \ldots, y_n) = \mathbf{Q}^T\mathbf{X}^{-1/2}\mathbf{z}$, and $1 + t\lambda_i > 0$ due to

$$
\begin{aligned}
\mathbf{dom}\, g &= \{t \mid \mathbf{X} + t\mathbf{V} \succ \mathbf{0}\} \\
&= \{t \mid 1 + t\lambda_i > 0, \; i = 1, \ldots, n\} \quad (\text{by (3.43)}).
\end{aligned}
$$

Since

$$g'(t) = \sum_{i=1}^{n} y_i^2 \frac{-1}{(1+t\lambda_i)^2} \lambda_i,$$

$$g''(t) = \sum_{i=1}^{n} y_i^2 \frac{2}{(1+t\lambda_i)^3} \lambda_i^2 \geq 0, \ \forall t \in \mathbf{dom} \ g,$$

$g(t)$ is convex on $\mathbf{dom} \ g$ for any $\mathbf{X}, \mathbf{V}$, and $\mathbf{z}$ (by second-order condition), by which we can confirm that $\mathbf{z}^T \mathbf{X}^{-1} \mathbf{z}$ is convex for every $\mathbf{z} \in \mathbb{R}^n$ by Fact 3.1, which in turn implies that $\mathbf{X}^{-1}$ is K-convex on $\mathbb{S}_{++}^n$ by Fact 3.4.   ∎ □

In addition to the definition of K-convex functions for verifying the K-convexity of a function $\boldsymbol{f}$, there are also other efficient means, including the convexity of $\boldsymbol{w}^T \boldsymbol{f}$ where $\boldsymbol{w} \in K^*$ and the first-order condition as discussed in the following two remarks.

Remark 3.39 A function $\boldsymbol{f} : \mathbb{R}^n \to \mathbb{R}^m$ is K-convex if and only if for every $\boldsymbol{w} \succeq_{K^*} \mathbf{0}$, the real-valued function $\boldsymbol{w}^T \boldsymbol{f}$ is convex (in the ordinary sense) by *(P1)* (cf. (2.121a)); $\boldsymbol{f}$ is strictly K-convex if and only if for every nonzero $\boldsymbol{w} \succeq_{K^*} \mathbf{0}$ the function $\boldsymbol{w}^T \boldsymbol{f}$ is strictly convex by *(P2)* (cf. (2.121b)).   □

Remark 3.40 A differentiable function $\boldsymbol{f} : \mathbb{R}^n \to \mathbb{R}^m$ is K-convex if and only if $\mathbf{dom} \ \boldsymbol{f}$ is convex, and for every $\mathbf{x}, \mathbf{y} \in \mathbf{dom} \ \boldsymbol{f}$,

$$\boldsymbol{f}(\mathbf{y}) \succeq_K \boldsymbol{f}(\mathbf{x}) + D\boldsymbol{f}(\mathbf{x})(\mathbf{y} - \mathbf{x}). \tag{3.134}$$

The function $\boldsymbol{f}$ is strictly K-convex if and only if for all $\mathbf{x}, \mathbf{y} \in \mathbf{dom} \ \boldsymbol{f}$ with $\mathbf{x} \neq \mathbf{y}$,

$$\boldsymbol{f}(\mathbf{y}) \succ_K \boldsymbol{f}(\mathbf{x}) + D\boldsymbol{f}(\mathbf{x})(\mathbf{y} - \mathbf{x}). \tag{3.135}$$

The above first-order condition (3.134) of a differentiable K-convex function $\boldsymbol{f}$ can be easily proven by the ordinary first-order condition given by (3.16) and Remark 3.39 above; so can (3.135). For the case of $m = 1$, the generalized inequalities (3.134) and (3.135) become ordinary inequalities (3.16) or (3.20). □

Example 3.14 In Example 3.13, $\boldsymbol{f}(\mathbf{X}) = \mathbf{X}^{-1}$ has been shown to be $\mathbb{S}_+^n$-convex on $\mathbb{S}_{++}^n$ for $n \geq 2$. In this example an alternative proof is done by Remark 3.39 above, which turns out to be a much simpler proof.

By Remark 3.39, we only need to prove that $\mathrm{Tr}(\mathbf{W}\mathbf{X}^{-1})$ is convex in $\mathbf{X} \in \mathbb{S}_{++}^n$ for any PSD matrix $\mathbf{W} \in \mathbb{S}_+^n$ since the proper cone $\mathbb{S}_+^n$ is self-dual. Equivalently, we need to prove the following ordinary first-order condition by (3.20):

$$\mathrm{Tr}(\mathbf{W}\mathbf{Y}^{-1}) \geq \mathrm{Tr}(\mathbf{W}\mathbf{X}^{-1}) + \mathrm{Tr}\left(D(\mathrm{Tr}(\mathbf{W}\mathbf{X}^{-1}))(\mathbf{Y} - \mathbf{X})\right), \tag{3.136}$$

where $\mathbf{X}, \mathbf{Y} \in \mathbb{S}_{++}^n$ and $\mathbf{W} \succeq \mathbf{0}$. By substituting [PP08]

$$D(\mathrm{Tr}(\mathbf{W}\mathbf{X}^{-1})) = -\mathbf{X}^{-1}\mathbf{W}\mathbf{X}^{-1} \tag{3.137}$$

into (3.136) yields

$$\mathrm{Tr}\big(\mathbf{W}(\mathbf{Y}^{-1} - 2\mathbf{X}^{-1} + \mathbf{X}^{-1}\mathbf{Y}\mathbf{X}^{-1})\big) = \mathrm{Tr}(\mathbf{W}\mathbf{A}\mathbf{A}^T) = \mathrm{Tr}(\mathbf{A}^T\mathbf{W}\mathbf{A}) \geq 0,$$

where $\mathbf{A} = \mathbf{Y}^{-1/2} - \mathbf{X}^{-1}\mathbf{Y}^{1/2}$, and $\mathbf{Y}^{1/2} \in \mathbb{S}^n_{++}$ such that $\mathbf{Y} = (\mathbf{Y}^{1/2})^2$. Thus we have completed the proof. $\square$

3.6   Summary and discussion

In this chapter, we have introduced the concepts and various properties and conditions associated with convex functions, quasiconvex functions, and K-convex functions. Some convexity preserving operations were also introduced together with many examples to illustrate how to verify the convexity and concavity of functions, and the ways to verify the convexity of functions are usually nonunique. The convexity of a set is conceptually totally different from that of a function, while the latter can be corroborated by showing that its epigraph is a convex set. The basics of the convex sets and the convex functions introduced in the previous chapter and the current chapter, respectively, are essential to solving convex and quasiconvex optimization problems, as we will see in the forthcoming chapters.

References for Chapter 3

[Ber09] D. P. Bertsekas, *Convex Optimization Theory*. Belmont, MA, USA: Athena Scientific, 2009.

[BV04] S. Boyd and L. Vandenberghe, *Convex Optimization*. Cambridge, UK: Cambridge University Press, 2004.

[LCC15] W.-C. Li, T.-H. Chang, and C.-Y. Chi, "Multicell coordinated beamforming with rate outage constraint–Part II: Efficient approximation algorithms," *IEEE Trans. Signal Process.*, vol. 63, no. 11, pp. 2763–2778, June 2015.

[PP08] K. B. Petersen and M. S. Pedersen, *The Matrix Cookbook*, Nov. 14, 2008.

[YCM+12] Y. Yang, T.-H. Chang, W.-K. Ma, J. Ge, C.-Y. Chi, and P.-C. Ching, "Noncoherent bit-interleaved coded OSTBC-OFDM with maximum spatial-frequency diversity," *IEEE Trans. Wireless Commun.*, vol. 11, no. 9, pp. 3335–3347, Sept. 2012.

4 Convex Optimization Problems

In the previous chapters we have introduced the notions of convex sets (Chapter 2) and convex functions (Chapter 3). Based on the foundation laid in the previous chapters, we now move forward to study the concepts of convex optimization. An optimization problem to be solved may appear to be a nonconvex problem due either to the nonconvexity of the objective function, to the the nonconvexity of the constraint set, or to both. However, it may be potentially a convex optimization problem. In this chapter, we will focus on formulation of such problems into a standard convex optimization problem so that the optimal solution can be obtained via either optimality conditions (to be presented partly in this chapter and partly in Chapter 9) or available convex problem solvers. The quasiconvex problem is also presented together with a sufficient optimality condition, and then the widely used bisection method for solving it is introduced.

However, when the problem under consideration cannot be reformulated as a convex or a quasiconvex problem, we may have to consider a stationary-point solution instead. To this end, we finally introduce an iterative *block successive upper bound minimization (BSUM)* method that can yield a stationary-point solution under some convergence conditions, followed by the widely used *successive convex approximation (SCA)* that may also yield a stationary-solution depending on the problem under consideration. In the subsequent four chapters, all the reformulation and approximation approaches introduced in this chapter will be extensively applied to solve various practical problems in signal processing and wireless communications.

For conciseness of presentation of the ensuing chapters, a function f is always defined as $f : \mathbb{R}^n \to \mathbb{R}$ implicitly, unless otherwise specified. The unknown variable vector is denoted by $\mathbf{x}$ and its ith element by x_i. The optimal solution of an optimization problem is denoted by either $\mathbf{x}^\star$ or $\widehat{\mathbf{x}}$. A standard convex optimization problem is simply termed as a convex problem. When the constraint set is $\mathbb{R}^n$ (i.e., an unconstrained optimization problem), the constraint set $\mathbb{R}^n$ does usually not show up explicitly in the problem for simplicity.

4.1   Optimization problems in a standard form

In general, an optimization problem has the following structure:

$$
\begin{aligned}
\min \quad & f_0(\mathbf{x}) \\
\text{s.t.} \quad & f_i(\mathbf{x}) \leq 0, \ i = 1, \ldots, m \\
& h_i(\mathbf{x}) = 0, \ i = 1, \ldots, p
\end{aligned}
\tag{4.1}
$$

where f_0 is the *objective function*, f_i, $i = 1, \ldots, m$, are *inequality constraint functions*, h_i, $i = 1, \ldots, p$, are *equality constraint functions*, and "s.t." stands for "subject to."

Before getting into the details of convex optimization, we need to introduce some terminologies which will be often encountered when discussing optimization problems.

4.1.1   Some terminologies

- The set

$$
\mathcal{D} = \left\{ \bigcap_{i=0}^{m} \mathbf{dom} \, f_i \right\} \cap \left\{ \bigcap_{i=1}^{p} \mathbf{dom} \, h_i \right\}
\tag{4.2}
$$

  is called the *problem domain* of (4.1).

- The set

$$
\mathcal{C} = \left\{ \mathbf{x} \mid \mathbf{x} \in \mathcal{D}, \ f_i(\mathbf{x}) \leq 0, \ i = 1, \ldots, m, \ h_i(\mathbf{x}) = 0, \ i = 1, \ldots, p \right\}
\tag{4.3}
$$

  is called the *feasible set*, or the *constraint set*.

- A point $\mathbf{x} \in \mathcal{D}$ is *feasible* if $\mathbf{x} \in \mathcal{C}$, and *infeasible* otherwise.

- A point $\mathbf{x}$ is *strictly feasible* if $f_i(\mathbf{x}) < 0$, $i = 1, \ldots, m$, $h_i(\mathbf{x}) = 0$, $i = 1, \ldots, p$, i.e., all the inequality constraints are inactive.

- The inequality constraint $f_i(\mathbf{x})$ is *active* at $\mathbf{x} \in \mathcal{C}$ if $f_i(\mathbf{x}) = 0$.

- A problem is *feasible* if there exists at least one $\mathbf{x} \in \mathcal{C}$; *infeasible* if $\mathcal{C} = \emptyset$ (empty set); *strictly feasible* if there exists a strictly feasible point.

- A problem is *unconstrained* if $\mathcal{C} = \mathbb{R}^n$.

4.1.2   Optimal value and solution

The optimal (objective) value $p^\star$ of the optimization problem (4.1) is defined as

$$
p^\star = \inf_{\mathbf{x} \in \mathcal{C}} f_0(\mathbf{x}) = \inf \left\{ f_0(\mathbf{x}) \mid \mathbf{x} \in \mathcal{C} \right\}.
\tag{4.4}
$$

- If the problem is infeasible (i.e., $\mathcal{C} = \emptyset$), we have $p^\star = +\infty$ since

$$
\inf \left\{ f_0(\mathbf{x}) \mid \mathbf{x} \in \mathcal{C} \right\} = \inf \emptyset = \infty.
$$

If $p^\star = -\infty$, the problem is *unbounded below*.

- A point $\mathbf{x}^\star$ is *globally optimal* or simply *optimal* if $\mathbf{x}^\star \in \mathcal{C}$ and $f_0(\mathbf{x}^\star) = p^\star$. If $\mathbf{x}^\star$ exists, it can also be represented as

$$\mathbf{x}^\star = \arg\min \{f_0(\mathbf{x}) \mid \mathbf{x} \in \mathcal{C}\}, \tag{4.5}$$

  which is also called a *global minimizer*.

- As f_0 is continuous and the feasible set $\mathcal{C} \subset \mathbb{R}^n$ is compact, there exists a global minimizer $\mathbf{x}^\star \in \mathcal{C}$ (by Theorem of Weierstrass for continuous functions [Ber09]), and so this is also true when f_0 is convex over a compact $\mathcal{C} \subset \mathbb{R}^n$, since f_0 must be continuous over the interior of $\mathcal{C}$ (cf. Remark 3.7).

- The problem is *solvable* if an optimal point $\mathbf{x}^\star$ exists. Figure 4.1 illustrates a convex, but unsolvable, optimization (unconstrained) problem, which, however, is feasible.

- A point $\boldsymbol{x}$ is *locally optimal* (or a *local minimizer*) if there is an $r > 0$ such that

$$f_0(\boldsymbol{x}) = \inf \left\{ f_0(\mathbf{x}) \mid \mathbf{x} \in \mathcal{C}, \|\mathbf{x} - \boldsymbol{x}\|_2 \le r \right\}. \tag{4.6}$$

- A feasible point $\mathbf{x}$ with $f_0(\mathbf{x}) \le p^\star + \epsilon$ (where $\epsilon > 0$) is called ϵ-*suboptimal*, and the set of all ϵ-suboptimal points is called the ϵ-suboptimal set for problem (4.1).

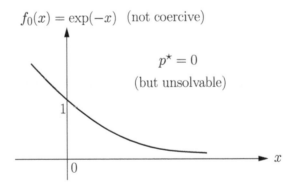

$f_0(x) = \exp(-x)$ (not coercive)

$p^\star = 0$

(but unsolvable)

Figure 4.1 An unsolvable unconstrained convex optimization problem with $\mathcal{D} = \mathbb{R}$ where the objective function f_0 is strictly convex but not coercive.

Some connections between the objective function's attributes and the existence and uniqueness of the optimal solution of problem (4.1) are as follows:

- f_0 is *coercive* if $\lim_{\|\mathbf{x}_k\| \to +\infty} f_0(\mathbf{x}_k) = +\infty$ for every sequence $\{\mathbf{x}_k\}$. For instance, $f_0(x) = e^{ax}$ is not coercive (as illustrated in Figure 4.1), $g(\mathbf{x}) = \|\mathbf{x}\|_1$ is coercive, and $h(\mathbf{x}) = \mathbf{a}^T\mathbf{x}$ is not coercive.

- If f_0 is coercive and continuous with unbounded domain, then the set of the global minimizers $\{\mathbf{x}^\star\}$ is nonempty. For instance, $f_0(x) = \log|x|, x \neq 0$ is coercive but not continuous, and no global minimizer $\mathbf{x}^\star$ exists for $\mathcal{C} = \mathbb{R}$.

- If f_0 is strictly convex and the interior of its domain is nonempty (and thus is continuous over the interior of its domain), then $\{\mathbf{x}^\star\}$ has at most one element; if f_0 is strictly convex and coercive or if f_0 is strictly convex and a global minimizer $\mathbf{x}^\star$ exists, then the optimal $\mathbf{x}^\star$ is unique; if f_0 is strictly convex with unbounded domain but not coercive, the solution set $\{\mathbf{x}^\star\} = \emptyset$ as illustrated in Figure 4.1.

4.1.3   Equivalent problems and feasibility problem

Maximizing an objective function $f_0(\mathbf{x})$ over a constraint set $\mathcal{C}$ is equivalent to a minimization problem. That is,

$$\max_{\mathbf{x}\in\mathcal{C}} f_0(\mathbf{x}) = -\min_{\mathbf{x}\in\mathcal{C}}\left\{-f_0(\mathbf{x})\right\} \equiv \min_{\mathbf{x}\in\mathcal{C}}\left\{-f_0(\mathbf{x})\right\}, \tag{4.7}$$

where "$\equiv$" means that the problems on both sides have identical solutions but the corresponding objective functions (and so the associated optimal values) are different. Moreover, if $f_0(\mathbf{x}) > 0 \ \forall \mathbf{x} \in \mathcal{C}$, it can be easily seen that

$$\max_{\mathbf{x}\in\mathcal{C}} f_0(\mathbf{x}) = \frac{1}{\min_{\mathbf{x}\in\mathcal{C}}\left(\frac{1}{f_0(\mathbf{x})}\right)} \equiv \min_{\mathbf{x}\in\mathcal{C}} \frac{1}{f_0(\mathbf{x})}, \tag{4.8}$$

$$\min_{\mathbf{x}\in\mathcal{C}} f_0(\mathbf{x}) = \frac{1}{\max_{\mathbf{x}\in\mathcal{C}}\left(\frac{1}{f_0(\mathbf{x})}\right)} \equiv \max_{\mathbf{x}\in\mathcal{C}} \frac{1}{f_0(\mathbf{x})}. \tag{4.9}$$

For an optimization problem under consideration with a feasible set $\mathcal{C}$, if one is only interested in whether the problem is feasible or not (i.e., whether $\mathcal{C} = \emptyset$ is true or not), one can consider a so-called feasibility problem instead, which is represented as

$$\begin{aligned}\text{find } \ &\mathbf{x}\\ \text{s.t. } \ &\mathbf{x} \in \mathcal{C}.\end{aligned} \tag{4.10}$$

This feasibility problem is actually a constrained optimization problem with a particular objective function $f_0(\mathbf{x}) = 0$. In this case,

$$p^\star = 0 \text{ if } \mathcal{C} \neq \emptyset, \text{ and } p^\star = +\infty \text{ if } \mathcal{C} = \emptyset.$$

The feasibility problem is very useful in initializing an iterative convex optimization algorithm that requires a feasible point as its input. The feasibility problem plays a central role in solving quasiconvex problems to be introduced later in this chapter.

4.2   Convex optimization problems

The optimization problem (4.1) is said to be a *convex optimization problem* if the objective function f_0 is a convex function and the constraint set $\mathcal{C}$ defined in (4.3) is a convex set. Note that $\mathcal{C}$ is convex if $f_1, \ldots, f_m$ are convex and $h_1, \ldots, h_p$ are affine. A standard convex problem is therefore defined as

$$\begin{aligned} \min \quad & f_0(\mathbf{x}) \\ \text{s.t.} \quad & \mathbf{Ax} = \mathbf{b}, \ f_i(\mathbf{x}) \leq 0, \ i = 1, \ldots, m, \end{aligned} \tag{4.11}$$

where $f_0, \ldots, f_m$ are convex, $\mathbf{A} \in \mathbb{R}^{p \times n}$, and $\mathbf{b} \in \mathbb{R}^p$. In other words, a standard convex problem requires the objective function f_0 to be a convex function along with a convex feasible set in the following form

$$\mathcal{C} = \left\{ \mathbf{x} \mid \mathbf{x} \in \mathcal{D}, \ f_i(\mathbf{x}) \leq 0, \ i = 1, \ldots, m, \ \mathbf{Ax} = \mathbf{b} \right\}. \tag{4.12}$$

Some examples of convex optimization problems are given next.

- Least-squares (LS) problem:

$$\min \ \|\mathbf{Ax} - \mathbf{b}\|_2^2 \tag{4.13}$$

is an unconstrained convex optimization problem because $\|\cdot\|_2^2$ is a convex function. A constrained LS problem, called a fully constrained (i.e., constrained on the unit simplex, not constrained (just) on the nonnegative orthant) least squares (FCLS), is defined as

$$\begin{aligned} \min \quad & \|\mathbf{Ax} - \mathbf{b}\|_2^2 \\ \text{s.t.} \quad & \mathbf{x} \succeq \mathbf{0}, \mathbf{1}_n^T \mathbf{x} = 1. \end{aligned} \tag{4.14}$$

FCLS has been widely applied in hyperspectral unmixing (in remote sensing) for abundance estimation, which will be introduced in Chapter 6. Note that FCLS is a convex optimization problem as the constraint set is a polyhedron.

- Unconstrained quadratic program (QP):

$$\min \ \mathbf{x}^T \mathbf{Px} + 2\mathbf{q}^T \mathbf{x} + r \tag{4.15}$$

which is convex if and only if $\mathbf{P} \succeq \mathbf{0}$. Note that the LS problem (cf. (4.13)) is an unconstrained QP with $\mathbf{P} = \mathbf{A}^T \mathbf{A} \succeq \mathbf{0}$. (Question: what happens when $\mathbf{P}$ is indefinite?)

- Linear program (LP): An LP has the following structure

$$\begin{aligned} \min \quad & \mathbf{c}^T \mathbf{x} \\ \text{s.t.} \quad & \mathbf{x} \succeq \mathbf{0}, \ \mathbf{Ax} = \mathbf{b}. \end{aligned} \tag{4.16}$$

This is a constrained convex problem since any linear function is a convex function, and the constraint set is a polyhedron.

- Minimum-norm approximation with bound constraints:

$$\begin{aligned} \min \quad & \|\mathbf{Ax} - \mathbf{b}\| \\ \text{s.t.} \quad & l_i \le x_i \le u_i, \quad i = 1, \ldots, n, \end{aligned} \tag{4.17}$$

for any norm $\|\cdot\|$, where x_i denotes the ith component of $\mathbf{x} \in \mathbb{R}^n$. This is a constrained convex problem since $\|\cdot\|$ is known to be a convex function, and the constraint set is a polyhedron.

Remark 4.1 Let

$$C_i = \{\mathbf{x} \mid \mathbf{x} \in \mathcal{D}, \ f_i(\mathbf{x}) \le 0\}, \ i = 1, \ldots, m, \tag{4.18}$$

$$H = \{\mathbf{x} \mid \mathbf{x} \in \mathcal{D}, \ \mathbf{Ax} = \mathbf{0}\}. \tag{4.19}$$

Then C_i and H are all convex sets for a convex problem, and the feasible set defined in (4.12) can expressed as

$$\mathcal{C} = H \cap \left(\cap_{i=1}^m C_i \right) \tag{4.20}$$

which is certainly convex. However, it is also true that if any f_i is quasiconvex because C_i is nothing but a sublevel set (a convex set) for this case, implying that the condition that f_i must be convex for all i is a sufficient condition for $\mathcal{C}$ to be a convex set. Nevertheless, for a standard convex problem, it is required to reformulate the nonconvex f_i into a convex function with C_i unchanged. How to perform the reformulation will be detailed later in this chapter. $\square$

4.2.1   Global optimality

Fact 4.1 For the convex optimization problem defined in (4.11), any locally optimal solution is globally optimal.

Proof: We will prove by contradiction. Let $\mathbf{x}^\star$ be a globally optimal solution and $\boldsymbol{x}$ be a locally optimal solution (i.e., $\boldsymbol{x}$ is a locally minimal point) of a convex optimization problem with $f_0(\mathbf{x}^\star) < f_0(\boldsymbol{x})$.

Let

$$C \triangleq \{t\mathbf{x}^\star + (1-t)\boldsymbol{x} = \boldsymbol{x} + t(\mathbf{x}^\star - \boldsymbol{x}), \ 0 \le t \le 1\} \tag{4.21}$$

which is the closed set of the line segment with the end points being $\mathbf{x}^\star$ and $\boldsymbol{x}$. Then by the definition of convex functions (cf. (3.3) and (3.4)) and the assumption that $f_0(\boldsymbol{x}) > f_0(\mathbf{x}^\star)$, we have

$$f_0(\mathbf{x}) \le \frac{\|\mathbf{x} - \mathbf{x}^\star\|_2}{\|\mathbf{x}^\star - \boldsymbol{x}\|_2} f_0(\boldsymbol{x}) + \frac{\|\mathbf{x} - \boldsymbol{x}\|_2}{\|\mathbf{x}^\star - \boldsymbol{x}\|_2} f_0(\mathbf{x}^\star) < f_0(\boldsymbol{x}) \ \forall \mathbf{x} \in C \setminus \{\boldsymbol{x}\}. \tag{4.22}$$

Let $B(\mathbf{y}, r) \subset \mathbf{dom} \ f$ denote the 2-norm ball with center $\mathbf{y}$ and radius r. Since $(C \setminus \{\boldsymbol{x}\}) \cap B(\boldsymbol{x}, r) \ne \emptyset$ for any $r > 0$, it can be inferred from (4.22) that

$$f_0(\mathbf{x}) < f_0(\boldsymbol{x}), \text{ for some } \mathbf{x} \in B(\boldsymbol{x}, r), \text{ for any } r > 0,$$

which contradicts with the assumption of locally minimal x by (4.6). Hence, we have completed the proof that for a convex optimization problem, any locally optimal point must be globally optimal. ∎

4.2.2   An optimality criterion

Assume that f_0 is differentiable, and that the associated optimization problem (4.11) with the constraint set $\mathcal{C}$ given by (4.12) is convex. Then a point $\mathbf{x} \in \mathcal{C}$ is optimal if and only if

$$\nabla f_0(\mathbf{x})^T(\mathbf{y} - \mathbf{x}) \geq 0, \ \forall \mathbf{y} \in \mathcal{C}. \tag{4.23}$$

Proof of (4.23): The proof of sufficiency of this optimality criterion is straightforward. From the first-order condition (given by (3.16)),

$$f_0(\mathbf{y}) \geq f_0(\mathbf{x}) + \nabla f_0(\mathbf{x})^T(\mathbf{y} - \mathbf{x}), \ \forall \mathbf{y} \in \mathcal{C}. \tag{4.24}$$

If $\mathbf{x}$ satisfies (4.23), then $f_0(\mathbf{y}) \geq f_0(\mathbf{x})$, for all $\mathbf{y} \in \mathcal{C}$.

The necessity of the optimality criterion can be proven by contradiction. Suppose that $\mathbf{x}$ is optimal but $\nabla f_0(\mathbf{x})^T(\mathbf{y} - \mathbf{x}) < 0$ for some $\mathbf{y} \in \mathcal{C}$. Let

$$\mathbf{z} = t\mathbf{y} + (1-t)\mathbf{x}, \ t \in [0, 1].$$

Since $\mathcal{C}$ is convex, $\mathbf{z} \in \mathcal{C}$, the objective function $f_0(\mathbf{z})$ is differentiable w.r.t. t and

$$\frac{df_0(\mathbf{z})}{dt} = \frac{d}{dt} f_0(\mathbf{x} + t(\mathbf{y} - \mathbf{x})) = \nabla f_0(\mathbf{x} + t(\mathbf{y} - \mathbf{x}))^T(\mathbf{y} - \mathbf{x}). \tag{4.25}$$

Then we have

$$\left.\frac{df_0(\mathbf{z})}{dt}\right|_{t=0} = \nabla f_0(\mathbf{x})^T(\mathbf{y} - \mathbf{x}) < 0, \tag{4.26}$$

which, together with $\mathbf{z} = \mathbf{x}$ for $t = 0$, implies that for small t, $f_0(\mathbf{z}) < f_0(\mathbf{x})$. This is a contradiction with the global optimality of $\mathbf{x}$. So (4.23) has been proved. ∎

When $f_0 : \mathbb{R}^{n \times m} \to \mathbb{R}$, i.e., f_0 is a function of matrices, a point $\mathbf{X} \in \mathcal{C}$ is optimal if and only if

$$\mathrm{Tr}\left(\nabla f_0(\mathbf{X})^T(\mathbf{Y} - \mathbf{X})\right) \geq 0, \forall \mathbf{Y} \in \mathcal{C}. \tag{4.27}$$

Two interesting perspectives can be inferred from the first-order optimality condition given by (4.23) and (4.27).

- For an unconstrained convex problem, $\mathbf{x} \in \mathcal{C} = \mathbb{R}^n$ is optimal if and only if

$$\nabla f_0(\mathbf{x}) = \mathbf{0}_n. \tag{4.28}$$

This can be easily shown from (4.23) because $\mathbf{y} - \mathbf{x}$ represents any arbitrary vector in $\mathbb{R}^n$. For this case, a closed-form optimal solution for $\mathbf{x}^\star$ may be obtained by solving (4.28). For the case when $f_0 : \mathbb{R}^{n \times m} \to \mathbb{R}$, the optimality condition given by (4.28) becomes

$$\nabla f_0(\mathbf{X}) = \mathbf{0}_{n \times m}. \tag{4.29}$$

- The optimality criterion given by (4.23) implies that an optimal $\mathbf{x} \in \mathcal{C}$ either yields $\nabla f_0(\mathbf{x}) = \mathbf{0}_n$ (see Case 1 in Figure 4.2) which is actually the same result as given by (4.28) associated with the unconstrained convex problem, or $-\nabla f_0(\mathbf{x}) \neq \mathbf{0}_n$ forming a supporting hyperplane of $\mathcal{C}$ at the optimal point $\mathbf{x} \in \mathbf{bd}\ \mathcal{C}$, such that

$$\mathcal{C} \subset \mathcal{H}_+(\mathbf{y}) = \left\{ \mathbf{y} \mid \nabla f_0(\mathbf{x})^T(\mathbf{y} - \mathbf{x}) \geq 0 \right\} \tag{4.30}$$

(see Case 2 in Figure 4.2). For this case, usually it may be non-trivial to find an optimal $\mathbf{x}$ when (4.23) turns out to be a complicated nonlinear inequality, depending on the applications. However, it is still possible to obtain a closed-form solution (which is always our favorite) rather than a numerical solution for the optimal $\mathbf{x}$.

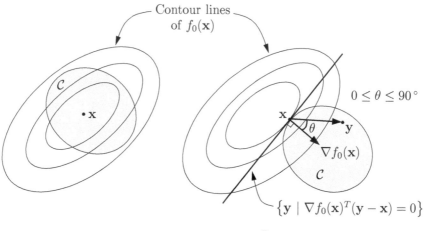

Case 1: $\nabla f_0(\mathbf{x}) = \mathbf{0}_n$ Case 2: $\nabla f_0(\mathbf{x})^T(\mathbf{y} - \mathbf{x}) \geq 0,\ \forall \mathbf{y} \in \mathcal{C}$

Figure 4.2 Geometric interpretation of the optimality conditions, where $\nabla f_0(\mathbf{x}) - \mathbf{0}_n$ in Case 1 and $\nabla f_0(\mathbf{x}) \neq \mathbf{0}_n$ in Case 2.

Next, let us give some examples to illustrate the effectiveness of the optimality condition in (4.23) and (4.28).

Example 4.1 The unconstrained QP

$$\min_{\mathbf{x} \in \mathbb{R}^n}\ \mathbf{x}^T \mathbf{P} \mathbf{x} + 2\mathbf{q}^T \mathbf{x} + r, \tag{4.31}$$

where $\mathbf{P} \in \mathbb{S}_+^n$, $\mathbf{q} \in \mathbb{R}^n$, and $r \in \mathbb{R}$. This problem is an unconstrained convex problem and the optimality criterion by (4.28) is

$$\nabla f_0(\mathbf{x}) = 2\mathbf{P}\mathbf{x} + 2\mathbf{q} = \mathbf{0}. \tag{4.32}$$

If a solution to (4.32) exists (i.e., $\mathbf{q} \in \mathcal{R}(\mathbf{P})$), then an optimal $\mathbf{x}^\star$ can surely be found; otherwise, one may need to reformulate the problem (4.31) to get the optimal solution. For instance, if $\mathbf{P}$ is invertible (or $\mathbf{P} \succ \mathbf{0}$), then

$$\mathbf{x}^\star = -\mathbf{P}^{-1}\mathbf{q} \tag{4.33}$$

is the optimal solution. Similarly, one can also prove the optimal solution of the unconstrained LS problem given by (4.13) (for which $\mathbf{P} = \mathbf{A}^T\mathbf{A}$) to be

$$\mathbf{x}_{\mathrm{LS}} = (\mathbf{A}^T\mathbf{A})^{-1}\mathbf{A}^T\mathbf{b} = \mathbf{A}^\dagger\mathbf{b} \tag{4.34}$$

(cf. (1.125)) provided that $\mathbf{A}$ is of full column rank. □

Example 4.2 *(Linear minimum mean-squared estimation of random vectors)* Given a measurement vector $\boldsymbol{x} \in \mathbb{R}^M$ with zero mean and positive definite correlation matrix $\boldsymbol{P}_{\mathrm{xx}} = \mathbb{E}\{\boldsymbol{x}\boldsymbol{x}^T\} \succ \mathbf{0}$, we would like to estimate a random vector $\boldsymbol{s} \in \mathbb{R}^N$ from $\boldsymbol{x}$, provided that $\boldsymbol{s}$ is zero-mean with autocorrelation matrix $\boldsymbol{P}_{\mathrm{ss}}$ and cross-correlation matrix $\boldsymbol{P}_{\mathrm{sx}} = \mathbb{E}\{\boldsymbol{s}\boldsymbol{x}^T\}$. With the given data vector $\boldsymbol{x}$, let $\widehat{\boldsymbol{s}}$ denote a linear estimate of $\boldsymbol{s}$, i.e.,

$$\widehat{\boldsymbol{s}} = \boldsymbol{F}^T\boldsymbol{x} \tag{4.35}$$

where $\boldsymbol{F} \in \mathbb{R}^{M \times N}$ is the linear estimator to be designed.

The widely used linear minimum mean-squared estimator (LMMSE) [Men95, Kay13] is designed by minimizing the following objective function, i.e., mean-squared error (MSE), as follows

$$\begin{aligned} J(\boldsymbol{F}) &= \mathbb{E}\left\{\|\widehat{\boldsymbol{s}} - \boldsymbol{s}\|_2^2\right\} = \mathbb{E}\left\{\|\boldsymbol{F}^T\boldsymbol{x} - \boldsymbol{s}\|_2^2\right\} \\ &= \mathbb{E}\left\{\mathrm{Tr}\left((\boldsymbol{F}^T\boldsymbol{x} - \boldsymbol{s})(\boldsymbol{F}^T\boldsymbol{x} - \boldsymbol{s})^T\right)\right\} \\ &= \mathrm{Tr}\left(\boldsymbol{F}^T\boldsymbol{P}_{\mathrm{xx}}\boldsymbol{F} - \boldsymbol{F}^T\boldsymbol{P}_{\mathrm{xs}} - \boldsymbol{P}_{\mathrm{sx}}\boldsymbol{F} + \boldsymbol{P}_{\mathrm{ss}}\right). \end{aligned} \tag{4.36}$$

Then we have

$$\begin{aligned} J(\boldsymbol{F} + \boldsymbol{G}) &= \mathrm{Tr}\left((\boldsymbol{F}^T + \boldsymbol{G}^T)\boldsymbol{P}_{\mathrm{xx}}(\boldsymbol{F} + \boldsymbol{G}) - (\boldsymbol{F}^T + \boldsymbol{G}^T)\boldsymbol{P}_{\mathrm{xs}} - \boldsymbol{P}_{\mathrm{sx}}(\boldsymbol{F} + \boldsymbol{G}) + \boldsymbol{P}_{\mathrm{ss}}\right) \\ &= J(\boldsymbol{F}) + \mathrm{Tr}\left(\nabla J(\boldsymbol{F})^T\boldsymbol{G}\right) + \mathrm{Tr}(\boldsymbol{G}^T\boldsymbol{P}_{\mathrm{xx}}\boldsymbol{G}) \quad \text{(cf. (1.56))} \\ &\geq J(\boldsymbol{F}) + \mathrm{Tr}\left(\nabla J(\boldsymbol{F})^T\boldsymbol{G}\right) \quad \text{(cf. (3.20))} \end{aligned}$$

where

$$\nabla J(\boldsymbol{F}) = -2\boldsymbol{P}_{\mathrm{xs}} + 2\boldsymbol{P}_{\mathrm{xx}}\boldsymbol{F}. \tag{4.37}$$

Hence, $J(\boldsymbol{F})$ is a differentiable convex function. Letting $\nabla J(\boldsymbol{F}) = \mathbf{0}_{M \times N}$ (according to (4.29)), one can obtain the optimal LMMSE

$$\boldsymbol{F}^\star \triangleq \arg \min_{\boldsymbol{F} \in \mathbb{R}^{M \times N}} J(\boldsymbol{F}) = \boldsymbol{P}_{\mathrm{xx}}^{-1}\boldsymbol{P}_{\mathrm{xs}} \tag{4.38}$$

as well as the minimum MSE

$$J(\boldsymbol{F}^\star) = \text{Tr}(\boldsymbol{P}_{\text{ss}}) - \text{Tr}(\boldsymbol{P}_{\text{sx}}\boldsymbol{P}_{\text{xx}}^{-1}\boldsymbol{P}_{\text{xs}}). \qquad (4.39)$$

This example is also an illustration that the LMMSE can be obtained via the first-order optimality condition (4.29).

Let us consider an application of LMMSE, where

$$\boldsymbol{x} = \boldsymbol{H}\boldsymbol{s} + \boldsymbol{n} \qquad (4.40)$$

in which $\boldsymbol{H}$ is a given MIMO channel, the channel input $\boldsymbol{s}$ is an unknown random vector to be estimated, and $\boldsymbol{n}$ is a random noise vector with zero mean and $\boldsymbol{P}_{\text{nn}} = \sigma_n^2 \boldsymbol{I}$, and uncorrelated with $\boldsymbol{s}$. For this case,

$$\boldsymbol{P}_{\text{xx}} = \boldsymbol{H}\boldsymbol{P}_{\text{ss}}\boldsymbol{H}^T + \sigma_n^2 \boldsymbol{I}_M, \quad \boldsymbol{P}_{\text{xs}} = \boldsymbol{H}\boldsymbol{P}_{\text{ss}} = \boldsymbol{P}_{\text{sx}}^T.$$

The minimum MSE can be shown to be

$$J(\boldsymbol{F}^\star) = \text{Tr}\left((\boldsymbol{P}_{\text{ss}}^{-1} + \boldsymbol{H}^T\boldsymbol{H}/\sigma_n^2)^{-1}\right) \quad \text{(by (4.39) and (1.79))}. \qquad (4.41)$$

As $N = 1$, $\boldsymbol{P}_{\text{ss}} = \sigma_s^2$ (i.e., single-input multiple-output (SIMO) in the linear model $\boldsymbol{x}$ in (4.40)), and $\boldsymbol{H}$ and $\boldsymbol{F}$ reduce to vectors $\boldsymbol{h} \in \mathbb{R}^M$ and $\boldsymbol{f} \in \mathbb{R}^M$, respectively. The optimal $\boldsymbol{f}$ and minimum MSE can be further expressed as

$$\boldsymbol{f}^\star = \frac{\sigma_s^2 \boldsymbol{h}}{\sigma_s^2 \|\boldsymbol{h}\|_2^2 + \sigma_n^2} \quad \text{(by (4.38) and (1.79))} \qquad (4.42)$$

$$J(\boldsymbol{f}^\star) = \min_{\boldsymbol{f}} \left\{ \sigma_s^2 |1 - \boldsymbol{f}^T\boldsymbol{h}|^2 + \sigma_n^2 \boldsymbol{f}^T\boldsymbol{f} \right\} \quad \text{(an unconstrained QP)}$$

$$= \frac{\sigma_s^2}{1 + \sigma_s^2 \|\boldsymbol{h}\|_2^2 / \sigma_n^2} = \frac{\sigma_s^2}{1 + \text{SNR}} \quad \text{(by (4.41))} \qquad (4.43)$$

where SNR denotes the signal-to-noise ratio in $\boldsymbol{x} \in \mathbb{R}^M$. It can been seen that for the SIMO case, the minimum MSE decreases with SNR, and the optimal LMMSE $\boldsymbol{f}^\star$, irrespective of SNR, is always matched with the channel $\boldsymbol{h}$.   □

Remark 4.2 Consider an optimization problem with the objective function $f(\boldsymbol{x} = (\boldsymbol{y}, \boldsymbol{z}))$ and constraint set $\mathcal{C} = \mathcal{C}_1 \times \mathcal{C}_2 = \{(\boldsymbol{y}, \boldsymbol{z}) \mid \boldsymbol{y} \in \mathcal{C}_1, \boldsymbol{z} \in \mathcal{C}_2\}$. Then we have

$$\inf_{\boldsymbol{x} \in \mathcal{C}} f(\boldsymbol{x}) = \inf_{\boldsymbol{y} \in \mathcal{C}_1} \left\{ \inf_{\boldsymbol{z} \in \mathcal{C}_2} f(\boldsymbol{y}, \boldsymbol{z}) \right\}. \qquad (4.44)$$

In other words, one can solve the inner optimization problem first (which is usually easier to solve) by treating $\boldsymbol{y}$ as a fixed parameter to get the inner optimal value $f(\boldsymbol{y}, \boldsymbol{z}^\star(\boldsymbol{y}))$ and the optimal $\boldsymbol{z}^\star$ in terms of $\boldsymbol{y}$, and then solve the outer optimization problem to obtain the optimal $\boldsymbol{y}^\star$. Thus the optimal $\boldsymbol{x}^\star = (\boldsymbol{y}^\star, \boldsymbol{z}^\star(\boldsymbol{y}^\star))$. This approach is quite useful when f is not convex in $\boldsymbol{x}$, while it is convex in $\boldsymbol{y}$ and in $\boldsymbol{z}$, respectively, or only convex in $\boldsymbol{z}$.   □

Example 4.3 *(Affine set fitting for dimension reduction and noise suppression [CMCW08, MCCW10])* Let us consider the widely known dimension reduction problem by the affine set fitting. Given a high-dimensional set of data $\{\mathbf{x}_1, \mathbf{x}_2, \ldots, \mathbf{x}_L\} \subset \mathbb{R}^M$, we would like to find an affine set $\{\mathbf{x} = \mathbf{C}\boldsymbol{\alpha} + \mathbf{d} \mid \boldsymbol{\alpha} \in \mathbb{R}^{N-1}\}$ of affine dimension being $N - 1 \ll M$ (cf. (2.7)), as an approximation to the given data set in the least-squares approximation error sense. In other words, we would like to find the projection of the given high-dimensional data set on the low-dimensional affine set with least-squares approximation error.

Let us define

$$\mathcal{X} = \{\mathbf{x}_1, \ldots, \mathbf{x}_L\} \subset \mathbb{R}^M \quad \text{(data set or data cloud)}$$
$$\mathbf{X} = [\mathbf{x}_1, \ldots, \mathbf{x}_L] \in \mathbb{R}^{M \times L} \quad \text{(data matrix)}$$
$$q(\mathbf{X}) = \text{affdim}(\mathcal{X}) \leq M \quad \text{(i.e., affine dimension of } \mathcal{X}).$$

This affine set fitting problem can be defined as follows:

$$p^\star = \min \|\mathbb{X} - \mathbf{X}\|_{\text{F}}^2$$
$$\text{s.t. } \mathbb{X} = [\boldsymbol{x}_1, \ldots, \boldsymbol{x}_L] \in \mathbb{R}^{M \times L}, \ q(\mathbb{X}) = N - 1$$

$$= \min_{\substack{\boldsymbol{\alpha}_i \in \mathbb{R}^{N-1}, \mathbf{d} \in \mathbb{R}^M, \\ \mathbf{C}^T\mathbf{C} = \mathbf{I}_{N-1}}} \sum_{i=1}^{L} \|(\mathbf{C}\boldsymbol{\alpha}_i + \mathbf{d}) - \mathbf{x}_i\|_2^2 \quad \text{(by (2.7))} \qquad (4.45)$$

where $L \gg N$ and $M \gg N$ and $\mathbf{C} \in \mathbb{R}^{M \times (N-1)}$ is a semi-unitary matrix. The optimal solution of the problem (4.45) in terms of the associated dimension reduced data set is denoted as

$$\mathcal{A} \triangleq \{\widehat{\boldsymbol{\alpha}}_1, \widehat{\boldsymbol{\alpha}}_2, \ldots, \widehat{\boldsymbol{\alpha}}_L\} \subset \mathbb{R}^{N-1}$$

where $\widehat{\boldsymbol{\alpha}}_i$ maps to $\mathbf{x}_i \in \mathcal{X}$ in a one-to-one manner. It can be easily inferred that as $q(\mathbf{X}) \leq (N-1)$, the optimal value $p^\star = 0$ and the optimal $\mathbb{X}^\star = \mathbf{X}$; otherwise, $p^\star > 0$ and the optimal $\mathbb{X}^\star \neq \mathbf{X}$.

This dimension reduction process for complexity and noise reduction of the ensuing processing (e.g., hyperspectral unmixing, endmember extraction, abundance estimation) has been successfully applied in hyperspectral imaging in remote sensing recently (to be introduced in Subsection 6.3.2), because high-dimensional hyperspectral image data $\mathcal{X}$ (which ideally (no noise) will lie in an affine hull with affine dimension equal to $N - 1$) are usually obtained with several hundreds of spectral bands (M) while the true number (N) of substances in a scene of interest is usually a value of several tens. It is remarkable that if **conv** $\mathcal{X} \subset \mathbb{R}^M$ is a simplex with N vertexes (provided that N is known or has been estimated), then **conv** $\mathcal{A} \subset \mathbb{R}^{N-1}$ is a simplest simplex with N vertexes (as illustrated in Figure 4.3). To extract the information embedded in the dimension reduced hyperspectral data set $\mathcal{A}$ is computationally much more efficient than that embedded in the original hyperspectral data set $\mathcal{X}$ with no loss of infor-

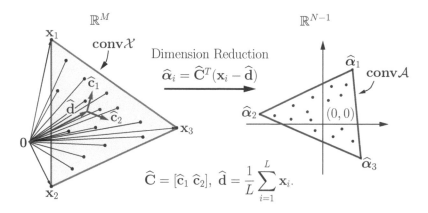

Figure 4.3 Dimension reduction illustration using affine set fitting for $N = 3$, where the geometric center $\widehat{\mathbf{d}}$ of the data cloud $\mathcal{X}$ in the M-dimensional space maps to the origin in the $(N-1)$-dimensional space which is also the geometric center of the dimension-reduced data cloud $\mathcal{A}$.

mation, hence having been an essential step in the signal processing algorithm design for hyperspectral unmixing.

The optimal solution to problem (4.45) is given by

$$\widehat{\mathbf{d}} = \frac{1}{L}\sum_{i=1}^{L}\mathbf{x}_i, \tag{4.46}$$

$$\widehat{\boldsymbol{\alpha}}_i = \widehat{\mathbf{C}}^T(\mathbf{x}_i - \widehat{\mathbf{d}}), \tag{4.47}$$

$$\widehat{\mathbf{C}} = [\widehat{\mathbf{c}}_1, \dots, \widehat{\mathbf{c}}_{N-1}], \tag{4.48}$$

where $\widehat{\mathbf{c}}_i$ is the unit-norm eigenvector associated with the ith principal eigenvalue of $\mathbf{U}\mathbf{U}^T$, denoted as $\lambda_i(\mathbf{U}\mathbf{U}^T)$, where $\mathbf{U}$ is the mean-removed data matrix

$$\mathbf{U} \triangleq \left[\mathbf{x}_1 - \widehat{\mathbf{d}}, \dots, \mathbf{x}_L - \widehat{\mathbf{d}}\right] \in \mathbb{R}^{M \times L}. \tag{4.49}$$

Note that the matrix

$$\frac{1}{L}\mathbf{U}\mathbf{U}^T = \frac{1}{L}\sum_{i=1}^{L}\left(\mathbf{x}_i - \widehat{\mathbf{d}}\right)\left(\mathbf{x}_i - \widehat{\mathbf{d}}\right)^T$$

is exactly the *sample covariance matrix* if the data in $\mathcal{X}$ are random with the same distribution function. The above optimal solution given by (4.46), (4.47), and (4.48) of the affine set fitting problem (4.45) is also the solution of the well-known *principal component analysis* (PCA) [And63, WAR+09]. We would like to emphasize that, in contrast to the PCA, the randomness of data is never a premise to the affine set fitting problem considered in this example. Next, let us derive the solution of problem (4.45) through the use of the optimality criterion (4.28), and prove its equivalence to the problem considered by the PCA.

The optimal solution of (4.45) can be derived as follows:

$$p^\star = \min_{\mathbf{C}^T\mathbf{C}=\mathbf{I}_{N-1}} \left\{ \min_{\mathbf{d}\in\mathbb{R}^M} \left(\sum_{i=1}^{L} \min_{\boldsymbol{\alpha}_i\in\mathbb{R}^{N-1}} \left\| \mathbf{C}\boldsymbol{\alpha}_i - (\mathbf{x}_i - \mathbf{d}) \right\|_2^2 \right) \right\}$$

$$= \min_{\mathbf{C}^T\mathbf{C}=\mathbf{I}_{N-1}} \left\{ \min_{\mathbf{d}\in\mathbb{R}^M} \sum_{i=1}^{L} \left\| \mathbf{C}\widehat{\boldsymbol{\alpha}}_i - (\mathbf{x}_i - \mathbf{d}) \right\|_2^2 \right\} \quad \text{(by (4.28))}$$

$$= \min_{\mathbf{C}^T\mathbf{C}=\mathbf{I}_{N-1}} \left\{ \sum_{i=1}^{L} \left(\mathbf{x}_i - \widehat{\mathbf{d}} \right)^T (\mathbf{I}_M - \mathbf{C}\mathbf{C}^T) \left(\mathbf{x}_i - \widehat{\mathbf{d}} \right) \right\} \quad \text{(by (4.28))}$$

$$= \min_{\mathbf{C}^T\mathbf{C}=\mathbf{I}_{N-1}} \text{Tr}\left([\mathbf{I}_M - \mathbf{C}\mathbf{C}^T]\mathbf{U}\mathbf{U}^T \right) \quad \text{(since (4.49))}. \tag{4.50}$$

Note that (4.50) is exactly the problem that the PCA, a statistical approach, tries to solve the dimension reduction problem in many signal processing applications. Moreover, (4.50) can be further simplified as

$$p^\star = \min_{\mathbf{C}^T\mathbf{C}=\mathbf{I}_{N-1}} \text{Tr}(\mathbf{U}\mathbf{U}^T) - \text{Tr}(\mathbf{C}^T(\mathbf{U}\mathbf{U}^T)\mathbf{C})$$

$$= \sum_{i=N}^{M} \lambda_i \left(\mathbf{U}\mathbf{U}^T \right) \quad \text{(by (1.94) and (1.96))}. \tag{4.51}$$

Hence the optimal solution of (4.45) can also be expressed as

$$\mathbb{X}^\star = \left[\boldsymbol{x}_1^\star, \ldots, \boldsymbol{x}_L^\star \right] \in \mathbb{R}^{M\times L},$$

where

$$\boldsymbol{x}_i^\star = \widehat{\mathbf{C}}\widehat{\boldsymbol{\alpha}}_i + \widehat{\mathbf{d}}, \ i = 1, \ldots, L, \tag{4.52}$$

in which $\widehat{\mathbf{d}}$ is given by (4.46) and $\widehat{\boldsymbol{\alpha}}_i$ is given by (4.47). If $\dim(\mathcal{R}([\widehat{\mathbf{C}}, \widehat{\mathbf{d}}])) = N$ (i.e., $\widehat{\mathbf{C}}\widehat{\mathbf{C}}^T\widehat{\mathbf{d}} \neq \widehat{\mathbf{d}}$), then $\text{rank}(\mathbb{X}^\star) = N$; otherwise $\text{rank}(\mathbb{X}^\star) = N - 1$. The associated "noise-reduced" data set in the original data space $\mathbb{R}^M$ is

$$\mathcal{X}^\star = \left\{ \boldsymbol{x}_i^\star, i = 1, \ldots, L \right\} \subset \mathbb{R}^M$$

with the affine dimension $q(\mathbb{X}^\star) = N - 1$.

If we are given a set of noise-contaminated data, $\mathbf{Y} = \mathbf{X} + \mathbf{V}$ instead of the noise-free data $\mathbf{X}$ with $q(\mathbf{X}) = N - 1$, where $\mathbf{V}$ is the noise matrix, thereby making $q(\mathbf{Y}) \gg N - 1$. The obtained $\mathcal{X}^\star$ is also a noise-reduced counterpart of $\mathcal{X}$. In the presence of additive noise in the data, the determination of N itself is a challenging problem in hyperspectral imaging. We will present some cutting-edge results on the estimation of N based on convex geometry and optimization in Section 6.5 later. This example also justifies that the affine setting fitting and the PCA are in fact equivalent, in spite of different philosophies behind them. □

Remark 4.3 *(Low rank matrix approximation)* The optimum low rank approximation to the data matrix $\mathbf{X} \in \mathbb{R}^{M \times L}$ (where $L \gg M$) can be expressed as

$$\widehat{\mathbf{Z}} = \arg \; \min \; \|\mathbf{Z} - \mathbf{X}\|_{\mathrm{F}}^2$$
$$\text{s.t. rank}(\mathbf{Z}) \le N, \; \mathbf{Z} \in \mathbb{R}^{M \times L} \tag{4.53}$$

which, however, is not a convex problem due to the nonconvex constraint set. Nevertheless, let us present how to solve it via the use of the first-order optimality condition (4.29). Because, by (1.63), any matrix $\mathbf{Z} \in \mathbb{R}^{M \times L}$ with $\text{rank}(\mathbf{Z}) \le N$ can be expressed as

$$\mathbf{Z} = \mathbf{CY} \in \mathbb{R}^{M \times L}, \; \mathbf{C}^T \mathbf{C} = \mathbf{I}_N, \; \mathbf{C} \in \mathbb{R}^{M \times N}, \; \mathbf{Y} \in \mathbb{R}^{N \times L},$$

the problem (4.53) is equivalent to

$$(\widehat{\mathbf{C}}, \widehat{\mathbf{Y}}) = \arg \min_{\mathbf{C}^T \mathbf{C} = \mathbf{I}_N} \left\{ \min_{\mathbf{Y} \in \mathbb{R}^{N \times L}} \|\mathbf{CY} - \mathbf{X}\|_{\mathrm{F}}^2 \right\}$$

$$= \arg \min_{\mathbf{C}^T \mathbf{C} = \mathbf{I}_N} \left\{ \min_{\mathbf{Y} \in \mathbb{R}^{N \times L}} \text{Tr}\left(\mathbf{Y}^T \mathbf{Y} - 2\mathbf{Y}^T \mathbf{C}^T \mathbf{X} + \mathbf{X}^T \mathbf{X}\right) \right\}$$

$$\Rightarrow \begin{cases} \widehat{\mathbf{C}} = \arg \min_{\mathbf{C}^T \mathbf{C} = \mathbf{I}_N} \text{Tr}\left([\mathbf{I}_M - \mathbf{CC}^T]\mathbf{XX}^T\right) \\ \widehat{\mathbf{Y}} = \widehat{\mathbf{C}}^T \mathbf{X} \end{cases} \quad \text{(by (4.29))}$$

(i.e., $\mathbf{U}$ in (4.50) and (4.51) replaced by $\mathbf{X}$ for obtaining $\widehat{\mathbf{C}}$)

where the inner minimization is an unconstrained convex QP w.r.t. $\mathbf{Y}$ and thus (4.29) can be applied, and $\widehat{\mathbf{C}} = [\mathbf{u}_1, \dots, \mathbf{u}_N]$ contains N principal orthonormal eigenvectors of $\mathbf{XX}^T$ (cf. (4.48)). Hence, we come up with the same result as the optimal low rank matrix approximation (1.121) by convex optimization as follows:

$$\widehat{\mathbf{Z}} = \widehat{\mathbf{C}}\widehat{\mathbf{Y}} = \widehat{\mathbf{C}}\widehat{\mathbf{C}}^T \mathbf{X} = \widehat{\mathbf{C}}\widehat{\mathbf{C}}^T \sum_{i=1}^{\text{rank}(\mathbf{X})} \sigma_i \mathbf{u}_i \mathbf{v}_i^T = \sum_{i=1}^{N} \sigma_i \mathbf{u}_i \mathbf{v}_i^T,$$

where σ_i is the ith principal singular value associated with the left singular vector $\mathbf{u}_i$ and right singular vector $\mathbf{v}_i$ of $\mathbf{X}$. Moreover,

$$\left\|\widehat{\mathbf{Z}} - \mathbf{X}\right\|_{\mathrm{F}}^2 = \sum_{i=N+1}^{\text{rank}(\mathbf{X})} \sigma_i^2 \le p^\star = \sum_{i=N}^{\text{rank}(\mathbf{X})} \lambda_i\left(\mathbf{UU}^T\right), \tag{4.54}$$

where $p^\star$ is the least-squares approximation error of the affine set fitting problem (4.45), simply because the feasible set for (4.45), i.e., $\mathbf{aff}\{\boldsymbol{x}_1, \dots, \boldsymbol{x}_L\} \subset \mathbb{R}^{M \times L}$ with affine dimension being $N - 1$, is a subset of the feasible set for problem (4.53), namely, $\{\mathbf{Z} \in \mathbb{R}^{M \times L} \mid \text{rank}(\mathbf{Z}) \le N\}$. $\square$

Example 4.4 Consider the following constrained optimization problem:

$$p^\star = \min_{\boldsymbol{\lambda} \in \mathbb{R}^n_+, \nu \in \mathbb{R}} \nu + \sum_{i=1}^n e^{\lambda_i - \nu - 1}$$

$$= \min_{\boldsymbol{\lambda} \succeq 0} \left\{ \min_{\nu \in \mathbb{R}} \nu + \sum_{i=1}^n e^{\lambda_i - \nu - 1} \right\} \quad \text{(by (4.44))} \tag{4.55}$$

The inner unconstrained minimization problem is convex and can be solved by applying the optimality condition given by (4.28), yielding the optimal

$$\nu^\star = \log \left(\sum_{i=1}^n e^{\lambda_i} \right) - 1.$$

Substituting the optimal $\nu^\star$ into (4.55) gives rise to

$$p^\star = \min_{\boldsymbol{\lambda} \succeq 0} \left(f_0(\boldsymbol{\lambda}) \triangleq \log \sum_{i=1}^n e^{\lambda_i} \right)$$

which can be easily seen to be a convex problem. Then it is straightforward to obtain

$$\nabla f_0(\boldsymbol{\lambda}) = \frac{1}{\sum_{i=1}^n e^{\lambda_i}} [e^{\lambda_1}, \ldots, e^{\lambda_n}]^T \succ \mathbf{0}_n, \ \forall \ \boldsymbol{\lambda} \in \mathbb{R}^n_+.$$

According to the first-order optimality condition given by (4.23), i.e.,

$$\nabla f_0(\boldsymbol{\lambda}^\star)^T (\boldsymbol{\lambda} - \boldsymbol{\lambda}^\star) = (\boldsymbol{\lambda} - \boldsymbol{\lambda}^\star)^T \nabla f_0(\boldsymbol{\lambda}^\star) \geq 0, \ \forall \ \boldsymbol{\lambda} \in \mathbb{R}^n_+, \tag{4.56}$$

and $\nabla f_0(\boldsymbol{\lambda}^\star) \in \mathbf{int}\, K$ (where $K = \mathbb{R}^n_+$ is a self-dual proper cone), it can be easily inferred by *(P2)* (cf. (2.121b)) that (4.56) must hold true with the strict inequality for any $(\boldsymbol{\lambda} - \boldsymbol{\lambda}^\star) \in K^* \setminus \{\mathbf{0}_n\}$, where $\boldsymbol{\lambda}, \boldsymbol{\lambda}^\star \in K^* = \mathbb{R}^n_+$, implying that the optimal $\boldsymbol{\lambda}^\star = \mathbf{0}_n$ (which is the unique minimal point in $\mathbb{R}^n_+$). Therefore, the optimal value and the solution are, respectively, given by

$$p^\star = f_0(\boldsymbol{\lambda}^\star = \mathbf{0}_n) = \log n \tag{4.57}$$
$$(\boldsymbol{\lambda}^\star, \nu^\star) = (\mathbf{0}_n, (\log n) - 1).$$

This example also illustrates that generalized inequalities are useful in the first-order optimization condition. $\quad\square$

Remark 4.4 *(Projected subgradient/gradient method)* Consider the following convex optimization problem

$$\min_{\mathbf{x} \in \mathbb{R}^n} f(\mathbf{x})$$
$$\text{s.t. } \mathbf{x} \in \mathcal{C} \tag{4.58}$$

where $f(\mathbf{x}) : \mathbb{R}^n \to \mathbb{R}$ is a nonsmooth convex function and the feasible set $\mathcal{C}$ is closed. Hence, all the optimality conditions that require f to be differentiable (e.g., first-order optimality condition and KKT conditions[1]) are no longer applicable to finding the optimal solution.

The iterative subgradient method can be applied to find a solution to problem (4.58). At each iteration, the unknown variable $\mathbf{x}$ is updated toward the direction of a negative subgradient and then is projected onto the feasible set $\mathcal{C}$, in order to reach a better solution of problem (4.58). Let $\mathbf{x}^{(k)}$ denote the kth iterate and $\bar{\nabla}f(\mathbf{x}^{(k)})$ a subgradient of f at $\mathbf{x}^{(k)}$. Then $\mathbf{x}^{(k+1)}$ is updated by

$$\mathbf{x}^{(k+1)} = \mathbb{P}_{\mathcal{C}}\left\{\mathbf{x}^{(k)} - s^{(k)}\bar{\nabla}f(\mathbf{x}^{(k)})\right\} \tag{4.59}$$

where $s^{(k)} > 0$ is a chosen step size, and

$$\mathbb{P}_{\mathcal{C}}\{\mathbf{x}\} = \arg\, \mathbf{dist}_C(\mathbf{x}) = \arg\min_{\mathbf{v}\in\mathcal{C}}\|\mathbf{x} - \mathbf{v}\|_2 \quad (\text{cf. } (3.91)) \tag{4.60}$$

(which is also a convex problem); namely, $\mathbb{P}_{\mathcal{C}}\{\cdot\}$ is a projection operator that maps a given point $\mathbf{x}$ to a point in $\mathcal{C}$ with least-squares approximation error if $\mathbf{x} \notin \mathcal{C}$, or $\|\mathbb{P}_{\mathcal{C}}\{\mathbf{x}\} - \mathbf{x}\|_2$ is the distance between the point $\mathbf{x}$ and the set $\mathcal{C}$. When $\mathbf{x} \in \mathcal{C}$, $\mathbb{P}_{\mathcal{C}}\{\mathbf{x}\} = \mathbf{x}$.

Suitable choices for the step size $s^{(k)}$ are required in order to guarantee the convergence of $\mathbf{x}^{(k)}$ to an ϵ-optimal solution. Two such choices are constant step size ($s^{(k)} = s$) and constant step length ($s^{(k)} = s/\|\bar{\nabla}f(\mathbf{x}^{(k)})\|_2$). For these two types of step size, the subgradient algorithm is guaranteed to converge and yield an ϵ-optimal solution, i.e.,

$$\lim_{k\to\infty} f(\mathbf{x}^{(k)}) - \min_{\mathbf{x}\in\mathcal{C}} f(\mathbf{x}) \le \epsilon,$$

where ϵ is a decreasing function of the step size parameter s, which is problem dependent. Furthermore, when the step size $s^{(k)}$ satisfies one of the following conditions

- $\sum_{k=1}^{\infty}(s^{(k)})^2 < \infty$ and $\sum_{k=1}^{\infty}s^{(k)} = \infty$,
- $\lim_{k\to\infty}s^{(k)} = 0$ and $\sum_{k=1}^{\infty}s^{(k)} = \infty$,

the subgradient algorithm is guaranteed to converge and to end up with an optimal solution of problem (4.58). Moreover, as f is differentiable, the subgradient method reduces to the *projected gradient method* that can be guaranteed to yield an optimal solution provided that a small enough constant step size is used. It has been known that the projected gradient method converges linearly if the objective function is Lipschitz continuous and strongly convex [Dun81]. □

---

[1]KKT conditions, a set of equalities and/or inequalities, for optimally solving a convex problem and its dual problem under strong duality when the objective function and all the constraint functions are differentiable will be introduced in detail in Chapter 9 later.

4.3   Equivalent representations and transforms

This section introduces how to convert the optimization problem under consideration into a convex problem (which is our goal), at the expense of more variables and more constraints involved, i.e., a larger problem size. This facilitates the access to all the available convex solvers for the performance evaluation before efficient algorithm design and implementation. To this end, let us introduce four widely used methods for the problem reformulation, including epigrapgh form, equality constraint elimination, function transformation, and change of variables in the subsequent subsections, respectively.

4.3.1   Equivalent problem: Epigraph form

It is often convenient to express a standard optimization problem (4.1) in epigraph form (introduced in Subsection 3.1.5).

$$
\begin{aligned}
\min \ \ & t \\
\text{s.t.} \ \ & f_0(\mathbf{x}) - t \leq 0, \\
& f_i(\mathbf{x}) \leq 0, \ i = 1, \ldots, m \\
& h_i(\mathbf{x}) = 0, \ i = 1, \ldots, p
\end{aligned}
\tag{4.61}
$$

where a new variable t is introduced, called an *auxiliary variable*. Now the minimization problem is w.r.t. $(\mathbf{x}, t)$ rather than $\mathbf{x}$ alone, i.e., both $\mathbf{x}$ and t are unknown variables. Note that if the original optimization problem (4.1) is convex as given by (4.11), the extra inequality constraint function $f_0(\mathbf{x}) - t$ is convex in $(\mathbf{x}, t)$ and the corresponding constraint set is exactly **epi** f_0 (convex), and hence the problem remains convex. It can be seen that for the optimal solution $(\mathbf{x}^\star, t^\star)$, the extra inequality must be active, i.e.,

$$
t^\star = f_0(\mathbf{x}^\star),
$$

implying that $(\mathbf{x}^\star, t^\star)$ is exactly the optimal solution and the optimal value of problem (4.11).

Example 4.5 The minimization of an unconstrained piecewise-linear objective function (which is convex) can be expressed in epigraph form as follows

$$
\min_{\mathbf{x}\in\mathbb{R}^n} \ \max_{i=1,\ldots,m} \{\mathbf{a}_i^T\mathbf{x} + b_i\} =
\begin{cases}
\min \ t \\
\text{s.t.} \ \max_{i=1,\ldots,m}\{\mathbf{a}_i^T\mathbf{x} + b_i\} \leq t, \ \mathbf{x} \in \mathbb{R}^n
\end{cases}
$$

$$
=
\begin{cases}
\min \ t \\
\text{s.t.} \ \mathbf{a}_i^T\mathbf{x} + b_i \leq t, \ i = 1, \ldots, m, \ \mathbf{x} \in \mathbb{R}^n.
\end{cases}
\tag{4.62}
$$

Obviously, this problem is also a constrained convex optimization problem since both the objective function and all the inequality constraint functions are affine functions, and actually it is an LP, which will be discussed in detail in Chapter 6. This example also indicates that a convex problem can be formulated in multiple equivalent forms. One may prefer one form for analysis and another form for algorithm development.   □

4.3.2   Equivalent problem: Equality constraint elimination

Linear equalities in the convex problem (4.11) can be equivalently expressed as follows:

$$\mathbf{Ax} = \mathbf{b} \Leftrightarrow \mathbf{x} = \mathbf{A}^\dagger \mathbf{b} + \mathbf{v}, \quad \mathbf{v} \in \mathcal{N}(\mathbf{A}) \tag{4.63}$$

$$\Leftrightarrow \mathbf{x} = \mathbf{A}^\dagger \mathbf{b} + \mathbf{Fz}, \quad \mathbf{z} \in \mathbb{R}^d$$

for some $\mathbf{F} \in \mathbb{R}^{n \times d}$ such that $\mathcal{R}(\mathbf{F}) = \mathcal{N}(\mathbf{A})$ and $d = \dim(\mathcal{N}(\mathbf{A})) = n - \operatorname{rank}(\mathbf{A})$ (cf. (1.71)). So the standard convex problem given by (4.11) can be rewritten as

$$\begin{aligned} \min \quad & f_0(\mathbf{A}^\dagger \mathbf{b} + \mathbf{Fz}) \\ \text{s.t.} \quad & f_i(\mathbf{A}^\dagger \mathbf{b} + \mathbf{Fz}) \le 0, \ i = 1, \ldots, m, \ \mathbf{z} \in \mathbb{R}^d. \end{aligned} \tag{4.64}$$

Note that both the objective function f_0 and all the constraint functions f_i in (4.11) are convex functions in $\mathbf{x}$, implying that the reformulated problem (4.64) is also a convex problem with the unknown variable $\mathbf{z}$ due to the affine mapping between $\mathbf{x}$ and $\mathbf{z}$. After solving problem (4.64) to obtain the optimal $\mathbf{z}^\star$, one can obtain the optimal solution of (4.11) as

$$\mathbf{x}^\star = \mathbf{A}^\dagger \mathbf{b} + \mathbf{Fz}^\star. \tag{4.65}$$

Example 4.6 Consider the convex problem

$$\begin{aligned} \min \quad & \mathbf{x}^T \mathbf{Rx} \\ \text{s.t.} \quad & \mathbf{Ax} = \mathbf{b}, \end{aligned} \tag{4.66}$$

where $\mathbf{R} \in \mathbb{S}^n_+$. It can be expressed as the following unconstrained convex optimization problem

$$\min_{\mathbf{z} \in \mathbb{R}^d} \ (\mathbf{x}_0 + \mathbf{Fz})^T \mathbf{R}(\mathbf{x}_0 + \mathbf{Fz}), \tag{4.67}$$

where $\mathbf{x}_0 = \mathbf{A}^\dagger \mathbf{b}$. Suppose that $\mathbf{F}^T \mathbf{RF} \succ \mathbf{0}$. By the optimality condition given by (4.28), the optimal solution to problem (4.67) can be easily obtained as

$$\mathbf{z}^\star = -(\mathbf{F}^T \mathbf{RF})^{-1} \mathbf{F}^T \mathbf{Rx}_0. \tag{4.68}$$

Hence, we have

$$\mathbf{x}^\star = \mathbf{x}_0 + \mathbf{Fz}^\star = \left(\mathbf{I} - \mathbf{F} \left(\mathbf{F}^T \mathbf{RF} \right)^{-1} \mathbf{F}^T \mathbf{R} \right) \mathbf{A}^\dagger \mathbf{b}, \tag{4.69}$$

which is the optimal solution to problem (4.66).   □

4.3.3   Equivalent problem: Function transformation

Suppose $\psi_0 : \mathbb{R} \to \mathbb{R}$ is monotone (i.e., strictly) increasing, $\psi_i : \mathbb{R} \to \mathbb{R}$, $i = 1, \ldots, m$ satisfy $\psi_i(u) \le 0$ if and only if $u \le 0$, and $\psi_i : \mathbb{R} \to \mathbb{R}$, $i = m +$

$1, \ldots, m + p$ satisfy $\psi_i(u) = 0$ if and only if $u = 0$. The standard optimization problem given by (4.1) is equivalent to

$$\begin{aligned} \min \quad & \psi_0(f_0(\mathbf{x})) \\ \text{s.t.} \quad & \psi_i(f_i(\mathbf{x})) \leq 0, \ i = 1, \ldots, m \\ & \psi_{m+i}(h_i(\mathbf{x})) = 0, \ i = 1, \ldots, p. \end{aligned} \quad (4.70)$$

By means of function transformation, we intend to convert the original optimization problem (4.1) into a solvable convex optimization problem (4.70). Such transformations are certainly not unique as illustrated in the following examples.

- The inequality constraint

$$\log x \leq 1 \quad (4.71)$$

  is nonconvex. However, via the exponential function transformation $\psi(u) = e(e^u - 1) \leq 0$ where $u \leq 0$, with u replaced by $(\log x) - 1$ we have

$$\log x \leq 1 \Leftrightarrow x \leq e, \quad (4.72)$$

  which leads to a convex constraint.

- Consider the following inequality constraints

$$1 - x_1 x_2 \leq 0, \ x_1 \geq 0, \ x_2 \geq 0, \quad (4.73)$$

  where the inequality constraint function $1 - x_1 x_2$ is nonconvex (which can be proved by using the second-order condition). However, through the function transformation $\psi(u) = -\log(1 - u) \leq 0$ where $u \leq 0$, with u replaced by $1 - x_1 x_2$, we can reformulate (4.73) as

$$-\log x_1 - \log x_2 \leq 0, \ x_1 \geq 0, \ x_2 \geq 0, \quad (4.74)$$

  leading to a convex constraint set. Alternatively, the above nonconvex constraints in (4.73) can also be rewritten as a convex PSD matrix constraint

$$\begin{bmatrix} x_1 & 1 \\ 1 & x_2 \end{bmatrix} = x_1 \begin{bmatrix} 1 & 0 \\ 0 & 0 \end{bmatrix} + x_2 \begin{bmatrix} 0 & 0 \\ 0 & 1 \end{bmatrix} + \begin{bmatrix} 0 & 1 \\ 1 & 0 \end{bmatrix} \succeq \mathbf{0}. \quad (4.75)$$

  One can easily verify that the 2×2 matrix on the left-hand side of the generalized inequality (4.75) is PSD if and only if all the three inequalities (4.73) are true. Note that the PSD constraint given by (4.75) is actually a linear matrix inequality (LMI), a matrix inequality with the unknown variables linearly involved in the constraint function. The LMI constraints that are convex constraints can be seen in many communication problems.

- The unconstrained 2-norm problem:

$$\min \ \|\mathbf{A}\mathbf{x} - \mathbf{b}\|_2 \quad (4.76)$$

  is equivalent to the LS problem $\min \|\mathbf{A}\mathbf{x} - \mathbf{b}\|_2^2$. In fact, we want to avoid (4.76), which exhibits less desirable differentiation properties, so that the optimality condition given in (4.28) can be applied to find the optimal solution.

- Suppose that $\mathbf{P} \succ \mathbf{0}$ and $\mathbf{x}^T\mathbf{P}\mathbf{x} > 0$ for all $\mathbf{x} \in \mathcal{C}$. The problem

$$\max_{\mathbf{x}\in\mathcal{C}} \frac{1}{\mathbf{x}^T\mathbf{P}\mathbf{x}}, \tag{4.77}$$

which is apparently a nonconvex problem and is equivalent to

$$\min_{\mathbf{x}\in\mathcal{C}} \mathbf{x}^T\mathbf{P}\mathbf{x} \quad (\text{cf. } (4.8)), \tag{4.78}$$

which is then a convex problem provided that $\mathcal{C}$ is a convex set.

- Maximum-likelihood (ML) estimation [Men95, Kay13] of mean:
  Consider the likelihood function of m Gaussian vector samples $\mathbf{x}_i \in \mathbb{R}^n$ (assumed to be independent and identically distributed) with mean vector $\boldsymbol{\mu}$ and covariance matrix $\boldsymbol{\Sigma}$, defined as their joint probability density function as follows:

$$L(\boldsymbol{\mu},\boldsymbol{\Sigma}) = \frac{1}{(2\pi)^{\frac{nm}{2}}(\det\boldsymbol{\Sigma})^{\frac{m}{2}}} \exp\left\{-\frac{1}{2}\sum_{i=1}^{m}(\mathbf{x}_i-\boldsymbol{\mu})^T\boldsymbol{\Sigma}^{-1}(\mathbf{x}_i-\boldsymbol{\mu})\right\}. \tag{4.79}$$

Given $\boldsymbol{\Sigma}$, the ML estimation of $\boldsymbol{\mu}$ is the maximization problem as follows

$$\max_{\boldsymbol{\mu}} L(\boldsymbol{\mu},\boldsymbol{\Sigma}). \tag{4.80}$$

We often put logarithm (which is a monotone increasing function) on $L(\boldsymbol{\mu},\boldsymbol{\Sigma})$ to obtain an equivalent problem

$$\max_{\boldsymbol{\mu}} \log L(\boldsymbol{\mu},\boldsymbol{\Sigma}) \equiv \max_{\boldsymbol{\mu}} -\sum_{i=1}^{m}(\mathbf{x}_i-\boldsymbol{\mu})^T\boldsymbol{\Sigma}^{-1}(\mathbf{x}_i-\boldsymbol{\mu})$$
$$= -\min_{\boldsymbol{\mu}} \sum_{i=1}^{m}(\mathbf{x}_i-\boldsymbol{\mu})^T\boldsymbol{\Sigma}^{-1}(\mathbf{x}_i-\boldsymbol{\mu}) \tag{4.81}$$

(where all the terms not involving $\boldsymbol{\mu}$ are redundant and so can be disposed of) which is an unconstrained convex optimization problem in $\boldsymbol{\mu}$. The optimal $\widehat{\boldsymbol{\mu}}$ can be easily shown (by the optimality condition (4.28)) to be

$$\widehat{\boldsymbol{\mu}} = \frac{1}{m}\sum_{i=1}^{m}\mathbf{x}_i, \tag{4.82}$$

which is also known as the sample mean.

Let us conclude this subsection with a simple practical example in wireless communications and networking.

Example 4.7 *(Optimal power assignment)* In this application example, consider the scenario in which there are K transmitters and K receivers (see Figure 4.4). Transmitter i sends signals to receiver i, and the other transmitters are interferers to Transmitter i.

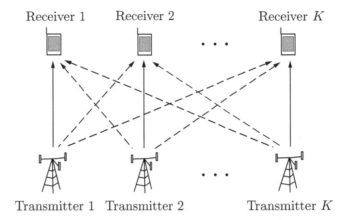

Figure 4.4 Illustration of power control: K transmitters and K receivers case.

The signal-to-interference-plus-noise ratio (SINR) at receiver i is given by

$$\gamma_i = \frac{G_{ii}p_i}{\sum_{j=1,j\neq i}^{K} G_{ij}p_j + \sigma_i^2},\qquad(4.83)$$

where

- p_i is the power of transmitter i;
- G_{ij} is the channel power gain from transmitter j to receiver i;
- σ_i^2 is the noise power at receiver i.

The problem of interest is a *power minimization problem*, that is, to minimize the average transmitter power, subject to a constraint that all SINRs are no less than a prespecified threshold γ_0, which can be defined as the following optimization problem:

$$
\begin{aligned}
\min\ & \sum_{i=1}^{K} p_i \\
\text{s.t.}\ & \frac{G_{ii}p_i}{\sum_{j=1,j\neq i}^{K} G_{ij}p_j + \sigma_i^2} \geq \gamma_0,\ i=1,\ldots,K \\
& p_i \geq 0,\ i=1,\ldots,K
\end{aligned}
\qquad(4.84)
$$

Note that the inequality constraint functions in the second line of (4.84) are nonconvex and thus this problem is not yet a convex optimization problem. By applying function transformation to these nonconvex constraints, the problem can be rewritten as

$$\min \ \sum_{i=1}^{K} p_i$$

$$\text{s.t.} \ -G_{ii}p_i + \gamma_0 \sum_{j=1, j \neq i}^{K} G_{ij}p_j + \gamma_0 \sigma_i^2 \leq 0, \ i = 1, \ldots, K \quad (4.85)$$

$$p_i \geq 0, \ i = 1, \ldots, K$$

which is obviously a convex optimization problem in p_i (more specifically an LP) and hence can be solved optimally. $\qquad\square$

4.3.4   Equivalent problem: Change of variables

Suppose $\phi : \mathbb{R}^n \to \mathbb{R}^n$ is one-to-one with image covering the problem domain. Define, by letting $\mathbf{x} = \phi(\mathbf{z})$ in $f_i(\mathbf{x})$ and $h_i(\mathbf{x})$ in (4.1),

$$\tilde{f}_i(\mathbf{z}) = f_i(\phi(\mathbf{z})), \ i = 0, \ldots, m, \ \tilde{h}_i(\mathbf{z}) = h_i(\phi(\mathbf{z})), \ i = 1, \ldots, p. \quad (4.86)$$

The standard optimization problem given by (4.1) is equivalent to

$$\begin{aligned}
\min \ & \tilde{f}_0(\mathbf{z}) \\
\text{s.t.} \ & \tilde{f}_i(\mathbf{z}) \leq 0, \ i = 1, \ldots, m \\
& \tilde{h}_i(\mathbf{z}) = 0, \ i = 1, \ldots, p.
\end{aligned} \quad (4.87)$$

If problem (4.87) is convex and thus can be optimally solved, the optimal solution $\mathbf{x}^\star = \phi(\mathbf{z}^\star)$ of problem (4.1) can be readily obtained, as illustrated in the following ML estimation examples.

- ML estimation of covariance matrix:
  Consider the logarithm of the likelihood function defined in (4.79) again. Given $\boldsymbol{\mu}$, the ML estimation of $\boldsymbol{\Sigma}$ is to solve the problem

$$\max_{\boldsymbol{\Sigma} \succ 0} \ -\frac{m}{2} \log(\det \boldsymbol{\Sigma}) - \frac{1}{2} \sum_{i=1}^{m} (\mathbf{x}_i - \boldsymbol{\mu})^T \boldsymbol{\Sigma}^{-1} (\mathbf{x}_i - \boldsymbol{\mu}), \quad (4.88)$$

  where the constant term $-\frac{nm}{2} \log 2\pi$ is ignored without change of ML estimation of $\boldsymbol{\Sigma}$. We have proved in Chapter 3 (see Example 3.13 in Section 3.5) that $\boldsymbol{\Sigma}^{-1}$ is $\mathbb{S}_+^n$-convex on $\mathbb{S}_{++}^n$, implying that $\sum_{i=1}^{m} (\mathbf{x}_i - \boldsymbol{\mu})^T \boldsymbol{\Sigma}^{-1} (\mathbf{x}_i - \boldsymbol{\mu})$ is convex in $\boldsymbol{\Sigma}$ as well (by Fact 3.4), and that $-\log(\det \boldsymbol{\Sigma})$ is convex on $\mathbb{S}_{++}^n$ (see Subsection 3.1.4). Therefore, the objective function in (4.88) is neither convex nor concave, and solving it seems non-trivial. However, change of variables can effectively solve the problem.

  Let $\boldsymbol{\Psi} = \boldsymbol{\Sigma}^{-1} \succ \mathbf{0}$. The ML estimation problem (4.88) then becomes

$$\max_{\boldsymbol{\Psi} \succ 0} \ \frac{m}{2} \log(\det \boldsymbol{\Psi}) - \frac{1}{2} \sum_{i=1}^{m} (\mathbf{x}_i - \boldsymbol{\mu})^T \boldsymbol{\Psi} (\mathbf{x}_i - \boldsymbol{\mu}) \equiv \min_{\boldsymbol{\Psi} \succ 0} \ f_0(\boldsymbol{\mu}, \boldsymbol{\Psi}), \quad (4.89)$$

(by (4.7)) where

$$f_0(\boldsymbol{\mu}, \boldsymbol{\Psi}) \triangleq -\frac{m}{2} \log(\det \boldsymbol{\Psi}) + \frac{1}{2} \sum_{i=1}^{m} (\mathbf{x}_i - \boldsymbol{\mu})^T \boldsymbol{\Psi}(\mathbf{x}_i - \boldsymbol{\mu}). \tag{4.90}$$

Note that $f_0(\boldsymbol{\mu}, \boldsymbol{\Psi})$ is convex in $\boldsymbol{\Psi} \in \mathbb{S}_{++}^n$ and in $\boldsymbol{\mu}$, but nonconvex in $(\boldsymbol{\mu}, \boldsymbol{\Psi})$. Then by deriving the gradient of $f_0(\boldsymbol{\mu}, \boldsymbol{\Psi})$ w.r.t. $\boldsymbol{\Psi}$, i.e., $\nabla_{\boldsymbol{\Psi}} f_0(\boldsymbol{\mu}, \boldsymbol{\Psi})$, and setting it to zero (i.e., by (4.29)), one can easily show that for a given $\boldsymbol{\mu}$, the optimal solution is

$$\widehat{\boldsymbol{\Sigma}} = \widehat{\boldsymbol{\Psi}}^{-1} = \frac{1}{m} \sum_{i=1}^{m} (\mathbf{x}_i - \boldsymbol{\mu})(\mathbf{x}_i - \boldsymbol{\mu})^T, \tag{4.91}$$

which is known as the sample covariance, where the derivation of $\nabla_{\boldsymbol{\Psi}} f_0$ is through the use of (3.40) and (3.46).

- ML estimation of mean and covariance matrix:
  In the previous examples we have considered the estimation of the mean $\boldsymbol{\mu}$ and the variance $\boldsymbol{\Sigma}$, separately. Now let us consider the joint ML estimation of $\boldsymbol{\mu}$ and $\boldsymbol{\Sigma}$. From (4.89), the joint ML estimation problem can be written as

$$\min_{\boldsymbol{\mu} \in \mathbb{R}^n, \boldsymbol{\Psi} \succ \mathbf{0}} f_0(\boldsymbol{\mu}, \boldsymbol{\Psi}) = \min_{\boldsymbol{\Psi} \succ \mathbf{0}} \left(\min_{\boldsymbol{\mu}} f_0(\boldsymbol{\mu}, \boldsymbol{\Psi}) \right) \quad \text{(cf. (4.44) and Remark 4.2)} \tag{4.92}$$

in spite of the nonconvexity of $f_0(\boldsymbol{\mu}, \boldsymbol{\Psi})$ w.r.t. $(\boldsymbol{\mu}, \boldsymbol{\Psi})$. We will first consider the inner minimization, given $\boldsymbol{\Psi}$. The optimality condition is $\nabla_{\boldsymbol{\mu}} f_0(\boldsymbol{\mu}, \boldsymbol{\Psi}) = \mathbf{0}$ (by (4.28)) since $f_0(\boldsymbol{\mu}, \boldsymbol{\Psi})$ is convex in $\boldsymbol{\mu}$. Hence we have

$$\nabla_{\boldsymbol{\mu}} f_0(\boldsymbol{\mu}, \boldsymbol{\Psi}) = \frac{1}{2} \sum_{i=1}^{m} (-2\boldsymbol{\Psi}\mathbf{x}_i + 2\boldsymbol{\Psi}\boldsymbol{\mu}) = \mathbf{0}. \tag{4.93}$$

Given $\boldsymbol{\Psi} \succ \mathbf{0}$, the optimal solution for $\boldsymbol{\mu}$ by solving (4.93) is exactly the sample mean $\widehat{\boldsymbol{\mu}}$ given by (4.82). Note that $\widehat{\boldsymbol{\mu}}$ does not depend on $\boldsymbol{\Psi}$. Now, let us focus on the outer minimization.

Putting the optimal $\widehat{\boldsymbol{\mu}}$ back to $f_0(\boldsymbol{\mu}, \boldsymbol{\Psi})$ in (4.92), we have

$$\min_{\boldsymbol{\Psi} \succ \mathbf{0}} f_0(\widehat{\boldsymbol{\mu}}, \boldsymbol{\Psi}) = \min_{\boldsymbol{\Psi} \succ \mathbf{0}} -\frac{m}{2} \log(\det \boldsymbol{\Psi}) + \frac{1}{2} \sum_{i=1}^{m} (\mathbf{x}_i - \widehat{\boldsymbol{\mu}})^T \boldsymbol{\Psi}(\mathbf{x}_i - \widehat{\boldsymbol{\mu}})$$

$$= \min_{\boldsymbol{\Psi} \succ \mathbf{0}} -\frac{m}{2} \log(\det \boldsymbol{\Psi}) + \frac{1}{2} \text{Tr}(\boldsymbol{\Psi}\mathbf{C}), \tag{4.94}$$

where

$$\mathbf{C} = \sum_{i=1}^{m} (\mathbf{x}_i - \widehat{\boldsymbol{\mu}})(\mathbf{x}_i - \widehat{\boldsymbol{\mu}})^T \succeq \mathbf{0}.$$

Applying the optimality condition (4.29) to $f_0(\widehat{\boldsymbol{\mu}}, \boldsymbol{\Psi})$, we get

$$\nabla_{\boldsymbol{\Psi}} f_0(\widehat{\boldsymbol{\mu}}, \boldsymbol{\Psi}) = -\frac{m}{2}\boldsymbol{\Psi}^{-1} + \frac{1}{2}\mathbf{C} = \mathbf{0} \quad \text{(by (3.40) and (3.46))} \qquad (4.95)$$

and thus the optimum ML estimate for $\boldsymbol{\Sigma}$ can be obtained as

$$\widehat{\boldsymbol{\Sigma}} = \widehat{\boldsymbol{\Psi}}^{-1} = \frac{1}{m}\mathbf{C} = \frac{1}{m}\sum_{i=1}^{m}(\mathbf{x}_i - \widehat{\boldsymbol{\mu}})(\mathbf{x}_i - \widehat{\boldsymbol{\mu}})^T. \qquad (4.96)$$

In summary, we have proved the widely known fact that the ML estimates $\widehat{\boldsymbol{\mu}}$ and $\widehat{\boldsymbol{\Sigma}}$ are given by (4.82) (the sample mean) and (4.96) (the sample covariance (4.91) with the true mean replaced by the sample mean).

4.3.5   Reformulation of complex-variable problems

Consider an optimization problem given in the form

$$\begin{aligned}
\min_{\mathbf{x}\in\mathbb{C}^n} \quad & f_0(\mathbf{x}) \\
\text{s.t.} \quad & f_i(\mathbf{x}) \leq 0, \ i = 1,\ldots,m \\
& h_i(\mathbf{x}) = 0, \ i = 1,\ldots,p
\end{aligned} \qquad (4.97)$$

with problem domain $\mathcal{D}$ and feasible set $\mathcal{C}$, where $f_i(\mathbf{x})$ and $h_j(\mathbf{x})$ are real-valued functions for all i and j. The first-order optimality condition for a convex problem is given in the following remark and the proof is similar to that for the real-variable case.

Remark 4.5 When $f_i(\mathbf{x})$ is convex for all i and $h_j(\mathbf{x})$ is affine for all j, the problem (4.97) is a convex problem though the unknown variable $\mathbf{x}$ is complex. The optimality condition for an optimal minimizer $\mathbf{x}^\star$ is given by

$$\begin{aligned}
\operatorname{Re}\left\{\nabla f_0(\mathbf{x}^\star)^H(\mathbf{y} - \mathbf{x}^\star)\right\} \geq 0, \ \forall \mathbf{y} \in \mathcal{C} \\
\nabla f_0(\mathbf{x}^\star) = \mathbf{0}, \ \text{if } \mathbf{x}^\star \in \mathbf{int}\ \mathcal{C}
\end{aligned} \qquad (4.98)$$

(the counterpart of (4.23)). When the decision variable is a complex matrix $\mathbf{X}$, the optimality condition for an optimal minimizer $\mathbf{X}^\star$ is given by

$$\begin{aligned}
\operatorname{Tr}\!\left(\operatorname{Re}\left\{\nabla f_0(\mathbf{X}^\star)^H(\mathbf{Y} - \mathbf{X}^\star)\right\}\right) \geq 0, \ \forall \mathbf{Y} \in \mathcal{C} \\
\nabla f_0(\mathbf{X}^\star) = \mathbf{0}, \ \text{if } \mathbf{X}^\star \in \mathbf{int}\ \mathcal{C}
\end{aligned} \qquad (4.99)$$

(the counterpart of (4.27)). $\qquad\qquad\qquad\qquad\qquad\qquad\qquad\qquad\qquad\square$

All the reformulation methods introduced previously in this section can be applied to make a reformulated convex problem [SBL12], and then it can be solved optimally using the first-order optimality condition or KKT conditions, etc. On the other hand, in view of the fact that a complex variable corresponds to the combination of two real variables (i.e., the real part and imaginary part), problem (4.97) with the variable $\mathbf{x} \in \mathbb{C}^n$ can surely be converted into an optimization problem with the associated variable $\boldsymbol{x} \in \mathbb{R}^{2n}$. Then we can proceed with the reformulation of this problem into a convex problem. Next, let us use a simple example to illustrate these two approaches.

Consider the following complex-variable QP:

$$\min_{\mathbf{w} \in \mathbb{C}^n} \ \{f(\mathbf{w}) \triangleq \mathbf{w}^H \mathbf{P} \mathbf{w}\}$$

$$\text{s.t. } h(\mathbf{w}) \triangleq \mathbf{w}^H \mathbf{a} - 1 = 0 \tag{4.100}$$

where $\mathbf{P} \in \mathbb{H}_+^n$. Next, let us show that (4.100) is a convex problem; namely, f is real-valued and convex in $\mathbf{w} \in \mathbb{C}^n$, and h (which is not real-valued in $\mathbf{w} \in \mathbb{C}^n$) can be re-expressed as a set of real-valued affine functions in $\mathbf{w} \in \mathbb{C}^n$.

Since f has been a real-valued function in $\mathbf{w} \in \mathbb{C}^n$, its gradient can be easily obtained as

$$\nabla_{\mathbf{w}} f(\mathbf{w}) = 2 \nabla_{\mathbf{w}^*} f(\mathbf{w}) = 2 \mathbf{P} \mathbf{w} \ \ (\text{cf. } (1.43)).$$

To prove whether f is convex or not, let us use the first-order condition (3.25) as follows:

$$\begin{aligned}
f(\mathbf{w} + \delta \mathbf{w}) &= (\mathbf{w} + \delta \mathbf{w})^H \mathbf{P} (\mathbf{w} + \delta \mathbf{w}) \\
&= \mathbf{w}^H \mathbf{P} \mathbf{w} + \mathrm{Re}\{(2\mathbf{P}\mathbf{w})^H \delta \mathbf{w}\} + (\delta \mathbf{w})^H \mathbf{P} \delta \mathbf{w} \\
&\geq \mathbf{w}^H \mathbf{P} \mathbf{w} + \mathrm{Re}\{(2\mathbf{P}\mathbf{w})^H \delta \mathbf{w}\} \ \ (\text{since } \mathbf{P} \succeq \mathbf{0}) \\
&= f(\mathbf{w}) + \mathrm{Re}\{\nabla_{\mathbf{w}} f(\mathbf{w})^H \delta \mathbf{w}\}
\end{aligned}$$

implying that f is convex in $\mathbf{w} \in \mathbb{C}^n$. Moreover, the equality constraint function h in problem (4.100) can be equivalently represented by two real-valued affine functions of $\mathbf{w} \in \mathbb{C}^n$ as follows:

$$\begin{cases} h_1(\mathbf{w}) = \mathrm{Re}\{h(\mathbf{w})\} = \mathrm{Re}\{\mathbf{w}^H \mathbf{a} - 1\} = \frac{1}{2}(\mathbf{w}^H \mathbf{a} + \mathbf{a}^H \mathbf{w}) - 1 \\ h_2(\mathbf{w}) = \mathrm{Im}\{h(\mathbf{w})\} = \mathrm{Im}\{\mathbf{w}^H \mathbf{a} - 1\} = \frac{1}{2}(-j\mathbf{w}^H \mathbf{a} + j\mathbf{a}^H \mathbf{w}). \end{cases} \tag{4.101}$$

It can be seen that $h(\mathbf{w}) = 0$ if and if only $h_1(\mathbf{w}) = 0$ and $h_2(\mathbf{w}) = 0$. Hence problem (4.100) is a convex problem (cf. Remark 4.5). However, to solve problem (4.100) using the first-order optimality condition given by (4.98) may not be easy, while the closed-form solution instead can be obtained by solving the associated KKT conditions (cf. Example 9.14).

Let $\mathbf{P}_R = \mathrm{Re}\{\mathbf{P}\}$ and $\mathbf{P}_I = \mathrm{Im}\{\mathbf{P}\}$. Then we have

$$\mathbf{P} = \mathbf{P}_R + j\mathbf{P}_I = \mathbf{P}^H = \mathbf{P}_R^T - j\mathbf{P}_I^T,$$

which implies that

$$\begin{aligned}
&\mathbf{P}_R = \mathbf{P}_R^T, \\
&\mathbf{P}_I = -\mathbf{P}_I^T \ \Rightarrow \ [\mathbf{P}_I]_{ij} = -[\mathbf{P}_I]_{ji} \ \Rightarrow \ [\mathbf{P}_I]_{ii} = 0 \ \forall i.
\end{aligned}$$

Let $\mathbf{w}_R = \text{Re}\{\mathbf{w}\}$ and $\mathbf{w}_I = \text{Im}\{\mathbf{w}\}$, and $\mathbf{w} = \mathbf{w}_R + j\mathbf{w}_I$. Then the objective function in (4.100) becomes

$$
\begin{aligned}
\mathbf{w}^H \mathbf{P} \mathbf{w} &= (\mathbf{w}_R^T - j\mathbf{w}_I^T)(\mathbf{P}_R + j\mathbf{P}_I)(\mathbf{w}_R + j\mathbf{w}_I) \\
&= \mathbf{w}_R^T \mathbf{P}_R \mathbf{w}_R - \mathbf{w}_R^T \mathbf{P}_I \mathbf{w}_I + \mathbf{w}_I^T \mathbf{P}_R \mathbf{w}_I + \mathbf{w}_I^T \mathbf{P}_I \mathbf{w}_R \\
&= \begin{bmatrix} \mathbf{w}_R^T & \mathbf{w}_I^T \end{bmatrix} \begin{bmatrix} \mathbf{P}_R & -\mathbf{P}_I \\ \mathbf{P}_I & \mathbf{P}_R \end{bmatrix} \begin{bmatrix} \mathbf{w}_R \\ \mathbf{w}_I \end{bmatrix} \geq 0 \quad (\text{since } \mathbf{P} \succeq \mathbf{0}),
\end{aligned}
$$

where we have used the fact that $\mathbf{w}_R^T \mathbf{P}_I \mathbf{w}_R = 0$ and $\mathbf{w}_I^T \mathbf{P}_I \mathbf{w}_I = 0$ due to $\mathbf{P}_I = -\mathbf{P}_I^T$. Coming to the constraints in (4.100), we have

$$
\begin{aligned}
\mathbf{w}^H \mathbf{a} &= (\mathbf{w}_R + j\mathbf{w}_I)^H (\text{Re}\{\mathbf{a}\} + j\text{Im}\{\mathbf{a}\}) = 1 \\
&\Leftrightarrow \text{Re}\{\mathbf{w}^H \mathbf{a}\} = 1, \ \text{Im}\{\mathbf{w}^H \mathbf{a}\} = 0 \\
&\Leftrightarrow \begin{bmatrix} \mathbf{w}_R^T & \mathbf{w}_I^T \end{bmatrix} \begin{bmatrix} \text{Re}\{\mathbf{a}\} \\ \text{Im}\{\mathbf{a}\} \end{bmatrix} = 1, \ \begin{bmatrix} \mathbf{w}_R^T & \mathbf{w}_I^T \end{bmatrix} \begin{bmatrix} \text{Im}\{\mathbf{a}\} \\ -\text{Re}\{\mathbf{a}\} \end{bmatrix} = 0
\end{aligned}
$$

(which, respectively, are real-variable counterparts of the complex-variable equality constraints $h_1(\mathbf{w}) = 0$ and $h_2(\mathbf{w}) = 0$ in (4.101)). Thus problem (4.100) can be converted to a real convex QP with the problem domain $\mathcal{D} = \mathbb{R}^{2n}$ and two linear inequality constraints as follows:

$$
\begin{aligned}
\min_{\boldsymbol{w} \in \mathbb{R}^{2n}} \ & \{f(\boldsymbol{w}) \triangleq \boldsymbol{w}^T \boldsymbol{P} \boldsymbol{w}\} \\
\text{s.t.} \ & \boldsymbol{w}^T \boldsymbol{a}_1 = 1, \ \boldsymbol{w}^T \boldsymbol{a}_2 = 0
\end{aligned}
\tag{4.102}
$$

where

$$
\boldsymbol{w} = \begin{bmatrix} \mathbf{w}_R \\ \mathbf{w}_I \end{bmatrix}, \ \boldsymbol{P} = \begin{bmatrix} \mathbf{P}_R & -\mathbf{P}_I \\ \mathbf{P}_I & \mathbf{P}_R \end{bmatrix} = \begin{bmatrix} \mathbf{P}_R & \mathbf{P}_I^T \\ \mathbf{P}_I & \mathbf{P}_R \end{bmatrix} \succeq \mathbf{0} \quad (\text{cf. Remark 1.20}),
$$

and

$$
\boldsymbol{a}_1 = \begin{bmatrix} \text{Re}\{\mathbf{a}\} \\ \text{Im}\{\mathbf{a}\} \end{bmatrix}, \ \boldsymbol{a}_2 = \begin{bmatrix} \text{Im}\{\mathbf{a}\} \\ -\text{Re}\{\mathbf{a}\} \end{bmatrix}.
$$

Problem (4.102) can be easily seen to be convex since Hessian of $f(\boldsymbol{w})$ is $\boldsymbol{P} \succeq \mathbf{0}$ and the two equalities' constraint functions are both affine. Though the problem conversion from the complex-variable form to the real-variable form is straightforward (as illustrated in this example), it could be quite complicated for a nonconvex problem.

Some more correspondences between the complex-variable form and the real-variable form that are useful in the reformulation of constraints of a complex-variable optimization problem are given as follows. Let

$$\mathbf{A} = \mathbf{A}_R + j\mathbf{A}_I \in \mathbb{C}^{m \times n}, \quad \mathcal{A} = \begin{bmatrix} \mathbf{A}_R & -\mathbf{A}_I \\ \mathbf{A}_I & \mathbf{A}_R \end{bmatrix} \in \mathbb{R}^{2m \times 2n}$$

$$\mathbf{X} = \mathbf{X}_R + j\mathbf{X}_I \in \mathbb{H}^n, \quad \mathcal{X} = \begin{bmatrix} \mathbf{X}_R & -\mathbf{X}_I \\ \mathbf{X}_I & \mathbf{X}_R \end{bmatrix} \in \mathbb{S}^{2n}$$

$$\mathbf{x} = \mathbf{x}_R + j\mathbf{x}_I \in \mathbb{C}^n, \quad \boldsymbol{x} = [\mathbf{x}_R^T, \mathbf{x}_I^T]^T \in \mathbb{R}^{2n}$$

$$\mathbf{b} = \mathbf{b}_R + j\mathbf{b}_I \in \mathbb{C}^m, \quad \boldsymbol{b} = [\mathbf{b}_R^T, \mathbf{b}_I^T]^T \in \mathbb{R}^{2m}$$

$$\mathbf{c} = \mathbf{c}_R + j\mathbf{c}_I \in \mathbb{C}^n, \quad \boldsymbol{c} = [\mathbf{c}_R^T, \mathbf{c}_I^T]^T \in \mathbb{R}^{2n}$$

$$d = d_R + jd_I \in \mathbb{C}.$$

Then following the same conversion as illustrated in the above complex-variable QP example, one can show that a complex SOC constraint and its counterpart are given by

$$\begin{cases} \|\mathbf{Ax} + \mathbf{b}\|_2 \leq \mathbf{c}^H\mathbf{x} + d \\ \text{Im}\{\mathbf{c}^H\mathbf{x} + d\} = 0 \end{cases} \iff \begin{cases} \|\mathcal{A}\boldsymbol{x} + \boldsymbol{b}\|_2 \leq \boldsymbol{c}^T\boldsymbol{x} + d_R \\ [-\mathbf{c}_I^T, \mathbf{c}_R^T]\,\boldsymbol{x} + d_I = 0 \end{cases} \quad (4.103)$$

and a complex LMI and a complex linear matrix equality and their real counterparts are as follows:

$$\begin{cases} \mathbf{AXA}^H + \mathbf{DYD}^H \succeq \mathbf{0} \iff \mathcal{A}\mathcal{X}\mathcal{A}^T + \mathcal{B}\mathcal{Y}\mathcal{B}^T \succeq \mathbf{0} \\ \text{Tr}(\mathbf{CX}) = t \in \mathbb{R}, \; \mathbf{C} \in \mathbb{H}^n \iff \text{Tr}(\mathcal{C}\mathcal{X}) = 2t \end{cases} \quad \text{(by Remark 1.20)}$$

$$(4.104)$$

where $\mathbf{X}$, $\mathbf{Y}$, and $\mathbf{x}$ are complex variables and $\mathbf{A}$, $\mathbf{B}$, $\mathbf{C}$, $\mathbf{b}$, $\mathbf{c}$, and d are fixed complex parameters, and the pairs of $(\mathbf{B}, \mathcal{B})$ and $(\mathbf{C}, \mathcal{C})$, and the pair of $(\mathbf{Y}, \mathcal{Y})$, are defined similarly as the pair of $(\mathbf{A}, \mathcal{A})$, and the pair of $(\mathbf{X}, \mathcal{X})$, respectively. Note that (4.103) exhibits a second-order cone and a hyperplane in complex-variable form and the corresponding real-variable form, and (4.104) exhibits a LMI and a linear equality in complex-variable form and the corresponding real-variable form. In general, real-variable and complex-variable mathematical expressions for constraints (such as (4.103) and (4.104)) and those for problems (such as the pair of (4.100) and (4.102)) are similar in both form and type. In other words, the reformulation of a complex-variable optimization problem can proceed in the same fashion as that of a real-variable optimization problem without need of conversion from the complex domain to the real domain ahead of time.

Next, let us present an interesting example of solving a convex problem of complex variable by the first-order optimality condition (4.98). Recall that Schur complement for the real case has been shown in Subsection 3.2.5. The proof for the corresponding complex case can be formulated as a complex-variable optimization problem as illustrated in the following example.

Example 4.8 *(Complex Schur complement)* Suppose that $\mathbf{C} \in \mathbb{H}_{++}^m$, $\mathbf{A} \in \mathbb{H}^n$, and $\mathbf{B} \in \mathbb{C}^{n \times m}$. Then

$$\mathbf{S} \triangleq \begin{bmatrix} \mathbf{A} & \mathbf{B} \\ \mathbf{B}^H & \mathbf{C} \end{bmatrix} \succeq \mathbf{0} \tag{4.105}$$

if and only if $\mathbf{S_C} \triangleq \mathbf{A} - \mathbf{BC}^{-1}\mathbf{B}^H \succeq \mathbf{0}$. Since the proof of sufficiency becomes trivial from the proof of necessity, we only prove the latter.

Proof of Necessity: Since $\mathbf{S} \succeq \mathbf{0}$, we have

$$f(\mathbf{x}, \mathbf{y}) \triangleq [\mathbf{x}^H \ \mathbf{y}^H] \, \mathbf{S} \begin{bmatrix} \mathbf{x} \\ \mathbf{y} \end{bmatrix}$$

$$= \mathbf{x}^H\mathbf{A}\mathbf{x} + \mathbf{x}^H\mathbf{B}\mathbf{y} + \mathbf{y}^H\mathbf{B}^H\mathbf{x} + \mathbf{y}^H\mathbf{C}\mathbf{y} \geq 0 \ \ \forall (\mathbf{x}, \mathbf{y}) \in \mathbb{C}^{n+m},$$

which is known to be convex in $(\mathbf{x}, \mathbf{y})$, and convex in $\mathbf{y}$ since $\mathbf{C} \succ \mathbf{0}$ (cf. (4.100)). Consider the complex-variable minimization problem

$$\mathbf{y}^\star = \arg \inf_{\mathbf{y} \in \mathbb{C}^m} f(\mathbf{x}, \mathbf{y}).$$

By the first-order optimization condition (4.98), we have

$$\nabla_{\mathbf{y}} f(\mathbf{x}, \mathbf{y}) = \ 2\nabla_{\mathbf{y}^*} f(\mathbf{x}, \mathbf{y}) = 2\mathbf{B}^H\mathbf{x} + 2\mathbf{C}\mathbf{y} = \mathbf{0}$$

$$\Rightarrow \ \mathbf{y}^\star = -\mathbf{C}^{-1}\mathbf{B}^H\mathbf{x}.$$

Then we have

$$\inf_{\mathbf{y} \in \mathbb{C}^m} f(\mathbf{x}, \mathbf{y}) = f(\mathbf{x}, \mathbf{y}^\star) = \mathbf{x}^H(\mathbf{A} - \mathbf{B}\mathbf{C}^{-1}\mathbf{B}^H)\mathbf{x}$$

$$= \mathbf{x}^H\mathbf{S_C}\mathbf{x} \geq 0 \ \forall \mathbf{x} \in \mathbb{C}^n,$$

implying that $\mathbf{S_C} \succeq \mathbf{0}$. ■ □

For applications in wireless communications and networks, not only real variables (e.g., transmission power and transmission rates) but also complex variables (e.g., beamforming vectors) are involved in the optimization problem of interest in general. Both reformulation approaches (i.e., with or without the conversion of the original problem into a real-variable problem beforehand) have been used, depending on the problem under consideration. Many examples and applications are presented in Chapters 7 and 8 for each reformulation approach. However, to obtain a numerical solution of the resulting convex problem (if no closed-form solution can be found), the conversion to the corresponding real-variable convex problem usually makes the algorithm design (such as primal-dual interior-point method to be introduced in Chapter 10) easier and more practical (especially when Hessian matrices are involved), such as off-the-shelf convex solvers CVX and SeDuMi that are also designed after this conversion.

4.4   Convex problems with generalized inequalities

This section introduces convex problems involving generalized inequalities, including the case that the feasible set involves generalized inequality constraints, and the case that the objective function, which itself is a vector or a matrix, involves generalized inequality. The problem for the second case is called a *vector optimization* problem.

4.4.1   Convex problems with generalized inequalitiy constraints

A convex problem with generalized inequalities can be defined as follows

$$\begin{aligned} \min\ & f_0(\mathbf{x}) \\ \text{s.t.}\ & \mathbf{f}_i(\mathbf{x}) \preceq_{K_i} \mathbf{0},\ i=1,\dots,m \\ & \mathbf{Ax} = \mathbf{b} \end{aligned} \tag{4.106}$$

where $\preceq_{K_i}$ is defined on a proper cone $K_i \subset \mathbb{R}^{n_i}$, f_0 is convex, and $\mathbf{f}_i$, $i = 1,2,\dots,m$ are K_i-convex. As the ordinary convex optimization problem defined in (4.11), the feasible set of (4.106)

$$\mathcal{C} = \{\mathbf{x} \mid \mathbf{x} \in \mathcal{D},\ \mathbf{f}_i(\mathbf{x}) \preceq_{K_i} \mathbf{0},\ i=1,\dots,m,\ \mathbf{Ax}=\mathbf{b}\}$$

is also convex, and any local optimal point of (4.106) is globally optimal, and the optimality condition presented in Subsection 4.2.2 also applies to problem (4.106). Some commonly known convex problems involving generalized inequality constraints are given below.

- Cone program: The standard form is defined as [BV04]

$$\begin{aligned} \min\ & \mathbf{c}^T\mathbf{x} \\ \text{s.t.}\ & \mathbf{x} \succeq_K \mathbf{0},\ \mathbf{Ax}=\mathbf{b} \end{aligned} \tag{4.107}$$

and the inequality form is defined as

$$\begin{aligned} \min\ & \mathbf{c}^T\mathbf{x} \\ \text{s.t.}\ & \mathbf{Fx}+\mathbf{g} \preceq_K \mathbf{0}. \end{aligned} \tag{4.108}$$

As the proper cone is the nonnegative orthant (i.e., $K = \mathbb{R}^n_+$), (4.107) and (4.108) are referred to as the standard form and the inequality form of LP, respectively.

- Second-order cone program (SOCP) (special case of (4.107)):

$$\begin{aligned} \min\ & \mathbf{c}_0^T\mathbf{x} \\ \text{s.t.}\ & (\mathbf{A}_i\mathbf{x}+\mathbf{b}_i, \mathbf{c}_i^T\mathbf{x}+d_i) \succeq_{K_i} \mathbf{0},\ i=1,\dots,m \\ & \mathbf{Fx}=\mathbf{g} \end{aligned} \tag{4.109}$$

where

$$K_i = \{(\mathbf{y},t) \in \mathbb{R}^{n_i+1} \mid \|\mathbf{y}\|_2 \le t\}$$

is a second-order cone. Note that the above generalized inequality constraint associated with K_i means that $(\mathbf{A}_i\mathbf{x} + \mathbf{b}_i, \mathbf{c}_i^T\mathbf{x} + d_i) \in K_i$ (cf. (2.78a)), i.e., $\|\mathbf{A}_i\mathbf{x} + \mathbf{b}_i\|_2 \le \mathbf{c}_i^T\mathbf{x} + d_i$.

- Semidefinite program (SDP): The standard form is defined as

$$
\begin{aligned}
\min \quad & \mathrm{Tr}(\mathbf{CX}) \\
\text{s.t.} \quad & \mathbf{X} \succeq_K \mathbf{0}, \\
& \mathrm{Tr}(\mathbf{A}_i\mathbf{X}) = b_i, \ i = 1, \ldots, p
\end{aligned}
\tag{4.110}
$$

where the variable $\mathbf{X} \in \mathbb{S}^n$, and $\mathbf{C}, \mathbf{A}_1, \ldots, \mathbf{A}_p \in \mathbb{S}^n$, and $K = \mathbb{S}_+^n$. The inequality form is defined as

$$
\begin{aligned}
\min \quad & \mathbf{c}^T\mathbf{x} \\
\text{s.t.} \quad & x_1\mathbf{A}_1 + \cdots + x_n\mathbf{A}_n \preceq \mathbf{B}
\end{aligned}
\tag{4.111}
$$

where the variable $\mathbf{x} \in \mathbb{R}^n$, and $\mathbf{c} \in \mathbb{R}^n$, and $\mathbf{B}, \mathbf{A}_1, \ldots, \mathbf{A}_n \in \mathbb{S}^k$.

How to formulate a problem into an LP or an SOCP or an SDP and their effective applications in communications and signal processing will be presented in Chapters 6, 7, and 8, respectively.

4.4.2   Vector optimization

A vector optimization problem is defined as

$$
\begin{aligned}
\min \ (\text{w.r.t. } K) \quad & \boldsymbol{f}_0(\mathbf{x}) \\
\text{subject to} \quad & f_i(\mathbf{x}) \le 0, \ i = 1, \ldots, m \\
& h_i(\mathbf{x}) = 0, \ i = 1, \ldots, p
\end{aligned}
\tag{4.112}
$$

where $\mathbf{x} \in \mathbb{R}^n$ is the optimization variable, $K \subset \mathbb{R}^q$ is a proper cone, the objective function is $\boldsymbol{f}_0 : \mathbb{R}^n \to \mathbb{R}^q$, $f_i : \mathbb{R}^n \to \mathbb{R}$, $i = 1, \ldots, m$, and $h_i(\mathbf{x})$, $i = 1, \ldots, p$ are inequality constraint functions and equality constraint functions, respectively.

Let $\mathcal{D}$ denote the problem domain, $\mathcal{C}$ denote the feasible set (cf. (4.3)), and $\mathcal{O}$ denote the set of $\boldsymbol{f}_0(\mathbf{x})$ ($q \times 1$ vectors) for all feasible points $\mathbf{x}$ in $\mathcal{C}$, i.e.,

$$
\begin{aligned}
\mathcal{O} &= \boldsymbol{f}_0(\mathcal{C}) \\
&= \{\boldsymbol{f}_0(\mathbf{x}) \mid \exists \mathbf{x} \in \mathcal{D}, \ f_i(\mathbf{x}) \le 0, \ i = 1, \ldots, m, \ h_i(\mathbf{x}) = 0, \ i = 1, \ldots, p\} \subset \mathbb{R}^q
\end{aligned}
\tag{4.113}
$$

which is also the set of achievable objective values (cf. Figure 4.5). A feasible point $\mathbf{x}^\star$ is optimal if $\boldsymbol{f}_0(\mathbf{x}^\star) \preceq_K \boldsymbol{f}_0(\mathbf{y})$ for all feasible $\mathbf{y}$, i.e., $\boldsymbol{f}_0(\mathbf{x}^\star)$ is a minimum element of the set $\mathcal{O}$ (cf. Figure 2.15), or equivalently

$$
\mathcal{O} \subseteq \{\boldsymbol{f}_0(\mathbf{x}^\star)\} + K \ \Leftrightarrow \ \mathcal{O} - \{\boldsymbol{f}_0(\mathbf{x}^\star)\} \subseteq K \quad (\text{by (2.83)}).
\tag{4.114}
$$

However, problem (4.112) may not have any optimal point and optimal value. For this case, a feasible point $\mathbf{x}^\star$ is *Pareto optimal* (or *Pareto efficient*) if $\boldsymbol{f}_0(\mathbf{x}^\star)$

is a minimal element (cf. Figure 2.16) of the set $\mathcal{O}$. Apparently, the set of Pareto optimal points,

$$\mathcal{C}_P \triangleq \{\mathbf{x} \in \mathcal{C} \mid \boldsymbol{f}_0(\mathbf{y}) \not\preceq \boldsymbol{f}_0(\mathbf{x}) \ \forall \mathbf{y} \in \mathcal{C}, \mathbf{y} \neq \mathbf{x}\} \quad \text{(cf. (2.86))}, \qquad (4.115)$$

is a subset of $\mathcal{C}$. For the case of $K = \mathbb{R}_+^q$, the Pareto optimality for $\mathbf{x}^\star \in \mathcal{C}_P$ can be interpreted as: it is impossible to decrease any element of the associated objective value $\boldsymbol{f}_0(\mathbf{x}^\star)$ (a $q \times 1$ vector) without increasing any value of the other $q - 1$ elements of $\boldsymbol{f}_0(\mathbf{x}^\star)$.

A point $\mathbf{x}^\star \in \mathcal{C}_P$ is Pareto optimal if and only if it is feasible and

$$(\{\boldsymbol{f}_0(\mathbf{x}^\star)\} - K) \cap \mathcal{O} = \{\boldsymbol{f}_0(\mathbf{x}^\star)\} \quad \text{(cf. (2.85) and Figure 2.16).} \qquad (4.116)$$

A vector optimization problem can have many Pareto optimal values and points. The set of Pareto optimal values, denoted by $\mathcal{P}$, must satisfy

$$\mathcal{P} = \boldsymbol{f}_0(\mathcal{C}_P) \subseteq (\mathcal{O} \cap \mathbf{bd}\ \mathcal{O}) \qquad (4.117)$$

(since the set $\mathcal{O}$ may be not closed and not open in the meantime); namely, every Pareto optimal value is an achievable objective vector that not only belongs to the set $\mathcal{O}$ but also lies on the boundary of the set $\mathcal{O}$. The set $\mathcal{P}$ is therefore called the *Pareto boundary* in some applications. An illustration for the Pareto boundary $\mathcal{P}$ and $K = \mathbb{R}_+^2$ is shown in the top plot of Figure 4.5.

Finding the set $\mathcal{C}_P$ and the associated set $\mathcal{P}$ is very important in some applications. A way to find a point of the set $\mathcal{C}_P$ is to solve the following scalar optimization problem

$$\mathbf{x}^\circ = \arg\min\ \{\boldsymbol{\lambda}^T \boldsymbol{f}_0(\mathbf{x}),\ \mathbf{x} \in \mathcal{C}\} \qquad (4.118)$$

where $\boldsymbol{\lambda} \succeq_{K^*} \mathbf{0}$ and $\boldsymbol{\lambda} \neq \mathbf{0}$ (cf. *(P1)* given by (2.121a)). Let

$$\boldsymbol{y}^\circ(\boldsymbol{\lambda}) \triangleq \boldsymbol{f}_0(\mathbf{x}^\circ) = \arg\min\ \{\boldsymbol{\lambda}^T \boldsymbol{y} \mid \boldsymbol{y} \in \mathcal{O}\}. \qquad (4.119)$$

Note that problem (4.119) may not be a convex problem with unknown variable $\boldsymbol{y}$ since the feasible set $\mathcal{O}$ may not be convex, and that $\boldsymbol{y}^\circ(\boldsymbol{\lambda})$ may not be unique. Nevertheless, in general, $\boldsymbol{y}^\circ(\boldsymbol{\lambda})$ is a minimal point of $\mathcal{O}$ (i.e., a Pareto optimal value of problem (4.112)) due to $\boldsymbol{\lambda} \succeq_{K^*} \mathbf{0}$ and $\boldsymbol{\lambda} \neq \mathbf{0}$ (cf. the bottom left plot in Figure 4.5 for $K = K^* = \mathbb{R}_+^2$).

Surely, to find the entire set $\mathcal{C}_P$, the approach of solving problem (4.118) by varying the weight vector $\boldsymbol{\lambda}$ over the set $K^* \setminus \{\mathbf{0}\}$ may not be very computationally efficient simply because $\mathcal{C}_P$ is usually constituted by continuous subsets of $\mathcal{C}$ in a high-dimensional domain. Effective and efficient algorithms for finding $\mathcal{C}_P$ are still a subject for challenging research in wireless communications and networking. Before presenting an illustrative example, two worthy remarks about finding Pareto optimal solutions via solving (4.118) are as follows.

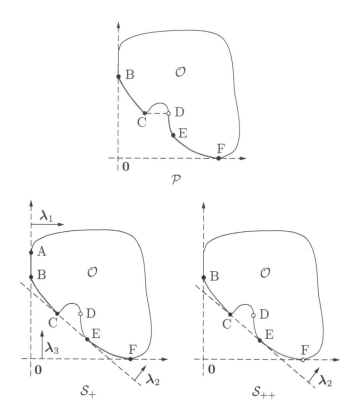

Figure 4.5 An illustration in $\mathbb{R}^2$ with the proper cone $K = \mathbb{R}^2_+$ in problem (4.112) for the set $\mathcal{O} = \boldsymbol{f}_0(\mathcal{C})$ (the achievable objective vectors) defined in (4.113), the Pareto boundary $\mathcal{P} = \boldsymbol{f}_0(\mathcal{C}_P)$ (top plot in blue color), the set $\mathcal{S}_+$ defined in (4.120) (bottom left plot in green color), and the set $\mathcal{S}_{++}$ defined in (4.121) (bottom right plot in red color excluding point F since $\mathcal{P}$ is smooth at F), where the line segment from point A to point B (at which $\mathcal{P}$ is nonsmooth) is the set of $\boldsymbol{y}^\circ(\boldsymbol{\lambda}_1)$, the set of $\boldsymbol{y}^\circ(\boldsymbol{\lambda}_2)$ contains only points C and E, and point F is exactly $\boldsymbol{y}^\circ(\boldsymbol{\lambda}_3)$.

Remark 4.6 Let

$$\mathcal{S}_+ = \{\boldsymbol{y}^\circ(\boldsymbol{\lambda}) \mid \boldsymbol{\lambda} \succeq_{K^\star} \boldsymbol{0}, \boldsymbol{\lambda} \neq \boldsymbol{0}\}, \tag{4.120}$$

$$\mathcal{S}_{++} = \{\boldsymbol{y}^\circ(\boldsymbol{\lambda}) \mid \boldsymbol{\lambda} \succ_{K^\star} \boldsymbol{0}\}, \tag{4.121}$$

where $\boldsymbol{y}^\circ(\boldsymbol{\lambda}) = \boldsymbol{f}_0(\mathbf{x}^\circ)$ (cf. (4.119)) in which $\mathbf{x}^\circ$ is an optimal solution of the preceding scalarized problem (4.118). Then

$$\begin{aligned} \mathcal{S}_{++} &\subseteq \mathcal{P}, \\ \mathcal{P} &\subseteq \mathcal{S}_+ \text{ if } \mathcal{O} \text{ is convex;} \end{aligned} \tag{4.122}$$

otherwise, $\mathcal{S}_+$ may only contain some points of $\mathcal{P}$. In other words, every solution of problem (4.118) with $\boldsymbol{\lambda} \succ_{K^\star} \boldsymbol{0}$ is Pareto optimal to the vector optimization problem (4.112), but it is not always true with $\boldsymbol{\lambda} \succeq_{K^\star} \boldsymbol{0}$ and $\boldsymbol{\lambda} \neq \boldsymbol{0}$. The mapping from $\boldsymbol{\lambda}$ to the point $\boldsymbol{y}^\circ(\boldsymbol{\lambda})$ in $\mathcal{S}_{++}$ and $\mathcal{S}_+$ may not be one-to-one; namely, it is possible that multiple and even infinitely many $y^\circ(\lambda)$ are associated with a certain λ depending on the problem under consideration. An illustration in $\mathbb{R}^2$ with the proper cone $K = \mathbb{R}^2_+$ for the sets $\mathcal{O}$ (achievable objective vectors of problem (4.112)), $\mathcal{P}$ (Pareto boundary), $\mathcal{S}_+$, and $\mathcal{S}_{++}$ is given in Figure 4.5, showing a case of $\mathcal{S}_{++} \subseteq \mathcal{P}$ but $\mathcal{P} \nsubseteq \mathcal{S}_+$. □

Remark 4.7 *(Convex vector optimization)* The vector optimization problem (4.112) is convex when $f_0(\mathbf{x})$ is K-convex and the feasible set $\mathcal{C}$ is a convex set. Then the associated scalar optimization problem (4.118) is also a convex problem (by Remark 3.39 in Section 3.5). For this case, it can be easily shown that the sum of the objective value set $\mathcal{O}$ and the proper cone K is a convex set given by

$$\mathcal{O} + K = \{t \in \mathbb{R}^q \mid \exists \mathbf{x} \in \mathcal{C}, \ f_0(\mathbf{x}) \preceq_K t\}, \tag{4.123}$$

implying that the two sets $\mathcal{O}$ and $\mathcal{O} + K$ have identical Pareto boundary set $\mathcal{P}$, and, thus,

$$\mathcal{S}_{++} \subseteq \mathcal{P} \subseteq \mathcal{S}_+, \tag{4.124}$$

though the set $\mathcal{O}$ is not necessarily a convex set. On the other hand, if the objective function f_0 is differentiable and the minimum point of $\mathcal{O}$ exists, the first-order optimality condition can be shown to be

$$D f_0(\mathbf{x}^\star)(\mathbf{y} - \mathbf{x}^\star) \succeq_K \mathbf{0}, \ \ \forall \mathbf{y} \in \mathcal{C} \tag{4.125}$$

by means of (3.134) and (4.114). □

Example 4.9 Consider the following convex vector optimization problem:

$$\min_{\mathbf{X} \in \mathbb{S}^n} (\text{w.r.t. } \mathbb{S}^n_+) \ \mathbf{X}$$
$$\text{subject to} \quad \mathbf{X} \succeq \mathbf{A}_i, \ i = 1,\ldots,m \tag{4.126}$$

where $\mathbf{A}_i \in \mathbb{S}^n$, $i = 1,\ldots,m$, are given, and we desire to find a Pareto optimal solution $\mathbf{X}^\star$. The corresponding convex scalar optimization problem is given by

$$\min_{\mathbf{X} \in \mathbb{S}^n} \text{Tr}(\mathbf{W}\mathbf{X})$$
$$\text{s.t. } \mathbf{X} \succeq \mathbf{A}_i, \ i = 1,\ldots,m \tag{4.127}$$

which is an SDP. Note that the objective function $\text{Tr}(\mathbf{W}\mathbf{X})$ in (4.127) is $\mathbb{S}^n_+$-nondecreasing for any nonzero weight matrix $\mathbf{W} \succeq \mathbf{0}$. One can obtain a Pareto optimal solution $\mathbf{X}^\star$ by solving (4.127).

Suppose that $\mathbf{A}_i \succ \mathbf{0}$, which uniquely defines an ellipsoid, i.e.,

$$\mathcal{E}_{\mathbf{A}_i} = \{\mathbf{u} \mid \mathbf{u}^T \mathbf{A}_i^{-1} \mathbf{u} \leq 1\} \ \ (\text{cf. (2.37)}).$$

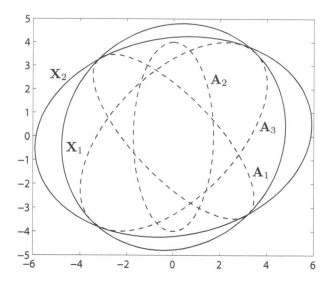

Figure 4.6 Geometric interpretation to problem (4.126). The two ellipsoids corresponding to Pareto optimal solutions $\mathbf{X}_1$ and $\mathbf{X}_2$ (obtained by solving the SDP (4.127) for two different weight matrices $\mathbf{W}_1$ and $\mathbf{W}_2$ given by (4.129), respectively) contain the three ellipsoids corresponding to $\mathbf{A}_1$, $\mathbf{A}_2$, and $\mathbf{A}_3$ given by (4.128).

In Figure 4.6, two Pareto optimal solutions, denoted as $\mathbf{X}_1$ and $\mathbf{X}_2$ for the case of $n = 2$, $m = 3$, and

$$\mathbf{A}_1 = \begin{bmatrix} 12 & -9 \\ -9 & 12 \end{bmatrix}, \ \mathbf{A}_2 = \begin{bmatrix} 3 & 0 \\ 0 & 16 \end{bmatrix}, \ \mathbf{A}_3 = \begin{bmatrix} 16 & 9 \\ 9 & 16 \end{bmatrix}, \tag{4.128}$$

$$\mathbf{W}_1 = \begin{bmatrix} 1 & 0 \\ 0 & 1 \end{bmatrix}, \ \mathbf{W}_2 = \begin{bmatrix} 1 & -1 \\ -1 & 5 \end{bmatrix} \tag{4.129}$$

are illustrated, where $\mathcal{E}_{\mathbf{A}_i} \subseteq \mathcal{E}_{\mathbf{X}_j}$, $i = 1, \ldots, m$, $j = 1, 2$ (due to $\mathbf{X}_j \succeq \mathbf{A}_i$ and by (2.79)). $\qquad\qquad\square$

In contrast to the standard vector optimization problem given by (4.112) which is a minimization problem, another vector optimization problem is a maximization problem given by

$$\begin{aligned} &\max \ (\text{w.r.t. } K) \ \ \boldsymbol{f}_0(\mathbf{x}) \\ &\text{subject to} \qquad f_i(\mathbf{x}) \leq 0, \ i = 1, \ldots, m \\ &\qquad\qquad\qquad h_i(\mathbf{x}) = 0, \ i = 1, \ldots, p \end{aligned} \tag{4.130}$$

which is convex when the objective function $\boldsymbol{f}_0 : \mathbb{R}^n \to \mathbb{R}^q$ is K-concave, i.e., $-\boldsymbol{f}_0$ is K-convex and the feasible set $\mathcal{C}$ is convex. Consider finding either the unique maximum element (if it exists) or maximal elements of the set $\mathcal{O}$ (i.e., the set of all the achievable objective values as defined in (4.113)). To find a maximal element of $\mathcal{O}$ and the associated Pareto optimal solution $\mathbf{x}^\star$, again by scalarization of problem (4.112), one needs to solve the corresponding scalar optimization problem given by

$$\max \ \boldsymbol{\lambda}^T \boldsymbol{f}_0(\mathbf{x})$$
$$\text{s.t.} \quad f_i(\mathbf{x}) \leq 0, \ i = 1, \ldots, m \tag{4.131}$$
$$h_i(\mathbf{x}) = 0, \ i = 1, \ldots, p$$

where $\boldsymbol{\lambda} \succeq_{K^*} \boldsymbol{0}$ and $\boldsymbol{\lambda} \neq \boldsymbol{0}$. Let us conclude this section with the following example for illustrating how to find Pareto optimal solutions of problem (4.130).

Example 4.10 Consider the following convex vector optimization problem:

$$\max \ (\text{w.r.t. } \mathbb{R}_+^n) \ \boldsymbol{R}(\boldsymbol{p}) \triangleq \left(R_1 = \log(1 + \frac{p_1}{\sigma_1^2}), \ldots, R_n = \log(1 + \frac{p_n}{\sigma_n^2}) \right)$$
$$\tag{4.132}$$
$$\text{subject to } \sum_{i=1}^{n} p_i \leq P, \ p_i \geq 0, \ i = 1, \ldots, n$$

where the objective function $\boldsymbol{R}(\boldsymbol{p})$ is $\mathbb{R}_+^n$-concave in $\boldsymbol{p} = (p_1, \ldots, p_n) \in \mathbb{R}^n$, and $\sigma_i^2 > 0, \ i = 1, \ldots, n$, are given parameters. Note that in multichannel wireless communications, R_i denotes the achievable transmission rate (bits/sec/Hz) over subchannel i as the base of the log function is 2, and it depends on the transmit power p_i and the associated channel noise variance σ_i^2 as defined in (4.132). This problem is also a resource allocation problem for the case of no interference between subchannels since each transmission rate R_i is only dependent upon its own channel state information (CSI) (characterized by σ_i^2). For instance, a practical scenario for this case is that all the subchannels are orthogonal, e.g., subchannels in orthogonal frequency division multiplexing (OFDM) systems.

The corresponding convex scalar optimization problem is given by

$$\max \ f(\boldsymbol{R}) \triangleq \boldsymbol{\lambda}^T \boldsymbol{R} = \lambda_1 R_1 + \cdots + \lambda_n R_n$$
$$\tag{4.133}$$
$$\text{s.t.} \ \sum_{i=1}^{n} p_i \leq P, \ p_i \geq 0, \ i = 1, \ldots, n$$

which is apparently a convex problem and the objective function $f(\boldsymbol{R})$ is K-increasing for $\boldsymbol{\lambda} \succ_{K^*} \boldsymbol{0}_n$ and K-nondecreasing for $\boldsymbol{\lambda} \succeq_{K^*} \boldsymbol{0}_n$ where $K^* = K = \mathbb{R}_+^n$. For this case, it can be shown that the set of achievable objective values $\mathcal{O}$ of problem (4.132) is convex, and so the Pareto boundary can be found by solving problem (4.133), i.e.,

$$\mathcal{P} = \{\boldsymbol{y}^\circ(\boldsymbol{\lambda}) = \boldsymbol{R}(\boldsymbol{p}^\circ) \mid \boldsymbol{\lambda} \succeq \boldsymbol{0}_n, \boldsymbol{\lambda} \neq \boldsymbol{0}_n\},$$

where $\boldsymbol{p}^\circ$ is an optimal solution of the convex problem (4.133), and

$$S_{\mathcal{O}}(\boldsymbol{\lambda}) = \max \ \{\boldsymbol{\lambda}^T \boldsymbol{R} \mid \boldsymbol{R} \in \mathcal{O}\} = f(\boldsymbol{y}^\circ(\boldsymbol{\lambda}))$$

is actually the support function (w.r.t. $\boldsymbol{\lambda}$) of the set $\mathcal{O}$ with domain $\{\boldsymbol{\lambda} \succeq \boldsymbol{0}_n, \boldsymbol{\lambda} \neq \boldsymbol{0}_n\}$ (cf. (3.82)).

An illustration for the case of $n = 2$ is shown in Figure 4.7, where each Pareto boundary (i.e., the set $\mathcal{P}$) consisting of all the maximal elements of $\mathcal{O}$ is shown together with a pair of maximal elements obtained for different weights λ_1 and λ_2. Note that each maximal element associated with $\lambda_1 = \lambda_2 > 0$ is also the maximum sum rate (i.e., $R_1 + R_2$ is maximized under the total power constraint $p_1 + p_2 \leq P$). For this case, the problem (4.133) corresponds to a water filling problem (9.128) with an analytical optimal power control solution $p_i^\star$ given by (9.133) in Chapter 9.   □

Remark 4.8 Though the set of achievable objective values for a vector optimization problem may be nonconvex, it can be shown that the corresponding set $\mathcal{O}$ of problem (4.132) is convex.

Before proceeding with the proof, we need to define *"normal set"* [Tuy00]. A set $\mathcal{A} \in \mathbb{R}^n$ is a normal set w.r.t. a proper cone $\mathcal{K}$, if

$$\boldsymbol{R}_1 \in \mathcal{A} \implies \boldsymbol{R}_2 \in \mathcal{A} \ \forall \boldsymbol{0} \preceq_\mathcal{K} \boldsymbol{R}_2 \preceq_\mathcal{K} \boldsymbol{R}_1. \tag{4.134}$$

For instance, the set of achievable objective values $\mathcal{O} = \boldsymbol{f}(\mathcal{C})$ of problem (4.132) where $\mathcal{K} = \mathbb{R}_+^n$ can be seen to be a normal set, which will be needed in its convexity proof below.

Let $\mathcal{C}$ denote the *convex* feasible set of (4.132) and

$$\boldsymbol{p}^{(i)} \triangleq (p_1^{(i)}, \ldots, p_n^{(i)}) \in \mathcal{C}, \quad i = 1, 2. \tag{4.135}$$

Then the rate tuple associated with $\boldsymbol{p}^{(i)}$ is given by

$$\boldsymbol{R}^{(i)} = (R_1^{(i)}, \ldots, R_n^{(i)}) \in \mathcal{O} \tag{4.136}$$

where

$$R_j^{(i)} = \log(1 + p_j^{(i)}/\sigma_j^2) \ \forall i = 1, 2 \text{ and } j = 1, \ldots, n. \tag{4.137}$$

Thus we have

$$\theta \boldsymbol{p}^{(1)} + (1 - \theta)\boldsymbol{p}^{(2)} \in \mathcal{C}, \quad \forall \theta \in [0, 1], \tag{4.138}$$

implying that $\mathcal{O}$ must contain the associated rate tuple, denoted as $\boldsymbol{R}^{(1,2)}$, i.e.,

$$\boldsymbol{R}^{(1,2)} \triangleq \begin{bmatrix} \log\left(1 + [\theta p_1^{(1)} + (1 - \theta)p_1^{(2)}]/\sigma_1^2\right) \\ \vdots \\ \log\left(1 + [\theta p_n^{(1)} + (1 - \theta)p_n^{(2)}]/\sigma_n^2\right) \end{bmatrix} \in \mathcal{O}. \tag{4.139}$$

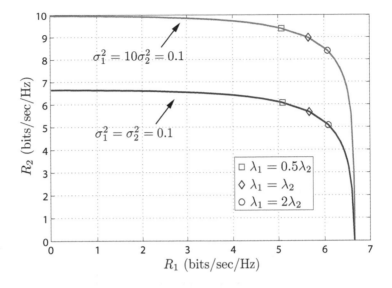

Figure 4.7 An illustration for showing Pareto boundaries of problem (4.132) for $P = 10$ dB and different values of parameters σ_1^2 and σ_2^2. One is denoted by the blue curve for $\sigma_1^2 = \sigma_2^2 = 0.1$ and the other by a green curve for $\sigma_1^2 = 10\sigma_2^2 = 0.1$ together with two sets of maximal elements (squares, diamonds, and circles) on each Pareto boundary with the corresponding weights $\lambda_1 = 0.5\lambda_2$, $\lambda_1 = \lambda_2$, and $\lambda_1 = 2\lambda_2$, respectively.

For any $\theta \in [0, 1]$, consider

$$\theta \boldsymbol{R}^{(1)} + (1-\theta)\boldsymbol{R}^{(2)} = \begin{bmatrix} \theta \log\left(1 + \frac{p_1^{(1)}}{\sigma_1^2}\right) + (1-\theta)\log\left(1 + \frac{p_1^{(2)}}{\sigma_1^2}\right) \\ \vdots \\ \theta \log\left(1 + \frac{p_n^{(1)}}{\sigma_n^2}\right) + (1-\theta)\log\left(1 + \frac{p_n^{(2)}}{\sigma_n^2}\right) \end{bmatrix} \tag{4.140a}$$

$$\preceq \begin{bmatrix} \log\left(1 + [\theta p_1^{(1)} + (1-\theta)p_1^{(2)}]/\sigma_1^2\right) \\ \vdots \\ \log\left(1 + [\theta p_n^{(1)} + (1-\theta)p_n^{(2)}]/\sigma_n^2\right) \end{bmatrix} = \boldsymbol{R}^{(1,2)} \in \mathcal{O} \tag{4.140b}$$

where the equality in (4.140b) is due to (4.139) and the inequality in (4.140b) holds since $\log(1 + x)$ is a concave function of $x \in \mathbb{R}_+$ (also showing that $\boldsymbol{R}$ is $\mathbb{R}_+^n$-concave in $\boldsymbol{p}$). By the fact that the rate region $\mathcal{O}$ is a normal set and by (4.140), $\theta \boldsymbol{R}^{(1)} + (1-\theta)\boldsymbol{R}^{(2)} \in \mathcal{O}$ must be true for any $\boldsymbol{R}^{(1)}, \boldsymbol{R}^{(2)} \in \mathcal{O}$, and $0 \le \theta \le 1$, and so the set $\mathcal{O}$ is convex. □

4.5   Quasiconvex optimization

A quasiconvex optimization problem has the following standard form

$$p^\star = \min\ f_0(\mathbf{x})$$
$$\text{s.t.}\ \ f_i(\mathbf{x}) \le 0,\ i = 1,\dots,m, \qquad\qquad (4.141)$$
$$\mathbf{Ax} = \mathbf{b},$$

where $f_0(\mathbf{x})$ is quasiconvex, and $f_i(\mathbf{x})$ is convex for all i. Though the feasible set is convex, a quasiconvex problem can have locally optimal points, and thus finding the optimal solution is usually not very computationally efficient. An optimal condition for (4.141) is given in the following fact.

Fact 4.2 Assume that $f_0(\mathbf{x})$ is differentiable. For the quasiconvex problem defined in (4.141), a point $\mathbf{x} \in \mathcal{C} = \{\mathbf{x} \mid f_i(\mathbf{x}) \le 0,\ i = 1,\dots,m, \mathbf{Ax} = \mathbf{b}\}$ is optimal if

$$\nabla f_0(\mathbf{x})^T(\mathbf{y} - \mathbf{x}) > 0,\ \forall \mathbf{y} \in \mathcal{C},\ \mathbf{y} \ne \mathbf{x} \qquad\qquad (4.142)$$

(sufficient condition), but the converse is not necessarily true.

Proof: Recall that f_0 is quasiconvex if and only if for all $\mathbf{x}, \mathbf{y} \in \mathbf{dom}\ f_0$,

$$f_0(\mathbf{y}) \le f_0(\mathbf{x}) \Rightarrow \nabla f_0(\mathbf{x})^T(\mathbf{y} - \mathbf{x}) \le 0$$

by the first-order condition of quasiconvex functions (see (3.110)). Here, it is given that $\nabla f_0(\mathbf{x})^T(\mathbf{y} - \mathbf{x}) > 0,\ \forall \mathbf{y} \in \mathcal{C}$, so we have $f_0(\mathbf{x}) < f_0(\mathbf{y})$, for all $\mathbf{y} \in \mathcal{C}$, which shows that $\mathbf{x}$ is optimal. ∎

Recall that the optimality condition of the convex problem given by (4.23) is both sufficient and necessary, while the optimality condition given by (4.142) of the quasiconvex problem is only sufficient, thus downgrading its usefulness in practical applications, although they look quite similar except that the former involves a closed halfspace while the latter involves the corresponding open halfspace.

By epigraph representation, problem (4.141) can be equivalently expressed as

$$p^\star = \min\ t$$
$$\text{s.t.}\ \ f_0(\mathbf{x}) \le t,\ f_i(\mathbf{x}) \le 0,\ i = 1,\dots,m, \qquad\qquad (4.143)$$
$$\mathbf{Ax} = \mathbf{b}.$$

Note that the inequality constraint set $\{(\mathbf{x}, t) \mid f_0(\mathbf{x}) \le t\} = \mathbf{epi}\ f_0$ is not convex in $(\mathbf{x}, t)$ since f_0 is not convex but quasiconvex, and so the problem (4.143) is nonconvex. However, by fixing t, the following feasibility problem is convex

$$\text{find}\ \ \mathbf{x}$$
$$\text{s.t.}\ \ f_0(\mathbf{x}) \le t,\ f_i(\mathbf{x}) \le 0,\ \ i = 1,\dots,m, \qquad\qquad (4.144)$$
$$\mathbf{Ax} = \mathbf{b}.$$

Note that the convex constraint set, denoted as $\mathcal{C}_t$, of the feasibility problem (4.144) can be expressed as

$$\mathcal{C}_t = \{\mathbf{x} \mid f_0(\mathbf{x}) \le t,\ f_i(\mathbf{x}) \le 0, i = 1, \dots, m,\ \mathbf{Ax} = \mathbf{b}\} \subseteq \mathcal{C}_s,\ \forall t \le s. \quad (4.145)$$

Moreover, $\mathcal{C}_t \ne \emptyset$ for $t \ge p^\star$ and $\mathcal{C}_t = \emptyset$ for $t < p^\star$ (cf. Figure 4.8).

An idea for solving (4.141) is to iteratively decrease t until the feasibility problem (4.144) is feasible for $t \in [p^\star, p^\star + \epsilon]$ where ϵ is a preassigned small positive real number. Assume that $p^\star$ is known to lie within $[\ell, u]$, where ℓ and u usually can be determined by the associated constraints (e.g., either maximum or minimum transmit power in wireless communications). Then the quasiconvex optimization problem (4.141) can be solved by the iterative method given in Algorithm 4.1, called the *bisection method*, for an ϵ-suboptimal solution.

Algorithm 4.1 Bisection method for solving quasiconvex problem (4.141)

1: Given bounds $\ell \le p^\star \le u$, and convergence tolerance ϵ.
2: **repeat**
3:    Update $t := (\ell + u)/2$.
4:    Solve the convex feasibility problem (4.144).
5:    If the problem is feasible ($\mathcal{C}_t \ne \emptyset$), update $u := t$; otherwise update $\ell := t$.
6: **until** $u - \ell \le \epsilon$.

The procedure of the bisection method is conceptually illustrated in Figure 4.8. Surely, the number of iterations needed depends on the preassigned ϵ and characteristics of the quasiconvex problem under consideration. However, the most preferable alternative to effectively solving a quasiconvex problem under consideration is to try to reformulate it into a convex problem (if possible) through various representations, function transformations, and variable changes introduced earlier in this chapter.

Next, let us use an example to demonstrate the bisection method to solve a quasiconvex optimization problem. Consider a linear-fractional function

$$f_0(\mathbf{x}) = \frac{\mathbf{c}^T\mathbf{x} + d}{\mathbf{f}^T\mathbf{x} + g}, \quad (4.146)$$

which is a quasiconvex function with $\mathbf{dom}\ f_0 = \{\mathbf{x} \mid \mathbf{f}^T\mathbf{x} + g > 0\}$. The optimization problem, called the *linear-fractional program*, is defined as

$$\begin{aligned} p^\star = \min\ &f_0(\mathbf{x}) \\ \text{s.t. } &\mathbf{f}^T\mathbf{x} + g > 0,\ \mathbf{Ax} \preceq \mathbf{b}. \end{aligned} \quad (4.147)$$

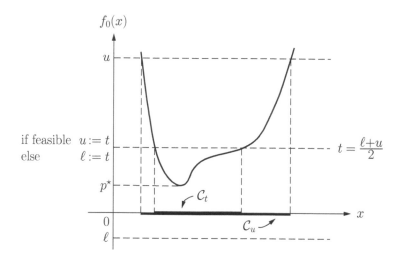

Figure 4.8 Illustration of the bisection method for solving a one-dimensional uncon-strained quasiconvex optimization problem.

By epigraph reformulation we have

$$\min \ t$$
$$\text{s.t.} \ \frac{\mathbf{c}^T\mathbf{x} + d}{\mathbf{f}^T\mathbf{x} + g} \le t \tag{4.148}$$
$$\mathbf{f}^T\mathbf{x} + g > 0, \ \mathbf{A}\mathbf{x} \preceq \mathbf{b}$$

which is exactly the same as

$$\min \ t$$
$$\text{s.t.} \ \mathbf{c}^T\mathbf{x} + d \le t(\mathbf{f}^T\mathbf{x} + g), \tag{4.149}$$
$$\mathbf{f}^T\mathbf{x} + g > 0, \ \mathbf{A}\mathbf{x} \preceq \mathbf{b}.$$

This problem can be solved by using the bisection method, in which the associ-ated feasibility problem is an LP as follows

$$\text{find} \ \mathbf{x}$$
$$\text{s.t.} \ \mathbf{c}^T\mathbf{x} + d \le t(\mathbf{f}^T\mathbf{x} + g), \tag{4.150}$$
$$\mathbf{f}^T\mathbf{x} + g > 0, \ \mathbf{A}\mathbf{x} \preceq \mathbf{b},$$

for a fixed value of t. If the feasibility problem is not feasible ($\mathcal{C}_t = \emptyset$), then $t < p^\star$. If the feasibility problem is feasible ($\mathcal{C}_t \ne \emptyset$), then $p^\star \le t$.

Similarly, we can easily verify that the *generalized linear-fractional program*, defined as

$$\min \ \max_{i=1,\ldots,K} \frac{\mathbf{c}_i^T\mathbf{x} + d_i}{\mathbf{f}_i^T\mathbf{x} + g_i} \tag{4.151}$$
$$\text{s.t.} \ \mathbf{A}\mathbf{x} \preceq \mathbf{b}, \ \mathbf{f}_i^T\mathbf{x} + g_i > 0, \ i = 1,\ldots,K,$$

is also a quasiconvex problem, and hence it can be solved using the iterative bisection method, where each feasibility problem involved is again an LP.

Though the quasiconvex problem (4.147) can be solved by using the bisection method, it is still preferred to solve it by reformulating the problem as a convex problem if possible. Next, let us show that problem (4.147) indeed can be reformulated as a convex problem by the change of variables so that it can be optimally solved.

If the feasible set $\{\mathbf{x} \mid \mathbf{A}\mathbf{x} \preceq \mathbf{b}, \mathbf{f}^T\mathbf{x} + g > 0\}$ is bounded, the linear-fractional program given by (4.147) can be transformed into an LP. By letting

$$z = \frac{1}{\mathbf{f}^T\mathbf{x} + g} \quad \text{(a.k.a. Charnes–Cooper transformation)},$$

$$\mathbf{y} = z\mathbf{x},$$

we get the following LP

$$\min_{\mathbf{y} \in \mathbb{R}^n, \ z \in \mathbb{R}} \quad \mathbf{c}^T\mathbf{y} + dz$$
$$\text{s.t. } \mathbf{A}\mathbf{y} - \mathbf{b}z \preceq \mathbf{0}, \ z \geq 0, \ \mathbf{f}^T\mathbf{y} + gz = 1. \tag{4.152}$$

It can be shown that if $\{\mathbf{x} \mid \mathbf{A}\mathbf{x} \preceq \mathbf{b}, \mathbf{f}^T\mathbf{x} + g > 0\}$ is bounded, then $z > 0$ for any feasible $(\mathbf{y}, z)$. If $(\mathbf{y}, z)$ is feasible to the LP (4.152), then $\mathbf{x} = \mathbf{y}/z$ is feasible to the linear-fractional program (4.147) [BV04].

Finally, let us conclude this section by the optimal power assignment example again, to illustrate the application of the bisection method.

Example 4.11 *(Optimal power assignment (revisited))* The problem of interest is called a *max-min-fair problem* to maximize the smallest γ_i defined in (4.83) subject to power constraints $0 \leq p_i \leq P_i$, where P_i is the maximum allowable power for transmitter i. The power allocation problem can be mathematically written as

$$\gamma^\star = \max_{p_i \in [0, P_i], i=1,\ldots,K} \ \min_{i=1,\ldots,K} \ \gamma_i$$
$$= \max_{p_i \in [0, P_i], i=1,\ldots,K} \ \min_{i=1,\ldots,K} \ \frac{G_{ii}p_i}{\sum_{j=1, j\neq i}^{K} G_{ij}p_j + \sigma_i^2}. \tag{4.153}$$

By either (4.8) or (4.9), the above problem can be reformulated as a generalized linear-fractional program:

$$\frac{1}{\gamma^\star} = \min_{p_i \in [0, P_i], i=1,\ldots,K} \ \max_{i=1,\ldots,K} \ \frac{1}{\gamma_i}$$
$$= \min_{p_i \in [0, P_i], i=1,\ldots,K} \ \max_{i=1,\ldots,K} \ \frac{\sum_{j=1, j\neq i}^{K} G_{ij}p_j + \sigma_i^2}{G_{ii}p_i}, \tag{4.154}$$

which can be solved by the bisection method as presented above. Notice that problem (4.153) and problem (4.154) share the same optimal solutions, but their

optimal values are the inverse of each other. At each iteration in running the
bisection algorithm, an LP feasibility problem is solved until convergence. □

4.6   Block successive upper bound minimization

So far, this chapter has been concentrating on some general approaches to reformulation of a nonconvex problem into a convex problem or a quasiconvex problem, so that any means for obtaining optimal solutions can be applied, such as
the first-order optimality condition (4.23) for solving the former and the bisection
method for solving the latter, which have also been introduced in this chapter.
Surely, there have been many other efficient methods for solving convex problems
that will be introduced later in Chapter 9, especially KKT conditions. However,
if it is formidable to convert the nonconvex problem under consideration into a
convex or quasiconvex problem, an approximate solution, referred to as a *station-*
ary point, is preferred. Next, let us introduce stationary points of a nonconvex
problem, followed by an efficient approach, called block successive upper bound
minimization (previously abbreviated by BSUM), to find a stationary-point solution of the nonconvex problem under some conditions.

4.6.1   Stationary point

Let $f: \mathcal{C} \to \mathbb{R}$ be a continuous nonconvex function, possibly nondifferentiable,
where $\mathcal{C} \subseteq \mathbb{R}^n$ is a closed convex set. Consider the following minimization problem:

$$\min_{\mathbf{x} \in \mathcal{C}} \ f(\mathbf{x}). \tag{4.155}$$

The *directional derivative* of f at a point $\mathbf{x}$ in direction $\mathbf{v}$ is defined as

$$f'(\mathbf{x}; \mathbf{v}) \triangleq \liminf_{\lambda \downarrow 0} \ \frac{f(\mathbf{x} + \lambda \mathbf{v}) - f(\mathbf{x})}{\lambda}$$

$$= \lim_{\lambda \to 0^+} \inf_{0 < \mu \leq \lambda} \frac{f(\mathbf{x} + \mu \mathbf{v}) - f(\mathbf{x})}{\mu}. \tag{4.156}$$

The point $\mathbf{x}$ is a *stationary point* [RHL13] of problem (4.155) if $f'(\mathbf{x}; \mathbf{v}) \geq 0$ for
all $\mathbf{v}$ such that $\mathbf{x} + \mathbf{v} \in \mathcal{C}$.

As an illustrative example for stationary points, let $\mathcal{C} = [a, b]$, where $a \in \mathbb{R}$,
$b \in \mathbb{R}$, and $a \leq b$, and let $f(x) = |x|$, which is convex but nondifferentiable. Then,
one can use (4.156) to readily obtain the directional derivative of f and the result
is given by

$$f'(x; v) = \begin{cases} |v|, & \text{if } x = 0, \\ |v|, & \text{if } x > 0, \, v > 0 \text{ or } x < 0, \, v < 0, \\ -|v|, & \text{if } x > 0, \, v < 0 \text{ or } x < 0, \, v > 0. \end{cases}$$

Consequently, the stationary point of problem (4.155) is $x = 0$ if $a \le 0 \le b$, $x = a$ if $a \ge 0$, and $x = b$ if $b \le 0$. Note that the stationary-point solution is also the optimal solution for this example.

As f is differentiable, the condition of stationarity defined in (4.156) can be simplified as

$$f'(\mathbf{x}; \mathbf{v}) = \lim_{\lambda \to 0^+} \frac{f(\mathbf{x} + \lambda \mathbf{v}) - f(\mathbf{x})}{\lambda} = \nabla f(\mathbf{x})^T \mathbf{v} \ge 0 \quad \forall \mathbf{x} + \mathbf{v} \in \mathcal{C} \qquad (4.157)$$

$$\Leftrightarrow \nabla f(\mathbf{x})^T (\mathbf{y} - \mathbf{x}) \ge 0 \quad \forall \mathbf{y} \in \mathcal{C};$$

namely, it is equivalent to the first-order optimality condition given by (4.23). Accordingly, $f'(\mathbf{x}; \mathbf{v})$ also reduces to $\nabla f(\mathbf{x}) = \mathbf{0}$ for $\mathbf{x} \in \text{int } \mathcal{C}$ (cf. (4.28)). In general, a stationary point can be a local minimum, local maximum, or a saddle point (cf. Figure 4.9), while for the case that f is convex, the stationary points are exactly the global minimizers to problem (4.155).

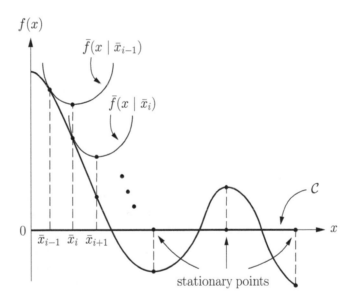

Figure 4.9 Illustration of the BSUM method (i.e., Algorithm 4.2) for $m = n = 1$ that can yield a stationary-point solution of problem (4.155) after convergence. Note that the subscript i in $\bar{x}_i$ here represents the iteration number rather than the block number since $m = 1$.

Remark 4.9 The convex problem (4.11) with the objective function f and all inequality constraint functions being differentiable can be viewed as a special case of problem (4.155). For this case, under Slater's condition (cf. Subsection 9.3.1), the stationary-point condition given by (4.157) will be shown to be equivalent to the KKT conditions of this problem in Chapter 9 (cf. Remark 9.14). Furthermore, the equivalence of the stationary points and the KKT points (that satisfy KKT conditions) of problem (4.155) (where f is nonconvex) is still valid under Slater's condition (cf. Remark 9.15). The BSUM method to be introduced below is efficient and effective for finding a stationary-point solution of the nonconvex problem (4.155). $\qquad\square$

With the stationary points defined above, let us introduce *regularity of a function* for ease of the introduction to the BSUM method. Let $f : \mathbb{R}^n \to \mathbb{R}$ and $\mathbf{x} = (\mathbf{x}_1, \ldots, \mathbf{x}_m) \in \mathbf{dom}\ f$, where $\mathbf{x}_i \in \mathbb{R}^{n_i}$ and $n_1 + \cdots + n_m = n$. Then $f : \mathbb{R}^n \to \mathbb{R}$ is regular at the point $\mathbf{x}$ if, for all

$$\begin{cases} \mathbf{v} \triangleq (\mathbf{v}_1, \ldots, \mathbf{v}_m) \in \mathbb{R}^{n_1} \times \cdots \times \mathbb{R}^{n_m} \\ \mathbf{v}_i \triangleq (\mathbf{0}_{n_1}, \ldots, \mathbf{0}_{n_{i-1}}, \mathbf{v}_i, \mathbf{0}_{n_{i+1}}, \ldots, \mathbf{0}_{n_m}),\ \mathbf{v}_i \in \mathbb{R}^{n_i}, \end{cases} \tag{4.158}$$

$f'(\mathbf{x}; \mathbf{v}_i) \geq 0$ for $i = 1, \ldots, m$ implies $f'(\mathbf{x}; \mathbf{v}) \geq 0$. It can be inferred that if f is differentiable at the point $\mathbf{x}$, then

$$f'(\mathbf{x}; \mathbf{v}) = \nabla f(\mathbf{x})^T \mathbf{v} = \nabla f(\mathbf{x})^T \left\{ \sum_{i=1}^m \mathbf{v}_i \right\} \quad (\text{by (4.157) and (4.158)})$$

$$= \sum_{i=1}^m f'(\mathbf{x}; \mathbf{v}_i) \geq 0 \quad \text{if } f'(\mathbf{x}; \mathbf{v}_i) \geq 0\ \forall i, \tag{4.159}$$

and so $\mathbf{x}$ must be a regular point of f. Next, we introduce BSUM for a stationary-point solution of the nonconvex problem (4.155).

4.6.2   BSUM

Suppose that $\mathcal{C} = \mathcal{C}_1 \times \cdots \times \mathcal{C}_m$, where $\mathcal{C}_i \subseteq \mathbb{R}^{n_i}$, $i = 1, \ldots, m$, are closed convex sets, and $\sum_{i=1}^m n_i = n$. By judiciously exploiting such block structure, the BSUM method can efficiently yield a stationary-point solution of problem (4.155) [RHL13] by iteratively updating the m blocks of variables in a round-robin manner (Gauss–Seidel update rule) [Ber99]. Specifically, given a feasible point $\bar{\mathbf{x}} = (\bar{\mathbf{x}}_1, \ldots, \bar{\mathbf{x}}_m) \in \mathcal{C}$ obtained at the $(r-1)$th iteration, the ith block $\bar{\mathbf{x}}_i$, where $i = ((r - 1) \bmod m) + 1$, is updated at the rth iteration by

$$\bar{\mathbf{x}}_i = \arg\min_{\mathbf{x}_i \in \mathcal{C}_i} \bar{f}_i(\mathbf{x}_i \mid \bar{\mathbf{x}}), \tag{4.160}$$

where $\bar{f}_i(\mathbf{x}_i \mid \bar{\mathbf{x}})$ is an upper bound approximation of $f(\mathbf{x})$ w.r.t. the ith block at the reference point $\mathbf{x} = \bar{\mathbf{x}} \in \mathcal{C}$. The steps of BSUM method are summarized in Algorithm 4.2. An illustration for the iterative operation of Algorithm 4.2 is given in Figure 4.9 for the case of $m = n = 1$ that will yield a stationary point of the nonconvex problem (4.155) under some convergence conditions to be presented next.

The convergence property of the BSUM method has been investigated in [RHL13], and is briefly summarized as follows. Suppose that either of the follow-

---

Algorithm 4.2 BSUM method

1: Given a feasible point $\bar{\mathbf{x}} = (\bar{\mathbf{x}}_1, \ldots, \bar{\mathbf{x}}_m) \in \mathcal{C}$ and set $r = 0$;
2: **repeat**
3:    Set $r := r + 1$ and $i = ((r-1) \bmod m) + 1$;
4:    Update $\bar{\mathbf{x}}_i$ by (4.160);
5: **until** Some convergence criterion is met;
6: Output $\bar{\mathbf{x}}$ as an approximate solution to problem (4.155).

---

ing two premises is true:

(a) $\bar{f}_i(\mathbf{x}_i \mid \bar{\mathbf{x}})$ is quasiconvex in $\mathbf{x}_i$ for $i = 1, \ldots, m$, and $f(\mathbf{x})$ is regular at every point $\mathbf{x} \in \mathcal{C}$. $\hfill (4.161\text{a})$

(b) There exists an $\mathbf{x}' \in \mathcal{C}$ such that the sublevel set $\mathcal{S} = \{\mathbf{x} \in \mathcal{C} \mid f(\mathbf{x}) \leq f(\mathbf{x}')\}$ is compact, and $f(\mathbf{x})$ is regular at every point $\mathbf{x} \in \mathcal{S}$. $\hfill (4.161\text{b})$

Then the sequence of the iterates $\bar{\mathbf{x}}$ generated by the BSUM algorithm converges to a point belonging to the set of stationary points of problem (4.155) as long as

$$\bar{f}_i(\bar{\mathbf{x}}_i \mid \bar{\mathbf{x}}) = f(\bar{\mathbf{x}}), \tag{4.162a}$$

$$\bar{f}_i(\mathbf{x}_i \mid \bar{\mathbf{x}}) > f(\bar{\mathbf{x}}_1, \ldots, \bar{\mathbf{x}}_{i-1}, \mathbf{x}_i, \bar{\mathbf{x}}_{i+1}, \ldots, \bar{\mathbf{x}}_m), \tag{4.162h}$$

$$\bar{f}_i'(\bar{\mathbf{x}}_i; \mathbf{v}_i \mid \bar{\mathbf{x}}) = f'(\bar{\mathbf{x}}; \boldsymbol{v}_i), \quad (\text{cf. (4.158) for } \mathbf{v}_i \text{ and } \boldsymbol{v}_i) \tag{4.162c}$$

$$\bar{f}_i(\mathbf{x}_i \mid \bar{\mathbf{x}}) \text{ is continuous in } (\mathbf{x}_i, \bar{\mathbf{x}}), \tag{4.162d}$$

$$\text{problem (4.160) has a unique solution,} \tag{4.162e}$$

for all $\mathbf{x}_i \in \mathcal{C}_i$, $\bar{\mathbf{x}} \in \mathcal{C}$, and for all $\mathbf{v}_i$ such that $\mathbf{x}_i + \mathbf{v}_i \in \mathcal{C}_i$, $\forall i$. Two noteworthy remarks are as follows:

Remark 4.10 As f and $\bar{f}_i$ are differentiable, it can be inferred from (4.162a), (4.162b), that

$$\bar{\mathbf{x}}_i = \arg\min \left\{ \bar{f}_i(\mathbf{x}_i \mid \bar{\mathbf{x}}) - f(\bar{\mathbf{x}}_1, \ldots, \bar{\mathbf{x}}_{i-1}, \mathbf{x}_i, \bar{\mathbf{x}}_{i+1}, \ldots, \bar{\mathbf{x}}_m) \right\}$$
$$\Rightarrow \nabla_{\mathbf{x}_i} \left(\bar{f}_i(\bar{\mathbf{x}}_i \mid \bar{\mathbf{x}}) - f(\bar{\mathbf{x}}) \right) = \mathbf{0}$$
$$\Rightarrow \nabla \bar{f}_i(\bar{\mathbf{x}}_i \mid \bar{\mathbf{x}}) = \nabla_{\mathbf{x}_i} f(\bar{\mathbf{x}})$$
$$\Rightarrow \nabla \bar{f}_i(\bar{\mathbf{x}}_i \mid \bar{\mathbf{x}})^T \mathbf{v}_i = \nabla_{\mathbf{x}_i} f(\bar{\mathbf{x}})^T \mathbf{v}_i = \nabla f(\bar{\mathbf{x}})^T \boldsymbol{v}_i$$
$$\Rightarrow \bar{f}_i'(\bar{\mathbf{x}}_i; \mathbf{v}_i \mid \bar{\mathbf{x}}) = f'(\bar{\mathbf{x}}; \boldsymbol{v}_i) \quad (\text{cf. (4.157)});$$

namely, (4.162c) must be true. So the convergence conditions of the BSUM method reduce to (4.162a), (4.162b), (4.162d), and (4.162e). $\hfill \square$

Remark 4.11 The convergence condition (4.162e), i.e., a unique solution to problem (4.160) is required only for $m > 1$. For $m = 1$, the BSUM method is simply called the SUM method which can yield a stationary-point solution to (4.155) without need of the convergence condition (4.162e). $\hfill \square$

Therefore, the key to solving problem (4.155) by the BSUM method is to properly design or find the approximate functions $\bar{f}_i(\mathbf{x}_i \mid \bar{\mathbf{x}})$, $i = 1, \ldots, m$, such that all the conditions in (4.162) are satisfied on one hand, and problem (4.160) can be solved efficiently on the other hand. Finally, let us illustrate how to apply the BSUM method by the following example.

Example 4.12 Consider a two-user single-input single-output (SISO) interference channel (which is a special case of the K-user multiple-input single-output (MISO) interference channel as illustrated in Figure 8.9 for $K = 3$), where two single-antenna transmitters communicate with their respective single-antenna receivers by using the same spectral band simultaneously. Therefore, these two pairs of transmitter and receiver interfere with each other in their received signals. The signal model of this system can be expressed as

$$y_1 = x_1 + h_{21}x_2 + n_1,$$
$$y_2 = h_{12}x_1 + x_2 + n_2,$$

where y_i is the signal received at receiver i, x_i is the signal transmitted by transmitter i, $h_{ki} \in \mathbb{C}$ is the crosslink channel gain between transmitter k and receiver i, and $n_i \sim \mathcal{CN}(0, \sigma_i^2)$ is the additive noise at receiver i for $i, k \in \{1, 2\}$. Note that the received signal y_i has been normalized by h_{ii}, and thus $h_{ii} = 1$ in the above signal model for simplicity. Assuming that the transmitted signals x_i are Gaussian encoded with zero mean and variance p_i, i.e., $x_i \sim \mathcal{CN}(0, p_i)$ for $i = 1, 2$, and that the receivers employ single-user detection schemes to decode the received signals y_i to obtain the desired signals x_i, the achievable data rates of the two transmitter-receiver pairs, which depend on the SINR in the respective received signal y_i, are given by

$$r_1(p_1, p_2) = \log_2\left(1 + \frac{\mathbb{E}\{|x_1|^2\}}{\mathbb{E}\{|h_{21}x_2 + n_1|^2\}}\right)$$

$$= \log_2\left(1 + \frac{p_1}{|h_{21}|^2 p_2 + \sigma_1^2}\right) \text{ bits/transmission,}$$

$$r_2(p_1, p_2) = \log_2\left(1 + \frac{\mathbb{E}\{|x_2|^2\}}{\mathbb{E}\{|h_{12}x_1 + n_2|^2\}}\right)$$

$$= \log_2\left(1 + \frac{p_2}{|h_{12}|^2 p_1 + \sigma_2^2}\right) \text{ bits/transmission.}$$

Consider the following power control problem to maximize the sum rate of the two transmitter-receiver pairs:

$$\max_{p_1, p_2} \ r_1(p_1, p_2) + r_2(p_1, p_2) \tag{4.163a}$$

$$\text{s.t. } 0 \le p_1 \le P_1, \tag{4.163b}$$

$$0 \le p_2 \le P_2, \tag{4.163c}$$

where P_1 and P_2 are the power budgets of transmitter 1 and transmitter 2, respectively. Note that with the feasible set being closed and convex (cf. (4.163b) and (4.163c)), this problem is nonconvex because the objective function is neither convex nor concave in (p_1, p_2). Moreover, it is almost impossible to convert this problem into a convex problem using the reformulation approaches previously introduced in this chapter.

Now let us present how to obtain a stationary-point solution to problem (4.163) using the BSUM method. First of all, let us express problem (4.163) as a standard optimization problem as follows:

$$\min_{p_1,p_2} \; f(p_1,p_2) \triangleq -r_1(p_1,p_2) - r_2(p_1,p_2) \qquad (4.164a)$$

$$\text{s.t. } 0 \leq p_1 \leq P_1, \qquad (4.164b)$$

$$0 \leq p_2 \leq P_2. \qquad (4.164c)$$

It can be shown by the second-order condition that $-r_1(p_1,p_2)$ is convex in p_1 but concave in p_2, and $-r_2(p_1,p_2)$ convex in p_2 but concave in p_1. The two desired approximate functions, denoted as $\bar{f}_1(p_1 \mid \bar{p}_1, \bar{p}_2)$ and $\bar{f}_2(p_2 \mid \bar{p}_1, \bar{p}_2)$, that satisfy the conditions in (4.162) can be obtained by the first-order approximation of concave functions. As a result, the two approximate functions for updating p_1, p_2 are, respectively, given by

$$\bar{f}_1(p_1 \mid \bar{p}_1, \bar{p}_2) \triangleq -r_1(p_1, \bar{p}_2) - r_2(\bar{p}_1, \bar{p}_2) + (p_1 - \bar{p}_1) \frac{\partial\{-r_2(p_1, \bar{p}_2)\}}{\partial p_1}\bigg|_{p_1 - \bar{p}_1}$$

$$= -r_1(p_1, \bar{p}_2) - r_2(\bar{p}_1, \bar{p}_2) + \frac{|h_{12}|^2 \bar{p}_2 (p_1 - \bar{p}_1)/\log 2}{(\bar{p}_2 + |h_{12}|^2 \bar{p}_1 + \sigma_2^2)(|h_{12}|^2 \bar{p}_1 + \sigma_2^2)}$$

$$\geq f(p_1, \bar{p}_2),$$

and

$$\bar{f}_2(p_2 \mid \bar{p}_1, \bar{p}_2) \triangleq -r_2(\bar{p}_1, p_2) - r_1(\bar{p}_1, \bar{p}_2) + (p_2 - \bar{p}_2) \frac{\partial\{-r_1(\bar{p}_1, p_2)\}}{\partial p_2}\bigg|_{p_2 = \bar{p}_2}$$

$$= -r_2(\bar{p}_2, p_2) - r_1(\bar{p}_1, \bar{p}_2) + \frac{|h_{21}|^2 \bar{p}_1 (p_2 - \bar{p}_2)/\log 2}{(\bar{p}_1 + |h_{21}|^2 \bar{p}_2 + \sigma_1^2)(|h_{21}|^2 \bar{p}_2 + \sigma_1^2)}$$

$$\geq f(\bar{p}_1, p_2),$$

where $\bar{p}_1$ and $\bar{p}_2$ can be any points satisfying the power constraints (4.164b) and (4.164c), respectively. It is not difficult to verify that conditions (4.162a)–(4.162d) are satisfied; moreover, each of the two corresponding subproblems (cf. (4.160))

$$\min_{0 \leq p_1 \leq P_1} \; \bar{f}_1(p_1 \mid \bar{p}_1, \bar{p}_2)$$

$$\min_{0 \leq p_2 \leq P_2} \; \bar{f}_2(p_2 \mid \bar{p}_1, \bar{p}_2)$$

is a convex problem with a unique solution (since both $\bar{f}_1$ and $\bar{f}_2$ are strictly convex in p_1 and in p_2, respectively) that can be obtained either by the first-order optimality condition (4.23) or by solving the associated KKT conditions as

$$
p_1^\star = \begin{cases} g_1(\bar{p}_1, \bar{p}_2), & \text{if } 0 \le g_1(\bar{p}_1, \bar{p}_2) \le P_1, \\ P_1, & \text{if } g_1(\bar{p}_1, \bar{p}_2) > P_1, \\ 0, & \text{if } g_1(\bar{p}_1, \bar{p}_2) < 0, \end{cases} \tag{4.165a}
$$

$$
p_2^\star = \begin{cases} g_2(\bar{p}_1, \bar{p}_2), & \text{if } 0 \le g_2(\bar{p}_1, \bar{p}_2) \le P_2, \\ P_2, & \text{if } g_2(\bar{p}_1, \bar{p}_2) > P_2, \\ 0, & \text{if } g_2(\bar{p}_1, \bar{p}_2) < 0, \end{cases} \tag{4.165b}
$$

where

$$
g_1(\bar{p}_1, \bar{p}_2) = \frac{(\bar{p}_2 + |h_{12}|^2 \bar{p}_1 + \sigma_2^2)(|h_{12}|^2 \bar{p}_1 + \sigma_2^2)}{|h_{12}|^2 \bar{p}_2} - (|h_{21}|^2 \bar{p}_2 + \sigma_1^2),
$$

$$
g_2(\bar{p}_1, \bar{p}_2) = \frac{(\bar{p}_1 + |h_{21}|^2 \bar{p}_2 + \sigma_1^2)(|h_{21}|^2 \bar{p}_2 + \sigma_1^2)}{|h_{21}|^2 \bar{p}_1} - (|h_{12}|^2 \bar{p}_1 + \sigma_2^2).
$$

The BSUM method for handling the sum rate maximization problem (4.164) is fulfilled by Algorithm 4.3. Note that since the objective function $f(p_1, p_2)$ is regular at any point (p_1, p_2) (since it is differentiable) and both $\bar{f}_1(p_1 \mid \bar{p}_1, \bar{p}_2)$ and $\bar{f}_2(p_2 \mid \bar{p}_1, \bar{p}_2)$ are strictly convex (implying that the premise (4.161a) holds true); furthermore, the feasible set $\{(p_1, p_2) \mid 0 \le p_1 \le P_1, 0 \le p_2 \le P_2\}$ is compact (implying that the premise (4.161b) also holds true). Hence Algorithm 4.3 guarantees to yield a stationary-point solution of problem (4.164). □

Algorithm 4.3 BSUM algorithm for problem (4.164)

1: Given a point $(\bar{p}_1, \bar{p}_2)$ satisfying (4.164b) and (4.164c);
2: **repeat**
3:    Update $\bar{p}_1$ by (4.165a);
4:    Update $\bar{p}_2$ by (4.165b);
5: **until** Some convergence criterion is met;
6: Output $(\bar{p}_1, \bar{p}_2)$ as an approximate solution to problem (4.164).

Some applications of the BSUM to outage constrained coordinated beamforming for MISO interference channel will be presented in Subsection 8.5.7 to show its effectiveness in improving the efficacy (performance and efficiency) of the designed transmit beamforming algorithms.

4.7   Successive convex approximation

In the previous section, we have introduced the BSUM method for iteratively finding a stationary-point solution to the nonconvex problem (4.155) where the feasible set is convex but the objective function is nonconvex. In this section, we consider a nonconvex problem as follows

$$p^{\star} = \min_{\mathbf{x} \in \mathcal{C}} \ f(\mathbf{x}) \tag{4.166}$$

where $f : \mathcal{C} \to \mathbb{R}$ is a convex function but $\mathcal{C} \subseteq \mathbb{R}^n$ is a closed nonconvex set. Next, we introduce the widely used successive convex approximation (previously abbreviated by SCA) [MW78] in wireless communications to obtain a stationary-point solution to (4.166).

The idea of the SCA is to successively find a convex restrictive approximation to problem (4.166) as follows:

$$\min_{\mathbf{x} \in C_i} \ f(\mathbf{x}), \tag{4.167}$$

where the feasible set $C_i \subset \mathcal{C}$ is convex. Let $\mathbf{x}_i^{\star} \in C_i$ denote an optimal solution to the convex problem (4.167). Next, we define the convex set C_{i+1} such that $\mathbf{x}_i^{\star} \in C_{i+1} \subset \mathcal{C}$. Then solve the problem (4.167) where C_i is replaced with C_{i+1} for an optimal solution $\mathbf{x}_{i+1}^{\star}$. It can be easily seen that

$$f(\mathbf{x}_i^{\star}) \geq f(\mathbf{x}_{i+1}^{\star}) = \min_{\mathbf{x} \in C_{i+1}} \ f(\mathbf{x}) \geq p^{\star}, \ C_{i+1} \triangleq \cup_{j=1}^{i+1} C_j \subset \mathcal{C}$$

$$C_i \cap C_{i+1} \neq \emptyset \ \forall i. \tag{4.168}$$

As the optimal value $f(\mathbf{x}_i^{\star})$ converges under some preassigned tolerance, a stationary-point solution may be obtained, depending on the problem under consideration and the choice of the convex set C_i. It is noticeable that not only $\mathbf{x}_i^{\star} \in C_i$ but also $\mathbf{x}_i^{\star} \in \mathbf{bd}\, C_i$ before convergence as illustrated in Figure 8.10 (in Chapter 8). The reason for this is that if $\mathbf{x}_i^{\star} \in \mathbf{int}\, C_i$, it must be a local optimum in $\mathcal{C}$ and thus a global optimum in $\mathcal{C}$ as well since f is convex (so we are lucky for this case), implying that the obtained solution using SCA should be theoretically provable to be globally optimal to the nonconvex problem (4.166) if this lucky case has also been demonstrated by simulation results. It is quite often, even after convergence, $\mathbf{x}_i^{\star} \in \mathbf{bd}\, C_i$ may still remain and thus $\mathbf{x}_i^{\star} \in \mathbf{bd}\, \mathcal{C}$.

The next question is how to find a suitable C_{i+1} that contains $\mathbf{x}_i^{\star}$. Let us answer this question with an illustrative example. Suppose that $g(\mathbf{x}) \leq 0$ is a nonconvex inequality constraint function of (4.166) and it is a continuous function, possibly nondifferentiable, given in the form

$$g(\mathbf{x}) = g_1(\mathbf{x}) + g_2(\mathbf{x}) \tag{4.169}$$

where $g_1(\mathbf{x})$ is convex but $g_2(\mathbf{x})$ is concave. By applying the first-order inequality of concave functions to $g_2(\mathbf{x})$ at $\mathbf{x} = \mathbf{x}_i^{\star}$, we have

$$\widehat{g}(\mathbf{x}) = g_1(\mathbf{x}) + g_2(\mathbf{x}_i^{\star}) + \bar{\nabla} g_2(\mathbf{x}_i^{\star})^T (\mathbf{x} - \mathbf{x}_i^{\star}) \geq g(\mathbf{x})$$

$$\implies C_{i+1} \triangleq \{\mathbf{x} \mid \widehat{g}(\mathbf{x}) \leq 0\} \subset \{\mathbf{x} \mid g(\mathbf{x}) \leq 0\}. \tag{4.170}$$

Note that $\widehat{g}(\mathbf{x})$ in (4.170) is convex and a tight upper bound to $g(\mathbf{x})$. Then a candidate of the feasible set C_{i+1} has been obtained as defined in (4.170), where the inequality constraint $g(\mathbf{x}) \leq 0$ is replaced by the restricted convex constraint $\widehat{g}(\mathbf{x}) \leq 0$.

Remark 4.12 The SCA is based on the convex restriction to a nonconvex constraint set since $C_i \subset C$ in (4.167) for all i. Sometimes if the nonconvex feasible set $\mathcal{C}$ of the considered problem neither has any closed-form nor tractable expressions (e.g., complicated probability functions involved in $\mathcal{C}$), one may face a challenge (or even a bottleneck) of finding a good restrictive well-defined convex set $C \subset \mathcal{C}$ such that the optimal solution of the reformulated convex problem will not be too conservative (i.e., C is a good convex approximation to $\mathcal{C}$). Various inequalities depending on the applications may be needed for finding a suitable C, and the performance analysis about the solution accuracy is of paramount importance to its applicability, but usually non-trivial as well. Some applications involving restrictive approximations to intractable probability inequality constraints are introduced in Subsections 8.5.8 and 8.5.9 (outage constrained robust transmit beamforming under single-cell MISO scenario for the former and multicell MISO scenario for the latter). □

Remark 4.13 In contrast to the convex restriction in SCA, the convex relaxation to a nonconvex constraint set $\mathcal{C}$ has also been used in wireless communications to find an approximate solution to a nonconvex problem of interest. For instance,

$$\mathcal{C} = \{-3, -1, +1, +3\} \text{ relaxed to } \mathbf{conv}\ \mathcal{C} = [-3, 3];$$
$$\mathcal{C} = \{\mathbf{X} \in \mathbb{S}_+^n \mid \text{rank}(\mathbf{X}) = 1\} \text{ relaxed to } \mathbf{conv}\ \mathcal{C} = \mathbb{S}_+^n \tag{4.171}$$

(called semidefinite relaxation (SDR) for the second case). However, the optimal solution, $\boldsymbol{x}^\star$, of the reformulated convex problem may not be feasible to the original nonconvex problem, and, consequently, one has to find a good approximate solution of the original problem based on $\boldsymbol{x}^\star$ through some problem-dependent procedures. The details will be addressed in Section 8.4. Nevertheless, the obtained solution can be a stationary-point solution under some certain conditions or scenarios. Actually, various restrictions and relaxations mixed with problem reformulations have been extensively employed to handle challenging nonconvex problems, such as in transmit beamforming for MIMO wireless communications, and some of them will be introduced in Chapter 8 later. □

4.8   Summary and discussion

In this chapter we have introduced the basics of convex optimization problems, the first-order optimality condition (which is both necessary and sufficient), and their equivalent forms. We have also presented many examples to illustrate convex optimization problems and the use of the first-order optimality condition for finding an optimal solution, by reformulation of the original optimization problem via equivalent representations, function transformations, and change of variables that are typical approaches used for problem reformulation. Some more optimality conditions for convex problems, so called KKT conditions (which are also both necessary and sufficient), will be introduced in Chapter 9. The quasi-convex optimization problem, the first-order optimality condition (which is only sufficient), and the widely used bisection method for solving the problem were also introduced. Note that once we come up with a convex problem, a global optimal numerical solution can be obtained using the off-the-shelf convex solvers.

However, when an optimization problem of interest cannot be reformulated into a convex problem via any preceding approaches introduced in this chapter, we instead may have to find convex approximations either to the objective function or to the constraint functions, or both (e.g., relaxation and restriction to the constraint set, first-order approximations to the objective function or constraint functions) so that we can still obtain an approximate solution to the optimization problem under consideration. Besides the SCA introduced in Subsection 4.7, the iterative *condensation method* to be introduced in Chapter 5 is an efficient restriction approach to reshaping the feasible set, especially suitable for the use of Geometric Programs. The BSUM approach introduced in this chapter is another powerful approach to obtaining a stationary-point solution. Subsequently, mathematical performance and complexity analysis to the obtained approximate solution or algorithm will be needed to pin down its essential characteristics and properties (e.g., a stationary-point solution, performance gap w.r.t. the optimal solution, convergence conditions, complexity order, and algorithm implementation) for its practical usefulness as shown in Figure 1.6. The procedure for the algorithm design shown in this figure has proven very useful in various applications, especially in wireless communications and networking to be introduced in Chapter 8 later.

References for Chapter 4

[And63]   T. W. Anderson, "Asymptotic theory for principal component analysis," *Ann. Math. Statist.*, vol. 34, no. 1, pp. 122–148, Mar. 1963.

[Ber99]   D. P. Bertsekas, *Nonlinear Programming*, 2nd ed. Belmont, MA:Athena Scientific, 1999.

[Ber09]   ——, *Convex Optimization Theory*. Belmont, MA, USA: Athena Scientific, 2009.

[BV04]   S. Boyd and L. Vandenberghe, *Convex Optimization*. Cambridge, UK: Cambridge University Press, 2004.

[CMCW08]   T.-H. Chan, W.-K. Ma, C.-Y. Chi, and Y. Wang, "A convex analysis framework for blind separation of non-negative sources," *IEEE Trans. Signal Processing*, vol. 56, no. 10, pp. 5120–5134, Oct. 2008.

[Dun81]   J. C. Dunn, "Global and asymptotic convergence rate estimates for a class of projected gradient processes," *SIAM J. Contr. Optimiz.*, vol. 19, no. 3, pp. 368–400, 1981.

[Kay13]   S. Kay, *Fundamentals of Statistical Signal Processing, Volume III: Practical Algorithm Development*. Pearson Education, 2013.

[MCCW10] W.-K. Ma, T.-H. Chan, C.-Y. Chi, and Y. Wang, "Convex analysis for non-negative blind source separation with application in imaging," in *Convex Optimization in Signal Processing and Communications*, D. P. Palomar and Y. C. Eldar, Eds. Cambridge University Press, 2010, ch. 7.

[Men95] J. M. Mendel, *Lessons in Estimation Theory for Signal Processing, Communications, and Control*. Pearson Education, 1995.

[MW78] B. R. Marks and G. P. Wright, "A general inner approximation algorithm for nonconvex mathematical programs," *Operations Research*, vol. 26, no. 4, pp. 681–683, 1978.

[RHL13] M. Razaviyayn, M. Hong, and Z.-Q. Luo, "A unified convergence analysis of block successive minimization methods for nonsmooth optimization," *SIAM J. Optimization*, vol. 23, no. 2, pp. 1126–1153, 2013.

[SBL12] L. Sorber, M. V. Barel, and L. Lathauwer, "Unconstrained optimzation of real function in complex variables," *SIAM J.Optim.*, vol. 22, no. 3, pp. 879–898, Jul. 2012.

[Tuy00] H. Tuy, "Monotonic optimization: Problems and solution approaches," *SIAM J. Optimization*, vol. 11, no. 2, pp. 464–49, 2000.

[WAR+09] J. Wright, G. Arvind, S. Rao, Y. Peng, and Y. Ma, "Robust principal component analysis: Exact recovery of corrupted low-rank matrices via convex optimization," in *Advances in Neural Information Processing Systems 22*. Curran Associates, Inc., 2009, pp. 2080–2088.

5   Geometric Programming

This chapter deals with another type of optimization problem called geometric program (GP). The GP has been found useful in some resource allocation problems in communication and networking, where all the unknowns are positive in nature, such as powers, achievable transmission rates, secrecy rates, etc. The GP itself is a nonconvex problem that, however, can be converted into a convex problem via the change of variables and the function transformation introduced in Chapter 4, thereby providing an exemplar of convex problem reformulation from a nonconvex problem (as detailed in Chapter 4). Then we introduce the condensation method that is effectively implemented using the GP via some conservative approximations to the constraints of a specific type problem, followed by its application in physical layer secret communications.

5.1   Some basics

- A function $f : \mathbb{R}^n \to \mathbb{R}$ with $\mathbf{dom}\, f = \mathbb{R}^n_{++}$, defined as

$$f(\mathbf{x}) = c x_1^{a_1} x_2^{a_2} \cdots x_n^{a_n}, \qquad (5.1)$$

where $c > 0$ and $a_i \in \mathbb{R}$ is called a *monomial function*. It is nonconvex in general.

- A sum of monomials

$$f(\mathbf{x}) = \sum_{k=1}^{K} c_k x_1^{a_{1k}} x_2^{a_{2k}} \cdots x_n^{a_{nk}}, \qquad (5.2)$$

where $c_k > 0$ and $a_{ik} \in \mathbb{R}$ is called a *posynomial function*. It is nonconvex in general.

- The log-sum-exp function which is convex on $\mathbb{R}^n$ has been defined in Subsection 3.1.4 as

$$f(\mathbf{x}) = \log(e^{x_1} + \cdots + e^{x_n}), \qquad (5.3)$$

where $\mathbf{dom}\, f \subseteq \mathbb{R}^n$. As mentioned in Subsection 3.1.4, its convexity can be proven by the second-order condition given by (3.27).

- The following is the extended log-sum-exp function

$$f(\mathbf{x}) = \log(e^{\mathbf{a}_1^T \mathbf{x} + b_1} + \cdots + e^{\mathbf{a}_m^T \mathbf{x} + b_m}), \tag{5.4}$$

where $\mathbf{a}_i \in \mathbb{R}^n$ and $b_i \in \mathbb{R}$. It is also convex on $\mathbb{R}^n$.

Proof: The convexity of (5.4) can be readily proved by the fact that the log-sum-exponential function is convex and its composition with any affine mapping preserves the convexity of the resulting function (as presented in Subsection 3.2.2). Let $g(\mathbf{y}) = \log(e^{y_1} + \cdots + e^{y_m})$, which is known to be convex on $\mathbb{R}^m$. Let $\mathbf{A} = [\mathbf{a}_1, \ldots, \mathbf{a}_m]^T \in \mathbb{R}^{m \times n}$, and $\mathbf{b} = [b_1, \ldots, b_m]^T \in \mathbb{R}^m$. Then $f(\mathbf{x}) = g(\mathbf{A}\mathbf{x} + \mathbf{b})$ is convex and given by (5.4). ∎

5.2   Geometric program (GP)

The general structure of a GP is as follows

$$
\begin{aligned}
\min \quad & \sum_{k=1}^{K_0} c_{0k} x_1^{a_{0,1k}} x_2^{a_{0,2k}} \cdots x_n^{a_{0,nk}} \\
\text{s.t.} \quad & \sum_{k=1}^{K_i} c_{ik} x_1^{a_{i,1k}} x_2^{a_{i,2k}} \cdots x_n^{a_{i,nk}} \le 1, \ i = 1, \ldots, m \\
& d_i x_1^{g_{i1}} \cdots x_n^{g_{in}} = 1, \ i = 1, \ldots, p
\end{aligned}
\tag{5.5}
$$

where $c_{ik} > 0$, $d_i > 0$, and $a_{i,jk}, g_{ij} \in \mathbb{R}$ for all i, j, k, and the problem domain is $\mathcal{D} = \mathbb{R}_{++}^n$. Some characteristics of this problem are as follows:

- The objective function and the inequality constraint functions are posynomials.
- The equality constraint functions are monomials.
- It is not a convex optimization problem in general.

The GP defined in (5.5) appears to be a nonconvex problem. Via change of variables and function transformation, it can be converted into a convex problem, as presented in the next section.

5.3   GP in a convex form

Let

$$
\begin{aligned}
x_i &= e^{y_i} \text{ (change of variables)} \\
\mathbf{y} &= [y_1, \ldots, y_n]^T.
\end{aligned}
$$

Then a monomial can be re-expressed as

$$cx_1^{a_1} x_2^{a_2} \cdots x_n^{a_n} = e^{\mathbf{a}^T \mathbf{y} + b}, \tag{5.6}$$

where $b = \log c$ and $\mathbf{a} = [a_1, \ldots, a_n]^T$. Likewise, a posynomial can be re-expressed as

$$\sum_{k=1}^{K} c_k x_1^{a_{1k}} x_2^{a_{2k}} \cdots x_n^{a_{nk}} = \sum_{k=1}^{K} e^{\mathbf{a}_k^T \mathbf{y} + b_k}, \tag{5.7}$$

where

$$b_k = \log c_k, \quad \mathbf{a}_k = [a_{1k}, \ldots, a_{nk}]^T.$$

By defining

$$\mathbf{g}_i = [g_{i1}, \ldots, g_{in}]^T, \quad h_i = \log d_i,$$

the GP defined in (5.5) can then be transformed to

$$
\begin{aligned}
\min \quad & \sum_{k=1}^{K_0} e^{\mathbf{a}_{0k}^T \mathbf{y} + b_{0k}} \\
\text{s.t.} \quad & \sum_{k=1}^{K_i} e^{\mathbf{a}_{ik}^T \mathbf{y} + b_{ik}} \le 1, \ i = 1, \ldots, m \\
& e^{\mathbf{g}_i^T \mathbf{y} + h_i} = 1, \ i = 1, \ldots, p
\end{aligned}
\tag{5.8}
$$

where $\mathbf{a}_{ik}$ and b_{ik} are defined similarly as $\mathbf{a}_k$ and b_k were defined above. Note that for notational simplicity, we have used $\mathbf{a}_k$ ($n \times 1$ vector) and b_k in (5.7), and $\mathbf{a}_{ik}$ ($n \times 1$ vector) and b_{ik} in (5.8) to denote the associated parameters, respectively. By applying a logarithm to the objective function and all the equality and inequality constraint functions in (5.8) (function transformation), we obtain an equivalent problem

$$
\begin{aligned}
\min \quad & \log \sum_{k=1}^{K_0} e^{\mathbf{a}_{0k}^T \mathbf{y} + b_{0k}} \\
\text{s.t.} \quad & \log \sum_{k=1}^{K_i} e^{\mathbf{a}_{ik}^T \mathbf{y} + b_{ik}} \le 0, \ i = 1, \ldots, m \\
& \mathbf{g}_i^T \mathbf{y} + h_i = 0, \ i = 1, \ldots, p
\end{aligned}
\tag{5.9}
$$

that is a convex optimization problem, where $\mathbf{y} \in \mathbb{R}^n$ is the unknown vector variable. Note that the objective function and the inequality constraints in both (5.8) and (5.9) are convex though (5.8) is not a convex problem. After the optimal solution $\mathbf{y}^\star$ of (5.9) is obtained, the associated optimal solution of the GP (5.5) is given by

$$\mathbf{x}^\star = (e^{y_1^\star}, \ldots, e^{y_n^\star}).$$

Example 5.1 *(Optimal power assignment—revisited)* Let us reconsider the max-min-fair power assignment problem (4.154) in Chapter 4:

$$
\min_{p_i, i=1,\ldots,K} \max_{i=1,\ldots,K} \frac{\sum_{j=1, j\neq i}^{K} G_{ij} p_j + \sigma_i^2}{G_{ii} p_i} \tag{5.10}
$$
$$
\text{s.t. } 0 < p_i \leq P_i, \ i = 1, \ldots, K.
$$

which is a quasiconvex problem with the problem domain $\mathcal{D} \subseteq \mathbb{R}_{++}^K$. In Section 4.5, the quasiconvex problem (5.10) was solved by the iterative bisection method, where the associated feasibility problem is an LP. Now, let us reformulate this problem into a GP and so the optimal solution can be obtained more efficiently.

By epigraph reformulation, we may rewrite problem (5.10) as

$$
\min_{t, p_i, i=1,\ldots,K} t
$$
$$
\text{s.t. } \frac{\sum_{j=1, j\neq i}^{K} G_{ij} p_j + \sigma_i^2}{G_{ii} p_i} \leq t, \ i = 1, \ldots, K \tag{5.11}
$$
$$
0 < p_i \leq P_i, \ i = 1, \ldots, K.
$$

The problem can be further reformulated as

$$
\min_{t, p_i, i=1,\ldots,K} t \tag{5.12a}
$$
$$
\text{s.t. } \sum_{j=1, j\neq i}^{K} G_{ij} G_{ii}^{-1} p_j p_i^{-1} t^{-1} + \sigma_i^2 G_{ii}^{-1} p_i^{-1} t^{-1} \leq 1, \ i = 1, \ldots, K \tag{5.12b}
$$
$$
P_i^{-1} p_i \leq 1, \ i = 1, \ldots, K. \tag{5.12c}
$$

Note that the objective function t in (5.12a) is a monomial, (5.12b) is a posynomial inequality for each i, and (5.12c) is also a monomial inequality for each i. The problem now is a GP with the problem domain $\mathcal{D} \subseteq \mathbb{R}_{++}^{K+1}$. Then this nonconvex GP can be converted into a convex problem as presented above. This example also illustrates that an optimization problem can be formulated into different types of convex problems, thereby providing different implementations for solving the same problem. □

5.4   Condensation method

In some applications (e.g., power allocation in wireless communications), one may face the following optimization problem:

$$
\min \ q(\mathbf{x})
$$
$$
\text{s.t. } \frac{f(\mathbf{x})}{g(\mathbf{x})} \leq 1 \tag{5.13}
$$
$$
h(\mathbf{x}) = 1
$$

where $q(\mathbf{x})$, $f(\mathbf{x})$, and $g(\mathbf{x})$ are posynomials and $h(\mathbf{x})$ is a monomial. This problem is not a standard GP. This is because the inequality constraint, which is the ratio of two posynomials, is not a posynomial. However, such a problem can be solved by a successive GP approximation method, known as the *condensation method* to be introduced next.

5.4.1   Successive GP approximation

In the condensation method, we approximate the posynomial function $g(\mathbf{x})$ with a monomial function, so that the left-hand side of the inequality constraint in (5.13) will be a posynomial (as ratio of a posynomial to a monomial is still a posynomial). Then we successively perform such a posynomial-by-monomial approximation to the modified problem (to be explained below), thereby yielding a GP for the resulting problem in a successive fashion.

A popular way of approximating the posynomial with a monomial is by the arithmetic-geometric mean inequality. Recall from Chapter 3 (see (3.65)) that for $\alpha_i > 0$ and $u_i > 0$, $i = 1, ..., n$, and $\sum_{i=1}^{n} \alpha_i = 1$,

$$\sum_{i=1}^{n} u_i \geq \prod_{i=1}^{n} \left(\frac{u_i}{\alpha_i} \right)^{\alpha_i}. \tag{5.14}$$

Let $\{u_i(\mathbf{x})\}$ be the monomial terms in the posynomial $g(\mathbf{x})$. Based on the above inequality, one can have

$$g(\mathbf{x}) = \sum_{i=1}^{n} u_i(\mathbf{x}) \geq \prod_{i=1}^{n} \left(\frac{u_i(\mathbf{x})}{\alpha_i} \right)^{\alpha_i}. \tag{5.15}$$

Note that the right-hand side of the inequality in (5.15) is clearly a monomial (as product of monomials is still a monomial). A common way to choose α_i, at a given feasible point $\mathbf{x}_0$, is

$$\alpha_i = \frac{u_i(\mathbf{x}_0)}{g(\mathbf{x}_0)}, \tag{5.16}$$

which satisfies $\alpha_i > 0$ and $\sum_{i=1}^{n} \alpha_i = 1$. Substituting α_i given by (5.16) into (5.15) gives rise to

$$\tilde{g}(\mathbf{x}, \mathbf{x}_0) = \prod_{i=1}^{n} \left(\frac{u_i(\mathbf{x})}{\alpha_i} \right)^{\alpha_i} \leq g(\mathbf{x}), \tag{5.17}$$

which is a monomial approximation of the posynomial function $g(\mathbf{x})$ at a given feasible point $\mathbf{x}_0$, and

$$\tilde{g}(\mathbf{x}_0, \mathbf{x}_0) = g(\mathbf{x}_0). \tag{5.18}$$

In other words, $\tilde{g}(\mathbf{x}_0, \mathbf{x}_0)$ is a tight lower bound of $g(\mathbf{x})$. Let $\mathcal{C}$ be the feasible set of problem (5.13) and $\mathbf{x}_0 \in \mathcal{C}$. Then it can be seen that

$$\mathcal{C}(\mathbf{x}_0) \triangleq \{\mathbf{x} \mid f(\mathbf{x})/\tilde{g}(\mathbf{x}, \mathbf{x}_0) \leq 1, h(\mathbf{x}) = 1\}$$
$$\subseteq \mathcal{C} \triangleq \{\mathbf{x} \mid f(\mathbf{x})/g(\mathbf{x}) \leq 1, h(\mathbf{x}) = 1\} \quad \text{(by (5.17))}. \qquad (5.19)$$

Now, we can approximate the inequality constraint in problem (5.13) with the following constraint:

$$\frac{f(\mathbf{x})}{\tilde{g}(\mathbf{x}, \mathbf{x}_0)} \leq 1. \qquad (5.20)$$

The resulting problem

$$\min\ q(\mathbf{x})$$
$$\text{s.t.}\ \frac{f(\mathbf{x})}{\tilde{g}(\mathbf{x}, \mathbf{x}_0)} \leq 1 \qquad (5.21)$$
$$h(\mathbf{x}) = 1$$

is a standard GP (with a condensed constraint set $\mathcal{C}(\mathbf{x}_0) \subseteq \mathcal{C}$ (cf. (5.19)). Then a solution to problem (5.13) can be obtained by the following successive approximation method (condensation method) given in Algorithm 5.1.

Algorithm 5.1 Pseudocode for Condensation Method

1: Given a feasible point $\mathbf{x}_0^{(0)} \in \mathcal{C}$.
2: Set $k := 0$.
3: **repeat**
4:    Approximate the posynomial function $g(\mathbf{x})$ by a monomial function $\tilde{g}(\mathbf{x}, \mathbf{x}_0^{(k)})$ in (5.17).
5:    Solve the approximated problem (5.21) and let the optimal solution be $\mathbf{x}_0^{(k+1)} \in \mathcal{C}(\mathbf{x}_0^{(k)}) \subseteq \mathcal{C}$.
6:    Set $k := k + 1$.
7: **until** A stopping condition on convergence of the optimal solution is achieved.

Although the objective value $q(\mathbf{x}_0^{(k)})$ monotonically decreases with the iteration number k until convergence (in N iterations), the solution $\mathbf{x}_0^{(N)}$ found by the above successive approximation method can be expressed as

$$\mathbf{x}_0^{(N)} = \arg\ \min\left\{q(\mathbf{x}) \mid \mathbf{x} \in \cup_{k=0}^{N-1} \mathcal{C}(\mathbf{x}_0^{(k)}) \subseteq \mathcal{C}\right\}, \qquad (5.22)$$

which is an approximate solution to (5.13), namely, a suboptimal solution. Let us conclude this subsection with the following remark:

Remark 5.1 For a nonconvex problem like problem (5.13), it may be almost formidable to be reformulated into a convex problem, and even finding an approximate solution is non-trivial. Through successive restrictive approximation to the constraint set of the nonconvex problem (5.13), the condensation method can efficiently yield a good approximation solution. On the other hand, the SCA introduced in Section 4.7 is another widely used restrictive approximation method to conservatively find some convex approximations to those nonconvex constraints of the original problem such that the original problem can be handled iteratively using advisable convex optimization tools or algorithms. Although the obtained solution is suboptimal, it may be provable to be a stationary point (cf. Section 4.7) of the original problem by the associated KKT conditions. If a stationary-point solution can be obtained by a computationally efficient algorithm with polynomial time and meanwhile the performance requirement is met, we actually have found a practical solution to the original problem instead. Some more interesting conservative approximations used in coordinated transmit beamforming for MISO interference channel will be presented in Subsection 8.5.6.   □

5.4.2   Physical-layer secret communications

An application example in physical-layer secret communications is the multi-stage training sequence design for discriminatory channel estimation in wireless MIMO systems [CCHC10]. The scenario considered is shown in Figure 5.1. This training signal design also considers artificial noise superimposed in the training signal, while it is added in the null space of the legitimate receiver's estimated channel. The design objective is to minimize the normalized mean-squared error of the legitimate receiver's channel estimation under the constraint on the total transmit energy and the constraint on the normalized mean-squared error of the unauthorized receiver's (e.g., eavesdropper) channel estimation. This problem can be formulated as (5.13), and therefore the above successive QP approximation method can be applied to obtain approximate solutions for the allocation of the training signal power and artificial noise power at each stage. The details can be found in [CCHC10].

5.5   Summary and discussion

In this chapter we have introduced a simple, yet powerful optimization problem called geometric program. The GP defined in terms of monomial functions and posynomial functions, in general, is nonconvex, and the transformation used to convert it into a convex problem was also elaborated. Finally, we briefly introduced the condensation method (a successive GP approximation method) that converts a nonconvex problem with the same form as (5.13) (also in terms of monomial functions and posynomial functions) into a GP through the use of a Jensen's inequality in the constraint set approximation introduced in Chapter 3, which has been regarded as a pragmatic approach to handling nonconvex problems like (5.13). In spite of guaranteed convergence of the condensation method together with a suboptimal solution, it has been effectively applied to power

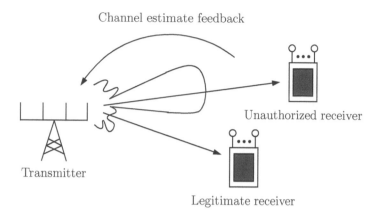

Channel estimate feedback

Unauthorized receiver

Transmitter

Legitimate receiver

Figure 5.1 A wireless MIMO system consisting of a multi-antenna transmitter, a multi-antenna legitimate receiver, and a multi-antenna unauthorized receiver.

assignment and allocation problems and physical-layer secret communications in wireless communications recently. More interesting convex approximations applied to handling nonconvex problems will be presented in Chapter 8, where the resulting convex SDP is obtained via various convex approximations for handling nonconvex MIMO communications problems.

References for Chapter 5

[CCHC10]   T.-H. Chang, W.-C. Chiang, Y.-W. P. Hong, and C.-Y. Chi, "Training sequence design for discriminatory channel estimation in wireless MIMO systems," *IEEE Trans. Signal Process.*, vol. 58, no. 12, pp. 6223–6237, Dec. 2010.

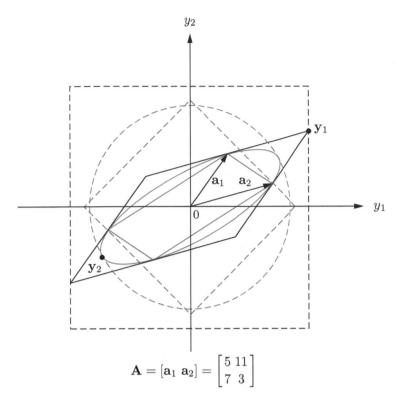

$$\mathbf{A} = [\mathbf{a}_1 \ \mathbf{a}_2] = \begin{bmatrix} 5 & 11 \\ 7 & 3 \end{bmatrix}$$

Figure 1.2 An illustration of the induced norm of $\mathbf{A} \in \mathbb{R}^{2 \times 2}$, $\|\mathbf{A}\|_1 = \|\mathbf{a}_2\|_1$, $\|\mathbf{A}\|_2 = \|\mathbf{y}_2\|_2$ (the largest semiaxis of the ellipsoid), and $\|\mathbf{A}\|_\infty = \|\mathbf{y}_1\|_\infty$, where the parallelograms and the ellipsoid (denoted by solid lines) are the images of the associated unit-radius 1-norm, 2-norm, and ∞-norm balls shown in Figure 1.1, via the linear transformation $\mathbf{Ax}$, together with the associated norm balls (denoted by dashed lines) with radii being $\|\mathbf{A}\|_1$, $\|\mathbf{A}\|_2$, and $\|\mathbf{A}\|_\infty$, respectively.

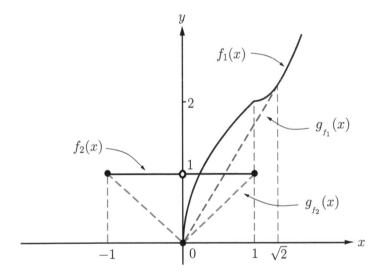

Figure 3.5 Convex envelopes $g_{f_1}(x)$ and $g_{f_2}(x)$ (cf. (3.57) and (3.58)) of two nonconvex functions $f_1(x)$ and $f_2(x)$ defined in (3.55) and (3.56), respectively.

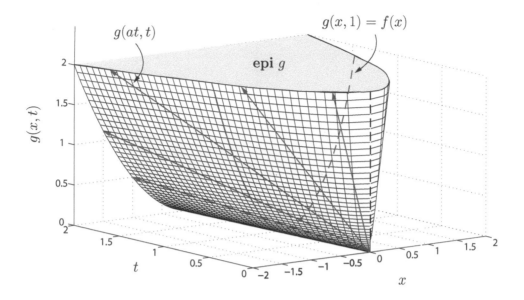

Figure 3.7 Epigraph of $g(x,t) = x^2/t$, the perspective of $f(x) = x^2$, where the red curve denotes $g(x,1) = f(x)$, and each ray in blue color is associated with $g(at,t) = a^2t$ for a different value of a.

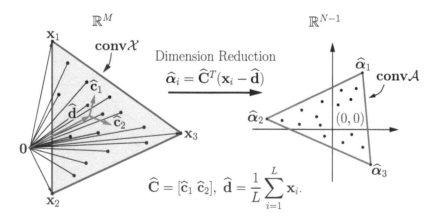

Figure 4.3 Dimension reduction illustration using affine set fitting for $N = 3$, where the geometric center $\widehat{\mathbf{d}}$ of the data cloud $\mathcal{X}$ in the M-dimensional space maps to the origin in the $(N-1)$-dimensional space which is also the geometric center of the dimension-reduced data cloud $\mathcal{A}$.

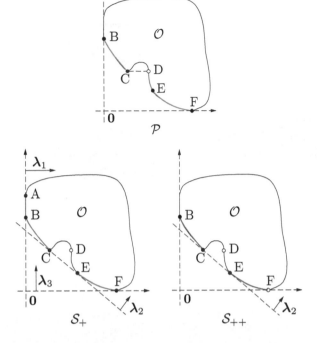

Figure 4.5 An illustration in $\mathbb{R}^2$ with the proper cone $K = \mathbb{R}_+^2$ in problem (4.112) for the set $\mathcal{O} = \boldsymbol{f}_0(\mathcal{C})$ (the achievable objective vectors) defined in (4.113), the Pareto boundary $\mathcal{P} = \boldsymbol{f}_0(\mathcal{C}_P)$ (top plot in blue color), the set $\mathcal{S}_+$ defined in (4.120) (bottom left plot in green color), and the set $\mathcal{S}_{++}$ defined in (4.121) (bottom right plot in red color excluding point F since $\mathcal{P}$ is smooth at F), where the line segment from point A to point B (at which $\mathcal{P}$ is nonsmooth) is the set of $\boldsymbol{y}^\circ(\boldsymbol{\lambda}_1)$, the set of $\boldsymbol{y}^\circ(\boldsymbol{\lambda}_2)$ contains only points C and E, and point F is exactly $\boldsymbol{y}^\circ(\boldsymbol{\lambda}_3)$.

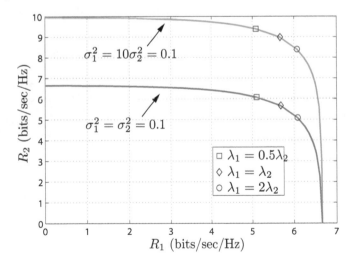

Figure 4.7 An illustration for showing Pareto boundaries of problem (4.132) for $P = 10$ dB and different values of parameters σ_1^2 and σ_2^2. One is denoted by the blue curve for $\sigma_1^2 = \sigma_2^2 = 0.1$ and the other by a green curve for $\sigma_1^2 = 10\sigma_2^2 = 0.1$ together with two sets of maximal elements (squares, diamonds, and circles) on each Pareto boundary with the corresponding weights $\lambda_1 = 0.5\lambda_2$, $\lambda_1 = \lambda_2$, and $\lambda_1 = 2\lambda_2$, respectively.

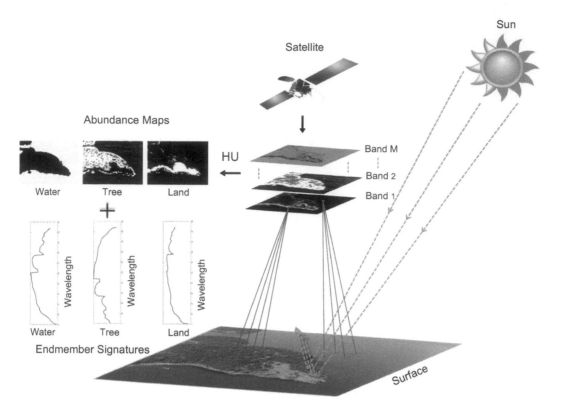

Figure 6.4 Illustration of hyperspectral unmixing.

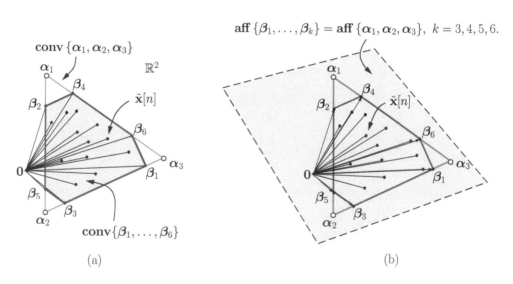

Figure 6.13 Illustration of GENE-CH algorithm, when no pure pixel is present in the noise-free hyperspectral data ($N = 3$ case). (a) The endmember estimates are denoted by $\boldsymbol{\beta}_i, i = 1, \ldots, N_{\max} = 6$, but $\mathbf{conv}\,\{\boldsymbol{\beta}_1, \ldots, \boldsymbol{\beta}_6\} \neq \mathbf{conv}\,\{\boldsymbol{\alpha}_1, \boldsymbol{\alpha}_2, \boldsymbol{\alpha}_3\}$ because the true endmembers $\boldsymbol{\alpha}_1, \boldsymbol{\alpha}_2, \boldsymbol{\alpha}_3$ are not present in the data cloud, whereas $\mathbf{aff}\,\{\boldsymbol{\beta}_1, \ldots, \boldsymbol{\beta}_k\} = \mathbf{aff}\,\{\boldsymbol{\alpha}_1, \boldsymbol{\alpha}_2, \boldsymbol{\alpha}_3\}$, $k = 3, 4, 5, 6$, as shown in (b).

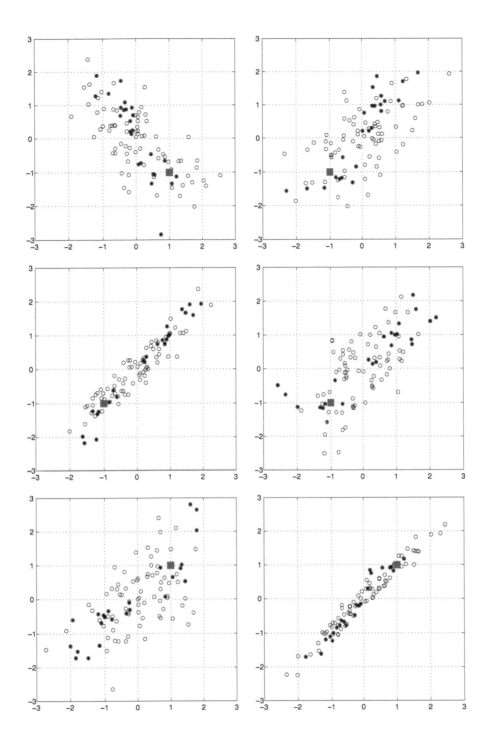

Figure 8.3 An illustration of Gaussian randomization for handling a BQP for $n = 12$ via SDR, where bullets and circles in the six subplots denote generated $L = 100$ random vectors $\boldsymbol{\xi} \in \mathbb{R}^n$ (with two components of $\boldsymbol{\xi}$ displayed in each subplot); the optimal solution $\mathbf{x}^\star = (1, -1, -1, -1, -1, -1, -1, -1, 1, 1, 1, 1)$ of the BQP is denoted by red bullets, and all the black bullets yield either $\widehat{\mathbf{x}} = \mathbf{x}^\star$ (denoted as blue squares) or $\widehat{\mathbf{x}} = -\mathbf{x}^\star$.

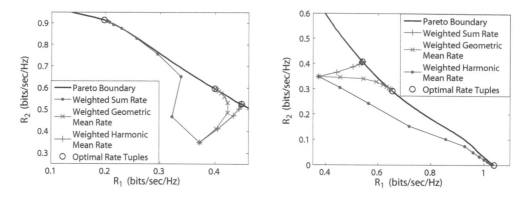

Figure 8.11 Convergence trajectories of the SCA algorithm by maximizing the weighted sum rate, weighted geometric mean rate, and weighted harmonic mean rate in a two-user MISO IFC with $N_t = 4$ and the user priority weights $(\alpha_1, \alpha_2) = (\frac{1}{2}, \frac{1}{2})$ for the left plot and $(\alpha_1, \alpha_2) = (\frac{2}{3}, \frac{1}{3})$ for the right plot. (© 2013 IEEE. Reprinted, with permission, from W.-C. Li, T.-H. Chang, C. Lin, and C.-Y. Chi, "Coordinated Beamforming for Multiuser MISO Interference Channel Under Rate Outage Constraints," Mar. 2013.)

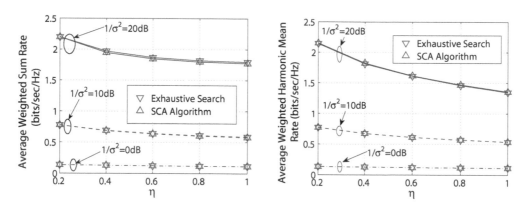

Figure 8.12 Average performance of the SCA algorithm in weighted sum rate (left plot) and weighted harmonic mean rate (right plot), for $K = 2$, $N_t = 4$, and the user priority weights $(\alpha_1, \alpha_2) = (\frac{1}{2}, \frac{1}{2})$, where each result is obtained by averaging over 500 realizations of $\{\mathbf{Q}_{ki}\}$. (© 2013 IEEE. Reprinted, with permission, from W.-C. Li, T.-H. Chang, C. Lin, and C.-Y. Chi, "Coordinated Beamforming for Multiuser MISO Interference Channel Under Rate Outage Constraints," Mar. 2013.)

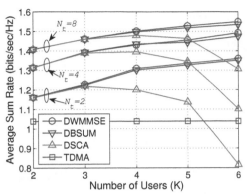

Figure 8.13 Performance comparison for Algorithm 8.2 (DBSUM), Algorithm 8.3 (DWMMSE), and the DSCA algorithm, where $1/\sigma^2 = 10$ dB, $\eta = 0.5$, $\alpha_1 = \cdots = \alpha_K = 1$, $\mathrm{rank}(\mathbf{Q}_{ki}) = N_t$ for all k, i. (© 2015 IEEE. Reprinted, with permission, from W.-C. Li, T.-H. Chang, and C.-Y. Chi, "Multicell Coordinated Beamforming with Rate Outage Constraint–Part II: Efficient Approximation Algorithms," Jun. 2015.)

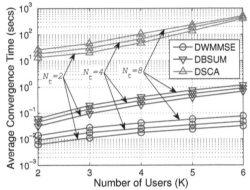

Figure 8.14 Complexity comparison for Algorithm 8.2 (DBSUM), Algorithm 8.3 (DWMMSE), and the DSCA algorithm, where $1/\sigma^2 = 10$ dB, $\eta = 0.5$, $\alpha_1 = \cdots = \alpha_K = 1$, $\mathrm{rank}(\mathbf{Q}_{ki}) = N_t$ for all k, i. (© 2015 IEEE. Reprinted, with permission, from W.-C. Li, T.-H. Chang, and C.-Y. Chi, "Multicell Coordinated Beamforming with Rate Outage Constraint–Part II: Efficient Approximation Algorithms," Jun. 2015.)

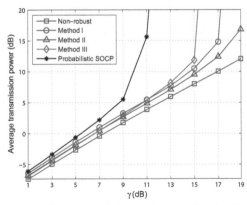

Figure 8.15 Transmit power performance of the various methods. $N_t = K = 3$; $\rho = 0.1$; spatially i.i.d. Gaussian CSI errors with $\sigma_e^2 = 0.002$. (© 2014 IEEE. Reprinted, with permission, from K.-Y. Wang, A. M.-C. So, T.-H. Chang, W.-K. Ma, and C.-Y. Chi, "Outage Constrained Robust Transmit Optimization for Multiuser MISO Downlinks: Tractable Approximations by Conic Optimization," Nov. 2014.)

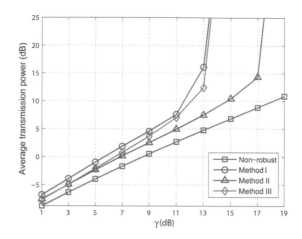

Figure 8.16 Transmit power performance under spatially correlated Gaussian CSI errors. $N_t = 8$; $K = 6$; $\rho = 0.01$; $\sigma_e^2 = 0.01$. (© 2014 IEEE. Reprinted, with permission, from K.-Y. Wang, A. M.-C. So, T.-H. Chang, W.-K. Ma, and C.-Y. Chi, "Outage Constrained Robust Transmit Optimization for Multiuser MISO Downlinks: Tractable Approximations by Conic Optimization," Nov. 2014.)

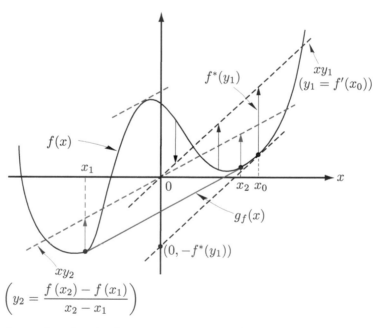

Figure 9.1 Conjugate function f^* of a nonconvex function $f : \mathbb{R} \to \mathbb{R}$, for $y_1, y_2 \in \mathbf{dom}\, f^* \subseteq \mathbb{R}$.

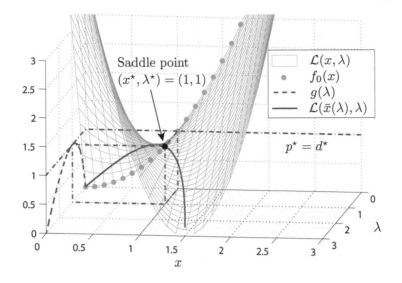

Figure 9.2 The objective function $f_0(x) = x^2$, Lagrangian $\mathcal{L}(x, \lambda)$, the curve $\mathcal{L}(\bar{x}(\lambda), \lambda)$, the dual function $g(\lambda) = \mathcal{L}(\bar{x}(\lambda) = 2\lambda/(1+\lambda), \lambda)$, and the primal-dual optimal solution $(x^\star, \lambda^\star) = (1, 1)$ of the convex problem $\min\{f_0(x) \mid (x-2)^2 \leq 1\}$ with strong duality.

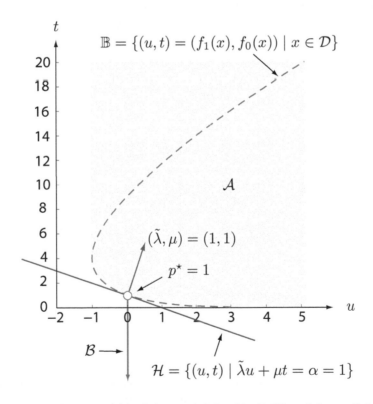

Figure 9.3 Illustration of the set $\mathcal{A}$ (shaded region) defined in (9.67) and the set $\mathcal{B}$ defined in (9.70) (a ray), and a separating hyperplane $\mathcal{H}$ between them, where $f_0(x) = x^2$, $f_1(x) = (x-2)^2 - 1$, and $\mathcal{D} = \mathbb{R}$.

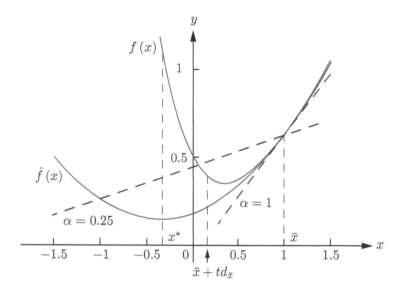

Figure 10.1 Illustration of backtracking line search by Algorithm 10.1, where $f(x)$ given by (10.14) (blue solid line), the associated quadratic convex function $\hat{f}$ given by (10.7) for $\bar{x} = 1$ (red solid line), and the associated linear approximations (dashed lines, one for $\alpha = 0.25$ and one for $\alpha = 1$) given by (10.15) are depicted. For this case, $x^* = -0.35$ (the optimal solution of $\hat{f}$) and $d_{\bar{x}} = x^* - \bar{x} = -1.35$; an admissible point $\bar{x} + td_{\bar{x}}$ for $t = 0.64$ is also indicated by an arrow.

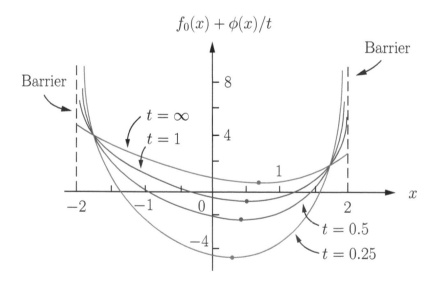

Figure 10.3 Illustration of problem (10.20) for different values of t, where $f_0(x)$ is given in (10.22) and the log barrier $\phi(x)$ is given in (10.23). The optimal solutions to the corresponding unconstrained convex problem (10.20) are $x = 0.310$ for $t = 0.25$, $x = 0.435$ for $t = 0.5$, $x = 0.539$ for $t = 1$, and $x = 0.7035$ for $t = \infty$, together with the associated optimal values of $f_0(x) + \phi(x)/t$ denoted by bullets. Note that the four associated curves intersect at $x = \pm\sqrt{3}$ due to $\phi(\pm\sqrt{3}) = 0$.

6 Linear Programming and Quadratic Programming

In this chapter, we will introduce two important types of convex optimization problems, LP and QP (including quadratic constrained QP (QCQP)), which have been widely applied in science and engineering problems. We will illustrate them via various examples and some cutting edge applications in signal processing (blind source separation in biomedical imaging and hyperspectral imaging) and in communications (power assignment, receive beamforming, and transmit beamforming). Through these examples and applications, the reformulation of the original optimization into an LP or a QP is clearly illustrated through change of variables, equivalent transformations, or representations introduced in Chapter 4, though closed-formed solutions and analytical solutions, which may exist, are usually hard to find.

6.1 Linear program (LP)

The linear program has the following general form:

$$\begin{aligned} \min \quad & \mathbf{c}^T \mathbf{x} \\ \text{s.t.} \quad & \mathbf{G}\mathbf{x} \preceq \mathbf{h}, \end{aligned} \tag{6.1}$$

which is also called the *inequality form* of an LP.

In a nutshell, we can say that LP is a problem of minimizing a linear objective function over a polyhedron (see Figure 6.1). One can observe, from Figure 6.1, one of the extreme points of the feasible set (polyhedron) will be an optimal point to both (6.1) and (6.4), if the feasible set is compact. Multiple solutions may exist as $\mathbf{c}$ is orthogonal to a facet of the polyhedron.

The general form of LP in (6.1) can be reformulated into the so-called *standard form* LP (cf. (6.4) below), by using a *slack variable*

$$\mathbf{s} \triangleq \mathbf{h} - \mathbf{G}\mathbf{x},$$

which plays the role as an auxiliary variable like a relay in a cooperative communication system or network. With the slack variable $\mathbf{s}$ defined above, problem

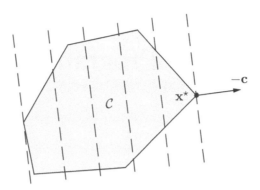

Figure 6.1 Illustration of an LP.

(6.1) can be written as follows:

$$\min \ \mathbf{c}^T \mathbf{x}$$
$$\text{s.t. } \mathbf{s} \succeq \mathbf{0}, \ \mathbf{h} - \mathbf{Gx} = \mathbf{s}. \tag{6.2}$$

Let $\mathbf{x} = \mathbf{x}_+ - \mathbf{x}_-$, where $\mathbf{x}_+, \mathbf{x}_- \succeq \mathbf{0}$. Problem (6.2) is equivalent to

$$\min \ \begin{bmatrix} \mathbf{c}^T & -\mathbf{c}^T & \mathbf{0}^T \end{bmatrix} \begin{bmatrix} \mathbf{x}_+ \\ \mathbf{x}_- \\ \mathbf{s} \end{bmatrix}$$
$$\text{s.t. } \begin{bmatrix} \mathbf{x}_+ \\ \mathbf{x}_- \\ \mathbf{s} \end{bmatrix} \succeq \mathbf{0}, \ \begin{bmatrix} \mathbf{G} & -\mathbf{G} & \mathbf{I} \end{bmatrix} \begin{bmatrix} \mathbf{x}_+ \\ \mathbf{x}_- \\ \mathbf{s} \end{bmatrix} = \mathbf{h}. \tag{6.3}$$

Problem (6.3) has been in the standard form of LP as follows:

$$\min \ \mathbf{c}^T \boldsymbol{x}$$
$$\text{s.t. } \boldsymbol{x} \succeq \mathbf{0}, \ \boldsymbol{Ax} = \boldsymbol{b} \tag{6.4}$$

where

$$\boldsymbol{A} = \begin{bmatrix} \mathbf{G} & -\mathbf{G} & \mathbf{I} \end{bmatrix}, \ \ \boldsymbol{c} = [\mathbf{c}^T \ -\mathbf{c}^T \ \mathbf{0}^T]^T, \ \ \boldsymbol{x} = [\mathbf{x}_+^T \ \mathbf{x}_-^T \ \mathbf{s}^T]^T,$$

while $\boldsymbol{x}$ is the unknown vector variable.

Note that maximization of an LP is equivalent to the minimization of an LP with the polarity of the objective function (linear function) reversed. So both maximization and minimization problems have been regarded as LP.

6.2   Examples using LP

An interesting example for LP is the piecewise linear minimization problem (defined in (4.62)) discussed in Subsection 4.3.1. More examples for LP will be given in this section.

6.2.1   Diet problem

Let us define the following variables and constraints:

- $x_i \geq 0$ is the quantity of food i, $i = 1, \ldots, n$.
- Each unit of food i has a cost of c_i.
- One unit of food i contains an amount a_{ji} of nutrient j, $j = 1, \ldots, m$.
- Nutrient i is required to be at least equal to b_i.

It is desired to find the cheapest diet such that the minimum nutrient requirements are fulfilled. This problem can be cast as an LP:

$$\begin{aligned} \min \quad & \mathbf{c}^T \mathbf{x} \\ \text{s.t.} \quad & \mathbf{A}\mathbf{x} \succeq \mathbf{b}, \ \mathbf{x} \succeq \mathbf{0}, \end{aligned} \tag{6.5}$$

where $\mathbf{x} \in \mathbb{R}^n$, $\mathbf{A} = \{a_{ji}\}_{m \times n}$, $\mathbf{c} = [c_1, \ldots, c_n]^T$, and $\mathbf{b} = [b_1, \ldots, b_m]^T$.

6.2.2   Chebyshev center

Consider a norm ball $B(\mathbf{x}_c, r) = \{\mathbf{x} \mid \|\mathbf{x} - \mathbf{x}_c\|_2 \leq r\}$ and a polyhedron $\mathcal{P} = \{\mathbf{x} \mid \mathbf{a}_i^T \mathbf{x} \leq b_i, \ i = 1, \ldots, m\}$.

♦ **Chebyshev Center Problem** Find the largest ball inside a polyhedron $\mathcal{P}$ (see Figure 6.2), which can be formulated as the following maximization problem:

$$\begin{aligned} \max_{\mathbf{x}_c, r} \quad & r \\ \text{s.t.} \quad & B(\mathbf{x}_c, r) \subseteq \mathcal{P} = \{\mathbf{x} \mid \mathbf{a}_i^T \mathbf{x} \leq b_i, \ i = 1, \ldots, m\}. \end{aligned} \tag{6.6}$$

At first glance, this problem is a nonconvex optimization problem, which, fortunately, can be formulated into an LP as presented next. The 2-norm ball $B(\mathbf{x}_c, r)$ can be represented as

$$B(\mathbf{x}_c, r) = \{\mathbf{x}_c + \mathbf{u} \mid \|\mathbf{u}\|_2 \leq r\} \quad \text{(by (2.36))}. \tag{6.7}$$

Recall that by Cauchy–Schwartz inequality,

$$\mathbf{a}_i^T \mathbf{u} \leq \|\mathbf{a}_i\|_2 \cdot \|\mathbf{u}\|_2 \tag{6.8}$$

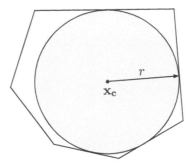

Figure 6.2 The maximal 2-norm ball with the center $\mathbf{x}_c$ and the radius r inside a polyhedron.

where the equality holds as $\mathbf{u} = \alpha \mathbf{a}_i$ for some $\alpha \geq 0$. Then the constraint set of (6.6) can be simplified as

$$B(\mathbf{x}_c, r) \subseteq \mathcal{P} \iff \sup \left\{ \mathbf{a}_i^T (\mathbf{x}_c + \mathbf{u}) \mid \|\mathbf{u}\|_2 \leq r \right\} \leq b_i, \quad \forall i$$
$$\iff \mathbf{a}_i^T \mathbf{x}_c + r \|\mathbf{a}_i\|_2 \leq b_i, \ \forall i \ \ (\text{i.e., } \mathbf{u} = r \frac{\mathbf{a}_i}{\|\mathbf{a}_i\|_2}), \tag{6.9}$$

where we have applied the Cauchy–Schwartz inequality in deriving (6.9). Hence, the Chebyshev center problem is equivalent to the following LP

$$\begin{aligned} \max_{\mathbf{x}_c, r} \ & r \\ \text{s.t. } & \mathbf{a}_i^T \mathbf{x}_c + r \|\mathbf{a}_i\|_2 \leq b_i, \ i = 1, \ldots, m. \end{aligned} \tag{6.10}$$

6.2.3 ℓ_∞-norm approximation

The problem is defined as

$$\min \ \|\mathbf{A}\mathbf{x} - \mathbf{b}\|_\infty \tag{6.11}$$

where $\mathbf{A} \in \mathbb{R}^{m \times n}$ and $\mathbf{b} \in \mathbb{R}^m$. By using the epigraph form, the ℓ_∞-norm approximation problem can be cast as an LP:

$$\begin{aligned} \min \ & t \\ \text{s.t. } & \max_{i=1,\ldots,m} |r_i| \leq t \\ & \mathbf{r} = \mathbf{A}\mathbf{x} - \mathbf{b} \end{aligned} \quad = \quad \begin{aligned} \min \ & t \\ \text{s.t. } & -t \mathbf{1}_m \preceq \mathbf{r} \preceq t \mathbf{1}_m \\ & \mathbf{r} = \mathbf{A}\mathbf{x} - \mathbf{b} \end{aligned} \tag{6.12}$$

where we have defined the auxiliary variable $\mathbf{r}$ in (6.12), and the unknown variables for this problem include $\mathbf{x}$, $\mathbf{r}$, and t.

6.2.4 ℓ_1-norm approximation

Similar to the above ℓ_∞-norm problem, the ℓ_1-norm approximation can be defined as

$$\min \ \|\mathbf{Ax} - \mathbf{b}\|_1, \tag{6.13}$$

where $\mathbf{A} \in \mathbb{R}^{m \times n}$ and $\mathbf{b} \in \mathbb{R}^m$. It can be rewritten as

$$\begin{aligned} &\min \ \sum_{i=1}^m |r_i| \\ &\text{s.t. } \mathbf{r} = \mathbf{Ax} - \mathbf{b} \end{aligned} \quad = \quad \begin{aligned} &\min \ \sum_{i=1}^m t_i \\ &\text{s.t. } -t_i \leq r_i \leq t_i, \ i = 1,\ldots,m \\ &\quad \mathbf{r} = \mathbf{Ax} - \mathbf{b} \end{aligned} \tag{6.14}$$

which is an LP. Again, we have defined the auxiliary variable $\mathbf{r}$ in (6.14), and the unknown variables for this problem include $\mathbf{x}$, $\mathbf{r}$, t_i, $i = 1,\ldots,m$. It can be easily inferred that the optimal $t_i^\star = |r_i^\star|$ for all i.

6.2.5 Maximization/minimization of matrix determinant

Often in blind source separation applications, the following optimization problem appears:

$$\max_{\mathbf{w}_j \in \mathcal{F}, \ j=1,\ldots,N} \ \det(\mathbf{W}), \tag{6.15}$$

where $\mathbf{w}_j$ is the jth column of matrix $\mathbf{W} = \{w_{ij}\}_{N \times N} \in \mathbb{R}^{N \times N}$, and $\mathcal{F}$ is a polyhedron (convex set). It can be seen that in problem (6.15), though the constraint set is convex, the nonconvexity of the objective function on $\mathbb{R}^{N \times N}$ makes the problem nonconvex.

However, problem (6.15) can be conveniently handled through an iterative maximization procedure using LP. The idea is motivated by the cofactor expansion of $\det(\mathbf{W})$ w.r.t. any one column or any row of $\mathbf{W}$ (cf. (1.76)). Here we consider the cofactor expansion along the jth column of $\mathbf{W}$, namely, $\mathbf{w}_j = [w_{1j}, w_{2j}, \ldots, w_{Nj}]^T$. It is given by

$$\det(\mathbf{W}) = \sum_{i=1}^N (-1)^{i+j} w_{ij} \det(\boldsymbol{\mathcal{W}}_{ij}), \tag{6.16}$$

where $\boldsymbol{\mathcal{W}}_{ij} \in \mathbb{R}^{(N-1) \times (N-1)}$ is the submatrix of $\mathbf{W}$ with the ith row and jth column removed, and it never depends upon $\mathbf{w}_j$ as j is fixed. It is easy to see that with $\boldsymbol{\mathcal{W}}_{ij}$ fixed (i.e., not depending on $\mathbf{w}_j$), $\det(\mathbf{W})$ becomes a linear function of $\mathbf{w}_j$. Therefore, by fixing all the column vectors of $\mathbf{W}$ other than $\mathbf{w}_j$, problem (6.15) can then be reduced to the following problem

$$\max_{\mathbf{w}_j \in \mathcal{F}} \ \sum_{i=1}^N (-1)^{i+j} w_{ij} \det(\boldsymbol{\mathcal{W}}_{ij}) = \max_{\mathbf{w}_j \in \mathcal{F}} \ \mathbf{b}_j^T \mathbf{w}_j, \tag{6.17}$$

where $\mathbf{b}_j \in \mathbb{R}^N$ with the ith component given by

$$[\mathbf{b}_j]_i = (-1)^{i+j}\det(\boldsymbol{\mathcal{W}}_{ij}), \quad i = 1, \ldots, N.$$

Thus, problem (6.15) can be handled iteratively by solving the LP (6.17) in a round-robin fashion (i.e., updating $\mathbf{W}$ either in column-by-column or row-by-row fashion) until the convergence of $\det(\mathbf{W})$. Though the obtained solution for $\mathbf{W}$ is an approximate solution to problem (6.15), it has proven useful in biomedical and hyperspectral image analysis to be introduced below in this chapter. Finally, it should be mentioned that the above procedure is also applicable to the minimization counterpart of (6.15).

6.3   Applications in blind source separation using LP/convex geometry

In this section, we will introduce some cutting-edge research that we recently accomplished by means of LP. The first one is nonnegative blind source separation (nBSS) [CZPA09, LS99, CMCW08] with a given set of linear mixtures of unknown source signals, aiming to extract the nonnegative source signals (e.g., biomedical image signals) and the other is hyperspectral image signals, where "blind" means no prior information of how they were mixed (which is a practical scenario during data acquisition). In contrast to the conventional independent component analysis (ICA) based approaches [Plu03], the introduced nBSS algorithms are unique with some desired characteristics, including no statistical assumption for the unknown sources (i.e., the source independence assumption relaxed), no training phase involved, no regularization and tuning parameters involved, insensitive to initial conditions, rigorous theoretical analysis and identifiability guarantee, and good computational efficiency.

6.3.1   nBSS of dependent sources using LP

In this application example, we will consider a linear mixing model with N inputs (sources) and $M \geq N$ outputs (observations). Here, each output observation vector $\boldsymbol{x}_i = [x_i[1], \ldots, x_i[L]] \in \mathbb{R}_+^L$ (e.g., an image of L pixels in biomedical or hyperspectral imaging) is given by

$$\boldsymbol{x}_i = \sum_{j=1}^{N} a_{ij}\boldsymbol{s}_j \succeq \mathbf{0}_L, \ i = 1, \ldots, M, \qquad (6.18)$$

where the sources $\boldsymbol{s}_j \in \mathbb{R}_+^L$ are nonnegative (and can be mutually statistically correlated, i.e., dependent sources) and the unknown mixing matrix $\mathbf{A} = \{a_{ij}\}_{M \times N} \in \mathbb{R}_+^{M \times N}$ is of full column rank with row-sum equal to unity, i.e.,

$$\mathbf{A}\mathbf{1}_N = \mathbf{1}_M. \qquad (6.19)$$

This unit row sum assumption seems very restrictive. It will be readily satisfied through a normalization w.r.t. all the observation vectors $\boldsymbol{x}_i$. In general, in biomedical imaging, the sources of interest are dependent. Therefore, accurate estimation of the source images is almost formidable for most of the existing independent component analysis based signal processing algorithms.

For the sake of simplicity, let us assume that the number of sources is equal to the number of observations (i.e., $M = N$). The problem now is to design an unmixing matrix $\mathbf{W} = \{w_{ij}\}_{N \times N} \in \mathbb{R}^{N \times N}$ such that

$$\{\boldsymbol{y}_i, i = 1, \ldots, N\} = \{\boldsymbol{s}_i, i = 1, \ldots N\}$$

(i.e., permutation ambiguity is allowed) without the information of the mixing matrix $\mathbf{A}$, where $\boldsymbol{y}_i$ is the ith separated or extracted source signal obtained as

$$\boldsymbol{y}_i = \sum_{j=1}^{N} w_{ij} \boldsymbol{x}_j = \sum_{j=1}^{N} w_{ij} \left(\sum_{k=1}^{N} a_{jk} \boldsymbol{s}_k \right) \quad \text{(by (6.18))}$$

$$= \sum_{k=1}^{N} \left(\sum_{j=1}^{N} w_{ij} a_{jk} \right) \boldsymbol{s}_k = \sum_{k=1}^{N} p_{ik} \boldsymbol{s}_k,$$

in which $p_{ik} = \sum_{j=1}^{N} w_{ij} a_{jk}$. In other words, it is desired that the product of the mixing matrix $\mathbf{A}$ and the unmixing matrix $\mathbf{W}$, $\mathbf{WA} = \mathbf{P} = \{p_{ik}\}_{N \times N}$ is an $N \times N$ permutation matrix (i.e., every row of $\mathbf{P}$ has only one nonzero entry equal to unity; so does every column of $\mathbf{P}$). The corresponding role of the unmixing matrix $\mathbf{W}$ in MIMO wireless communications is known as the *blind linear zero-forcing equalizer*.

The unmixing matrix $\mathbf{W}$ can be designed by maximizing the volume of the simplex $\mathbf{conv}\{\mathbf{0}, \boldsymbol{y}_1, \ldots, \boldsymbol{y}_N\}$ (after unmixing), which is related to the volume of the simplex $\mathbf{conv}\{\mathbf{0}, \boldsymbol{x}_1, \ldots, \boldsymbol{x}_N\}$ (before unmixing) by

$$\text{vol}\big(\mathbf{conv}\{\mathbf{0}, \boldsymbol{y}_1, \ldots, \boldsymbol{y}_N\}\big) = |\det(\mathbf{W})| \cdot \text{vol}\big(\mathbf{conv}\{\mathbf{0}, \boldsymbol{x}_1, \ldots, \boldsymbol{x}_N\}\big). \quad (6.20)$$

It has been shown in [WCCW10] that solving the volume maximization problem yields the optimum

$$\{\boldsymbol{y}_i, i = 1, \ldots, N\} = \{\boldsymbol{s}_i, i = 1, \ldots, N\} \quad (6.21)$$

(i.e., source identifiability can be achieved), provided that a pure pixel l_i exists for each source image $\boldsymbol{s}_i$, namely, $s_i[l_i] > 0$ and $s_j[l_i] = 0$ for all $j \neq i$. Hence, the source separation problem can be cast as the following nonconvex optimization problem

$$\max_{\mathbf{W} \in \mathbb{R}^{N \times N}} \; |\det(\mathbf{W})|$$

$$\text{s.t. } \mathbf{W} \mathbf{1}_N = \mathbf{1}_N, \; \boldsymbol{y}_i = \sum_{j=1}^{N} w_{ij} \boldsymbol{x}_j \succeq \mathbf{0}. \quad (6.22)$$

Note that problem (6.22) is NP-hard in general. Next, we present how to find a good approximate solution to this problem, through convex problem approximation of LP.

Closed-form solution for $M = N = 2$: Although problem (6.22) is a nonconvex problem, fortunately we can find a closed-form solution for the case $M = N = 2$. By incorporating the equality constraint in (6.22) into its objective function, we have

$$\det(\mathbf{W}) = \det \begin{bmatrix} w_{11} & 1 - w_{11} \\ w_{21} & 1 - w_{21} \end{bmatrix} = w_{11} - w_{21}.$$

Consider the case that $\det(\mathbf{W}) \geq 0$. Then (6.22) becomes

$$\max \quad w_{11} - w_{21} \tag{6.23a}$$
$$\text{s.t.} \quad w_{11}x_1[n] + (1 - w_{11})x_2[n] \geq 0, \tag{6.23b}$$
$$\qquad w_{21}x_1[n] + (1 - w_{21})x_2[n] \geq 0, \ \forall n = 1, \ldots, L. \tag{6.23c}$$

The constraints (6.23b) and (6.23c) can be equivalently written as

$$\beta \leq w_{11}, w_{21} \leq \alpha \tag{6.24}$$

where

$$\alpha = \min_n \left\{ \frac{-x_2[n]}{x_1[n] - x_2[n]} \ \middle| \ x_1[n] < x_2[n] \right\} > 0 \tag{6.25}$$

and

$$\beta = \max_n \left\{ \frac{-x_2[n]}{x_1[n] - x_2[n]} \ \middle| \ x_1[n] > x_2[n] \right\} < 0. \tag{6.26}$$

Thus, problem (6.23) becomes

$$\max_{\beta \leq w_{11} \leq \alpha} w_{11} - \min_{\beta \leq w_{21} \leq \alpha} w_{21}, \tag{6.27}$$

for which the optimal solutions are $w_{11}^\star = \alpha$ and $w_{21}^\star = \beta$, and so the optimal value is $\det(\mathbf{W}^\star) = \alpha - \beta > 0$. Similarly, one can prove that for the case of $\det(\mathbf{W}^\star) \leq 0$, the optimal (minimum) value is $\det(\mathbf{W}^\star) = \beta - \alpha < 0$.

Closed-form solutions for optimization problems are in general not easy to obtain. However, for some optimization problems it is possible to obtain closed-form solutions depending on some conditions inherent in the optimization problem. This will be addressed in Chapter 9, where KKT conditions can be utilized to find closed-form or analytical solutions to the convex optimization problem under consideration, even though only numerical solutions of the problem can be found most of the time in the sequel.

In order to solve (6.22) for a general case ($M = N > 2$), the procedure illustrated in Subsection 6.2.5 is employed. Here, we consider the cofactor expansion of $\det(\mathbf{W})$ w.r.t. any one row of $\mathbf{W}$, say the ith row of $\mathbf{W}$, denoted as

$\mathbf{w}_i^T = [w_{i1}, w_{i2}, \ldots, w_{iN}]$. It is given by

$$\det(\mathbf{W}) = \sum_{j=1}^{N} (-1)^{i+j} w_{ij} \det \boldsymbol{\mathcal{W}}_{ij}, \qquad (6.28)$$

where $\boldsymbol{\mathcal{W}}_{ij}$ is the submatrix of $\mathbf{W}$ with the ith row and jth column removed. It can be seen that with $\boldsymbol{\mathcal{W}}_{ij}$ fixed for $j = 1, 2, \ldots, N$, $\det(\mathbf{W})$ becomes a linear function of $\mathbf{w}_i$. By fixing all the other rows of $\mathbf{W}$, problem (6.22) can then be reduced to a partial nonconvex maximization problem

$$\max_{\mathbf{w}_i} \left| \sum_{j=1}^{N} (-1)^{i+j} w_{ij} \det(\boldsymbol{\mathcal{W}}_{ij}) \right| \qquad (6.29)$$

$$\text{s.t. } \mathbf{w}_i^T \mathbf{1}_N = 1, \ \boldsymbol{y}_i = \sum_{j=1}^{N} w_{ij} \boldsymbol{x}_j \succeq \mathbf{0}.$$

Note that the objective function in (6.29) is still nonconvex. Fortunately, the partial maximization problem (6.29) can be solved in a globally optimal manner by solving the following two LPs:

$$p^\star = \max_{\mathbf{w}_i} \ \sum_{j=1}^{N} (-1)^{i+j} w_{ij} \det(\boldsymbol{\mathcal{W}}_{ij}) \qquad (6.30)$$

$$\text{s.t. } \mathbf{w}_i^T \mathbf{1}_N = 1, \ \boldsymbol{y}_i = \sum_{j=1}^{N} w_{ij} \boldsymbol{x}_j \succeq \mathbf{0}$$

and

$$q^\star = \min_{\mathbf{w}_i} \ \sum_{j=1}^{N} (-1)^{i+j} w_{ij} \det(\boldsymbol{\mathcal{W}}_{ij}) \qquad (6.31)$$

$$\text{s.t. } \mathbf{w}_i^T \mathbf{1}_N = 1, \ \boldsymbol{y}_i = \sum_{j=1}^{N} w_{ij} \boldsymbol{x}_j \succeq \mathbf{0}.$$

The iterative nBSS algorithm (called nLCA-IVM algorithm) given in Algorithm 6.1 can be used to find the desired unmixing matrix $\mathbf{W}^\star$, where the above partial maximization procedure is conducted in updating $\mathbf{W}$ in a row-by-row manner at each iteration until convergence.

Note that "Iter" in the preceding nLCA-IVM algorithm indicates the number of iterations spent in obtaining $\mathbf{W}^\star$, where one iteration means that all the N row vectors of $\mathbf{W}$ are updated once. In each iteration, either the solution of (6.30) if $|p^\star| > |q^\star|$, or the solution of (6.31) if $|q^\star| > |p^\star|$, is chosen as the solution of problem (6.29) (for maximizing $|\det(\mathbf{W})|$). Thus an nBSS problem can be effectively solved using LP.

Algorithm 6.1 Pseudocode for nLCA-IVM Algorithm

1: Set a convergence tolerance $\varepsilon > 0$, $\mathbf{W} := \mathbf{I}_N$, and Iter := 0.
2: **repeat**
3:    Set Iter:=Iter+1, and $\upsilon := |\det(\mathbf{W})|$.
4:    **for** $i = 1, \ldots, N$ **do**
5:       Solve the two LPs (6.30) and (6.31) to obtain their optimal solutions, denoted by $(p^\star, \bar{\mathbf{w}}_i)$ and $(q^\star, \widehat{\mathbf{w}}_i)$, respectively.
6:       If $|p^\star| > |q^\star|$, then update $\mathbf{w}_i := \bar{\mathbf{w}}_i$. Otherwise, update $\mathbf{w}_i := \widehat{\mathbf{w}}_i$.
7:    **end for**
8: **until** $|\max\{|p^\star|, |q^\star|\} - \upsilon|/\upsilon < \varepsilon$
9: $\mathbf{W}^\star = \mathbf{W}$ is obtained.

Actually, the nLCA-IVM algorithm can be speeded up by replacing the objective function $|\det(\mathbf{W})|$ in problem (6.22) by $\det(\mathbf{W})$ because

$$|\det(\mathbf{W})| = \det(\mathbf{PW})$$

where $\mathbf{P} \in \mathbb{R}^{N \times N}$ is a permutation matrix with $\det(\mathbf{P})$ taking the value 1 or -1. Moreover, some redundant inequalities in (6.22) can be removed with the constraint set (polyhedron) unchanged before solving it. With the above two computational reduction strategies, together with customized implementation for LP using interior-point method (to be introduced in Chapter 10), the resulting algorithm (which yields the same solution as the nLCA-IVM algorithm) can run significantly faster than using the available general purpose convex solvers such as SeDuMi and CVX. Such an approach has been discussed in [CSA$^+$11].

We now show an application of the nLCA-IVM algorithm in multispectral image analysis. Fluorescence microscopy uses an optical sensor array (e.g., CCD camera) to produce multispectral images in which the targets of a specimen are labeled with different fluorescence probes. In an attempt to dissect the multispectral images, the ability of identifying spectral biomarkers is limited due to the spectral-overlapped problem among the probes, leading to information leakthrough from one spectral channel to another. Resolving such problems by separating the fluorescence microscopy into individual maps associated with specific biomarkers can be generally formulated as an nBSS problem. The nLCA-IVM algorithm is employed to analyze a set of dividing newt lung cell images taken from http://publications.nigms.nih.gov/insidethecell/chapter1.html. In Figure 6.3(a), the observed images of intermediate filaments, spindle fibers, and chromosomes are displayed from top to bottom, respectively. Because of spectral overlapping of the fluorescence probes, the spectral biomarkers are indeed overlapped as shown in Figure 6.3(a). The regions of interest (ROI) of the observed images are shown in Figure 6.3(b), and the obtained results are shown in Figure 6.3(c), where the unmixed spectral images of intermediate filaments, chromosomes, and spindle fibers are visually much clearer than the ROIs of the observed images before unmixing. Specifically, the rope shape of the extracted chromosomes and the spindle shape of the extracted spindle fibers can be clearly observed, and these results exhibit a good agreement with biological expectation. (Figure 6.3 is taken from [WCCW10].)

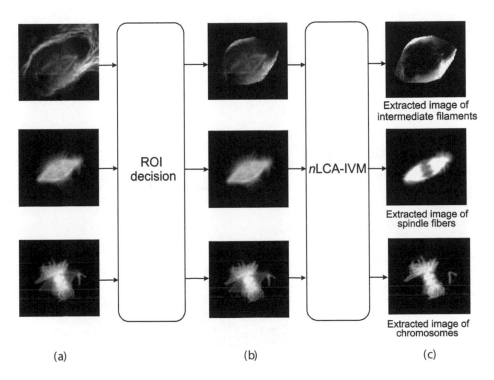

(a)   (b)   (c)

Figure 6.3 Fluorescence microscopy images: (a) the measured newt lung cell images, (b) the ROI images of (a), and (c) the unmixed images of intermediate filaments (top), chromosomes (middle), and spindle fibers (bottom) obtained by the proposed nLCA-IVM.

6.3.2   Hyperspectral unmixing using LP

Hyperspectral images consist of the electromagnetic scattering patterns of materials present in an area, several hundreds of spectral bands that range from visible to shortwave near-infrared wavelength region. The limited spatial resolution of the sensor used for hyperspectral imaging demands an effective hyperspectral unmixing (HU) scheme to extract the underlying endmember signatures (or simply endmembers) and the associated abundance maps (or abundance fractions) distributed over a scene of interest [KM02, BDPCV+13, MBDCG14]. The endmember signature corresponds to the reflection pattern of a mineral (or substance) in different wavelengths and the abundance fraction is the fractional distribution of a mineral (or substance) over the given scene. Figure 6.4 [ACMC11] shows a scenario to illustrate the notion of pure pixels and mixed pixels, where the red pixel corresponds to a mixed pixel (contributed by land, vegetation, and water), and the blue pixel corresponds to a pure pixel (contributed only by water).

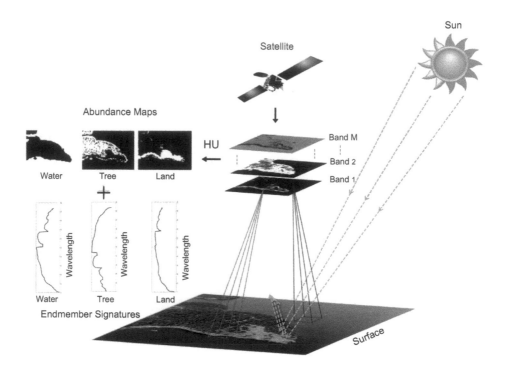

Figure 6.4 Illustration of hyperspectral unmixing.

Each pixel vector of the measured hyperspectral image cube can be described by an $M \times N$ linear mixing model:

$$\mathbf{x}[n] = \mathbf{A}\mathbf{s}[n] = \sum_{i=1}^{N} s_i[n]\mathbf{a}_i, \ \forall n = 1, \dots, L, \qquad (6.32)$$

where M is the number of spectral bands and N is the number of endmembers present in the scene. Further, $\mathbf{x}[n] = [\ x_1[n], \dots, x_M[n]\ ]^T$ is the nth pixel vector in the hyperspectral data, $\mathbf{A} = [\ \mathbf{a}_1, \dots, \mathbf{a}_N\ ] \in \mathbb{R}^{M \times N}$ denotes the endmember signature matrix whose ith column vector $\mathbf{a}_i$ is the ith endmember signature (or simply endmember), $\mathbf{s}[n] = [\ s_1[n], \dots, s_N[n]\ ]^T \in \mathbb{R}^N$ is the nth abundance vector comprising N fractional abundances and L is the total number of observed pixel vectors. In this section, we focus on the estimation of endmember signatures and abundances with a given hyperspectral data. The estimation of the number N of materials present in the hyperspectral scene corresponds to a model-order determination problem, which will be introduced in Section 6.5 using QP and convex geometry.

Assuming that N is known *a priori*, the aim of HU is now to find $\mathbf{A}$ and $\mathbf{s}[n]$, $\forall n = 1, \ldots, L$, from the hyperspectral observations, under the following general assumptions [CCHM09]:

(A1) Source (i.e., abundance) nonnegativity: For each abundance map $\mathbf{s}_j = [s_j[1], s_j[2], \ldots, s_j[L]]^T \succeq \mathbf{0}_L$, $j \in \{1, \ldots, N\}$.

(A2) $\mathbf{A} = [\, \mathbf{a}_1, \ldots, \mathbf{a}_N \,] \in \mathbb{R}^{M \times N}$ is of full column rank and $\mathbf{a}_j \succeq \mathbf{0}_M$ for all j.

(A3) Abundance fractions are proportionally distributed in each observed pixel, i.e., $\sum_{j=1}^{N} s_j[n] = \mathbf{1}_N^T \mathbf{s}[n] = 1$, for all n.

(A4) Pure pixel assumption: For each $i \in \{1, \ldots, N\}$, there exists an (unknown) index l_i such that $s_i[l_i] = 1$ and $s_j[l_i] = 0$, $\forall j \neq i$ (e.g., the blue pixel in Figure 6.4 is a pure pixel of water).

Since M is usually much larger than N, reducing the dimension of the observed hyperspectral data cube is often considered as the pre-processing step. The affine set fitting given in Example 4.3 in Subsection 4.2.2 is employed here for this purpose. That is, the dimension-reduced (DR) data $\tilde{\mathbf{x}}[n]$ are obtained by the following affine transformation:

$$\tilde{\mathbf{x}}[n] = \mathbf{C}^T(\mathbf{x}[n] - \mathbf{d}) = \sum_{j=1}^{N} s_j[n] \boldsymbol{\alpha}_j \in \mathbb{R}^{N-1}, \tag{6.33}$$

where $\mathbf{d}$ (mean of the data $\mathbf{x}[n], n = 1, \ldots, L$) and the $M \times (N-1)$ semi-unitary matrix $\mathbf{C}$ are given by (4.46) and (4.48), respectively, and

$$\boldsymbol{\alpha}_j = \mathbf{C}^T(\mathbf{a}_j - \mathbf{d}) \in \mathbb{R}^{N-1}, \; j = 1, \ldots, N, \;\; (\text{by } (4.47)) \tag{6.34}$$

is the jth DR endmember. Note that the DR hyperspectral data $\tilde{\mathbf{x}}[n]$ contain the same information of endmembers $\mathbf{a}_i, i = 1, \ldots, N$ (via the affine mapping from $\boldsymbol{\alpha}_i$ given by (6.34)) and the abundance maps $\mathbf{s}_i, i = 1, \ldots, N$ as the original high-dimensional hyperspectral data $\mathbf{x}[n]$ under the above assumptions (cf. Figure 4.3).

A well-known criterion for HU is the Craig's criterion, which states that the vertices of the minimum volume simplex enclosing the hyperspectral data should yield high fidelity estimates of the endmember signatures associated with the data cloud. An illustration of Craig's criterion is given in Figure 6.5. Note that the mean $\mathbf{d}$ (not a data pixel) of the original data $\mathbf{x}[n]$ maps to the origin in the DR space (cf. (6.33)).

Based on Craig's criterion, the unmixing problem of finding a minimum volume simplex enclosing all the DR data can be written as the following optimization problem:

$$\min_{\boldsymbol{\beta}_1, \ldots, \boldsymbol{\beta}_N \in \mathbb{R}^{N-1}} \quad V(\boldsymbol{\beta}_1, \ldots, \boldsymbol{\beta}_N)$$
$$\text{s.t.} \quad \tilde{\mathbf{x}}[n] \in \mathbf{conv}\{\boldsymbol{\beta}_1, \ldots, \boldsymbol{\beta}_N\}, \; \forall n, \tag{6.35}$$

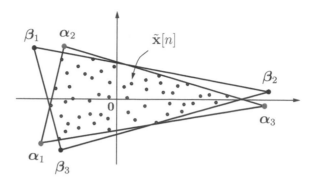

Figure 6.5 Scatter plot of the DR pixels $\tilde{\mathbf{x}}[n]$ for $N = 3$, illustrating the Craig's unmixing criterion (that can be formulated as the simplex volume minimization problem (6.35)) for hyperspectral unmixing, where the simplex $\mathbf{conv}\{\boldsymbol{\beta}_1, \boldsymbol{\beta}_2, \boldsymbol{\beta}_3\}$ is an approximate solution and the true simplex $\mathbf{conv}\{\boldsymbol{\alpha}_1, \boldsymbol{\alpha}_2, \boldsymbol{\alpha}_3\}$ is the optimal solution.

where $V(\boldsymbol{\beta}_1, \ldots, \boldsymbol{\beta}_N)$ is the volume of the simplex $\mathbf{conv}\{\boldsymbol{\beta}_1, \ldots, \boldsymbol{\beta}_N\}$ [Str06] and is given by

$$V(\boldsymbol{\beta}_1, \ldots, \boldsymbol{\beta}_N) = \frac{|\det(\mathbf{B})|}{(N-1)!}, \tag{6.36}$$

in which

$$\mathbf{B} = [\ \boldsymbol{\beta}_1 - \boldsymbol{\beta}_N, \ldots, \boldsymbol{\beta}_{N-1} - \boldsymbol{\beta}_N \ ] \in \mathbb{R}^{(N-1)\times(N-1)}. \tag{6.37}$$

The optimal solution for $\{\boldsymbol{\beta}_1, \ldots, \boldsymbol{\beta}_N\}$ can be shown to be identical to $\{\boldsymbol{\alpha}_1, \ldots, \boldsymbol{\alpha}_N\}$. The problem (6.35) is nonconvex because $V(\boldsymbol{\beta}_1, \ldots, \boldsymbol{\beta}_N)$ is nonconvex in spite of its convex constraint set. Next, let us present how to reformulate this problem into a problem that can be solved using LP.

By (2.22), the constraint of (6.35) can be rewritten in terms of $\mathbf{B}$ as

$$\tilde{\mathbf{x}}[n] = \boldsymbol{\beta}_N + \mathbf{B}\boldsymbol{\theta}_n, \tag{6.38}$$

where

$$\boldsymbol{\theta}_n = [s_1[n], \ldots, s_{N-1}[n]]^T \succeq \mathbf{0}_{N-1}, \tag{6.39}$$

$$\mathbf{1}_{N-1}^T \boldsymbol{\theta}_n \leq 1. \tag{6.40}$$

Hence, problem (6.35) can be equivalently written as

$$\begin{aligned} \min_{\substack{\mathbf{B} \in \mathbb{R}^{(N-1)\times(N-1)}, \\ \boldsymbol{\beta}_N, \boldsymbol{\theta}_1, \ldots, \boldsymbol{\theta}_L \in \mathbb{R}^{N-1}}} \quad & |\det(\mathbf{B})| \\ \text{s.t.} \quad & \boldsymbol{\theta}_n \succeq \mathbf{0}_{N-1}, \ \mathbf{1}_{N-1}^T \boldsymbol{\theta}_n \leq 1 \\ & \tilde{\mathbf{x}}[n] = \boldsymbol{\beta}_N + \mathbf{B}\boldsymbol{\theta}_n, \ \forall n. \end{aligned} \tag{6.41}$$

The inequality constraints in (6.41) are convex, but the equality constraints in (6.41) are nonconvex which, however, can be made convex as follows.

By the following variable changes

$$\mathbf{H} = \mathbf{B}^{-1}, \tag{6.42}$$

$$\mathbf{g} = \mathbf{B}^{-1}\boldsymbol{\beta}_N, \tag{6.43}$$

$$\boldsymbol{\theta}_n = \mathbf{H}\tilde{\mathbf{x}}[n] - \mathbf{g}, \tag{6.44}$$

one can eliminate the variables $\boldsymbol{\theta}_n$ for all n in (6.41) and come up with

$$\max_{\mathbf{H} \in \mathbb{R}^{(N-1)\times(N-1)}, \, \mathbf{g} \in \mathbb{R}^{N-1}} \quad |\det(\mathbf{H})|$$
$$\text{s.t.} \quad \mathbf{1}_{N-1}^T(\mathbf{H}\tilde{\mathbf{x}}[n] - \mathbf{g}) \leq 1 \tag{6.45}$$
$$\mathbf{H}\tilde{\mathbf{x}}[n] - \mathbf{g} \succeq \mathbf{0}_{N-1}, \; \forall n$$

where the constraint set is a polyhedron (convex).

Problem (6.45) now becomes a problem similar to (6.22). The iterative partial maximization strategy explained in Subsection 6.3.1 can be employed here to estimate the optimal $\mathbf{H}^\star$ and $\mathbf{g}^\star$. The resulting unmixing algorithm is called the minimum-volume enclosing simplex (MVES) algorithm [CCHM09]. Then the endmember signatures can be recovered by

$$\hat{\mathbf{a}}_i = \mathbf{C}\hat{\boldsymbol{\alpha}}_i + \mathbf{d}, \; i = 1,\ldots,N \quad \text{(by (6.34))}, \tag{6.46}$$

where $\hat{\boldsymbol{\alpha}}_i$ denotes the optimal solution of (6.35) and can be obtained from the above variable changes as

$$\hat{\boldsymbol{\alpha}}_N = (\mathbf{H}^\star)^{-1}\mathbf{g}^\star \quad \text{(by (6.42) and (6.43))}, \tag{6.47}$$

$$[\hat{\boldsymbol{\alpha}}_1,\ldots,\hat{\boldsymbol{\alpha}}_{N-1}] = \hat{\boldsymbol{\alpha}}_N \mathbf{1}_{N-1}^T + (\mathbf{H}^\star)^{-1} \quad \text{(by (6.37) and (6.42))}. \tag{6.48}$$

Finally, the abundance vectors can be estimated as

$$\hat{\mathbf{s}}[n] = [\; \boldsymbol{\theta}_n^T \quad 1 - \mathbf{1}_{N-1}^T\boldsymbol{\theta}_n \;]^T \quad \text{(by (6.39), (6.40), and ($A3$))}$$
$$= [\; (\mathbf{H}^\star\tilde{\mathbf{x}}[n] - \mathbf{g}^\star)^T \quad 1 - \mathbf{1}_{N-1}^T(\mathbf{H}^\star\tilde{\mathbf{x}}[n] - \mathbf{g}^\star) \;]^T \quad \text{(by (6.44))} \tag{6.49}$$

for all $n = 1,\ldots,L$. The MVES algorithm has been shown to perfectly identify the true endmembers (i.e., $\{\hat{\boldsymbol{\alpha}}_1,\ldots,\hat{\boldsymbol{\alpha}}_N\} = \{\boldsymbol{\alpha}_1,\ldots,\boldsymbol{\alpha}_N\}$) under the sufficient conditions ($A1$) through ($A4$).

Note that, in the presence of noise, the abundance estimate $\hat{\mathbf{s}}[n]$ given by (6.49) and the endmember estimate $\hat{\mathbf{a}}_i$ given by (6.46) may not satisfy the nonnegativity assumptions in ($A1$) and ($A2$), respectively. For those negative entries in $\hat{\mathbf{s}}[n]$ and $\hat{\mathbf{a}}_i$, they are set to zero by the MVES algorithm.

Let us conclude this subsection with real data experiment results obtained by applying the MVES algorithm to the AVIRIS[1] hyperspectral data taken over the Cuprite Nevada site. A 200×200 subimage of the hyperspectral data is chosen as the ROI, with 224 spectral bands. The bands 1-2, 104-113, 148-167, and 221-224 are less significant (due to strong noise or dense water-vapor content) and

---

[1] http://speclab.cr.usgs.gov/PAPERS/cuprite.gr.truth.1992/swayze.1992.html.

thus are excluded. A total of 188 bands are therefore considered. The subimage of the 100th band is shown in Figure 6.6. It has been reported that there are $N = 14$ endmembers present in the scene.

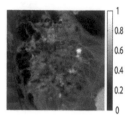

Figure 6.6 The 200×200 subimage of the AVIRIS hyperspectral image data for the 100th band.

The endmember signatures estimated by MVES and the corresponding ground-truth endmember signatures obtained from the United States Geological Survey (USGS) library are shown in Figure 6.7 (taken from [CCHM09]). The figure reveals high correlation between the true and the estimated signatures. The abundance maps $s_i, i = 1, \ldots, N$ of the minerals estimated by the MVES algorithm are shown in Figure 6.8 (taken from [CCHM09]), which shows high agreement with the available groundtruth.

6.3.3   Hyperspectral unmixing by simplex geometry

As discussed in Subsection 6.3.2, HU is a crucial signal processing procedure to identify the underlying materials (or endmembers) and their corresponding proportions (or abundances) from an observed hyperspectral scene, and Craig's criterion is an important blind HU criterion as it has been found effective even in the scenario of no pure pixels. However, most Craig criterion-based algorithms directly solve problem (6.35) for the vertices of Craig's simplex (e.g., the MVES algorithm introduced in Subsection 6.3.2), thus suffering from heavy simplex volume computations and making these Craig criterion-based algorithms very computationally expensive.

Thanks to the observation that abundance maps (i.e., $s[n]$) often show large sparseness in practice, quite many pixels lie on the boundary hyperplanes of the Craig's simplex. According to this observation and the fact that a boundary hyperplane of a simplest simplex with N extreme points can be uniquely determined by any $N - 1$ affinely independent points on the hyperplane, the high-fidelity identification of Craig's simplex can be efficiently fulfilled by accurate estimation of its boundary hyperplanes. Next, based on this creative idea, a fast blind HU algorithm is introduced, referred to as the hyperplane-based Craig

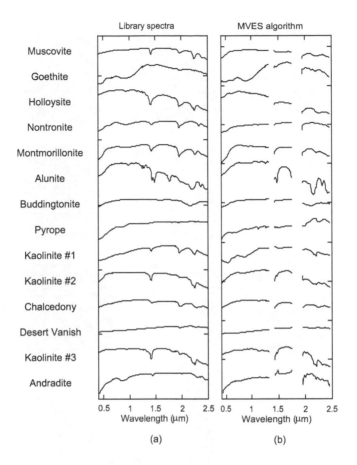

Figure 6.7 (a) The endmember signatures provided by the USGS library and (b) the endmember estimates obtained by the MVES algorithm.

simplex identification (HyperCSI) algorithm that no longer involves any simplex volume computations.

We reformulate the problem (6.35) into N hyperplane estimation subproblems, based on the simplex geometry fact stated in Property 2.7 in Chapter 2. Without resorting to numerical optimization, the HyperCSI algorithm [LCWC16] bypasses cumbersome simplex volume computations, and only needs to search for a specific set of *"active"* data pixels $\mathbf{x}[n]$ via simple linear algebraic computations, thereby accounting for its computational efficiency.

We begin with noticing from ($A2$) that the set of DR endmembers $\{\boldsymbol{\alpha}_1, \ldots, \boldsymbol{\alpha}_N\} \subseteq \mathbb{R}^{N-1}$ is affinely independent. Hence, according to Property 2.7, the endmembers' simplex $\mathbf{conv}\{\boldsymbol{\alpha}_1, \ldots, \boldsymbol{\alpha}_N\} \subseteq \mathbb{R}^{N-1}$ is a simplest simplex in the DR space $\mathbb{R}^{N-1}$ and can be reconstructed from the N hyperplanes $\{\mathcal{H}_1, \ldots, \mathcal{H}_N\}$,

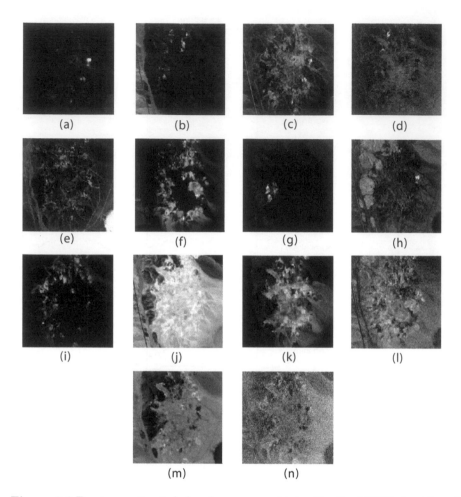

Figure 6.8 Fourteen estimated abundance maps obtained by the MVES algorithm: (a) Muscovite, (b) Goethite, (c) Holloysite, (d) Nontronite, (e) Montmorillonite, (f) Alunite, (g) Buddingtonite, (h) Pyrope, (i) Kaolinite #1, (j) Kaolinite #2, (k) Chalcedony, (l) Desert Vanish, (m) Kaolinite #3, and (n) Andradite.

where

$$\mathcal{H}_i \triangleq \mathbf{aff}(\{\boldsymbol{\alpha}_1,\dots,\boldsymbol{\alpha}_N\} \setminus \{\boldsymbol{\alpha}_i\}) \equiv \{ \mathbf{x} \in \mathbb{R}^{N-1} \mid \mathbf{b}_i^T \mathbf{x} = h_i \}. \tag{6.50}$$

Then, one can decouple the problem (6.35) into N subproblems of hyperplane estimation, or, equivalently, estimation of the N parameter vectors $(\mathbf{b}_i, h_i)$. Next, let us present how to blindly estimate the normal vector $\mathbf{b}_i$ and the inner product constant h_i, respectively, from the DR data set

$$\mathcal{X} \triangleq \{ \tilde{\mathbf{x}}[1],\dots,\tilde{\mathbf{x}}[L] \}. \tag{6.51}$$

♦ Estimation of normal vectors

First of all, we introduce how the HyperCSI algorithm [LCWC16] estimates the normal vector $\mathbf{b}_i$. To this end, we note that the normal vector $\mathbf{b}_i$ can be represented by any affinely independent set $\{\mathbf{p}_1^{(i)}, \ldots, \mathbf{p}_{N-1}^{(i)}\} \subseteq \mathcal{H}_i$. Specifically, as it can be proven true that $\mathbf{0}_{N-1}$ is the data mean in the DR space $\mathbb{R}^{N-1}$ (by (6.33)), we have [LCWC16, Proposition 2]

$$\mathbf{b}_i = \boldsymbol{v}_i(\mathbf{p}_1^{(i)}, \ldots, \mathbf{p}_{i-1}^{(i)}, \mathbf{0}_{N-1}, \mathbf{p}_i^{(i)}, \ldots, \mathbf{p}_{N-1}^{(i)}), \tag{6.52}$$

where $\boldsymbol{v}_i(\cdot)$ is defined in (2.52), and

$$\mathcal{P}_i \triangleq \{\mathbf{p}_1^{(i)}, \ldots, \mathbf{p}_{N-1}^{(i)}\} \subseteq \mathcal{H}_i \tag{6.53}$$

is any given "affinely independent" subset of the $(N-2)$-dimensional hyperplane $\mathcal{H}_i$. Therefore, we need to search for $N-1$ affinely independent pixels (referred to as *active* pixels in $\mathcal{X}$) which are as close to the associated hyperplane as possible.

Assume that we are given N purest pixels $\{\tilde{\boldsymbol{\alpha}}_1, \ldots, \tilde{\boldsymbol{\alpha}}_N\} \subseteq \mathcal{X}$ (identified by successive projection algorithm (SPA) [AGH+12]) that define N disjoint proper regions, each centered at a different purest pixel (cf. Figure 6.9). Then, for each hyperplane $\mathcal{H}_i$ of the Craig's simplex, the desired $N-1$ active pixels in $\mathcal{P}_i$, which are as close to the hyperplane as possible, are respectively sifted from $N-1$ subsets of $\mathcal{X}$, each from the data set associated with one different proper region (cf. Figure 6.9). Specifically, the affinely independent points $\{\mathbf{p}_1^{(i)}, \ldots, \mathbf{p}_{N-1}^{(i)}\}$ in $\mathcal{P}_i$ are sifted by

$$\mathbf{p}_k^{(i)} \in \arg\max \{\tilde{\mathbf{b}}_i^T \mathbf{p} \mid \mathbf{p} \in \mathcal{X} \cap \mathcal{R}_k^{(i)}\}, \quad \forall k \in \mathcal{I}_{N-1}, \tag{6.54}$$

where

$$\begin{cases} \mathcal{I}_n \triangleq \{1, \ldots, n\} \\ \tilde{\mathbf{b}}_i \triangleq \boldsymbol{v}_i(\tilde{\boldsymbol{\alpha}}_1, \ldots, \tilde{\boldsymbol{\alpha}}_N) \quad (\text{cf. } (2.52)) \end{cases} \tag{6.55}$$

and $\mathcal{R}_1^{(i)}, \ldots, \mathcal{R}_{N-1}^{(i)}$ are $N-1$ disjoint sets defined as

$$\mathcal{R}_k^{(i)} \equiv \mathcal{R}_k^{(i)}(\tilde{\boldsymbol{\alpha}}_1, \ldots, \tilde{\boldsymbol{\alpha}}_N) \triangleq \begin{cases} \mathcal{B}(\tilde{\boldsymbol{\alpha}}_k, r), & k < i, \\ \mathcal{B}(\tilde{\boldsymbol{\alpha}}_{k+1}, r), & k \geq i, \end{cases} \tag{6.56}$$

in which $r \triangleq (1/2) \cdot \min\{\|\tilde{\boldsymbol{\alpha}}_i - \tilde{\boldsymbol{\alpha}}_j\|_2 \mid 1 \leq i < j \leq N\} > 0$ is the radius of the "open" Euclidean norm ball $\mathcal{B}(\tilde{\boldsymbol{\alpha}}_k, r)$. Note that the choice of the radius r is to guarantee that $\mathcal{R}_1^{(i)}, \ldots, \mathcal{R}_{N-1}^{(i)}$ are $N-1$ non-overlapping regions, thereby guaranteeing that the $N-1$ points extracted by (6.54) are distinct and hence always affinely independent (cf. Theorem 6.1 below). Moreover, each hyperball $\mathcal{R}_k^{(i)}$ must contain at least one pixel as it contains either the purest pixel $\tilde{\boldsymbol{\alpha}}_k$ or $\tilde{\boldsymbol{\alpha}}_{k+1}$ (cf. (6.56)), i.e., $\mathcal{X} \cap \mathcal{R}_k^{(i)} \neq \emptyset$, and hence problem (6.54) must be a feasible problem.

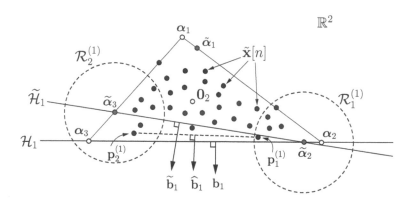

Figure 6.9 An illustration of hyperplanes and DR data in $\mathbb{R}^2$ for the case of $N = 3$, where $\widetilde{\alpha}_3$ is a purest pixel in $\mathcal{X}$ (a purest pixel $\widetilde{\alpha}_i$ can be considered as the pixel closest to α_i) but not necessarily very close to hyperplane $\mathcal{H}_1 = \mathbf{aff}\{\alpha_2, \alpha_3\}$, leading to non-trivial orientation difference between $\widetilde{\mathbf{b}}_1$ and $\mathbf{b}_1$. However, the active pixels $\mathbf{p}_1^{(1)}$ and $\mathbf{p}_2^{(1)}$ identified by (6.54) will be very close to $\mathcal{H}_1$ (especially, for large L), and hence the orientations of $\widehat{\mathbf{b}}_1$ and $\mathbf{b}_1$ will be almost the same.

If the $N - 1$ points extracted by (6.54) are affinely independent, then the estimated normal vector associated with $\mathcal{H}_i$ can be uniquely (up to a positive scale factor) determined by (6.52):

$$\widehat{\mathbf{b}}_i = \boldsymbol{v}_i(\mathbf{p}_1^{(i)}, \ldots, \mathbf{p}_{i-1}^{(i)}, \mathbf{0}_{N-1}, \mathbf{p}_i^{(i)}, \ldots, \mathbf{p}_{N-1}^{(i)}). \tag{6.57}$$

The obtained $\mathcal{P}_i \triangleq \{\mathbf{p}_1^{(i)}, \ldots, \mathbf{p}_{N-1}^{(i)}\}$ by (6.54) can be proved (in Theorem 6.1 below) to be always affinely independent with one more assumption.

(A5) The abundance vectors $\{\mathbf{s}[n]\} \subseteq \mathbb{R}^N$ (defined in (6.32)) are independent and identically distributed (i.i.d.) following *Dirichlet distribution* with parameter $\boldsymbol{\gamma} = [\gamma_1, \ldots, \gamma_N] \in \mathbb{R}_{++}^N$ whose probability density function (p.d.f.) is given by

$$f(\mathbf{s}) = \begin{cases} \dfrac{\Gamma(\gamma_0)}{\prod_{i=1}^N \Gamma(\gamma_i)} \cdot \prod_{i=1}^N s_i^{\gamma_i - 1}, & \mathbf{s} \in \mathbf{dom}\, f \\ 0, & \text{otherwise} \end{cases} \tag{6.58}$$

where $\mathbf{s} = (s_1, \ldots, s_N) \in \mathbb{R}^N$, $\gamma_0 = \sum_{i=1}^N \gamma_i$, $\mathbf{dom}\, f = \{\mathbf{s} \in \mathbb{R}_{++}^N \mid \mathbf{1}_N^T \mathbf{s} = 1\}$, and

$$\Gamma(\gamma) = \int_0^\infty x^{\gamma - 1} e^{-x} \, dx \tag{6.59}$$

is the Gamma function.

Theorem 6.1. *Assume (A1)–(A3) and (A5) hold true. Let* $\mathbf{p}_k^{(i)}$ *be a solution to (6.54) with* $\mathcal{R}_k^{(i)}$ *defined in (6.56), for all* $i \in \mathcal{I}_N$ *and* $k \in \mathcal{I}_{N-1}$. *Then, the set* $\mathcal{P}_i \triangleq \{\mathbf{p}_1, \ldots, \mathbf{p}_{N-1}\}$ *is affinely independent with probability 1 (w.p.1).*

Proof: For a fixed $i \in \mathcal{I}_N$, one can see from (6.56) that $\mathcal{R}_k^{(i)} \cap \mathcal{R}_\ell^{(i)} = \emptyset$, $\forall k \neq \ell$, implying that the $N-1$ pixels $\mathbf{p}_k^{(i)}$, $\forall k \in \mathcal{I}_{N-1}$, identified by solving (6.54) must be distinct. Hence, it suffices to show that $\mathcal{P}$ is affinely independent w.p.1 for any $\mathcal{P} \triangleq \{\mathbf{p}_1, \ldots, \mathbf{p}_{N-1}\} \subseteq \mathcal{X}$ that satisfies

$$\mathbf{p}_k \neq \mathbf{p}_\ell, \text{ for all } 1 \leq k < \ell \leq N - 1. \tag{6.60}$$

Then, as $\mathbf{p}_k \in \mathcal{X}$, $\forall k \in \mathcal{I}_{N-1}$, we have from *(A4)* and (6.60) that there exist i.i.d. Dirichlet distributed random vectors $\{\mathbf{s}_1, \ldots, \mathbf{s}_{N-1}\} \subseteq \mathbf{dom}\, f$ such that

$$\mathbf{p}_k = [\boldsymbol{\alpha}_1 \cdots \boldsymbol{\alpha}_N]\, \mathbf{s}_k, \quad \text{for all } k \in \mathcal{I}_{N-1}. \tag{6.61}$$

For ease of the ensuing presentation, we define the following events:

E1	The set $\mathcal{P}$ is affinely dependent.
E2	The set $\{\mathbf{s}_1, \ldots, \mathbf{s}_{N-1}\}$ is affinely dependent.
E3$^{(k)}$	$\mathbf{s}_k \in \mathbf{aff}\,\{\{\mathbf{s}_1, \ldots, \mathbf{s}_{N-1}\} \setminus \{\mathbf{s}_k\}\}$, $\forall k \in \mathcal{I}_{N-1}$.

Then, to prove that $\mathcal{P}_i$ is affinely independent w.p.1, it suffices to prove $\mathrm{Prob}\{\mathsf{E1}\} = 0$ because $\mathcal{P}_i$ is just a special case of $\mathcal{P}$.

Next, let us show that E1 implies E2. Assume E1 is true. Then $\mathbf{p}_k \in \mathbf{aff}\{\mathcal{P} \setminus \{\mathbf{p}_k\}\}$ for some $k \in \mathcal{I}_{N-1}$. Without loss of generality, let us assume $k = 1$. Then,

$$\mathbf{p}_1 = \theta_2 \cdot \mathbf{p}_2 + \cdots + \theta_{N-1} \cdot \mathbf{p}_{N-1}, \tag{6.62}$$

for some $\theta_i, i = 2, \ldots, N - 1$, satisfying

$$\theta_2 + \cdots + \theta_{N-1} = 1. \tag{6.63}$$

By substituting (6.61) into (6.62), we have

$$[\boldsymbol{\alpha}_1, \ldots, \boldsymbol{\alpha}_N]\, \mathbf{s}_1 = [\boldsymbol{\alpha}_1, \ldots, \boldsymbol{\alpha}_N]\, \mathbf{t}, \tag{6.64}$$

where

$$\mathbf{t} \triangleq \sum_{m=2}^{N-1} \theta_m \cdot \mathbf{s}_m.$$

Then, from the facts of $\mathbf{1}_N^T \mathbf{t} = 1$ (by (6.63)) and $\mathbf{1}_N^T \mathbf{s}_1 = 1$, we come up with $\mathbf{s}_1 = \mathbf{t} = \sum_{m=2}^{N-1} \theta_m \cdot \mathbf{s}_m$ by Property 2.2, or, equivalently, $\mathbf{s}_1 \in \mathbf{aff}\{\mathbf{s}_2, \ldots, \mathbf{s}_{N-1}\}$ (by (6.63)), implying that E2 is true. Thus we have proved that E1 implies E2, and hence

$$\mathrm{Prob}\{\mathsf{E1}\} \leq \mathrm{Prob}\{\mathsf{E2}\}. \tag{6.65}$$

As Dirichlet distribution is a continuous multivariate distribution for a random vector $\mathbf{s} \in \mathbb{R}^N$ to satisfy *(A1)–(A2)* with an $(N-1)$-dimensional domain, any given affine hull $\mathcal{A} \subseteq \mathbb{R}^N$ with affine dimension P must satisfy

$$\text{Prob}\{\, \mathbf{s} \in \mathcal{A} \,\} = 0, \text{ if } P < N - 1. \tag{6.66}$$

Moreover, as $\{\mathbf{s}_1, \ldots, \mathbf{s}_{N-1}\}$ are i.i.d. random vectors and the affine hull $\mathbf{aff}\,\{\{\mathbf{s}_1, \ldots, \mathbf{s}_{N-1}\} \setminus \{\mathbf{s}_k\}\}$ must have affine dimension $P < N - 1$, we have from (6.66) that

$$\text{Prob}\{\mathsf{E}3^{(k)}\} = 0, \text{ for all } k \in \mathcal{I}_{N-1}. \tag{6.67}$$

Then we have the following inferences:

$$0 \le \text{Prob}\{\mathsf{E}1\} \le \text{Prob}\{\mathsf{E}2\} \quad \text{(by (6.65))}$$
$$= \text{Prob}\{\cup_{k=1}^{N-1} \mathsf{E}3^{(k)}\} \quad \text{(by the definitions of E2 and } \mathsf{E}3^{(k)}\text{)}$$
$$\le \sum_{k=1}^{N-1} \text{Prob}\{\mathsf{E}3^{(k)}\} = 0 \quad \text{(by the union bound and (6.67))}.$$

i.e., $\text{Prob}\{\mathsf{E}1\} = 0$. Therefore, the proof is completed.   ∎

By Theorem 6.1, the normal vector estimate $\widehat{\mathbf{b}}_i$ given by (6.57) exists uniquely (up to a positive scale factor) w.p.1. Its estimation accuracy will be further discussed in Remark 6.1 below.

♦ Estimation of inner product constants

Next, let us present how the HyperCSI algorithm estimates the inner product constant h_i. For Craig's simplex (the minimum-volume data-enclosing simplex), all the data in $\mathcal{X}$ should lie on the same side of $\mathcal{H}_i$ (otherwise, it is not data-enclosing), and $\mathcal{H}_i$ should be as tightly close to the data cloud $\mathcal{X}$ as possible (otherwise, it is not minimum-volume); the only possibility is when the hyper-plane $\mathcal{H}_i$ is externally tangent to the data cloud. In other words, $\mathcal{H}_i$ will incorporate the pixel that has maximum inner product with $\widehat{\mathbf{b}}_i$, and hence it can be determined as $\mathcal{H}_i(\widehat{\mathbf{b}}_i, \widehat{h}_i)$, where $\widehat{h}_i$ is obtained by solving

$$\widehat{h}_i = \max \{\, \widehat{\mathbf{b}}_i^T \mathbf{p} \mid \mathbf{p} \in \mathcal{X} \,\}. \tag{6.68}$$

However, it has been reported that when the observed data pixels are noise-corrupted, the random noise may expand the data cloud, thereby inflating the volume of the Craig's data-enclosing simplex [ACMC11]. As a result, the estimated hyperplanes are pushed away from the origin (i.e., the data mean in the DR space) due to noise effect, and hence the estimated inner product constant in (6.68) would be larger than that of the ground truth. To mitigate this effect, the estimated hyperplanes need to be properly shifted closer to the origin, so instead, $\mathcal{H}_i(\widehat{\mathbf{b}}_i, \widehat{h}_i/c)$, $\forall i \in \mathcal{I}_N$, are the desired hyperplane estimates for some $c \ge 1$. Therefore, the corresponding DR endmember estimates are obtained by (cf. (2.50) in the proof of Property 2.7)

$$\widehat{\boldsymbol{\alpha}}_i = \widehat{\mathbf{B}}_{-i}^{-1} \cdot \frac{\widehat{\mathbf{h}}_{-i}}{c}, \quad \forall i \in \mathcal{I}_N, \tag{6.69}$$

where $\widehat{\mathbf{B}}_{-i}$ and $\widehat{\mathbf{h}}_{-i}$ are given by (2.49) with $\mathbf{b}_j$ and h_j replaced by $\widehat{\mathbf{b}}_j$ and $\widehat{h}_j$, $\forall j \neq i$, respectively. Moreover, it is necessary to choose c such that the associated endmember estimates in the original space are nonnegative, i.e.,

$$\widehat{\mathbf{a}}_i = \mathbf{C}\,\widehat{\boldsymbol{\alpha}}_i + \mathbf{d} \succeq \mathbf{0}_M, \ \forall i \in \mathcal{I}_N \ \text{(cf. (6.34))}. \tag{6.70}$$

By (6.69) and (6.70), the hyperplanes should be shifted closer to the origin with $c = c'$ at least, where

$$c' \triangleq \min_{c'' \geq 1} \left\{ c'' \mid \mathbf{C}\,(\widehat{\mathbf{B}}_{-i}^{-1} \cdot \widehat{\mathbf{h}}_{-i}) + c'' \cdot \mathbf{d} \succeq \mathbf{0}_M, \ \forall i \right\} \tag{6.71}$$

which can be further shown to have a closed-form solution:

$$c' = \max\left\{1, \max\{-v_{ij}/d_j \mid i \in \mathcal{I}_N, \ j \in \mathcal{I}_M\}\right\}, \tag{6.72}$$

where v_{ij} is the jth component of $\mathbf{C}\,(\widehat{\mathbf{B}}_{-i}^{-1} \cdot \widehat{\mathbf{h}}_{-i}) \in \mathbb{R}^M$ and d_j is the jth component of $\mathbf{d}$.

Note that c' is just the minimum value for c to yield nonnegative endmember estimates. Thus, we can generally set $c = c'/\eta \geq c'$ for some $\eta \in (0, 1]$. Moreover, the value of $\eta = 0.9$ is empirically found to be a good choice.

The asymptotic identifiability of the proposed HyperCSI algorithm can be guaranteed as stated in the following theorem.

Theorem 6.2. *Under (A1)–(A3) and (A5), the noiseless assumption and $L \to \infty$, the simplex identified by HyperCSI algorithm with $c = 1$ is exactly the Craig's minimum-volume simplex (i.e., solution of (6.35)) and the true endmembers' simplex $\mathbf{conv}\{\boldsymbol{\alpha}_1, \ldots, \boldsymbol{\alpha}_N\}$ w.p.1.*

The proof of Theorem 6.2 can be found in [LCWC16]. Here, we provide the philosophies and intuitions behind the proof of this theorem in the following two remarks.

Remark 6.1 With the abundance vector distribution stated in (A5), the $N - 1$ pixels in $\mathcal{P}_i$ can be shown to be arbitrarily close to $\mathcal{H}_i$ as the pixel number $L \to \infty$, and they are affinely independent w.p.1 (cf. Theorem 6.1). Therefore, $\widehat{\mathbf{b}}_i$ can be uniquely obtained by (6.57), and its orientation approaches to that of $\mathbf{b}_i$ w.p.1. ☐

Remark 6.2 Remark 6.1 together with (6.51) implies that $\widehat{h}_i$ is upper bounded by h_i w.p.1 (assuming without loss of generality that $\|\widehat{\mathbf{b}}_i\| = \|\mathbf{b}_i\|$), and this upper bound can be shown to be achievable w.p.1 as $L \to \infty$. Thus, as $c = 1$, we have that $\widehat{h}_i/c = h_i$ w.p.1. ☐

It can be further inferred, from the above two remarks, that $\widehat{\alpha}_i$ is exactly the true α_i w.p.1 (cf. (6.69)) as $L \to \infty$ in the absence of noise. Although the identifiability analysis in Theorem 6.2 is conducted for the noiseless case and $L \to \infty$, it has been empirically found that the HyperCSI algorithm [LCWC16] can yield high-fidelity endmember estimates for a moderate L and finite SNR, to be demonstrated by simulation results later.

Remark 6.3 The well-known Dirichlet distribution in $(A5)$ for the abundance vector $\mathbf{s}[n]$, under which both the nonnegativity and full-additivity of $\mathbf{s}[n]$ are automatically satisfied, has been widely used in the performance evaluation of various HU algorithms. However, the statistical assumption $(A5)$ is only for the analysis purpose of the HyperCSI algorithm but not for the algorithm development. So even if abundance vectors are neither i.i.d. nor Dirichlet distributed, the HyperCSI algorithm can still work well (cf. the Monte Carlo simulation in the end of this section). □

♦ Estimation of abundances

As the normal vector and inner product constant is handy, the HyperCSI algorithm [LCWC16] can compute the abundances $s_i[n]$ using the following property.

Property 6.1 Assume $(A1)$–$(A3)$ hold true. Then $\mathbf{s}[n] = [s_1[n] \cdots s_N[n]]^T$ has the following closed-form expression:

$$s_i[n] = \frac{h_i - \mathbf{b}_i^T \tilde{\mathbf{x}}[n]}{h_i - \mathbf{b}_i^T \alpha_i}, \quad \forall i \in \mathcal{I}_N, \ \forall n \in \mathcal{I}_L. \tag{6.73}$$

Proof: Given $i \in \mathcal{I}_N$ and $n \in \mathcal{I}_L$. For the case of $s_i[n] = 1$, we have from $(A1)$–$(A2)$ that $\tilde{\mathbf{x}}[n] = \alpha_i$, and hence (6.73) holds true. For the case of $s_i[n] < 1$, we define

$$\mathbf{q}_i[n] \triangleq \sum_{j=1, j\neq i}^{N} \frac{s_j[n]}{1 - s_i[n]} \cdot \alpha_j = \sum_{j=1, j\neq i}^{N} s_j'[n]\alpha_j, \tag{6.74}$$

where the coefficients $s_j'[n] \triangleq \frac{s_j[n]}{1 - s_i[n]}$ clearly satisfy $\sum_{j\neq i} s_j'[n] = 1$ (by $(A2)$), and thus

$$\mathbf{q}_i[n] \in \mathrm{aff}(\ \{\alpha_1, \ldots, \alpha_N\} \setminus \{\alpha_i\}\ ) = \mathcal{H}_i. \tag{6.75}$$

Then we can see from (6.74) that

$$\tilde{\mathbf{x}}[n] = \sum_{j=1}^{N} s_j[n]\alpha_j = s_i[n]\alpha_i + (1 - s_i[n])\mathbf{q}_i[n]. \tag{6.76}$$

By computing the inner product of $\tilde{\mathbf{x}}[n]$ and $\mathbf{b}_i$ on both sides of (6.76), we have from (2.45) and (6.75) that

$$\mathbf{b}_i^T \tilde{\mathbf{x}}[n] = s_i[n]\mathbf{b}_i^T \boldsymbol{\alpha}_i + (1 - s_i[n])\mathbf{b}_i^T \mathbf{q}_i[n],$$
$$= s_i[n]\mathbf{b}_i^T \boldsymbol{\alpha}_i + (1 - s_i[n])h_i,$$

which is equivalent to

$$s_i[n](h_i - \mathbf{b}_i^T \boldsymbol{\alpha}_i) = h_i - \mathbf{b}_i^T \tilde{\mathbf{x}}[n]. \tag{6.77}$$

Finally, we note that $h_i - \mathbf{b}_i^T \boldsymbol{\alpha}_i \neq 0$; otherwise, $\boldsymbol{\alpha}_i \in \mathbf{aff}(\, \{\boldsymbol{\alpha}_1,\ldots,\boldsymbol{\alpha}_N\} \setminus \{\boldsymbol{\alpha}_i\}\,)$ (cf. (2.45)), which violates ($A3$). Therefore, (6.77) yields (6.73), and Theorem 6.1 is proven. ∎

Based on (6.73), the HyperCSI algorithm [LCWC16] estimates the abundance vector $\mathbf{s}[n]$ as

$$\widehat{s}_i[n] = \left[\frac{\widehat{h}_i - \widehat{\mathbf{b}}_i^T \tilde{\mathbf{x}}[n]}{\widehat{h}_i - \widehat{\mathbf{b}}_i^T \widehat{\boldsymbol{\alpha}}_i} \right]^+, \quad \forall i \in \mathcal{I}_N, \ \forall n \in \mathcal{I}_L, \tag{6.78}$$

where $[y]^+ \triangleq \max\{y, 0\}$ is to enforce the nonnegativity of abundance $\mathbf{s}[n]$.

♦ Computational complexity of HyperCSI Algorithm

Next, we discuss the computational complexity of the HyperCSI algorithm that is summarized in Table 6.2. The computational complexity of Hyper-CSI is primarily dominated by the computations of the feasible sets $\mathcal{X} \cap \mathcal{R}_k^{(i)}$ (which costs $\mathcal{O}(N(N+1)L)$) in Step 4, the active pixels in $\mathcal{P}_i$ (which costs $\mathcal{O}(N(N+1)L)$) in Step 5, and the abundances $\widehat{s}_i[n]$ (which costs $\mathcal{O}(N^2L)$) in Step 8 [LCWC16]. Therefore, the overall computational complexity of HyperCSI is $\mathcal{O}(2N(N+1)L + N^2L) = \mathcal{O}(N^2L)$ [LCWC16].

Algorithm 6.2 Pseudocode for the HyperCSI Algorithm

1: **Given** Hyperspectral data $\{\mathbf{x}[1],\ldots,\mathbf{x}[L]\}$, number of endmembers N, and $\eta = 0.9$.
2: Calculate the DR dataset $\mathcal{X} = \{\tilde{\mathbf{x}}[1],\ldots,\tilde{\mathbf{x}}[L]\}$ using (6.33).
3: Obtain purest pixels $\{\tilde{\boldsymbol{\alpha}}_1,\ldots,\tilde{\boldsymbol{\alpha}}_N\}$ [AGH$^+$12].
4: Obtain $\tilde{\mathbf{b}}_i, \forall i$, and $\mathcal{X} \cap \mathcal{R}_k^{(i)}, \forall i, k$, using (6.56).
5: Obtain $(\mathcal{P}_i, \widehat{\mathbf{b}}_i, \widehat{h}_i)$ by (6.54), (6.57), and (6.68), $\forall i$.
6: Obtain c' by (6.72), and set $c = c'/\eta$.
7: $\widehat{\boldsymbol{\alpha}}_i$ by (6.69) and $\widehat{\mathbf{a}}_i = \mathbf{C}\, \widehat{\boldsymbol{\alpha}}_i + \mathbf{d}$ by (6.70), $\forall i$.
8: Calculate $\widehat{\mathbf{s}}[n] = [\widehat{s}_1[n] \cdots \widehat{s}_N[n]]^T$ by (6.78), $\forall n$.
9: **Output** The endmember estimates $\{\widehat{\mathbf{a}}_1,\ldots,\widehat{\mathbf{a}}_N\}$ and abundance estimates $\{\widehat{\mathbf{s}}[1],\ldots,\widehat{\mathbf{s}}[L]\}$.

Surprisingly, the complexity order $\mathcal{O}(N^2L)$ (including both endmember estimation and abundance estimation) of the proposed HyperCSI algorithm is the

same as (rather than much higher than) that of some pure-pixel-based HU algorithms (i.e., HU algorithms requiring the pure pixel assumption ($A4$)). Moreover,
to the best of our knowledge, the MVES algorithm [CCHM09] is the existing
Craig criterion-based algorithm with lowest complexity order $\mathcal{O}(\tau N^2 L^{1.5})$, where
τ is the number of iterations [CCHM09]. Hence, the introduced "hyperplane identification approach" (without involving simplex volume computations) indeed
yields significantly lower complexity than the "vertex identification approach"
exploited by other Craig criterion-based algorithms.

♦ Monte Carlo simulations

To demonstrate the superior efficacy of the HyperCSI algorithm [LCWC16],
we evaluate the performance of the proposed HyperCSI algorithm, along with a
performance comparison with five state-of-the-art Craig criterion-based algorithms, including MVC-NMF [MQ07], MVSA [LBD08], MVES [CCHM09],
SISAL [BD09], and ipMVSA [LAZ$^+$15]. The root-mean-square (rms) spectral
angle error between the true endmembers $\{\mathbf{a}_1, \dots, \mathbf{a}_N\}$ and their estimates
$\{\widehat{\mathbf{a}}_1, \dots, \widehat{\mathbf{a}}_N\}$ defined as

$$\phi_{en} = \min_{\boldsymbol{\pi} \in \Pi_N} \sqrt{\frac{1}{N} \sum_{i=1}^{N} \left[\arccos\left(\frac{\mathbf{a}_i^T \widehat{\mathbf{a}}_{\pi_i}}{\|\mathbf{a}_i\| \cdot \|\widehat{\mathbf{a}}_{\pi_i}\|} \right) \right]^2} \tag{6.79}$$

is used as the performance measure of endmember estimation, where $\Pi_N = \{\boldsymbol{\pi} = (\pi_1, \dots, \pi_N) \in \mathbb{R}^N \mid \pi_i \in \{1, \dots, N\}, \ \pi_i \neq \pi_j \text{ for } i \neq j\}$ is the set of all permu-
tations of $\{1, \dots, N\}$. All the HU algorithms under test are implemented using
Mathworks Matlab R2013a running on a desktop computer equipped with Core-
i7-4790K CPU with 4.00 GHz speed and 16 GB random access memory, and
all the performance results in terms of ϕ_{en}, ϕ_{ab}, and computational time T are
averaged over 100 independent realizations.

A. Abundances satisfying ($A5$)

In the first Monte Carlo simulation, $N = 6$ endmembers (i.e., Jarsoite, Pyrope,
Dumortierite, Buddingtonite, Muscovite, and Goethite) with $M = 224$ spectral
bands randomly selected from the USGS library are used to generate $L = 10,000$
synthetic hyperspectral data $\mathbf{x}[n]$ based on ($A5$), following a standard data generation procedure as described in [LCWC16]. The data are generated for different
SNR and data purity level $\rho \in [1/\sqrt{N}, 1]$ [CCHM09]. The simulation results for
ϕ_{en}, ϕ_{ab}, and computational time T are displayed in Table 6.1, where boldface
numbers correspond to the best performance (i.e., the smallest ϕ_{en} and T) of all
the HU algorithms under test for a specific (ρ, SNR).

Some general observations from Table 6.1 are as follows. For fixed purity level
ρ, all the algorithms under test perform better for larger SNR. As expected,
the HyperCSI algorithm performs better for higher data purity level ρ, but this
performance behavior does not apply to the other five Craig criterion-based

Table 6.1. Performance comparison, in terms of ϕ_{en} (degrees), ϕ_{ab} (degrees), and average running time T (seconds), of various Craig criterion-based algorithms for different data purity levels ρ and SNRs, where abundances are i.i.d. and Dirichlet distributed.

Methods	ρ	ϕ_{en} (degrees) SNR (dB)			ϕ_{ab} (degrees) SNR (dB)			T (seconds)
		20	30	40	20	30	40	
MVC-NMF	0.8	2.87	1.63	1.14	13.18	7.14	5.04	1.68E+2
	0.9	2.98	0.98	0.40	12.67	4.64	2.16	
	1	3.25	1.00	**0.21**	12.30	4.14	**1.11**	
MVSA	0.8	11.08	3.41	1.03	21.78	8.71	2.85	3.54E+0
	0.9	11.55	3.48	1.05	21.89	8.63	2.82	
	1	11.64	3.54	1.06	21.67	8.49	2.72	
MVES	0.8	10.66	3.39	1.16	21.04	9.04	3.33	2.80E+1
	0.9	10.17	3.48	1.12	21.51	9.28	3.45	
	1	9.95	3.55	1.30	22.50	10.32	4.49	
SISAL	0.8	3.97	1.59	0.53	13.70	5.22	1.80	2.59E+0
	0.9	4.18	1.64	0.54	13.55	5.11	1.75	
	1	4.49	1.73	0.54	13.40	5.03	1.66	
ipMVSA	0.8	12.03	4.04	1.16	21.81	9.58	2.23	9.86E-1
	0.9	12.63	4.04	1.25	22.33	9.37	3.31	
	1	12.89	4.00	1.28	22.16	9.06	3.28	
HyperCSI	0.8	**1.65**	**0.79**	**0.37**	**11.17**	**4.32**	**1.64**	**5.39E-2**
	0.9	**1.37**	**0.64**	**0.32**	**10.08**	**3.62**	**1.38**	
	1	**1.21**	**0.57**	0.27	**9.28**	**3.23**	1.15	

algorithms, perhaps because the involved nonconvexity of the complicated simplex volume makes their performance behaviors more intractable w.r.t. different data purities. The HyperCSI algorithm outperforms all the other five algorithms when the data are heavily mixed (i.e., $\rho = 0.8$) or moderately mixed (i.e., $\rho = 0.9$). As for high data purity $\rho = 1$ (corresponding to the existence of pure pixels), the HyperCSI algorithm also performs best except for the case of $(\rho, \text{SNR}) = (1, 40 \text{ dB})$. On the other hand, the computational efficiency of the HyperCSI algorithm is about 1 to 4 orders of magnitude faster than the other five HU algorithms under test.

B. Abundances violating ($A5$)

In view of the sparseness of abundance maps in practical applications rather than Dirichlet distribution, two sets of sparse and spatially correlated abundance maps displayed in Figure 6.10 are used to generate two synthetic hyperspectral images, denoted as SYN1 ($L = 100 \times 100$) and SYN2 ($L = 130 \times 130$). Then all the algorithms listed in Table 6.1 are tested again with these two synthetic data sets for which the abundance vectors obviously violate ($A5$) (i.e., neither i.i.d. nor Dirichlet distributed).

The simulation results are shown in Table 6.2, where boldface numbers correspond to the best performance among the algorithms under test. As expected, for both data sets, all the algorithms perform better for larger SNR. One can see

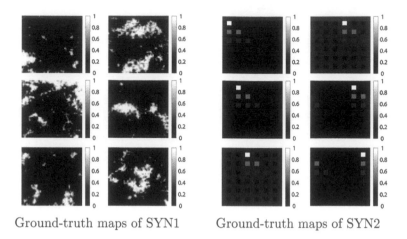

Ground-truth maps of SYN1   Ground-truth maps of SYN2

Figure 6.10 Two sets of sparse and spatially correlated abundance maps.

from Table 6.2 that for both data sets, HyperCSI yields more accurate endmember estimates than the other algorithms, except for the case of SNR= 40 (dB). As for abundance estimation, HyperCSI performs best for SYN1, while MVC-NMF performs best for SYN2. However, in both data sets, the computational efficiency of the HyperCSI algorithm is at least one order of magnitude faster than the other five algorithms. These simulation results have demonstrated the superior efficacy of the HyperCSI algorithm over the other Craig criterion-based algorithms under test in both estimation accuracy and computational efficiency.

Let us conclude this subsection with a list of some remarkable features of the HyperCSI algorithm as follows:

- It only requires the Assumptions $(A1)$–$(A3)$, and can work very well even when $(A4)$ and $(A5)$ are both violated.

- Without involving any simplex volume computations, the Craig's simplex is reconstructed from N hyperplane estimates, i.e., the N estimates $(\widehat{\mathbf{b}}_i, \widehat{h}_i)$, which can be obtained in parallel by searching for $N(N-1)$ active pixels from the data set $\mathcal{X}$.

- All the processing steps can be carried out either by simple linear algebraic formulations or by closed-form expressions. Its computational complexity (without using parallel implementation) is $\mathcal{O}(N^2 L)$, which is also the complexity of some state-of-the-art pure-pixel-based HU algorithms.

- The abundance estimation is readily fulfilled by a closed-form expression, and thus is also very computationally efficient.

- The estimated endmembers are guaranteed nonnegative, and the identified simplex was proven to be both Craig's simplex and true endmembers' simplex w.p.1 as $L \to \infty$ for the noiseless case. Moreover, it is reproducible without involving random initialization.

Table 6.2. Performance comparison, in terms of ϕ_{en} (degrees), ϕ_{ab} (degrees), and running time T (seconds), of various Craig criterion-based algorithms using synthetic data SYN1 and SYN2 for different SNRs, where abundances are non-i.i.d., non-Dirichlet, and sparse (see Figure 6.10).

	Methods	ϕ_{en} (degrees) SNR (dB)			ϕ_{ab} (degrees) SNR (dB)			T (seconds)
		20	30	40	20	30	40	
SYN1	MVC-NMF	3.23	1.05	**0.25**	13.87	4.79	**1.34**	1.74E+2
	MVSA	10.65	3.38	1.05	22.93	9.34	3.19	3.53E+0
	MVES	9.55	3.60	1.22	23.89	14.49	5.66	3.42E+1
	SISAL	4.43	1.81	0.86	15.85	6.89	4.65	2.66E+0
	ipMVSA	11.62	3.38	1.05	24.05	9.34	3.19	1.65E+0
	HyperCSI	**1.55**	**0.79**	0.35	**12.03**	**4.16**	1.46	**5.56E-2**
SYN2	MVC-NMF	2.86	0.97	**0.23**	22.86	**9.39**	**2.67**	2.48E+2
	MVSA	10.21	3.08	0.95	29.86	15.57	5.83	5.65E+0
	MVES	10.12	3.15	3.77	29.43	15.66	13.17	2.22E+1
	SISAL	3.25	1.48	0.63	24.79	11.51	4.21	4.45E+0
	ipMVSA	11.34	3.34	1.01	30.23	16.29	6.39	8.14E-1
	HyperCSI	**1.48**	**0.71**	0.31	**22.64**	11.10	4.40	**7.48E-2**

6.4   Quadratic program (QP)

The second class of optimization problem is the quadratic program and it has the following structure:

$$\min \frac{1}{2}\mathbf{x}^T\mathbf{P}\mathbf{x} + \mathbf{q}^T\mathbf{x} + r \tag{6.80}$$
$$\text{s.t. } \mathbf{Ax} = \mathbf{b}, \ \mathbf{Gx} \preceq \mathbf{h}$$

where $\mathbf{P} \in \mathbb{S}^n$, $\mathbf{G} \in \mathbb{R}^{m \times n}$, and $\mathbf{A} \in \mathbb{R}^{p \times n}$. A QP is convex if and only if $\mathbf{P} \succeq \mathbf{0}$. Simply speaking, QP finds the minimum of a quadratic function over a polyhedral set (see Figure 6.11). From Figure 6.11, it can be observed that the optimal point $\mathbf{x}^\star$ may be located at an extreme point or on a facet of the polyhedron if the unconstrained global minimum of the objective function lies outside the feasible set (the polyhedron) of problem (6.80).

Let us consider the following unconstrained QP:

$$\min \frac{1}{2}\mathbf{x}^T\mathbf{P}\mathbf{x} + \mathbf{q}^T\mathbf{x} + r. \tag{6.81}$$

When $\mathbf{P}$ is a PSD matrix, the optimality condition given by (4.28) for this problem yields

$$\mathbf{Px} = -\mathbf{q},$$

from which one can try to find the optimal solution $\mathbf{x}^\star$ if it exists, as discussed in the following cases:

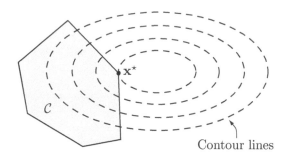

Figure 6.11 Illustration of a convex QP.

- If $\mathbf{P} \succ \mathbf{0}$, then $\mathbf{x}^\star = -\mathbf{P}^{-1}\mathbf{q}$, which is also the unique solution.

- If $\mathbf{P} \succeq \mathbf{0}$ but $\mathbf{q} \notin \mathcal{R}(\mathbf{P})$, then there is no solution for $\mathbf{Px} = -\mathbf{q}$. For this case,

$$\widehat{\mathbf{q}} \triangleq \mathbf{q} - \mathbf{P}\mathbf{P}^\dagger\mathbf{q} \neq \mathbf{0}, \quad \widehat{\mathbf{q}} \in \mathcal{R}(\mathbf{P})^\perp = \mathcal{N}(\mathbf{P}) \quad (\text{by}(1.120) \text{ and } (1.90)).$$

There exists an $\tilde{\mathbf{x}} \in \mathcal{R}(\mathbf{P})^\perp$, such that $\widehat{\mathbf{q}}^T\tilde{\mathbf{x}} \neq 0$. However, $\mathbf{P}^T\tilde{\mathbf{x}} = \mathbf{P}\tilde{\mathbf{x}} = \mathbf{0}$. Then the objective function in (6.81) for $\mathbf{x} = \tilde{\mathbf{x}}$ reduces to $\mathbf{q}^T\tilde{\mathbf{x}} + r = \widehat{\mathbf{q}}^T\tilde{\mathbf{x}} + r$, implying that the objective function in (6.81) is unbounded simply because $\widehat{\mathbf{q}}^T(\alpha\tilde{\mathbf{x}}) + r$ approaches $-\infty$, as $\alpha \to \infty$ if $\widehat{\mathbf{q}}^T\tilde{\mathbf{x}} < 0$ and as $\alpha \to -\infty$ if $\widehat{\mathbf{q}}^T\tilde{\mathbf{x}} > 0$. Hence, $p^\star = -\infty$ and the optimal $\mathbf{x}^\star$ is not existent, or the problem is not solvable for this case.

- If $\mathbf{P} \succeq \mathbf{0}$ with $\lambda_{\min}(\mathbf{P}) = 0$ and $\mathbf{q} \in \mathcal{R}(\mathbf{P})$, then

$$\mathbf{x}^\star = -\mathbf{P}^\dagger\mathbf{q} + \boldsymbol{\nu}, \quad \boldsymbol{\nu} \in \mathcal{N}(\mathbf{P}) \quad (\text{by } (1.119)),$$

which will be nonunique since $\dim(\mathcal{N}(\mathbf{P})) \geq 1$.

Some more examples are given below to further illustrate QP.

- LS with bound constraints:

$$
\begin{aligned}
\min \ &\|\mathbf{Ax} - \mathbf{b}\|_2^2 \\
\text{s.t. } &\boldsymbol{\ell} \preceq \mathbf{x} \preceq \mathbf{u}
\end{aligned}
\tag{6.82}
$$

which is a QP because

$$\|\mathbf{Ax} - \mathbf{b}\|_2^2 = \mathbf{x}^T\left(\mathbf{A}^T\mathbf{A}\right)\mathbf{x} - 2\mathbf{b}^T\mathbf{Ax} + \mathbf{b}^T\mathbf{b}$$

is a quadratic function of $\mathbf{x}$.

- Distance between polyhedra:

$$
\begin{aligned}
\min \ &\|\mathbf{x}_1 - \mathbf{x}_2\|_2^2 \\
\text{s.t. } &\mathbf{A}_1\mathbf{x}_1 \preceq \mathbf{b}_1, \ \mathbf{A}_2\mathbf{x}_2 \preceq \mathbf{b}_2
\end{aligned}
\tag{6.83}
$$

where $\mathbf{x}_1 \in \mathbb{R}^n$ and $\mathbf{x}_2 \in \mathbb{R}^n$. Let $\mathbf{x} = [\mathbf{x}_1^T, \mathbf{x}_2^T]^T \in \mathbb{R}^{2n}$. Then problem (6.83) becomes

$$
\begin{aligned}
\min \ & \left\| [\mathbf{I}_n \ -\mathbf{I}_n] \mathbf{x} \right\|_2^2 \\
\text{s.t.} \ & \begin{bmatrix} \mathbf{A}_1 & \mathbf{0} \\ \mathbf{0} & \mathbf{A}_2 \end{bmatrix} \mathbf{x} \preceq \begin{bmatrix} \mathbf{b}_1 \\ \mathbf{b}_2 \end{bmatrix}
\end{aligned}
\tag{6.84}
$$

which is a QP, because

$$
\left\| [\mathbf{I}_n \ -\mathbf{I}_n] \mathbf{x} \right\|_2^2 = \mathbf{x}^T \begin{bmatrix} \mathbf{I}_n & -\mathbf{I}_n \\ -\mathbf{I}_n & \mathbf{I}_n \end{bmatrix} \mathbf{x}
$$

is a quadratic function of $\mathbf{x}$.

- The FCLS problem (cf. (4.14)) is a QP as follows.

$$
\begin{aligned}
\min \ & \left\| \mathbf{A}\mathbf{x} - \mathbf{b} \right\|_2^2 \\
\text{s.t.} \ & \mathbf{x} \succeq \mathbf{0}, \mathbf{1}_n^T \mathbf{x} = 1.
\end{aligned}
\tag{6.85}
$$

This problem has been widely applied in hyperspectral unmixing based on the linear mixture model given by (6.32), where the endmember signatures $\mathbf{a}_i, i = 1, \ldots, N$, of N unknown materials are first estimated using an endmember extraction algorithm and then the abundance maps $\mathbf{s}_i, i = 1, \ldots, N$, of the N materials are obtained by solving (6.85), where $\mathbf{A} = [\hat{\mathbf{a}}_1, \ldots, \hat{\mathbf{a}}_N]$. Actually, N needs to be estimated prior to estimation of the endmember signatures and abundances, and the estimation of N still remains to be a challenging research in hyperspectral image analysis.

6.5 Applications of QP and convex geometry in hyperspectral image analysis

The notion of HU has been introduced in Subsection 6.3.2, where it was assumed that the number of endmembers is known *a priori*. However, in practice, estimating the number of endmembers (also known as model order selection) [BA02, KMB07] is a daunting task that confronts researchers not only in HU, but also in various signal/image processing areas. Interestingly, based on the linear mixing model given by (6.32) and under the standard assumptions (A1)–(A4) mentioned in Subsection 6.3.2, the geometry of the hyperspectral data can be judiciously used for estimating the number of endmembers. Two such hyperspectral data geometry-based algorithms, namely geometry-based estimation of number of endmembers convex hull (GENE-CH) algorithm and affine hull (GENE-AH) algorithm, have been proposed [ACCK13]. The GENE algorithms (GENE-CH and GENE-AH) exploit a successive estimation property of a reliable and reproducible pure-pixel-based endmember extraction algorithm (EEA) (for instance, the TRI-P algorithm in [ACCK11]), and aim to decide when the

EEA should stop estimating the next endmember signature via the widely known binary hypothesis testing.

The GENE-CH and GENE-AH algorithms are devised based on the data geometry fact that all the observed pixel vectors should lie in the convex hull (CH) and affine hull (AH) of the endmember signatures, respectively. Since the EEAs identify endmember estimates from the set of observed pixel vectors, the fact pertaining to the data geometry also implies that the current endmember estimate should lie in the CH/AH of the previously found endmembers when the current endmember estimate is obtained for an overly estimated number of endmembers. In the noisy scenario, each observed pixel vector $\mathbf{x}[n]$ (given by (6.32)) is corrupted by additive Gaussian noise vector $\mathbf{w}[n]$ with the distribution $\mathcal{N}(\mathbf{0}, \mathbf{D})$, i.e.,

$$\mathbf{x}[n] = \mathbf{A}\mathbf{s}[n] + \mathbf{w}[n], \ n = 1, \ldots, L. \tag{6.86}$$

In such a scenario, the decision of whether the current endmember estimate is in the CH/AH of the previously found endmembers can be formulated as a binary hypothesis testing problem, which can then be handled by Neyman-Pearson detection theory. In both of the GENE algorithms, the decision parameters for the hypothesis testing are suitably obtained by solving a QP, which can be formulated based on the data geometry. The details of the GENE algorithms are as follows.

As in Subsection 6.3.2, we first begin with the dimension reduction of the observed data. Similar to (6.33), by assuming $N_{\max}$ to be an upper bound on the number of endmembers, where $N \le N_{\max} \le M$, the DR noisy observed pixel vectors $\tilde{\mathbf{x}}[n]$ can be obtained by the following affine transformation of $\mathbf{x}[n]$

$$\tilde{\mathbf{x}}[n] = \widehat{\mathcal{C}}^T (\mathbf{x}[n] - \widehat{\mathbf{d}}) \in \mathbb{R}^{N_{\max}-1}, \tag{6.87}$$

where $\widehat{\mathcal{C}}$ and $\widehat{\mathbf{d}}$ are derived in [ACMC11], and are given by

$$\widehat{\mathbf{d}} = \frac{1}{L} \sum_{n=1}^{L} \mathbf{x}[n] = \frac{1}{L} \sum_{n=1}^{L} \mathbf{A}\mathbf{s}[n] + \frac{1}{L} \sum_{n=1}^{L} \mathbf{w}[n], \tag{6.88}$$

$$\widehat{\mathcal{C}} = [\ \boldsymbol{q}_1(\mathbf{U}\mathbf{U}^T - L\widehat{\mathbf{D}}), \ldots, \boldsymbol{q}_{N_{\max}-1}(\mathbf{U}\mathbf{U}^T - L\widehat{\mathbf{D}})\ ], \tag{6.89}$$

where $\mathbf{U}$ is the mean removed data matrix defined in (4.49), $\boldsymbol{q}_i(\mathbf{R})$ denotes the orthonormal eigenvector associated with the ith principal eigenvalue of the matrix $\mathbf{R}$, and $\widehat{\mathbf{D}}$ is a given estimate of the noise covariance matrix $\mathbf{D}$.

Further, due to ($A3$), (6.32), and (6.86),

$$\tilde{\mathbf{x}}[n] = \sum_{i=1}^{N} s_i[n]\boldsymbol{\alpha}_i + \tilde{\mathbf{w}}[n], \ n = 1, \ldots, L, \tag{6.90}$$

where

$$\boldsymbol{\alpha}_i = \widehat{\mathcal{C}}^T (\mathbf{a}_i - \widehat{\mathbf{d}}) \in \mathbb{R}^{N_{\max}-1}, \ i = 1, \ldots, N, \tag{6.91}$$

is the ith dimension-reduced endmember, and $\tilde{\mathbf{w}}[n] \triangleq \widehat{\boldsymbol{C}}^T \mathbf{w}[n] \sim \mathcal{N}(\mathbf{0}, \boldsymbol{\Sigma})$, in which

$$\boldsymbol{\Sigma} = \widehat{\boldsymbol{C}}^T \mathbf{D} \widehat{\boldsymbol{C}} \in \mathbb{R}^{(N_{\max}-1) \times (N_{\max}-1)}. \tag{6.92}$$

Also from (6.90) and under the pure pixel assumption $(A4)$,

$$\tilde{\mathbf{x}}[l_i] = \boldsymbol{\alpha}_i + \tilde{\mathbf{w}}[l_i], \ \forall i = 1, \ldots, N, \tag{6.93}$$

which will be made use of in the GENE algorithms.

The GENE algorithms stem from the following important geometrical facts pertaining to the noise-free hyperspectral data.

(F1) In the noise-free case, by $(A1)$–$(A4)$, any DR pixel vectors $\tilde{\mathbf{x}}[n]$ lie in the convex hull of the DR endmember signatures, and

$$\mathbf{conv}\,\mathcal{X} = \mathbf{conv}\,\{\tilde{\mathbf{x}}[n], n = 1, \ldots, L\} = \mathbf{conv}\,\{\boldsymbol{\alpha}_1, \ldots, \boldsymbol{\alpha}_N\}, \tag{6.94}$$

in which $\mathbf{conv}\,\{\boldsymbol{\alpha}_1, \ldots, \boldsymbol{\alpha}_N\}$ is a simplex with N extreme points being $\boldsymbol{\alpha}_1, \ldots, \boldsymbol{\alpha}_N$.

(F2) In the noise-free case, by $(A2)$ and $(A3)$, (namely, $(A1)$ and $(A4)$ are relaxed)) any DR pixel vectors $\tilde{\mathbf{x}}[n]$ lie in the affine hull of the DR endmember signatures, and

$$\mathbf{aff}\,\mathcal{X} = \mathbf{aff}\,\{\boldsymbol{\alpha}_1, \ldots, \boldsymbol{\alpha}_N\}. \tag{6.95}$$

A simple illustration of (F1) and (F2), for the $N = 3$ case, is shown in Figure 6.12. Moreover, it is true that

$$\mathrm{affdim}(\mathcal{X}) = N - 1 \tag{6.96}$$

as long as either (F1) or (F2) holds true.

6.5.1   GENE-CH algorithm for endmember number estimation

GENE-CH is based on (F1) which assumes that the pure pixel assumption $(A4)$ holds true. Suppose that a reliable, successive EEA has found the pixel indices $l_1, \ldots, l_N, l_{N+1}, \ldots, l_{k-1}, l_k$, in which $l_1, \ldots, l_N$ are pure pixel indices and the rest are not. Here, l_k is the current pixel index estimate and $\{l_1, l_2, \ldots, l_{k-1}\}$ are the previously found pixel index estimates, and $k \le N_{\max}$. For ease of later use, let

$$\widetilde{X}_k = \{\tilde{\mathbf{x}}[l_1], \ldots, \tilde{\mathbf{x}}[l_k]\}.$$

Then by (6.90) and (6.93), it can be readily inferred that

$$\tilde{\mathbf{x}}[l_i] = \boldsymbol{\beta}_i + \tilde{\mathbf{w}}[l_i], \ i = 1, \ldots, k, \tag{6.97}$$

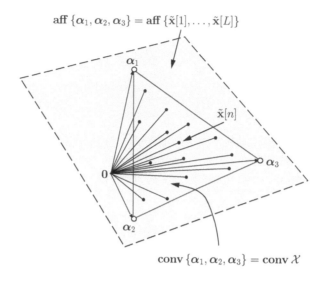

Figure 6.12 Illustration of (F1) and (F2), for $N = 3$ case.

where

$$
\boldsymbol{\beta}_i = \begin{cases} \boldsymbol{\alpha}_i, & i = 1, \ldots, N, \\ \sum_{j=1}^{N} s_j[l_i]\boldsymbol{\alpha}_j, & i = N+1, \ldots, k. \end{cases} \tag{6.98}
$$

The idea behind GENE-CH can be well explained by first considering the noise-free scenario, i.e., $\tilde{\mathbf{w}}[l_i] = \mathbf{0}$, $\forall i = 1, \ldots, k$, in (6.97). It is true from (F1) that the total number of extreme points of **conv** $\mathcal{X}$ is N. That is to say, if $\tilde{\mathbf{x}}[l_k] = \boldsymbol{\beta}_k$ cannot contribute a new extreme point to the **conv** $\widetilde{X}_k$, then

$$
\mathbf{conv}\ \widetilde{X}_k = \mathbf{conv}\ \widetilde{X}_{k-1} = \mathbf{conv}\ \{\boldsymbol{\beta}_1, \ldots, \boldsymbol{\beta}_{k-1}\}, \tag{6.99}
$$

which together with (6.98) implies that all the DR endmembers have already been found, surely belonging to the convex set **conv** $\widetilde{X}_{k-1}$. Therefore, the smallest k such that **conv** $\widetilde{X}_k = $ **conv** $\widetilde{X}_{k-1}$ (a simplex of $k - 1 = N$ vertexes) must take the value of $N + 1$, and thus N can be estimated as $k - 1$, provided that the smallest k can be reliably estimated.

However, in a real scenario, since only noisy $\widetilde{X}_k$ are available rather than $\boldsymbol{\beta}_1, \ldots, \boldsymbol{\beta}_k$ (cf. (6.97)), in the process of estimating the number of endmembers, the presence of noise in the $\widetilde{X}_k$ must be taken into account. In other words, the problem now is to determine whether $\boldsymbol{\beta}_k \in \mathbf{conv}\ \{\boldsymbol{\beta}_1, \ldots, \boldsymbol{\beta}_{k-1}\}$, or not, based on noisy $\widetilde{X}_k$. Hence, the following constrained least-squares problem is considered to measure how far $\tilde{\mathbf{x}}[l_k]$ is away from the **conv** $\widetilde{X}_{k-1}$, that is:

$$
\boldsymbol{\theta}^\star = \arg \left\{ \min_{\boldsymbol{\theta} \succeq \mathbf{0}, \mathbf{1}_{k-1}^T \boldsymbol{\theta} = 1} \left\| \tilde{\mathbf{x}}[l_k] - \widehat{\mathbf{A}}_{k-1}\boldsymbol{\theta} \right\|_2^2 \right\}, \tag{6.100}
$$

where

$$\widehat{\mathbf{A}}_{k-1} = \left[\tilde{\mathbf{x}}[l_1], \ldots, \tilde{\mathbf{x}}[l_{k-1}] \right] \in \mathbb{R}^{(N_{\max}-1) \times (k-1)}. \tag{6.101}$$

The optimization problem in (6.100) is a convex QP and can be solved by using available convex optimization solvers such as SeDuMi and CVX. Once (6.100) is solved, the fitting error vector e can be defined as

$$e = \tilde{\mathbf{x}}[l_k] - \widehat{\mathbf{A}}_{k-1}\theta^\star. \tag{6.102}$$

It can be shown that when $\beta_k \in \mathbf{conv}\{\beta_1, \ldots, \beta_{k-1}\}$, the error vector e can be approximated as a zero-mean Gaussian random vector, i.e.,

$$e \sim \mathcal{N}(\mathbf{0}, \xi^\star \mathbf{\Sigma}),$$
$$\xi^\star = 1 + \theta_1^{\star 2} + \theta_2^{\star 2} + \cdots + \theta_{k-1}^{\star 2},$$

and $\mathbf{\Sigma}$ is given by (6.92). When $\beta_k \notin \mathbf{conv}\{\beta_1, \ldots, \beta_{k-1}\}$, the error vector $e \sim \mathcal{N}(\mu_k, \xi^\star \mathbf{\Sigma})$, where the mean μ_k is unknown. Then a decision statistic r can be defined as

$$r = e^T (\xi^\star \mathbf{\Sigma})^{-1} e. \tag{6.103}$$

The decision statistic r has a central chi-square distribution if $\beta_k \in \mathbf{conv}\{\beta_1, \ldots, \beta_{k-1}\}$, and noncentral chi-square distribution, otherwise. Hence, the Neyman–Pearson binary hypothesis testing methodology with a given false alarm probability P_{FA}, can be suitably employed to estimate the true number of endmembers present in the given hyperspectral data (refer to [ACCK13] for details). The entire procedure for GENE-CH (along with the procedure of the ensuing GENE-AH) is summarized in Algorithm 6.3, where Step 8 to Step 11 corresponds to the Neyman–Pearson binary hypothesis testing methodology.

6.5.2   GENE-AH algorithm for endmember number estimation

The GENE-CH algorithm is based on the assumption that the pure pixels are present in the data (i.e., ($A4$) holds true). However, for practical hyperspectral data the presence of pure pixels cannot be guaranteed. In this case, the DR endmembers estimated by an EEA can be expressed in general as in (6.97), where

$$\beta_i = \sum_{j=1}^{N} s_j[l_i]\alpha_j, \ \forall i = 1, \ldots, k. \tag{6.104}$$

Therefore, the GENE-CH algorithm may not provide an accurate estimate of the number of endmembers especially when the observations are highly mixed. A pictorial illustration is given in Figure 6.13, where $N = 3$ endmembers $\alpha_1, \alpha_2, \alpha_3$ are not present in the noise-free hyperspectral data. For this case, the endmember estimates, denoted by $\beta_i, i = 1, \ldots, N_{\max} = 6$, obtained by an EEA are shown

---

Algorithm 6.3 Pseudocode for GENE-CH and GENE-AH algorithms.

1: Given noisy hyperspectral data $\mathbf{x}[n]$, maximum number of endmembers $N \leq N_{\max} \leq M$, false alarm probability P_{FA}, and estimate of noise covariance matrix $\widehat{\mathbf{D}}$.

2: Compute $(\widehat{\mathcal{C}}, \widehat{\mathbf{d}})$ given by (6.88) and (6.89).

3: Obtain the first pixel index l_1 by a successive EEA and compute $\tilde{\mathbf{x}}[l_1] = \widehat{\mathcal{C}}^T (\mathbf{x}[l_1] - \widehat{\mathbf{d}}) \in \mathbb{R}^{N_{\max}-1}$.

4: **repeat**

5:    Set $k := k + 1$.

6:    Obtain the kth pixel index l_k using the successive EEA and compute $\tilde{\mathbf{x}}[l_k] = \widehat{\mathcal{C}}^T (\mathbf{x}[l_k] - \widehat{\mathbf{d}}) \in \mathbb{R}^{N_{\max}-1}$ and form $\widehat{\mathbf{A}}_{k-1} = [\tilde{\mathbf{x}}[l_1], \dots, \tilde{\mathbf{x}}[l_{k-1}]] \in \mathbb{R}^{(N_{\max}-1)\times(k-1)}$.

7:    Solve the following QPs:

$$\mathrm{GENE-CH} : \boldsymbol{\theta}^\star = \arg \min_{\boldsymbol{\theta} \succeq 0, \mathbf{1}_{k-1}^T \boldsymbol{\theta} = 1} \|\tilde{\mathbf{x}}[l_k] - \widehat{\mathbf{A}}_{k-1} \boldsymbol{\theta}\|_2^2,$$

$$\mathrm{GENE-AH} : \boldsymbol{\theta}^\star = \arg \min_{\mathbf{1}_{k-1}^T \boldsymbol{\theta} = 1} \|\tilde{\mathbf{x}}[l_k] - \widehat{\mathbf{A}}_{k-1} \boldsymbol{\theta}\|_2^2,$$

and calculate $e = \tilde{\mathbf{x}}[l_k] - \widehat{\mathbf{A}}_{k-1} \boldsymbol{\theta}^\star$.

8:    Compute $r = e^T (\xi^\star \boldsymbol{\Sigma})^{-1} e$, where $\xi^\star = 1 + \boldsymbol{\theta}^{\star T} \boldsymbol{\theta}^\star$ and $\boldsymbol{\Sigma} = \widehat{\mathcal{C}}^T \widehat{\mathbf{D}} \widehat{\mathcal{C}}$.

9:    Calculate

$$\psi = 1 - \frac{\gamma(r/2, (N_{\max} - 1)/2)}{\Gamma((N_{\max} - 1)/2)},$$

where $\Gamma(x)$ is the Gamma function (cf. (6.59)) and

$$\gamma(s, x) = \int_0^x t^{s-1} e^{-t} dt$$

is the lower incomplete Gamma function.

10: **until** $\psi > P_{\mathrm{FA}}$ or $k = N_{\max}$.

11: If $\psi > P_{\mathrm{FA}}$, output $k - 1$ as the estimate for number of endmembers; otherwise output k as the estimate for number of endmembers.

---

in Figure 6.13(a) and can be expressed as

$$\boldsymbol{\beta}_i = \tilde{\mathbf{x}}[l_i] = \sum_{j=1}^3 s_j[l_i] \boldsymbol{\alpha}_j, \ i = 1, \dots, N_{\max} = 6 \ \ \text{(by (6.97) and (6.104))}, \quad (6.105)$$

where $l_1, \dots, l_6$ are the pixel indices provided by the EEA under considera-tion. Then, as can be inferred from Figure 6.13(a), for the **conv** $\{\boldsymbol{\beta}_1, \dots, \boldsymbol{\beta}_6\} =$ **conv** $\mathcal{X}$, there can be more than 3 extreme points, which, in fact, is 6 in this case. Hence, using the GENE-CH algorithm under (F2) will obviously result in an overestimation of the number of endmembers for this case. However, from

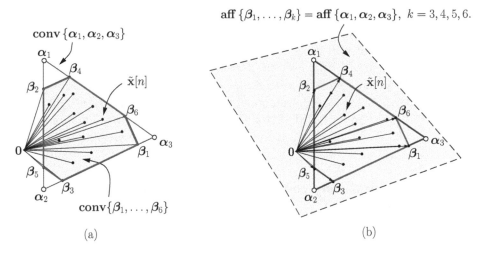

(a)   (b)

Figure 6.13 Illustration of GENE-CH algorithm, when no pure pixel is present in the noise-free hyperspectral data ($N = 3$ case). (a) The endmember estimates are denoted by $\boldsymbol{\beta}_i, i = 1, \ldots, N_{\max} = 6$, but $\mathbf{conv}\{\boldsymbol{\beta}_1, \ldots, \boldsymbol{\beta}_6\} \neq \mathbf{conv}\{\boldsymbol{\alpha}_1, \boldsymbol{\alpha}_2, \boldsymbol{\alpha}_3\}$ because the true endmembers $\boldsymbol{\alpha}_1, \boldsymbol{\alpha}_2, \boldsymbol{\alpha}_3$ are not present in the data cloud, whereas $\mathbf{aff}\{\boldsymbol{\beta}_1, \ldots, \boldsymbol{\beta}_k\} = \mathbf{aff}\{\boldsymbol{\alpha}_1, \boldsymbol{\alpha}_2, \boldsymbol{\alpha}_3\}, \ k = 3, 4, 5, 6$, as shown in (b).

Figure 6.13(b), it can be readily inferred that

$$\boldsymbol{\beta}_k \notin \mathbf{aff}\{\boldsymbol{\beta}_1, \ldots, \boldsymbol{\beta}_{k-1}\}, \ k = 2, 3, \ \text{and}$$
$$\boldsymbol{\beta}_k \in \mathbf{aff}\{\boldsymbol{\beta}_1, \ldots, \boldsymbol{\beta}_{k-1}\}, \ k = 4, 5, 6.$$

The following GENE-AH algorithm is motivated by the above illustration.

The GENE-AH algorithm uses the fact (F2), which implies that in the noise-free case, if $\tilde{\mathbf{x}}[l_k] = \boldsymbol{\beta}_k$ cannot contribute an increment to the affine dimension of $\mathbf{aff}\,\widetilde{X}_k$, i.e.,

$$\mathbf{aff}\,\widetilde{X}_k = \mathbf{aff}\,\widetilde{X}_{k-1} = \mathbf{aff}\{\boldsymbol{\beta}_1, \ldots, \boldsymbol{\beta}_{k-1}\}, \tag{6.106}$$

(i.e., $\boldsymbol{\beta}_k \in \mathbf{aff}\{\boldsymbol{\beta}_1, \ldots, \boldsymbol{\beta}_{k-1}\}$), then

$$\mathrm{affdim}(\widetilde{X}_k) = \mathrm{affdim}(\widetilde{X}_{k-1}) = \mathrm{affdim}(\mathcal{X}) = N - 1 \ \ (\text{by (6.96)}),$$

namely, $\widetilde{X}_{k-1}$ is affinely independent with the dimension equal to $k - 2 = N - 1$, but $\widetilde{X}_k$ is affinely dependent. Therefore, the smallest k such that $\mathbf{aff}\,\widetilde{X}_k = \mathbf{aff}\,\widetilde{X}_{k-1}$ must take the value of $N + 1$, and thus N can be estimated as $k - 1$.

Owing to the presence of noise, the problem now is to determine whether $\boldsymbol{\beta}_k \in \mathbf{aff}\{\boldsymbol{\beta}_1, \ldots, \boldsymbol{\beta}_{k-1}\}$, or not, based on noisy $\widetilde{X}_k$. Therefore, the following constrained least-squares problem is considered to measure how far $\tilde{\mathbf{x}}[l_k]$ is away from the $\mathbf{aff}\,\widetilde{X}_{k-1}$, that is:

$$\boldsymbol{\theta}^\star = \arg\left\{\min_{\mathbf{1}_{k-1}^T \boldsymbol{\theta} = 1} \ \left\|\tilde{\mathbf{x}}[l_k] - \widehat{\mathbf{A}}_{k-1}\boldsymbol{\theta}\right\|_2^2\right\}, \tag{6.107}$$

where $\widehat{\mathbf{A}}_{k-1}$ is defined in (6.101). Since (6.107) is again a convex QP, $\boldsymbol{\theta}^\star$ can be obtained by available convex optimization solvers such as SeDuMi and CVX. As in GENE-CH, by computing the fitting error vector $\boldsymbol{e}$ defined in (6.102) with $\boldsymbol{\theta}^\star$ given by (6.107), and the decision statistic r as in (6.103), a similar Neyman-Pearson hypothesis testing procedure can be devised for GENE-AH to estimate the number of endmembers present in the data (see [ACCK13] for details). The procedure for GENE-AH is also given in Algorithm 6.3 and is similar to that of GENE-CH, except that in Step 7, the optimal $\boldsymbol{\theta}^\star$ is obtained by solving (6.107).

6.6   Quadratically constrained QP (QCQP)

The general structure of a QCQP is as follows

$$
\begin{aligned}
\min \quad & \frac{1}{2}\mathbf{x}^T\mathbf{P}_0\mathbf{x} + \mathbf{q}_0^T\mathbf{x} + r_0 \\
\text{s.t.} \quad & \frac{1}{2}\mathbf{x}^T\mathbf{P}_i\mathbf{x} + \mathbf{q}_i^T\mathbf{x} + r_i \leq 0, \ i = 1,\ldots,m \\
& \mathbf{A}\mathbf{x} = \mathbf{b}
\end{aligned}
\tag{6.108}
$$

where $\mathbf{P}_i \in \mathbb{S}^n$, $i = 0,1,\ldots,m$, and $\mathbf{A} \in \mathbb{R}^{p \times n}$. Some special cases and characteristics of the above problem are as follows:

- QCQP is convex if $\mathbf{P}_i \succeq \mathbf{0}$ for all i.
- When $\mathbf{P}_i \succ \mathbf{0}$ for $i = 1,\ldots,m$, QCQP is a quadratic minimization problem over an intersection of m ellipsoids (cf. (2.37)), and the affine set $\{\mathbf{x} \mid \mathbf{A}\mathbf{x} = \mathbf{b}\}$.
- If $\mathbf{P}_i = \mathbf{0}$ for $i = 1,\ldots,m$, then QCQP reduces to QP.
- If $\mathbf{P}_i = \mathbf{0}$ for $i = 0,1,\ldots,m$, then QCQP reduces to LP.
- As

$$
\frac{1}{2}\mathbf{x}^T\mathbf{P}_i\mathbf{x} + \mathbf{q}_i^T\mathbf{x} + r_i = \left\|\mathbf{A}_i\mathbf{x} + \mathbf{b}_i\right\|_2^2 - \left(\mathbf{f}_i^T\mathbf{x} + d_i\right)^2,
$$

for all $i = 0,\ldots,m$, with $\mathbf{f}_0 = \mathbf{0}_n$ and $\mathbf{f}_i^T\mathbf{x} + d_i > 0$ for all $\mathbf{x}$ in the problem domain, problem (6.108) is equivalent to

$$
\begin{aligned}
\min \quad & \left\|\mathbf{A}_0\mathbf{x} + \mathbf{b}_0\right\|_2 \\
\text{s.t.} \quad & \left\|\mathbf{A}_i\mathbf{x} + \mathbf{b}_i\right\|_2 \leq \mathbf{f}_i^T\mathbf{x} + d_i, \ i = 1,\ldots,m \\
& \mathbf{A}\mathbf{x} = \mathbf{b}
\end{aligned}
\tag{6.109}
$$

or in epigraph form

$$
\begin{aligned}
\min \quad & t \\
\text{s.t.} \quad & \left\|\mathbf{A}_0\mathbf{x} + \mathbf{b}_0\right\|_2 \leq t \\
& \left\|\mathbf{A}_i\mathbf{x} + \mathbf{b}_i\right\|_2 \leq \mathbf{f}_i^T\mathbf{x} + d_i, \ i = 1,\ldots,m \\
& \mathbf{A}\mathbf{x} = \mathbf{b}
\end{aligned}
\tag{6.110}
$$

which is an SOCP as defined in (4.109).

6.7 Applications of QP and QCQP in beamformer design

In this section, we present how QP and QCQP can be used to design an optimal beamformer for a uniform linear antenna array. Here, we consider far-field situations so that source waves are planar and narrowband source signals and the received signal of the ith sensor, $x_i(t)$, is a phase shifted version of $x_j(t)$, i.e.,

$$x_i(t) = x_j(t - (i - j)d\sin\theta/c),$$

where θ and c are the direction of arrival and the propagation speed of the source wave, respectively, and d is the interspacing of antenna elements of a uniform linear array (see Figure 6.14).

If a source signal $s(t) \in \mathbb{C}$ comes from a direction of θ, then the received signal vector by a P-element linear sensor array, denoted as $\mathbf{y}(t) = [y_1(t), \ldots, y_P(t)]^T$, is given by

$$\mathbf{y}(t) = \mathbf{a}(\theta)s(t), \tag{6.111}$$

where

$$\mathbf{a}(\theta) = [1, e^{-j2\pi d\sin(\theta)/\lambda}, \ldots, e^{-j2\pi d(P-1)\sin(\theta)/\lambda}]^T \in \mathbb{C}^P \tag{6.112}$$

is the steering vector, and λ is the signal wavelength.

The estimated source signal is obtained as

$$\hat{s}(t) = \mathbf{w}^H \mathbf{y}(t) = \mathbf{w}^H \mathbf{a}(\theta)s(t) \tag{6.113}$$

where $\mathbf{w} \in \mathbb{C}^P$ is a beamformer weight vector. Let $\theta_{\text{des}} \in [-\pi/2, \pi/2]$ be the desired direction which is assumed to be perfectly known. A simple (conventional) beamformer is $\mathbf{w} = \mathbf{a}(\theta_{\text{des}})$, but it does not provide good sidelobe suppression (see Figure 6.15).

6.7.1 Receive beamforming: Average sidelobe energy minimization

Let

$$\Omega = [-\pi/2, \theta_\ell] \cup [\theta_u, \pi/2]$$

denote the sidelobe band, for some θ_ℓ, θ_u. We would like to find a $\mathbf{w}$ that minimizes sidelobe energy subject to a pass response constraint $\mathbf{w}^H \mathbf{a}(\theta_{\text{des}}) = 1$ where $\theta_{\text{des}} \in [\theta_\ell, \theta_u]$. Two receive beamforming designs are introduced next to illustrate applications of QP.

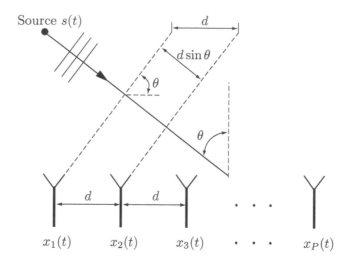

Figure 6.14 Uniform linear array.

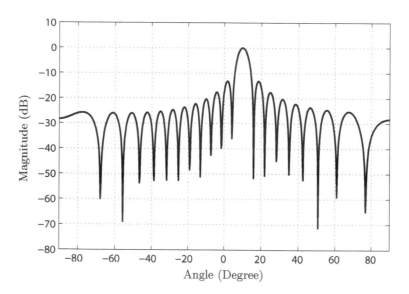

Figure 6.15 Direction pattern of a conventional beamformer with the number of antenna elements $P = 20$, the ratio of antenna elements interspace to signal wavelength $d/\lambda = 0.5$, and $\theta_{\text{des}} = 10°$.

The problem for minimizing the total energy of $\hat{s}(t)$ over the sidelobe band is then

$$\min_{\mathbf{w} \in \mathbb{C}^P} \int_{\Omega} \left| \mathbf{w}^H \mathbf{a}(\theta) \right|^2 d\theta$$
$$\text{s.t. } \mathbf{w}^H \mathbf{a}(\theta_{\text{des}}) = 1. \tag{6.114}$$

The designed beamformer output $\hat{s}(t)$ will be exactly $s(t)$ for $\theta = \theta_{\text{des}}$, but nonzero residuals remain for $\theta \neq \theta_{\text{des}}$, though we try to minimize the total residual power over $\boldsymbol{\Omega}$. The problem is equivalent to an equality constrained QP:

$$\min_{\mathbf{w} \in \mathbb{C}^P} \{f(\mathbf{w}) \triangleq \mathbf{w}^H \mathbf{P} \mathbf{w}\} \qquad (6.115)$$

$$\text{s.t. } \mathbf{w}^H \mathbf{a}(\theta_{\text{des}}) = 1$$

where

$$\mathbf{P} = \int_{\boldsymbol{\Omega}} \mathbf{a}(\theta) \mathbf{a}^H(\theta) d\theta = \mathbf{P}^H \succeq \mathbf{0}$$

(which can be computed by numerical integration). This problem has been proven to be a convex problem (cf. (4.100)).

6.7.2 Receive beamforming: Worst-case sidelobe energy minimization

The worst-case sidelobe energy minimization problem is defined as

$$\min_{\mathbf{w} \in \mathbb{C}^P} \max_{\theta \in \boldsymbol{\Omega}} \left| \mathbf{w}^H \mathbf{a}(\theta) \right|^2 \qquad (6.116)$$

$$\text{s.t. } \mathbf{w}^H \mathbf{a}(\theta_{\text{des}}) = 1.$$

It can be easily reformulated as

$$\min_{\mathbf{w} \in \mathbb{C}^P, \, t \in \mathbb{R}} t$$

$$\text{s.t. } \left| \mathbf{w}^H \mathbf{a}(\theta) \right|^2 \leq t, \, \forall \theta \in \boldsymbol{\Omega} \qquad (6.117)$$

$$\mathbf{w}^H \mathbf{a}(\theta_{\text{des}}) = 1$$

(epigraph form) which is a QCQP with infinitely many constraints. The inequality constraint function in (6.117) can be expressed as

$$\left| \mathbf{w}^H \mathbf{a}(\theta) \right|^2 - t = \mathbf{w}^H \mathbf{P}(\theta) \mathbf{w} - t \qquad (6.118)$$

where $\mathbf{P}(\theta) = \mathbf{a}(\theta) \mathbf{a}^H(\theta) \succeq \mathbf{0}$. It can easily be seen that the constraint function in (6.118) is the sum of a quadratic convex function in $\mathbf{w} \in \mathbb{C}^P$ and a linear function in $t \in \mathbb{R}$, and so it is convex in $(\mathbf{w}, t)$. Hence problem (6.117) is convex in $(\mathbf{w}, t)$, which can also be equivalently written as the standard form of QCQP as follows

$$\min_{\mathbf{w}, t} \begin{bmatrix} \mathbf{0}_P^T & 1 \end{bmatrix} \begin{bmatrix} \mathbf{w} \\ t \end{bmatrix}$$

$$\text{s.t. } \begin{bmatrix} \mathbf{w}^H & t \end{bmatrix} \begin{bmatrix} \mathbf{P}(\theta) & \mathbf{0}_P \\ \mathbf{0}_P^T & 0 \end{bmatrix} \begin{bmatrix} \mathbf{w} \\ t \end{bmatrix} + \begin{bmatrix} \mathbf{0}_P^T & -1 \end{bmatrix} \begin{bmatrix} \mathbf{w} \\ t \end{bmatrix} \leq 0, \, \forall \theta \in \boldsymbol{\Omega} \qquad (6.119)$$

$$\begin{bmatrix} \mathbf{w}^H & t \end{bmatrix} \begin{bmatrix} \mathbf{a}(\theta_{\text{des}}) \\ 0 \end{bmatrix} = 1.$$

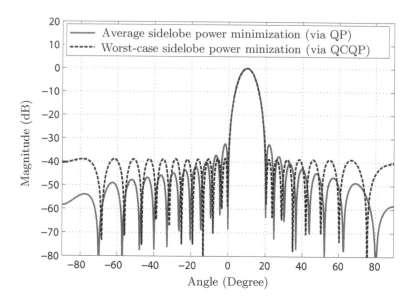

Figure 6.16 Direction patterns of the two beamformer designs with the number of antenna elements $P = 20$, $\theta_{\text{des}} = 10°$. Sidelobe suppression was applied to directions outside $[0°,\ 20°]$.

Note that this is also a *semi-infinite optimization problem* (a problem with a finite number of variables and infinite number of constraints) although it is convex. Next, let us present how to handle the case of infinitely many constraints.

The worst-case sidelobe energy minimization problem can be approximated by discretization. Let $\theta_1, \theta_2, \ldots, \theta_L$ be a set of sample points in Ω. We approximate the problem (6.117) by

$$\min_{\mathbf{w},t}\ t$$
$$\text{s.t.}\ \left|\mathbf{w}^H\mathbf{a}(\theta_i)\right|^2 \leq t,\ i = 1,\ldots,L \qquad (6.120)$$
$$\mathbf{w}^H\mathbf{a}(\theta_{\text{des}}) = 1.$$

Some simulation results for the average sidelobe power minimization problem and the worst-case sidelobe power minimization problem are shown in Figure 6.16. Though discretization in one-dimension parameter space such as Ω in (6.117) may be effective in converting a semi-infinite optimization problem into an approximate optimization problem with finite constraints, it may not be very practical as the parameter space is a high-dimensional space. We will address this issue in Chapter 8 again.

6.7.3   Transmit beamforming in cognitive radio using QCQP

In this subsection, let us consider a cognitive radio network [ALLP12]. The scenario considered is that there is one secondary user who wants to use the same spectrum used by K primary users. Assume that the kth primary transmitter has N_k antennas and the secondary transmitter has N_S antennas, and all the intended receivers have a single antenna. We use $\mathbf{h}_{SS} \in \mathbb{C}^{N_S}$, $\mathbf{h}_{Sk} \in \mathbb{C}^{N_S}$, and $\mathbf{h}_{kS} \in \mathbb{C}^{N_k}$ (which are assumed to be perfectly known) to denote the channel from the secondary transmitter to the secondary receiver, from the secondary transmitter to the kth primary receiver, and from the kth primary transmitter to the secondary receiver, respectively. Assuming that the information bearing signal to be transmitted by each transmitter is of unit power for simplicity, the SINR of the secondary receiver can be expressed as

$$\gamma_S = \frac{|\mathbf{h}_{SS}^H \mathbf{w}_S|^2}{\sum_{k=1}^{K} |\mathbf{h}_{kS}^H \mathbf{w}_k|^2 + \sigma_S^2}, \tag{6.121}$$

where $\mathbf{w}_k \in \mathbb{C}^{N_k}$ and $\mathbf{w}_S \in \mathbb{R}^{N_S}$ represent the beamforming vector of the kth primary transmitter and that of the secondary transmitter, respectively, and σ_S^2 is the noise power at the secondary receiver. The interference power at the kth primary receiver caused by the secondary transmitter is $|\mathbf{h}_{Sk}^H \mathbf{w}_S|^2$. Our goal is to design the secondary transmit beamforming vector $\mathbf{w}_S$ such that γ_S is maximized, while the interference power to the kth primary receiver is below a preassigned threshold ϵ_k.

Mathematically, the problem can be formulated as

$$\begin{aligned} \max_{\mathbf{w}_S} \ & \gamma_S \\ \text{s.t. } & \left|\mathbf{h}_{Sk}^H \mathbf{w}_S\right|^2 \le \epsilon_k, \ k = 1, \dots, K \\ & \left\|\mathbf{w}_S\right\|_2^2 \le P_S \end{aligned} \tag{6.122}$$

where P_S is the maximum transmission power of the secondary link. Define

$$\begin{aligned} \mathbf{A} &= \mathbf{h}_{SS} \mathbf{h}_{SS}^H, \\ \mathbf{B}_k &= \mathbf{h}_{Sk} \mathbf{h}_{Sk}^H, \ k = 1, \dots, K, \end{aligned}$$

which are all rank-one PSD matrices. Problem (6.122) can then be equivalently represented as

$$\begin{aligned} \max_{\mathbf{w}_S} \ & \mathbf{w}_S^H \mathbf{A} \mathbf{w}_S \\ \text{s.t. } & \mathbf{w}_S^H \mathbf{B}_k \mathbf{w}_S \le \epsilon_k, \ k = 1, \dots, K \\ & \mathbf{w}_S^H \mathbf{w}_S \le P_S \end{aligned} \tag{6.123}$$

or it can be equivalently expressed as the following QCQP:

$$\min_{\mathbf{w}_S} \ - \mathbf{w}_S^H \mathbf{A} \mathbf{w}_S$$

$$\text{s.t. } \mathbf{w}_S^H \mathbf{B}_k \mathbf{w}_S \leq \epsilon_k, \ k = 1, \ldots, K \qquad (6.124)$$

$$\mathbf{w}_S^H \mathbf{w}_S \leq P_S.$$

One should note that the problem is not convex as the objective function is not convex. Therefore, it cannot be optimally solved without further reformulation. To further reformulate the problem into a convex problem, we can apply the SDR method (see for instance, [ZS11]), which will be introduced in Subsection 8.5.2.

6.8   Summary and discussion

Two important types of convex optimization problems, namely LP and QP (which are special cases of QCQP), have been discussed in this chapter. Again, via various problem reformulations, the nBSS problems in biomedical image analysis and the hyperspectral unmixing were presented to illustrate applications of LP and QP, and some beamformer design problems were also introduced to illustrate applications of QP and QCQP. However, a QCQP is not necessarily a convex problem. Nevertheless, it can still be solvable if it can be formulated into other types of convex problems such as SDP. The optimal solutions in these applications may have closed-form solutions by solving the optimality condition given by (4.23). In practical applications, closed-form or analytical solutions are surely preferred over the solutions obtained using any convex optimization solvers or algorithms, if they can be obtained through tractable and sometimes involved analysis. We will address this issue in Chapter 9 where in some practical applications, closed-form solutions can be obtained by solving KKT conditions of convex problems.

References for Chapter 6

[ACCK11]   A. Ambikapathi, T.-H. Chan, C.-Y. Chi, and K. Keizer, "Two effective and computationally efficient pure-pixel based algorithms for hyperspectral endmember extraction," in *Proc. 2011 IEEE ICASSP*, Prague, Czech Republic, May 22–27, 2011, pp. 1369–1372.

[ACCK13]   ——, "Hyperspectral data geometry based estimation of number of endmembers using p-norm based pure pixel identification," *IEEE Trans. Geoscience and Remote Sensing*, vol. 51, no. 5, pp. 2753–2769, May 2013.

[ACMC11]   A. Ambikapathi, T.-H. Chan, W.-K. Ma, and C.-Y. Chi, "Chance constrained robust minimum volume enclosing simplex algorithm for hyperspectral unmixing," *IEEE Trans. Geoscience and Remote Sensing*, vol. 49, no. 11, pp. 4194–4209, Nov. 2011.

[AGH+12]   S. Arora, R. Ge, Y. Halpern, D. Mimno, A. Moitra, D. Sontag, Y. Wu, and M. Zhu, "A practical algorithm for topic modeling with provable guarantees," *arXiv preprint arXiv:1212.4777*, 2012.

[ALLP12]   E. Axell, G. Leus, E. G. Larsson, and H. V. Poor, "Spectrum sensing for cognitive radio: State-of-the-art and recent advances," *IEEE Signal Process. Mag.*, vol. 29, no. 3, pp. 101–116, May 2012.

[BA02]   K. P. Burnham and D. R. Anderson, *Model Selection and Multimodel Inference: A Practical Information Theoretic Approach*. New York, USA: Springer-Verlag, 2002.

[BD09]   J. M. Bioucas-Dias, "A variable splitting augmented Lagrangian approach to linear spectral unmixing," in *Proc. IEEE WHISPERS*, Grenoble, France, Aug. 26-28, 2009, pp. 1–4.

[BDPCV+13] J. M. Bioucas-Dias, A. Plaza, G. Camps-Valls, P. Scheunders, N. Nasrabadi, and J. Chanussot, "Hyperspectral remote sensing data analysis and future challenges," *IEEE Geosci. Remote Sens. Mag.*, vol. 1, no. 2, pp. 6–36, Jun. 2013.

[CCHM09]   T.-H. Chan, C.-Y. Chi, Y.-M. Huang, and W.-K. Ma, "A convex analysis based minimum-volume enclosing simplex algorithm for hyperspectral unmixing," *IEEE Trans. Signal Process.*, vol. 57, no. 11, pp. 4418–4432, Nov. 2009.

[CMCW08]   T.-H. Chan, W.-K. Ma, C.-Y. Chi, and Y. Wang, "A convex analysis framework for blind separation of non-negative sources," *IEEE Trans. Signal Processing*, vol. 56, no. 10, pp. 5120–5134, Oct. 2008.

[CSA+11]   T.-H. Chan, C.-J. Song, A. Ambikapathi, C.-Y. Chi, and W.-K. Ma, "Fast alternating volume maximization algorithm for blind separation of non-negative sources," in *Proc. 2011 IEEE International Workshop on Machine Learning for Signal Processing (MLSP)*, Beijing, China, Sept. 18–21, 2011.

[CZPA09]   A. Cichocki, R. Zdunek, A. H. Phan, and S. I. Amari, *Nonnegative Matrix and Tensor Factorizations: Applications to Exploratory Multi-Way Data Analysis and Blind Source Separation*. United Kingdom: John Wiley & Son, 2009.

[KM02]   N. Keshava and J. F. Mustard, "Spectral unmixing," *IEEE Signal Processing Magazine*, vol. 19, no. 1, pp. 44–57, Jan 2002.

[KMB07]   O. Kuybeda, D. Malah, and M. Barzohar, "Rank estimation and redundancy reduction of high-dimensional noisy signals with preservation of rare vectors," *IEEE Trans. Signal Processing*, vol. 55, no. 12, pp. 5579–5592, 2007.

[LAZ+15]   J. Li, A. Agathos, D. Zaharie, J. M. Bioucas-Dias, A. Plaza, and X. Li, "Minimum volume simplex analysis: A fast algorithm for linear hyperspectral unmixing," *IEEE Trans. Geosci. Remote Sens.*, vol. 53, no. 9, pp. 5067–5082, Apr. 2015.

[LBD08]   J. Li and J. M. Bioucas-Dias, "Minimum volume simplex analysis: A fast algorithm to unmix hyperspectral data," in *Proc. IEEE IGARSS*, vol. 4, Boston, MA, Aug. 8–12, 2008, pp. 2369–2371.

[LCWC16]   C.-H. Lin, C.-Y. Chi, Y.-H. Wang, and T.-H. Chan, "A fast hyperplane-based minimum-volume enclosing simplex algorithm for blind hyperspectral unmixing," *IEEE Trans. Signal Processing*, vol. 64, no. 8, pp. 1946–1961, Apr. 2016.

[LS99]   D. Lee and H. Seung, "Learning the parts of objects by nonnegative matrix factorization," *Nature*, vol. 401, pp. 788–791, Oct. 1999.

[MBDCG14]   W.-K. Ma, J. M. Bioucas-Dias, J. Chanussot, and P. Gader, "Special issue on signal and image processing in hyperspectral remote sensing," *IEEE Signal Process. Mag.*, vol. 31, no. 1, Jan. 2014.

[MQ07]   L. Miao and H. Qi, "Endmember extraction from highly mixed data using minimum volume constrained nonnegative matrix factorization," *IEEE Trans. Geosci. Remote Sens.*, vol. 45, no. 3, pp. 765–777, 2007.

[Plu03]   M. D. Plumbley, "Algorithms for non-negative independent component analysis," *IEEE Trans. Neural Netw.*, vol. 14, no. 3, pp. 534–543, 2003.

[Str06]   G. Strang, *Linear Algebra and Its Applications*, 4th ed. San Diego, CA: Thomson, 2006.

[WCCW10]   F.-Y. Wang, C.-Y. Chi, T.-H. Chan, and Y. Wang, "Nonnegative least correlated component analysis for separation of dependent sources by volume maximization," *IEEE Trans. Pattern Analysis and Machine Intelligence*, vol. 32, no. 5, pp. 875–888, May 2010.

[ZS11]   Y.-J. Zhang and A. M.-C. So, "Optimal spectrum sharing in MIMO cognitive radio networks via semidefinite programming," *IEEE J. Sel. Area Commun.*, vol. 29, no. 2, pp. 362–373, Feb. 2011.

7   Second-order Cone Programming

In this chapter, we will introduce another convex optimization problem called second-order cone program (previously abbreviated by SOCP). SOCP has been widely applied in communications and signal processing problems, such as various robust algorithm designs against some system information uncertainty, e.g., robust receive beamforming, transmit beamforming, etc., which will be used to illustrate SOCP after some proper problem formulations in this chapter.

7.1   Second-order cone program (SOCP)

As introduced in Subsection 4.4.1, the SOCP has the following structure

$$
\begin{aligned}
\min \;\; & \mathbf{c}^T \mathbf{x} \\
\text{s.t.} \;\; & \left\| \mathbf{A}_i \mathbf{x} + \mathbf{b}_i \right\|_2 \leq \mathbf{f}_i^T \mathbf{x} + d_i, \; i = 1, \ldots, m \\
& \mathbf{F}\mathbf{x} = \mathbf{g}
\end{aligned}
\tag{7.1}
$$

where $\mathbf{A}_i \in \mathbb{R}^{n_i \times n}$. That is, each inequality constraint in (7.1) involves a generalized inequality defined by a second-order cone $K_i = \{ (\mathbf{y}, t) \in \mathbb{R}^{n_i + 1} \mid \| \mathbf{y} \|_2 \leq t \}$, or

$$
(\mathbf{A}_i \mathbf{x} + \mathbf{b}_i, \mathbf{f}_i^T \mathbf{x} + d_i) \succeq_{K_i} \mathbf{0} \Longleftrightarrow \begin{bmatrix} \mathbf{A}_i \mathbf{x} + \mathbf{b}_i \\ \mathbf{f}_i^T \mathbf{x} + d_i \end{bmatrix} \in K_i \quad (\text{cf. (2.78a)}). \tag{7.2}
$$

Some QCQPs may be regarded as special cases of the SOCP. For example, in Section 6.6 we have introduced a QCQP in the form of

$$
\begin{aligned}
\min \;\; & \left\| \mathbf{A}_0 \mathbf{x} + \mathbf{b}_0 \right\|_2^2 \quad \left(\equiv \min \left\| \mathbf{A}_0 \mathbf{x} + \mathbf{b}_0 \right\|_2 \right) \\
\text{s.t.} \;\; & \left\| \mathbf{A}_i \mathbf{x} + \mathbf{b}_i \right\|_2^2 \leq r_i, \; i = 1, \ldots, L.
\end{aligned}
\tag{7.3}
$$

The problem can be equivalently expressed as

$$
\begin{aligned}
\min \;\; & t \\
\text{s.t.} \;\; & \left\| \mathbf{A}_0 \mathbf{x} + \mathbf{b}_0 \right\|_2 \leq t \\
& \left\| \mathbf{A}_i \mathbf{x} + \mathbf{b}_i \right\|_2 \leq \sqrt{r_i}, \; i = 1, \ldots, L
\end{aligned}
\tag{7.4}
$$

which is an SOCP.

7.2   Robust linear program

When there is uncertainty in any parameter in a system of interest, we need a robust version of the associated optimization problem. In this section we will introduce the robust linear program, which turns out to be an SOCP through reformulation.

Consider the LP

$$\begin{aligned} &\min \ \mathbf{c}^T \mathbf{x} \\ &\text{s.t. } \mathbf{a}_i^T \mathbf{x} \le b_i, \ i = 1, \dots, m. \end{aligned} \qquad (7.5)$$

So the optimal $\mathbf{x}^\star(\mathbf{a}_i)$ and the associated objective value $\mathbf{c}^T \mathbf{x}^\star$ is determined by the actually used $\mathbf{a}_i$.

Suppose that $\mathbf{c}$ is exactly known, but there is some uncertainty in $\mathbf{a}_i$ as follows:

$$\mathbf{a}_i \in \Upsilon_i \triangleq \{\bar{\mathbf{a}}_i + \mathbf{P}_i \mathbf{u} \mid \|\mathbf{u}\|_2 \le 1\} \qquad (7.6)$$

where $\bar{\mathbf{a}}_i$ and $\mathbf{P}_i$ (perturbation matrix) are only known system parameters rather than the true system vector $\mathbf{a}_i$. However, the linear inequality constraints associated with the true but unknown $\mathbf{a}_i$ may be very different from those associated with the nominal system parameter vector $\bar{\mathbf{a}}_i$. Hence the nominal optimal solution $\mathbf{x}^\star(\bar{\mathbf{a}}_i)$, which may be very different from the true optimal solution $\mathbf{x}^\star(\mathbf{a}_i)$, may fail to meet the system constraints during operation if treated as the true solution.

♦ Robust LP formulation

It is desired to find the optimal $\mathbf{x}^\star$ with all the inequality constraints satisfied over all possible uncertainties in the system parameters $\mathbf{a}_i$. The resulting optimization problem can be seen to be

$$\begin{aligned} &\min \ \mathbf{c}^T \mathbf{x} \\ &\text{s.t. } \mathbf{a}_i^T \mathbf{x} \le b_i, \ \forall \mathbf{a}_i \in \Upsilon_i, \ i = 1, \dots, m. \end{aligned} \qquad (7.7)$$

This is also a semi-infinite optimization problem. Since

$$\mathbf{a}_i^T \mathbf{x} \le b_i, \ \forall \mathbf{a}_i \in \Upsilon_i \iff \sup_{\|\mathbf{u}\|_2 \le 1} \left(\bar{\mathbf{a}}_i + \mathbf{P}_i \mathbf{u}\right)^T \mathbf{x} = \bar{\mathbf{a}}_i^T \mathbf{x} + \left\|\mathbf{P}_i^T \mathbf{x}\right\|_2 \le b_i, \quad (7.8)$$

where we used $\mathbf{u} = \mathbf{P}_i^T \mathbf{x} / \|\mathbf{P}_i^T \mathbf{x}\|_2$ (by Schwartz inequality) in the derivation of (7.8), the robust LP is equivalent to

$$\begin{aligned} &\min \ \mathbf{c}^T \mathbf{x} \\ &\text{s.t. } \bar{\mathbf{a}}_i^T \mathbf{x} + \left\|\mathbf{P}_i^T \mathbf{x}\right\|_2 \le b_i, \ i = 1, \dots, m \end{aligned} \qquad (7.9)$$

which is an SOCP.

7.3   Chance constrained linear program

Another way of dealing with the uncertainty in a system is by using chance constraints. We again consider the optimization problem in (7.5). Suppose that the parameters $\mathbf{a}_i$ in the inequality constraints are independent Gaussian random vectors, with known mean vector $\bar{\mathbf{a}}_i$ and covariance matrix $\mathbf{\Sigma}_i$. Since there is some randomness involved in the inequality constraints, we require that those constraints be satisfied with some probability (or confidence level) no less than η. In other words, the probability

$$\text{Prob}\{\mathbf{a}_i^T\mathbf{x} \leq b_i\} \geq \eta, \tag{7.10}$$

where $0 \leq \eta \leq 1$ is a design parameter. The constraint given by (7.10) can be thought of as a soft constraint, while that in (7.8) is a hard constraint.

An optimization problem that involves chance constraints is termed as a *chance constrained* optimization problem. The chance constrained optimization problem corresponding to (7.5) can be therefore written as

$$\begin{aligned} \min \ & \mathbf{c}^T\mathbf{x} \\ \text{s.t. } & \text{Prob}\{\mathbf{a}_i^T\mathbf{x} \leq b_i\} \geq \eta, \ i = 1, \ldots, m. \end{aligned} \tag{7.11}$$

Since $\mathbf{a}_i^T\mathbf{x}$ is a Gaussian random variable with mean $\bar{\mathbf{a}}_i^T\mathbf{x}$ and variance $\mathbf{x}^T\mathbf{\Sigma}_i\mathbf{x}$, (7.11) can be equivalently expressed as

$$\begin{aligned} \min \ & \mathbf{c}^T\mathbf{x} \\ \text{s.t. } & \Phi^{-1}(\eta)\big\|\mathbf{\Sigma}_i^{1/2}\mathbf{x}\big\|_2 + \bar{\mathbf{a}}_i^T\mathbf{x} \leq b_i, \ i = 1, \ldots, m \end{aligned} \tag{7.12}$$

where $\Phi(\cdot)$ is the cumulative distribution function of a zero-mean unit-variance Gaussian random variable, i.e.,

$$\Phi(v) = \frac{1}{\sqrt{2\pi}} \int_{-\infty}^{v} e^{-x^2/2} dx, \tag{7.13}$$

and $\Phi^{-1}(\cdot)$ is its inverse. It can be observed that (7.12) is an SOCP when $\eta > 0.5$ (as $\Phi^{-1}(\eta) > 0$), is a nonconvex problem when $\eta < 0.5$ (as $\Phi^{-1}(\eta) < 0$), and reduces to the original problem (7.5) (with $\mathbf{a}_i$ replaced by $\bar{\mathbf{a}}_i$) when $\eta = 0.5$ (as $\Phi^{-1}(\eta) = 0$).

Chance constrained optimization has been currently applied to transmit beamforming, specifically, robust beamforming in wireless communications, where the essential channel state information needed for the transmit beamforming is never perfect either due to the finite-length training signal used or due to the finite number of bits feedback from the receiver, and has been applied to hyperspectral unmixing in remote sensing [ACMC11], where the noise effects can be effectively taken into account to improve the performance of the endmember extraction.

7.4   Robust least-squares approximation

Recall the standard LS problem:

$$\min_{\mathbf{x}} \|\mathbf{Ax} - \mathbf{b}\|_2^2 \equiv \min_{\mathbf{x}} \|\mathbf{Ax} - \mathbf{b}\|_2. \tag{7.14}$$

Consider that there is uncertainty in $\mathbf{A}$:

$$\mathbf{A} \in \mathcal{A} \triangleq \{\bar{\mathbf{A}} + \mathbf{U} \mid \|\mathbf{U}\|_2 \leq \alpha\} \tag{7.15}$$

and we only have knowledge of $\bar{\mathbf{A}}$ and α.

♦ **Worst-case robust LS formulation**

The LS problem with the uncertainty in the system matrix $\mathbf{A}$ taken into account is defined as

$$\min_{\mathbf{x}} \ \sup_{\mathbf{A} \in \mathcal{A}} \|\mathbf{Ax} - \mathbf{b}\|_2. \tag{7.16}$$

For $\mathbf{A} = \bar{\mathbf{A}} + \mathbf{U}$, $\|\mathbf{U}\|_2 = \sup\{\|\mathbf{Uu}\|_2 \mid \|\mathbf{u}\|_2 \leq 1\} \leq \alpha$ (see (1.7)),

$$\begin{aligned}
\|\mathbf{Ax} - \mathbf{b}\|_2 &= \|\bar{\mathbf{A}}\mathbf{x} - \mathbf{b} + \mathbf{Ux}\|_2 \\
&\leq \|\bar{\mathbf{A}}\mathbf{x} - \mathbf{b}\|_2 + \|\mathbf{Ux}\|_2 \quad \text{(by triangle inequality)} \\
&\leq \|\bar{\mathbf{A}}\mathbf{x} - \mathbf{b}\|_2 + \|\mathbf{U}\|_2 \cdot \|\mathbf{x}\|_2 \quad \text{(by (1.15))} \\
&\leq \|\bar{\mathbf{A}}\mathbf{x} - \mathbf{b}\|_2 + \alpha\|\mathbf{x}\|_2,
\end{aligned} \tag{7.17}$$

where all the inequalities will hold with equality for some $\|\mathbf{U}\|_2 \leq \alpha$. The robust LS problem (7.16) then becomes

$$\min \ \|\bar{\mathbf{A}}\mathbf{x} - \mathbf{b}\|_2 + \alpha\|\mathbf{x}\|_2 \ = \ \begin{array}{l} \min \ t_1 + \alpha t_2 \\ \text{s.t. } \|\bar{\mathbf{A}}\mathbf{x} - \mathbf{b}\|_2 \leq t_1, \ \|\mathbf{x}\|_2 \leq t_2 \end{array} \tag{7.18}$$

which is an SOCP.

7.5   Robust receive beamforming via SOCP

The average sidelobe energy minimization design problem given by (6.115) is repeated here for convenience

$$\begin{aligned}
\min \ &\mathbf{w}^H \mathbf{Pw} \\
\text{s.t. } &\mathbf{w}^H \mathbf{a}(\theta_{\text{des}}) = 1
\end{aligned} \tag{7.19}$$

where $\mathbf{P} = \sum_{i=1}^K \mathbf{a}(\theta_i)\mathbf{a}^H(\theta_i)$, in which θ_is are directions that are not of interest. However, in practice, those interfering directions θ_is are not known. Instead, we can minimize the total energy of the beamformer output under the constraint of $\mathbf{w}^H \mathbf{a}(\theta_{\text{des}}) = 1$ (i.e., signal energy is fixed in the beamformer output) without need of the information of θ_is. In other words, the desired beamformer will suppress any interfering signal coming from the direction $\theta \neq \theta_{\text{des}}$, provided that θ_{des} is perfectly known.

7.5.1   Minimum-variance beamformer

The received signal model is given by

$$\mathbf{y}(t) = \mathbf{a}(\theta_{\text{des}})s(t) + \sum_{i=1}^{K} \mathbf{a}(\theta_i)u_i(t) + \mathbf{v}(t), \tag{7.20}$$

where the source signal $s(t)$ and the interfering signals $u_i(t)$ are assumed to be mutually uncorrelated and wide-sense stationary with zero mean and variances, σ_s^2 and $\sigma_{u_i}^2$, respectively, and $\mathbf{v}(t)$ is spatially white with zero mean and variance σ_v^2. The covariance matrix of the received signal $\mathbf{y}(t)$ can be easily shown to be

$$\mathbf{R} = \mathbb{E}\{\mathbf{y}(t)\mathbf{y}^H(t)\}$$

$$= \sigma_s^2 \mathbf{a}(\theta_{\text{des}})\mathbf{a}^H(\theta_{\text{des}}) + \sum_{i=1}^{K} \sigma_{u_i}^2 \mathbf{a}(\theta_i)\mathbf{a}^H(\theta_i) + \sigma_v^2 \mathbf{I}. \tag{7.21}$$

A widely used method for the estimation of $\mathbf{R}$ is the time-average from $\mathbf{y}(t)$ as follows:

$$\widehat{\mathbf{R}} = \frac{1}{N} \sum_{t=1}^{N} \mathbf{y}(t)\mathbf{y}^H(t). \tag{7.22}$$

The beamformer output $\hat{s}(t) = \mathbf{w}^H\mathbf{y}(t)$ has the average power $\mathbb{E}[|\hat{s}(t)|^2] = \mathbf{w}^H\mathbf{R}\mathbf{w}$, in which the true signal power is equal to σ_s^2 under the equality constraint in (7.19). As the covariance of $\mathbf{y}(t)$, $\mathbf{R}$ is given or estimated in advance, and the energy minimization beamforming problem becomes a QP as follows

$$\min \ \mathbf{w}^H\mathbf{R}\mathbf{w}$$
$$\text{s.t. } \mathbf{w}^H\mathbf{a}(\theta_{\text{des}}) = 1 \tag{7.23}$$

which is equivalent to

$$\min \ \sum_{i=1}^{K} \sigma_{u_i}^2 \left|\mathbf{w}^H\mathbf{a}(\theta_i)\right|^2 + \sigma_v^2 \|\mathbf{w}\|_2^2 \quad \text{(by (7.21))}$$
$$\text{s.t. } \mathbf{w}^H\mathbf{a}(\theta_{\text{des}}) = 1 \tag{7.24}$$

where we actually minimize the output interference plus noise power. In the signal processing literature, (7.23) is known as the *minimum variance beamformer* design.

The closed-form solution for the QP (7.23) also exists, whereas we will find it through KKT conditions to be introduced in Chapter 9. Let

$$\mathbf{R} = \mathbf{V}^H\mathbf{V} \tag{7.25}$$

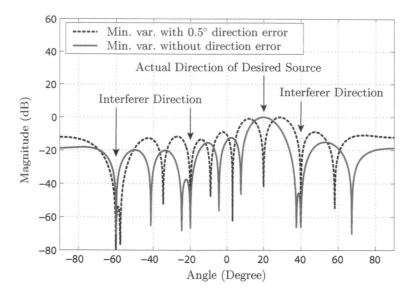

Figure 7.1 Direction patterns of minimum variance beamformers with the number of antenna elements $P = 10$. The solid line represents the beam pattern of minimum variance beamformer without direction error and the dotted line represents the beam pattern of minimum variance beamformer designed with a direction error of $0.5°$.

where $\mathbf{V}$ is the square root of $\mathbf{R}$. Then problem (7.23) can also be converted to an SOCP as follows (cf. (4.103)):

$$
\begin{aligned}
\min \ & t \\
\text{s.t.} \ & \left\| \mathbf{V}\mathbf{w} \right\|_2 \leq t \\
& \mathbf{w}^H \mathbf{a}(\theta_{\text{des}}) = 1.
\end{aligned}
\tag{7.26}
$$

7.5.2   Robust beamforming via SOCP

Consider situations where there is uncertainty with the true direction θ_{des}, or the corresponding true steering vector $\mathbf{a}(\theta_{\text{des}})$ is imperfectly known. Suppose that $\bar{\theta}_{\text{des}}$ is the given nominal arrival direction of the signal of interest and $\mathbf{a}(\bar{\theta}_{\text{des}})$ is the associated steering vector. The uncertainty effect can be modeled as [VGL03]

$$
\mathbf{a}(\theta_{\text{des}}) = \mathbf{a}(\bar{\theta}_{\text{des}}) + \mathbf{u}
\tag{7.27}
$$

where $\mathbf{u}$ is the uncertainty vector. The minimum variance beamformer design based on $\mathbf{a}(\bar{\theta}_{\text{des}})$ can be very sensitive to uncertainty in the given $\mathbf{a}(\bar{\theta}_{\text{des}})$ (see Figure 7.1).

The robust beamforming problem formulation is given by:

$$\min \ \mathbf{w}^H \mathbf{R} \mathbf{w}$$
$$\text{s.t.} \ \left| \mathbf{w}^H (a + \mathbf{u}) \right| \geq 1, \ \forall \ \|\mathbf{u}\|_2 \leq \epsilon \tag{7.28}$$

(which is also a semi-infinite optimization problem) where $a = \mathbf{a}(\bar{\theta}_{\text{des}})$ for notational simplicity. This problem can also be written as

$$\min \ \mathbf{w}^H \mathbf{R} \mathbf{w}$$
$$\text{s.t.} \ \inf_{\|\mathbf{u}\|_2 \leq \epsilon} \left| \mathbf{w}^H (a + \mathbf{u}) \right| \geq 1. \tag{7.29}$$

With a first look at the above problem, it is a nonconvex problem, where the nonconvex constraint itself is a minimization problem. By triangular inequality:

$$\left| \mathbf{w}^H (a + \mathbf{u}) \right| \geq \left| \mathbf{w}^H a \right| - \left| \mathbf{w}^H \mathbf{u} \right|$$
$$\geq \left| \mathbf{w}^H a \right| - \epsilon \|\mathbf{w}\|_2, \ \forall \ \|\mathbf{u}\|_2 \leq \epsilon \tag{7.30}$$

where in the second inequality we have used Cauchy–Schwartz inequality and also assumed $|\mathbf{w}^H a| > \epsilon \|\mathbf{w}\|_2$. (What happens if $|\mathbf{w}^H a| \leq \epsilon \|\mathbf{w}\|_2$? Ans: The problem will become infeasible, i.e., the constraint set will be empty, as $\inf_{\|\mathbf{u}\|_2 \leq \epsilon} |\mathbf{w}^H (a + \mathbf{u})| = 0 \not\geq 1$). Now choose

$$\mathbf{u} = -\frac{\epsilon e^{j \angle (\mathbf{w}^H a)}}{\|\mathbf{w}\|_2} \mathbf{w}, \tag{7.31}$$

so that the equality in (7.30) is achieved. Thus

$$\inf_{\|\mathbf{u}\|_2 \leq \epsilon} \left| \mathbf{w}^H (a + \mathbf{u}) \right| = \left| \mathbf{w}^H a \right| - \epsilon \|\mathbf{w}\|_2. \tag{7.32}$$

The robust beamforming problem can be rewritten as

$$\min \ \mathbf{w}^H \mathbf{R} \mathbf{w}$$
$$\text{s.t.} \ \left| \mathbf{w}^H a \right| - \epsilon \|\mathbf{w}\|_2 \geq 1 \tag{7.33}$$

which is still nonconvex due to the nonconvex inequality constraint. Note that if $\mathbf{w}^\star$ is a solution, then $e^{j\psi} \mathbf{w}^\star$ is also a solution for any phase shift ψ. This fact can lead us to the reformulation of (7.33) into a convex optimization problem.

Without loss of optimality, let us add two extra constraints to problem (7.33):

$$\text{Re}\{\mathbf{w}^H a\} \geq 0, \ \text{Im}\{\mathbf{w}^H a\} = 0. \tag{7.34}$$

Then, problem (7.33) becomes

$$\min \ \|\mathbf{V} \mathbf{w}\|_2 \qquad \text{(by (7.25))}$$
$$\text{s.t.} \ \mathbf{w}^H a \geq 1 + \epsilon \|\mathbf{w}\|_2 \tag{7.35}$$
$$\text{Im}\{\mathbf{w}^H a\} = 0.$$

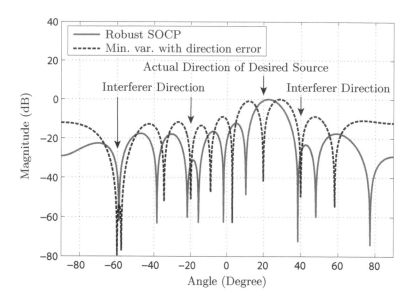

Figure 7.2 Direction patterns of the robust beamformer with the number of antenna elements $P = 10$, $\bar{\theta}_{\mathrm{des}} = 20.5°$, $\theta_{\mathrm{des}} = 20°$, and $\epsilon = 0.2\|\boldsymbol{a}\|_2$.

Finally, by the epigraph reformulation, the robust beamforming problem can be rewritten as an SOCP (cf. (4.103)):

$$\min \ t$$
$$\text{s.t. } \|\mathbf{V}\mathbf{w}\|_2 \leq t, \ \epsilon\|\mathbf{w}\|_2 \leq \mathbf{w}^H \boldsymbol{a} - 1 \qquad (7.36)$$
$$\operatorname{Im}\{\mathbf{w}^H \boldsymbol{a}\} = 0.$$

Some simulation results for a performance comparison of the robust beamformer and the minimum variance beamformer are shown in Figure 7.2 for $\bar{\theta}_{\mathrm{des}} = 20.5°$, $\theta_{\mathrm{des}} = 20°$, and $\epsilon = 0.2\|\boldsymbol{a}\|_2$. It can be observed that the former performs well while the latter fails to extract the signal with the DOA of $20°$.

7.6   Transmit downlink beamforming via SOCP

Consider the following scenario as depicted in Figure 7.3:

- The base station (BS) has m antennas.
- The BS sends data to n mobile stations (MSs), each of which has one antenna.
- The BS uses transmit beamforming to simultaneously transmit signals to the n MSs, over the same channel.

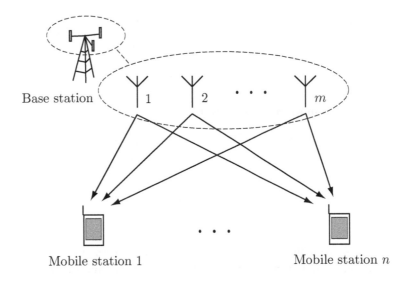

Figure 7.3 Transmit downlink beamforming.

Assuming frequency flat channel fading, the received signal at MS i at each time instant may be expressed as

$$y_i = \mathbf{h}_i^T \mathbf{x} + v_i, \tag{7.37}$$

where

- $\mathbf{h}_i \in \mathbb{C}^m$ is an MISO channel for MS i,
- $v_i \in \mathbb{C}$ is the additive white Gaussian noise (AWGN) with zero mean and variance σ_i^2, and
- $\mathbf{x} \in \mathbb{C}^m$ is the BS transmitted signal vector, with the ith component x_i denoting the transmitted signal of the ith antenna of the BS.

Note that the received signal model in (7.37) can also be expressed as $y_i = \mathbf{h}_i^H \mathbf{x} + v_i$. We may use any of them hereafter freely.

The BS transmitted signal is given by

$$\mathbf{x} = \sum_{i=1}^{n} \mathbf{f}_i s_i = \mathbf{Fs}, \tag{7.38}$$

where

- $s_i \in \mathbb{C}$ is the information carrying signal for MS i,
- $\mathbf{f}_i \in \mathbb{C}^m$ is the corresponding transmit beamforming vector for MS i,
- $\mathbf{F} = [\mathbf{f}_1, \mathbf{f}_2, \ldots, \mathbf{f}_n] \in \mathbb{C}^{m \times n}$, and
- $\mathbf{s} = [s_1, \ldots, s_n]^T \in \mathbb{C}^n$.

Assume that $\mathbb{E}\{|s_i|^2\} = 1$ for all i. The SINR of MS i is given by

$$\gamma_i(\mathbf{F}) = \frac{\left|\mathbf{h}_i^T \mathbf{f}_i\right|^2}{\sum_{j=1, j\neq i}^{n} \left|\mathbf{h}_i^T \mathbf{f}_j\right|^2 + \sigma_i^2}. \tag{7.39}$$

7.6.1   Power minimization beamforming

Given a minimum SINR requirement γ_0, we would like to find a beamformer matrix $\mathbf{F}$ that minimizes the total transmit power [BO02, WES06]:

$$\min \sum_{i=1}^{n} \|\mathbf{f}_i\|_2^2 \tag{7.40}$$

$$\text{s.t. } \gamma_i(\mathbf{F}) \geq \gamma_0, \ i = 1, \ldots, n.$$

This problem in the present form is nonconvex since the inequality constraint function apparently is nonconvex. Next, let us reformulate it into an SOCP.

The inequality constraints in (7.40) can be re-expressed as

$$\left(1 + \frac{1}{\gamma_0}\right) \left|\mathbf{h}_i^T \mathbf{f}_i\right|^2 \geq \sum_{j=1}^{n} \left|\mathbf{h}_i^T \mathbf{f}_j\right|^2 + \sigma_i^2, \ i = 1, \ldots, n. \tag{7.41}$$

With no loss of optimality, we can add the following extra constraints to problem (7.40),

$$\text{Re}\{\mathbf{h}_i^T \mathbf{f}_i\} \geq 0, \ \text{Im}\{\mathbf{h}_i^T \mathbf{f}_i\} = 0, \ i = 1, \ldots, n, \tag{7.42}$$

by virtue of which (7.41) can be re-expressed as

$$\|\mathbf{H}_i \mathbf{f} + \mathbf{b}_i\|_2 \leq \sqrt{1 + \frac{1}{\gamma_0}} \mathbf{h}_i^T \mathbf{f}_i, \tag{7.43}$$

where

$$\mathbf{f} = [\mathbf{f}_1^T, \mathbf{f}_2^T, \ldots, \mathbf{f}_n^T]^T \in \mathbb{C}^{nm}, \tag{7.44}$$

$$\mathbf{b}_i = [\mathbf{0}_n^T, \sigma_i]^T \in \mathbb{R}^{n+1}, \tag{7.45}$$

$$\mathbf{H}_i = \begin{bmatrix} \mathbf{DIAG}\left(\mathbf{h}_i^T, \mathbf{h}_i^T, \ldots, \mathbf{h}_i^T\right) \\ \mathbf{0}_{nm}^T \end{bmatrix} \in \mathbb{C}^{(n+1)\times nm}. \tag{7.46}$$

Thus, the transmit beamformer design problem (7.40) can now be cast as an SOCP as follows (cf. (4.103))

$$\min \ t$$

$$\text{s.t. } \|\mathbf{f}\|_2 \leq t$$

$$\text{Im}\{\mathbf{h}_i^T \mathbf{f}_i\} = 0, \ i = 1, \ldots, n \tag{7.47}$$

$$\|\mathbf{H}_i \mathbf{f} + \mathbf{b}_i\|_2 \leq \sqrt{1 + \frac{1}{\gamma_0}} \mathbf{h}_i^T \mathbf{f}_i, \ i = 1, \ldots, n.$$

Though problem (7.40) can be formulated as the SOCP (7.47), it can also be formulated as an SDP to be presented in Subsection 9.8.3. Moreover, the worst-case robust design to (7.40) for the single-cell MISO scenario will be presented in Subsection 8.5.4 and the corresponding rate outage constrained robust design will be addressed in Subsection 8.5.8.

7.6.2　Max-Min-Fair beamforming

Given a power P_0, we would like to find a beamformer matrix $\mathbf{F}$ that maximizes the smallest (or worst-case) SINR (cf. (4.153) and (4.154)):

$$
\begin{array}{ll}
\max & \min_{i=1,\ldots,n} \gamma_i(\mathbf{F}) \\
\text{s.t.} & \sum_{i=1}^{n} \|\mathbf{f}_i\|_2^2 \le P_0
\end{array}
\equiv
\begin{array}{ll}
\min & \max_{i=1,\ldots,n} \left\{ \frac{1}{\gamma_i(\mathbf{F})} \right\} \\
\text{s.t.} & \sum_{i=1}^{n} \|\mathbf{f}_i\|_2^2 \le P_0.
\end{array}
\tag{7.48}
$$

This problem in its current form is nonconvex since the objective function is nonconvex. It is not a quasiconvex problem either. Next, we present how to formulate this problem into a solvable problem using the bisection method as used to solve a quasiconvex problem.

By using the epigraph reformulation, the above problem can be cast as

$$
\begin{aligned}
\min \quad & t \\
\text{s.t.} \quad & \frac{1}{\gamma_i(\mathbf{F})} \le t, \ i = 1, 2, \ldots, n \\
& \sum_{i=1}^{n} \|\mathbf{f}_i\|_2^2 \le P_0.
\end{aligned}
\tag{7.49}
$$

As in the previous subsection, the first n inequality constraints in (7.49) can be written as

$$
\frac{1}{\gamma_i(\mathbf{F})} = \frac{\sum_{j=1, j\neq i}^{n} \left|\mathbf{h}_i^T \mathbf{f}_j\right|^2 + \sigma_i^2}{\left|\mathbf{h}_i^T \mathbf{f}_i\right|^2} \le t, \ i = 1, 2, \ldots, n
\tag{7.50}
$$

$$
\Rightarrow \sum_{j=1}^{n} \left|\mathbf{h}_i^T \mathbf{f}_j\right|^2 + \sigma_i^2 \le (1+t) \cdot \left|\mathbf{h}_i^T \mathbf{f}_i\right|^2, \ i = 1, 2, \ldots, n
\tag{7.51}
$$

which has the same form as (7.41) with $1/\gamma_0$ replaced by t. Let us emphasize that γ_0 is a preassigned SINR threshold (a constant) but t is an unknown variable here. Moreover, the last inequality constraint in (7.49) is equivalent to

$$
\|\mathbf{f}\|_2 \le \sqrt{P_0}.
\tag{7.52}
$$

Finally, the problem (7.49) can be written as

$$\min \ t$$

$$\text{s.t.} \ \sum_{j=1}^{n} \left| \mathbf{h}_i^T \mathbf{f}_j \right|^2 + \sigma_i^2 \leq (1+t) \cdot \left| \mathbf{h}_i^T \mathbf{f}_i \right|^2, \ i = 1, 2, \ldots, n \qquad (7.53)$$

$$\|\mathbf{f}\|_2 \leq \sqrt{P_0}.$$

Problem (7.53) is nonconvex because the n inequality constraint functions involving t are nonconvex. However, it can be solved under a quasiconvex optimization framework. Note that each of the inequalities in the second line of (7.53) can be further transformed to a second-order cone constraint (similar to the one shown in (7.43) with $1/\gamma_0$ replaced by t). The resultant problem is an SOCP for any fixed t as follows:

$$\min \ t$$

$$\text{s.t.} \ \left\| \mathbf{H}_i \mathbf{f} + \mathbf{b}_i \right\|_2 \leq \sqrt{1+t} \cdot \mathbf{h}_i^T \mathbf{f}_i, \ i = 1, 2, \ldots, n$$

$$\text{Im}\left\{ \mathbf{h}_i^T \mathbf{f}_i \right\} = 0, \ i = 1, \ldots, n \qquad (7.54)$$

$$\|\mathbf{f}\|_2 \leq \sqrt{P_0}.$$

Problem (7.54) can be solved using the bisection method (cf. Algorithm 4.1) introduced in Section 4.5, except that in Step 4, the convex feasibility problem is an SOCP as follows (cf. (4.103))

$$\text{find} \ \mathbf{f}$$

$$\text{s.t.} \ \left\| \mathbf{H}_i \mathbf{f} + \mathbf{b}_i \right\|_2 \leq \sqrt{1+t} \cdot \mathbf{h}_i^T \mathbf{f}_i, \ i = 1, 2, \ldots, n$$

$$\text{Im}\left\{ \mathbf{h}_i^T \mathbf{f}_i \right\} = 0, \ i = 1, \ldots, n \qquad (7.55)$$

$$\|\mathbf{f}\|_2 \leq \sqrt{P_0}.$$

7.6.3   Multicell beamforming

Consider a multicell wireless system with N_c cells as shown in Figure 7.4, where each cell consists of a BS sharing the same frequency band and equipped with N_t antennas and K single-antenna mobile stations (MSs). Let $s_{ik}(t)$ denote the information signal for MS k in the ith cell (denoted as MS_{ik}) with $\mathbb{E}\{|s_{ik}(t)|^2\} = 1$, and let $\mathbf{w}_{ik} \in \mathbb{C}^{N_t}$ denote the associated beamforming vector. Then the transmitted signal by the ith BS can be expressed as

$$\mathbf{x}_i(t) = \sum_{k=1}^{K} \mathbf{w}_{ik} s_{ik}(t) \qquad (7.56)$$

for $i = 1, \ldots, N_c$. Let $\mathbf{h}_{jik} \in \mathbb{C}^{N_t}$ denote the channel vector from the jth BS to MS_{ik}. The received signal of MS_{ik} can be expressed as

$$
\begin{aligned}
y_{ik}(t) &= \sum_{j=1}^{N_c} \mathbf{h}_{jik}^T \mathbf{x}_j(t) + z_{ik}(t) \\
&= \sum_{j=1}^{N_c} \mathbf{h}_{jik}^T \left(\sum_{\ell=1}^{K} \mathbf{w}_{j\ell} s_{j\ell}(t) \right) + z_{ik}(t) \\
&= \mathbf{h}_{iik}^T \mathbf{w}_{ik} s_{ik}(t) + \sum_{\ell \neq k}^{K} \mathbf{h}_{iik}^T \mathbf{w}_{i\ell} s_{i\ell}(t) + \sum_{j \neq i}^{N_c} \mathbf{h}_{jik}^T \sum_{\ell=1}^{K} \mathbf{w}_{j\ell} s_{j\ell}(t) + z_{ik}(t),
\end{aligned}
\tag{7.57}
$$

where the first term in (7.57) is the signal of interest, the second and third terms are the *intra-cell interference* and *inter-cell interference*, respectively, and $z_{ik}(t)$ is the additive noise with zero mean and variance $\sigma_{ik}^2 > 0$. From (7.57), the SINR of MS_{ik} can be shown to be

$$
\mathrm{SINR}_{ik}\left(\mathbf{w}_{11}, \ldots, \mathbf{w}_{N_cK}\right) = \frac{\left|\mathbf{h}_{iik}^T \mathbf{w}_{ik}\right|^2}{\sum\limits_{\ell \neq k}^{K} \left|\mathbf{h}_{iik}^T \mathbf{w}_{i\ell}\right|^2 + \sum\limits_{j \neq i}^{N_c} \sum\limits_{\ell=1}^{K} \left|\mathbf{h}_{jik}^T \mathbf{w}_{j\ell}\right|^2 + \sigma_{ik}^2}.
\tag{7.58}
$$

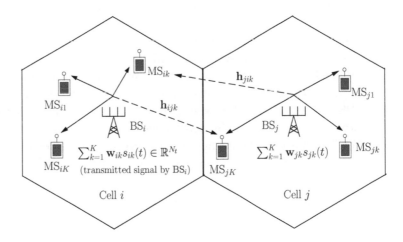

Figure 7.4 An illustration of a multicell system.

With the SINR in (7.58) used as the MSs' quality of service (QoS) measure, a multicell beamforming (MCBF) design problem is defined as

$$
\min \sum_{i=1}^{N_c} \sum_{k=1}^{K} \left\| \mathbf{w}_{ik} \right\|_2^2
\tag{7.59}
$$

$$
\text{s.t. } \mathrm{SINR}_{ik}\left(\mathbf{w}_{11}, \ldots, \mathbf{w}_{N_cK}\right) \geq \gamma_{ik}, \quad k = 1, \ldots, K, \ i = 1, \ldots, N_c
$$

where $\gamma_{ik} > 0$ is the target SINR for MS_{ik}. One can see from (7.59) that the N_c BSs jointly design their beamforming vectors such that the sum power of BSs is minimized, and meanwhile each of the MSs can be served with SINR no less than γ_{ik}. Although problem (7.59) is not convex, it can be reformulated as a convex SOCP through the similar reformulation procedure as presented in Subsection 7.6.1. The resulting SOCP is given by

$$
\begin{aligned}
\min ~ & t \\
\text{s.t.} ~ & \left(\sum_{i=1}^{N_c} \sum_{k=1}^{K} \left\| \mathbf{w}_{ik} \right\|_2^2 \right)^{1/2} \leq t \\
& \mathbf{h}_{iik}^T \mathbf{w}_{ik} \geq \sqrt{ \gamma_{ik} \left(\sum_{\ell \neq k}^{K} \left| \mathbf{h}_{iik}^T \mathbf{w}_{i\ell} \right|^2 + \sum_{j \neq i}^{N_c} \sum_{\ell=1}^{K} \left| \mathbf{h}_{jik}^T \mathbf{w}_{j\ell} \right|^2 + \sigma_{ik}^2 \right) }, ~ \forall~ k, i \\
& \text{Im}\{ \mathbf{h}_{iik}^T \mathbf{w}_{ik} \} = 0, ~ k = 1, \ldots, K, ~ i = 1, \ldots, N_c.
\end{aligned}
\tag{7.60}
$$

The MCBF design of problem (7.59) is also called the non-robust design. The associated worst-case robust design will be addressed in Subsection 8.5.5.

7.6.4   Femtocell beamforming

Consider a two-tier heterogeneous network that consists of a macrocell coverage and a localized femtocell as shown in Fig. 7.5, where the macrocell base station (MBS), equipped with N_M antennas, communicates with a single-antenna macrocell user equipment (MUE), and the femtocell base station (FBS), equipped with N_F antennas, serves a single-antenna femtocell user equipment (FUE) which shares the MUE spectrum during the downlink transmission of MBS and FBS. The MUE belongs to tier-1 with a higher priority for QoS guarantee while the FUE is in tier-2 whose service can be characterized as "best-effort" in nature. The channel from MBS to FUE is denoted by $\mathbf{h}_{FM} \in \mathbb{C}^{N_M}$; that from FBS to FUE is denoted by $\mathbf{h}_{FF} \in \mathbb{C}^{N_F}$; and that from FBS to MUE is denoted by $\mathbf{h}_{MF} \in \mathbb{C}^{N_F}$.

The transmit signal at the FBS is given by

$$
\mathbf{x}(t) = \mathbf{w}_F \cdot s_F(t),
\tag{7.61}
$$

where $s_F(t) \in \mathbb{C}$ is the information-bearing signal intended for the FUE, which is generated from a Gaussian random codebook with zero mean, and $\mathbf{w}_F \in \mathbb{C}^{N_F}$ is the associated beamforming vector. Let the transmit signal at the MBS be $\mathbf{w}_M \cdot s_M(t)$. Therefore, the received signal at the FUE can be expressed as

$$
y(t) = \mathbf{h}_{FF}^H \mathbf{w}_F s_F(t) + \mathbf{h}_{FM}^H \mathbf{w}_M s_M(t) + n_F(t),
\tag{7.62}
$$

where the first term is the intended signal for FUE, the second term is the interference from the macrocell, and $n_F(t) \in \mathbb{C}$ is additive noise at FUE with power $\sigma_F^2 > 0$. The QoS of the FUE is measured in terms of its SINR. Without loss of generality, suppose that $\mathbb{E}[|s_F(t)|^2] = 1$ and $\mathbb{E}[|s_M(t)|^2] = 1$. Then the SINR at FUE can be represented as

$$\text{SINR}_F = \frac{\left|\mathbf{h}_{FF}^H \mathbf{w}_F\right|^2}{\left|\mathbf{h}_{FM}^H \mathbf{w}_M\right|^2 + \sigma_F^2}, \tag{7.63}$$

and the interference power at the MUE from the femtocell can be seen to be $|\mathbf{h}_{MF}^H \mathbf{w}_F|^2$.

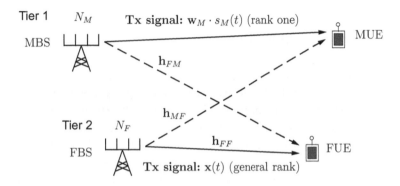

Figure 7.5 System model.

The FBS aims to design its beamforming vector $\mathbf{w}_F$ such that the transmit power is minimized, subject to an SINR constraint no smaller than $\gamma_F \geq 0$ on the FUE and an interference power constraint no larger than $\epsilon_M \geq 0$ on the MUE. Mathematically, the problem can be formulated as

$$\min_{\mathbf{w}_F \in \mathbb{C}^{N_F}} \ \|\mathbf{w}_F\|_2^2$$

$$\text{s.t.} \ \frac{\left|\mathbf{h}_{FF}^H \mathbf{w}_F\right|^2}{\left|\mathbf{h}_{FM}^H \mathbf{w}_M\right|^2 + \sigma_F^2} \geq \gamma_F \tag{7.64}$$

$$\left|\mathbf{h}_{MF}^H \mathbf{w}_F\right|^2 \leq \epsilon_M.$$

Although problem (7.64) is not convex, it can be equivalently reformulated into a convex SOCP through the similar reformulation procedure as presented in Subsection 7.6.1. The resulting SOCP is given by

$$\min \ t$$

$$\text{s.t.} \ \|\mathbf{w}_F\|_2 \leq t$$

$$\mathbf{h}_{FF}^H \mathbf{w}_F \geq \sqrt{\gamma_F\left(\left|\mathbf{h}_{FM}^H \mathbf{w}_M\right|^2 + \sigma_F^2\right)} \tag{7.65}$$

$$\left|\mathbf{h}_{MF}^H \mathbf{w}_F\right| \leq \sqrt{\epsilon_M}$$

$$\text{Im}\left\{\mathbf{h}_{FF}^H \mathbf{w}_F\right\} = 0.$$

Note that the first and the third inequality constraints in problem (7.65) are SOC constraints, while the second inequality constraint involving $\mathbf{w}_M$ is a second-order cone constraint when $\mathbf{w}_M$ is an unknown variable. However, usually it is known to FBS in advance because it can be provided by MBS via a broadband backhaul link. Then this inequality constraint actually reduces to a linear inequality constraint.

The rate outage constrained robust transmit beamforming design to problem (7.64) will be addressed in Subsection 8.5.9 where it is formulated as an SDP.

7.7   Summary and discussion

In this chapter we have presented an important class of convex optimization problems called SOCP together with some applications in communications. Through a variety of problem reformulations, the effectiveness of SOCP has been demonstrated by robust receive beamforming, and by various transmit downlink beamforming in single cell and multicell scenarios. Closed-form solutions for SOCP problems may be hard to obtain in general, and so available convex optimization solvers are usually used to obtain optimal numerical solutions instead. However, more cutting-edge challenging research in wireless communications and networking have appeared recently, such as robust multicell downlink beamforming due to uncertainty of the channel state information, and downlink interference management and resource allocation for heterogenous networks, where SDP turns out to be more effective and powerful than SOCP and will be introduced in the next chapter.

References for Chapter 7

[ACMC11] A. Ambikapathi, T.-H. Chan, W.-K. Ma, and C.-Y. Chi, "Chance constrained robust minimum volume enclosing simplex algorithm for hyperspectral unmixing," *IEEE Trans. Geoscience and Remote Sensing*, vol. 49, no. 11, pp. 4194–4209, Nov. 2011.

[BO02] M. Bengtsson and B. Ottersten, "Optimal and suboptimal transmit beamforming," in *Handbook of Antennas in Wireless Communications*, L. C. Godara, Ed. CRC Press, 2002.

[VGL03] S. Vorobyov, A. Gershman, and Z.-Q. Luo, "Robust adaptive beamforming using worst-case performance optimization: A solution to the signal mismatch problem," *IEEE Trans. Signal Process.*, vol. 52, no. 2, pp. 313–324, Feb. 2003.

[WES06] A. Wiesel, Y. C. Eldar, and S. Shamai, "Linear precoding via conic optimization for fixed MIMO receivers," *IEEE Trans. Signal Process.*, vol. 54, no. 1, pp. 161–176, Jan. 2006.

8 Semidefinite Programming

In this chapter we will introduce a widely known optimization problem called semidefinite program (previously abbreviated by SDP), which has been applied in many signal processing problems in MIMO wireless communications. However, the optimization problem of interest may often be NP-hard, and hence we may have to resort to various problem reformulations and convex approximations to either the objective function or the constraint functions or both of the original problem, that either were presented in the previous chapters (e.g., the SCA (cf. Section 4.7)) or are introduced in this chapter (e.g., the widely used semidefinite relaxation). Consequently, we may come up with a solvable SDP together with a suboptimal solution to the original problem instead. Nevertheless, the obtained solution may be analytically proven to be an optimal solution or a stationary point of the original problem for some practical scenarios. Though the SDP is the focus of this chapter, the BSUM introduced in Section 4.7 (that may not involve SDP at all) may be potentially a more powerful alternative than the SDP for finding a stationary-point solution of the same problem. Hence some applications using both the SDP and the BSUM will also be presented in this chapter for the comparison of the two methods in terms of complexity and solution accuracy.

Since the ways of solving an optimization problem are always nonunique, conversion from LP, QP, QCQP, SOCP to SDP is briefly presented first, and then the *S-procedure* is presented that is effective to convert infinitely many quadratic constraints into an LMI constraint, together with some practical examples to illustrate the effectiveness of SDP, via the use of the semidefinite relaxation. Then we present some cutting edge research results in MIMO wireless communications through the use of SDP, including coherent/noncoherent detection, physical-layer secret communications, worst-case robust transmit beamforming for both single-cell and multicell scenarios, outage constrained multicell coordinated beamforming, and outage constrained robust transmit beamforming for both single-cell and multicell scenarios. For notational simplicity in this chapter, $\mathbf{X} \succeq \mathbf{0}$ means that the matrix $\mathbf{X}$ is PSD; $\mathbf{x} \succeq \mathbf{0}$ means that all the components of the column vector $\mathbf{x}$ are greater than or equal to zero.

8.1   Semidefinite program (SDP)

The SDP has the following structures:

- Inequality form:

$$\begin{aligned} \min \ & \mathbf{c}^T\mathbf{x} \\ \text{s.t. } & \mathbf{F}(\mathbf{x}) \preceq \mathbf{0} \end{aligned} \tag{8.1}$$

with variable $\mathbf{x} \in \mathbb{R}^n$ where $\mathbf{c} \in \mathbb{R}^n$;

$$\mathbf{F}(\mathbf{x}) = \mathbf{F}_0 + x_1\mathbf{F}_1 + \cdots + x_n\mathbf{F}_n \tag{8.2}$$

is an LMI in which $\mathbf{F}_i \in \mathbb{S}^k$.

- Standard form:

$$\begin{aligned} \min \ & \mathrm{Tr}(\mathbf{CX}) \\ \text{s.t. } & \mathbf{X} \succeq \mathbf{0} \\ & \mathrm{Tr}(\mathbf{A}_i\mathbf{X}) = b_i \in \mathbb{R}, \ i = 1,\ldots,m \end{aligned} \tag{8.3}$$

with variable $\mathbf{X} \in \mathbb{S}^n$, where $\mathbf{A}_i \in \mathbb{S}^n$, and $\mathbf{C} \in \mathbb{S}^n$.

The inequality form and the standard form can be shown to be equivalent (in Chapter 9). An SDP with multiple LMIs

$$\begin{aligned} \min \ & \mathbf{c}^T\mathbf{x} \\ \text{s.t. } & \mathbf{F}_i(\mathbf{x}) \preceq \mathbf{0}, \ i = 1,\ldots,m \end{aligned} \tag{8.4}$$

can be reduced to an SDP with one LMI since $\mathbf{F}_i(\mathbf{x}) \preceq \mathbf{0} \ \forall i = 1,\ldots,m$, are equivalent to

$$\mathbf{DIAG}(\mathbf{F}_1(\mathbf{x}),\ldots,\mathbf{F}_m(\mathbf{x})) \preceq \mathbf{0}. \tag{8.5}$$

Example 8.1 *(Maximum eigenvalue minimization)* Let $\lambda(\mathbf{X})$ $(\lambda_{\max}(\mathbf{X}))$ denote any eigenvalue (the maximum eigenvalue) of a matrix $\mathbf{X} \in \mathbb{S}^m$. The maximum eigenvalue minimization problem is defined as

$$\min_{\mathbf{x}} \ \lambda_{\max}(\mathbf{A}(\mathbf{x})) \tag{8.6}$$

where

$$\mathbf{A}(\mathbf{x}) = \mathbf{A}_0 + x_1\mathbf{A}_1 + \cdots + x_n\mathbf{A}_n \in \mathbb{S}^m.$$

This unconstrained problem is convex since $\lambda_{\max}(\mathbf{A}(\mathbf{x}))$ is convex in $\mathbf{A}(\mathbf{x})$, while $\mathbf{A}(\mathbf{x})$ is affine in $\mathbf{x}$.

Because of the fact that

$$\lambda(\mathbf{A}(\mathbf{x}) - t\mathbf{I}_m) = \lambda(\mathbf{A}(\mathbf{x})) - t$$

for any fixed $\mathbf{x}$, we have

$$\lambda_{\max}(\mathbf{A}(\mathbf{x})) \le t \iff \mathbf{A}(\mathbf{x}) - t\mathbf{I}_m \preceq \mathbf{0}. \tag{8.7}$$

Hence, the problem (8.6) is equivalent to (by epigraph reformulation)

$$\min_{\mathbf{x},t} \ t$$
$$\text{s.t. } \mathbf{A}(\mathbf{x}) - t\mathbf{I}_m \preceq 0 \tag{8.8}$$

which is an SDP. □

Remark 8.1 Note that the matrix function $\mathbf{F}(\mathbf{x})$ given by (8.2) is affine in $\mathbf{x}$, i.e., every component of $\mathbf{F}(\mathbf{x})$ is affine in $\mathbf{x}$. As all the coefficient matrices $\mathbf{F}_i$ in $\mathbf{F}(\mathbf{x})$ are diagonal, the SDP (8.1) reduces to an LP; namely, LP is actually a special case of SDP. Surely, an LP can also be reformulated as an SDP. Recall the standard form LP:

$$\min \ \mathbf{c}^T\mathbf{x}$$
$$\text{s.t. } \mathbf{x} \succeq 0 \tag{8.9}$$
$$\mathbf{a}_i^T\mathbf{x} = b_i, \ i = 1,\dots,m.$$

Let $\mathbf{C} = \mathbf{Diag}(\mathbf{c}) \in \mathbb{S}^n$, and $\mathbf{A}_i = \mathbf{Diag}(\mathbf{a}_i) \in \mathbb{S}^n$. Then problem (8.9) is equivalent to the standard form SDP given by

$$\min \ \mathrm{Tr}(\mathbf{CX}) \ \ (\text{since } [\mathbf{X}]_{ii} = x_i)$$
$$\text{s.t. } \mathbf{X} \succeq 0 \tag{8.10}$$
$$\mathrm{Tr}(\mathbf{A}_i\mathbf{X}) = b_i, \ i = 1,\dots,m$$

since $\mathbf{X} \succeq 0$ implies $[\mathbf{X}]_{ii} = x_i \geq 0$ for all i. □

8.2 QCQP and SOCP as SDP via Schur complement

The Schur complement, introduced in Subsection 3.2.5 (cf. (3.92) and (3.96)) for the real-variable case and in Example 4.8 (cf. (4.105)) for the complex-variable case, has been widely used in formulating quadratic inequality constraints into equivalent LMI constraints so that an optimization problem under consideration can be transformed to an SDP. In this section, we only consider the conversion of QCQP and SOCP into SDP for the real-variable case, because that for the complex-variable case can be similarly performed.

First of all, by the Schur complement (3.96), it can be easily inferred that the convex quadratic inequality

$$(\mathbf{Ax} + \mathbf{b})^T(\mathbf{Ax} + \mathbf{b}) - \mathbf{c}^T\mathbf{x} - d \leq 0, \ \forall \mathbf{x} \in \mathbb{R}^n \tag{8.11}$$

is true if and only if the LMI

$$\begin{bmatrix} \mathbf{I} & \mathbf{Ax} + \mathbf{b} \\ (\mathbf{Ax} + \mathbf{b})^T & \mathbf{c}^T\mathbf{x} + d \end{bmatrix} \succeq 0, \ \forall \mathbf{x} \in \mathbb{R}^n \tag{8.12}$$

is true. In other words, the constraint set under (8.11) and that under (8.12) are exactly the same.

Next, let us show how a QCQP can be converted to an SDP. Recall that a convex QCQP can be written as

$$
\begin{aligned}
\min_{\mathbf{x} \in \mathbb{R}^n} \quad & \left\| \mathbf{A}_0 \mathbf{x} + \mathbf{b}_0 \right\|_2^2 - \mathbf{c}_0^T \mathbf{x} - d_0 \\
\text{s.t.} \quad & \left\| \mathbf{A}_i \mathbf{x} + \mathbf{b}_i \right\|_2^2 - \mathbf{c}_i^T \mathbf{x} - d_i \leq 0, \ i = 1, \ldots, m.
\end{aligned}
\tag{8.13}
$$

By (3.96) and the epigraph representation to the problem, the QCQP can be equivalently written as

$$
\begin{aligned}
\min_{\mathbf{x} \in \mathbb{R}^n, t \in \mathbb{R}} \quad & t \\
\text{s.t.} \quad & \begin{bmatrix} \mathbf{I} & \mathbf{A}_0 \mathbf{x} + \mathbf{b}_0 \\ (\mathbf{A}_0 \mathbf{x} + \mathbf{b}_0)^T & \mathbf{c}_0^T \mathbf{x} + d_0 + t \end{bmatrix} \succeq \mathbf{0} \\
& \begin{bmatrix} \mathbf{I} & \mathbf{A}_i \mathbf{x} + \mathbf{b}_i \\ (\mathbf{A}_i \mathbf{x} + \mathbf{b}_i)^T & \mathbf{c}_i^T \mathbf{x} + d_i \end{bmatrix} \succeq \mathbf{0}, \ i = 1, \ldots, m
\end{aligned}
\tag{8.14}
$$

which is an SDP.

Finally, consider the second-order cone inequality:

$$
\|\mathbf{A}\mathbf{x} + \mathbf{b}\|_2 \leq \mathbf{f}^T \mathbf{x} + d, \ \mathbf{x} \in \mathbb{R}^n
\tag{8.15}
$$

where $\mathbf{A} \in \mathbb{R}^{m \times n}$ and $\mathbf{b} \in \mathbb{R}^m$. If the problem domain is such that $\mathbf{f}^T \mathbf{x} + d > 0$, we have $\|\mathbf{A}\mathbf{x} + \mathbf{b}\|_2^2 \leq (\mathbf{f}^T \mathbf{x} + d)^2$, and then the inequality can be re-expressed as

$$
\mathbf{f}^T \mathbf{x} + d - \frac{1}{\mathbf{f}^T \mathbf{x} + d} (\mathbf{A}\mathbf{x} + \mathbf{b})^T (\mathbf{A}\mathbf{x} + \mathbf{b}) \geq 0, \ \mathbf{x} \in \mathbb{R}^n.
\tag{8.16}
$$

By (3.96) again, the constraint subject to this inequality is identical to that subject to the LMI

$$
\begin{bmatrix} (\mathbf{f}^T \mathbf{x} + d) \mathbf{I}_m & \mathbf{A}\mathbf{x} + \mathbf{b} \\ (\mathbf{A}\mathbf{x} + \mathbf{b})^T & \mathbf{f}^T \mathbf{x} + d \end{bmatrix} \succeq \mathbf{0}, \ \mathbf{x} \in \mathbb{R}^n
\tag{8.17}
$$

implying that an SOCP can also be converted to an SDP.

8.3   S-Procedure

Schur complement is an effective method to convert a single quadratic inequality into a single semidefinite matrix inequality. The S-procedure is an effective method that converts a set of infinitely many quadratic constraints into a single semidefinite matrix inequality with an extra unknown nonnegative parameter as stated next.

S-procedure: Let $\mathbf{F}_1$, $\mathbf{F}_2 \in \mathbb{S}^n$, $\mathbf{g}_1$, $\mathbf{g}_2 \in \mathbb{R}^n$, h_1, $h_2 \in \mathbb{R}$. Then the following implication

$$\mathbf{x}^T \mathbf{F}_1 \mathbf{x} + 2\mathbf{g}_1^T \mathbf{x} + h_1 \leq 0 \tag{8.18a}$$

$$\implies \mathbf{x}^T \mathbf{F}_2 \mathbf{x} + 2\mathbf{g}_2^T \mathbf{x} + h_2 \leq 0 \tag{8.18b}$$

i.e., $\{\mathbf{x} \in \mathbb{R}^n \mid \mathbf{x}^T \mathbf{F}_1 \mathbf{x} + 2\mathbf{g}_1^T \mathbf{x} + h_1 \leq 0\} \subseteq \{\mathbf{x} \in \mathbb{R}^n \mid \mathbf{x}^T \mathbf{F}_2 \mathbf{x} + 2\mathbf{g}_2^T \mathbf{x} + h_2 \leq 0\}$, holds true if and only if there exists a $\lambda \geq 0$ such that

$$\begin{bmatrix} \mathbf{F}_2 & \mathbf{g}_2 \\ \mathbf{g}_2^T & h_2 \end{bmatrix} \preceq \lambda \begin{bmatrix} \mathbf{F}_1 & \mathbf{g}_1 \\ \mathbf{g}_1^T & h_1 \end{bmatrix}, \tag{8.19}$$

provided that there exists a point $\widehat{\mathbf{x}}$ with $\widehat{\mathbf{x}}^T \mathbf{F}_1 \widehat{\mathbf{x}} + 2\mathbf{g}_1^T \widehat{\mathbf{x}} + h_1 < 0$.

One proof of the S-procedure will be presented using theorems of alternatives to be introduced in Subsection 9.9.3 later. A mathematically equivalent lemma for the S-procedure will be presented and proven via strong duality of a convex problem in Subsection 9.3.2. An alternative interpretation for the S-procedure, which is instrumental in how to practically apply the S-procedure, is given in the following remark:

Remark 8.2 Suppose that there exists a point $\widehat{\mathbf{x}}$ satisfying $\widehat{\mathbf{x}}^T \mathbf{F}_1 \widehat{\mathbf{x}} + 2\mathbf{g}_1^T \widehat{\mathbf{x}} + h_1 < 0$. The statement, that the second-order inequality (8.18b) is true for all $\mathbf{x}$ satisfying the second-order inequality (8.18a), is true if and only if (8.19) is true for some $\lambda \geq 0$. $\square$

In contrast to the equivalence of the convex quadratic inequality given by (8.11) and the semidefinite matrix inequality given by (8.12), the equivalent semidefinite matrix inequality given by (8.19) never involves $\mathbf{x}$. Note that if the inequalities in (8.18) are parameterized by $\mathbf{x}$ (i.e., $\mathbf{x}$ does not stand for the unknown variable while $\mathbf{F}_i$, $\mathbf{g}_i$, and h_i for $i = 1, 2$ contain unknown variables instead) in a semi-infinite optimization problem, i.e., there are infinitely many inequality constraints. The equivalent semidefinite matrix constraint given by (8.19) only introduces an extra unknown variable λ under nonnegative constraint in the reformulated optimization problem. Currently, the S-procedure has been widely used in robust transmit beamforming in MIMO wireless communications. The complex version of the S-procedure is given in the following remark:

Remark 8.3 *(S-procedure for the complex case)* Let $\mathbf{F}_1, \mathbf{F}_2 \in \mathbb{H}^n$, $\mathbf{g}_1, \mathbf{g}_2 \in \mathbb{C}^n$, $h_1, h_2 \in \mathbb{R}$. The following implication

$$\mathbf{x}^H \mathbf{F}_1 \mathbf{x} + 2\mathrm{Re}\{\mathbf{g}_1^H \mathbf{x}\} + h_1 \leq 0 \implies \mathbf{x}^H \mathbf{F}_2 \mathbf{x} + 2\mathrm{Re}\{\mathbf{g}_2^H \mathbf{x}\} + h_2 \leq 0 \tag{8.20}$$

holds true if and only if there exists a $\lambda \geq 0$ such that

$$\begin{bmatrix} \mathbf{F}_2 & \mathbf{g}_2 \\ \mathbf{g}_2^H & h_2 \end{bmatrix} \preceq \lambda \begin{bmatrix} \mathbf{F}_1 & \mathbf{g}_1 \\ \mathbf{g}_1^H & h_1 \end{bmatrix}, \tag{8.21}$$

provided that there exists a point $\widehat{\mathbf{x}}$ with $\widehat{\mathbf{x}}^H \mathbf{F}_1 \widehat{\mathbf{x}} + 2\mathrm{Re}\{\mathbf{g}_1^H \widehat{\mathbf{x}}\} + h_1 < 0$. $\square$

The S-procedure for the complex case (i.e., (8.20) and (8.21)) can be converted to that for the real case (i.e., (8.18) and (8.19)) with the corresponding real $\mathbf{F}_i$, $\mathbf{g}_i$, and $\mathbf{x}$ defined as follows:

$$\mathbf{F}_i = \mathbf{F}_{iR} + j\mathbf{F}_{iI} \in \mathbb{H}^n, \quad \boldsymbol{F}_i = \begin{bmatrix} \mathbf{F}_{iR} & -\mathbf{F}_{iI} \\ \mathbf{F}_{iI} & \mathbf{F}_{iR} \end{bmatrix} \in \mathbb{S}^{2n}, \quad i = 1, 2$$

$$\mathbf{x} = \mathbf{x}_R + j\mathbf{x}_I \in \mathbb{C}^n, \quad \boldsymbol{x} = [\mathbf{x}_R^T, \mathbf{x}_I^T]^T \in \mathbb{R}^{2n}$$

$$\mathbf{g}_i = \mathbf{g}_{iR} + j\mathbf{g}_{iI} \in \mathbb{C}^n, \quad \boldsymbol{g}_i = [\mathbf{g}_{iR}^T, \mathbf{g}_{iI}^T]^T \in \mathbb{R}^{2n}, \quad i = 1, 2.$$

Hence, the S-procedure is true for the real case implies that it is true for the complex case.

8.4   Applications in combinatorial optimization

8.4.1   Boolean quadratic program (BQP)

Consider the following BQP:

$$\begin{aligned} \max \ & \mathbf{x}^T \mathbf{C} \mathbf{x} \\ \text{s.t. } & x_i \in \{-1, +1\}, \ i = 1, \ldots, n, \ (\text{i.e., } \mathbf{x} \in \{-1, +1\}^n) \end{aligned} \tag{8.22}$$

where $\mathbf{C} \in \mathbb{S}^n$. The quadratic objective function is convex only when $\mathbf{C} \succeq \mathbf{0}$. Moreover, the feasible set $\mathbf{x} \in \{-1, +1\}^n$ is equivalent to the set $x_i^2 = 1, i = 1, \ldots, n$ that all are equality constraints but not affine. Hence the BQP is non-convex even when $\mathbf{C}$ is PSD. The BQP is an exhaustive search problem with complexity of 2^n. In fact the BQP is NP-hard in general. Loosely speaking, for an NP-hard problem, the worst-case complexity order is at least $\mathcal{O}(r^n) \geq \mathcal{O}(n^k)$ for n large, where $r > 1$, k are positive real numbers, and n denotes the number of decision variables. Next, let us introduce some examples of BQPs prior to the presentation of how to find approximate solutions with good accuracy that can be obtained efficiently by solving a polynomial-time solvable convex problem via a reformulation and relaxation procedure from the BQP.

8.4.2   Practical example I: MAXCUT

Consider a graph $G = (V, E)$ where V represents a set of nodes and E is a set of edges (or equivalently the set of all the ordered pairs (i, j) of the connected nodes i and j in V) with weights $w_{ij} \geq 0$ for $(i, j) \in E$. Assume $w_{ij} = 0$ if $(i, j) \notin E$ (cf. Figure 8.1). A cut is to divide all the nodes into two disjoint sets K and $V \setminus K$. Let $\mathcal{C}(K)$ denote the cut determined by K, that is,

$$\mathcal{C}(K) = \{(i, j) \in E \mid i \in K, j \in V \setminus K\}.$$

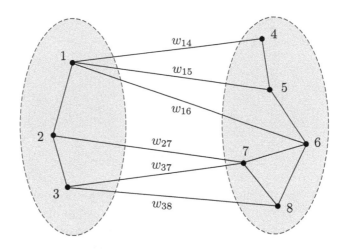

Figure 8.1 The MAXCUT problem.

The total weight of the cut is defined as

$$\mathsf{w}(\mathcal{C}(K)) = \sum_{(i,j)\in\,\mathcal{C}(K)} w_{ij}.$$

The MAXCUT problem is the problem to maximize the total weight of the edges over all possible cuts $\mathcal{C}(K)$. Hence, this problem can be formulated as the following optimization problem

$$\max \; \mathsf{w}(\mathcal{C}(K)) \tag{8.23}$$
$$\text{s.t. } K \subset V.$$

Let $V = \{1, 2, \ldots, n\}$ (cf. Figure 8.1 for $n = 8$). By introducing variables x_i, $i = 1, \ldots, n$ and $x_i = +1$ if $i \in K$, $x_i = -1$ if $i \in V \setminus K$. Then we come up with

$$1 - x_i x_j = \begin{cases} 2, & \text{if } i \in K \text{ and } j \in V \setminus K \text{ or } j \in K \text{ and } i \in V \setminus K \\ 0, & \text{otherwise}, \end{cases}$$

and hence

$$\mathsf{w}(\mathcal{C}(K)) = \sum_{i=1}^{n} \sum_{j=i+1}^{n} w_{ij} \frac{1 - x_i x_j}{2}. \tag{8.24}$$

By (8.23) and (8.24), the above MAXCUT problem takes the following form precisely

$$\max \; \sum_{i=1}^{n} \sum_{j=i+1}^{n} w_{ij} \frac{1 - x_i x_j}{2} \tag{8.25}$$
$$\text{s.t. } x_i \in \{-1, +1\}, \; i = 1, \ldots, n$$

which can be formulated as a BQP as follows

$$\max \ \mathbf{x}^T \mathbf{C} \mathbf{x}$$
$$\text{s.t. } x_i \in \{-1, +1\}, \ i = 1, \ldots, n \tag{8.26}$$

where $\mathbf{x} = (x_1, \ldots, x_n)$, $\mathbf{C} = \{c_{ij}\}_{n \times n} \in \mathbb{S}^n$,

$$c_{ij} = c_{ji} = -\frac{1}{4} w_{ij} \leq 0, \ i \neq j,$$

$$c_{ii} = \frac{1}{2} \sum_{j=i+1}^{n} w_{ij} \geq 0.$$

Thus the optimal solution must comprise a group of nodes (i.e., nodes of a set K) associated with all $x_i = 1$ and a group of nodes (i.e., nodes of the set $V \setminus K$) associated with all $x_i = -1$, and meanwhile the two disjoint groups are complements of each other with the maximum sum of the edge weights between them.

8.4.3   Practical example II: ML MIMO detection

Consider the scenario that the transmitter and the receiver have n and m antennas, respectively (see Figure 8.2). We consider a spatial multiplexing system, where each transmitter antenna transmits its own sequence of symbols. Assume frequency flat fading and antipodal modulation. The received signal model can be expressed as

$$\mathbf{y} = \mathbf{H}\mathbf{s} + \mathbf{v}, \tag{8.27}$$

where

- $\mathbf{y} \in \mathbb{C}^m$ is the multireceiver (channel output) vector,
- $\mathbf{s} \in \{-1, +1\}^n$ is the signal vector consisting of n transmitted symbols,
- $\mathbf{H} = \{h_{ij}\}_{m \times n} \in \mathbb{C}^{m \times n}$ is the MIMO (or multiantenna) channel, and
- $\mathbf{v} \in \mathbb{C}^m$ is Gaussian noise with zero mean and covariance $\sigma^2 \mathbf{I}_m$.

The corresponding real signal model can be easily seen to be

$$y = \begin{bmatrix} \text{Re}\{\mathbf{y}\} \\ \text{Im}\{\mathbf{y}\} \end{bmatrix} = \mathcal{H}s + v, \tag{8.28}$$

where

$$\mathcal{H} = \begin{bmatrix} \text{Re}\{\mathbf{H}\} \\ \text{Im}\{\mathbf{H}\} \end{bmatrix} \in \mathbb{R}^{2m \times n}, \quad v = \begin{bmatrix} \text{Re}\{\mathbf{v}\} \\ \text{Im}\{\mathbf{v}\} \end{bmatrix} \in \mathbb{R}^{2m}.$$

ML detection of $\mathbf{s}$ can be formulated as the following problem

$$\min_{\mathbf{s} \in \{\pm 1\}^n} \|y - \mathcal{H}s\|_2^2 \equiv \min_{\mathbf{s} \in \{\pm 1\}^n} s^T \mathcal{H}^T \mathcal{H}s - 2s^T \mathcal{H}^T y, \tag{8.29}$$

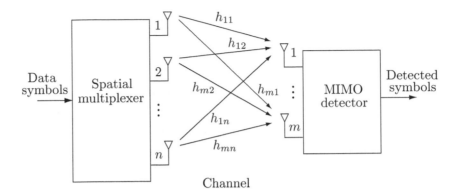

Figure 8.2 A MIMO spatial multiplexing system with n transmit antennas and m receive antennas, where h_{ij} denotes the channel from transmit antenna j to receive antenna i.

which obviously is a nonhomogeneous BQP. It can be homogenized as follows:

$$
\begin{aligned}
& \min_{\mathbf{s} \in \{\pm 1\}^n} \mathbf{s}^T \mathcal{H}^T \mathcal{H} \mathbf{s} - 2\mathbf{s}^T \mathcal{H}^T \mathbf{y} \\
&= \min_{\tilde{\mathbf{s}} \in \{\pm 1\}^n, c \in \{\pm 1\}} (c\tilde{\mathbf{s}})^T \mathcal{H}^T \mathcal{H}(c\tilde{\mathbf{s}}) - 2(c\tilde{\mathbf{s}})^T \mathcal{H}^T \mathbf{y} && (\mathbf{s} = c\tilde{\mathbf{s}}) \\
&= \min_{(\tilde{\mathbf{s}}, c) \in \{\pm 1\}^{n+1}} \tilde{\mathbf{s}}^T \mathcal{H}^T \mathcal{H} \tilde{\mathbf{s}} - 2c\tilde{\mathbf{s}}^T \mathcal{H}^T \mathbf{y} && (c^2 = 1) \\
&= \min_{(\tilde{\mathbf{s}}, c) \in \{\pm 1\}^{n+1}} \begin{bmatrix} \tilde{\mathbf{s}}^T & c \end{bmatrix} \begin{bmatrix} \mathcal{H}^T \mathcal{H} & -\mathcal{H}^T \mathbf{y} \\ -\mathbf{y}^T \mathcal{H} & 0 \end{bmatrix} \begin{bmatrix} \tilde{\mathbf{s}} \\ c \end{bmatrix}.
\end{aligned}
\tag{8.30}
$$

Thus the ML detection problem reduces to a BQP. Next, let us present how to solve the nonconvex BQP by finding an approximate solution where solving an SDP problem is needed.

8.4.4 BQP approximation by semidefinite relaxation

Now let us get back to the standard BQP (8.22). To handle (8.22), we first formulate a convex problem (which is never an equivalent reformulation of (8.22)) through relaxation of the nonconvex constraints of (8.22), so-called *semidefinite relaxation* [LMS+10]. Then we address how to find an approximate solution of (8.22) from the optimal solution of the formulated convex problem.

By $\mathbf{x}^T \mathbf{C} \mathbf{x} = \mathrm{Tr}(\mathbf{C} \mathbf{x} \mathbf{x}^T)$ and defining the auxiliary variable

$$
\mathbf{X} = \mathbf{x} \mathbf{x}^T,
$$

the BQP in (8.22) can be reformulated as

$$\max_{\mathbf{x},\mathbf{X}} \ \text{Tr}(\mathbf{C}\mathbf{X})$$

$$\text{s.t. } \mathbf{X} = \mathbf{x}\mathbf{x}^T \qquad\qquad\qquad (8.31)$$

$$[\mathbf{X}]_{ii} = 1, \ i = 1, \ldots, n.$$

Now the objective function is linear (and hence convex) in $\mathbf{X}$ for any $\mathbf{C}$. The nonconvex constraint set associated with the $\{\pm 1\}$ constraint on $\mathbf{x}$ becomes a convex set of $\mathbf{X}$ (i.e., all the diagonal elements equal to unity), but the equality constraint $\mathbf{X} = \mathbf{x}\mathbf{x}^T$ (i.e., $\mathbf{X}$ must be a rank-one PSD matrix) is nonconvex; namely, the set $\{(\mathbf{x}, \mathbf{X}) \mid \mathbf{X} = \mathbf{x}\mathbf{x}^T\}$ is nonconvex.

In spite of the fact that

$$\mathbf{X} = \mathbf{x}\mathbf{x}^T \iff \mathbf{X} \succeq \mathbf{0} \text{ and } \text{rank}(\mathbf{X}) = 1, \qquad\qquad (8.32)$$

by relaxing the nonconvex constraint $\mathbf{X} = \mathbf{x}\mathbf{x}^T$ in (8.31) to $\mathbf{X} \succeq \mathbf{0}$ (i.e., removal of the rank-one constraint on $\mathbf{X}$), we come up with the following SDP:

$$\max \ \text{Tr}(\mathbf{C}\mathbf{X})$$

$$\text{s.t. } \mathbf{X} \succeq \mathbf{0} \qquad\qquad\qquad (8.33)$$

$$[\mathbf{X}]_{ii} = 1, \ i = 1, \ldots, n.$$

This relaxation is called the SDR. Problem (8.33) is called the SDR of problem (8.31), and they have the identical objective function $\text{Tr}(\mathbf{C}\mathbf{X})$ with the feasible set of the former containing that of the latter. It is noticeable that the optimal value of problem (8.33) will be no less than that of problem (8.31) (or (8.22)), and any feasible point of the latter is feasible to the former, implying that it is possible to find a good approximate solution of the latter, properly guided by the optimal solution of the former.

Once the SDP (8.33) is solved, its solution $\mathbf{X}^\star$ is used to find an approximation to the solution of the BQP (8.22). If the optimal $\mathbf{X}^\star$ is indeed of rank one, decomposing $\mathbf{X}^\star = \mathbf{x}^\star \mathbf{x}^{\star T}$ is straightforward to obtain the optimal solution $\mathbf{x}^\star$ of (8.22). If $\mathbf{X}^\star$ is not a rank-one matrix, two commonly used approximation methods [LMS+10] are as follows.

1. *Rank-one approximation*: One can simply choose the principal eigenvector $\tilde{\mathbf{x}}$ of $\mathbf{X}^\star$ and then quantize it into $\hat{\mathbf{x}} \in \{\pm 1\}^n$ by $[\hat{\mathbf{x}}]_i = \text{sgn}([\tilde{\mathbf{x}}]_i)$ for all i. The obtained $\hat{\mathbf{x}}$ is used as an approximate solution to the original BQP (8.22).

2. *Gaussian randomization*: This method generates L random vectors $\{\boldsymbol{\xi}^{(\ell)}, \ell = 1, \ldots, L\}$ with Gaussian distribution of zero mean and covariance matrix $\mathbf{X}^\star$ and then quantizes $\boldsymbol{\xi}^{(\ell)}$ into a vector in the set $\{\pm 1\}^n$, denoted as $\hat{\mathbf{x}}^{(\ell)}$, by

$$[\hat{\mathbf{x}}^{(\ell)}]_i = \text{sgn}([\boldsymbol{\xi}^{(\ell)}]_i) \ \forall i, \qquad\qquad (8.34)$$

and finally chooses $\widehat{\mathbf{x}} = \widehat{\mathbf{x}}^{(\ell^*)}$ as an approximate solution to problem (8.22) where

$$\ell^* = \arg\max_{\ell=1,\ldots,L} \left(\widehat{\mathbf{x}}^{(\ell)}\right)^T \mathbf{C}\widehat{\mathbf{x}}^{(\ell)}. \tag{8.35}$$

In general, the Gaussian randomization yields a better approximation solution than the rank-one approximation in many wireless communication applications. Let us conclude this subsection with the following remarks.

Remark 8.4 The philosophy of Gaussian randomization is that the SDR problem (8.33) and the following stochastic optimization problem share the identical solution $\mathbf{X}^*$:

$$\begin{aligned} \max \ & \mathbb{E}\left\{\boldsymbol{\xi}^T \mathbf{C}\boldsymbol{\xi}\right\} \\ \text{s.t. } & \boldsymbol{\xi} \sim \mathcal{N}(\mathbf{0}, \mathbf{X}) \\ & \mathbf{X} = \mathbb{E}\{\boldsymbol{\xi}\boldsymbol{\xi}^T\} \succeq \mathbf{0}, \ [\mathbf{X}]_{ii} = 1, \ i = 1,\ldots,n. \end{aligned} \tag{8.36}$$

The Gaussian randomization for finding an approximate solution $\widehat{\mathbf{x}}$ to the BQP (8.22) can be summarized by the following inequalities:

$$\begin{aligned} \widehat{\mathbf{x}}^T \mathbf{C}\widehat{\mathbf{x}} = \max_{\ell=1,\ldots,L} & \ \left(\widehat{\mathbf{x}}^{(\ell)}\right)^T \mathbf{C}\widehat{\mathbf{x}}^{(\ell)} \quad \text{(cf. (8.34) and (8.35))} \\ & \leq (\mathbf{x}^*)^T \mathbf{C}\mathbf{x}^* \quad \text{(where } \mathbf{x}^* \text{ is optimal to (8.22))} \\ & \leq \text{Tr}(\mathbf{C}\mathbf{X}^*) = \mathbb{E}\left\{\boldsymbol{\xi}^T \mathbf{C}\boldsymbol{\xi}\right\} \quad \text{(optimal value of (8.33) and (8.36))} \end{aligned}$$

where $\boldsymbol{\xi} \sim \mathcal{N}(\mathbf{0}, \mathbf{X}^*)$. When $\boldsymbol{\xi} \sim \mathcal{N}(\mathbf{0}, \mathbf{X}^*)$, a "good opportunity" (e.g., with a moderate L) can be expected, that is, the obtained approximate solution $\widehat{\mathbf{x}}$ well approximates to $\mathbf{x}^*$ though the nonlinear quantization is performed due to $\boldsymbol{\xi} \notin \{\pm 1\}^n$ with probability one. (Note that the second inequality holds with equality if $\mathbf{X}^*$ is of rank one, and that if no quantization is performed, $\sum_{\ell=1}^{L} (\widehat{\mathbf{x}}^{(\ell)})^T \mathbf{C}\widehat{\mathbf{x}}^{(\ell)}/L \to \text{Tr}(\mathbf{C}\mathbf{X}^*)$ as L increases.) Surely the larger the value of L, the better the solution accuracy. An empirically good choice is $L = 100$ (not a large number), consistent with the "good opportunity" mentioned above. An illustration for Gaussian randomization is given in Figure 8.3 for $n = 12$ and $L = 100$, where either $\widehat{\mathbf{x}} = \mathbf{x}^*$ or $\widehat{\mathbf{x}} = -\mathbf{x}^*$ since the optimal value of problem (8.22) remains identical for $\mathbf{x} = \pm\mathbf{x}^*$. $\square$

Remark 8.5 *(Convex-hull relaxation)* The SDR is to relax the nonconvex constraint of $\{\mathbf{X} = \mathbf{x}\mathbf{x}^T \mid \mathbf{x} \in \mathbb{R}^n\}$ (cf. (8.31)) by its convex hull

$$\text{conv}\,\{\mathbf{X} = \mathbf{x}\mathbf{x}^T \mid \mathbf{x} \in \mathbb{R}^n\} = \mathbb{S}_+^n \quad \text{(cf. (8.32)).} \tag{8.37}$$

The widely used Gaussian randomization for SDR turns out to be an effective follow-up procedure for finding a good approximate solution to the original nonconvex problem (8.22). This implies that the relaxation of a nonconvex constraint by its convex hull may help for finding an approximate solution to the original

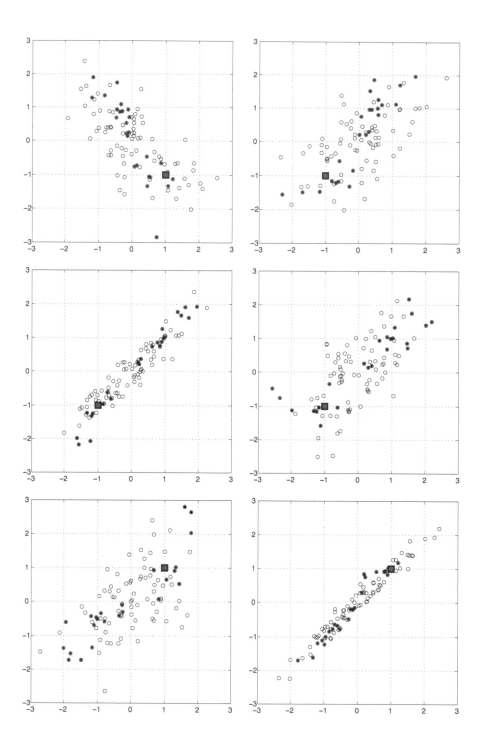

Figure 8.3 An illustration of Gaussian randomization for handling a BQP for $n = 12$ via SDR, where bullets and circles in the six subplots denote generated $L = 100$ random vectors $\boldsymbol{\xi} \in \mathbb{R}^n$ (with two components of $\boldsymbol{\xi}$ displayed in each subplot); the optimal solution $\mathbf{x}^\star = (1, -1, -1, -1, -1, -1, -1, -1, 1, 1, 1, 1)$ of the BQP is denoted by red bullets, and all the black bullets yield either $\widehat{\mathbf{x}} = \mathbf{x}^\star$ (denoted as blue squares) or $\widehat{\mathbf{x}} = -\mathbf{x}^\star$.

problem, while the SDR is merely a special case of convex-hull relaxation. In practical applications, not only SDR but also relaxation of other nonconvex constraints may be needed simultaneously in order to relax a nonconvex problem of interest into a convex problem. ☐

Remark 8.6 *(Complex SDR)* For the complex case of BQP, where the quadratic objective function is $\mathbf{x}^H \mathbf{C} \mathbf{x}$ with $\mathbf{C} \in \mathbb{H}^n$ and $\mathbf{x} \in \{\alpha \pm j\beta \in \mathbb{C}\}^n$ with $\alpha, \beta \in \mathbb{R}$, the SDR is to replace the equality constraint $\mathbf{X} = \mathbf{x}\mathbf{x}^H$ by the PSD constraint $\mathbf{X} = \mathbf{X}^H \succeq \mathbf{0}$ as well as $[\mathbf{X}]_{ii} = \alpha^2 + \beta^2$ for all i. Surely, the extension of SDR and Gaussian randomization to the case that $\mathbf{x} \in \mathcal{A}^n$, where $\mathcal{A}$ is a finite set of complex or real numbers with nonconstant modulus (e.g., the symbol constellation set in digital communications), is straightforward. An application in noncoherent detection involving both SDR and convex-hull relaxation of other nonconvex constraints will be introduced in the next subsection. ☐

Remark 8.7 Prior to the application of SDR, the nonconvex problem under consideration needs to be reformulated in the homogeneous quadratic form, as illustrated in the practical example of ML MIMO detection in the previous subsection. The SDR technique has been widely applied to many nonconvex problems in wireless communications involving quadratic constraints and/or quadratic functions embedded in the objective function. A rank-one solution may exist and can be analytically proven true, depending on the problem under consideration, and thus provides an optimal solution to the original problem without need of Gaussian randomization. As no rank-one solution exists, the preceding Gaussian randomization together with the way of obtaining a good rank-one approximate solution becomes problem dependent, and so advisable modifications are needed. Some applications in transmit beamforming will be introduced later in the next section. ☐

8.4.5 Practical example III: Linear fractional SDR (LFSDR) approach to noncoherent ML detection of higher-order QAM OSTBC

We consider a MIMO orthogonal space-time block code (OSTBC) system with N_t transmit antennas and N_r receive antennas as shown in Figure 8.2 wherein $n = N_t$ and $m = N_r$. It is assumed that the channel is frequency flat and it remains static for P consecutive code blocks. The respective received signal model is given by [CHMC10]

$$\mathbf{Y}_p = \mathbf{H}\mathbf{C}(\mathbf{u}_p) + \mathbf{W}_p, \ p = 1, \dots, P. \tag{8.38}$$

Here,

$\mathbf{Y}_p \in \mathbb{C}^{N_r \times T}$ received code matrix at block p, with T denoting the block length of the OSTBCs;

$\mathbf{u}_p \in \mathcal{U}^K$ transmitted symbol vector at block p, with $\mathcal{U} \subset \mathbb{C}$ denoting the symbol constellation set, K denoting the number of symbols per block;

$\mathbf{C} : \mathbb{C}^K \to \mathbb{C}^{N_t \times T}$ function that maps the given symbol vector to an OSTBC block;

$\mathbf{H} \in \mathbb{C}^{N_r \times N_t}$ MIMO channel matrix;

$\mathbf{W}_p \in \mathbb{C}^{N_r \times T}$ additive white Gaussian noise matrix with σ_w^2 denoting the average power per entry.

An OSTBC mapping function $\mathbf{C}(\cdot)$ can always be expressed in a linear dispersion form as

$$\mathbf{C}(\mathbf{u}_p) = \sum_{k=1}^{K} \mathrm{Re}\{[\mathbf{u}_p]_k\}\mathbf{A}_k + j \sum_{k=1}^{K} \mathrm{Im}\{[\mathbf{u}_p]_k\}\mathbf{B}_k, \qquad (8.39)$$

where $[\mathbf{u}_p]_k$ denotes the kth entry of $\mathbf{u}_p$, and $\mathbf{A}_k, \mathbf{B}_k \in \mathbb{R}^{N_t \times T}$ are the code basis matrices. The basis matrices are specially designed such that, for any $\mathbf{u}_p \in \mathcal{U}^K$, the orthogonality condition is satisfied [TJC99, GS01]:

$$\mathbf{C}(\mathbf{u}_p)\mathbf{C}^H(\mathbf{u}_p) = \|\mathbf{u}_p\|_2^2 \cdot \mathbf{I}_{N_t}. \qquad (8.40)$$

Here we are interested in detecting $\{\mathbf{u}_p\}_{p=1}^P$ from $\{\mathbf{Y}_p\}_{p=1}^P$ without knowing $\mathbf{H}$ (as known as noncoherent OSTBC detection). One can show that the blind ML OSTBC detection problem for the 4^q-QAM ($q \geq 2$) signaling case can be simplified to a real discrete maximization problem as follows

$$\mathbf{s}^\star = \arg \max_{\mathbf{s} \in \mathcal{U}^{2PK}} \frac{\mathbf{s}^T \mathbf{F} \mathbf{s}}{\mathbf{s}^T \mathbf{s}}, \qquad (8.41)$$

where

$$\mathcal{U} = \{\pm 1, \pm 3, \ldots, \pm(2^q - 1)\} \qquad (8.42)$$

and the real $2PK \times 1$ vector $\mathbf{s}$ contains all the real parts and the imaginary parts of $\{\mathbf{u}_p\}_{p=1}^P$, and the matrix $\mathbf{F} \in \mathbb{S}^{2PK}$ is a function of $\mathbf{Y}_p$ and basis matrices $\mathbf{A}_k$ and $\mathbf{B}_k$. Note that the problem (8.41) to be solved has a Rayleigh quotient objective function.

Let us partition $\mathbf{F}$ and $\mathbf{s}$ in the following forms

$$\mathbf{F} = \begin{bmatrix} u & \mathbf{v}^T \\ \mathbf{v} & \mathbf{R} \end{bmatrix}, \quad \mathbf{s} = \begin{bmatrix} s_1 \\ \tilde{\mathbf{x}} \end{bmatrix}, \qquad (8.43)$$

where $u \in \mathbb{R}$, $\mathbf{v} \in \mathbb{R}^{2PK-1}$, $\mathbf{R} \in \mathbb{R}^{(2PK-1) \times (2PK-1)}$, and $\tilde{\mathbf{x}} \in \mathcal{U}^{2PK-1}$. Define $n = 2PK$, and

$$\mathbf{G} = \begin{bmatrix} \mathbf{R} & s_1\mathbf{v} \\ s_1\mathbf{v}^T & s_1^2 u \end{bmatrix}, \quad \mathbf{D} = \begin{bmatrix} \mathbf{I}_{n-1} & \mathbf{0} \\ \mathbf{0}^T & s_1^2 \end{bmatrix}. \qquad (8.44)$$

With the assumption that s_1 (i.e., a pilot symbol) is known to the receiver in order to get rid of the scalar ambiguity of the Rayleigh quotient objective function in (8.41), the problem (8.41) can be reformulated as

$$\max_{\mathbf{x} \in \mathbb{R}^n} \frac{\mathbf{x}^T \mathbf{G} \mathbf{x}}{\mathbf{x}^T \mathbf{D} \mathbf{x}} \tag{8.45a}$$

$$\text{s.t.}\ \ x_k \in \mathcal{U},\ k = 1, \ldots, n-1, \tag{8.45b}$$

$$x_n \in \{\pm 1\}. \tag{8.45c}$$

If $\mathbf{x}^\star = [x_1^\star, \ldots, x_{n-1}^\star, x_n^\star]^T$ is a solution of (8.45), then

$$\tilde{\mathbf{x}}^\star = [x_1^\star x_n^\star, \ldots, x_{n-1}^\star x_n^\star]^T$$

is a solution of (8.41) via (8.43).

Let us now introduce the LFSDR approach to (8.45). By defining $\mathbf{X} = \mathbf{x}\mathbf{x}^T$, one can rewrite (8.45) in terms of $\mathbf{X}$ as follows:

$$\max_{\mathbf{X} \in \mathbb{R}^{n \times n}} \frac{\text{Tr}(\mathbf{G}\mathbf{X})}{\text{Tr}(\mathbf{D}\mathbf{X})} \tag{8.46a}$$

$$\text{s.t.}\ \ [\mathbf{X}]_{k,k} \in \{1, 9, \ldots, (2^q - 1)^2\},\ k = 1, \ldots, n-1, \tag{8.46b}$$

$$[\mathbf{X}]_{n,n} = 1, \tag{8.46c}$$

$$\mathbf{X} \succ \mathbf{0}, \tag{8.46d}$$

$$\text{rank}(\mathbf{X}) = 1, \tag{8.46e}$$

which is a nonconvex problem. Via SDR (i.e., removal of (8.46e) and convex hull relaxation to (8.46b) (cf. Remark 8.5)), we come up with the following LFSDR problem

$$\mathbf{X}^\star = \arg \max_{\mathbf{X} \in \mathbb{R}^{n \times n}} \frac{\text{Tr}(\mathbf{G}\mathbf{X})}{\text{Tr}(\mathbf{D}\mathbf{X})} \tag{8.47a}$$

$$\text{s.t.}\ \ 1 \leq [\mathbf{X}]_{k,k} \leq (2^q - 1)^2,\ k = 1, \ldots, n-1, \tag{8.47b}$$

$$[\mathbf{X}]_{n,n} = 1, \tag{8.47c}$$

$$\mathbf{X} \succeq \mathbf{0}, \tag{8.47d}$$

which is a quasiconvex problem and can be solved using the bisection method presented in Chapter 4. Of course, if one can further convert this quasiconvex problem into a convex problem, it is certainly preferred from the computational efficiency viewpoint. This is possible as presented next.

Through the change of variables (Charnes–Cooper transformation) [CC62]

$$\mathbf{Z} = \frac{\mathbf{X}}{\text{Tr}(\mathbf{D}\mathbf{X})} \tag{8.48}$$

(thus implying that $[\mathbf{Z}]_{n,n} = 1/\mathrm{Tr}(\mathbf{DX})$), the quasiconvex problem (8.47) can be further converted into a (convex) SDP as follows:

$$\mathbf{Z}^\star = \arg \max_{\mathbf{Z} \in \mathbb{R}^{n \times n}} \mathrm{Tr}(\mathbf{GZ}) \tag{8.49a}$$

$$\text{s.t. } \mathrm{Tr}(\mathbf{DZ}) = 1, \tag{8.49b}$$

$$[\mathbf{Z}]_{n,n} \leq [\mathbf{Z}]_{k,k} \leq (2^q - 1)^2 [\mathbf{Z}]_{n,n}, \quad k = 1, \ldots, n-1, \tag{8.49c}$$

$$\mathbf{Z} \succeq \mathbf{0}. \tag{8.49d}$$

Once the SDP is solved, its solution

$$\mathbf{X}^\star = \mathbf{Z}^\star \cdot \mathrm{Tr}(\mathbf{DX}^\star) = \frac{\mathbf{Z}^\star}{[\mathbf{Z}^\star]_{n,n}} \qquad \text{(by (8.48))}$$

is used to find an approximation to the solution for (8.45) (e.g., rank-one approximation and Gaussian randomization).

Some simulation results are provided to demonstrate the performance of the LFSDR-based higher-order QAM blind ML OSTBC detector. The channel coefficients in $\mathbf{H}$ were i.i.d. circular complex Gaussian random variables with zero mean and unity variance. The SNR was defined as

$$\mathrm{SNR} = \frac{\mathbb{E}\{\|\mathbf{HC}(\mathbf{s}_p)\|_{\mathrm{F}}^2\}}{\mathbb{E}\{\|\mathbf{W}_p\|_{\mathrm{F}}^2\}} = \frac{\gamma N_t K}{T \sigma_w^2},$$

where $\gamma = 10$ for 16-QAM and $\gamma = 42$ for 64-QAM. The complex 3×4 OSTBC ($N_t = 3$, $T = 4$, $K = 3$) [TJC99]

$$\mathbf{C}(\mathbf{s}) = \begin{bmatrix} s_1 + js_2 & -s_3 + js_4 & -s_5 + js_6 & 0 \\ s_3 + js_4 & s_1 - js_2 & 0 & -s_5 + js_6 \\ s_5 + js_6 & 0 & s_1 - js_2 & s_3 - js_4 \end{bmatrix} \tag{8.50}$$

was used in the simulation, and the LFSDR problem (8.49) was solved by a specialized interior-point method. An approximate solution of problem (8.41) was obtained either by quantizing the principal eigenvector of $\mathbf{X}^\star$ or by the Gaussian randomization procedure in Subsection 8.4.4 with 100 random vectors ($L = 100$) generated. The detector performance was evaluated using average symbol error rate (SER), and at least 10,000 trials were performed for each simulation result. Simulations results (SER versus SNR) are shown in Figure 8.4 for the case of 16-QAM OSTBC and for the 64-QAM OSTBC, respectively.

Figure 8.4 shows the SER performance for the proposed LFSDR blind ML detector, the norm relaxed blind ML detector (cf. Remark 8.8 below), the blind subspace channel estimator by Shahbazpanahi *et al.* [SGM05], the cyclic ML method [LSL03] (initialized by the norm relaxed blind ML detector), and the coherent ML detector (which assumes perfect channel state information (CSI)). These results illustrate that in the case of $N_r = 4$ and either for 16-QAM or 64-QAM OSTBCs, the LFSDR approach is accurate in the approximation of the

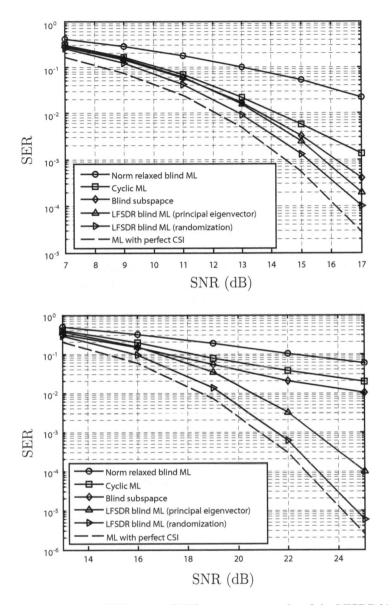

Figure 8.4 Performance (SER versus SNR) comparison results of the LFSDR blind ML detector and some existing methods for the complex 3×4 OSTBC for $N_r = 4$, $P = 8$, and (a) 16-QAM (top plot); (b) 64-QAM (bottom plot). (© 2010 IEEE. Reprinted, with permission, from T.-H. Chang, C.-W. Hsin, W.-K. Ma, and C.-Y. Chi, "A Linear Fractional Semidefinite Relaxation Approach to Maximum-Likelihood Detection of Higher-Order QAM OSTBC in Unknown Channels," Apr. 2010.)

true blind ML solution, and indicates that the Gaussian randomization procedure is a better approximation method than the principal eigenvector procedure.

Remark 8.8 Let $v^\star \in \mathbb{R}^{2PK}$ denote the principal eigenvector of $\mathbf{F}$, which is actually the optimal solution of the original problem (8.41) with the nonconvex feasible set $\mathcal{U}^{2PK}$ (cf. (8.42)) replaced by $\mathbb{R}^{2PK}$. The norm relaxed blind ML detector obtains an approximate solution of (8.41) by

$$\widehat{\mathbf{s}}_{\mathrm{NR}} = \sigma_{\mathrm{PAM}} \left(\frac{s_1}{v_1^\star} v^\star \right), \tag{8.51}$$

where $\sigma_{\mathrm{PAM}} : \mathbb{R}^{2PK} \to \mathcal{U}^{2PK}$ is a function in which the ith element x_i of $\sigma_{\mathrm{PAM}}(x)$ is obtained by rounding it to an integer in the symbol constellation set $\mathcal{U}$ given by (8.42) for all i. □

8.5   Applications in transmit beamforming

In this section, we will introduce various applications of SDP in transmit beamforming designs [GSS+10], where each design is an optimization problem of complex variables. Without converting the problem under consideration into a problem of real variables, all the complex-variable reformulations are directly applied so that the inherent structure of the original problem can be maximally maintained in the reformulated problem. The fundamental theory introduced in early chapters will be referred to advisably during the design stage. Some necessary but not complete simulation results are presented to demonstrate the efficacy of the designed beamformers. This section can be thought of as a set of some existing practical designs of how the fundamental theory introduced in early chapters are applied in various combination forms.

8.5.1   Downlink beamforming for broadcasting

The problem of downlink transmit beamforming for broadcasting considered here is the same as the one defined in Chapter 7 (Section 7.6), except that the transmitter (i.e., BS) sends common information to all the n receivers (i.e., MS). The received signal in form remains the same, namely,

$$y_i = \mathbf{h}_i^T \mathbf{x} + v_i, \ i = 1, \ldots, n, \quad (\text{cf. (7.37)}) \tag{8.52}$$

where $\mathbf{h}_i$ is the channel vector from the transmitter to the receiver i and v_i is noise with zero mean and variance σ_i^2 at the receiver i, but the transmitted signal is now given by

$$\mathbf{x} = \mathbf{f}s, \tag{8.53}$$

where $\mathbf{f} \in \mathbb{C}^m$ is the transmit beamformer vector and $s \in \mathbb{C}$ is the information bearing signal with $\mathbb{E}\{|s|^2\} = 1$.

♦ **Power Minimization Problem**

The problem is to minimize the transmit power, subject to the constraint that the SNR for each receiver is no less than a pre-specified threshold γ_0, i.e.,

$$\gamma_i = \frac{|\mathbf{h}_i^T \mathbf{f}|^2}{\sigma_i^2} \geq \gamma_0, \ i = 1, \ldots, n. \tag{8.54}$$

Note that $|\mathbf{h}_i^T \mathbf{f}|^2 = (\mathbf{h}_i^T \mathbf{f}) \mathbf{f}^H \mathbf{h}_i^* = \text{Tr}(\mathbf{f}\mathbf{f}^H \mathbf{h}_i^* \mathbf{h}_i^T)$. Now let $\mathbf{Q}_i = \mathbf{h}_i^* \mathbf{h}_i^T / (\gamma_0 \sigma_i^2)$. Then the problem can be written as

$$\begin{aligned} &\min \ \|\mathbf{f}\|_2^2 \\ &\text{s.t. } \text{Tr}(\mathbf{f}\mathbf{f}^H \mathbf{Q}_i) \geq 1, \ i = 1, \ldots, n. \end{aligned} \tag{8.55}$$

Unfortunately, the transmit beamformer design problem here is nonconvex. Even worse, the problem is shown to be NP-hard [SDL06]. However, one can approximate the problem using SDR. An equivalent form of the problem is as follows:

$$\begin{aligned} &\min_{\mathbf{f}\in\mathbb{C}^m, \mathbf{F}\in\mathbb{H}^m} \ \text{Tr}(\mathbf{F}) \\ &\text{s.t. } \mathbf{F} = \mathbf{f}\mathbf{f}^H, \ \text{Tr}(\mathbf{F}\mathbf{Q}_i) \geq 1, \ i = 1, \ldots, n. \end{aligned} \tag{8.56}$$

The SDR approximation to the above problem is:

$$\begin{aligned} &\min \ \text{Tr}(\mathbf{F}) \\ &\text{s.t. } \mathbf{F} \succeq \mathbf{0}, \ \text{Tr}(\mathbf{F}\mathbf{Q}_i) \geq 1, \ i = 1, \ldots, n. \end{aligned} \tag{8.57}$$

♦ **Max-Min-Fair problem**

We would like to maximize the weakest SNR over all the receivers, subject to the constraint that the transmit power is no greater than a threshold P_0:

$$\begin{aligned} &\max \ \min_{i=1,\ldots,n} \ \frac{|\mathbf{h}_i^T \mathbf{f}|^2}{\sigma_i^2} \\ &\text{s.t. } \|\mathbf{f}\|_2^2 \leq P_0. \end{aligned} \tag{8.58}$$

The associated SDR approximation is then

$$\begin{aligned} &\max \ t \\ &\text{s.t. } \frac{1}{\sigma_i^2} \text{Tr}(\mathbf{h}_i^* \mathbf{h}_i^T \mathbf{F}) \geq t, \ i = 1, \ldots, n \\ &\mathbf{F} \succeq \mathbf{0}, \ \text{Tr}(\mathbf{F}) \leq P_0. \end{aligned} \tag{8.59}$$

Again, let us mention that an approximate solution to the original nonconvex problem (either problem (8.55) or problem (8.58)) can then be obtained from the solution of the associated SDP problem (via SDR) by rank-one approximation or Gaussian randomization.

8.5.2   Transmit beamforming in cognitive radio

In this subsection, we revisit the example in Subsection 6.7.3, and reformulate the problem (6.124) into a convex problem by using SDR. By letting $\mathbf{W}_S = \mathbf{w}_S \mathbf{w}_S^H$ and removing the rank-one constraint, the problem (6.124) can be relaxed as

$$\begin{aligned} \min \quad & - \operatorname{Tr}(\mathbf{A}\mathbf{W}_S) \\ \text{s.t.} \quad & \operatorname{Tr}(\mathbf{B}_k \mathbf{W}_S) \leq \epsilon_k, \ k = 1, \dots, K \\ & \operatorname{Tr}(\mathbf{W}_S) \leq P_S, \ \mathbf{W}_S \succeq \mathbf{0} \end{aligned} \tag{8.60}$$

which is an SDP. The optimal solution $\mathbf{W}_S^\star$ of the SDP given by (8.60), which may not be of rank one, is then used to find a rank-one approximate solution for the original nonconvex QCQP given by (6.124). Surely, if the $\mathbf{W}_S^\star$ is of rank one, the corresponding rank-one solution is also the optimal solution of (6.124). Otherwise, finding rank-one approximation to the optimal (6.124) is needed.

8.5.3   Transmit beamforming in secrecy communication: Artificial noise (AN) aided approach

Under open wireless media, information security using cryptographic encryption (in the network layer) may be subject to vulnerabilities, such as problems with secret key distribution and management. Motivated by this fact, information secrecy using physical-layer transmit designs, commonly known as physical layer secrecy in the present literature, has received growing attention recently. Physical-layer secrecy may serve as an alternative to, or a complement to, cryptographic encryption.

Consider a commercial wireless downlink scenario where a transmitter (Alice) with N_t antennas sends a data stream to a legitimate single-antenna receiver (Bob) in the presence of M non-colluding single-antenna eavesdroppers (Eves). A simple diagram is depicted in Figure 8.5 to illustrate the scenario. The base station has eavesdroppers' full or partial channel state information (CSI), which would be true for active eavesdroppers or participating system users, where some participating users of the system attempt to access service requiring additional charges (e.g., high-definition video) by overhearing. With transmit beamforming, one can utilize the spatial degree of freedom (DoF) to cripple eavesdroppers' interceptions.

The transmit signal depicted in Figure 8.5 is given by

$$\mathbf{x}(t) = \boldsymbol{w}s(t) + \mathbf{z}(t) \in \mathbb{C}^{N_t}. \tag{8.61}$$

Here, $s(t) \in \mathbb{C}$ is the data stream intended for Bob only, where we assume $\mathbb{E}\{|s(t)|^2\} = 1$; $\boldsymbol{w} \in \mathbb{C}^{N_t}$ is the transmit beamforming vector corresponding to $s(t)$; $\mathbf{z}(t) \in \mathbb{C}^{N_t}$ is a noise vector artificially generated by Alice to interfere with Eves; i.e., the so-called AN. Assume that

$$\mathbf{z}(t) \sim \mathcal{CN}(\mathbf{0}, \boldsymbol{\Sigma}), \tag{8.62}$$

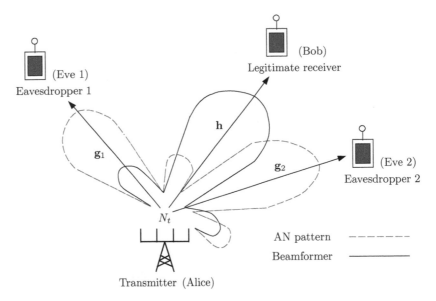

Figure 8.5 Illustration of AN-aided transmit beamforming.

where $\boldsymbol{\Sigma} \succeq \mathbf{0}$ denotes the AN spatial covariance. The received signals at Bob and Eves are, respectively, modeled as:

$$y_b(t) = \mathbf{h}^H \mathbf{x}(t) + n(t), \tag{8.63}$$

$$y_{e,m}(t) = \mathbf{g}_m^H \mathbf{x}(t) + v_m(t), \quad m = 1, \ldots, M, \tag{8.64}$$

where $\mathbf{h} \in \mathbb{C}^{N_t}$ is the channel from Alice to Bob, $\mathbf{g}_m \in \mathbb{C}^{N_t}$ is the channel from Alice to the mth Eve, and $n(t)$ and $v_m(t)$ are i.i.d. complex circular Gaussian noises with variances $\sigma_n^2 > 0$ and $\sigma_{v,m}^2 > 0$, respectively.

Suppose that the Alice-to-Bob channel $\mathbf{h}$ is random with mean $\bar{\mathbf{h}}$ and covariance $\mathbf{C}_h$. From Bob's model (8.63) and the AN-aided transmit structure (8.61), the SINR of Bob w.r.t. $(\boldsymbol{w}, \boldsymbol{\Sigma})$ can be defined as

$$\mathrm{SINR}_b(\boldsymbol{w}, \boldsymbol{\Sigma}) = \frac{\mathbb{E}\{|\mathbf{h}^H \boldsymbol{w} s(t)|^2\}}{\mathbb{E}\{|\mathbf{h}^H \mathbf{z}(t)|^2\} + \sigma_n^2} = \frac{\boldsymbol{w}^H \mathbf{R}_h \boldsymbol{w}}{\mathrm{Tr}(\boldsymbol{\Sigma} \mathbf{R}_h) + \sigma_n^2}, \tag{8.65}$$

where

$$\mathbf{R}_h = \mathbb{E}\{\mathbf{h}\mathbf{h}^H\} = \bar{\mathbf{h}}\bar{\mathbf{h}}^H + \mathbf{C}_h \tag{8.66}$$

is the correlation matrix of $\mathbf{h}$. Assume that $\mathbf{R}_h$ is known to Alice, while for the case where the instantaneous channel realization of Bob is known to Alice, it is defined as $\mathbf{R}_h = \mathbf{h}\mathbf{h}^H$. Likewise, the SINRs of Eves under the model (8.64) are given by

$$\mathrm{SINR}_{e,m}(\boldsymbol{w}, \boldsymbol{\Sigma}) = \frac{\mathbb{E}\{|\mathbf{g}_m^H \boldsymbol{w} s(t)|^2\}}{\mathbb{E}\{|\mathbf{g}_m^H \mathbf{z}(t)|^2\} + \sigma_{v,m}^2} = \frac{\boldsymbol{w}^H \mathbf{R}_{g,m} \boldsymbol{w}}{\mathrm{Tr}(\boldsymbol{\Sigma} \mathbf{R}_{g,m}) + \sigma_{v,m}^2}, \tag{8.67}$$

where, for the case where the instantaneous channel realizations of Eves are known to Alice, we define $\mathbf{R}_{g,m} = \mathbf{g}_m \mathbf{g}_m^H$; and, for the case where only Eves' channel correlation matrices are available to Alice, we define

$$\mathbf{R}_{g,m} = \mathbb{E}\{\mathbf{g}_m \mathbf{g}_m^H\} = \bar{\mathbf{g}}_m \bar{\mathbf{g}}_m^H + \mathbf{C}_{g,m} \tag{8.68}$$

with $\bar{\mathbf{g}}_m$ and $\mathbf{C}_{g,m}$ denoting the mean and covariance of $\mathbf{g}_m$, respectively. Again, $\mathbf{R}_{g,1}, \ldots, \mathbf{R}_{g,M}$ are assumed to be available to Alice.

The goal of the transmitter is to jointly optimize the transmit beamforming vector $\boldsymbol{w}$ and AN spatial covariance $\boldsymbol{\Sigma}$ such that Bob's and Eves' SINRs are enhanced and degraded, respectively. To this end, consider a power minimization formulation as follows:

$$\min_{\boldsymbol{w} \in \mathbb{C}^{N_t}, \boldsymbol{\Sigma} \in \mathbb{H}^{N_t}} \|\boldsymbol{w}\|_2^2 + \mathrm{Tr}(\boldsymbol{\Sigma})$$

$$\text{s.t. } \mathrm{SINR}_b(\boldsymbol{w}, \boldsymbol{\Sigma}) \geq \gamma_b, \tag{8.69}$$

$$\mathrm{SINR}_{e,m}(\boldsymbol{w}, \boldsymbol{\Sigma}) \leq \gamma_e, \ m = 1, \ldots, M,$$

$$\boldsymbol{\Sigma} \succeq \mathbf{0},$$

where $\gamma_b > 0$ and $\gamma_e > 0$ are a preset minimum SINR requirement on Bob and a preset maximum allowable SINR threshold on Eves, respectively. Let us explicitly express the power minimization design (8.69) as

$$\min_{\boldsymbol{w}, \boldsymbol{\Sigma}} \|\boldsymbol{w}\|_2^2 + \mathrm{Tr}(\boldsymbol{\Sigma}) \tag{8.70a}$$

$$\text{s.t. } \frac{1}{\gamma_b} \boldsymbol{w}^H \mathbf{R}_h \boldsymbol{w} \geq \mathrm{Tr}(\boldsymbol{\Sigma} \mathbf{R}_h) + \sigma_n^2, \tag{8.70b}$$

$$\frac{1}{\gamma_e} \boldsymbol{w}^H \mathbf{R}_{g,m} \boldsymbol{w} \leq \mathrm{Tr}(\boldsymbol{\Sigma} \mathbf{R}_{g,m}) + \sigma_{v,m}^2, \ m = 1, \ldots, M, \tag{8.70c}$$

$$\boldsymbol{\Sigma} \succeq \mathbf{0}. \tag{8.70d}$$

The AN-aided design (8.70) is a nonconvex quadratic optimization problem, only because of Bob's nonconvex SINR constraint in (8.70b), and thus is intractable. Furthermore, it has been shown to be an NP-hard problem. However, the optimal power of problem (8.70) will be smaller than that without using AN (i.e., $\boldsymbol{\Sigma} = \mathbf{0}$) and than that using isotropic AN.

Applying SDR to problem (8.70) gives rise to

$$\min_{\mathbf{W}, \boldsymbol{\Sigma}} \mathrm{Tr}(\mathbf{W}) + \mathrm{Tr}(\boldsymbol{\Sigma}) \tag{8.71a}$$

$$\text{s.t. } \frac{1}{\gamma_b} \mathrm{Tr}(\mathbf{W} \mathbf{R}_h) - \mathrm{Tr}(\mathbf{R}_h \boldsymbol{\Sigma}) \geq \sigma_n^2, \tag{8.71b}$$

$$\frac{1}{\gamma_e} \mathrm{Tr}(\mathbf{W} \mathbf{R}_{g,m}) - \mathrm{Tr}(\mathbf{R}_{g,m} \boldsymbol{\Sigma}) \leq \sigma_{v,m}^2, \ m = 1, \ldots, M, \tag{8.71c}$$

$$\boldsymbol{\Sigma} \succeq \mathbf{0}, \ \mathbf{W} \succeq \mathbf{0}, \tag{8.71d}$$

which is an SDP and hence can be efficiently solved.

The SDR problem (8.71) is generally an approximation to the AN-aided design (8.70), because the former does not guarantee a rank-one optimal $\mathbf{W}$ given an arbitrary problem instance $\mathbf{R}_h, \mathbf{R}_{g,1}, \ldots, \mathbf{R}_{g,M} \succeq \mathbf{0}, \gamma_b, \gamma_e > 0$. Some conditions

[LCMC11] under which the SDR problem (8.71) has been proven, by the associated KKT conditions, to yield rank-one solutions are given below:

($C1$) (Instantaneous CSI on Bob) $\mathbf{R}_h = \mathbf{h}\mathbf{h}^H$, while $\mathbf{R}_{g,1}, \ldots, \mathbf{R}_{g,M} \succeq \mathbf{0}$ are arbitrary PSD matrices;

($C2$) (Correlation-based CSI on Bob and Eves, with white channel covariance) $N_t > M$, and the channel correlation matrices take the form

$$\mathbf{R}_h = \bar{\mathbf{h}}\bar{\mathbf{h}}^H + \sigma_h^2 \mathbf{I}_{N_t},$$
$$\mathbf{R}_{g,m} = \bar{\mathbf{g}}_m \bar{\mathbf{g}}_m^H + \sigma_{g,m}^2 \mathbf{I}_{N_t}, \quad m = 1, \ldots, M,$$

where $\sigma_h^2, \sigma_{g,1}^2, \ldots, \sigma_{g,M}^2 \geq 0$, and $\bar{\mathbf{h}}, \bar{\mathbf{g}}_1, \ldots, \bar{\mathbf{g}}_M \in \mathbb{C}^{N_t}$ are such that $\bar{\mathbf{h}} \notin$ span$[\bar{\mathbf{g}}_1, \ldots, \bar{\mathbf{g}}_M]$;

($C3$) (The number of Eves is no greater than two) $M \leq 2$, and $\mathbf{R}_h, \mathbf{R}_{g,1}, \ldots, \mathbf{R}_{g,M}$ are arbitrary.

If any one of the above three conditions is satisfied, there exists an optimal SDR solution, denoted as $(\mathbf{W}^\star, \boldsymbol{\Sigma}^\star)$, such that $\mathbf{W}^\star$ is of rank one, i.e., $\mathbf{W}^\star = \boldsymbol{w}^\star(\boldsymbol{w}^\star)^H$, and hence is also an optimal solution to the original problem (8.69). Note that the above three conditions are sufficient but not necessary conditions, which can be proven through the KKT conditions of the SDP (8.71). There may exist other conditions under which the SDP (8.71) has rank-one solutions.

Let us show some simulation results to demonstrate the performance of the AN-aided design under the power minimization formulation in (8.70). In the simulation, $\mathbf{R}_h = \mathbf{h}\mathbf{h}^H$ and $\mathbf{R}_{g,m} = \mathbf{g}_m\mathbf{g}_m^H$ for all m (i.e., the instantaneous CSI case as stated in ($C1$)). The channel realizations $\mathbf{h}, \mathbf{g}_1, \ldots, \mathbf{g}_M$ generated are i.i.d. complex Gaussian distributed, with zero mean and covariance matrix $\mathbf{I}_{N_t}/N_t$. Let Bob's noise power be $\sigma_n^2 = 0$ dB and all Eves have identical noise power $\sigma_{v,1}^2 = \cdots = \sigma_{v,M}^2 \triangleq \sigma_v^2$.

Besides the design obtained by solving the SDR problem, i.e., problem (8.71), the no-AN design and the isotropic AN design are also evaluated in the simulation for performance comparison, which are also under the power minimization formulation (8.70) but with some restrictions on the transmit structure. Specifically, the no-AN design sets $\boldsymbol{\Sigma} = \mathbf{0}$ in (8.70), which can be reformulated as a convex second-order cone program. The isotropic AN design fixes the structure of $(\boldsymbol{w}, \boldsymbol{\Sigma})$ to be [LCMC11]

$$\boldsymbol{w} = \sqrt{\rho} \cdot \mathbf{h} \quad \text{and} \quad \boldsymbol{\Sigma} = \beta \mathbf{P}_{\mathbf{h}}^{\perp}, \tag{8.72}$$

where $\mathbf{P}_{\mathbf{h}}^{\perp} = \mathbf{I} - \mathbf{h}\mathbf{h}^H/\|\mathbf{h}\|_2^2$ (which is a PSD matrix with rank equal to $N_t - 1$ and all nonzero eigenvalues equal to unity) is the orthogonal complement projector of $\mathbf{h}$, and

$$\rho = \frac{\sigma_n^2 \gamma_b}{\|\mathbf{h}\|_2^4}, \quad \beta = \max\left\{0, \max_{m=1,\ldots,M} \frac{(\rho/\gamma_e)\mathbf{h}^H \mathbf{R}_{g,m}\mathbf{h} - \sigma_{v,m}^2}{\text{Tr}(\mathbf{P}_{\mathbf{h}}^{\perp} \mathbf{R}_{g,m})}\right\}.$$

The above power allocation for (ρ, β) can be shown to yield the smallest total transmit power under the isotropic AN structure (8.72). The other simulation settings are $\gamma_e = 0$ dB, $\gamma_b = 10$ dB. Simulation results are shown in Figures 8.6 and 8.7, which are obtained over 1000 independent trials and all the involved optimization problems are solved by SeDuMi (http://sedumi.ie.lehigh.edu/) [Stu99].

In Figure 8.6, the average transmit powers of the various designs for $N_t = 4$ and $M = 3$ are plotted over a wide range of values of $1/\sigma_v^2$. Note that large $1/\sigma_v^2$ physically means strong (clean) overhearing ability for Eves, while small $1/\sigma_v^2$ means weak (noisy) overhearing ability. One can observe, from the top plot of Figure 8.6, that for $1/\sigma_v^2 < -10$ dB, the obtained powers by all the designs are quite similar, simply because of no need of spending more resources to deal with weak Eves. However, for $1/\sigma_v^2 > 0$ dB, the AN-aided design in problem (8.71) yields the smallest powers among the three designs, and its performance gaps relative to the other two designs are wider as $1/\sigma_v^2$ increases. For example, at $1/\sigma_v^2 = 20$ dB, the performance gap between the AN-aided and no-AN designs is 12 dB in power, while that between the AN-aided and isotropic AN designs is 4 dB. This figure also reveals that for the strong Eves regime (say, $1/\sigma_v^2 > 0$ dB), using AN, even in an isotropic manner, would give better performance than not using AN.

To get more performance insights, in the bottom plot of Figure 8.6, we separately show the powers of the transmit beamforming and AN. It can be seen that the power allocated to AN increases with $1/\sigma_v^2$, confirming that using AN is the reason behind the good power saving performance of the AN-aided design in problem (8.71). Moreover, the power allocated to AN in this design is much lower than that in the isotropic AN design.

It is interesting to demonstrate how the transmit powers of the various designs change with the number of Eves, with the same simulation settings associated with Figure 8.6 except $N_t = 20$ and $1/\sigma_v^2 = 15$ dB. Again, from the top plot of Figure 8.7, the AN-aided design in problem (8.71) can be seen to exhibit the best performance. Moreover, the no-AN design performs better than the isotropic AN design, except at $M = 18$. We should recall that the no-AN design focuses on manipulating the transmit DoF to deal with Eves, while the isotropic AN design does not. The corresponding allocations of transmit beamforming power and AN power are shown in the bottom plot of Figure 8.7. One can see from this plot that the AN-aided design tends to use less AN power for small M. Our interpretation with the results is that for small numbers of Eves, using transmit DoF to degrade Eves would be more effective than using AN. The fact that, as previously mentioned, solutions of the AN-aided design are of rank-one for the instantaneous CSI case (cf. $(C1)$) has also been validated by the above simulation results shown in Figures 8.6 and 8.7.

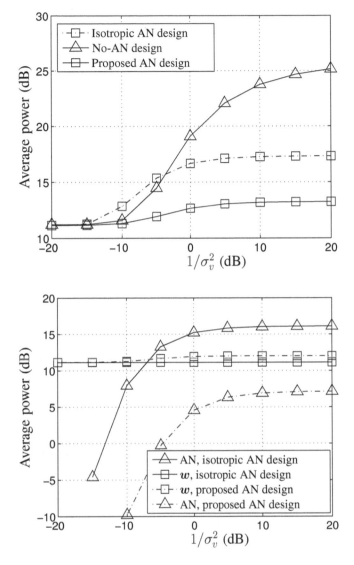

Figure 8.6 Performance of the various transmit designs under the power minimization formulation for $N_t = 4$ and $M = 3$. (a) (top plot) Transmit powers versus the reciprocal of Eves' noise power; (b) (bottom plot) the power allocations of the isotropic design and the AN-aided design in problem (8.71) of the results in (a). (© 2011 IEEE. Reprinted, with permission, from W.-C. Liao, T.-H. Chang, W.-K. Ma, and C.-Y. Chi, "Qos-Based Transmit Beamforming in the Presence of Eavesdroppers: An Optimized Artificial-Noise-Aided Approach," Mar. 2011.)

8.5.4   Worst-case robust transmit beamforming: Single-cell MISO scenario

In this subsection, we consider the robust transmit beamforming design of a multiuser MISO communication system. The scenario considered here consists

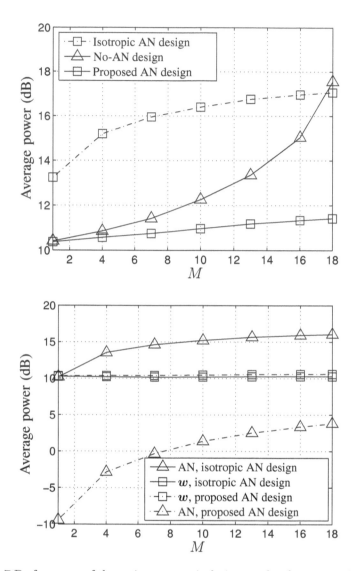

Figure 8.7 Performance of the various transmit designs under the power minimization formulation for $N_t = 20$ and $1/\sigma_v^2 = 15$ dB. (a) (top plot) transmit powers versus the number of Eves; and (b) (bottom plot) the power allocations of the isotropic design and the AN-aided design in problem (8.71) of the results in (a). (© 2011 IEEE. Reprinted, with permission, from W.-C. Liao, T.-H. Chang, W.-K. Ma, and C.-Y. Chi, "Qos-Based Transmit Beamforming in the Presence of Eavesdroppers: An Optimized Artificial-Noise-Aided Approach," Mar. 2011.)

of a single transmitter equipped with N_t antennas, which wants to transmit K independent information signals $s_i(t) \in \mathbb{C}$, $i = 1, \ldots, K$, to the K single-antenna receivers. The received signal at the ith receiver is

$$y_i(t) = \mathbf{h}_i^H \sum_{k=1}^{K} \mathbf{w}_k s_k(t) + n_i(t),$$

where $\mathbf{h}_i \in \mathbb{C}^{N_t}$ denotes the channel between the transmitter and the ith receiver; $\mathbf{w}_i \in \mathbb{C}^{N_t}$ denotes the beamforming vector for $s_i(t)$; and $n_i(t) \in \mathbb{C}$ is the additive noise at receiver i with zero mean and variance σ_i^2. For simplicity, we assume that each of the information signals $s_i(t)$ is of unit power. The SINR of the ith receiver can be expressed as

$$\gamma_i(\{\mathbf{w}_i\}_{i=1}^{K}) = \frac{|\mathbf{h}_i^H \mathbf{w}_i|^2}{\sum_{k=1,k\neq i}^{K} |\mathbf{h}_i^H \mathbf{w}_k|^2 + \sigma_i^2} \quad \text{(cf. (7.39))}.$$

In Subsection 7.6.1, under the assumption that the CSI is perfectly known to the transmitter, we have solved the following transmit beamforming problem

$$\min \ \sum_{i=1}^{K} \|\mathbf{w}_i\|_2^2 \quad \text{(cf. (7.40))}$$

$$\text{s.t. } \gamma_i(\{\mathbf{w}_i\}_{i=1}^{K}) \geq \gamma_0, \ i = 1, \ldots, K$$

by reformulating it into a SOCP. Furthermore, let us present a robust design against the CSI uncertainty next, which is suitable for practical applications.

In wireless systems, CSI at the transmitter is either estimated via uplink training or acquired through receivers' feedback. Hence it is inevitable to have CSI estimation errors in practice due to finite-length training data or have CSI quantization errors due to limited uplink feedback. Let $\bar{\mathbf{h}}_i \in \mathbb{C}^{N_t}$, $i = 1, \ldots, K$, denote the available channel estimates at the transmitter. The true channel vectors $\{\mathbf{h}_i\}_{i=1}^{K}$ can be modeled as

$$\mathbf{h}_i = \bar{\mathbf{h}}_i + \mathbf{e}_i, \ i = 1, \ldots, K,$$

where $\mathbf{e}_i \in \mathbb{C}^{N_t}$ represents the channel error. Under the assumption that the channel error $\mathbf{e}_i$ is ball-bounded with radius r_i for all i, consider the following worst-case robust transmit beamforming design problem [ZWN08]:

$$\min \ \sum_{k=1}^{K} \|\mathbf{w}_k\|_2^2 \tag{8.73a}$$

$$\text{s.t. } \frac{|(\bar{\mathbf{h}}_i + \mathbf{e}_i)^H \mathbf{w}_i|^2}{\sum_{k=1,k\neq i}^{K} |(\bar{\mathbf{h}}_i + \mathbf{e}_i)^H \mathbf{w}_k|^2 + \sigma_i^2} \geq \gamma_0, \ \forall \|\mathbf{e}_i\|_2^2 \leq r_i^2, \ i = 1, \ldots, K, \tag{8.73b}$$

where γ_0 is a preset target SINR value. The problem (8.73) is not convex due to the nonconvex constraint set. To handle this problem, we can apply SDR, i.e., let $\mathbf{W}_i = \mathbf{w}_i \mathbf{w}_i^H$ and remove the rank-one constraint, and rewrite the constraints in

(8.73b) as

$$
\mathbf{e}_i^H \left(-[1 + \frac{1}{\gamma_0}] \mathbf{W}_i + \sum_{k=1}^{K} \mathbf{W}_k \right) \mathbf{e}_i + 2\mathrm{Re}\left\{ \bar{\mathbf{h}}_i^H \left(-[1 + \frac{1}{\gamma_0}] \mathbf{W}_i + \sum_{k=1}^{K} \mathbf{W}_k \right) \mathbf{e}_i \right\}
$$

$$
+ \bar{\mathbf{h}}_i^H \left(-[1 + \frac{1}{\gamma_0}] \mathbf{W}_i + \sum_{k=1}^{K} \mathbf{W}_k \right) \bar{\mathbf{h}}_i + \sigma_i^2 \leq 0, \ \forall \|\mathbf{e}_i\|_2^2 \leq r_i^2, \ i = 1, \ldots, K.
$$

$$
(8.74)
$$

Since there are infinitely many constraints in (8.74), problem (8.73) is difficult to solve. Fortunately, the constraint in (8.74) actually involves two second-order inequalities in $\mathbf{e}_i$ and one implies the other, so by applying S-procedure (introduced in Section 8.3), the constraints in (8.74) can be equivalently reformulated as a PSD constraint. Specifically, the quadratic inequality

$$
\|\mathbf{e}_i\|_2^2 \leq r_i^2 \iff \mathbf{e}_i^H \mathbf{I}_{N_t} \mathbf{e}_i - r_i^2 \leq 0
$$

corresponds to the first inequality in (8.20) with

$$
\mathbf{F}_1 = \mathbf{I}_{N_t}, \quad \mathbf{g}_1 = \mathbf{0}, \quad h_1 = -r_i^2.
$$

The other quadratic inequality in (8.74) corresponds to the second inequality in (8.20) with

$$
\mathbf{F}_2 = -\left(1 + \frac{1}{\gamma_0}\right) \mathbf{W}_i + \sum_{k=1}^{K} \mathbf{W}_k,
$$

$$
\mathbf{g}_2 = \mathbf{F}_2 \bar{\mathbf{h}}_i,
$$

$$
h_2 = \bar{\mathbf{h}}_i^H \mathbf{F}_2 \bar{\mathbf{h}}_i + \sigma_i^2.
$$

Therefore, substituting the above corresponding parameters into (8.21) yields the following LMI (cf. (4.104)):

$$
\begin{bmatrix} \mathbf{I}_{N_t} \\ \bar{\mathbf{h}}_i^H \end{bmatrix} \left\{ \left(1 + \frac{1}{\gamma_0}\right) \mathbf{W}_i - \sum_{k=1}^{K} \mathbf{W}_k \right\} \begin{bmatrix} \mathbf{I}_{N_t} \\ \bar{\mathbf{h}}_i^H \end{bmatrix}^H + \begin{bmatrix} \lambda_i \mathbf{I}_{N_t} & \mathbf{0}_{N_t} \\ \mathbf{0}_{N_t}^H & -\sigma_i^2 - \lambda_i r_i^2 \end{bmatrix} \succeq \mathbf{0},
$$

$$
\lambda_i \geq 0, \ i = 1, \ldots, K.
$$

Therefore, the SDR approximation of problem (8.73) can be represented by

$$
\min \ \sum_{k=1}^{K} \mathrm{Tr}(\mathbf{W}_k)
$$

$$
\text{s.t.} \ \begin{bmatrix} \mathbf{I}_{N_t} \\ \bar{\mathbf{h}}_i^H \end{bmatrix} \left\{ \left(1 + \frac{1}{\gamma_0}\right) \mathbf{W}_i - \sum_{k=1}^{K} \mathbf{W}_k \right\} \begin{bmatrix} \mathbf{I}_{N_t} \\ \bar{\mathbf{h}}_i^H \end{bmatrix}^H + \begin{bmatrix} \lambda_i \mathbf{I}_{N_t} & \mathbf{0}_{N_t} \\ \mathbf{0}_{N_t}^H & -\sigma_i^2 - \lambda_i r_i^2 \end{bmatrix} \succeq \mathbf{0}
$$

$$
\lambda_i \geq 0, \ \mathbf{W}_i \succeq \mathbf{0}, \ i = 1, \ldots, K
$$

$$
(8.75)
$$

which is an SDP and can be efficiently solved. However, the optimal $\{\mathbf{W}_i^\star\}_{i=1}^K$ may not be of rank one, so rank-one approximation should be considered to obtain a rank-one suboptimal solution set of $\{\mathbf{w}_i^\star\}_{i=1}^K$.

The Gaussian randomization procedure in this case will be as follows (see, e.g., [WSC$^+$14]). Firstly, a set of beamformer directions $\mathbf{u}_1, \ldots, \mathbf{u}_K$, are randomly generated according to $\mathbf{u}_i \sim \mathcal{CN}(\mathbf{0}, \mathbf{W}_i^\star)$ and normalized by $\|\mathbf{u}_i\|_2^2 = 1$. Secondly, by substituting $\mathbf{W}_i = p_i \mathbf{u}_i \mathbf{u}_i^H$ in problem (8.75), where p_i denotes the transmission power for receiver i, then the resulting SDP is solved for $p_1, \ldots, p_K \geq 0$. If the problem is feasible, then the obtained powers $p_i^\star$ and the beamformer directions $\mathbf{u}_i$ are combined to form a feasible approximate solution $\{\mathbf{w}_i = \sqrt{p_i^\star}\mathbf{u}_i\}_{i=1}^K$ of problem (8.73). Finally, the above steps are repeated for multiple sets of beamformer directions and the solution, denoted as $\{\mathbf{w}_i^\star\}_{i=1}^K$, that yields the lowest total power $\sum_{i=1}^K p_i^\star$ is the desired rank-one solution.

8.5.5 Worst-case robust transmit beamforming: Multicell MISO scenario

The preceding SOCP muti-cell beamforming formulation given by (7.60) assumes that all the coordinating BSs know the perfect CSI $\mathbf{h}_{jik}$ for all i, j, and k. In the presence of CSI errors, the optimal solution to (7.60) can no longer guarantee the desired SINR requirements. To resolve this problem, again we can consider the worst-case robust design formulation [SCW$^+$12].

Supposing that $\bar{\mathbf{h}}_{jik}$ is the given channel estimate and the CSI error vector $\mathbf{e}_{jik}$ is elliptically bounded, a worst-case SINR constraint on MS$_{ik}$ is defined as

$$\mathrm{SINR}_{ik}\left(\{\mathbf{w}_{j\ell}\}, \{\bar{\mathbf{h}}_{jik} + \mathbf{e}_{jik}\}_{j=1}^{N_c}\right)$$

$$\triangleq \frac{\left|(\bar{\mathbf{h}}_{iik} + \mathbf{e}_{iik})^H \mathbf{w}_{ik}\right|^2}{\sum_{\ell \neq k}^{K}\left|(\bar{\mathbf{h}}_{iik} + \mathbf{e}_{iik})^H \mathbf{w}_{i\ell}\right|^2 + \sum_{j \neq i}^{N_c}\sum_{\ell=1}^{K}\left|(\bar{\mathbf{h}}_{jik} + \mathbf{e}_{jik})^H \mathbf{w}_{j\ell}\right|^2 + \sigma_{ik}^2} \geq \gamma_{ik},$$

$$\forall \mathbf{e}_{jik}^H \mathbf{C}_{jik} \mathbf{e}_{jik} \leq 1, \quad j = 1, \ldots, N_c, \quad (8.76)$$

where $\mathbf{C}_{jik} \succ \mathbf{0}$ determines the size and the shape of the error ellipsoid. Note from (8.76) that the SINR specification γ_{ik} must be satisfied for all possible CSI errors. Under the worst-case SINR constraints in (8.76), we come up with the following robust MCBF design problem:

$$\min \sum_{i=1}^{N_c}\sum_{k=1}^{K}\|\mathbf{w}_{ik}\|_2^2 \qquad (8.77a)$$

$$\text{s.t. } \mathrm{SINR}_{ik}\left(\{\mathbf{w}_{j\ell}\}, \{\bar{\mathbf{h}}_{jik} + \mathbf{e}_{jik}\}_{j=1}^{N_c}\right) \geq \gamma_{ik}, \quad \forall \mathbf{e}_{jik}^H \mathbf{C}_{jik} \mathbf{e}_{jik} \leq 1,$$

$$i, j = 1, \ldots, N_c, \ k = 1, \ldots, K, \qquad (8.77b)$$

which is nothing but a worst-case robust counterpart of problem (7.60). Solving the optimization problem (8.77) is challenging due to the infinitely many nonconvex SINR constraints in (8.77b). To handle this problem, let us present a suboptimal method via SDR and S-procedure.

Expressing the objective function of problem (8.77) as $\sum_{i=1}^{N_c}\sum_{k=1}^{K}\text{Tr}(\mathbf{w}_{ik}\mathbf{w}_{ik}^H)$, and the worst-case SINR constraint on MS_{ik} in (8.76) as

$$
(\bar{\mathbf{h}}_{iik}^H + \mathbf{e}_{iik}^H)\left(\frac{1}{\gamma_{ik}}\mathbf{w}_{ik}\mathbf{w}_{ik}^H - \sum_{\ell\neq k}^{K}\mathbf{w}_{i\ell}\mathbf{w}_{i\ell}^H\right)(\bar{\mathbf{h}}_{iik} + \mathbf{e}_{iik})
$$
$$
\geq \sum_{j\neq i}^{N_c}(\bar{\mathbf{h}}_{jik}^H + \mathbf{e}_{jik}^H)\left(\sum_{\ell=1}^{K}\mathbf{w}_{j\ell}\mathbf{w}_{j\ell}^H\right)(\bar{\mathbf{h}}_{jik} + \mathbf{e}_{jik}) + \sigma_{ik}^2,
$$
$$
\forall \mathbf{e}_{jik}^H\mathbf{C}_{jik}\mathbf{e}_{jik} \leq 1,\ j = 1,\ldots,N_c, \tag{8.78}
$$

and then applying SDR to (8.77) gives rise to the following problem:

$$
\min \sum_{i=1}^{N_c}\sum_{k=1}^{K}\text{Tr}(\mathbf{W}_{ik}) \tag{8.79a}
$$

$$
\text{s.t. } (\bar{\mathbf{h}}_{iik}^H + \mathbf{e}_{iik}^H)\left(\frac{1}{\gamma_{ik}}\mathbf{W}_{ik} - \sum_{\ell\neq k}^{K}\mathbf{W}_{i\ell}\right)(\bar{\mathbf{h}}_{iik} + \mathbf{e}_{iik})
$$
$$
\geq \sum_{j\neq i}^{N_c}(\bar{\mathbf{h}}_{jik}^H + \mathbf{e}_{jik}^H)\left(\sum_{\ell=1}^{K}\mathbf{W}_{j\ell}\right)(\bar{\mathbf{h}}_{jik} + \mathbf{e}_{jik}) + \sigma_{ik}^2, \tag{8.79b}
$$
$$
\forall \mathbf{e}_{jik}^H\mathbf{C}_{jik}\mathbf{e}_{jik} \leq 1,\ i,j = 1,\ldots,N_c,\ k = 1,\ldots,K,
$$

which again involves infinitely many convex constraints. Though the right-hand side of the inequality constraint (8.79b) involves coupling CSI errors, it can be nicely decoupled into the following N_c constraints:

$$
(\bar{\mathbf{h}}_{iik}^H + \mathbf{e}_{iik}^H)\left(\frac{1}{\gamma_{ik}}\mathbf{W}_{ik} - \sum_{\ell\neq k}^{K}\mathbf{W}_{i\ell}\right)(\bar{\mathbf{h}}_{iik} + \mathbf{e}_{iik})
$$
$$
\geq \sum_{j\neq i}^{N_c}t_{jik} + \sigma_{ik}^2,\quad \forall \mathbf{e}_{iik}^H\mathbf{C}_{iik}\mathbf{e}_{iik} \leq 1, \tag{8.80}
$$

$$
(\bar{\mathbf{h}}_{jik}^H + \mathbf{e}_{jik}^H)\left(\sum_{\ell=1}^{K}\mathbf{W}_{j\ell}\right)(\bar{\mathbf{h}}_{jik} + \mathbf{e}_{jik}) \leq t_{jik},\quad \forall \mathbf{e}_{jik}^H\mathbf{C}_{jik}\mathbf{e}_{jik} \leq 1,\ j\neq i,
$$
$$
\tag{8.81}
$$

where $\{t_{jik}\}_{j\neq i}$ are auxiliary variables. Note that (8.80) involves only the CSI error $\mathbf{e}_{iik}$ and each of the constraints in (8.81) involves only one CSI error $\mathbf{e}_{jik}$. Furthermore, by applying the S-procedure, (8.80) and (8.81) can be, respectively, replaced by the following finite LMIs (cf. (4.104)):

$$\boldsymbol{\Phi}_{ik}\left(\{\mathbf{W}_{i\ell}\}_{\ell=1}^{K}, \{t_{jik}\}_{j\neq i}, \lambda_{iik}\right) \triangleq$$

$$\begin{bmatrix} \mathbf{I}_{N_t} \\ \bar{\mathbf{h}}_{iik}^H \end{bmatrix} \left(\frac{1}{\gamma_{ik}}\mathbf{W}_{ik} - \sum_{\ell\neq k}^{K} \mathbf{W}_{i\ell}\right) \begin{bmatrix} \mathbf{I}_{N_t} \\ \bar{\mathbf{h}}_{iik}^H \end{bmatrix}^H + \begin{bmatrix} \lambda_{iik}\mathbf{C}_{iik} & \mathbf{0}_{N_t} \\ \mathbf{0}_{N_t}^H & -\sum_{j\neq i}^{N_c} t_{jik} - \sigma_{ik}^2 - \lambda_{iik} \end{bmatrix} \succeq \mathbf{0},$$

$$(8.82)$$

and

$$\boldsymbol{\Psi}_{jik}\left(\{\mathbf{W}_{j\ell}\}_{\ell=1}^{K}, t_{jik}, \lambda_{jik}\right) \triangleq$$

$$\begin{bmatrix} \mathbf{I}_{N_t} \\ \bar{\mathbf{h}}_{jik}^H \end{bmatrix} \left(-\sum_{\ell=1}^{K} \mathbf{W}_{j\ell}\right) \begin{bmatrix} \mathbf{I}_{N_t} \\ \bar{\mathbf{h}}_{jik}^H \end{bmatrix}^H + \begin{bmatrix} \lambda_{jik}\mathbf{C}_{jik} & \mathbf{0}_{N_t} \\ \mathbf{0}_{N_t}^H & t_{jik} - \lambda_{jik} \end{bmatrix} \succeq \mathbf{0}, \quad j\neq i, \quad (8.83)$$

where $\lambda_{jik} \geq 0$ for all $i, j = 1, \ldots, N_c$, and $k = 1, \ldots, K$.

Replacing (8.79b) with (8.82) and (8.83) leads to the following SDR problem

$$\min \sum_{i=1}^{N_c} \sum_{k=1}^{K} \mathrm{Tr}(\mathbf{W}_{ik})$$

$$\text{s.t. } \boldsymbol{\Phi}_{ik}\left(\{\mathbf{W}_{i\ell}\}_{\ell=1}^{K}, \{t_{jik}\}_{j\neq i}, \lambda_{iik}\right) \succeq \mathbf{0}$$

$$\boldsymbol{\Psi}_{jik}\left(\{\mathbf{W}_{j\ell}\}_{\ell=1}^{K}, t_{jik}, \lambda_{jik}\right) \succeq \mathbf{0}, \; j\neq i$$

$$t_{jik} \geq 0, \; j\neq i \qquad\qquad (8.84)$$

$$\mathbf{W}_{ik} \succeq \mathbf{0}, \; \lambda_{jik} \geq 0, \; j = 1, \ldots, N_c$$

$$i = 1, \ldots, N_c, \; k = 1, \ldots, K.$$

Problem (8.84) is a convex SDP; hence it can be efficiently solved.

What remains is to analyze whether the SDR problem (8.84) can yield a rank-one solution, i.e., whether an optimal solution $\mathbf{W}_{ik}^\star = \mathbf{w}_{ik}^\star(\mathbf{w}_{ik}^\star)^H$ exists for some $\mathbf{w}_{ik}^\star \in \mathbb{C}^{N_t}$, for all i, k. If this is true, then $\{\mathbf{w}_{ik}^\star\}$ is an optimal solution of the original robust MCBF problem (8.77). Suppose that the SDR problem (8.84) is feasible. Some conditions [SCW⁺12] under which the SDR problem (8.84) has been proven to yield rank-one solutions are given below:

(C1) $K = 1$, i.e., there is only one MS in each cell;

(C2) $\mathbf{C}_{iik} = \infty \mathbf{I}_{N_t}$ for all i, k, i.e., $\mathbf{e}_{iik} = \mathbf{0}$ for all i, k, i.e., perfect intra-cell CSI $\{\mathbf{h}_{iik}\}$;

(C3) For the spherical model, i.e., $\|\mathbf{e}_{jik}\|_2^2 \leq \varepsilon_{jik}^2$ for all i, j, k, the CSI error bounds $\{\varepsilon_{jik}\}$ satisfy

$$\varepsilon_{jik} \leq \bar{\varepsilon}_{jik} \text{ and } \varepsilon_{iik} < \sqrt{\sigma_{ik}^2 \gamma_{ik}/f^\star} \qquad (8.85)$$

for all i, j, k, where $\{\bar{\varepsilon}_{jik}\}$ are some CSI error bounds under which problem (8.84) is feasible, with $f^\star > 0$ denoting the associated optimal value.

If any one of the above three conditions is satisfied, then the SDR problem (8.84) must yield a rank-one solution. Nevertheless, the above three conditions are sufficient conditions which can be proven through some KKT conditions of problem (8.84). To find other sufficient conditions for the rank-one solutions is still a challenging research.

Let us show some simulation results to demonstrate the performance of the SDR formulation in (8.84) for handling the robust MCBF problem in (8.77), together with the convex restrictive approximation method proposed in [TPW11] for performance comparison, and the non-robust MCBF design (7.59) as a performance benchmark. All the design formulations are solved by SeDuMi.

In the simulations, we consider the following channel model [DY10]:

$$\mathbf{h}_{ijk} = 10^{-(128.1+37.6 \cdot \log_{10} d_{ijk})/20} \cdot \psi_{ijk} \cdot \varphi_{ijk}$$
$$\cdot \left[\widehat{\mathbf{h}}_{ijk} + \mathbf{e}_{ijk} \right], \tag{8.86}$$

where the exponential term is due to the path loss depending on the distance between the ith BS and MS$_{jk}$ (denoted by d_{ijk} in kilometers), ψ_{ijk} reflects the shadowing effect, and φ_{ijk} represents the transmit-receive antenna gain. The term inside the brackets in (8.86) denotes the small scale fading which consists of the preassumed CSI $\widehat{\mathbf{h}}_{ijk}$ and the CSI error $\mathbf{e}_{ijk}$. As seen from (8.86), it is assumed that the BSs can accurately track the large scale fading, and suffers only from the small scale CSI errors.

The inter-BS distance is 500 meters, and the locations of the MSs in each cell are randomly determined with the distance to the serving BS at least 35 meters, i.e., $d_{iik} \geq 0.035$ for all i, k. The shadowing coefficient ψ_{ijk} follows the log-normal distribution with zero mean and standard deviation equal to 8. The elements of the presumed CSI $\{\widehat{\mathbf{h}}_{ijk}\}$ are i.i.d. complex Gaussian random variables with zero mean and unity variance. We also assume that all MSs have the same noise power spectral density equal to -162 dBm/Hz (i.e., $\sigma_{ik}^2 = -92$ dBm for all i, k over a 10 MHz bandwidth), and each BS has a maximum power limit 46 dBm (as an extra power constraint on all the designs under test for practical consideration). The SINR requirements of MSs are the same, i.e., $\gamma_{ik} \triangleq \gamma$, and each link has the same antenna gain $\varphi_{ijk} = 15$ dBi. For the CSI errors, the spherical error model is considered, i.e., $\mathbf{C}_{ijk} = (1/\epsilon^2)\mathbf{I}_{N_t}$ for all i, j, and k, where ϵ denotes the radius of the error sphere.

Figure 8.8 shows some simulation results of average sum power (dBm) versus the SINR requirement γ (the top plot) and versus the CSI error radius ϵ (the bottom plot), respectively. Note that, over the generated seven thousand channel realizations, each of the results in Figure 8.8 is obtained by averaging over those channel realizations for which the three methods under test are all feasible. We should emphasize that the SDR problem (8.84) yields rank-one solutions for all the results shown in this figure. One can observe from both plots that, as a price for worst-case performance guarantee, the robust MCBF designs require higher

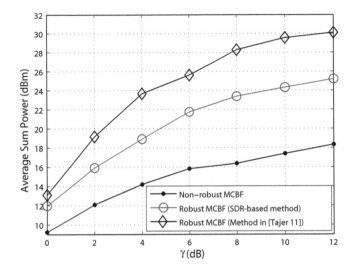

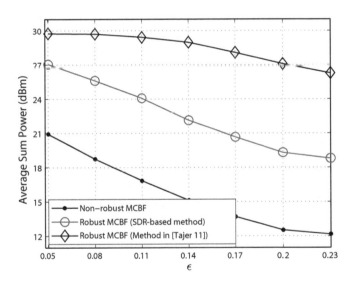

Figure 8.8 Average transmission sum power (dBm) of various methods for $N_c = 2$, $K = 4$, $N_t = 6$, and (a) (top plot) $\epsilon = 0.1$ and (b) (bottom plot) $\gamma = 10$ dB. (© 2012 IEEE. Reprinted, with permission, from C. Shen, T.-H. Chang, K.-Y. Wang, Z. Qiu, and C.-Y. Chi, "Distributed Robust Multi-Cell Coordinated Beamforming with Imperfect CSI: An ADMM Approach," Jun. 2012.)

average transmission powers than the non-robust design, while the SDR-based method is more power efficient than the method in [TPW11]. For example, for $\gamma = 10$ dB in the top plot of Figure 8.8, the SDR-based method consumes around 24 dBm while the method in [TPW11] requires 29 dBm. On the other hand, it is noticeable from the bottom plot of Figure 8.8 that the average powers of all the three methods decrease with ϵ, which seems counterintuitive. The reason for this is that the set and the total number of feasible channel realizations used for evaluating the average powers significantly vary with ϵ, and thus the obtained average powers turn out not necessarily to increase with ϵ.

Remark 8.9 Using a distributed convex optimization technique known as alternating direction method of multipliers [BT89b, BPC+10], which will be introduced in Section 9.7, two decentralized implementations for obtaining the centralized beamforming solutions of problem (8.84) only with local CSI used at each BS and limited backhaul information exchange between BSs have also been proposed in [SCW+12]. □

8.5.6   Outage constrained coordinated beamforming for MISO interference channel: Part I (centralized algorithm)

The interference channel has recently drawn extensive study for multiple communication systems sharing the common resource, e.g., multiple base stations sharing the same frequency band for higher spectrum efficiency in cellular wireless communication systems. However, the common frequency band (resource) used by multiple base stations simultaneously will cause serious inter-cell interference, inevitably leading to system performance (e.g., achievable transmission rate, bit error rate or QoS) degradation. This naturally motivates joint beamforming over the coordinated base stations so as to maximize the spectrum efficiency.

Consider the K-user MISO interference channel (IFC) (see Figure 8.9 for an illustrative example of 3-user MISO IFC, where each transmitter is equipped with N_t antennas and each receiver with a single antenna). It is assumed that all transmitters employ transmit beamforming to communicate with their respective receivers. Let $s_i(t)$ denote the information signal sent from the ith transmitter, and let $\boldsymbol{w}_i \in \mathbb{C}^{N_t}$ be the corresponding beamforming vector. The received signal at receiver i can be expressed as

$$x_i(t) = \mathbf{h}_{ii}^H \boldsymbol{w}_i s_i(t) + \sum_{k=1,k\neq i}^{K} \mathbf{h}_{ki}^H \boldsymbol{w}_k s_k(t) + n_i(t), \qquad (8.87)$$

where $\mathbf{h}_{ki} \in \mathbb{C}^{N_t}$ denotes the channel vector from transmitter k to receiver i, and $n_i(t) \sim \mathcal{CN}(0, \sigma_i^2)$ is the additive noise of receiver i with the noise variance $\sigma_i^2 > 0$. Assume that all receivers employ single-user detection where the cross-link interference is simply treated as background noise. Under Gaussian signaling, i.e., $s_i(t) \sim \mathcal{CN}(0,1)$, the instantaneous achievable rate of the ith transmitter-

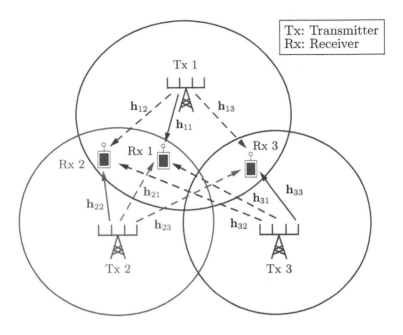

Figure 8.9 3-user MISO IFC.

receiver pair can be written as

$$r_i\left(\{\mathbf{h}_{ki}\}_k, \{\boldsymbol{w}_k\}\right) = \log_2\left(1 + \mathrm{SINR}_i\right)$$

$$= \log_2\left(1 + \frac{\left|\mathbf{h}_{ii}^H \boldsymbol{w}_i\right|^2}{\sum_{k \neq i}\left|\mathbf{h}_{ki}^H \boldsymbol{w}_k\right|^2 + \sigma_i^2}\right) \text{ bits/transmission,}$$

$$(8.88)$$

where $\{\mathbf{h}_{ki}\}_k$ denotes the set $\{\mathbf{h}_{1i}, \ldots, \mathbf{h}_{Ki}\}$, $\{\boldsymbol{w}_k\}$ denotes the set $\{\boldsymbol{w}_1, \ldots, \boldsymbol{w}_K\}$, and SINR_i denotes the signal-to-interference-plus-noise ratio for user i (cf. (8.87)), as given in the second term inside the parentheses.

Assume that the channel coefficients $\mathbf{h}_{ki}$ are block-faded, and that the transmitters have only the statistical information of the channels, i.e., the channel distribution information (CDI). In particular, it is assumed that $\mathbf{h}_{ki} \sim \mathcal{CN}(\mathbf{0}, \mathbf{Q}_{ki})$ for all $k, i = 1, \ldots, K$, where $\mathbf{Q}_{ki} \succeq \mathbf{0}$ denotes the channel covariance matrix and is known to all the transmitters. Since the transmission rate R_i cannot be adapted without channel state information, the communication would be in outage whenever the transmission rate R_i is higher than the instantaneous capacity that the channel can support. For a given outage probability requirement $(\epsilon_1, \ldots, \epsilon_K)$, the set of beamforming vectors $\{\boldsymbol{w}_i\}$ is required to satisfy $\mathrm{Prob}\{r_i(\{\mathbf{h}_{ki}\}_k, \{\boldsymbol{w}_i\}) < R_i\} \leq \epsilon_i$ to maintain the system in reliable operation.

Given an outage requirement $(\epsilon_1, \ldots, \epsilon_K)$ and an individual power budget $(P_1, \ldots, P_K)$, it is desired to optimize $\{w_k\}$ such that the predefined system utility function $U(R_1, \ldots, R_K)$ is maximized. To this end, let us consider the following *outage constrained coordinated beamforming design problem*

$$\max_{\substack{w_i \in \mathbb{C}^{N_t}, R_i \geq 0, \\ i=1,\ldots,K}} \quad U(R_1, \ldots, R_K) \tag{8.89a}$$

$$\text{s.t. } \text{Prob}\,\{r_i(\{\mathbf{h}_{ki}\}_k, \{w_k\}) < R_i\} \leq \epsilon_i, \ i = 1, \ldots, K, \tag{8.89b}$$

$$\|w_i\|_2^2 \leq P_i, \quad i = 1, \ldots, K. \tag{8.89c}$$

Note that, as each user would prefer a higher transmission rate, a sensible system utility function $U(R_1, \ldots, R_K)$ should be monotonically increasing w.r.t. $R_1, \ldots, R_K$, respectively. On the other hand, when the user fairness is taken into consideration, the utility function $U(R_1, \ldots, R_K)$ should be jointly concave w.r.t. $R_1, \ldots, R_K$. We hence make the assumption that $U(R_1, \ldots, R_K)$ is $\mathbb{R}_+^K$-increasing and concave. In this case, the difficulty in solving problem (8.89) lies in the probability constraint (8.89b). As for the system utility function, the following three have been extensively used in wireless communications and networking:

- Weighted sum rate:

$$U_S(R_1, \ldots, R_K) = \sum_{i=1}^{K} \alpha_i R_i;$$

- Weighted geometric mean rate:

$$U_G(R_1, \ldots, R_K) = \prod_{i=1}^{K} R_i^{\alpha_i};$$

- Weighted harmonic mean rate:

$$U_H(R_1, \ldots, R_K) = \frac{1}{\sum_{i=1}^{K} \alpha_i R_i^{-1}}.$$

The coefficients $\alpha_1, \ldots, \alpha_K$ represent the user priority, which satisfy $\alpha_i \in [0, 1]$ for $i = 1, \ldots, K$ and $\sum_{i=1}^{K} \alpha_i = 1$. It can be seen that U_S, U_G, and U_H are all strictly increasing w.r.t. R_i for all i (i.e., $\mathbb{R}_+^K$-increasing) provided that $\alpha_i > 0$ for all i; moreover, U_S is a concave function, and maximizing U_G and U_H is equivalent to maximizing the concave functions $\sum_{i=1}^{K} \alpha_i \ln R_i$ and $-\sum_{i=1}^{K} \alpha_i R_i^{-1}$, respectively. Also notice that maximizing U_S, U_G, and U_H is, respectively, equivalent to achieving the maximal throughput, proportional fairness, and minimal potential delay of the users [BM01].

According to [KB02] that provides a closed-form expression for the probability function in (8.89b), problem (8.89) can be expressed as

$$\max_{\substack{\boldsymbol{w}_i \in \mathbb{C}^{N_t}, R_i \geq 0, \\ i=1,\ldots,K}} U(R_1, \ldots, R_K) \tag{8.90a}$$

$$\text{s.t. } \rho_i \exp\left(\frac{(2^{R_i}-1)\sigma_i^2}{\boldsymbol{w}_i^H \mathbf{Q}_{ii}\boldsymbol{w}_i}\right) \prod_{k \neq i}\left(1 + \frac{(2^{R_i}-1)\boldsymbol{w}_k^H \mathbf{Q}_{ki}\boldsymbol{w}_k}{\boldsymbol{w}_i^H \mathbf{Q}_{ii}\boldsymbol{w}_i}\right) \leq 1, \tag{8.90b}$$

$$\|\boldsymbol{w}_i\|_2^2 \leq P_i, \ i = 1, \ldots, K, \tag{8.90c}$$

where $\rho_i \triangleq 1 - \epsilon_i$. As (8.90b) is a complicated nonconvex constraint, problem (8.90) is almost formidable to solve and so far no effective solutions were reported until an approximate solution proposed by Wei-Chiang Li, et al. [LCLC11, LCLC13]. Next, let us introduce how an approximate solution is obtained through the use of SDR and some other convexity approximations to the constraints in (8.90b) as reported in [LCLC11, LCLC13].

The idea is to find a convex approximation of (8.90b) such that problem (8.90) can be approximated by a convex problem. First of all, by SDR, (8.90) can be approximated by

$$\max_{\substack{\mathbf{W}_i \in \mathbb{H}^{N_t}, R_i \geq 0, \\ i=1,\ldots,K}} U(R_1, \ldots, R_K) \tag{8.91a}$$

$$\text{s.t. } \rho_i \exp\left(\frac{(2^{R_i}-1)\sigma_i^2}{\text{Tr}(\mathbf{W}_i \mathbf{Q}_{ii})}\right) \prod_{k \neq i}\left(1 + \frac{(2^{R_i}-1)\text{Tr}(\mathbf{W}_k \mathbf{Q}_{ki})}{\text{Tr}(\mathbf{W}_i \mathbf{Q}_{ii})}\right) \leq 1, \tag{8.91b}$$

$$\text{Tr}(\mathbf{W}_i) \leq P_i, \tag{8.91c}$$

$$\mathbf{W}_i \succeq \mathbf{0}, \ i = 1, \ldots, K. \tag{8.91d}$$

However, problem (8.91) is still not convex due to the nonconvex constraints in (8.91b). Therefore, further approximations are needed for problem (8.91).

In contrast to SDR that essentially results in a larger feasible set, the second approximation is restrictive, in the sense that the obtained solution must also be feasible to problem (8.91). To obtain approximate convex constraints, let us define the following auxiliary variables:

$e^{x_{ki}} \triangleq \text{Tr}(\mathbf{W}_k \mathbf{Q}_{ki})$ (interference power from transmitter k to user i),

$e^{y_i} \triangleq 2^{R_i} - 1$ (SINR$_i$ corresponding to rate R_i),

$z_i \triangleq \dfrac{2^{R_i} - 1}{\text{Tr}(\mathbf{W}_i \mathbf{Q}_{ii})} = e^{y_i - x_{ii}}$ (inverse of interference-plus-noise power at user i),

for $i, k = 1, \ldots, K$. The above auxiliary variables x_{ki}, y_i, and z_i will convert (8.91b) into a convex constraint, but they also introduce a set of nonconvex equality constraints that need to be somehow handled by convex approximations to be presented below. Problem (8.91) can be equivalently expressed as

$$\max_{\substack{(\mathbf{W}_1,\ldots,\mathbf{W}_K)\in\mathcal{S}, \\ R_i\geq 0, x_{ki}, y_i, z_i\in\mathbb{R}, \\ k,i=1,\ldots,K}} U(R_1,\ldots,R_K) \tag{8.92a}$$

$$\text{s.t.} \quad \rho_i e^{\sigma_i^2 z_i} \prod_{k\neq i}\left(1+e^{-x_{ii}+x_{ki}+y_i}\right)\leq 1, \tag{8.92b}$$

$$\text{Tr}(\mathbf{W}_k\mathbf{Q}_{ki})\leq e^{x_{ki}}, \quad k\in\mathcal{K}_i^c, \tag{8.92c}$$

$$\text{Tr}(\mathbf{W}_i\mathbf{Q}_{ii})\geq e^{x_{ii}}, \tag{8.92d}$$

$$R_i\leq\log_2(1+e^{y_i}), \tag{8.92e}$$

$$e^{y_i-x_{ii}}\leq z_i, \tag{8.92f}$$

where $\mathcal{K}_i^c\triangleq\{1,\ldots,i-1,i+1,\ldots,K\}$, and $\mathcal{S}$ is defined below in (8.93). Note that the optimal solution of (8.92) occurs only when the inequality constraints (8.92b)–(8.92f) hold with equality, and thus guaranteeing the same optimal solution with (8.91). For instance, if the inequality (8.92e) is not active, the utility function $U(R_1,\ldots,R_K)$ can be further increased by increasing R_i until the equality holds true; if the optimal solution is the one such that the inequality (8.92f) is not active, one can find another feasible point with smaller z_i and larger y_i, R_i, which is still feasible to (8.92b), (8.92e), and (8.92f), and achieves a larger value of $U(R_1,\ldots,R_K)$.

We also note that if the optimal solution satisfies $\text{Tr}(\mathbf{W}_i\mathbf{Q}_{ik})=0$ in (8.92c) and (8.92d), then the optimal x_{ik} will be minus infinity which is not attainable, thereby possibly leading to numerical problems in solving the resulting optimization problem (8.96) below. In view of this, in (8.92) we have enforced $\mathbf{W}_1,\ldots,\mathbf{W}_K$ to lie in the subset

$$\mathcal{S}\triangleq\left\{(\mathbf{W}_1,\ldots,\mathbf{W}_K)\ \middle|\ \begin{array}{l}\mathbf{W}_i\succeq\mathbf{0}, \text{Tr}(\mathbf{W}_i)\leq P_i, \\ \text{Tr}(\mathbf{W}_i\mathbf{Q}_{ik})\geq\delta\ \forall i,k=1,\ldots,K\end{array}\right\}, \tag{8.93}$$

where δ is a small positive number. It can be observed that, with this reformulation, constraint (8.92b) is now convex. Furthermore, constraints (8.92d) and (8.92f) are also convex. Although constraints (8.92c) and (8.92e) are still nonconvex, they are relatively easy to handle in comparison with the original constraint (8.91b).

Let $\{\bar{\boldsymbol{w}}_i,\bar{R}_i\}_{i=1}^K$ be a feasible point of (8.90). Define

$$\bar{x}_{ki}\triangleq\ln(\bar{\boldsymbol{w}}_k^H\mathbf{Q}_{ki}\bar{\boldsymbol{w}}_k), \quad k=1,\ldots,K, \tag{8.94a}$$

$$\bar{y}_i\triangleq\ln(2^{\bar{R}_i}-1), \quad i=1,\ldots,K, \tag{8.94b}$$

which are nothing but the associated values of auxiliary variables x_{ki} and y_i at this feasible point, respectively. According to the first-order condition of the convex functions $e^{x_{ki}}$ and $\log_2(1+e^{y_i})$ (see (3.16)), one can obtain the following lower bounds for $e^{x_{ki}}$ in (8.92c) and $\log_2(1 + e^{y_i})$ in (8.92e), respectively,

$$e^{\bar{x}_{ki}}\left(x_{ki} - \bar{x}_{ki} + 1\right) \le e^{x_{ki}}, \tag{8.95a}$$

$$\frac{1}{\ln 2}\left(\ln(1 + e^{\bar{y}_i}) + \frac{e^{\bar{y}_i}(y_i - \bar{y}_i)}{1 + e^{\bar{y}_i}}\right) \le \log_2(1 + e^{y_i}). \tag{8.95b}$$

By (8.95), one can obtain a conservative approximation of (8.92)

$$\max_{\substack{(\mathbf{W}_1,\dots,\mathbf{W}_K)\in\mathcal{S},R_i\ge 0, \\ x_{ki},y_i,z_i\in\mathbb{R}, \\ k,i=1,\dots,K}} U(R_1,\dots,R_K), \tag{8.96a}$$

$$\text{s.t.}\quad \rho_i e^{\sigma_i^2 z_i} \prod_{k\ne i}\left(1 + e^{-x_{ii}+x_{ki}+y_i}\right) \le 1, \tag{8.96b}$$

$$\text{Tr}(\mathbf{W}_k\mathbf{Q}_{ki}) \le e^{\bar{x}_{ki}}(x_{ki} - \bar{x}_{ki} + 1), \quad k \in \mathcal{K}_i^c, \tag{8.96c}$$

$$\text{Tr}(\mathbf{W}_i\mathbf{Q}_{ii}) \ge e^{x_{ii}}, \tag{8.96d}$$

$$R_i \le \frac{1}{\ln 2}\left(\ln(1 + e^{\bar{y}_i}) + \frac{e^{\bar{y}_i}}{1 + e^{\bar{y}_i}}(y_i - \bar{y}_i)\right), \tag{8.96e}$$

$$e^{y_i - x_{ii}} \le z_i, \tag{8.96f}$$

which now is convex and the optimal solution can be obtained.

The preceding formulation (8.96) is obtained by approximating problem (8.92) at the given feasible point $\{\bar{\boldsymbol{w}}_i, \bar{R}_i\}_{i=1}^K$ as described in (8.94). The approximations can be successively improved by approximate problem (8.92) based on the optimal solution $(\{\mathbf{W}_i\}, \{R_i\})$ obtained by solving (8.96). The resulting algorithm, exactly an SCA algorithm (cf. Section 4.7), is named Algorithm 8.1. Moreover, a graphical illustration of the SCA algorithm running from one iteration to another is provided in Figure 8.10.

The Pareto boundary of problem (8.90) is constituted by all the maximal elements of the achievable set of $(R_1,\dots,R_K)$ of the following vector maximization problem with the same feasible set of problem (8.90):

$$\underset{\substack{\boldsymbol{w}_i\in\mathbb{C}^{N_t},R_i\ge 0, \\ i=1,\dots,K}}{\text{maximize (w.r.t. } \mathcal{K}=\mathbb{R}_+^K)}\ \boldsymbol{R} \triangleq (R_1,\dots,R_K)$$

$$\text{subject to}\ \ \rho_i\, \exp\left(\frac{(2^{R_i}-1)\sigma_i^2}{\boldsymbol{w}_i^H\mathbf{Q}_{ii}\boldsymbol{w}_i}\right)\prod_{k\ne i}\left(1 + \frac{(2^{R_i}-1)\boldsymbol{w}_k^H\mathbf{Q}_{ki}\boldsymbol{w}_k}{\boldsymbol{w}_i^H\mathbf{Q}_{ii}\boldsymbol{w}_i}\right) \le 1 \tag{8.97}$$

$$\|\boldsymbol{w}_i\|_2^2 \le P_i,\ i=1,\dots,K.$$

Note that the objective function $\boldsymbol{R} = (R_1,\dots,R_K)$ is $\mathcal{K}$-concave, but the problem is nonconvex due to the nonconvex inequality constraints.

Some interesting properties can be inferred from the outage probability constraints in problem (8.97) about the achievable rate region (i.e., the set of achievable objective values), denoted as $\mathcal{O}$, and Pareto boundary $\mathcal{P}$ (i.e., the set of maximal elements of $\mathcal{O}$). The first one is that $\mathcal{O}$ is a normal set (cf. (4.134)), for which if $\boldsymbol{R}_1 \in \mathcal{O}$, then $\boldsymbol{R}_2 \in \mathcal{O}$ for all $\boldsymbol{0} \preceq_\mathcal{K} \boldsymbol{R}_2 \preceq_\mathcal{K} \boldsymbol{R}_1$. The second one is that $\mathcal{O}$

Algorithm 8.1 SCA algorithm for solving problem (8.89)

1: Given $(\widehat{\boldsymbol{w}}_1\widehat{\boldsymbol{w}}_1^H, \ldots, \widehat{\boldsymbol{w}}_K\widehat{\boldsymbol{w}}_K^H) \in \mathcal{S}$ and $(\widehat{R}_1, \ldots, \widehat{R}_K)$ that are feasible to (8.91);
2: Set $\widehat{\mathbf{W}}_i = \widehat{\boldsymbol{w}}_i\widehat{\boldsymbol{w}}_i^H$ for all $i = 1, \ldots, K$;
3: **repeat**
4:     Set $\bar{x}_{ki} = \ln(\widehat{\mathbf{W}}_k \mathbf{Q}_{ki})$, for all $k \in \mathcal{K}_i^c$, and $\bar{y}_i = \ln(2^{\widehat{R}_i} - 1)$ for $i = 1, \ldots, K$;
5:     Obtain an optimal solution $(\{\widehat{\mathbf{W}}_i\}, \{\widehat{R}_i\}, \{\hat{x}_{ik}\}, \{\hat{y}_i\}, \{\hat{z}_i\})$ of problem (8.96).
6: **until** the predefined stopping criterion is met.
7: Obtain $\boldsymbol{w}_i^\star$ by decomposition of $\widehat{\mathbf{W}}_i = \boldsymbol{w}_i^\star(\boldsymbol{w}_i^\star)^H$ for all i, if $\widehat{\mathbf{W}}_i$ are all of rank one; otherwise perform Gaussian randomization to obtain a rank-one feasible approximate solution of (8.89).

---

may not be a convex set, while both $\mathcal{O}$ and $\mathcal{P}$ are closed continuous sets. Hence any optimal $\boldsymbol{R}$ of problem (8.90) will be a point on the Pareto boundary as the system utility function $U(\boldsymbol{R})$ is $\mathcal{K}$-increasing. These properties can be clearly observed in the simulation results below.

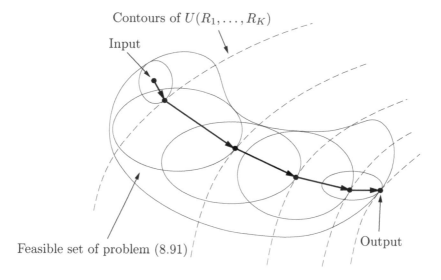

Figure 8.10 Illustration of SCA algorithm.

Next, we provide some simulation results to demonstrate the efficacy of the SCA algorithm. In the simulations, all receivers are assumed to have the same noise power, i.e., $\sigma_1^2 = \cdots = \sigma_K^2 = \sigma^2$, and all power constraints are set to one, i.e., $P_1 = \cdots = P_K = 1$. The parameter δ in (8.93) is set to 10^{-5} and the channel covariance matrices $\mathbf{Q}_{ki}$ are randomly generated.

For a randomly generated two-user MISO IFC with four antennas at each transmitter (i.e., $K = 2$ and $N_t = 4$), the rate tuples (R_1, R_2) achieved in each

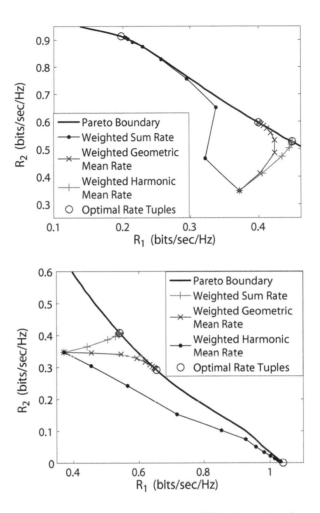

Figure 8.11 Convergence trajectories of the SCA algorithm by maximizing the weighted sum rate, weighted geometric mean rate, and weighted harmonic mean rate in a two-user MISO IFC with $N_t = 4$ and the user priority weights $(\alpha_1, \alpha_2) = (\frac{1}{2}, \frac{1}{2})$ for the top plot and $(\alpha_1, \alpha_2) = (\frac{2}{3}, \frac{1}{3})$ for the bottom plot. (© 2013 IEEE. Reprinted, with permission, from W.-C. Li, T.-H. Chang, C. Lin, and C.-Y. Chi, "Coordinated Beamforming for Multiuser MISO Interference Channel Under Rate Outage Constraints," Mar. 2013.)

iteration of the SCA algorithm are shown in Figure 8.11, wherein the "Pareto Boundary" represents the right-upper boundary of the set of all achievable rate tuples which was obtained by exhaustive search. One can see, from Figure 8.11, that the SCA algorithm converges to the global optimum for each of the three utility functions and for the user priority weights $(\alpha_1, \alpha_2) = (1/2, 1/2)$ (top plot) and $(\alpha_1, \alpha_2) = (2/3, 1/3)$ (bottom plot).

We also compare the average performance of the SCA algorithm with the global optimum obtained by an exhaustive search method, for different relative cross-link interference levels, defined as

$$\eta \triangleq \frac{\lambda_{\max}(\mathbf{Q}_{ki})}{\lambda_{\max}(\mathbf{Q}_{ii})}$$

for all $i = 1, \ldots, K$, $k \neq i$. From the top plot of Figure 8.12, one can see that the SCA algorithm achieves almost the same average weighted sum rates as the exhaustive search method for different values of $1/\sigma^2$ and interference levels. Similarly, one can observe from the bottom plot of Figure 8.12 that the SCA algorithm also performs near-optimally for maximizing the weighted harmonic mean rate.

Let us conclude this subsection with some remarks on the SCA algorithm as follows:

Remark 8.10 It can be proved that problems (8.91) and (8.96) can yield rank-one optimal $\{\mathbf{W}_i\}_{i=1}^K$ for $K \leq 3$. For this case, an approximate solution to (8.90) can be directly obtained by rank-one decomposition of the optimal $\{\mathbf{W}_i\}_{i=1}^K$. $\square$

Remark 8.11 Let $\{\widehat{\mathbf{W}}_i[n]\}$, $\{\widehat{R}_i[n]\}$, $\{\widehat{x}_{ik}[n]\}$, $\{\widehat{y}_i[n]\}$, and $\{\widehat{z}_i[n]\}$ denote the optimal solution obtained at the nth iteration of the SCA algorithm. Then $(\{\widehat{\mathbf{W}}_i[n]\}, \{\widehat{R}_i[n]\})$ can be proven to be a stationary point of problem (8.91), i.e., the SDR of problem (8.90), with the extra constraint (8.93) as $n \to \infty$. This also implies that as $n \to \infty$, if $\{\widehat{\mathbf{W}}_i[n]\}$ is of rank one, $(\{\widehat{\mathbf{W}}_i[n]\}, \{\widehat{R}_i[n]\})$ must also be a stationary point of problem (8.90). $\square$

Remark 8.12 The exhaustive search method, used in obtaining those results of optimal solutions of problem (8.90) shown in Figure 8.12, is only viable for $K \leq 2$ due to the computational complexity. For $K \geq 3$, the SCA algorithm performs better than the maximum ratio transmission algorithm (by only maximizing individual receiving power $\mathbf{w}_i^H \mathbf{Q}_{ii} \mathbf{w}_i$) and the zero-forcing beamformer (as channel covariance matrices $\mathbf{Q}_{ki}$ are sufficiently rank deficient) based on simulation experience. $\square$

Remark 8.13 The SCA algorithm is actually a centralized coordinated beamforming solution, via solving a sequence of subproblems (cf. (8.96)) with problem size $\mathcal{O}(K^2)$. For the system scalability, decentralized implementation with only local CDI used by each transmitter for its own beamforming design can also achieve a stationary point of problem (8.91) as detailed in [LCLC13]. The resulting algorithm, referred to as the DSCA algorithm, involves solving subproblems with problem size $\mathcal{O}(K)$ at each iteration. $\square$

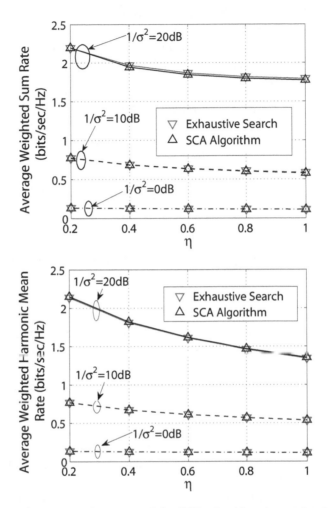

Figure 8.12 Average performance of the SCA algorithm in weighted sum rate (top plot) and weighted harmonic mean rate (bottom plot), for $K = 2$, $N_t = 4$, and the user priority weights $(\alpha_1, \alpha_2) = (\frac{1}{2}, \frac{1}{2})$, where each result is obtained by averaging over 500 realizations of $\{\mathbf{Q}_{ki}\}$. (© 2013 IEEE. Reprinted, with permission, from W.-C. Li, T.-H. Chang, C. Lin, and C.-Y. Chi, "Coordinated Beamforming for Multiuser MISO Interference Channel Under Rate Outage Constraints," Mar. 2013.)

8.5.7   Outage constrained coordinated beamforming for MISO interference channel: Part II (efficient algorithms using BSUM)

Although the SCA algorithm (DSCA algorithm) only involves solving a sequence of convex SDPs (cf. (8.96)), the structure of the subproblems is complicated; moreover, the problem size increases with $\mathcal{O}(K^2)$ ($\mathcal{O}(K)$). Therefore, the computational load need be further reduced for practical applications, especially for large K. In this subsection, we introduce two efficient algorithms based on the

BSUM method (cf. Section 4.7) without resorting to any SDP formulations. As the SCA algorithm, they can also yield a stationary point of the outage constrained coordinated beamforming design problem (8.89) [LCC15b].

The key ingredient of the two efficient algorithms is to equivalently reformulate problem (8.89) as

$$\max_{\boldsymbol{w}_i \in \mathbb{C}^{N_t}, \; i=1,\ldots,K} \; \mathcal{U}(\{\boldsymbol{w}_i\}) \triangleq U\big(R_1(\{\boldsymbol{w}_i\}), \ldots, R_K(\{\boldsymbol{w}_i\})\big) \tag{8.98a}$$

$$\text{s.t. } \|\boldsymbol{w}_i\|_2^2 \leq P_i, \; i=1,\ldots,K, \tag{8.98b}$$

where

$$R_i(\{\boldsymbol{w}_k\}) \triangleq \log_2\big(1 + \xi_i(\{\boldsymbol{w}_k\}_{k\neq i})\boldsymbol{w}_i^H \boldsymbol{Q}_{ii}\boldsymbol{w}_i\big), \tag{8.99}$$

and $\xi_i(\{\boldsymbol{w}_k\}_{k\neq i})$ is a continuously differentiable function of $\{\boldsymbol{w}_k\}_{k\neq i}$ and is a unique solution to the equation

$$\Phi_i(\xi_i, \{\boldsymbol{w}_k\}_{k\neq i}) \triangleq \ln \rho_i + \sigma_i^2 \xi_i + \sum_{k\neq i} \ln\big(1 + (\boldsymbol{w}_k^H \boldsymbol{Q}_{ki}\boldsymbol{w}_k)\cdot\xi_i\big) = 0, \tag{8.100}$$

for $i=1,\ldots,K$. As the outage inequality constraint (8.90b) must hold with equality for the optimal solution of problem (8.90), (8.100) stems from taking the logarithm on both sides of the resulting equality constraint. Here we purposely rewrite the outage constraint function in this form for ease of later derivations. The equivalence between problem (8.89) and problem (8.98) can be proved by exploiting the fact that the left-hand side function in (8.90b) is monotonic in $(2^{R_i} - 1)/\boldsymbol{w}_i^H \boldsymbol{Q}_{ii}\boldsymbol{w}_i$. The details of the proof can be found in [LCC15a].

A. BSUM method

Observing the block structure of the constraint set (8.98b), the BSUM method can be conveniently adopted to implement distributed and low-complexity algorithms for handling problem (8.98). Specifically, a set of approximation functions satisfying the BSUM conditions (4.162) can be obtained by exploiting the monotonicity and convexity/concavity of $R_i(\{\boldsymbol{w}_k\})$ (cf. (8.99)) w.r.t. $\{\boldsymbol{w}_k^H \boldsymbol{Q}_{ki}\boldsymbol{w}_k\}_k$. It is shown in [LCC15b, Lemma 1] that the function $R_i(\{\boldsymbol{w}_k\})$ defined in (8.99) is strictly increasing and strictly concave w.r.t. $\boldsymbol{w}_i^H \boldsymbol{Q}_{ii}\boldsymbol{w}_i$, while it is nonincreasing and convex w.r.t. each $\boldsymbol{w}_k^H \boldsymbol{Q}_{ki}\boldsymbol{w}_k$, $k \neq i$. For ease of later use, let

$$\bar{R}_j^{(i)}(\boldsymbol{w}_i \mid \{\bar{\boldsymbol{w}}_k\}) \triangleq$$
$$\begin{cases} \log_2\big(1 + \xi_i(\{\bar{\boldsymbol{w}}_k\}_{k\neq i})(2\mathrm{Re}\{\bar{\boldsymbol{w}}_i^H \boldsymbol{Q}_{ii}\boldsymbol{w}_i\} - \bar{\boldsymbol{w}}_i^H \boldsymbol{Q}_{ii}\bar{\boldsymbol{w}}_i)\big), \; j = i, \\ R_j(\{\bar{\boldsymbol{w}}_k\}) + \dfrac{\partial R_j(\{\bar{\boldsymbol{w}}_k\})}{\partial \boldsymbol{w}_i^H \boldsymbol{Q}_{ij}\boldsymbol{w}_i}(\boldsymbol{w}_i^H \boldsymbol{Q}_{ij}\boldsymbol{w}_i - \bar{\boldsymbol{w}}_i^H \boldsymbol{Q}_{ij}\bar{\boldsymbol{w}}_i), \; j \neq i, \end{cases} \tag{8.101}$$

which is a concave function in $\boldsymbol{w}_i$ for all i when $\partial R_j(\{\bar{\boldsymbol{w}}_k\})/\partial \boldsymbol{w}_i^H \boldsymbol{Q}_{ij}\boldsymbol{w}_i \leq 0$ for all $j \neq i$. Note that $\bar{R}_i^{(i)}(\boldsymbol{w}_i \mid \{\bar{\boldsymbol{w}}_k\})$ defined in (8.101) is obtained by linearizing the average signal power $\boldsymbol{w}_i^H \boldsymbol{Q}_{ii}\boldsymbol{w}_i$ in $R_i(\{\boldsymbol{w}_k\})$ w.r.t. $\boldsymbol{w}_i$ (cf. (3.25)), and that $\bar{R}_j^{(i)}(\boldsymbol{w}_i \mid \{\bar{\boldsymbol{w}}_k\})$ for $j \neq i$ is obtained by linearizing the outage-constrained achievable rate $R_j(\{\boldsymbol{w}_k\})$ w.r.t. the average interference power $\boldsymbol{w}_i^H \boldsymbol{Q}_{ij}\boldsymbol{w}_i$.

Furthermore, $\partial R_j(\{\bar{\boldsymbol{w}}_k\})/\partial \boldsymbol{w}_i^H \mathbf{Q}_{ij} \boldsymbol{w}_i$ in (8.101) can be expressed as

$$
\frac{\partial R_j(\{\bar{\boldsymbol{w}}_k\})}{\partial \boldsymbol{w}_i^H \mathbf{Q}_{ij} \boldsymbol{w}_i} = \left. \frac{\partial \log_2(1 + \xi_j \cdot \bar{\boldsymbol{w}}_j^H \mathbf{Q}_{jj} \bar{\boldsymbol{w}}_j)}{\partial \xi_j} \right|_{\xi_j = \xi_j(\{\bar{\boldsymbol{w}}_k\}_{k \neq j})} \qquad \text{(cf. (8.99))}
$$

$$
\times \left. \frac{\partial \xi_j(\{\boldsymbol{w}_k\}_{k \neq j})}{\partial \boldsymbol{w}_i^H \mathbf{Q}_{ij} \boldsymbol{w}_i} \right|_{\boldsymbol{w}_k = \bar{\boldsymbol{w}}_k, \forall k \neq i} \qquad (8.102)
$$

where the second term $\partial \xi_j(\{\boldsymbol{w}_k\}_{k \neq j})/\partial \boldsymbol{w}_i^H \mathbf{Q}_{ij} \boldsymbol{w}_i$, though non-trivial to compute, can be obtained by applying the implicit function theorem to (8.100), thereby leading to $\partial R_j(\{\bar{\boldsymbol{w}}_k\})/\partial \boldsymbol{w}_i^H \mathbf{Q}_{ij} \boldsymbol{w}_i \leq 0$ along with a closed-form expression for all $j \neq i$ [LCC15b]. Hence $\bar{R}_j^{(i)}(\boldsymbol{w}_i \mid \{\bar{\boldsymbol{w}}_k\})$ defined in (8.101) is indeed concave for all i.

Therefore, by (cf. (3.76c)), the following approximations $\bar{\mathcal{U}}^{(i)}(\boldsymbol{w}_i \mid \{\bar{\boldsymbol{w}}_k\})$ of the differentiable objective function $\mathcal{U}(\{\boldsymbol{w}_i\})$ in (8.98) are differentiable concave functions of $\boldsymbol{w}_i$:

$$
\bar{\mathcal{U}}^{(i)}(\boldsymbol{w}_i \mid \{\bar{\boldsymbol{w}}_k\}) \triangleq U\left(\bar{R}_1^{(i)}(\boldsymbol{w}_i | \{\bar{\boldsymbol{w}}_k\}), \dots, \bar{R}_K^{(i)}(\boldsymbol{w}_i | \{\bar{\boldsymbol{w}}_k\}) \right)
$$
$$
- \frac{c}{2} \left\| \boldsymbol{w}_i - \bar{\boldsymbol{w}}_i \right\|_2^2, \; i = 1, \dots, K, \qquad (8.103)
$$

where $c > 0$ is a penalty parameter, meanwhile satisfying the BSUM conditions (4.162) (except that $\geq$ in (4.162b) is replaced by $\leq$ since the problem (8.98) under consideration is a maximization problem). Note that the second term in (8.103) is for $\bar{\mathcal{U}}^{(i)}(\boldsymbol{w}_i \mid \{\bar{\boldsymbol{w}}_k\})$ to be strictly concave, guaranteeing the unique solution of problem (8.104) below (required by the convergence conditions of the BSUM method).

Due to the fact that the constraint set (8.98b) is compact and convex and the fact that the objective function of (8.98) is regular at every point of the constraint set (thus satisfying the promise (4.161b)), a stationary point of problem (8.98) can be efficiently obtained by the BSUM method as summarized in Algorithm 8.2. It is noteworthy that the optimization problem

$$
\max_{\|\boldsymbol{w}_i\|_2^2 \leq P_i} \bar{\mathcal{U}}^{(i)}(\boldsymbol{w}_i \mid \{\bar{\boldsymbol{w}}_k\}) \qquad (8.104)
$$

involved in each iteration of Algorithm 8.2 is a convex problem that can be efficiently solved by the projected gradient method (cf. Remark 4.4) instead of by the interior-point method due to only a simple 2-norm constraint involved. Moreover, the problem size of (8.104) no longer increases with K. It will be demonstrated by simulations later in this subsection that Algorithm 8.2 outperforms the DSCA algorithm not only in computational efficiency but also in the utility performance.

B. SUM method with parallel structure

Algorithm 8.2 is a Gauss–Seidel algorithm that updates the beamformers sequentially. Next, we introduce a Jacobi-type algorithm, which updates all the

---

Algorithm 8.2 BSUM method for handling problem (8.98)

1: **Input** a set of beamformers $\{\bar{w}_i\}$ satisfying (8.98b), and set $n := 0$;
2: **repeat**
3:     Set $n := n + 1$ and $i := (n - 1 \bmod K) + 1$;
4:     Update $\bar{w}_i = \arg \max\limits_{\|w_i\|_2^2 \leq P_i} \bar{\mathcal{U}}^{(i)}(w_i \mid \{\bar{w}_k\})$ (i.e., problem (8.104));
5: **until** some convergence criterion is met;
6: **Output** $\{\bar{w}_i\}$ as an approximate solution of problem (8.98).

---

beamformers in parallel at each iteration, particularly for handling the weighted sum rate (WSR) utility

$$\mathcal{U}_{wsr}(\{w_k\}) \triangleq \sum_{i=1}^{K} \alpha_i \log_2 \left(1 + \xi_i(\{w_k\}_{k \neq i}) w_i^H Q_{ii} w_i \right).$$

This algorithm is based on the BSUM method with only one block, referred to as the SUM method (cf. Remark 4.11), for which we approximate $\mathcal{U}_{wsr}(\{w_k\})$ by a surrogate function that can be decoupled over the K variable blocks, $w_1, \ldots, w_K$ so that the K variable blocks can be updated in parallel at each iteration.

Given a feasible point $\{\bar{w}_k\}$ satisfying (8.98b), an upper bound of $\Phi_i(\zeta_i, \{w_k\}_{k \neq i})$ is needed as follows

$$\Phi_i(\zeta_i, \{w_k\}_{k \neq i}) = \ln \rho_i + \sigma_i^2 \zeta_i + \sum_{k \neq i} \ln \left(1 + w_k^H Q_{ki} w_k \zeta_i\right)$$

$$\leq \ln \rho_i + \sigma_i^2 \zeta_i + \sum_{k \neq i} \ln \left(1 + \bar{w}_k^H Q_{ki} \bar{w}_k \bar{\zeta}_i\right) + \sum_{k \neq i} \frac{w_k^H Q_{ki} w_k \zeta_i - \bar{w}_k^H Q_{ki} \bar{w}_k \bar{\zeta}_i}{1 + \bar{w}_k^H Q_{ki} \bar{w}_k \bar{\zeta}_i}$$

$$= \ln \rho_i + \sum_{k \neq i} \ln \left(1 + \bar{w}_k^H Q_{ki} \bar{w}_k \bar{\zeta}_i\right) - \sum_{k \neq i} \frac{\bar{w}_k^H Q_{ki} \bar{w}_k \bar{\zeta}_i}{1 + \bar{w}_k^H Q_{ki} \bar{w}_k \bar{\zeta}_i}$$

$$+ \left(\sigma_i^2 + \sum_{k \neq i} \frac{w_k^H Q_{ki} w_k}{1 + \bar{w}_k^H Q_{ki} \bar{w}_k \bar{\zeta}_i}\right) \cdot \zeta_i$$

$$\triangleq \Psi_i(\zeta_i, \{w_k\}_{k \neq i} \mid \{\bar{w}_k\}_{k \neq i}) \tag{8.105}$$

where $\bar{\zeta}_i = \xi_i(\{\bar{w}_k\}_{k \neq i})$, and the inequality is due to the first-order approximation of the concave logarithm function, i.e., $\ln(y) \leq \ln(x) + (y - x)/x$, $\forall x, y > 0$. Similar to $\Phi_i(\xi_i, \{w_k\}_{k \neq i})$, there exists a unique continuously differentiable function, denoted by $\zeta_i(\{w_k\}_{k \neq i} \mid \{\bar{w}_k\}_{k \neq i})$, such that

$$\Psi_i\big(\zeta_i(\{w_k\}_{k \neq i} \mid \{\bar{w}_k\}_{k \neq i}), \{w_k\}_{k \neq i} \mid \{\bar{w}_k\}_{k \neq i}\big) = 0,$$

for all $\{w_k\}_{k \neq i}$. As a result, $\zeta_i(\{w_k\}_{k \neq i} \mid \{\bar{w}_k\}_{k \neq i})$ has a closed-form expression given by

$$\zeta_i(\{\boldsymbol{w}_k\}_{k\neq i} \mid \{\bar{\boldsymbol{w}}_k\}_{k\neq i}) = \frac{\gamma_i(\{\bar{\boldsymbol{w}}_k\}_{k\neq i})}{\sigma_i^2 + \sum_{j\neq i} \dfrac{\boldsymbol{w}_j^H \mathbf{Q}_{ji} \boldsymbol{w}_j}{1 + \bar{\boldsymbol{w}}_j^H \mathbf{Q}_{ji} \bar{\boldsymbol{w}}_j \xi_i(\{\bar{\boldsymbol{w}}_k\}_{k\neq i})}} \quad (8.106)$$

where

$$\begin{aligned}
\gamma_i(\{\bar{\boldsymbol{w}}_k\}_{k\neq i}) &\triangleq \sum_{j\neq i} \frac{\bar{\boldsymbol{w}}_j^H \mathbf{Q}_{ji} \bar{\boldsymbol{w}}_j \cdot \xi_i(\{\bar{\boldsymbol{w}}_k\}_{k\neq i})}{1 + \bar{\boldsymbol{w}}_j^H \mathbf{Q}_{ji} \bar{\boldsymbol{w}}_j \cdot \xi_i(\{\bar{\boldsymbol{w}}_k\}_{k\neq i})} - \ln \rho_i \\
&\quad - \sum_{j\neq i} \ln\left(1 + \bar{\boldsymbol{w}}_j^H \mathbf{Q}_{ji} \bar{\boldsymbol{w}}_j \cdot \xi_i(\{\bar{\boldsymbol{w}}_k\}_{k\neq i})\right) \\
&= \sum_{j\neq i} \frac{\bar{\boldsymbol{w}}_j^H \mathbf{Q}_{ji} \bar{\boldsymbol{w}}_j \cdot \xi_i(\{\bar{\boldsymbol{w}}_k\}_{k\neq i})}{1 + \bar{\boldsymbol{w}}_j^H \mathbf{Q}_{ji} \bar{\boldsymbol{w}}_j \cdot \xi_i(\{\bar{\boldsymbol{w}}_k\}_{k\neq i})} + \sigma_i^2 \xi_i(\{\bar{\boldsymbol{w}}_k\}_{k\neq i}) \quad \text{(by (8.100))} \\
&> 0, \ \forall \{\bar{\boldsymbol{w}}_k\}_{k\neq i}. \quad (8.107)
\end{aligned}$$

From (8.105) and (8.106), one can see that $\zeta_i(\{\boldsymbol{w}_k\}_{k\neq i} \mid \{\bar{\boldsymbol{w}}_k\}_{k\neq i})$ is a locally tight lower bound of $\xi_i(\{\boldsymbol{w}_k\}_{k\neq i})$ because $\Phi_i(x, \{\boldsymbol{w}_k\}_{k\neq i})$ is strictly increasing in x, implying that the following function

$$\widetilde{\mathcal{U}}_{wsr}(\{\boldsymbol{w}_k\} \mid \{\bar{\boldsymbol{w}}_k\}) \triangleq \sum_{i=1}^K \alpha_i \log_2(1 + \zeta_i(\{\boldsymbol{w}_k\}_{k\neq i} \mid \{\bar{\boldsymbol{w}}_k\}_{k\neq i}) \cdot \boldsymbol{w}_i^H \mathbf{Q}_{ii} \boldsymbol{w}_i) \quad (8.108)$$

is also a locally tight lower bound of the WSR utility $\mathcal{U}_{wsr}(\{\boldsymbol{w}_k\})$. By defining

$$\bar{\mathbf{Q}}_{ii}(\{\bar{\boldsymbol{w}}_k\}_{k\neq i}) \triangleq \gamma_i(\{\bar{\boldsymbol{w}}_k\}_{k\neq i}) \cdot \mathbf{Q}_{ii}, \quad (8.109a)$$

$$\bar{\mathbf{Q}}_{ji}(\{\bar{\boldsymbol{w}}_k\}_{k\neq i}) \triangleq \left(1 + \bar{\boldsymbol{w}}_j^H \mathbf{Q}_{ji} \bar{\boldsymbol{w}}_j \xi_i(\{\bar{\boldsymbol{w}}_k\}_{k\neq i})\right)^{-1} \mathbf{Q}_{ji}, \quad (8.109b)$$

for $i, j = 1, \ldots, K$, $j \neq i$, and by (8.106) one can further express $\widetilde{\mathcal{U}}_{wsr}(\cdot \mid \cdot)$ as

$$\widetilde{\mathcal{U}}_{wsr}(\{\boldsymbol{w}_k\} \mid \{\bar{\boldsymbol{w}}_k\}) = \sum_{i=1}^K -\alpha_i \log_2\left(1 + \frac{\boldsymbol{w}_i^H \bar{\mathbf{Q}}_{ii} \boldsymbol{w}_i}{\sigma_i^2 + \sum_{j\neq i} \boldsymbol{w}_j^H \bar{\mathbf{Q}}_{ji} \boldsymbol{w}_j}\right)^{-1}, \quad (8.110)$$

where $\bar{\mathbf{Q}}_{ji}(\{\bar{\boldsymbol{w}}_k\}_{k\neq i})$ are denoted by $\bar{\mathbf{Q}}_{ji}$ for all $i, j = 1, \ldots, K$, for notational simplicity. Furthermore, the relationship between LMMSE and SINR given by (4.42) and (4.43) with

$$\sigma_s^2 = 1, \ \sigma_n^2 = \sigma_i^2 + \sum_{j\neq i} \boldsymbol{w}_j^H \bar{\mathbf{Q}}_{ji} \boldsymbol{w}_j, \ \boldsymbol{h} = \bar{\mathbf{Q}}_{ii}^{1/2} \boldsymbol{w}_i \quad (8.111)$$

can be applied to the reformulation of (8.110) as

$$\begin{aligned}
&\widetilde{\mathcal{U}}_{wsr}(\{\boldsymbol{w}_k\} \mid \{\bar{\boldsymbol{w}}_k\}) \\
&= \sum_{i=1}^K -\alpha_i \log_2\left(\min_{\boldsymbol{y}_i \in \mathbb{C}^{N_t}} \left|1 - \boldsymbol{y}_i^H \bar{\mathbf{Q}}_{ii}^{1/2} \boldsymbol{w}_i\right|^2 + \left(\sigma_i^2 + \sum_{j\neq i} \boldsymbol{w}_j^H \bar{\mathbf{Q}}_{ji} \boldsymbol{w}_j\right) \boldsymbol{y}_i^H \boldsymbol{y}_i\right). \quad (8.112)
\end{aligned}$$

Then, by the first-order approximation of the convex function $-\ln(x) \geq -\ln \bar{x} + (1 - x/\bar{x})$, we further obtain the following locally tight lower bound of $\tilde{\mathcal{U}}_{wsr}(\cdot \mid \cdot)$:

$$
\begin{aligned}
\bar{\mathcal{U}}_{wsr}(\{\boldsymbol{w}_k\} \mid \{\bar{\boldsymbol{w}}_k\}) &\triangleq \\
& \sum_{i=1}^{K} -\alpha_i \log_2 \left(\left| 1 - \bar{\boldsymbol{y}}_i^H \bar{\mathbf{Q}}_{ii}^{1/2} \bar{\boldsymbol{w}}_i \right|^2 + \left(\sigma_i^2 + \sum_{j \neq i} \bar{\boldsymbol{w}}_j^H \bar{\mathbf{Q}}_{ji} \bar{\boldsymbol{w}}_j \right) \bar{\boldsymbol{y}}_i^H \bar{\boldsymbol{y}}_i \right) \\
& + \frac{\alpha_i}{\ln 2} \left(1 - \frac{|1 - \bar{\boldsymbol{y}}_i^H \bar{\mathbf{Q}}_{ii}^{1/2} \boldsymbol{w}_i|^2 + (\sigma_i^2 + \sum_{j \neq i} \boldsymbol{w}_j^H \bar{\mathbf{Q}}_{ji} \boldsymbol{w}_j) \bar{\boldsymbol{y}}_i^H \bar{\boldsymbol{y}}_i}{|1 - \bar{\boldsymbol{y}}_i^H \bar{\mathbf{Q}}_{ii}^{1/2} \bar{\boldsymbol{w}}_i|^2 + (\sigma_i^2 + \sum_{j \neq i} \bar{\boldsymbol{w}}_j^H \bar{\mathbf{Q}}_{ji} \bar{\boldsymbol{w}}_j) \bar{\boldsymbol{y}}_i^H \bar{\boldsymbol{y}}_i} \right),
\end{aligned} \tag{8.113}
$$

where

$$
\begin{aligned}
\bar{\boldsymbol{y}}_i &= \arg \min_{\boldsymbol{y}_i \in \mathbb{C}^{N_t}} \left| 1 - \boldsymbol{y}_i^H \bar{\mathbf{Q}}_{ii}^{1/2} \bar{\boldsymbol{w}}_i \right|^2 + \left(\sigma_i^2 + \sum_{j \neq i} \bar{\boldsymbol{w}}_j^H \bar{\mathbf{Q}}_{ji} \bar{\boldsymbol{w}}_j \right) \boldsymbol{y}_i^H \boldsymbol{y}_i \\
&= \frac{\bar{\mathbf{Q}}_{ii}^{1/2} \bar{\boldsymbol{w}}_i}{\sigma_i^2 + \sum_{j=1}^{K} \bar{\boldsymbol{w}}_j^H \bar{\mathbf{Q}}_{ji} \bar{\boldsymbol{w}}_j}, \quad i = 1, \ldots, K \quad \text{(by (4.42) and (8.111)).}
\end{aligned}
$$

Thus we have already shown that $\bar{\mathcal{U}}_{wsr}(\{\boldsymbol{w}_k\} \mid \{\bar{\boldsymbol{w}}_k\})$ defined by (8.113) is a tight lower bound of $\mathcal{U}_{wsr}(\{\boldsymbol{w}_k\})$ that satisfies the convergence conditions of the SUM method [RHL13][1] (cf. Remark 4.11); moreover, $\bar{\mathcal{U}}_{wsr}(\{\boldsymbol{w}_k\} \mid \{\bar{\boldsymbol{w}}_k\})$ can be equivalently (with all the constant terms removed and the indices i and j interchanged in the double summation term) re-expressed as

$$
\bar{\mathcal{U}}_{wsr}(\{\boldsymbol{w}_k\} \mid \{\bar{\boldsymbol{w}}_k\}) = \sum_{i=1}^{K} -\bar{\mathcal{U}}^{(i)}(\boldsymbol{w}_i \mid \{\bar{\boldsymbol{w}}_k\}) \tag{8.114}
$$

where

$$
\bar{\mathcal{U}}^{(i)}(\boldsymbol{w}_i \mid \{\bar{\boldsymbol{w}}_k\}) = \eta_i |1 - \bar{\boldsymbol{y}}_i^H \bar{\mathbf{Q}}_{ii}^{1/2} \boldsymbol{w}_i|^2 + \sum_{j \neq i} \eta_j \left(\boldsymbol{w}_i^H \bar{\mathbf{Q}}_{ij} \boldsymbol{w}_i \right) \bar{\boldsymbol{y}}_j^H \bar{\boldsymbol{y}}_j
$$

$$
\eta_j \triangleq \frac{\alpha_j}{\ln 2} \left[|1 - \bar{\boldsymbol{y}}_j^H \bar{\mathbf{Q}}_{jj}^{1/2} \bar{\boldsymbol{w}}_j|^2 + \left(\sigma_j^2 + \sum_{k \neq j} \bar{\boldsymbol{w}}_k^H \bar{\mathbf{Q}}_{kj} \bar{\boldsymbol{w}}_k \right) \bar{\boldsymbol{y}}_j^H \bar{\boldsymbol{y}}_j \right]^{-1}.
$$

Note that $\bar{\mathcal{U}}^{(i)}(\boldsymbol{w}_i \mid \{\bar{\boldsymbol{w}}_k\})$ depends only on $\boldsymbol{w}_i$, implying that the decision variables $\boldsymbol{w}_1, \ldots, \boldsymbol{w}_K$ are fully decoupled in $\bar{\mathcal{U}}_{wsr}(\{\boldsymbol{w}_k\} \mid \{\bar{\boldsymbol{w}}_k\})$, and hence maximizing $\bar{\mathcal{U}}_{wsr}(\{\boldsymbol{w}_k\} \mid \{\bar{\boldsymbol{w}}_k\})$ subject to $\{\boldsymbol{w}_k \mid \|\boldsymbol{w}_k\|_2^2 \leq P_k\}, k = 1, \ldots, K$ can be decomposed into K convex subproblems of QCQP:

$$
\min_{\|\boldsymbol{w}_i\|_2^2 \leq P_i} \bar{\mathcal{U}}^{(i)}(\boldsymbol{w}_i \mid \{\bar{\boldsymbol{w}}_k\}), \quad i = 1, \ldots, K. \tag{8.115}
$$

As problem (8.115) is simply to minimize a convex quadratic function under a 2-norm constraint, it can be solved very efficiently by the projected gradient

---

[1] The convergence conditions of the SUM method are almost the same as those of the BSUM method except that the approximation function $\bar{\mathcal{U}}_{wsr}(\cdot \mid \cdot)$ need not have a unique maximizer.

Algorithm 8.3 SUM method (parallel algorithm) for handling problem (8.98) with WSR utility

---

1: **Input** a set of beamformers $\{\bar{w}_i\}$ satisfying (8.98b);
2: **repeat**
3:   Update $\{\bar{w}_i\}$ by solving problem (8.115) for $i = 1, \ldots, K$ in parallel;
4: **until** Some convergence criterion is met;
5: **Output** $\{\bar{w}_i\}$ as an approximate solution of problem (8.98).

---

method or the Lagrange dual method (cf. (9.151)). The resulting parallel algorithm is summarized in Algorithm 8.3. As a final remark, the problem size of (8.115) does not increase with K either.

Finally, let us demonstrate the efficacy and efficiency of Algorithm 8.2 and Algorithm 8.3 by comparing with the decentralized version of the SCA algorithm (referred to as the DSCA algorithm) since their performances are comparable. The simulation setting is the same as that in the previous subsection.

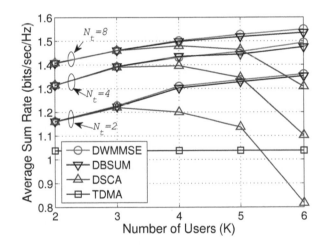

Figure 8.13 Performance comparison for Algorithm 8.2 (DBSUM), Algorithm 8.3 (DWMMSE), and the DSCA algorithm, where $1/\sigma^2 = 10$ dB, $\eta = 0.5$, $\alpha_1 = \cdots = \alpha_K = 1$, rank$(\mathbf{Q}_{ki}) = N_t$ for all k, i. (© 2015 IEEE. Reprinted, with permission, from W.-C. Li, T.-H. Chang, and C.-Y. Chi, "Multicell Coordinated Beamforming with Rate Outage Constraint–Part II: Efficient Approximation Algorithms," Jun. 2015.)

It can be observed from Figure 8.13 that Algorithm 8.2 (denoted as DBSUM) and Algorithm 8.3 (denoted as DWMMSE) yield nearly the same average sum rates, which increase with the number of users and the number of transmit antennas, and they outperform the DSCA algorithm when the number of users increases. The reason for this might be that the DSCA algorithm is relatively easier to get trapped in some local maximum when $K \geq 4$, because it involves SDR and thus Gaussian randomization is needed for obtaining rank-1 solutions for $\boldsymbol{w}_i$. The curve denoted by TDMA represents the achieved sum rate by time division multiple access. Apparently, allowing all the users to access the spectrum simultaneously attains higher spectral efficiency than TDMA even when only CDI of each user is available to the associated transmitter; moreover, the performance gain also increases with the number of users.

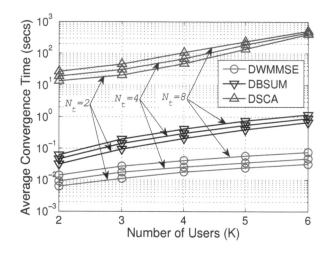

Figure 8.14 Complexity comparison for Algorithm 8.2 (DBSUM), Algorithm 8.3 (DWMMSE), and the DSCA algorithm, where $1/\sigma^2 = 10$ dB, $\eta = 0.5$, $\alpha_1 = \cdots = \alpha_K = 1$, $\mathrm{rank}(\mathbf{Q}_{ki}) = N_t$ for all k, i. (© 2015 IEEE. Reprinted, with permission, from W.-C. Li, T.-H. Chang, and C.-Y. Chi, "Multicell Coordinated Beamforming with Rate Outage Constraint–Part II: Efficient Approximation Algorithms," Jun. 2015.)

The average computation times per problem instance (in seconds) for Algorithm 8.2, Algorithm 8.3 (without parallel processing applied), and the DSCA algorithm are shown in Figure 8.14. In the simulations, the convex subproblems involved in Algorithm 8.2, Algorithm 8.3, and the DSCA algorithm are solved by the projected gradient method, the Lagrange dual method, and CVX, respectively. It is observed that the average computation time of Algorithm 8.2 is $10^2 \sim 10^3$ faster than the DSCA algorithm, and Algorithm 8.3 is about ten times faster than Algorithm 8.2. These simulation results have justified the efficiency of Algorithm 8.2 and Algorithm 8.3 much more superiorly to the DSCA algorithm.

8.5.8   Outage constrained robust transmit beamforming: Single-cell MISO scenario

Consider a MISO system, wherein each channel error $\mathbf{e}_i \in \mathbb{C}^{N_t}$ is modeled as an i.i.d. random vector with Gaussian distribution $\mathcal{CN}(\mathbf{0}, \mathbf{C}_i)$ (i.e., zero mean and covariance matrix $\mathbf{C}_i$). Equivalently, the complex Gaussian CSI error $\mathbf{e}_i \in \mathbb{C}^{N_t}$

can be expressed as

$$\mathbf{e}_i = \mathbf{C}_i^{1/2}\mathbf{v}_i, \ i = 1,\ldots,K, \tag{8.116}$$

where $\mathbf{C}_i^{1/2} \succeq \mathbf{0}$ is the PSD square root of $\mathbf{C}_i$, and $\mathbf{v}_i \in \mathbb{C}^{N_t}$ is a normalized random vector with Gaussian distribution, i.e., $\mathbf{v}_i \sim \mathcal{CN}(\mathbf{0},\mathbf{I}_{N_t})$.

Let $\rho_i \in (0,1]$ denote the maximum tolerable SINR outage probability of receiver i, i.e., SINR satisfaction probability is equal to $1 - \rho_i$. The following chance constrained robust transmit beamforming design problem [SD08] has been a cutting-edge research problem under extensive study:

$$\min \ \sum_{k=1}^K \|\mathbf{w}_k\|_2^2 \tag{8.117a}$$

$$\text{s.t. Prob}\left\{ \frac{|(\bar{\mathbf{h}}_i + \mathbf{e}_i)^H\mathbf{w}_i|^2}{\sum_{k\neq i}^K |(\bar{\mathbf{h}}_i + \mathbf{e}_i)^H\mathbf{w}_k|^2 + \sigma_i^2} \geq \gamma_i \right\} \geq 1 - \rho_i, \ i = 1,\ldots,K. \tag{8.117b}$$

This problem tries to find the most power efficient design under the constraint on SINR satisfaction probability. Solving problem (8.117) is a challenging task because each probabilistic SINR constraint in (8.117b) is nonconvex without any tractable closed-form expression. To solve this problem, this chance constraint must be reformulated into a tractable deterministic constraint in terms of $\mathbf{C}_i$ and unknown variables.

By applying SDR, the probabilistic constraints in (8.117b) can be expressed as

$$\text{Prob}\left\{\mathbf{v}_i^H\mathbf{Q}_i(\mathbf{W}_1,\ldots,\mathbf{W}_K)\mathbf{v}_i + 2\text{Re}\left\{\mathbf{v}_i^H\mathbf{u}_i(\mathbf{W}_1,\ldots,\mathbf{W}_K)\right\} \geq c_i(\mathbf{W}_1,\ldots,\mathbf{W}_K)\right\}$$
$$> 1 - \rho_i \tag{8.118}$$

for $i = 1,\ldots,K$, where

$$\mathbf{Q}_i(\mathbf{W}_1,\ldots,\mathbf{W}_K) \triangleq \mathbf{C}_i^{1/2}\left(\frac{1}{\gamma_i}\mathbf{W}_i - \sum_{k\neq i}\mathbf{W}_k\right)\mathbf{C}_i^{1/2}, \tag{8.119a}$$

$$\mathbf{u}_i(\mathbf{W}_1,\ldots,\mathbf{W}_K) \triangleq \mathbf{C}_i^{1/2}\left(\frac{1}{\gamma_i}\mathbf{W}_i - \sum_{k\neq i}\mathbf{W}_k\right)\bar{\mathbf{h}}_i, \tag{8.119b}$$

$$c_i(\mathbf{W}_1,\ldots,\mathbf{W}_K) \triangleq \sigma_i^2 - \bar{\mathbf{h}}_i^H\left(\frac{1}{\gamma_i}\mathbf{W}_i - \sum_{k\neq i}\mathbf{W}_k\right)\bar{\mathbf{h}}_i, \tag{8.119c}$$

which all are affine functions of $\mathbf{W}_i$. Note that the argument of the probability function in (8.118) is a quadratic function of random vector $\mathbf{v}_i \sim \mathcal{CN}(\mathbf{0},\mathbf{I}_{N_t})$, in spite of a linear function of the non-random matrices $\mathbf{W}_i$. Nevertheless, the probabilistic constraint (8.118) is still intangible because it is almost prohibitive to obtain a tractable or analytical closed-form expression for the probability function in (8.118).

Instead, the constraint (8.118) can be handled in a conservative manner (cf. Remark 4.12). The idea is to find another probability constraint that is sufficient for (8.118) to hold true, and more importantly is convex. Then replace the probability constraint in problem (8.117) by this probability constraint. The obtained solutions (supposing that it is of rank one) can be thought of as conservative solutions to problem (8.117), meaning that the consequent probability SINR satisfaction probability will be larger and may be much larger than the specified value of $1 - \rho_i$, and meanwhile the total transmit power will also be higher than the optimal transmit power of problem (8.117).

Two conservative approaches are introduced in the subsequent two subsections, respectively, that can give rise to a conservative convex constraint approximation to (8.118), meaning that if the resulting convex constraint holds true, then the constraint (8.118) holds true.

A. Sphere bounding approach

The sphere bounding approach is based on the following simple lemma [WSC⁺14]:

Lemma 8.1 *Let $\mathbf{v} \in \mathbb{C}^{N_t}$ be a continuous random vector following certain statistical distribution and let $G(\mathbf{v}) : \mathbb{C}^{N_t} \to \mathbb{R}$ be a function of $\mathbf{v}$. Let $r > 0$ be the radius of the ball $\{\mathbf{v} \mid \|\mathbf{v}\|_2^2 \le r^2\}$ such that $\mathrm{Prob}\{\|\mathbf{v}\|_2^2 \le r^2\} \ge 1 - \rho$ where $\rho \in (0, 1]$. Then*

$$G(\mathbf{v}) \ge 0 \quad \forall \|\mathbf{v}\|_2^2 \le r^2 \tag{8.120}$$

implies $\mathrm{Prob}\,\{G(\mathbf{v}) \ge 0\} \ge 1 - \rho$.

Let

$$\begin{aligned} G(\mathbf{v}_i) =& \mathbf{v}_i^H \mathbf{Q}_i(\mathbf{W}_1, \dots, \mathbf{W}_K)\mathbf{v}_i + 2\mathrm{Re}\{\mathbf{v}_i^H \mathbf{u}_i(\mathbf{W}_1, \dots, \mathbf{W}_K)\} \\ & - c_i(\mathbf{W}_1, \dots, \mathbf{W}_K). \end{aligned} \tag{8.121}$$

By applying Lemma 8.1 to (8.118), we can see that (8.118) is satisfied whenever

$$G(\mathbf{v}_i) \ge 0, \quad \forall \|\mathbf{v}_i\|_2^2 \le r_i^2, \tag{8.122}$$

where $r_i > 0$ is the radius of the ball $\{\mathbf{v}_i \mid \|\mathbf{v}_i\|_2^2 \le r_i^2\}$ such that $\mathrm{Prob}\{\|\mathbf{v}_i\|_2^2 \le r_i^2\} \ge 1 - \rho_i$. However, for the case of $\mathbf{v}_i \sim \mathcal{CN}(\mathbf{0}, \mathbf{I}_{N_t})$, $2\|\mathbf{v}_i\|_2^2$ is a chi-square random variable with $2N_t$ degrees of freedom. Let $\mathrm{ICDF}(\cdot)$ be the inverse cumulative distribution function of $2\|\mathbf{v}_i\|_2^2$. Then it can be shown that the choice of r_i given by

$$r_i = \sqrt{\frac{\mathrm{ICDF}(1 - \rho_i)}{2}} \tag{8.123}$$

is sufficient to guarantee $\mathrm{Prob}\{\|\mathbf{v}_i\|_2^2 \le r_i^2\} \ge 1 - \rho_i$ for $i = 1, \dots, K$. Thus, it is true, by Lemma 8.1, that $\mathrm{Prob}\,\{G(\mathbf{v}_i) \ge 0\} \ge 1 - \rho_i$, i.e., the constraint (8.118) is true.

However, the constraint (8.122) has infinitely many constraints, and is difficult to solve. As discussed in Subsection 8.5.4, we can apply the S-procedure to equivalently reformulate the constraint (8.122) into a PSD constraint and a linear constraint:

$$\begin{bmatrix} \mathbf{Q}_i(\mathbf{W}_1,\ldots,\mathbf{W}_K) + \lambda_i \mathbf{I}_{N_t} & \mathbf{u}_i(\mathbf{W}_1,\ldots,\mathbf{W}_K) \\ \mathbf{u}_i(\mathbf{W}_1,\ldots,\mathbf{W}_K)^H & -c_i(\mathbf{W}_1,\ldots,\mathbf{W}_K) - \lambda_i r_i^2 \end{bmatrix} \succeq \mathbf{0}, \qquad (8.124a)$$

$$\lambda_i \geq 0. \qquad (8.124b)$$

As a result, through the preceding SDR, followed by conservative sphere bounding approximation to (8.118), we end up with the following convex problem (as an approximation to the original problem (8.117))

$$\min \sum_{k=1}^{K} \mathrm{Tr}(\mathbf{W}_k)$$

$$\text{s.t.} \begin{bmatrix} \mathbf{Q}_i(\mathbf{W}_1,\ldots,\mathbf{W}_K) + \lambda_i \mathbf{I}_{N_t} & \mathbf{u}_i(\mathbf{W}_1,\ldots,\mathbf{W}_K) \\ \mathbf{u}_i(\mathbf{W}_1,\ldots,\mathbf{W}_K)^H & -c_i(\mathbf{W}_1,\ldots,\mathbf{W}_K) - \lambda_i r_i^2 \end{bmatrix} \succeq \mathbf{0}, \ i = 1,\ldots,K$$

$$\lambda_i \geq 0, \ \mathbf{W}_i \succeq \mathbf{0}, \ i = 1,\ldots,K$$

$$(8.125)$$

which is an SDP and can be efficiently solved. It is not difficult to see that if the optimal $\{\mathbf{W}_i^\star\}_{i=1}^{K}$ is of rank one, i.e., $\mathbf{W}_i^\star = \mathbf{w}_i^\star(\mathbf{w}_i^\star)^H$, it is also a conservative solution to the original problem (8.117). Otherwise rank-one approximation is needed to find a rank-one approximate solution.

B. Bernstein-type inequality approach

Now we introduce another conservative probability inequality, which is computationally feasible, to replace the intractable probability constraint (8.118). Surely the truth of the former must be a sufficient condition for the truth of the latter. This conservative probability inequality is based on the following lemma [Bec09] (that provides a probability inequality involving a quadratic function of Gaussian random vector as in (8.118)).

Lemma 8.2 Let $G = \mathbf{v}^H \mathbf{Q} \mathbf{v} + 2\mathrm{Re}\{\mathbf{v}^H \mathbf{u}\}$ where $\mathbf{Q} \in \mathbb{H}^{N_t}$, $\mathbf{u} \in \mathbb{C}^{N_t}$, and $\mathbf{v} \sim \mathcal{CN}(\mathbf{0}, \mathbf{I}_{N_t})$. Then for any $\delta > 0$, we have

$$\mathrm{Prob}\left\{ G \geq \mathrm{Tr}(\mathbf{Q}) - \sqrt{2\delta}\sqrt{\|\mathbf{Q}\|_{\mathrm{F}}^2 + 2\|\mathbf{u}\|_2^2} - \delta s^+(\mathbf{Q}) \right\} \geq 1 - e^{-\delta}, \qquad (8.126)$$

where $s^+(\mathbf{Q}) = \max\{\lambda_{\max}(-\mathbf{Q}), 0\}$ in which $\lambda_{\max}(-\mathbf{Q})$ denotes the maximum eigenvalue of matrix $-\mathbf{Q}$.

Let $\delta \triangleq -\ln(\rho)$ where $\rho \in (0, 1]$. Lemma 8.2 implies that the inequality

$$\mathrm{Prob}\left\{ \mathbf{v}^H \mathbf{Q} \mathbf{v} + 2\mathrm{Re}\{\mathbf{v}^H \mathbf{u}\} \geq c \right\} \geq 1 - \rho \qquad (8.127)$$

holds true if the following inequality is satisfied

$$\text{Tr}(\mathbf{Q}) - \sqrt{2\delta}\sqrt{\|\mathbf{Q}\|_F^2 + 2\|\mathbf{u}\|_2^2} - \delta s^+(\mathbf{Q}) \geq c. \tag{8.128}$$

Equation (8.128) thus serves as a conservative formulation for (8.127). Now, a key observation is that (8.128) can be represented by

$$\text{Tr}(\mathbf{Q}) - \sqrt{2\delta}x - \delta y \geq c, \tag{8.129a}$$

$$\sqrt{\|\mathbf{Q}\|_F^2 + 2\|\mathbf{u}\|_2^2} \leq x, \tag{8.129b}$$

$$y\mathbf{I}_{N_t} + \mathbf{Q} \succeq \mathbf{0}, \tag{8.129c}$$

$$y \geq 0, \tag{8.129d}$$

where $x, y \in \mathbb{R}$ are auxiliary variables. By defining

$$\delta_i \triangleq -\ln \rho_i, \quad i = 1, \ldots, K, \tag{8.130}$$

and applying (8.129) to (8.118), we obtain the following problem

$$\min \sum_{k=1}^{K} \text{Tr}(\mathbf{W}_k) \tag{8.131a}$$

$$\text{s.t. } \text{Tr}(\mathbf{Q}_i(\mathbf{W}_1, \ldots, \mathbf{W}_K)) - \sqrt{2\delta_i}x_i - \delta_i y_i \geq c_i(\mathbf{W}_1, \ldots, \mathbf{W}_K), \quad i = 1, \ldots, K, \tag{8.131b}$$

$$\left\| \begin{bmatrix} \text{vec}(\mathbf{Q}_i(\mathbf{W}_1, \ldots, \mathbf{W}_K)) \\ \sqrt{2}\mathbf{u}_i(\mathbf{W}_1, \ldots, \mathbf{W}_K) \end{bmatrix} \right\|_2 \leq x_i, \quad i = 1, \ldots, K, \tag{8.131c}$$

$$y_i\mathbf{I}_{N_t} + \mathbf{Q}_i(\mathbf{W}_1, \ldots, \mathbf{W}_K) \succeq \mathbf{0}, \quad i = 1, \ldots, K, \tag{8.131d}$$

$$y_i \geq 0, \quad \mathbf{W}_i \succeq \mathbf{0}, \quad i = 1, \ldots, K. \tag{8.131e}$$

Note that the constraints in (8.131b), (8.131c), and (8.131d) are, respectively, a linear inequality constraint, a convex SOC constraint, and a convex PSD constraint. Hence problem (8.131) is a convex problem and can be efficiently solved.

As the convex problem (8.125), problem (8.131) is also obtained through SDR. If the optimal solution $\{\mathbf{W}_i^\star\}_{i=1}^K$ is not of rank one, again one has to find a rank-one approximation solution that provides a conservative solution to the original problem (8.117). However, as reported in [WSC+14], the optimal solution of problem (8.131) turns out to be less conservative than that of problem (8.125). Finding other less conservative and implementation-efficient solutions to the original problem (8.117) still remains to be a challenging research in chance constrained robust transmit beamforming.

Let us show some simulation results to demonstrate the performance of the approximation formulations in (8.125) (called Method I) and (8.131) (called Method II) for handling the chance constrained robust transmit beamforming design problem in (8.117). In the simulations, $\sigma_1^2 = \cdots = \sigma_K^2 = 0.1$ (identical users' noise power), $\rho_1 = \cdots = \rho_K = \rho$ (identical users' outage probability requirement), and $\gamma_1 = \cdots = \gamma_K \triangleq \gamma$ (identical users' QoS requirement). Five

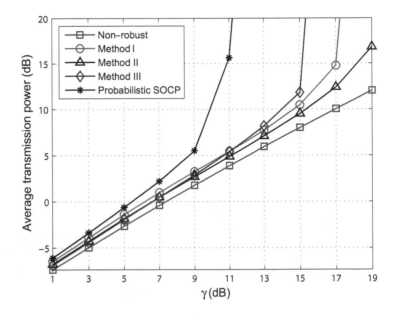

Figure 8.15 Transmit power performance of the various methods. $N_t = K = 3$; $\rho = 0.1$; spatially i.i.d. Gaussian CSI errors with $\sigma_e^2 = 0.002$. (© 2014 IEEE. Reprinted, with permission, from K.-Y. Wang, A. M.-C. So, T.-H. Chang, W.-K. Ma, and C.-Y. Chi, "Outage Constrained Robust Transmit Optimization for Multiuser MISO Downlinks: Tractable Approximations by Conic Optimization," Nov. 2014.)

hundred realizations of the presumed channels $\{\bar{\mathbf{h}}_i\}_{i=1}^K$ are generated in the simulations. In each simulation trial, the presumed channels $\{\bar{\mathbf{h}}_i\}_{i=1}^K$ are randomly and independently generated according to the standard circularly symmetric complex Gaussian distribution $\mathcal{CN}(\mathbf{0}, \mathbf{I}_{N_t})$. The CSI errors are spatially i.i.d. complex Gaussian with $\mathbf{C}_1 = \cdots = \mathbf{C}_K = \sigma_e^2 \mathbf{I}_{N_t}$, where the error variance $\sigma_e^2 = 0.002$.

In addition to Methods I and II, we also evaluate the performance of Method III proposed in [WSC⁺14] (which is developed using a decomposition-based large derivation inequality) and the existing probabilistic SOCP method proposed in [SD08] because both of them are also approximation methods for handling problem (8.117). Moreover, the conventional perfect-CSI-based SINR constrained design (7.40), called the "non-robust method" for convenience, is also tested as if the presumed channels $\{\bar{\mathbf{h}}_i\}_{i=1}^K$ were perfect CSI. All the involved convex formulations in the simulations are solved by the conic optimization solver SeDuMi, implemented through the parser software CVX.

Figure 8.15 shows some results of the average transmit powers of the various methods under test, where each result is obtained from those realizations of $\{\bar{\mathbf{h}}_i\}_{i=1}^K$ (181 out of 500 realizations) for which all the methods yield feasible solutions at $\gamma = 11$ dB. As can be seen from this figure, Method II yields the

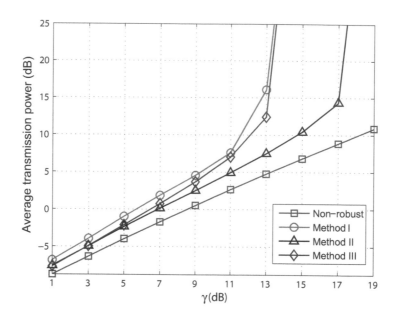

Figure 8.16 Transmit power performance under spatially correlated Gaussian CSI errors. $N_t = 8$; $K = 6$; $\rho = 0.01$; $\sigma_e^2 = 0.01$. (© 2014 IEEE. Reprinted, with permission, from K.-Y. Wang, A. M.-C. So, T.-H. Chang, W.-K. Ma, and C.-Y. Chi, "Outage Constrained Robust Transmit Optimization for Multiuser MISO Downlinks: Tractable Approximations by Conic Optimization," Nov. 2014.)

best average transmit power performance, followed by Methods I and III (with Method I exhibiting noticeably better performance for $\gamma > 15$ dB), while the probabilistic SOCP method performs worst. As expected, the non-robust method yields the least transmit power, also indicating how much more transmit power would be needed for the robust methods to accommodate the target outage probability. It can be observed that for $\gamma \leq 11$ dB, the transmit power difference between any one of Methods I–III and the non-robust method is within 1.5 dB, while the transmit power gap between the probabilistic SOCP and the non-robust method for $\gamma > 11$ dB is larger and increases much faster than Methods I–III. These results also illustrate that imperfect CSI effects are more difficult to cope with when users' target SINRs are too high.

Next, we show some simulation results for the case of $\mathbf{C}_1 = \cdots = \mathbf{C}_K = \mathbf{C}_e$, where

$$[\mathbf{C}_e]_{m,n} = \sigma_e^2 \times 0.9^{|m-n|}$$

with $\sigma_e^2 = 0.01$, and the other parameters used are $N_t = 8$, $K = 6$, and $\rho = 0.01$. However, the preceding probabilistic SOCP method is not applicable to this case because of formidable computational load. Through the same simulation procedure associated with those results shown in Figure 8.16 except that $\gamma = 13$

dB is used as the pick-up point of feasible channel realizations of all the methods, the corresponding results are shown in Figure 8.16. Again, it can be seen from this figure that Method II demonstrates the superior performance over the other two robust designs, although Method III performs slightly better than Method I for this case. All the obtained optimal solutions $\mathbf{W}^\star$ of Methods I–III in the simulation results shown in Figures 8.15 and 8.16 are of rank one.

8.5.9   Outage constrained robust transmit beamforming: Multicell MISO scenario

Let us revisit the multicell example in Subsection 7.6.4, where the scenario is shown in Fig. 7.5. In contrast to the assumption of the perfect CSI $\mathbf{h}_{FF}$, $\mathbf{h}_{FM}$, and $\mathbf{h}_{MF}$ made in the conventional formulation in (7.64), two cases based on the available channel information at the FBS are considered in [WJDC13]:

1. *No CSI feedback:* The FBS has no instantaneous channel estimate sent back from the FUE, and the FBS knows only the statistical information of the channels:

$$\begin{cases} \mathbf{h}_{FF} \sim \mathcal{CN}(\mathbf{0}, \mathbf{C}_{h,FF}) \\ \mathbf{h}_{FM} \sim \mathcal{CN}(\mathbf{0}, \mathbf{C}_{h,FM}) \\ \mathbf{h}_{MF} \sim \mathcal{CN}(\mathbf{0}, \mathbf{C}_{h,MF}) \end{cases} \tag{8.132}$$

where channel covariance matrices $\mathbf{C}_{h,FF}$, $\mathbf{C}_{h,FM}$, and $\mathbf{C}_{h,MF}$ are positive definite. This model is more advisable for a fast fading system, where the feedback channel is unable to provide a reliable estimate of the current CSI, and thus no instantaneous CSI estimate is fed back to FBS.

2. *Partial CSI feedback:* The FBS receives estimated CSI $\mathbf{h}_{FF}$ and $\mathbf{h}_{FM}$ from FUE, but knows only the statistical information of $\mathbf{h}_{MF} \sim \mathcal{CN}(\mathbf{0}, \mathbf{C}_{h,MF})$ (due to the lack of feedback link from MUE). Specifically, the true $\mathbf{h}_{FF}$ and $\mathbf{h}_{FM}$ are modeled as

$$\mathbf{h}_{FF} = \widehat{\mathbf{h}}_{FF} + \mathbf{e}_{FF} \quad \text{and} \quad \mathbf{h}_{FM} = \widehat{\mathbf{h}}_{FM} + \mathbf{e}_{FM}, \tag{8.133}$$

where $\widehat{\mathbf{h}}_{FF} \in \mathbb{C}^{N_F}$ and $\widehat{\mathbf{h}}_{FM} \in \mathbb{C}^{N_M}$ are the given channel estimates of $\mathbf{h}_{FF}$ and $\mathbf{h}_{FM}$, respectively, and $\mathbf{e}_{FF} \in \mathbb{C}^{N_F}$ and $\mathbf{e}_{FM} \in \mathbb{C}^{N_M}$ denote the corresponding estimation error vectors. Assume that

$$\mathbf{e}_{FF} \sim \mathcal{CN}(\mathbf{0}, \mathbf{C}_{e,FF}) \quad \text{and} \quad \mathbf{e}_{FM} \sim \mathcal{CN}(\mathbf{0}, \mathbf{C}_{e,FM}). \tag{8.134}$$

The model given by (8.133) is suitable for slow fading channels.

Typically, femtocells are connected to a macrocell network via a wired broadband backhaul link such as a digital subscriber line (DSL), and thus the beamforming vector at the MBS, $\mathbf{w}_M$, is assumed to be perfectly known to the FBS.

In the presence of CSI uncertainty, the multicell coordinated transmit beam-forming problem presented in Subsection 7.6.4

$$\min_{\mathbf{w}_F \in \mathbb{C}^{N_F}} \|\mathbf{w}_F\|_2^2 \qquad \text{(cf. (7.64))}$$

$$\text{s.t.} \quad \frac{\left|\mathbf{h}_{FF}^H \mathbf{w}_F\right|^2}{\left|\mathbf{h}_{FM}^H \mathbf{w}_M\right|^2 + \sigma_F^2} \geq \gamma_F$$

$$\left|\mathbf{h}_{MF}^H \mathbf{w}_F\right|^2 \leq \epsilon_M$$

has been solved by reformulating it into a SOCP, provided that the CSI is per-fectly known to the FBS. However, the beamforming solution to problem (7.64) will no longer guarantee the QoS requirement in (7.64) universally. To mitigate this QoS outage, the following chance constrained robust FBS transmit beam-forming design problem [WJDC13] has been a cutting-edge research problem under extensive study:

$$\min_{\mathbf{w}_F \in \mathbb{C}^{N_F}} \|\mathbf{w}_F\|_2^2 \tag{8.135a}$$

$$\text{s.t. Prob}\left\{\frac{\left|\mathbf{h}_{FF}^H \mathbf{w}_F\right|^2}{\left|\mathbf{h}_{FM}^H \mathbf{w}_M\right|^2 + \sigma_F^2} \geq \gamma_F\right\} \geq 1 - \rho_F, \tag{8.135b}$$

$$\text{Prob}\left\{|\mathbf{h}_{MF}^H \mathbf{w}_F|^2 \leq \epsilon_M\right\} \geq 1 - \rho_M, \tag{8.135c}$$

where ρ_F and ρ_M denote the preset maximum tolerable outage probabilities for SINR and interference power constraints, respectively. This problem tries to find the most power efficient design for femtocell under the constraints on the associated QoS satisfaction probabilities of $1 - \rho_F$ and $1 - \rho_M$. Problem (8.135) is almost intractable since constraints (8.135b) and (8.135c) may not be convex and have no closed-form expressions in general. Next, we discuss how to solve problem (8.135) for the *No CSI feedback* and *Partial CSI feedback* cases, respectively.

A. No CSI feedback scenario

For the probability function in (8.135b), one can observe that the random variables $|\mathbf{h}_{FF}^H \mathbf{w}_F|^2$ and $|\mathbf{h}_{FM}^H \mathbf{w}_M|^2$ are independently exponential distributed with parameters $1/(\mathbf{w}_F^H \mathbf{C}_{h,FF} \mathbf{w}_F)$ and $1/(\mathbf{w}_M^H \mathbf{C}_{h,FM} \mathbf{w}_M)$, respectively, and the closed-form expression has been shown in [KB02] to be

$$\exp\left(\frac{-\gamma_F \sigma_F^2}{\mathbf{w}_F^H \mathbf{C}_{h,FF} \mathbf{w}_F}\right) \frac{\mathbf{w}_F^H \mathbf{C}_{h,FF} \mathbf{w}_F}{\mathbf{w}_F^H \mathbf{C}_{h,FF} \mathbf{w}_F + \gamma_F \mathbf{w}_M^H \mathbf{C}_{h,FM} \mathbf{w}_M} \geq 1 - \rho_F. \tag{8.136}$$

The probability function in (8.135c) is in fact a cumulative distribution function of an exponential random variable with parameter $1/(\mathbf{w}_F^H \mathbf{C}_{h,MF} \mathbf{w}_F)$. Therefore, constraint (8.135c) can be represented as

$$\mathbf{w}_F^H \mathbf{C}_{h,MF} \mathbf{w}_F \leq \frac{\epsilon_M}{\ln(1/\rho_M)}. \tag{8.137}$$

As a result, problem (8.135) can be equivalently expressed as

$$\min_{\mathbf{w}_F \in \mathbb{C}^{N_F}} \ \|\mathbf{w}_F\|_2^2 \tag{8.138a}$$

$$\text{s.t.} \ \exp\left(\frac{-\gamma_F \sigma_F^2}{\mathbf{w}_F^H \mathbf{C}_{h,FF} \mathbf{w}_F}\right) \frac{\mathbf{w}_F^H \mathbf{C}_{h,FF} \mathbf{w}_F}{\mathbf{w}_F^H \mathbf{C}_{h,FF} \mathbf{w}_F + \gamma_F \mathbf{w}_M^H \mathbf{C}_{h,FM} \mathbf{w}_M} \geq 1 - \rho_F, \tag{8.138b}$$

$$\mathbf{w}_F^H \mathbf{C}_{h,MF} \mathbf{w}_F \leq \epsilon_M / \ln(1/\rho_M). \tag{8.138c}$$

Although problem (8.138) is still not convex due to nonconvex constraint (8.138b), the problem has a more tractable form than problem (8.135).

By applying SDR to problem (8.138), the relaxed problem can be expressed as

$$\min_{\mathbf{W}_F \in \mathbb{H}^{N_F}} \ \text{Tr}(\mathbf{W}_F) \tag{8.139a}$$

$$\text{s.t.} \ \text{Tr}(\mathbf{C}_{h,FF} \mathbf{W}_F) \geq (1 - \rho_F) \left\{ \text{Tr}(\mathbf{C}_{h,FF} \mathbf{W}_F) \exp\left(\frac{\gamma_F \sigma_F^2}{\text{Tr}(\mathbf{C}_{h,FF} \mathbf{W}_F)}\right) \right.$$

$$\left. + \gamma_F \mathbf{w}_M^H \mathbf{C}_{h,FM} \mathbf{w}_M \exp\left(\frac{\gamma_F \sigma_F^2}{\text{Tr}(\mathbf{C}_{h,FF} \mathbf{W}_F)}\right) \right\}, \tag{8.139b}$$

$$\text{Tr}(\mathbf{C}_{h,MF} \mathbf{W}_F) \leq \epsilon_M / \ln(1/\rho_M), \tag{8.139c}$$

$$\mathbf{W}_F \succeq \mathbf{0}. \tag{8.139d}$$

The relaxed optimization problem in (8.139) is now convex (by the second-order condition of convexity), and can be efficiently solved using off-the-shelf optimization software.

If the obtained solution of problem (8.139), denoted by $\mathbf{W}_F^\star$, is of rank one, then we can simply perform the rank-one decomposition $\mathbf{W}_F^\star = \mathbf{w}_F^\star (\mathbf{w}_F^\star)^H$, and output $\mathbf{w}_F^\star$ as the optimal beamforming solution to problem (8.135), or equivalently problem (8.138). If $\mathbf{W}_F^\star$ has rank higher than one, one can apply some rank-one approximation procedure [LMS+10] to obtain a feasible beamforming solution for problem (8.138).

B. Partial CSI feedback scenario

Let us discuss the transmit power minimization problem for the partial CSI feedback case that the FBS acquires imperfect channel estimates of $\mathbf{h}_{FF}$ and $\mathbf{h}_{FM}$ modeled by (8.133), but knows only the statistical information of $\mathbf{h}_{MF} \sim \mathcal{CN}(\mathbf{0}, \mathbf{C}_{h,MF})$ (due to the lack of feedback link from MUE). For this case, problem (8.135) can be written as

$$\min_{\mathbf{w}_F \in \mathbb{C}^{N_F}} \ \|\mathbf{w}_F\|_2^2 \tag{8.140a}$$

$$\text{s.t.} \ \text{Prob}\left\{\frac{|(\widehat{\mathbf{h}}_{FF} + \mathbf{e}_{FF})^H \mathbf{w}_F|^2}{|(\widehat{\mathbf{h}}_{FM} + \mathbf{e}_{FM})^H \mathbf{w}_M|^2 + \sigma_F^2} \geq \gamma_F\right\} \geq 1 - \rho_F, \tag{8.140b}$$

$$\text{Prob}\left\{|\mathbf{h}_{MF}^H \mathbf{w}_F|^2 \leq \epsilon_M\right\} \geq 1 - \rho_M. \tag{8.140c}$$

Again, problem (8.140) is difficult to solve since the probability function in (8.140b) has no closed-form expression and may not be convex in general. To proceed, the Bernstein-type inequality approach presented in Subsection 8.5.8 will be used to handle problem (8.140) again.

Let us express

$$\mathbf{e}_{FF} = \mathbf{C}_{e,FF}^{1/2}\mathbf{v}_{FF} \ \ \text{and} \ \ \mathbf{e}_{FM} = \mathbf{C}_{e,FM}^{1/2}\mathbf{v}_{FM}, \tag{8.141}$$

where $\mathbf{C}_{e,FF}^{1/2} \succeq \mathbf{0}$ and $\mathbf{C}_{e,FM}^{1/2} \succeq \mathbf{0}$ are the positive semidefinite square roots of $\mathbf{C}_{e,FF}$ and $\mathbf{C}_{e,FM}$, respectively, and $\mathbf{v}_{FF} \sim \mathcal{CN}(\mathbf{0},\mathbf{I}_{N_F})$ and $\mathbf{v}_{FM} \sim \mathcal{CN}(\mathbf{0},\mathbf{I}_{N_M})$. Let

$$\widehat{\mathbf{h}} = \begin{bmatrix} \widehat{\mathbf{h}}_{FF} \\ \widehat{\mathbf{h}}_{FM} \end{bmatrix}, \ \ \mathbf{C}^{1/2} = \begin{bmatrix} \mathbf{C}_{e,FF}^{1/2} & \mathbf{0} \\ \mathbf{0} & \mathbf{C}_{e,FM}^{1/2} \end{bmatrix}, \ \ \mathbf{v} = \begin{bmatrix} \mathbf{v}_{FF} \\ \mathbf{v}_{FM} \end{bmatrix}, \tag{8.142}$$

where $\mathbf{v} \sim \mathcal{CN}(\mathbf{0},\mathbf{I}_{N_F+N_M})$ since $\mathbf{v}_{FF}$ and $\mathbf{v}_{FM}$ are statistically independent. Applying SDR to problem (8.140) yields

$$\min_{\mathbf{W}_F \in \mathbb{H}^{N_F}} \ \text{Tr}(\mathbf{W}_F) \tag{8.143a}$$

$$\text{s.t.} \ \text{Prob}\big\{\mathbf{v}^H\boldsymbol{\Phi}(\mathbf{W}_F)\mathbf{v}+2\text{Re}\big\{\mathbf{v}^H\boldsymbol{\eta}(\mathbf{W}_F)\big\}\geq s(\mathbf{W}_F)\big\} \geq 1-\rho_F, \tag{8.143b}$$

$$\text{Tr}(\mathbf{C}_{h,MF}\mathbf{W}_F) \leq \frac{\epsilon_M}{\ln(1/\rho_M)}, \tag{8.143c}$$

$$\mathbf{W}_F \succeq \mathbf{0}, \tag{8.143d}$$

where

$$\begin{cases} \boldsymbol{\Phi}(\mathbf{W}_F) \triangleq \mathbf{C}^{1/2}\mathbf{W}\mathbf{C}^{1/2} \\ \boldsymbol{\eta}(\mathbf{W}_F) \triangleq \mathbf{C}^{1/2}\mathbf{W}\widehat{\mathbf{h}} \\ s(\mathbf{W}_F) \triangleq \sigma_F^2 - \widehat{\mathbf{h}}^H\mathbf{W}\widehat{\mathbf{h}} \end{cases} \tag{8.144}$$

(all affine functions in $\mathbf{W}_F$) in which

$$\mathbf{W} \triangleq \begin{bmatrix} \frac{1}{\gamma_F}\mathbf{W}_F & \mathbf{0} \\ \mathbf{0} & -\mathbf{w}_M\mathbf{w}_M^H \end{bmatrix}. \tag{8.145}$$

However, problem (8.143) is still intractable since the probability function in (8.143b) does not have a closed-form expression in general due to indefinite $\mathbf{W}$. By Lemma 8.2 and following the discussions in Subsection 8.5.8, the constraint in the form (8.143b) can be conservatively approximated by tractable convex constraints as follows:

$$\text{Tr}\left(\boldsymbol{\Phi}(\mathbf{W}_F)\right) - \sqrt{2\delta}x - \delta y \geq s(\mathbf{W}_F), \tag{8.146a}$$

$$\sqrt{\|\boldsymbol{\Phi}(\mathbf{W}_F)\|_{\text{F}}^2 + 2\|\boldsymbol{\eta}(\mathbf{W}_F)\|_2^2} \leq x, \tag{8.146b}$$

$$y\mathbf{I}_{N_F+N_M} + \boldsymbol{\Phi}(\mathbf{W}_F) \succeq \mathbf{0}, \tag{8.146c}$$

$$y \geq 0, \tag{8.146d}$$

where $\delta \triangleq -\ln(\rho_F)$. Specifically, the constraints in (8.146) can be represented as

$$\frac{1}{\gamma_F}\mathrm{Tr}\left(\left(\mathbf{C}_{e,FF}+\widehat{\mathbf{h}}_{FF}\widehat{\mathbf{h}}_{FF}^H\right)\mathbf{W}_F\right)-\sqrt{2\delta}x-\delta y$$
$$\geq \sigma_F^2+\mathbf{w}_M^H(\mathbf{C}_{e,FM}+\widehat{\mathbf{h}}_{FM}\widehat{\mathbf{h}}_{FM}^H)\mathbf{w}_M, \quad (8.147\mathrm{a})$$

$$\frac{1}{\gamma_F}\left\|\begin{bmatrix}\mathbf{vec}\left(\mathbf{C}_{e,FF}^{1/2}\mathbf{W}_F\mathbf{C}_{e,FF}^{1/2}\right)\\\sqrt{2}\mathbf{vec}\left(\mathbf{C}_{e,FF}^{1/2}\mathbf{W}_F\widehat{\mathbf{h}}_{FF}\right)\\\xi_{FM}\end{bmatrix}\right\|_2 \leq x, \quad (8.147\mathrm{b})$$

$$y\mathbf{I}_{N_F+N_M}+\begin{bmatrix}\frac{1}{\gamma_F}\mathbf{C}_{e,FF}^{1/2}\mathbf{W}_F\mathbf{C}_{e,FF}^{1/2} & \mathbf{0}\\\mathbf{0} & -\mathbf{C}_{e,FM}^{1/2}\mathbf{w}_M\mathbf{w}_M^H\mathbf{C}_{e,FM}^{1/2}\end{bmatrix}\succeq\mathbf{0}, \quad (8.147\mathrm{c})$$

$$y\geq 0, \quad (8.147\mathrm{d})$$

which is a convex constraint set w.r.t. $(\mathbf{W}_F,x,y)$, and the constant ξ_{FM} in (8.147b) is defined as

$$\xi_{FM}\triangleq\gamma_F\sqrt{\left\|\mathbf{C}_{e,FM}^{1/2}\mathbf{w}_M\mathbf{w}_M^H\mathbf{C}_{e,FM}^{1/2}\right\|_F^2+2\left\|\mathbf{C}_{e,FM}^{1/2}\mathbf{w}_M\mathbf{w}_M^H\widehat{\mathbf{h}}_{FM}\right\|_2^2}. \quad (8.148)$$

To minimize the transmit power $\mathrm{Tr}(\mathbf{W}_F)$, one can infer from (8.147) that y must be the principal eigenvalue of $\mathbf{C}_{e,FM}^{1/2}\mathbf{w}_M\mathbf{w}_M^H\mathbf{C}_{e,FM}^{1/2}$, i.e.,

$$y=\left\|\mathbf{C}_{e,FM}^{1/2}\mathbf{w}_M\right\|_2^2,$$

when the optimal $\mathbf{W}_F$ is achieved. As a result, by (8.143) and (8.147) with y replaced by $\|\mathbf{C}_{e,FM}^{1/2}\mathbf{w}_M\|_2^2$, a tractable approximation to problem (8.140) is given by

$$\min_{\substack{\mathbf{W}_F\in\mathbb{H}^{N_F},\\x\in\mathbb{R}}}\ \mathrm{Tr}(\mathbf{W}_F) \quad (8.149\mathrm{a})$$

$$\text{s.t.}\ \frac{1}{\gamma_F}\mathrm{Tr}\left(\left(\mathbf{C}_{e,FF}+\widehat{\mathbf{h}}_{FF}\widehat{\mathbf{h}}_{FF}^H\right)\mathbf{W}_F\right)-\sqrt{2\delta}x$$
$$\geq \sigma_F^2+\mathbf{w}_M^H\left((1+\delta)\mathbf{C}_{e,FM}+\widehat{\mathbf{h}}_{FM}\widehat{\mathbf{h}}_{FM}^H\right)\mathbf{w}_M, \quad (8.149\mathrm{b})$$

$$\frac{1}{\gamma_F}\left\|\begin{bmatrix}\mathbf{vec}(\mathbf{C}_{e,FF}^{1/2}\mathbf{W}_F\mathbf{C}_{e,FF}^{1/2})\\\sqrt{2}\mathbf{vec}(\mathbf{C}_{e,FF}^{1/2}\mathbf{W}_F\widehat{\mathbf{h}}_{FF})\\\xi_{FM}\end{bmatrix}\right\|_2 \leq x, \quad (8.149\mathrm{c})$$

$$\mathrm{Tr}(\mathbf{C}_{h,MF}\mathbf{W}_F)\leq \epsilon_M/\ln(1/\rho_M), \quad (8.149\mathrm{d})$$

$$\mathbf{W}_F\succeq\mathbf{0}, \quad (8.149\mathrm{e})$$

where $\delta\triangleq-\ln(\rho_F)$ and ξ_{FM} is defined in (8.148). Problem (8.149) is convex, and can be efficiently solved to yield a global optimal $\mathbf{W}_F$. If the obtained solution $\mathbf{W}_F^\star$ is not of rank one, then the rank-one approximation procedure [LMS+10] can be applied to obtain a feasible (conservative) beamforming solution to problem (8.140), where "conservative" is due to that the constraint set of problem (8.149) is a subset of the constraint set of problem (8.143).

Let us show some simulation results to justify the efficacy of the robust beamformers, respectively, designed by solving (8.139) (for the No CSI case), and (8.149) (for the Partial CSI case), and the efficacy of the non-robust design by using (7.65) (for the Naive CSI case). In the simulation, we consider FBS and MBS each equipped with four transmit antennas, i.e., $N_F = N_M = 4$, with tolerable outage probabilities $\rho_F = \rho_M = 0.1$, noise variance $\sigma_F^2 = 0.01$, and maximum interference power $\epsilon_M = -3$ dB. The beamforming vector $\mathbf{w}_M$ at MBS is randomly generated uniformly on the unit-norm sphere $\|\mathbf{w}_M\|_2 = 1$.

Let the channel covariance matrices in (8.136) be

$$\mathbf{C}_{h,FF} = \sigma_{h,FF}^2 \mathbf{C}_h, \ \mathbf{C}_{h,MF} = \sigma_{h,MF}^2 \mathbf{C}_h, \ \mathbf{C}_{h,FM} = \sigma_{h,FM}^2 \mathbf{C}_h,$$
$$[\mathbf{C}_h]_{m,n} = \varrho^{|m-n|},$$

where $\sigma_{h,FF}^2 = 1$ and $\sigma_{h,MF}^2 = \sigma_{h,FM}^2 = 0.01$, and ϱ denotes the channel correlation coefficient. On the other hand, let the channel error covariance matrices in (8.134) be

$$\mathbf{C}_{e,FF} = \sigma_e^2 \mathbf{I}_{N_F}, \ \mathbf{C}_{e,FM} = \sigma_e^2 \mathbf{I}_{N_M},$$

respectively, where $\sigma_e^2 = 0.002$. For the Partial CSI and Naive CSI cases, presumed CSI realizations are generated following $\widehat{\mathbf{h}}_{FF} \sim \mathcal{CN}(\mathbf{0}, \mathbf{C}_{h,FF} - \sigma_e^2 \mathbf{I}_{N_F})$, $\widehat{\mathbf{h}}_{MF} \sim \mathcal{CN}(\mathbf{0}, \mathbf{C}_{h,MF} - \sigma_e^2 \mathbf{I}_{N_F})$, and $\widehat{\mathbf{h}}_{FM} \sim \mathcal{CN}(\mathbf{0}, \mathbf{C}_{h,FM} - \sigma_e^2 \mathbf{I}_{N_M})$. The associated optimal beamforming vectors $\mathbf{w}_F$ are then obtained by solving (8.149) for the former, and (7.65) for the latter (for which the presumed CSI is treated as if they were true channel vectors), respectively. As for the No CSI case, the optimum beamforming vector is obtained by solving (8.139) (with no need of any channel estimates).

Figures 8.17(a) and 8.17(b) show the obtained average transmit power for $\varrho = 0.9$ and $\varrho = 0.01$, respectively, where each result for both cases of Partial CSI and Naive CSI is obtained over 500 independent CSI realizations. One can observe from these two figures that the transmit power (in dB) of each design linearly increases with γ_F (in dB). The transmit power performance for the Naive CSI case (non-robust design) is the best, indicating that more power must be consumed for each robust design. The transmit power performance for the Partial CSI case is better than that for the No CSI case. The performance gap between the No CSI case and the Naive CSI case (around 6 dB for $\varrho = 0.9$ and 15 dB for $\varrho = 0.01$) is much larger for smaller ϱ, while that between the Partial CSI case and the Naive CSI case (around 3 dB for $\varrho = 0.9$ and 4 dB for $\varrho = 0.01$) increases mildly. This also indicates that the more CSI information is available, the less transmit power is needed. In all the simulation instances, the solutions to the SDR-based problems (8.139) and (8.149) are always of rank one.

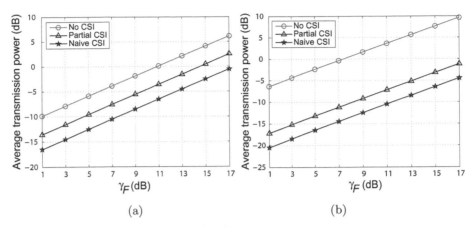

Figure 8.17 Average transmit power for (a) $\varrho = 0.9$ and (b) $\varrho = 0.01$ obtained by solving problem (8.139) for the No CSI case, problem (8.149) for the Partial CSI case, and problem (7.65) for Naive CSI case. (© 2013 IEEE. Reprinted, with permission, from K.-Y. Wang, N. Jacklin, Z. Ding, and C.-Y. Chi, "Robust MISO Transmit Optimization Under Outage-Based QoS Constraints in Two-Tier Heterogeneous Networks," Apr. 2013.)

In contrast to the two robust beamforming designs which guarantee the outage probability requirements, the non-robust design is more power efficient but it seriously fails to meet the outage probability requirements. Figure 8.18 including the top plot for $\varrho = 0.9$ and the bottom plot for $\varrho = 0.01$ shows the corresponding distribution (i.e., histogram) of achievable SINR of FUE (i.e., the value of SINR_F in (7.63)) for target SINR $\gamma_F = 15$ dB, where the results for the Partial CSI case and the Naive CSI case are obtained by adding 10^5 randomly generated realizations of the CSI errors ($\mathbf{e}_{FF} \sim \mathcal{CN}(\mathbf{0}, \sigma_e^2 \mathbf{I}_{N_F})$ and $\mathbf{e}_{FM} \sim \mathcal{CN}(\mathbf{0}, \sigma_e^2 \mathbf{I}_{N_M})$) to a single presumed CSI $\widehat{\mathbf{h}}_{FF} \sim \mathcal{CN}(\mathbf{0}, \mathbf{C}_{h,FF} - \sigma_e^2 \mathbf{I}_{N_F})$ and $\widehat{\mathbf{h}}_{FM} \sim \mathcal{CN}(\mathbf{0}, \mathbf{C}_{h,FM} - \sigma_e^2 \mathbf{I}_{N_M})$; those for the No CSI case are obtained over 10^5 randomly generated realizations of true channels $\mathbf{h}_{FF} \sim \mathcal{CN}(\mathbf{0}, \mathbf{C}_{h,FF})$, $\mathbf{h}_{MF} \sim \mathcal{CN}(\mathbf{0}, \mathbf{C}_{h,MF})$, and $\mathbf{h}_{FM} \sim \mathcal{CN}(\mathbf{0}, \mathbf{C}_{h,FM})$. One can see that both of the robust beamforming designs can meet the 10% outage probability requirement, while the non-robust beamforming design can only achieve 55% outage probability, seriously violating the outage probability requirement of 10%. This corroborates that the conventional non-robust beamforming design is quite sensitive to CSI errors. One can also see, from Figures 8.17 and 8.18, that the performance of the robust beamforming design for the Partial CSI case can be further improved (i.e., lower transmit power) through the bisection method because the achievable SINR outage probability is far below the required 10%. The bisection method is to iteratively reset $\rho_F = \rho_M$ to a value larger than the target value of 0.1, such that the resulting SINR outage probability increases but maintains within 10%, at the expense of increasing the complexity of the beamforming algorithm for the partial CSI case.

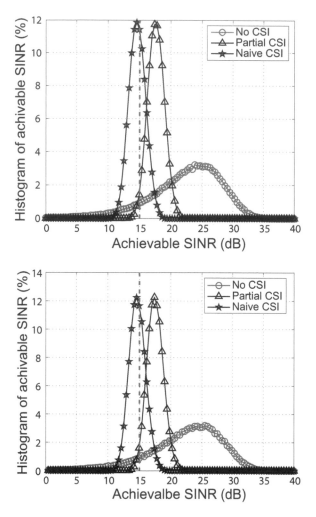

Figure 8.18 Distributions of achievable SINR values of FUE at $\gamma_F = 15$ dB and $\rho_F = \rho_M = 10\%$ for $\varrho = 0.9$ (top plot) and $\varrho = 0.01$ (bottom plot), indicating that the SINR outage probabilities of the No CSI case, the Partial CSI case, and the Naive CSI case are 10%, 1.24%, and 55.1%, respectively, for $\varrho = 0.9$, and 10%, 1.28%, and 54.5%, respectively, for $\varrho = 0.01$.

8.6   Summary and discussion

In this chapter we have presented the concept of SDP and their applications in combinatorial optimization and transmit beamforming. We have explained SDP approximations through SDR to solve BQP, specifically for coherent and noncoherent ML detection. Then we presented various transmit beamforming designs in MIMO wireless communications, where SDR (together with the S-procedure and convex constraint approximations in some design examples) is employed to convert the nonconvex problem of interest into an SDP, but without detailed performance analysis about the obtained suboptimal solution via the use of KKT conditions (e.g., its optimality to the original problem under some practical scenarios). However, almost all of SDP involved problems do not have close-form solutions and thus we may need to pragmatically employ off-the-shelf convex solvers such as CVX and SeDuMi instead, for obtaining optimal numerical solutions (without consideration for the computational efficiency for the time being). Moreover, the analysis about the rank-one solutions of the SDR problems relies on the associated KKT conditions to be introduced in the next chapter. We would like to mention that SDP has been regarded as a powerful optimization tool with lots of available complexity analysis [BTN01, Ye97, HRVW96, NN94].

On the other hand, the BSUM method (that may or may not involve SDP) also shows its strength for obtaining a stationary-point solution of nonconvex problems under some certain conditions that can be taken into account during the algorithm design stage, and thus the prospective complexity of the designed algorithm may well be under control in advance, though the solution accuracy of the obtained stationary-point solution compared to the unknown optimal solution still needs to be analyzed.

Overall, in Chapters 5 to 8, we have introduced various convex optimization problems and their equivalent forms or representations, and approximations, that are essential in reformulation of a given problem into a problem that can be handled to obtain a reasonable solution. Many examples and applications in signal processing and communications have been illustrated via problem reformulations, from the problem definition, formulation, approximation via SDR, to the resulting convex optimization problem. Surely, we merely presented some application examples for illustrating the effectiveness of convex optimization, and we believe that more successful and exciting applications can be found or continue to emerge not only in the signal processing and communications but also in other science and engineering areas [PE10].

References for Chapter 8

[Bec09]   I. Bechar, "A Bernstein-type inequality for stochastic processes of quadratic forms of Gaussian variables," *arXiv preprint arXiv:0909.3595*, 2009. [Online]. Available: http://arxiv.org/abs/0909.3595.

[BM01]   T. Bonald and L. Massoulié, "Impact of fairness on internet performance," in *Proc. ACM SIGMETRICS*, Cambridge, MA, June 16–20, 2001, pp. 82–91.

[BPC⁺10]   S. Boyd, N. Parikh, E. Chu, B. Peleato, and J. Eckstein, "Distributed optimization and statistical learning via the alternating direction method of multipliers," *Foundations and Trends in Machine Learning*, vol. 3, no. 1, pp. 1–122, 2010.

[BT89]   D. P. Bertsekas and J. N. Tsitsiklis, *Parallel and Distributed Computation: Numerical Methods*. Upper Saddle River, NJ: Prentice-Hall, 1989.

[BTN01]   A. Ben-Tal and A. Nemirovski, *Lectures on Modern Convex Optimization: Analysis, Algorithms, and Engineering Applications*. Philadelphia, PA, USA: MPSSIAM Series on Optimization, 2001.

[CC62]   A. Charnes and W. W. Cooper, "Programming with linear fractional functionals," *Naval Research Logistics Quarterly*, vol. 9, no. 3–4, pp. 181–186, Sept–Dec. 1962.

[CHMC10]   T.-H. Chang, C.-W. Hsin, W.-K. Ma, and C.-Y. Chi, "A linear fractional semidefinite relaxation approach to maximum-likelihood detection of higher order QAM OSTBC in unknown channels," *IEEE Trans. Signal Process.*, vol. 58, no. 4, pp. 2315–2326, Apr. 2010.

[DY10]   H. Dahrouj and W. Yu, "Coordinated beamforming for the multicell multi-antenna wireless system," *IEEE Trans. Wireless Commun.*, vol. 9, no. 5, pp. 1748–1759, May 2010.

[GS01]   G. Ganasan and P. Stoica, "Space-time block codes: A maximum SNR approach," *IEEE Trans. Inf. Theory*, vol. 47, no. 4, pp. 1650–1656, May 2001.

[GSS+10]   A. B. Gershman, N. D. Sidiropoulos, S. Shahbazpanahi, M. Bengtsson, and B. Otter-sten, "Convex optimization-based beamforming," *IEEE Signal Process. Mag.*, vol. 27, no. 3, pp. 62–75, May 2010.

[HRVW96]   C. Helmberg, F. Rendl, R. Vanderbei, and H. Wolkowicz, "An interior point method for semidefinite programming," *SIAM J. Optim.*, vol. 6, no. 2, pp. 342–361, 1996.

[KB02]   S. Kandukuri and S. Boyd, "Optimal power control in interference-limited fading wire-less channels with outage-probability specifications," *IEEE Trans. Wireless Commun.*, vol. 1, pp. 46–55, Jan. 2002.

[LCC15a]   W.-C. Li, T.-H. Chang, and C.-Y. Chi, "Multicell coordinated beamforming with rate outage constraint–Part I: Complexity analysis," *IEEE Trans. Signal Process.*, vol. 63, no. 11, pp. 2749–2762, June 2015.

[LCC15b]   ——, "Multicell coordinated beamforming with rate outage constraint–Part II: Efficient approximation algorithms," *IEEE Trans. Signal Process.*, vol. 63, no. 11, pp. 2763–2778, June 2015.

[LCLC11]   W.-C. Li, T.-H. Chang, C. Lin, and C.-Y. Chi, "A convex approximation approach to weighted sum rate maximization of multiuser MISO interference channel under outage constraints," in *Proc. 2011 IEEE ICASSP*, Prague, Czech, May 22–27, 2011, pp. 3368–3371.

[LCLC13]   ——, "Coordinated beamforming for multiuser MISO interference channel under rate outage constraints," *IEEE Trans. Signal Process.*, vol. 61, no. 5, pp. 1087–1103, Mar. 2013.

[LCMC11]   W.-C. Liao, T.-H. Chang, W.-K. Ma, and C.-Y. Chi, "QoS-based transmit beamforming in the presence of eavesdroppers: An optimized artificial-noise-aided approach," *IEEE Trans. Signal Process.*, vol. 59, no. 3, pp. 1202–1216, Mar. 2011.

[LMS+10]   Z.-Q. Luo, W.-K. Ma, A. M.-C. So, Y. Ye, and S. Zhang, "Semidefinite relaxation of quadratic optimization problems," *IEEE Signal Process. Mag.*, vol. 27, no. 3, pp. 20–34, May 2010.

[LSL03]   E. G. Larsson, P. Stoica, and J. Li, "Orthogonal space-time block codes: Maximum like-lihood detection for unknown channels and unstructured interferences," *IEEE Trans. Signal Process.*, vol. 51, no. 2, pp. 362–372, Feb. 2003.

[NN94]   Y. Nesterov and A. Nemirovskii, *Interior-Point Polynomial Algorithms in Convex Programming*. Philadelphia, US: SIAM, 1994.

[PE10]   D. P. Palomar and Y. C. Eldar, *Convex Optimization in Signal Processing and Communications*. Cambridge, UK: Cambridge University Press, 2010.

[RHL13]   M. Razaviyayn, M. Hong, and Z.-Q. Luo, "A unified convergence analysis of block successive minimization methods for nonsmooth optimization," *SIAM J. Optimization*, vol. 23, no. 2, pp. 1126–1153, 2013.

[SCW+12]   C. Shen, T.-H. Chang, K.-Y. Wang, Z. Qiu, and C.-Y. Chi, "Distributed robust multi-cell coordinated beamforming with imperfect CSI: An ADMM approach," *IEEE Trans. Signal Process.*, vol. 60, no. 6, pp. 2988–3003, June 2012.

[SD08]   M. B. Shenouda and T. N. Davidson, "Probabilistically-constrained approaches to the design of the multiple antenna downlink," in *Proc. 42nd Asilomar Conference*, Pacific Grove, Oct. 26–29, 2008, pp. 1120–1124.

[SDL06]   N. D. Sidiropoulos, T. N. Davidson, and Z.-Q. Luo, "Transmit beamforming for physical layer multicasting," *IEEE Trans. Signal Process.*, vol. 54, no. 6, pp. 2239–2251, June 2006.

[SGM05]   S. Shahbazpanahi, A. Gershman, and J. Manton, "Closed-form blind mimo channel estimation for orthogonal space-time block codes," *IEEE Trans. Signal Process.*, vol. 53, no. 12, pp. 4506–4517, Dec. 2005.

[Stu99]   J. F. Sturm, "Using SeDuMi 1.02, a Matlab toolbox for optimization over symmetric cones," *Optimization Methods and Software*, vol. 11, no. 1–4, pp. 625–653, 1999.

[TJC99]   V. Tarokh, H. Jafarkhani, and A. R. Calderbank, "Space-time block codes from orthogonal designs," *IEEE Trans. Inf. Theory*, vol. 45, no. 5, pp. 1456–1467, Jul. 1999.

[TPW11]   A. Tajer, N. Prasad, and X.-D. Wang, "Robust linear precoder design for multi-cell downlink transmission," *IEEE Trans. Signal Process.*, vol. 59, no. 1, pp. 235–251, Jan. 2011.

[WJDC13]   K.-Y. Wang, N. Jacklin, Z. Ding, and C.-Y. Chi, "Robust MISO transmit optimization under outage-based QoS constraints in two-tier heterogeneous networks," *IEEE Trans. Wireless Commun.*, vol. 12, no. 4, pp. 1883–1897, Apr. 2013.

[WSC+14]   K.-Y. Wang, A. M.-C. So, T.-H. Chang, W.-K. Ma, and C.-Y. Chi, "Outage constrained robust transmit optimization for multiuser MISO downlinks: Tractable approximations by conic optimization," *IEEE Trans. Signal Process.*, vol. 62, no. 21, pp. 5690–5705, Nov. 2014.

[Ye97]   Y. Ye, *Interior Point Algorithms: Theory and Analysis*. New York, US: John Wiley & Sons, 1997.

[ZWN08]   G. Zheng, K.-K. Wong, and T.-S. Ng, "Robust linear MIMO in the downlink: A worst-case optimization with ellipsoidal uncertainty regions," *EURASIP Journal on Advances in Signal Processing*, vol. 2008, pp. 1–15, June 2008.

9 Duality

In this chapter we will introduce the concept of duality, a perspective of the primal (i.e., the original) minimization problem from its dual maximization problem and vice versa. Like the analysis of a time-domain signal from its frequency-domain representation, the duality plays a vital role in understanding and solving an optimization problem.

The Slater's condition for *strong duality* (i.e., the same optimal value for the primal problem and the dual problem) and KKT conditions are also elaborated for the convex optimization problems with ordinary inequalities. For convex problems with the objective function and constraint functions being differentiable, the KKT conditions and the first-order optimality condition can be shown equivalent, whereas the former can yield both primal and dual optimal solution simultaneously but the latter can only yield the primal optimal solution. The S-procedure (or called S-lemma) introduced in Chapter 8 can be efficiently proven true via strong duality of a convex problem with strong duality.

Two alternative methods for solving a convex problem with strong duality will be introduced in this chapter. One is to solve the dual maximization problem to get the optimal dual solution first, and then obtain the optimal solution of the primal problem via the associated KKT conditions and the obtained optimal dual solution. The other, so-called Lagrange dual optimization or dual decomposition, is to iteratively obtain the primal solution and the dual solution, particularly useful for the solution in distributed fashion. This optimization method will be introduced in detail together with the well-known alternating direction method of multipliers (which is also a dual decomposition method) widely used in wireless communications.

Then the duality for the optimization problems with strong duality and generalized inequalities (defined by proper cones) is introduced. Many illustrative examples are provided to show the usefulness of duality and KKT conditions in solving various convex optimization problems, whereas their utilization in developing interior-point methods for finding numerical solutions of convex problems will be addressed in the next chapter.

Finally, we introduce alternatives for systems composed of equalities and inequalities based on either of strong duality and weak duality of advisable optimization problems, and conclude the chapter with the proof of the S-procedure via the use of theorems of alternatives.

9.1   Lagrange dual function and conjugate function

Consider a standard (not necessarily convex) optimization problem as defined in (4.1)

$$p^* = \min \ f_0(\mathbf{x})$$
$$\text{s.t. } f_i(\mathbf{x}) \leq 0, \ i = 1,\ldots,m \qquad (9.1)$$
$$h_i(\mathbf{x}) = 0, \ i = 1,\ldots,p$$

with the problem domain

$$\mathcal{D} = \left(\bigcap_{i=0}^{m} \mathbf{dom} \ f_i \right) \cap \left(\bigcap_{i=1}^{p} \mathbf{dom} \ h_i \right). \qquad (9.2)$$

This problem is termed as the *primal problem* and the unknown variable $\mathbf{x}$ is termed as the *primal variable*. In this section, let us present the Lagrange dual function of problem (9.1) which will be the objective function of the dual problem of (9.1). Then we present the conjugate function of $f_0(\mathbf{x})$ and its relationship with the dual function of problem (9.1).

9.1.1   Lagrange dual function

The Lagrangian associated with problem (9.1) is defined as

$$\mathcal{L}(\mathbf{x},\boldsymbol{\lambda},\boldsymbol{\nu}) = f_0(\mathbf{x}) + \sum_{i=1}^{m} \lambda_i f_i(\mathbf{x}) + \sum_{i=1}^{p} \nu_i h_i(\mathbf{x}), \qquad (9.3)$$

with $\mathbf{dom} \ \mathcal{L} = \mathcal{D} \times \mathbb{R}^m \times \mathbb{R}^p$, where $\boldsymbol{\lambda} = [\lambda_1,...,\lambda_m]^T$ and $\boldsymbol{\nu} = [\nu_1,...,\nu_p]^T$ are called the *dual variables* or *Lagrange multipliers*, associated with the m inequality constraints and p equality constraints, respectively. Note that $\mathcal{L}(\mathbf{x},\boldsymbol{\lambda},\boldsymbol{\nu})$ is an affine function of $(\boldsymbol{\lambda},\boldsymbol{\nu})$ for any fixed $\mathbf{x}$, and it provides an essential linkage of the primal problem given by (9.1) and the dual problem to be introduced in Section 9.2.

The associated Lagrange dual function, or simply called the dual function, is defined as

$$g(\boldsymbol{\lambda},\boldsymbol{\nu}) \triangleq \inf_{\mathbf{x}\in\mathcal{D}} \mathcal{L}(\mathbf{x},\boldsymbol{\lambda},\boldsymbol{\nu}), \qquad (9.4)$$

with

$$\mathbf{dom}\, g = \{(\boldsymbol{\lambda}, \boldsymbol{\nu}) \mid g(\boldsymbol{\lambda}, \boldsymbol{\nu}) > -\infty\}. \tag{9.5}$$

Since the dual function is the pointwise infimum of a family of affine functions (also concave functions) of $(\boldsymbol{\lambda}, \boldsymbol{\nu})$, the dual function g is concave in $(\boldsymbol{\lambda}, \boldsymbol{\nu})$ by (3.81), even when the original or primal problem is nonconvex. For a given optimization problem, to find $g(\boldsymbol{\lambda}, \boldsymbol{\nu})$ itself is basically an unconstrained optimization problem, while it is a convex problem for $\boldsymbol{\lambda} \succeq \mathbf{0}$ if the primal problem (9.1) is a convex problem.

Let

$$\mathcal{S}(\boldsymbol{\lambda}, \boldsymbol{\nu}) \triangleq \{\bar{\mathbf{x}}(\boldsymbol{\lambda}, \boldsymbol{\nu}) \mid g(\boldsymbol{\lambda}, \boldsymbol{\nu}) = \mathcal{L}(\bar{\mathbf{x}}(\boldsymbol{\lambda}, \boldsymbol{\nu}), \boldsymbol{\lambda}, \boldsymbol{\nu})\}, \tag{9.6}$$

which is nothing but the solution set of the unconstrained problem (9.4). It is easy to see that if $\mathcal{S}(\boldsymbol{\lambda}, \boldsymbol{\nu}) \neq \emptyset$, then

$$g(\boldsymbol{\lambda}, \boldsymbol{\nu}) = \mathcal{L}(\bar{\mathbf{x}}(\boldsymbol{\lambda}, \boldsymbol{\nu}), \boldsymbol{\lambda}, \boldsymbol{\nu}) \;\; \forall \bar{\mathbf{x}}(\boldsymbol{\lambda}, \boldsymbol{\nu}) \in \mathcal{S}(\boldsymbol{\lambda}, \boldsymbol{\nu}). \tag{9.7}$$

Note that $\bar{\mathbf{x}}(\boldsymbol{\lambda}, \boldsymbol{\nu}) \in \mathcal{S}(\boldsymbol{\lambda}, \boldsymbol{\nu})$ may not be a feasible point of the primal problem. However, when $\mathcal{S}(\boldsymbol{\lambda}, \boldsymbol{\nu}) = \emptyset$, the Lagrange dual function $g(\boldsymbol{\lambda}, \boldsymbol{\nu})$ itself is not a solvable problem, but it may still be well defined. For instance, consider the case of $f_0(x) = e^{-|x|}$, $f_1(x) = -x$, and $\mathcal{D} = \mathbb{R}$. Then the Lagrangian is given by

$$\mathcal{L}(x, \lambda) = e^{-|x|} - \lambda x,$$

and the Lagrange dual function $g(\lambda) = 0$, only for $\lambda = 0$ (i.e., $\mathbf{dom}\, g = \{0\}$), is a concave function, but $\mathcal{S}(\lambda = 0) = \emptyset$ and $\min_{x \in \mathbb{R}} \mathcal{L}(x, \lambda = 0)$ does not exist. Two examples are presented next to illustrate how to obtain the concave Lagrange dual function $g(\boldsymbol{\lambda}, \boldsymbol{\nu})$ defined in (9.4) where $\mathcal{S}(\boldsymbol{\lambda}, \boldsymbol{\nu}) \neq \emptyset$.

Example 9.1 *(Minimum-norm solution for linear equations)* Consider the following problem

$$\begin{aligned} \min \;\; & \|\mathbf{x}\|_2^2 \\ \text{s.t.} \;\; & \mathbf{Ax} = \mathbf{b} \end{aligned} \tag{9.8}$$

where $\mathbf{A} \in \mathbb{R}^{p \times n}$ and $\mathrm{rank}(\mathbf{A}) = p < n$ (full row rank). Then its Lagrangian is

$$\mathcal{L}(\mathbf{x}, \boldsymbol{\nu}) = \mathbf{x}^T \mathbf{x} + \boldsymbol{\nu}^T (\mathbf{Ax} - \mathbf{b}) = \mathbf{x}^T \mathbf{x} + (\mathbf{A}^T \boldsymbol{\nu})^T \mathbf{x} - \boldsymbol{\nu}^T \mathbf{b}. \tag{9.9}$$

Note that the dual function $g(\boldsymbol{\nu}) = \inf_{\mathbf{x}} \mathcal{L}(\mathbf{x}, \boldsymbol{\nu})$ is an unconstrained convex QP. Since $\mathcal{L}(\mathbf{x}, \boldsymbol{\nu})$ is a convex quadratic function of $\mathbf{x}$, we can find the optimum $\mathbf{x}$ from the optimality condition given by (4.28)

$$\nabla_{\mathbf{x}} \mathcal{L}(\mathbf{x}, \boldsymbol{\nu}) = 2\mathbf{x} + \mathbf{A}^T \boldsymbol{\nu} = \mathbf{0}, \tag{9.10}$$

which yields the optimal

$$\bar{\mathbf{x}}(\boldsymbol{\nu}) = -\frac{1}{2} \mathbf{A}^T \boldsymbol{\nu} \in \mathcal{S}(\boldsymbol{\nu}) \tag{9.11}$$

and is unique, i.e., $\mathcal{S}(\nu)$ is a singleton set. Thus, the dual function defined in (9.4) can be obtained as

$$g(\nu) = \mathcal{L}\left(\bar{\mathbf{x}}(\nu) = -\frac{1}{2}\mathbf{A}^T\nu, \nu\right) = -\frac{1}{4}\nu^T\mathbf{A}\mathbf{A}^T\nu - \mathbf{b}^T\nu, \qquad (9.12)$$

which is indeed concave with $\mathbf{dom}\ g = \mathbb{R}^p$.   □

Example 9.2 *(Standard form LP)* For the standard form LP given by (6.4)

$$\min\ \mathbf{c}^T\mathbf{x} \qquad (9.13)$$
$$\text{s.t. } \mathbf{x} \succeq 0,\ \mathbf{A}\mathbf{x} = \mathbf{b}$$

the associated Lagrangian can be written as

$$\mathcal{L}(\mathbf{x}, \lambda, \nu) = \mathbf{c}^T\mathbf{x} - \lambda^T\mathbf{x} + \nu^T(\mathbf{A}\mathbf{x} - \mathbf{b}) = (\mathbf{c} - \lambda + \mathbf{A}^T\nu)^T\mathbf{x} - \mathbf{b}^T\nu.$$

Note that $\mathcal{L}(\mathbf{x}, \lambda, \nu)$ is affine in $\mathbf{x}$, which is unbounded below unless $\nabla_{\mathbf{x}}\mathcal{L}(\mathbf{x}, \lambda, \nu) = \mathbf{c} - \lambda + \mathbf{A}^T\nu = \mathbf{0}$. So the set $\mathcal{S}(\lambda, \nu) = \mathcal{D} = \mathbb{R}^n$ for those (λ, ν) satisfying $\mathbf{c} - \lambda + \mathbf{A}^T\nu = \mathbf{0}$, and thus the dual function is given by

$$g(\lambda, \nu) = \inf_{\mathbf{x}} \mathcal{L}(\mathbf{x}, \nu, \lambda)$$
$$= \begin{cases} -\mathbf{b}^T\nu, & \mathbf{c} - \lambda + \mathbf{A}^T\nu = \mathbf{0} \\ -\infty, & \text{otherwise,} \end{cases} \qquad (9.14)$$

with $\mathbf{dom}\ g = \{(\lambda, \nu) \mid \mathbf{c} - \lambda + \mathbf{A}^T\nu = \mathbf{0}\}$, and it is concave.   □

Fact 9.1 For any $\lambda \succeq \mathbf{0}$ and $\nu \in \mathbb{R}^p$,

$$g(\lambda, \nu) \le p^\star, \qquad (9.15)$$

i.e., the optimal value of the primal problem is an upper bound of the dual function.

Proof: Suppose that $\tilde{\mathbf{x}}$ is feasible and so $f_i(\tilde{\mathbf{x}}) \le 0$ and $h_i(\tilde{\mathbf{x}}) = 0$. For any $\lambda \succeq \mathbf{0}$ and ν,

$$\mathcal{L}(\tilde{\mathbf{x}}, \lambda, \nu) = f_0(\tilde{\mathbf{x}}) + \sum_{i=1}^{m}\lambda_i f_i(\tilde{\mathbf{x}}) + \sum_{i=1}^{p}\nu_i h_i(\tilde{\mathbf{x}}) \le f_0(\tilde{\mathbf{x}})$$
$$\implies f_0(\tilde{\mathbf{x}}) \ge \mathcal{L}(\tilde{\mathbf{x}}, \lambda, \nu) \ge \inf_{\mathbf{x} \in \mathcal{D}} \mathcal{L}(\mathbf{x}, \lambda, \nu) = g(\lambda, \nu). \qquad (9.16)$$

Therefore, $p^\star \ge g(\lambda, \nu)$, for any $\lambda \succeq \mathbf{0}$ and ν.   ■

Due to Fact 9.1, the concave function $g(\lambda, \nu)$ defined in (9.4) will be used as the objective function of the dual problem which is a maximization problem to be introduced in the next section. In the subsequent subsections, let us introduce an alternative approach for the computation of $g(\lambda, \nu)$ via the conjugate of the objective function $f_0(\mathbf{x})$ of the primal problem.

9.1.2   Conjugate function

The conjugate function, denoted as f^*, of a function $f : \mathbb{R}^n \to \mathbb{R}$, is a function from $\mathbb{R}^n \to \mathbb{R}$ and is defined as

$$f^*(\mathbf{y}) = \sup_{\mathbf{x} \in \mathbf{dom}\, f} (\mathbf{y}^T \mathbf{x} - f(\mathbf{x})), \tag{9.17}$$

where $\mathbf{dom}\, f^* = \{\mathbf{y} \mid f^*(\mathbf{y}) < \infty\}$ or, in other words, $\mathbf{y}^T\mathbf{x} - f(\mathbf{x})$ is bounded above on $\mathbf{dom}\, f$. It could be readily inferred from (9.17) that $f^*(\mathbf{y})$ is a convex function in $\mathbf{y}$ (even when $f(\mathbf{x})$ is not convex), as it is the pointwise supremum of a family of affine functions in $\mathbf{y}$ by (3.80). Two instances about the conjugate function are illustrated in the following example.

Example 9.3 Consider the instance that f is a general norm, i.e.,

$$f(\mathbf{x}) = \|\mathbf{x}\|, \quad \mathbf{dom}\, f = \mathbb{R}^n, \tag{9.18}$$

which is a convex function but not necessarily differentiable. It can be inferred that

$$f^*(\mathbf{y}) = \max \left\{ \sup_{\|\mathbf{x}\| \neq 0} \{\|\mathbf{x}\| \sup_{\|\mathbf{u}\|=1} (\mathbf{y}^T\mathbf{u} - 1)\}, 0 \right\} \quad (\text{where } \mathbf{u} = \mathbf{x}/\|\mathbf{x}\|)$$

$$= \max \left\{ \sup_{\|\mathbf{x}\| \neq 0} \{\|\mathbf{x}\|(\|\mathbf{y}\|_* - 1)\}, 0 \right\} \quad (\text{cf. (2.88)})$$

$$= I_B(\mathbf{y}) \triangleq \begin{cases} 0, & \mathbf{y} \in B \triangleq \{\mathbf{y} \in \mathbb{R}^n \mid \|\mathbf{y}\|_* \leq 1\} \\ \infty, & \mathbf{y} \notin B \end{cases} \tag{9.19}$$

is an indicator function (a convex function).

Another instance is

$$f(\mathbf{x}) = \|\mathbf{x}\|_0, \quad \mathbf{dom}\, f = \{\mathbf{x} \in \mathbb{R}^n \mid \|\mathbf{x}\|_1 \leq 1\}, \tag{9.20}$$

which is lower semi-continuous (cf. (1.31)) and nonconvex. According to (9.17), we have

$$f^*(\mathbf{y}) = \max \left\{ \sup_{0 < \|\mathbf{x}\|_1 \leq 1} \{ \sup_{\|\mathbf{u}\|_1=1} (\|\mathbf{x}\|_1 \mathbf{y}^T\mathbf{u} - \|\mathbf{u}\|_0) \}, 0 \right\} \quad (\text{where } \mathbf{x} = \mathbf{u}\|\mathbf{x}\|_1)$$

$$= \max \left\{ \sup_{0 < \|\mathbf{x}\|_1 \leq 1} \{\|\mathbf{x}\|_1 \|\mathbf{y}\|_\infty - 1\}, 0 \right\}, \quad \mathbf{y} \in \mathbb{R}^n$$

$$= \max \{\|\mathbf{y}\|_\infty - 1, 0\}, \quad \mathbf{y} \in \mathbb{R}^n \tag{9.21}$$

which is also convex but nonsmooth. Note that we have used the fact that the dual of 1-norm is ∞-norm and $\|\mathbf{y}\|_\infty = \mathbf{y}^T(\mathbf{e}_i \mathrm{sgn}(y_i))$ (cf. (2.90)), where $i = \arg_j \max\{|y_j|, j = 1, \ldots, n\}$, in the derivation (the second line) of (9.21). $\square$

Note that f^* may not exist. For instance, considering the concave function $f(x) = -x^2$ with $\mathbf{dom}\, f = \mathbb{R}$, its conjugate f^* does not exist due to $\mathbf{dom}\, f^* = \emptyset$. Provided that f^* exists, to find $f^*(\mathbf{y})$ itself is also an optimization problem.

Moreover, as $f(\mathbf{x})$ is differentiable, if the maximum difference between the linear function $\mathbf{y}^T\mathbf{x}$ and the function $f(\mathbf{x})$ must occur at a point $\mathbf{x} \in \mathbf{dom}\ f$ obtained from $\nabla_\mathbf{x}(\mathbf{y}^T\mathbf{x} - f(\mathbf{x})) = \mathbf{0}$, i.e.,

$$\mathbf{y} = \nabla_\mathbf{x}f(\mathbf{x}^\star) \in \mathbf{dom}\ f^*, \tag{9.22}$$

which, however, is a necessary condition for obtaining $f^*(\mathbf{y})$. As $f(\mathbf{x})$ is convex $\nabla_\mathbf{x}^2(\mathbf{y}^T\mathbf{x} - f(\mathbf{x})) = -\nabla_\mathbf{x}^2 f(\mathbf{x}) \preceq \mathbf{0}$ (i.e., the second-order condition satisfied), it will also be a sufficient condition, implying that, as $f(\mathbf{x})$ is differentiable and convex,

$$f^*(\mathbf{y}) = (\mathbf{x}^\star)^T\nabla_\mathbf{x}f(\mathbf{x}^\star) - f(\mathbf{x}^\star), \tag{9.23}$$

where $\mathbf{x}^\star$ can be solved from (9.22) in terms of $\mathbf{y}$. Furthermore, by (9.17), it can be inferred that

$$\begin{aligned}
f^*(\nabla_\mathbf{x}f(\boldsymbol{x})) &= \boldsymbol{x}^T\nabla_\mathbf{x}f(\boldsymbol{x}) - f(\boldsymbol{x}) \quad \text{(by (9.23))}\\
&= \sup_{\mathbf{x}\in\mathbf{dom}\ f}\ \{\mathbf{x}^T\nabla_\mathbf{x}f(\boldsymbol{x}) - f(\mathbf{x})\}\\
&\geq \mathbf{x}^T\nabla_\mathbf{x}f(\boldsymbol{x}) - f(\mathbf{x}) \quad \forall \mathbf{x}\in\mathbf{dom}\ f\\
&\Leftrightarrow\ f(\mathbf{x}) \geq f(\boldsymbol{x}) + \nabla_\mathbf{x}f(\boldsymbol{x})^T(\mathbf{x}-\boldsymbol{x}) \quad \forall \mathbf{x}\in\mathbf{dom}\ f
\end{aligned} \tag{9.24}$$

provided that $\nabla_\mathbf{x}f(\boldsymbol{x}) \in \mathbf{dom}\ f^*$. It is noticeable from (9.24) that $f(\mathbf{x})$ is tightly bounded below by an affine function passing $\boldsymbol{x}$ no matter if $f(\mathbf{x})$ is convex or not. An illustration of how to obtain a conjugate of a one-dimensional nonconvex differentiable function is shown in Figure 9.1 together with some instances of the tight lower bound given by (9.24). Next, some conjugate function examples for convex functions are presented.

- Affine function (convex): $f(x) = ax + b$. Since $f(x)$ is differentiable, $y = f'(x) = a$ (by (9.22)), a constant. Therefore, $\mathbf{dom}\ f^* = \{a\}$ (a singleton set). Thus by (9.23), we have the pair

$$f(x) = ax + b\ \leftrightarrow\ f^*(y) = -b,\ y \in \{a\} \tag{9.25}$$

where $f^*(y)$ is indeed convex.

- Negative logarithm (convex): $f(x) = -\log x$, with $\mathbf{dom}\ f = \mathbb{R}_{++}$. Then $y = f'(x^\star) = -1/x^\star$ (by (9.22)) and $\mathbf{dom}\ f^* = -\mathbb{R}_{++}$. Therefore, by (9.23) and $x^\star = -1/y$, we obtain

$$f(x) = -\log x\ \leftrightarrow\ f^*(y) = -1 - \log(-y) \tag{9.26}$$

where $f^*(y) = -1 - \log(-y)$ is indeed convex.

- Negative entropy (convex): $f(x) = x\log x$, with $f(0) = 0$ and $\mathbf{dom}\ f = \mathbb{R}_+$. Then $y = f'(x^\star) = 1 + \log x^\star$ (by (9.22)) and $\mathbf{dom}\ f^* = \mathbb{R}$. Therefore, by (9.23) and $x^\star = e^{y-1}$ we obtain

$$f(x) = x\log x\ \leftrightarrow\ f^*(y) = ye^{y-1} - e^{y-1}(y-1) = e^{y-1} \tag{9.27}$$

where $f^*(y) = e^{y-1}$ is indeed convex.

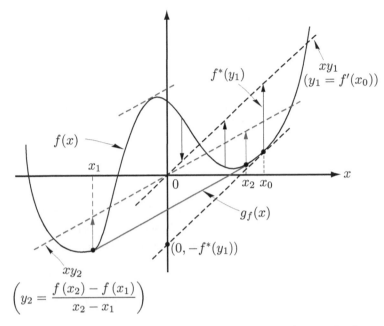

Figure 9.1 Conjugate function f^* of a nonconvex function $f : \mathbb{R} \to \mathbb{R}$, for $y_1, y_2 \in$ **dom** $f^* \subseteq \mathbb{R}$.

- Strictly convex quadratic function (convex): Let $\mathbf{Q} \in \mathbb{S}_{++}^n$. Using (9.22) and (9.23) one can similarly prove the following pair of a quadratic convex function and its convex conjugate:

$$f(\mathbf{x}) = \frac{1}{2}\mathbf{x}^T\mathbf{Q}\mathbf{x} \;\leftrightarrow\; f^*(\mathbf{y}) = \frac{1}{2}\mathbf{y}^T\mathbf{Q}^{-1}\mathbf{y}, \qquad (9.28)$$

with **dom** $f = $ **dom** $f^* = \mathbb{R}^n$.

Though the conjugate f^* of a convex function f is a convex function, the conjugate of f^*, i.e., f^{**} is certainly convex provided that f^* exists. Apparently, $f^{**} \neq f$ when f is nonconvex. Are they the same when f is convex? The relationship between f^{**} and f is presented for f to be convex and nonconvex in the following two remarks, respectively.

Remark 9.1 [Ber03, Proposition 7.1.1] If the function f is convex and meanwhile closed (cf. Remark 3.23), then $f^{**} = f$, i.e., the biconjugate of f is f again. In all the above examples, it is true that $f^{**} = f$ since they are convex and closed.

Proof: For simplicity, assume that f is differentiable. Let $\mathbf{y} = \nabla f(\mathbf{x}_0) \in$ **dom** f^*. Then one can obtain the conjugate of $f^*(\mathbf{y})$ as follows:

$$
\begin{aligned}
f^{**}(\mathbf{x}) &= \sup_{\mathbf{y} \in \mathbf{dom}\ f^*} \left\{ \mathbf{x}^T \mathbf{y} - f^*(\mathbf{y}) \right\} \\
&= \sup_{\mathbf{x}_0 \in \mathbf{dom}\ f} \left\{ \mathbf{x}^T \nabla f(\mathbf{x}_0) - \nabla f(\mathbf{x}_0)^T \mathbf{x}_0 + f(\mathbf{x}_0) \right\} \quad \text{(by (9.23))} \\
&= \sup_{\mathbf{x}_0 \in \mathbf{dom}\ f} \left\{ f(\mathbf{x}_0) + \nabla f(\mathbf{x}_0)^T (\mathbf{x} - \mathbf{x}_0) \right\} \\
&= f(\mathbf{x}) \quad \text{(by (3.17))}.
\end{aligned}
$$

Thus the proof is completed.   ■ □

Remark 9.2 As mentioned above, even if f is nonconvex, its conjugate f^* is still a convex function. Hence the convex $f^{**} \neq f$. What will be f^* and f^{**}? Let us consider the nonconvex differentiable function $f(x)$ defined on $\mathbb{R}$ illustrated in Figure 9.1. Suppose that g_f is the convex envelope of f. It can be easily seen that

$$
g_f(x) = \begin{cases} f(x), & x \notin (x_1, x_2) \\ f'(x_1)x + \dfrac{f(x_1)x_2 - f(x_2)x_1}{x_2 - x_1}, & x \in (x_1, x_2) \end{cases}
$$

where $f'(x_1) = \dfrac{f(x_2) - f(x_1)}{x_2 - x_1} = f'(x_2)$. By (9.24), if $f'(x_0) \in \mathbf{dom}\ f^*$, then

$$
f(x) \geq f(x_0) + f'(x_0)(x - x_0)\ \forall x
$$

that, however, holds true only for $x_0 \notin (x_1, x_2)$, implying that $f^*(f'(x_0)) = g_f^*(f'(x_0))$ for $x_0 \notin (x_1, x_2)$. Moreover, $g_f^*(g_f'(x_0)) = f^*(f'(x_1)) = f^*(f'(x_2))$ for all $x_0 \in (x_1, x_2)$. Hence, $f^* = g_f^*$ and

$$
\mathbf{dom}\ g_f^* = \left\{ g_f'(x) \mid x \in \mathbb{R} \right\} = \left\{ f'(x) \mid x \in \mathbb{R} \setminus (x_1, x_2) \right\} = \mathbf{dom}\ f^*.
$$

So one can conclude that $f^{**} = g_f^{**} = g_f$ for this case.

If the **dom** f is compact and f is lower semi-continuous (cf. (1.31)), it can be shown [Fal69] that $f^{**} = g_f$. Again, consider the nonsmooth nonconvex zero-norm function defined in (9.20) (whose convex envelope has been presented in Remark 3.25 without proof). Let us prove it by deriving the conjugate of its conjugate given by (9.21). According to (9.17), we have

$$
\begin{aligned}
g_f(\mathbf{x}) = f^{**}(\mathbf{x}) &= \sup_{\|\mathbf{y}\|_\infty \neq 0} \left\{ \|\mathbf{y}\|_\infty \sup_{\|\mathbf{u}\|_\infty = 1} \left(\mathbf{x}^T \mathbf{u} - \max \left\{ 1 - \frac{1}{\|\mathbf{y}\|_\infty}, 0 \right\} \right) \right\} \\
&= \sup_{\|\mathbf{y}\|_\infty \neq 0} \left\{ \|\mathbf{y}\|_\infty \|\mathbf{x}\|_1 - \max \left\{ \|\mathbf{y}\|_\infty - 1, 0 \right\} \right\} \\
&= \sup_{\|\mathbf{y}\|_\infty \neq 0} \min \left\{ \|\mathbf{y}\|_\infty (\|\mathbf{x}\|_1 - 1) + 1, \|\mathbf{y}\|_\infty \|\mathbf{x}\|_1 \right\} \\
&= \begin{cases} \|\mathbf{x}\|_1, & \text{if } \|\mathbf{x}\|_1 \leq 1 \\ \infty, & \text{if } \|\mathbf{x}\|_1 > 1 \end{cases}
\end{aligned} \tag{9.29}
$$

where we have used the variable change $\mathbf{u} = \mathbf{y}/\|\mathbf{y}\|_\infty$, the fact that the dual of ∞-norm is 1-norm, and the fact that $\|\mathbf{y}\|_\infty(\|\mathbf{x}\|_1 - 1) + 1$ is a nonincreasing affine function of $\|\mathbf{y}\|_\infty$ for $\|\mathbf{x}\|_1 \leq 1$ and $\|\mathbf{y}\|_\infty\|\mathbf{x}\|_1$ is a nondecreasing linear function of $\|\mathbf{y}\|_\infty$, and their intersection takes the value $\|\mathbf{x}\|_1$ when $\|\mathbf{y}\|_\infty = 1$. Note that $\|\mathbf{y}\|_\infty = 0$ was left out in the derivation of (9.29) without harm because of the obtained $g_f(\mathbf{x}) \geq 0$ for all $\|\mathbf{x}\|_1 \leq 1$. ☐

9.1.3   Relationship between Lagrange dual function and conjugate function

To show the close relationship between Lagrange dual function and the conjugate function of the objective function when all the constraint functions are affine (i.e., the feasible set of the primal problem is a polyhedron), consider the following simple problem

$$\begin{aligned} \min \ & f_0(\mathbf{x}) \\ \text{s.t. } & \mathbf{x} = \mathbf{0}. \end{aligned} \tag{9.30}$$

The Lagrangian of (9.30) is $\mathcal{L}(\mathbf{x}, \boldsymbol{\nu}) = f_0(\mathbf{x}) + \boldsymbol{\nu}^T\mathbf{x}$, and the corresponding dual function is

$$\begin{aligned} g(\boldsymbol{\nu}) &= \inf_{\mathbf{x}} \ \{f_0(\mathbf{x}) + \boldsymbol{\nu}^T\mathbf{x}\}, \\ &= -\sup_{\mathbf{x}} \ \{(-\boldsymbol{\nu})^T\mathbf{x} - f_0(\mathbf{x})\} = -f_0^*(-\boldsymbol{\nu}), \end{aligned}$$

with $\mathbf{dom}\ g = -\mathbf{dom}\ f_0^*$, which is concave since f_0^* is convex. Similarly, let us consider the following more general optimization problem

$$\begin{aligned} \min \ & f_0(\mathbf{x}) \\ \text{s.t. } & \mathbf{A}\mathbf{x} \preceq \mathbf{b}, \ \mathbf{C}\mathbf{x} = \mathbf{d} \end{aligned} \tag{9.31}$$

(with the feasible set being a polyhedron). The dual function of (9.31) can be written in terms of the conjugate function of the objective function, as shown below

$$\begin{aligned} g(\boldsymbol{\lambda}, \boldsymbol{\nu}) &= \inf_{\mathbf{x}} \ \{f_0(\mathbf{x}) + \boldsymbol{\lambda}^T(\mathbf{A}\mathbf{x} - \mathbf{b}) + \boldsymbol{\nu}^T(\mathbf{C}\mathbf{x} - \mathbf{d})\} \\ &= -\mathbf{b}^T\boldsymbol{\lambda} - \mathbf{d}^T\boldsymbol{\nu} + \inf_{\mathbf{x}} \ \{f_0(\mathbf{x}) + (\mathbf{A}^T\boldsymbol{\lambda} + \mathbf{C}^T\boldsymbol{\nu})^T\mathbf{x}\} \\ &= -\mathbf{b}^T\boldsymbol{\lambda} - \mathbf{d}^T\boldsymbol{\nu} - \sup_{\mathbf{x}} \ \{-(\mathbf{A}^T\boldsymbol{\lambda} + \mathbf{C}^T\boldsymbol{\nu})^T\mathbf{x} - f_0(\mathbf{x})\} \\ &= -\mathbf{b}^T\boldsymbol{\lambda} - \mathbf{d}^T\boldsymbol{\nu} - f_0^*(-\mathbf{A}^T\boldsymbol{\lambda} - \mathbf{C}^T\boldsymbol{\nu}), \end{aligned} \tag{9.32}$$

which is concave since f_0^* is convex. From the definition of conjugate function and (9.5), the domain of g can then be written as

$$\mathbf{dom}\ g = \{(\boldsymbol{\lambda}, \boldsymbol{\nu}) \mid -(\mathbf{A}^T\boldsymbol{\lambda} + \mathbf{C}^T\boldsymbol{\nu}) \in \mathbf{dom}\ f_0^*\}. \tag{9.33}$$

Let us conclude this subsection with the following example in which the considered problem takes the form of (9.31).

Example 9.4 *(Entropy maximization)* In this example we will show how the conjugate function is employed to find the Lagrange dual function of an entropy maximization problem defined as

$$
\max \left\{ \sum_{i=1}^{n} x_i \log \frac{1}{x_i} \right\} \quad \equiv \quad \min \left\{ f_0(\mathbf{x}) \triangleq \sum_{i=1}^{n} x_i \log x_i \right\} \tag{9.34}
$$
$$
\text{s.t. } \mathbf{x} \in \mathbb{R}_{+}^{n}, \ \mathbf{1}_{n}^{T}\mathbf{x} = 1 \qquad\qquad \text{s.t. } \mathbf{x} \in \mathbb{R}_{+}^{n}, \ \mathbf{1}_{n}^{T}\mathbf{x} = 1.
$$

Note that $f_0(\mathbf{x})$ in (9.34) is a sum of negative entropy functions, where each $x_i \log x_i$ is only a function of one scalar variable x_i, and $x_1, \ldots, x_n$ are independent variables. The conjugate function of negative entropy has been discussed in Subsection 9.1.2, and therefore the conjugate function of f_0 can be readily seen, from (9.27), to be

$$
f_0^*(\mathbf{y}) = \sup_{\mathbf{x} \in \mathbb{R}_{+}^{n}} \left\{ \mathbf{y}^{T}\mathbf{x} - f_0(\mathbf{x}) \right\}
$$
$$
= \sum_{i=1}^{n} \sup_{x_i \in \mathbb{R}_{+}} \left\{ y_i x_i - x_i \log x_i \right\} = \sum_{i=1}^{n} e^{y_i - 1}, \tag{9.35}
$$

with $\mathbf{dom}\, f_0^* = \mathbb{R}^n$.

By (9.32) with $\mathbf{A} = -\mathbf{I}_n$, $\mathbf{b} = \mathbf{0}$, $\mathbf{C} = \mathbf{1}_n^T$, and $d = 1$, the Lagrange dual function of problem (9.34) can be written as

$$
g(\boldsymbol{\lambda}, \nu) = -\nu - f_0^*(\boldsymbol{\lambda} - \mathbf{1}_n \nu)
$$
$$
= -\nu - \sum_{i=1}^{n} e^{\lambda_i - \nu - 1} = -\nu - e^{-\nu - 1} \sum_{i=1}^{n} e^{\lambda_i}. \tag{9.36}
$$

Thus, the Lagrange dual function of an optimization problem (not necessarily convex) can be readily obtained via the conjugate function of the objective function. $\qquad\qquad\square$

9.2   Lagrange dual problem

The dual problem of the primal problem (9.1) is defined as

$$
\max_{(\boldsymbol{\lambda}, \boldsymbol{\nu}) \in \mathbf{dom}\, g} \quad g(\boldsymbol{\lambda}, \boldsymbol{\nu}) \tag{9.37}
$$
$$
\text{s.t. } \boldsymbol{\lambda} \succeq \mathbf{0}
$$

where $\mathbf{dom}\, g$ has been defined in (9.5). Let us emphasize that the dual problem is convex, even when the primal problem is nonconvex. A pair $(\boldsymbol{\lambda}, \boldsymbol{\nu})$ is said to be dual feasible if $\boldsymbol{\lambda} \succeq \mathbf{0}$ and $(\boldsymbol{\lambda}, \boldsymbol{\nu}) \in \mathbf{dom}\, g$, i.e., $g(\boldsymbol{\lambda}, \boldsymbol{\nu}) > -\infty$, where the former is obviously an inequality constraint, but the latter may induce some equality constraints implicitly imposed in the dual problem (9.37), as can be clearly observed via illustrative examples in the subsequent sections of this chapter.

For a general case that $\mathcal{L}(\mathbf{x}, \boldsymbol{\lambda}, \boldsymbol{\nu})$ is differentiable in $\mathbf{x}$, we often need to ascertain if a $(\boldsymbol{\lambda} \succeq \mathbf{0}, \boldsymbol{\nu})$ is dual feasible. The following fact provides a sufficient condition for a given $(\boldsymbol{\lambda} \succeq \mathbf{0}, \boldsymbol{\nu})$ to be dual feasible.

Fact 9.2 For a given dual variable $(\boldsymbol{\lambda}, \boldsymbol{\nu})$, if $\mathcal{L}(\mathbf{x}, \boldsymbol{\lambda}, \boldsymbol{\nu})$ is convex in $\mathbf{x}$ and $\nabla_{\mathbf{x}}\mathcal{L}(\bar{\mathbf{x}}, \boldsymbol{\lambda}, \boldsymbol{\nu}) = \mathbf{0}$, for some $\bar{\mathbf{x}} \in \mathcal{D}$ (problem domain of (9.1)) which implies $g(\boldsymbol{\lambda}, \boldsymbol{\nu}) = \mathcal{L}(\bar{\mathbf{x}}, \boldsymbol{\lambda}, \boldsymbol{\nu}) > -\infty$, then $(\boldsymbol{\lambda}, \boldsymbol{\nu}) \in \mathbf{dom}\ g$. Furthermore, if $\boldsymbol{\lambda} \succeq \mathbf{0}$, then the $(\boldsymbol{\lambda}, \boldsymbol{\nu})$ is a feasible point of the dual problem (9.37).

The optimal value of the dual problem (9.37) is

$$d^\star = \sup\ \{g(\boldsymbol{\lambda}, \boldsymbol{\nu}) \mid \boldsymbol{\lambda} \succeq \mathbf{0}, \boldsymbol{\nu} \in \mathbb{R}^p\}, \tag{9.38}$$

which serves as a lower bound of the primal optimal value $p^\star$ by Fact 9.1. The result

$$d^\star \leq p^\star \tag{9.39}$$

is called *weak duality*. If the problem is such that

$$d^\star = p^\star, \tag{9.40}$$

then we say that *strong duality* holds.

For a primal dual feasible pair $\mathbf{x}$, $(\boldsymbol{\lambda}, \boldsymbol{\nu})$, the quantity

$$\eta(\mathbf{x}, \boldsymbol{\lambda}, \boldsymbol{\nu}) = f_0(\mathbf{x}) - g(\boldsymbol{\lambda}, \boldsymbol{\nu}) \tag{9.41}$$

is called the *duality gap*. Note that the dual function $g(\boldsymbol{\lambda}, \boldsymbol{\nu})$ is determined by the set $\mathcal{S}(\boldsymbol{\lambda}, \boldsymbol{\nu})$ defined in (9.6) when it is nonempty, while $\mathbf{x}$ in (9.41) is independent of $(\boldsymbol{\lambda}, \boldsymbol{\nu})$. For notational simplicity, η is also used in this chapter to denote $\eta(\mathbf{x}, \boldsymbol{\lambda}, \boldsymbol{\nu})$ without confusion.

It is easy to see that

$$g(\boldsymbol{\lambda}, \boldsymbol{\nu}) \leq d^\star \leq p^\star \leq f_0(\mathbf{x}), \tag{9.42}$$

implying that $\eta \geq p^\star - d^\star$ and $\eta \geq f_0(\mathbf{x}) - p^\star$, and $\mathbf{x}$ is η-suboptimal (because a feasible point $\mathbf{x}$ is ϵ-suboptimal if $f_0(\mathbf{x}) \leq p^\star + \epsilon$). If the duality gap $\eta = 0$, i.e., $p^\star = d^\star$, then $\mathbf{x}$ is a primal optimal solution and $(\boldsymbol{\lambda}, \boldsymbol{\nu})$ is a dual optimal solution. In other words, as $\eta(\mathbf{x}^\star, \boldsymbol{\lambda}^\star, \boldsymbol{\nu}^\star) = 0$ for a primal-dual feasible pair of $\mathbf{x}^\star$ and $(\boldsymbol{\lambda}^\star, \boldsymbol{\nu}^\star)$, it must be true that $f(\mathbf{x}^\star) < \infty$, $g(\boldsymbol{\lambda}^\star, \boldsymbol{\nu}^\star) > -\infty$, and

$$\begin{aligned}
f_0(\mathbf{x}^\star) &= g(\boldsymbol{\lambda}^\star, \boldsymbol{\nu}^\star) \quad (\text{since } \eta(\mathbf{x}^\star, \boldsymbol{\lambda}^\star, \boldsymbol{\nu}^\star) = 0) \\
&= \inf_{\mathbf{x} \in \mathcal{D}} \mathcal{L}(\mathbf{x}, \boldsymbol{\lambda}^\star, \boldsymbol{\nu}^\star) \\
&\leq \mathcal{L}(\mathbf{x}^\star, \boldsymbol{\lambda}^\star, \boldsymbol{\nu}^\star) \leq f_0(\mathbf{x}^\star) \quad (\text{by (9.16)}) \\
&\Rightarrow g(\boldsymbol{\lambda}^\star, \boldsymbol{\nu}^\star) = \mathcal{L}(\mathbf{x}^\star, \boldsymbol{\lambda}^\star, \boldsymbol{\nu}^\star) = f_0(\mathbf{x}^\star) \tag{9.43}
\end{aligned}$$

which is an essential link for the pair of a primal optimal point and a dual optimal point of an optimization problem with strong duality. The important relationship (9.43) between the primal and dual optimal points also further leads to the complementary slackness of strong duality to be addressed in Subsection 9.4.3 later.

In order to further illustrate the preceding relationship among the objective function, the Lagrangian, the dual function, and the primal and dual optimal points, let us consider a simple convex QCQP defined as

$$\min_x \ \{f_0(x) \triangleq x^2\}$$
$$\text{s.t. } (x-2)^2 \leq 1. \tag{9.44}$$

The optimal solution of this problem can be readily seen to be $x^* = 1$, and the primal optimal value $p^* = f_0(x^*) = 1$.

The Lagrangian of this problem can be easily seen to be

$$\mathcal{L}(x,\lambda) = x^2 + \lambda[(x-2)^2 - 1] = (1+\lambda)x^2 - 4\lambda x + 3\lambda,$$

which is convex in x for $\lambda > -1$, but unbounded below for $\lambda \leq -1$. The dual function $g(\lambda) > -\infty$ is actually determined by (4.28), yielding

$$\bar{x}(\lambda) = \arg\min_x \mathcal{L}(x,\lambda) = \frac{2\lambda}{1+\lambda}, \quad \text{for } \lambda > -1,$$

and the dual function $g(\lambda)$ (which is differentiable and concave) can be easily obtained as

$$g(\lambda) = \mathcal{L}(\bar{x}(\lambda),\lambda) = \frac{4\lambda}{1+\lambda} - \lambda, \quad \mathbf{dom} \ g = (-1,\infty).$$

Then it can be shown by (4.28) that

$$\lambda^* = \arg\max\{g(\lambda), \lambda \geq 0\} = 1,$$

and the optimal dual value $d^* = g(\lambda^*) = 1$. So the strong duality holds (due to $p^* = d^* = 1$) with the primal-dual optimal solution $(x^* = \bar{x}(\lambda^*), \lambda^*) = (1,1)$. This example has also illustrated the key result given by (9.43) due to strong duality.

Figure 9.2 shows the objective function $f_0(x)$, the 2-dimensional Lagrangian $\mathcal{L}(x,\lambda)$, $\mathcal{L}(\bar{x}(\lambda),\lambda)$ (a curve on the surface of $\mathcal{L}(x,\lambda)$), and the dual function $g(\lambda)$. From this figure, one can see that $g(\lambda) = \mathcal{L}(\bar{x}(\lambda),\lambda)$ is a curve determined by $\mathcal{S}(\lambda) = \{\bar{x}(\lambda)\}$, that $g(\lambda) \leq f_0(x)$ for all $\lambda \geq 0$ and $(x-2)^2 \leq 1$, and that

$$g(\lambda^*) = \mathcal{L}(x^*,\lambda^*) = f_0(x^*) = 1.$$

Moreover, the primal-dual optimal solution $(x^*,\lambda^*) = (1,1)$ is also a saddle point of $\mathcal{L}(x,\lambda)$. This can be proven true by deriving the gradient and Hessian of $\mathcal{L}(x,\lambda)$ for $(x,\lambda) = (1,1)$, which are a zero vector for the former and an indefinite matrix for the latter (see Remark 9.3 below). The coincidence between the saddle point of $\mathcal{L}(x,\lambda)$ and the primal-dual optimal solution is actually due to the max-min characteristic of strong duality to be addressed in Subsection 9.4.1 later.

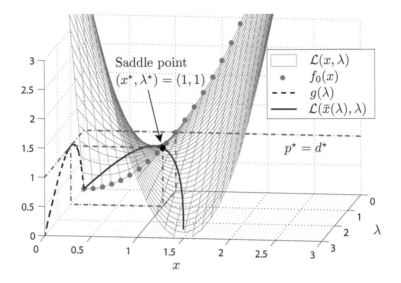

Figure 9.2 The objective function $f_0(x) = x^2$, Lagrangian $\mathcal{L}(x, \lambda)$, the curve $\mathcal{L}(\bar{x}(\lambda), \lambda)$, the dual function $g(\lambda) = \mathcal{L}(\bar{x}(\lambda) = 2\lambda/(1 + \lambda), \lambda)$, and the primal-dual optimal solution $(x^\star, \lambda^\star) = (1, 1)$ of the convex problem $\min\{f_0(x) \mid (x - 2)^2 \leq 1\}$ with strong duality.

Remark 9.3 Consider a function $f : W \times Z \to \mathbb{R}$, where $W \subseteq \mathbb{R}^m$ and $Z \subseteq \mathbb{R}^n$. A pair $(\widetilde{w}, \widetilde{z}) \in W \times Z$ is said to be a saddle point of f if

$$f(\widetilde{w}, z) \leq f(\widetilde{w}, \widetilde{z}) \leq f(w, \widetilde{z}), \ \forall w \in W, \ \forall z \in Z. \qquad (9.45)$$

For the case that f is twice differentiable and $m = n = 1$, $(\widetilde{w}, \widetilde{z})$ can be shown to be a saddle point if $(\widetilde{w}, \widetilde{z})$ is a stationary point (i.e., $\nabla f(\widetilde{w}, \widetilde{z}) = (0, 0)$) and its Hessian matrix $\nabla^2 f(\widetilde{w}, \widetilde{z})$ is indefinite. $\qquad \square$

Remark 9.4 *(Bidual of problem* (9.1)*)* The dual problem defined in (9.37) can be equivalently expressed as

$$- \min_{(\boldsymbol{\lambda}, \boldsymbol{\nu}) \in \mathbf{dom} \ g} \ - g(\boldsymbol{\lambda}, \boldsymbol{\nu}) \qquad (9.46)$$
$$\text{s.t.} \ \boldsymbol{\lambda} \succeq \mathbf{0}.$$

The dual of problem (9.46) (i.e.,(9.37)), which is the bidual of the primal problem (9.1), can be seen to be

$$- \max_{\boldsymbol{w} \succeq \mathbf{0}} \ \inf \left\{ -g(\boldsymbol{\lambda}, \boldsymbol{\nu}) - \boldsymbol{w}^T \boldsymbol{\lambda} \mid (\boldsymbol{\lambda}, \boldsymbol{\nu}) \in \mathbf{dom} \ g \right\} \quad \text{(by (9.4))}$$
$$= \min_{\boldsymbol{w} \succeq \mathbf{0}} \ - \inf \left\{ -g(\boldsymbol{\lambda}, \boldsymbol{\nu}) - \boldsymbol{w}^T \boldsymbol{\lambda} \mid (\boldsymbol{\lambda}, \boldsymbol{\nu}) \in \mathbf{dom} \ g \right\} \qquad (9.47)$$
$$= \min_{\boldsymbol{w} \succeq \mathbf{0}} \ \left\{ h(\boldsymbol{w}) \triangleq \sup \left\{ g(\boldsymbol{\lambda}, \boldsymbol{\nu}) + \boldsymbol{w}^T \boldsymbol{\lambda} \mid (\boldsymbol{\lambda}, \boldsymbol{\nu}) \in \mathbf{dom} \ g \right\} \right\}$$

where $\mathbf{dom}\ h = \{\boldsymbol{w} \mid h(\boldsymbol{w}) < \infty\}$. Let us discuss when problem (9.47) will be the same as the primal problem (9.1) in the following examples. $\square$

Example 9.5 *(Duals of standard form LP and inequality form LP)* By the dual function given by (9.14) for the standard form LP given by (9.13), the dual problem is

$$\begin{aligned} \max \quad & -\mathbf{b}^T \boldsymbol{\nu} \\ \text{s.t.} \quad & \boldsymbol{\lambda} \succeq \mathbf{0} \\ & \mathbf{c} - \boldsymbol{\lambda} + \mathbf{A}^T \boldsymbol{\nu} = \mathbf{0} \end{aligned} \tag{9.48}$$

which forms an equivalent problem as (by letting $\mathbf{y} = -\boldsymbol{\nu}$)

$$\begin{aligned} \max \quad & \mathbf{b}^T \mathbf{y} \\ \text{s.t.} \quad & \mathbf{c} - \mathbf{A}^T \mathbf{y} \succeq \mathbf{0} \end{aligned} \tag{9.49}$$

which is the inequality form of LP. Next, let us prove that the dual of (9.49) is (9.13).

By (9.47) and (9.48), the bidual of (9.13) can be obtained as

$$\begin{aligned} & \min_{\boldsymbol{w} \succeq \mathbf{0}} \ \sup_{\boldsymbol{\lambda}, \boldsymbol{\nu}} \left\{ -\mathbf{b}^T \boldsymbol{\nu} + \boldsymbol{w}^T \boldsymbol{\lambda} \mid \boldsymbol{\lambda} = \mathbf{A}^T \boldsymbol{\nu} + \mathbf{c} \right\} \\ & = \min_{\boldsymbol{w} \succeq \mathbf{0}} \ \sup_{\boldsymbol{\nu}} \left\{ (\mathbf{A} \boldsymbol{w} - \mathbf{b})^T \boldsymbol{\nu} + \mathbf{c}^T \boldsymbol{w} \right\} \\ & = \min_{\boldsymbol{w} \succeq \mathbf{0}, \mathbf{A}\boldsymbol{w} = \mathbf{b}} \ \mathbf{c}^T \boldsymbol{w}. \end{aligned} \tag{9.50}$$

Hence, the bidual of an LP is exactly the original LP given by (9.13). $\square$

Example 9.6 *(Dual and bidual of a general convex problem)* Consider, again, the optimization problem (9.31). Suppose that the objective function $f_0(\mathbf{x})$ is a closed convex function with $\mathbf{dom}\ f_0 = \mathbb{R}^n$ and $f_0^*(\mathbf{x})$ denotes its conjugate. Thus $f_0^{**}(\mathbf{x}) = f_0(\mathbf{x})$ by Remark 9.1. Since the dual function of problem (9.31) is given by (9.32), the dual of problem (9.31) can be seen to be

$$\max_{\boldsymbol{\lambda} \succeq \mathbf{0}, \boldsymbol{\nu}} \left\{ -\mathbf{b}^T \boldsymbol{\lambda} - \mathbf{d}^T \boldsymbol{\nu} - f_0^*(-\mathbf{A}^T \boldsymbol{\lambda} - \mathbf{C}^T \boldsymbol{\nu}) \right\} \tag{9.51}$$

which surely is a convex problem. Next, let us find the dual of problem (9.51), i.e., bidual of problem (9.31).

Let

$$\mathbf{y} \triangleq -[\mathbf{A}^T\ \mathbf{C}^T] \begin{bmatrix} \boldsymbol{\lambda} \\ \boldsymbol{\nu} \end{bmatrix} = -\mathbf{B}^T \begin{bmatrix} \boldsymbol{\lambda} \\ \boldsymbol{\nu} \end{bmatrix}$$

$$\Rightarrow \begin{bmatrix} \boldsymbol{\lambda} \\ \boldsymbol{\nu} \end{bmatrix} = -(\mathbf{B}^T)^\dagger \mathbf{y} + \boldsymbol{v}, \ \boldsymbol{v} \in \mathcal{N}(\mathbf{B}^T)$$

where $\mathbf{B}^T = [\mathbf{A}^T \; \mathbf{C}^T]$. By (9.47) and (9.51), the bidual of problem (9.31) is given by

$$\min_{\boldsymbol{w} \succeq 0} \; \sup_{\boldsymbol{\lambda}, \boldsymbol{\nu}} \left\{ (\boldsymbol{w} - \mathbf{b})^T \boldsymbol{\lambda} - \mathbf{d}^T \boldsymbol{\nu} - f_0^*(\boldsymbol{y}) \mid \boldsymbol{y} = -\mathbf{B}^T \begin{bmatrix} \boldsymbol{\lambda} \\ \boldsymbol{\nu} \end{bmatrix} \in \mathbf{dom} \; f_0^* \right\}$$

$$= \min_{\boldsymbol{w} \succeq 0} \; \sup_{\boldsymbol{y} \in \mathbf{dom} \; f_0^*, \; \boldsymbol{v} \in \mathcal{N}(\mathbf{B}^T)} \left\{ [(\boldsymbol{w} - \mathbf{b})^T \; -\mathbf{d}^T] \left(-(\mathbf{B}^T)^\dagger \boldsymbol{y} + \boldsymbol{v} \right) - f_0^*(\boldsymbol{y}) \right\}$$

$$= \min_{\boldsymbol{w} \succeq 0} \; f_0 \left(-\mathbf{B}^\dagger \begin{bmatrix} \boldsymbol{w} - \mathbf{b} \\ -\mathbf{d} \end{bmatrix} \right) \tag{9.52}$$

where the condition of $\sup\{\cdot\} < \infty$ has been applied in the derivations, thus requiring

$$\begin{bmatrix} \boldsymbol{w} - \mathbf{b} \\ -\mathbf{d} \end{bmatrix} \perp \mathcal{N}(\mathbf{B}^T) = \mathcal{R}(\mathbf{B})^\perp \implies \begin{bmatrix} \boldsymbol{w} - \mathbf{b} \\ -\mathbf{d} \end{bmatrix} \in \mathcal{R}(\mathbf{B}) \; \text{(cf. (1.68))}.$$

By change of variable of

$$\mathbf{x} = -\mathbf{B}^\dagger \begin{bmatrix} \boldsymbol{w} - \mathbf{b} \\ -\mathbf{d} \end{bmatrix} \implies -\mathbf{Bx} = \begin{bmatrix} -\mathbf{Ax} \\ -\mathbf{Cx} \end{bmatrix} = \begin{bmatrix} \boldsymbol{w} - \mathbf{b} \\ -\mathbf{d} \end{bmatrix} \in \mathcal{R}(\mathbf{B})$$

$$\implies \boldsymbol{w} = -\mathbf{Ax} + \mathbf{b} \succeq 0, \; \mathbf{Cx} = \mathbf{d},$$

problem (9.52) can be equivalently expressed as

$$\min_{\mathbf{Ax} \preceq \mathbf{b}, \mathbf{Cx} = \mathbf{d}} \; f_0(\mathbf{x}) \tag{9.53}$$

which is exactly the original problem (9.31). In other words, the bidual of problem (9.31) is itself, and this also applies to its dual problem (9.51).

A general convex problem corresponds to the problem (9.31) with $f_0(x)$ being closed and convex, and the finite inequality constraints $\mathbf{Ax} \preceq \mathbf{b}$ replaced by $f_i(\mathbf{x}) \leq 0, i = 1, \dots, m$. Without loss of generality, assume that $m = 1$, and $f_1(\mathbf{x})$ is nonsmooth or differentiable, and convex. By the supporting hyperplane theorem, the closed convex set associated with $f_1(x) \leq 0$ can be expressed as

$$C \triangleq \{\mathbf{x} \mid f_1(\mathbf{x}) \leq 0\}$$
$$= \bigcap_{\mathbf{x}_0 \in \mathbf{bd} \, C} \mathcal{A}(\mathbf{x}_0) \triangleq \{\mathbf{x} \mid \bar{\nabla} f_1(\mathbf{x}_0)^T (\mathbf{x} - \mathbf{x}_0) \leq 0\} \; \text{(cf. (3.23) and (3.24))}$$

(provided that $\mathbf{int} \, C$ is nonempty) which actually corresponds to a constraint set of (possibly infinitely many) linear inequalities. Let $\{A_i\}$ be a set sequence that satisfies

$$|A_i| < \infty, \; A_i \subseteq A_{i+1} \subseteq \mathbf{bd} \, C, \; \forall i \in \mathbb{Z}_{++},$$
$$\lim_{i \to \infty} \mathbf{conv} \, A_i = \mathbf{conv} \, \mathbf{bd} \, C = C,$$

where $\lim_{i \to \infty} A_i$ is *countable* and *dense* on $\mathbf{bd} \, C$ [Apo07]. Then

$$C \subset \bigcap_{\mathbf{x}_0 \in A_i} \mathcal{A}(\mathbf{x}_0) \; \forall i \in \mathbb{Z}_{++}, \; \text{and} \; \lim_{i \to \infty} \bigcap_{\mathbf{x}_0 \in A_i} \mathcal{A}(\mathbf{x}_0) = C.$$

Therefore, the bidual of the following convex problem with finite linear inequalities and equalities

$$\min\ f_0(\mathbf{x})$$
$$\text{s.t.}\ \ \begin{aligned}\mathbf{x} &\in \mathcal{A}(\mathbf{x}_0)\ \forall \mathbf{x}_0 \in A_i \\ \mathbf{C}\mathbf{x} &= \mathbf{d}\end{aligned} \tag{9.54}$$

is the same as itself, and furthermore as $i \to \infty$, problem (9.54) becomes the original convex problem

$$\min\ f_0(\mathbf{x})$$
$$\text{s.t.}\ f_1(\mathbf{x}) \le 0,\ \mathbf{C}\mathbf{x} = \mathbf{d} \tag{9.55}$$

since the feasible set of the former converges to that of the latter. Thus we have finished the proof that the bidual of a convex problem is the same as the original problem if the objective function is closed and all the inequality constraint functions are nonsmooth or differentiable.

Furthermore, consider the following minimization problem:

$$\min_{\mathbf{x} \in \mathcal{C}}\ f(\mathbf{x}) \tag{9.56}$$

where $f: \mathcal{C} \to \mathbb{R}$ is a nonconvex function, but $\mathcal{C}$ is a closed convex set. As introduced in Section 4.7, a stationary-point solution to this problem can be obtained by applying the BSUM method under some convergence conditions. On the other hand, it can be inferred that when f^* exists, the bidual of the nonconvex problem (9.56) is a convex problem given by

$$\min_{\mathbf{x} \in \mathcal{C}}\ f^{**}(\mathbf{x}). \tag{9.57}$$

Note that $f^{**}(\mathbf{x}) = g_f(\mathbf{x})$ under some mild conditions (cf. Remark 9.2). When $f^{**}(\mathbf{x}) = g_f(\mathbf{x})$, any optimal solution of problem (9.57) will also be an optimal solution of problem (9.56), if the set of all global minimizers of problem (9.56) is a convex set. $\qquad\square$

Example 9.7 *(Dual and bidual of BQP)* The nonconvex BQP has been introduced in Chapter 8, and it takes the following form (see (8.22)):

$$\min\ \mathbf{x}^T \mathbf{C}\mathbf{x}$$
$$\text{s.t.}\ x_i^2 = 1,\ i = 1,\ldots,n \tag{9.58}$$

where $\mathbf{C} \in \mathbb{S}^n$. The Lagrangian is

$$\begin{aligned}\mathcal{L}(\mathbf{x}, \boldsymbol{\nu}) =\ & \mathbf{x}^T \mathbf{C}\mathbf{x} + \sum_{i=1}^{n} \nu_i(x_i^2 - 1) \\ =\ & \mathbf{x}^T \mathbf{C}\mathbf{x} + \mathbf{x}^T \mathbf{Diag}(\boldsymbol{\nu})\mathbf{x} - \sum_{i=1}^{n} \nu_i \\ =\ & \mathbf{x}^T\big(\mathbf{C} + \mathbf{Diag}(\boldsymbol{\nu})\big)\mathbf{x} - \boldsymbol{\nu}^T \mathbf{1}_n. \end{aligned} \tag{9.59}$$

Then we have

$$g(\nu) = \inf_{\mathbf{x}} \mathcal{L}(\mathbf{x}, \nu) = \begin{cases} -\nu^T \mathbf{1}_n, & \mathbf{C} + \mathbf{Diag}(\nu) \succeq \mathbf{0} \\ -\infty, & \text{otherwise.} \end{cases} \tag{9.60}$$

The dual problem can be expressed as

$$\begin{array}{ll} \max \ -\nu^T \mathbf{1}_n & \quad -\min \ \nu^T \mathbf{1}_n \\ \text{s.t. } \mathbf{C} + \mathbf{Diag}(\nu) \succeq \mathbf{0} & = \quad \text{s.t. } \mathbf{C} + \mathbf{Diag}(\nu) \succeq \mathbf{0} \end{array} \tag{9.61}$$

which is an SDP.

Next, let us find the dual of the minimization problem on the right-hand side in (9.61). Its Lagrangian is

$$\begin{aligned} \mathcal{L}(\nu, \mathbf{Z}) &= \nu^T \mathbf{1}_n - \mathrm{Tr}\big(\mathbf{Z}[\mathbf{C} + \mathbf{Diag}(\nu)]\big), \ \mathbf{Z} \in \mathbb{S}^n \quad \text{(see Remark 9.5 below)} \\ &= \nu^T \mathbf{1}_n - \mathrm{Tr}(\mathbf{Z}\mathbf{C}) - \nu^T \mathbf{vecdiag}(\mathbf{Z}) \\ &= \nu^T (\mathbf{1}_n - \mathbf{vecdiag}(\mathbf{Z})) - \mathrm{Tr}(\mathbf{Z}\mathbf{C}), \end{aligned} \tag{9.62}$$

where the Lagrange multiplier $\mathbf{Z}$ is a symmetric matrix instead of a column vector to be discussed in the Remark 9.5 below. The dual function is

$$g(\mathbf{Z}) = \inf_{\nu} \mathcal{L}(\nu, \mathbf{Z}) = \begin{cases} -\mathrm{Tr}(\mathbf{Z}\mathbf{C}), & \mathbf{1}_n - \mathbf{vecdiag}(\mathbf{Z}) = \mathbf{0}_n \\ -\infty, & \text{otherwise;} \end{cases} \tag{9.63}$$

so the dual of problem (9.61) can be expressed as

$$\begin{array}{ll} -\max \ -\mathrm{Tr}(\mathbf{Z}\mathbf{C}) & \quad \min \ \mathrm{Tr}(\mathbf{Z}\mathbf{C}) \\ \text{s.t. } \mathbf{Z} \succeq \mathbf{0}, & = \quad \text{s.t. } \mathbf{Z} \succeq \mathbf{0}, \\ \quad [\mathbf{Z}]_{ii} = 1, \ i = 1, \dots, n & \qquad [\mathbf{Z}]_{ii} = 1, \ i = 1, \dots, n \end{array} \tag{9.64}$$

which is the SDP obtained through SDR of the BQP (see (8.33)).

In summary, the original BQP (9.58) has the dual problem given by the left-hand side of problem (9.61), whose dual is given by the right-hand side of the SDP (9.64), the bidual of the original BQP (9.58). It is reminiscent of the corresponding property of Fourier transform pair $(x(t), X(f))$:

$$X(f) = \mathcal{F}\{x(t)\} = \int_{-\infty}^{\infty} x(t) e^{-j2\pi f t} dt;$$

$$\mathcal{F}\{X(f)\} = \mathcal{F}\{\mathcal{F}\{x(t)\}\} = x(-t).$$

The time function $x(t)$ and its Fourier transform $X(f)$ appears horizontally (left-right) related to each other since they carry exactly the same information in the time domain and the frequency domain, respectively. Double Fourier transform of $x(t)$ is identical to $x(t)$ when it is symmetric. Analogous situations are also existent for a primal problem and its dual as discussed next.

In contrast with the Fourier transform, the pair of a primal problem and the associated dual problem appears vertically (top-down) related to each other since

$d^\star \le p^\star$; namely,

Primal Problem:

$$p^\star = \min \{f_0(\mathbf{x}) \mid f_i(\mathbf{x}) \le 0, h_j(\mathbf{x}) = 0, i = 1, \dots, m, j = 1, \dots, p, \mathbf{x} \in \mathcal{D}\} \ge d^\star$$

($\mathcal{D}$ is the primal problem domain)

$$\Downarrow \text{ Dual transformation } \Downarrow$$

Dual Problem:

$$d^\star = \max \{g(\boldsymbol{\lambda}, \boldsymbol{\nu}) = \inf_{\mathbf{x} \in \mathcal{D}} \mathcal{L}(\mathbf{x}, \boldsymbol{\lambda}, \boldsymbol{\nu}) > -\infty \mid \boldsymbol{\lambda} \succeq \mathbf{0}, \boldsymbol{\lambda} \in \mathbb{R}^m, \boldsymbol{\nu} \in \mathbb{R}^p\} \le p^\star$$

$$\left(\mathcal{L}(\mathbf{x}, \boldsymbol{\lambda}, \boldsymbol{\nu}) = f_0(\mathbf{x}) + \sum_{i=1}^{m} \lambda_i f_i(\mathbf{x}) + \sum_{i=1}^{p} \nu_i h_i(\mathbf{x})\right).$$

The dual problem is a convex problem regardless of if the primal problem is convex or not, and when the primal problem is convex, its bidual is also identical to the original problem if $f_0(\mathbf{x})$ is closed and all $f_i(\mathbf{x})$ are nonsmooth or differentiable, as illustrated in Example 9.6. □

Remark 9.5 The Lagrange multiplier associated with a matrix inequality constraint is very different from that associated with a scalar inequality in the Lagrangian. Consider the semidefinite matrix inequality constraint $\mathbf{X} \succeq \mathbf{0}$, which holds true if and only if the following infinitely many scalar inequality constraints hold true:

$$\mathbf{a}^T \mathbf{X} \mathbf{a} \ge 0, \ \forall \mathbf{a} \in \mathbb{R}^n \Leftrightarrow -\mathrm{Tr}(\mathbf{a}^T \mathbf{X} \mathbf{a}) \le 0, \ \forall \mathbf{a} \in \mathbb{R}^n.$$

Now, the associated term in the Lagrangian involving Lagrange multipliers $\lambda_{\mathbf{a}}$ for all $\mathbf{a} \in \mathbb{R}^n$ is then

$$- \sum_{\mathbf{a} \in \mathbb{R}^n} \lambda_{\mathbf{a}} \mathrm{Tr}(\mathbf{a}^T \mathbf{X} \mathbf{a}) = -\mathrm{Tr}\left(\mathbf{X} \sum_{\mathbf{a} \in \mathbb{R}^n} \lambda_{\mathbf{a}} \mathbf{a} \mathbf{a}^T\right) = -\mathrm{Tr}(\mathbf{X}\mathbf{Z}),$$

where

$$\mathbf{Z} = \sum_{\mathbf{a} \in \mathbb{R}^n} \lambda_{\mathbf{a}} \mathbf{a} \mathbf{a}^T \in \mathbb{S}^n,$$

and $\mathbf{Z} \succeq \mathbf{0}$ if $\lambda_{\mathbf{a}} \ge 0$ for all $\mathbf{a} \in \mathbb{R}^n$. Therefore, there will be a term $-\mathrm{Tr}(\mathbf{X}\mathbf{Z}) = -\mathrm{Tr}(\mathbf{Z}\mathbf{X})$ in the Lagrangian of an optimization problem with the matrix inequality constraint $\mathbf{X} \succeq \mathbf{0}$. For instance, it is $-\mathrm{Tr}(\mathbf{Z}[\mathbf{C} + \mathbf{Diag}(\boldsymbol{\nu})])$ in (9.62) above due to the inequality constraint $[\mathbf{C} + \mathbf{Diag}(\boldsymbol{\nu})] \succeq \mathbf{0}$ in (9.61). Furthermore, in the associated dual problem (9.64), the dual variable is $\mathbf{Z}$ with the inequality constraint $\mathbf{Z} \succeq \mathbf{0}$ (corresponding to $\lambda_{\mathbf{a}} \ge 0$ for all $\mathbf{a} \in \mathbb{R}^n$). □

9.3   Strong duality

9.3.1   Slater's condition

Strong duality usually holds for convex problems. Consider the standard convex optimization problem given by (4.11) which is repeated here for convenience:

$$\min \ f_0(\mathbf{x})$$
$$\text{s.t. } \mathbf{Ax} = \mathbf{b}, \ f_i(\mathbf{x}) \le 0, \ i = 1, \ldots, m \tag{9.65}$$

where $f_i(\mathbf{x})$, $i = 0, 1, \ldots, m$ are convex, and

$$\mathbf{A} = [\mathbf{a}_1, \ldots, \mathbf{a}_p]^T \in \mathbb{R}^{p \times n}, \ \mathbf{b} = [b_1, \ldots, b_p]^T \in \mathbb{R}^p.$$

A sufficient condition for strong duality to hold for this convex problem is described as follows:

Slater's condition (or ***Slater's constraint qualification***): If there exists an $\mathbf{x} \in \mathbf{relint} \ \mathcal{D}$ with $f_i(\mathbf{x}) < 0$ for $i = 1, \ldots, m$ and

$$h_i(\mathbf{x}) = \mathbf{a}_i^T \mathbf{x} - b_i = 0, \quad i = 1, \ldots, p \tag{9.66}$$

(i.e., at least one strictly feasible point exists or $\mathbf{relint} \ \mathcal{C} \neq \emptyset$ where $\mathcal{C}$ is the feasible set of the convex problem (9.65)), then strong duality holds for the convex problem (9.65).

Let us first present the proof outline of the strong duality under Slater's condition so that the proof to be presented below is easier to follow. The Lagrangian associated with problem (9.65) is given by

$$\mathcal{L}(\mathbf{x}, \boldsymbol{\lambda}, \boldsymbol{\nu}) = f_0(\mathbf{x}) + \sum_{i=1}^{m} \lambda_i f_i(\mathbf{x}) + \sum_{i=1}^{p} \nu_i h_i(\mathbf{x})$$
$$= (\boldsymbol{\lambda}, \boldsymbol{\nu}, 1)^T (f_1(\mathbf{x}), \ldots, f_m(\mathbf{x}), h_1(\mathbf{x}), \ldots, h_p(\mathbf{x}), f_0(\mathbf{x}))$$
$$\ge p^\star, \text{ if } f_i(\mathbf{x}) = 0, i = 1, \ldots, m, h_j(\mathbf{x}) = 0, j = 1, \ldots, p,$$

implying that the nonconvex set $\mathbb{B} \subset \mathbb{R}^{m+p+1}$ (defined in (9.69) below) that is constituted by $(f_1(\mathbf{x}), \ldots, f_m(\mathbf{x}), h_1(\mathbf{x}), \ldots, h_p(\mathbf{x}), f_0(\mathbf{x}))$ for all $\mathbf{x} \in \mathcal{D}$ and the convex set $\mathcal{B} \subset \mathbb{R}^{m+p+1}$ (defined in (9.70) below) are disjoint. So, by extending the nonconvex set $\mathbb{B}$, we will define a closed convex set $\mathcal{A} \subset \mathbb{R}^{m+p+1}$ such that $\mathcal{A} \cap \mathcal{B} = \emptyset$ besides $\mathbb{B} \subset \mathcal{A}$, and thus a separating hyperplane, which separates $\mathcal{A}$ and $\mathcal{B}$, exists. Specifically, under the Slater's condition, there exists a $(\boldsymbol{\lambda} \succeq \mathbf{0}, \boldsymbol{\nu}, 1)$ such that

$$\mathcal{L}(\mathbf{x}, \boldsymbol{\lambda}, \boldsymbol{\nu}) = (\boldsymbol{\lambda}, \boldsymbol{\nu}, 1)^T (\boldsymbol{u}, \boldsymbol{v}, t) \ge p^\star, \ \forall (\boldsymbol{u}, \boldsymbol{v}, t) \in \mathbb{B} \subset \mathcal{A}.$$

Consequently, $d^\star \ge g(\boldsymbol{\lambda}, \boldsymbol{\nu}) = \inf_{\mathbf{x} \in \mathcal{D}} \mathcal{L}(\mathbf{x}, \boldsymbol{\lambda}, \boldsymbol{\nu}) \ge p^\star$, implying that $d^\star = p^\star$ must be true by Fact 9.1.

Proof. For simplicity, let us make the following assumptions:

- The problem domain $\mathcal{D}$ (cf. (9.2)) has nonempty interior, i.e., **relint** $\mathcal{D} =$ **int** $\mathcal{D} \neq \emptyset$.
- $\text{rank}(\mathbf{A}) = p \leq n$ (full row rank).
- $|p^\star| < \infty$.

Let $\mathcal{A}$ be a closed set defined as

$$\mathcal{A} \triangleq \{(\mathbf{u}, \mathbf{v}, t) \in \mathbb{R}^m \times \mathbb{R}^p \times \mathbb{R} \mid (\mathbf{x}, \mathbf{u}, \mathbf{v}, t) \in \mathbb{A}\}$$
$$= \{(\mathbf{u}, \mathbf{v}, t) \in \mathbb{R}^m \times \mathbb{R}^p \times \mathbb{R} \mid \mathbf{u} \succeq \mathbf{u}, \ \mathbf{v} = \mathbf{v}, \ t \geq t_0, \ (\mathbf{u}, \mathbf{v}, t_0) \in \mathbb{B}\} \quad (9.67)$$

where

$$\mathbb{A} = \{(\mathbf{x}, \mathbf{u}, \mathbf{v}, t) \in \mathbb{R}^n \times \mathbb{R}^m \times \mathbb{R}^p \times \mathbb{R} \mid \mathbf{x} \in \mathcal{D}, \ f_i(\mathbf{x}) \leq u_i, \ i = 1, \ldots, m,$$
$$h_i(\mathbf{x}) = v_i, \ i = 1, \ldots, p, \ f_0(\mathbf{x}) \leq t\}, \quad (9.68)$$
$$\mathbb{B} = \{(\mathbf{u}, \mathbf{v}, t) \in \mathbb{R}^m \times \mathbb{R}^p \times \mathbb{R} \mid \exists \mathbf{x} \in \mathcal{D}, \ \mathbf{u} = (f_1(\mathbf{x}), \ldots, f_m(\mathbf{x})),$$
$$\mathbf{v} = (h_1(\mathbf{x}), \ldots, h_p(\mathbf{x})), \ t = f_0(\mathbf{x})\} \subset \mathcal{A}. \quad (9.69)$$

It is easy to show that $\mathbb{A}$ is a convex set, and so $\mathcal{A}$ defined in (9.67) is a convex set by (2.63). Moreover, $(\mathbf{0}_m, \mathbf{0}_p, t) \in \mathcal{A}$ for all $t \geq p^\star$. In order to illustrate the set $\mathcal{A}$, let us consider the same convex QCQP given by (9.44) in the previous section again for which $n = m = 1$, $p = 0$, $f_0(x) = x^2$, and $f_1(x) = (x-2)^2 - 1$. Figure 9.3 shows the set $\mathcal{A}$, the set $\mathbb{B}$ and the separating hyperplane that separates the set $\mathcal{A}$ given by (9.67) and the set $\mathcal{B}$ (a ray) to be defined in (9.70) next.

Define

$$\mathcal{B} \triangleq \{(\mathbf{0}_m, \mathbf{0}_p, s) = [\mathbf{0}_{m+p}^T, s]^T \mid s < p^\star\}, \quad (9.70)$$

which can be easily shown to be convex but not closed. Note that $(\mathbf{0}_m, \mathbf{0}_p, p^\star) \in$ **relbd** $\mathcal{B}$ in spite of $(\mathbf{0}_m, \mathbf{0}_p, p^\star) \in \mathcal{A}$. The sets $\mathcal{A}$ and $\mathcal{B}$ are disjoint, i.e., $\mathcal{A} \cap \mathcal{B} = \emptyset$. This can be proved as follows.

Consider a point $(\mathbf{u}, \mathbf{v}, t) \in \mathcal{B}$ implying that $\mathbf{u} = \mathbf{0}_m$, $\mathbf{v} = \mathbf{0}_p$, $t < p^\star$. Now, if this point $(\mathbf{u}, \mathbf{v}, t) \in \mathcal{A}$, there must exist an $\mathbf{x}$ satisfying $f_i(\mathbf{x}) \leq 0$, $i = 1, \ldots, m$, $\mathbf{Ax} - \mathbf{b} = \mathbf{0}_p$, and $f_0(\mathbf{x}) \leq t < p^\star$, which is impossible since $p^\star$ is the optimal value of the primal problem. Hence the sets $\mathcal{A}$ and $\mathcal{B}$ are disjoint.

By the separating hyperplane theorem, there exist $(\tilde{\boldsymbol{\lambda}}, \tilde{\boldsymbol{\nu}}, \mu) \neq \mathbf{0}_{m+p+1}$ and α such that

$$(\tilde{\boldsymbol{\lambda}}, \tilde{\boldsymbol{\nu}}, \mu)^T (\mathbf{u}, \mathbf{v}, t) \geq \alpha, \ \forall (\mathbf{u}, \mathbf{v}, t) \in \mathcal{A}, \quad (9.71)$$
$$(\tilde{\boldsymbol{\lambda}}, \tilde{\boldsymbol{\nu}}, \mu)^T (\mathbf{u}, \mathbf{v}, t) \leq \alpha, \ \forall (\mathbf{u}, \mathbf{v}, t) \in \mathcal{B}. \quad (9.72)$$

Then (9.71) can be rewritten as

$$\tilde{\boldsymbol{\lambda}}^T \mathbf{u} + \tilde{\boldsymbol{\nu}}^T \mathbf{v} + \mu t \geq \alpha, \ \forall (\mathbf{u}, \mathbf{v}, t) \in \mathcal{A},$$
$$\Rightarrow \mu p^\star \geq \alpha \ (\text{since } (\mathbf{0}_m, \mathbf{0}_p, p^\star) \in \mathcal{A}). \quad (9.73)$$

Next let us prove that $\mu \geq 0$ and $\tilde{\boldsymbol{\lambda}} \succeq \mathbf{0}_m$ (cf. Figure 9.3).

Consider the point $(\mathbf{0}_m, \mathbf{0}_p, t) \in \mathcal{A}$ for all $t \geq p^\star$. Then $\mu t \geq \alpha$, $\forall t \geq p^\star$ (by (9.71)). However, the term μt for large positive t will be unbounded below for $\mu < 0$. Therefore, $\mu \geq 0$. Similarly, one can show that $\tilde{\boldsymbol{\lambda}} \succeq \mathbf{0}$ by considering the point $(\mathbf{u}, \mathbf{0}_p, p^\star) \in \mathcal{A}$ for all $\mathbf{u} \succeq \mathbf{0}$ in (9.71).

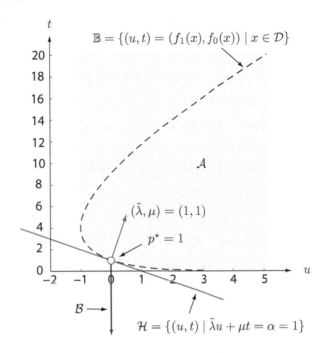

Figure 9.3 Illustration of the set $\mathcal{A}$ (shaded region) defined in (9.67) and the set $\mathcal{B}$ defined in (9.70) (a ray), and a separating hyperplane $\mathcal{H}$ between them, where $f_0(x) = x^2$, $f_1(x) = (x-2)^2 - 1$, and $\mathcal{D} = \mathbb{R}$.

Moreover, from (9.72), $\mu t \leq \alpha$, $\forall t < p^\star$ implies $\mu p^\star \leq \alpha$ (due to $(\mathbf{0}_m, \mathbf{0}_p, p^\star) \in$ **relbd** $\mathcal{B}$ and by Remark 2.23 in Subsection 2.6.1) which together with (9.73) leads to $\mu p^\star = \alpha$. Let us summarize the above necessary results as follows:

$$\tilde{\boldsymbol{\lambda}} \succeq \mathbf{0}_m, \ \mu \geq 0, \ \text{and} \ \alpha = \mu p^\star. \tag{9.74}$$

From (9.71) and (9.74), we have

$$\mathcal{L}(\mathbf{x}, \tilde{\boldsymbol{\lambda}}, \tilde{\boldsymbol{\nu}}) = (\tilde{\boldsymbol{\lambda}}, \tilde{\boldsymbol{\nu}}, \mu)^T (\mathbf{u}, \mathbf{v}, t) \geq \mu p^\star, \ \forall (\mathbf{u}, \mathbf{v}, t) \in \mathbb{B} \subset \mathcal{A} \tag{9.75}$$

where $(\mathbf{u}, \mathbf{v}, t)$, a function of $\mathbf{x} \in \mathcal{D}$, has been defined in (9.70). Since $\mu \geq 0$, we have two cases of μ to consider. One is when $\mu > 0$ and the other is when $\mu = 0$.

Case 1 $(\mu > 0)$: For $\mu > 0$, let $\boldsymbol{\lambda} = \tilde{\boldsymbol{\lambda}}/\mu \succeq \mathbf{0}$, $\boldsymbol{\nu} = \tilde{\boldsymbol{\nu}}/\mu$. Then (9.75) reduces to

$$\mathcal{L}(\mathbf{x}, \boldsymbol{\lambda}, \boldsymbol{\nu}) = (\boldsymbol{\lambda}, \boldsymbol{\nu}, 1)^T (\mathbf{u}, \mathbf{v}, t) \geq p^\star, \ \forall (\mathbf{u}, \mathbf{v}, t) \in \mathbb{B}$$

$$\Rightarrow \sum_{i=1}^m \lambda_i f_i(\mathbf{x}) + \boldsymbol{\nu}^T (\mathbf{A}\mathbf{x} - \mathbf{b}) + f_0(\mathbf{x}) \geq p^\star, \quad \forall \mathbf{x} \in \mathcal{D} \qquad (9.76)$$

$$\Rightarrow \mathcal{L}(\mathbf{x}, \boldsymbol{\lambda}, \boldsymbol{\nu}) \geq p^\star, \quad \forall \mathbf{x} \in \mathcal{D}$$

$$\Rightarrow d^\star \geq g(\boldsymbol{\lambda}, \boldsymbol{\nu}) = \inf_{\mathbf{x} \in \mathcal{D}} \mathcal{L}(\mathbf{x}, \boldsymbol{\lambda}, \boldsymbol{\nu}) \geq p^\star \quad (\text{by Fact 9.1})$$

$$\Rightarrow d^\star = p^\star \quad (\text{by } (9.39)),$$

i.e., the strong duality holds for $\mu > 0$.

Case 2 $(\mu = 0)$: We will prove that $\mu = 0$ is not possible under Slater's condition. Suppose $\mu = 0$. Then (9.75) becomes

$$\sum_{i=1}^m \tilde{\lambda}_i f_i(\mathbf{x}) + \tilde{\boldsymbol{\nu}}^T (\mathbf{A}\mathbf{x} - \mathbf{b}) \geq 0 \text{ for any } \mathbf{x} \in \mathcal{D}. \qquad (9.77)$$

Let $\tilde{\mathbf{x}} \in \mathcal{D}$ be a point satisfying Slater's condition. Then we have from (9.77)

$$\sum_{i=1}^m \tilde{\lambda}_i f_i(\tilde{\mathbf{x}}) \geq 0 \Longrightarrow \tilde{\lambda}_i = 0 \ \forall i \qquad (9.78)$$

since $f_i(\tilde{\mathbf{x}}) < 0$ and $\tilde{\lambda}_i \geq 0$ for all i. Then (9.77) can be further reduced to a halfspace in $\mathcal{D}$ as follows

$$\tilde{\boldsymbol{\nu}}^T (\mathbf{A}\mathbf{x} - \mathbf{b}) = (\mathbf{A}^T \tilde{\boldsymbol{\nu}})^T \mathbf{x} - \tilde{\boldsymbol{\nu}}^T \mathbf{b} \geq 0 \text{ for any } \mathbf{x} \in \mathcal{D}. \qquad (9.79)$$

Since $\tilde{\mathbf{x}} \in \mathbf{int}\ \mathcal{D}$, there exists an $r > 0$ such that $\tilde{\mathbf{x}} + r\mathbf{u} \in \mathcal{D}$, $\forall \|\mathbf{u}\|_2 \leq 1$. Then $\tilde{\boldsymbol{\nu}}^T (\mathbf{A}\tilde{\mathbf{x}} - \mathbf{b}) = 0$ implies that there exists an $\mathbf{x} \in \mathcal{D}$ with $\tilde{\boldsymbol{\nu}}^T (\mathbf{A}\mathbf{x} - \mathbf{b}) < 0$, thereby violating (9.79), unless $\mathbf{A}^T \tilde{\boldsymbol{\nu}} = \mathbf{0}_n$. On the other hand, because of the aforementioned $\tilde{\boldsymbol{\lambda}} = \mathbf{0}_m$ by (9.78) and $\mu = 0$ by the assumption for Case 2, we have $(\tilde{\boldsymbol{\lambda}}, \tilde{\boldsymbol{\nu}}, \mu) = (\mathbf{0}_m, \tilde{\boldsymbol{\nu}}, 0) \neq \mathbf{0}_{p+m+1}$ leading to $\tilde{\boldsymbol{\nu}} \neq \mathbf{0}_p$ (otherwise contradiction with $(\tilde{\boldsymbol{\lambda}}, \tilde{\boldsymbol{\nu}}, \mu) \neq \mathbf{0}_{p+m+1}$). Hence $\mathbf{A}^T \tilde{\boldsymbol{\nu}} = \mathbf{0}_n$ together with $\tilde{\boldsymbol{\nu}} \neq \mathbf{0}_p$ contradicts with the $\text{rank}(\mathbf{A}) = p$ assumption. Thus we conclude that the case $\mu = 0$ is impossible, thereby having completed the proof of strong duality for the convex problem (9.65) under Slater's condition. ∎

Remark 9.6 A refined version of Slater's condition: If $f_1, \ldots, f_k$ are affine, then strong duality for the convex problem given by (9.65) holds when there exists an $\mathbf{x} \in \mathbf{relint}\ \mathcal{D}$ such that

$$f_i(\mathbf{x}) \leq 0, \ i = 1, \ldots, k, \qquad (9.80a)$$

$$f_i(\mathbf{x}) < 0, \ i = k+1, \ldots, m, \qquad (9.80b)$$

$$\mathbf{A}\mathbf{x} = \mathbf{b}, \qquad (9.80c)$$

i.e., strict linear inequalities can be relaxed in Slater's condition. □

Strong duality generally does not hold for nonconvex problems, but it may hold true for some cases such as the ones in the following examples.

Example 9.8 *(The smallest eigenvalue problem)* Consider the problem

$$\min \ \mathbf{x}^T \mathbf{C} \mathbf{x} \qquad \text{(cf. (1.91))} \tag{9.81}$$
$$\text{s.t.} \ \mathbf{x}^T \mathbf{x} = 1$$

for any $\mathbf{C} \in \mathbb{S}^n$. The problem is nonconvex and it has been known that the optimal value of this problem is $p^\star = \lambda_{\min}(\mathbf{C})$. The Lagrangian of the above problem is

$$\mathcal{L}(\mathbf{x}, \nu) = \mathbf{x}^T \mathbf{C} \mathbf{x} + \nu(\mathbf{x}^T \mathbf{x} - 1) = \mathbf{x}^T (\mathbf{C} + \nu \mathbf{I}_n) \mathbf{x} - \nu; \tag{9.82}$$

so its Lagrange dual function is

$$g(\nu) = \begin{cases} -\nu, & \mathbf{C} + \nu \mathbf{I}_n \succeq \mathbf{0} \\ -\infty, & \text{otherwise.} \end{cases} \tag{9.83}$$

Then its dual problem is

$$\max \ -\nu \tag{9.84}$$
$$\text{s.t.} \ \mathbf{C} + \nu \mathbf{I}_n \succeq \mathbf{0}$$

which is equivalent to

$$\max \ -\nu \tag{9.85}$$
$$\text{s.t.} \ \lambda_i(\mathbf{C}) + \nu \geq 0, \ i = 1, \ldots, n$$

where $\lambda_i(\mathbf{C})$ denotes the ith largest eigenvalue of $\mathbf{C}$. It is trivial to see that the optimal value of the above problem is

$$d^\star = -\nu^\star = \lambda_{\min}(\mathbf{C}) = p^\star,$$

which is equal to the optimal value. Hence strong duality holds.

Though strong duality holds true for the nonconvex problem (9.81), the dual of its dual given by (9.84) can be easily shown by (9.47) to be

$$\min \ \mathrm{Tr}(\mathbf{C} \mathbf{X}) \tag{9.86}$$
$$\text{s.t.} \ \mathbf{X} \succeq \mathbf{0}, \ \mathrm{Tr}(\mathbf{X}) = 1$$

which is a convex SDP, and the SDR of (9.81). Next, let us show that the optimal value of (9.86) is $\mathrm{Tr}(\mathbf{C} \mathbf{X}^\star) = \lambda_{\min}(\mathbf{C})$ and $\mathbf{X}^\star = \mathbf{q}_{\min} \mathbf{q}_{\min}^T$ must be an optimal solution, where $\mathbf{q}_{\min}$ is an orthonormal eigenvector associated with $\lambda_{\min}(\mathbf{C})$.

The associated KKT conditions (to be introduced later in this chapter) for solving (9.86) and its dual (9.84) can be shown to be

$$\begin{cases} \mathbf{X} \succeq \mathbf{0}, \ \mathrm{Tr}(\mathbf{X}) = 1 \\ \mathbf{C} + \nu \mathbf{I}_n \succeq \mathbf{0}, \ (\mathbf{C} + \nu \mathbf{I}_n)\mathbf{X} = \mathbf{0}. \end{cases} \tag{9.87}$$

Let ℓ denote the multiplicity of the minimum eigenvalue $\lambda_{\min}(\mathbf{C})$, and $\boldsymbol{q}_{\min,1}$, $\boldsymbol{q}_{\min,2}, \ldots, \boldsymbol{q}_{\min,\ell}$ be the associated orthonormal eigenvectors, implying

$$\dim(V_{\min}) = \ell \text{ where } V_{\min} \triangleq \mathrm{span}[\boldsymbol{q}_{\min,1}, \ldots, \boldsymbol{q}_{\min,\ell}].$$

By (9.87), it can be inferred that $\mathrm{rank}(\mathbf{C} + \nu\mathbf{I}_n) \le n - 1$ must be true, otherwise $\mathbf{X} = \mathbf{0}$ which contracts with $\mathrm{Tr}(\mathbf{X}) = 1$. Furthermore, $\mathrm{rank}(\mathbf{C} + \nu\mathbf{I}_n) \le n - 1$, $\mathbf{C} + \nu\mathbf{I}_n \succeq \mathbf{0}$, and $(\mathbf{C} + \nu\mathbf{I}_n)\mathbf{X} = \mathbf{0}$ imply that

$$\lambda_{\min}(\mathbf{C} + \nu\mathbf{I}_n) = \lambda_{\min}(\mathbf{C}) + \nu = 0 \ \Rightarrow\ \nu^\star = -\lambda_{\min}(\mathbf{C})$$

and that

$$\mathcal{R}(\mathbf{X}) \subseteq \mathcal{N}(\mathbf{C} + \nu^\star\mathbf{I}_n) = V_{\min} \quad (\text{cf. Remark 1.18 and (1.90)}).$$

It can be seen that $\mathbf{X}^\star = \boldsymbol{q}\boldsymbol{q}^T$ must be an optimal rank-one solution, where $\boldsymbol{q} \in V_{\min}$ and $\|\boldsymbol{q}\|_2 = 1$, and $\mathrm{Tr}(\mathbf{C}\mathbf{X}^\star) = \lambda_{\min}(\mathbf{C})$. For the case of $\ell = 1$, $\mathbf{X}^\star$ is unique and meanwhile problem (9.86) reduces to problem (9.81); for the case of $\ell > 1$, $\mathbf{X}^\star$ is nonunique (e.g., $\mathbf{X}^\star = \theta\boldsymbol{q}_{\min,1}\boldsymbol{q}_{\min,1}^T + (1 - \theta)\boldsymbol{q}_{\min,2}\boldsymbol{q}_{\min,2}^T$ are rank-two solutions for all $0 < \theta < 1$) simply due to $\dim(V_{\min}) = \ell > 1$. $\qquad \square$

Example 9.9 *(Trust region problem)* Consider the trust region problem, defined as

$$\begin{aligned} \min\ &f(\mathbf{x}) \triangleq \mathbf{x}^T\mathbf{A}\mathbf{x} + 2\mathbf{b}^T\mathbf{x} \\ \text{s.t.}\ &\mathbf{x}^T\mathbf{x} \le 1 \end{aligned} \tag{9.88}$$

where $\mathbf{A} \in \mathbb{S}^n$ and $\mathbf{b} \in \mathbb{R}^n$. When $\mathbf{A} \not\succeq \mathbf{0}$, this is not a convex problem. The Lagrangian of (9.88) can be seen to be

$$\mathcal{L}(\mathbf{x}, \lambda) = \mathbf{x}^T(\mathbf{A} + \lambda\mathbf{I})\mathbf{x} + 2\mathbf{b}^T\mathbf{x} - \lambda. \tag{9.89}$$

Then the dual function can be proven, by using (4.28) and (1.119), to be

$$\begin{aligned} g(\lambda) &= \inf_{\mathbf{x}}\mathcal{L}(\mathbf{x}, \lambda) \\ &= \begin{cases} -\lambda - \mathbf{b}^T(\mathbf{A} + \lambda\mathbf{I})^\dagger\mathbf{b}, & \mathbf{A} + \lambda\mathbf{I} \succeq \mathbf{0},\ \mathbf{b} \in \mathcal{R}(\mathbf{A} + \lambda\mathbf{I}) \\ -\infty, & \text{otherwise.} \end{cases} \end{aligned} \tag{9.90}$$

Then the dual of problem (9.88) is given by

$$\begin{aligned} d^\star = \max\ &-\lambda - \mathbf{b}^T(\mathbf{A} + \lambda\mathbf{I})^\dagger\mathbf{b} \\ \text{s.t.}\ &\mathbf{A} + \lambda\mathbf{I} \succeq \mathbf{0},\ \lambda \ge 0,\ \mathbf{b} \in \mathcal{R}(\mathbf{A} + \lambda\mathbf{I}) \end{aligned} \tag{9.91}$$

which, by epigraph representation followed by applying Schur complement (cf. (3.96) and Remark 3.29) to the resulting generalized inequality constraint, can be equivalently expressed as the following SDP

$$\begin{aligned} \max\ &t \\ \text{s.t.}\ &\begin{bmatrix} \mathbf{A} + \lambda\mathbf{I} & \mathbf{b} \\ \mathbf{b}^T & -\lambda - t \end{bmatrix} \succeq \mathbf{0},\ \lambda \ge 0,\ \mathbf{b} \in \mathcal{R}(\mathbf{A} + \lambda\mathbf{I}). \end{aligned} \tag{9.92}$$

Note that the two constraints $\lambda \geq 0$ and $\mathbf{b} \in \mathcal{R}(\mathbf{A} + \lambda\mathbf{I})$ in (9.92) actually correspond to $\lambda - a \geq 0$ for some $a \geq 0$. So problem (9.92) must be a convex problem. It can be seen by Slater's condition that strong duality holds for problems (9.88) and (9.92) if $\mathbf{A} \succeq \mathbf{0}$. Next, assuming that $a = 0$ without loss of generality, let us find the dual of the SDP (9.92) using (9.47), because the resulting dual of (9.92) is not dependent upon the value of a.

Let

$$
\begin{aligned}
\tilde{h}(\lambda, t) &= t + y\lambda + \mathrm{Tr}\left(\begin{bmatrix} \mathbf{X} & \mathbf{x} \\ \mathbf{x}^T & x \end{bmatrix} \begin{bmatrix} \mathbf{A} + \lambda\mathbf{I} & \mathbf{b} \\ \mathbf{b}^T & -\lambda - t \end{bmatrix} \right) \\
&= t + y\lambda + \mathrm{Tr}\left(\begin{bmatrix} \mathbf{X}(\mathbf{A}+\lambda\mathbf{I}) + \mathbf{x}\mathbf{b}^T & \mathbf{X}\mathbf{b} - (\lambda + t)\mathbf{x} \\ \mathbf{x}^T(\mathbf{A}+\lambda\mathbf{I}) + x\mathbf{b}^T & \mathbf{x}^T\mathbf{b} - (\lambda + t)x \end{bmatrix} \right) \\
&= (-x + 1)t + \big(\mathrm{Tr}(\mathbf{X}) - x + y\big)\lambda + \mathrm{Tr}(\mathbf{X}\mathbf{A}) + 2\mathbf{b}^T\mathbf{x}. \quad (9.93)
\end{aligned}
$$

Then we have

$$
\sup_{\lambda, t} \tilde{h}(\lambda, t) = \begin{cases} \mathrm{Tr}(\mathbf{X}\mathbf{A}) + 2\mathbf{b}^T\mathbf{x}, & -x + 1 = 0, \ \mathrm{Tr}(\mathbf{X}) - x + y = 0 \\ \infty, & \text{otherwise.} \end{cases} \quad (9.94)
$$

Then by (9.47) we come up with the dual of the SDP (9.92) as follows

$$
\begin{array}{ccc}
\begin{aligned} &\min \ \sup_{\lambda,t} \tilde{h}(\lambda,t) \\ &\text{s.t.} \ \begin{bmatrix} \mathbf{X} & \mathbf{x} \\ \mathbf{x}^T & x \end{bmatrix} \succeq \mathbf{0}, \ y \geq 0 \end{aligned} & = & \begin{aligned} &\min \ \mathrm{Tr}(\mathbf{A}\mathbf{X}) + 2\mathbf{b}^T\mathbf{x} \\ &\text{s.t.} \ \begin{vmatrix} \mathbf{X} & \mathbf{x} \\ \mathbf{x}^T & 1 \end{vmatrix} \succeq \mathbf{0}, \ \mathrm{Tr}(\mathbf{X}) - 1 \leq 0 \end{aligned}
\end{array} \quad (9.95)
$$

which, by applying Schur complement (cf. (3.92) and Remark 3.29) to the generalized inequality constraint, can be alternatively expressed as

$$
\begin{aligned}
p^\star = &\min \ \mathrm{Tr}(\mathbf{A}\mathbf{X}) + 2\mathbf{b}^T\mathbf{x} \\
&\text{s.t.} \ \mathrm{Tr}(\mathbf{X}) - 1 \leq 0, \ \mathbf{X} \succeq \mathbf{x}\mathbf{x}^T
\end{aligned} \quad (9.96)
$$

which corresponds to a convex problem obtained by replacing $\mathbf{X} = \mathbf{x}\mathbf{x}^T$ and then relaxing it to the convex constraint $\mathbf{X} \succeq \mathbf{x}\mathbf{x}^T$ in problem (9.88). Next, let us prove the strong duality of problem (9.88) when the minimum eigenvalue of $\mathbf{A}$ is strictly less than zero, i.e., $\lambda_{\min}(\mathbf{A}) < 0$.

Let $\lambda^\star$ denote an optimal solution to (9.91), and $(\mathbf{X}^\star, \mathbf{x}^\star)$ an optimal solution to (9.96). By the KKT conditions (cf. (9.205)) associated with the problem (9.96) and its dual (9.91), it can be shown (cf. Example 9.17) that

$$
\begin{cases}
\lambda^\star \geq -\lambda_{\min}(\mathbf{A}) > 0 \\
\mathbf{x}^\star = -(\mathbf{A} + \lambda^\star\mathbf{I})^\dagger \mathbf{b}, \ \mathbf{b} \in \mathcal{R}(\mathbf{A} + \lambda^\star\mathbf{I}) \\
\mathbf{X}^\star = \mathbf{x}^\star(\mathbf{x}^\star)^T + \mathbf{v}\mathbf{v}^T, \ \mathbf{v} \in \mathcal{N}(\mathbf{A} + \lambda^\star\mathbf{I}) \\
\mathrm{Tr}(\mathbf{X}^\star) = \|\mathbf{x}^\star\|_2^2 + \|\mathbf{v}\|_2^2 = 1 \\
p^\star = d^\star = -\lambda^\star - \mathbf{b}^T(\mathbf{A} + \lambda^\star\mathbf{I})^\dagger \mathbf{b}
\end{cases} \quad (9.97)
$$

provided that $\dim\left(\mathcal{N}(\mathbf{A} + \lambda^\star \mathbf{I})\right) \geq 1$. Note that $\mathbf{X}^\star$ given in (9.97) is a rank-2 solution but not unique; otherwise $\mathbf{X}^\star = \mathbf{x}\mathbf{x}^\star$ will be a unique rank-1 solution if $\mathbf{A} + \lambda^\star \mathbf{I} \succ \mathbf{0}$.

Let $\boldsymbol{x}^\star = \mathbf{x}^\star + \mathbf{v}$, which is a feasible point of problem (9.88) due to $\|\boldsymbol{x}^\star\|_2^2 = \|\mathbf{x}^\star\|_2^2 + \|\mathbf{v}\|_2^2 = 1$. Then by (9.97), substituting $\boldsymbol{x}^\star$ into the objective function of problem (9.88) gives rise to

$$
\begin{aligned}
f(\boldsymbol{x}^\star) &= (\boldsymbol{x}^\star)^T \mathbf{A}\boldsymbol{x}^\star + 2\mathbf{b}^T\boldsymbol{x}^\star \\
&= (\boldsymbol{x}^\star)^T(\mathbf{A} + \lambda^\star \mathbf{I})\boldsymbol{x}^\star + 2\mathbf{b}^T\boldsymbol{x}^\star - \lambda^\star (\boldsymbol{x}^\star)^T \boldsymbol{x}^\star \\
&= (\mathbf{x}^\star + \mathbf{v})^T(\mathbf{A} + \lambda^\star \mathbf{I})(\mathbf{x}^\star + \mathbf{v}) + 2\mathbf{b}^T(\mathbf{x}^\star + \mathbf{v}) - \lambda^\star \\
&= -\lambda^\star - \mathbf{b}^T(\mathbf{A} + \lambda^\star \mathbf{I})^\dagger \mathbf{b} = d^\star.
\end{aligned}
$$

Hence, the strong duality holds true for the trust region problem even though $\mathbf{A} \not\succeq \mathbf{0}$ as long as Slater's condition holds true and $\mathbf{b} \in \mathcal{R}(\mathbf{A} + \lambda^\star \mathbf{I})$. $\square$

Remark 9.7 Note that, as expected, when $\mathbf{A} \succeq \mathbf{0}$, problem (9.96) reduces to problem (9.88) due to $\mathrm{Tr}(\mathbf{A}\mathbf{X}) \geq \mathrm{Tr}(\mathbf{A}\mathbf{x}\mathbf{x}^T) = \mathbf{x}^T\mathbf{A}\mathbf{x}$. When $\mathbf{A} \not\succeq \mathbf{0}$, it can be easily shown that f^* does not exist even though the strong duality holds. Although problem (9.88) is nothing but a special case of the nonconvex problem (9.56), (9.57) cannot be employed to find the bidual of problem (9.88) due to inexistence of f^*. $\square$

9.3.2   S-Lemma

The strong duality of convex problems plays an important role not only in solving various convex problems, but also in obtaining many useful results, such as the S-procedure (cf. (8.18)) (which has been extensively applied to robust transmit beamforming design against CSI uncertainty in wireless communication systems as presented in Chapter 8), and alternatives of two systems consisting of equalities and inequalities to be introduced in Section 9.9. An alternative to the S-procedure (referred to as the S-lemma) that can be proven via strong duality is presented next.

Let

$$
\mathbf{A} = -\begin{bmatrix} \mathbf{F}_1 & \mathbf{g}_1 \\ \mathbf{g}_1^T & h_1 \end{bmatrix}, \quad \mathbf{B} = -\begin{bmatrix} \mathbf{F}_2 & \mathbf{g}_2 \\ \mathbf{g}_2^T & h_2 \end{bmatrix}, \tag{9.98}
$$

where $\mathbf{F}_i$, $\mathbf{g}_i$, and h_i are the corresponding parameters mentioned in the S-procedure. Then the implication involving the two quadratic inequalities in (8.18), i.e.,

$$
\mathbf{x}^T\mathbf{F}_1\mathbf{x} + 2\mathbf{g}_1^T\mathbf{x} + h_1 \leq 0 \implies \mathbf{x}^T\mathbf{F}_2\mathbf{x} + 2\mathbf{g}_2^T\mathbf{x} + h_2 \leq 0, \tag{9.99}
$$

corresponds to the implication given in (9.101) below with $\mathbf{z} = (\mathbf{x}, 1)$; the premise $\widehat{\mathbf{x}}^T\mathbf{F}_1\widehat{\mathbf{x}} + 2\mathbf{g}_1^T\widehat{\mathbf{x}} + h_1 < 0$ corresponds to $\mathbf{y}^T\mathbf{A}\mathbf{y} > 0$ with $\mathbf{y} = (\widehat{\mathbf{x}}, 1)$; the LMI given by (8.19), i.e.,

$$\begin{bmatrix} \mathbf{F}_2 & \mathbf{g}_2 \\ \mathbf{g}_2^T & h_2 \end{bmatrix} \preceq \lambda \begin{bmatrix} \mathbf{F}_1 & \mathbf{g}_1 \\ \mathbf{g}_1^T & h_1 \end{bmatrix}, \quad \text{for some } \lambda \geq 0 \tag{9.100}$$

corresponds to (9.102) below. With the above correspondences, the S-procedure can be alternatively stated in the following lemma, which, as an illustration of the essential role of the strong duality, will be proven next via the use of Lemma 9.1 below.

S-Lemma: *Let* $\mathbf{A}, \mathbf{B} \in \mathbb{S}^n$ *and assume that* $\mathbf{y}^T \mathbf{A} \mathbf{y} > 0$ *for some vector* $\mathbf{y} \in \mathbb{R}^n$. *Then, the implication*

$$\mathbf{z}^T \mathbf{A} \mathbf{z} \geq 0 \;\Rightarrow\; \mathbf{z}^T \mathbf{B} \mathbf{z} \geq 0 \tag{9.101}$$

is valid if and only if

$$\mathbf{B} \succeq \lambda \mathbf{A} \;\text{ for some } \lambda \geq 0. \tag{9.102}$$

Proof: Let us first prove the sufficient condition [i.e., (9.102) ⇒ (9.101)]. Due to $\mathbf{B} - \lambda \mathbf{A} \succeq \mathbf{0}$ for some $\lambda \geq 0$ and $\mathbf{z}^T \mathbf{A} \mathbf{z} \geq 0$, we have

$$\mathbf{z}^T (\mathbf{B} - \lambda \mathbf{A}) \mathbf{z} \geq 0 \text{ and } \mathbf{z}^T \mathbf{A} \mathbf{z} \geq 0 \;\Rightarrow\; \mathbf{z}^T \mathbf{B} \mathbf{z} \geq \lambda \mathbf{z}^T \mathbf{A} \mathbf{z} \geq 0. \tag{9.103}$$

Therefore, the sufficient condition is proved.

Next, we show the necessary condition [i.e., (9.101) ⇒ (9.102)]. To this end, consider the following convex problem:

$$s^\star = \max_{s, \lambda \in \mathbb{R}} s \tag{9.104a}$$

$$\text{s.t. } \mathbf{B} - \lambda \mathbf{A} \succeq s \mathbf{I}_n, \tag{9.104b}$$

$$\lambda \geq 0. \tag{9.104c}$$

One can see that if $s^\star \geq 0$, then the statement of (9.102) is true. Hence, it suffices to show that $s^\star$ is nonnegative, provided that $\mathbf{z}^T \mathbf{A} \mathbf{z} \geq 0 \;\Rightarrow\; \mathbf{z}^T \mathbf{B} \mathbf{z} \geq 0$ holds true. Since there exists some vector $\mathbf{y} \in \mathbb{R}^n$ such that $\mathbf{y}^T \mathbf{A} \mathbf{y} > 0$, the matrix $\mathbf{A}$ must have a positive eigenvalue. In other words, there exists $a > 0$ and $\mathbf{v} \in \mathbb{R}^n$ with $\|\mathbf{v}\|_2 = 1$ such that $\mathbf{A} \mathbf{v} = a \mathbf{v}$. Then, $\mathbf{v}^T (\mathbf{B} - \lambda \mathbf{A}) \mathbf{v} = \mathbf{v}^T \mathbf{B} \mathbf{v} - \lambda a \leq \mathbf{v}^T \mathbf{B} \mathbf{v}$, and from (9.104b), one can show that $s \leq \mathbf{v}^T \mathbf{B} \mathbf{v}$, implying that the optimal value of problem (9.104) is bounded above, i.e., the optimal value $s^\star < \infty$.

Because problem (9.104) is convex and satisfies Slater's condition (implying that $s^\star > -\infty$), strong duality holds true. So the optimal value $s^\star$ of problem (9.104) is finite and the same as the optimal value of the associated dual problem

(9.105), which is feasible and can be readily shown by (9.47) to be

$$s^\star = \min_{\mathbf{X} \in \mathbb{S}^n} \mathrm{Tr}(\mathbf{BX}) \tag{9.105a}$$

$$\text{s.t. } \mathrm{Tr}(\mathbf{AX}) \geq 0, \tag{9.105b}$$

$$\mathrm{Tr}(\mathbf{X}) = 1, \tag{9.105c}$$

$$\mathbf{X} \succeq \mathbf{0}. \tag{9.105d}$$

Let $\mathbf{X}^\star = \mathbf{DD}^T$ (due to $\mathbf{X}^\star \succeq \mathbf{0}$) be an optimal solution to problem (9.105). Then, the associated constraint (9.105b) and the optimal value can be, respectively, written as

$$0 \leq \mathrm{Tr}(\mathbf{AX}^\star) = \mathrm{Tr}(\mathbf{D}^T\mathbf{AD}) \tag{9.106a}$$

$$s^\star = \mathrm{Tr}(\mathbf{BX}^\star) = \mathrm{Tr}(\mathbf{D}^T\mathbf{BD}). \tag{9.106b}$$

Now, we show $s^\star \geq 0$ by contradiction. Supposing $s^\star < 0$ and denoting

$$\mathbf{P} = \mathbf{D}^T\mathbf{AD} \text{ and } \mathbf{Q} = \mathbf{D}^T\mathbf{BD},$$

we have $\mathrm{Tr}(\mathbf{P}) \geq 0$ and $\mathrm{Tr}(\mathbf{Q}) = s^\star < 0$ [by (9.106a) and (9.106b), respectively]. According to Lemma 9.1, there exists a vector $\mathbf{x}$ such that

$$\mathbf{x}^T\mathbf{Px} = (\mathbf{Dx})^T\mathbf{A}(\mathbf{Dx}) \geq 0 \text{ and } \mathbf{x}^T\mathbf{Qx} = (\mathbf{Dx})^T\mathbf{B}(\mathbf{Dx}) < 0,$$

which contradicts the premise $\mathbf{z}^T\mathbf{Az} \geq 0 \; \Rightarrow \; \mathbf{z}^T\mathbf{Bz} \geq 0$ (in which $\mathbf{z} = \mathbf{Dx}$). As a result, we have shown $s^\star \geq 0$, and the proof of S-lemma is complete. ∎

Lemma 9.1. *Let* $\mathbf{P}, \mathbf{Q} \in \mathbb{S}^n$ *with* $\mathrm{Tr}(\mathbf{P}) \geq 0$ *and* $\mathrm{Tr}(\mathbf{Q}) < 0$. *Then, there exists a vector* $\mathbf{x} \in \mathbb{R}^n$ *such that* $\mathbf{x}^T\mathbf{Px} \geq 0$ *and* $\mathbf{x}^T\mathbf{Qx} < 0$.

Proof: Let $\mathbf{Q} = \boldsymbol{U}\boldsymbol{\Lambda}\boldsymbol{U}^T$ (EVD of $\mathbf{Q}$) where $\boldsymbol{U}$ is an orthogonal matrix (i.e., $\boldsymbol{U}^T\boldsymbol{U} = \boldsymbol{U}\boldsymbol{U}^T = \mathbf{I}_n$) and $\boldsymbol{\Lambda} = \mathbf{Diag}(\lambda_1, \dots, \lambda_n)$ is a diagonal matrix formed by its eigenvalues. Let $\mathbf{w} \in \mathbb{R}^n$ be a discrete random vector with i.i.d. entries taking on values of 1 or -1 with equal probability $(1/2)$. Then,

$$(\boldsymbol{U}\mathbf{w})^T\mathbf{Q}(\boldsymbol{U}\mathbf{w}) = (\boldsymbol{U}\mathbf{w})^T\boldsymbol{U}\boldsymbol{\Lambda}\boldsymbol{U}^T(\boldsymbol{U}\mathbf{w}) = \mathbf{w}^T\boldsymbol{\Lambda}\mathbf{w} = \mathrm{Tr}(\mathbf{Q}) < 0 \tag{9.107}$$

$$(\boldsymbol{U}\mathbf{w})^T\mathbf{P}(\boldsymbol{U}\mathbf{w}) = \mathbf{w}^T(\boldsymbol{U}^T\mathbf{P}\boldsymbol{U})\mathbf{w}. \tag{9.108}$$

Taking expectation of the quadratic function of $\mathbf{w}$ in (9.108) yields

$$\mathbb{E}\{(\boldsymbol{U}\mathbf{w})^T\mathbf{P}(\boldsymbol{U}\mathbf{w})\} = \mathrm{Tr}(\boldsymbol{U}^T\mathbf{P}\boldsymbol{U}\mathbb{E}\{\mathbf{ww}^T\}) = \mathrm{Tr}(\mathbf{P}) \geq 0, \tag{9.109}$$

where we have used the fact of $\mathbb{E}\{\mathbf{ww}^T\} = \mathbf{I}_n$. By (9.107) and (9.109), there exists at least one vector $\boldsymbol{w} \in \mathbb{R}^n$ such that $(\boldsymbol{U}\boldsymbol{w})^T\mathbf{Q}(\boldsymbol{U}\boldsymbol{w}) < 0$ and $(\boldsymbol{U}\boldsymbol{w})^T\mathbf{P}(\boldsymbol{U}\boldsymbol{w}) \geq 0$, i.e., there exists an $\mathbf{x} = \boldsymbol{U}\boldsymbol{w}$ such that $\mathbf{x}^T\mathbf{Qx} < 0$ and $\mathbf{x}^T\mathbf{Px} \geq 0$. The proof is thus complete. ∎

Remark 9.8 The S-lemma above is presented in terms of the matrices $\mathbf{A} \in \mathbb{S}^n$ and $\mathbf{B} \in \mathbb{S}^n$ defined in (9.98), and $\mathbf{A}$ has been shown to have a positive eigenvalue in the preceding proof. Note that S-lemma and S-procedure have the identical sufficient condition (cf. (9.100) and (9.102)). However, their premise and necessary condition look similar in form with the following implications:

$$\text{S-lemma} \quad \text{versus} \quad \text{S-procedure}$$
$$\text{Premise: } \exists \mathbf{y}, \mathbf{y}^T \mathbf{A} \mathbf{y} > 0 \implies \widehat{\mathbf{x}}^T \mathbf{F}_1 \widehat{\mathbf{x}} + 2\mathbf{g}_1^T \widehat{\mathbf{x}} + h_1 < 0 \ \ (\text{e.g., } \mathbf{y} = (\widehat{\mathbf{x}}, 1))$$
$$\text{Necessary condition: } (9.101) \implies (9.99) \ \ (\text{e.g., } \mathbf{z} = (\mathbf{x}, 1)).$$

The reverse implications are also true and the corresponding proof is relegated to Remark 9.22 (for the proof of the feasibility equivalence of (9.258) and (9.262)) in Subsection 9.9.3, where $\mathbf{A}_1$ and $\mathbf{A}_2$ (cf. (9.259) and (9.260)) that define (9.258) correspond to $-\mathbf{A}$ and $\mathbf{B}$ (cf. (9.98)) in the preceding S-lemma, respectively. $\square$

9.4   Implications of strong duality

Strong duality provides some analytical insights into an optimization problem, alternatives to solving the problem, necessary conditions for optimal solutions, and certificate of suboptimality if the primal problem given by (9.1) together with the dual problem given by (9.37) is solved iteratively. These characteristics implied from strong duality are discussed in the following subsections.

9.4.1   Max-min characterization of weak and strong duality

Suppose that there are no equality constraints for simplicity. Then,

$$\sup_{\boldsymbol{\lambda} \succeq 0} \mathcal{L}(\mathbf{x}, \boldsymbol{\lambda}) = \sup_{\boldsymbol{\lambda} \succeq 0} \left\{ f_0(\mathbf{x}) + \sum_{i=1}^m \lambda_i f_i(\mathbf{x}) \right\}$$
$$= \begin{cases} f_0(\mathbf{x}), & \text{if } f_i(\mathbf{x}) \le 0, \ i = 1, \dots, m \\ \infty, & \text{otherwise.} \end{cases} \quad (9.110)$$

Then

$$p^\star = \inf_{\mathbf{x}} \sup_{\boldsymbol{\lambda} \succeq 0} \mathcal{L}(\mathbf{x}, \boldsymbol{\lambda}). \quad (9.111)$$

By the definition of the dual problem, we have

$$d^\star = \sup_{\boldsymbol{\lambda} \succeq 0} \inf_{\mathbf{x}} \mathcal{L}(\mathbf{x}, \boldsymbol{\lambda}). \quad (9.112)$$

Thus, weak duality can be expressed as the inequality

$$d^\star = \sup_{\boldsymbol{\lambda} \succeq 0} \inf_{\mathbf{x}} \mathcal{L}(\mathbf{x}, \boldsymbol{\lambda}) \le \inf_{\mathbf{x}} \sup_{\boldsymbol{\lambda} \succeq 0} \mathcal{L}(\mathbf{x}, \boldsymbol{\lambda}) = p^\star. \quad (9.113)$$

When strong duality holds,

$$d^\star = \sup_{\boldsymbol{\lambda} \succeq 0} \inf_{\mathbf{x}} \mathcal{L}(\mathbf{x}, \boldsymbol{\lambda}) = \mathcal{L}(\mathbf{x}^\star, \boldsymbol{\lambda}^\star) = \inf_{\mathbf{x}} \sup_{\boldsymbol{\lambda} \succeq 0} \mathcal{L}(\mathbf{x}, \boldsymbol{\lambda}) = p^\star, \quad (9.114)$$

(cf. (9.43)) which means the primal-dual optimal $(\mathbf{x}^\star, \boldsymbol{\lambda}^\star)$ is a saddle point of the Lagrangian (cf. Figure 9.2), i.e.,

$$\mathcal{L}(\mathbf{x}^\star, \boldsymbol{\lambda}) \leq \sup_{\boldsymbol{\lambda} \succeq 0} \mathcal{L}(\mathbf{x}^\star, \boldsymbol{\lambda}) = \mathcal{L}(\mathbf{x}^\star, \boldsymbol{\lambda}^\star) = \inf_{\mathbf{x}} \mathcal{L}(\mathbf{x}, \boldsymbol{\lambda}^\star) \leq \mathcal{L}(\mathbf{x}, \boldsymbol{\lambda}^\star), \qquad (9.115)$$

Hence, supposing that there exist a primal optimal $\mathbf{x}^\star$ and a dual optimal $\boldsymbol{\lambda}^\star$ to a convex optimization problem with strong duality (say (9.112)), it is possible to obtain the primal-dual optimal solution $(\mathbf{x}^\star, \boldsymbol{\lambda}^\star)$ (a saddle point of $\mathcal{L}(\mathbf{x}, \boldsymbol{\lambda})$) by alternatively handling the inner minimization (i.e., updating the primal variable $\mathbf{x}$) and the outer maximization (i.e., updating the dual variable $\boldsymbol{\lambda}$).

When equality constraints exist, the saddle point implication from the strong duality for the primal-dual optimal solution $(\mathbf{x}^\star, \boldsymbol{\lambda}^\star, \boldsymbol{\nu}^\star)$ is also true, (i.e., (9.115) still holds true with $\boldsymbol{\lambda}$ replaced by $\boldsymbol{\lambda}, \boldsymbol{\nu}$, and $\boldsymbol{\lambda}^\star$ replaced by $\boldsymbol{\lambda}^\star, \boldsymbol{\nu}^\star$). We end up with the following remark:

Remark 9.9 [BSS06, Theorem 6.2.5] $(\mathbf{x}^\star, \boldsymbol{\lambda}^\star, \boldsymbol{\nu}^\star)$ is a saddle point of the Lagrangian (9.3) if and only if $\mathbf{x}^\star$ and $(\boldsymbol{\lambda}^\star, \boldsymbol{\nu}^\star)$ are, respectively, optimal solutions to the primal problem (9.1) and the dual problem (9.37) with zero duality gap (i.e., $p^\star = d^\star$). ☐

Remark 9.10 *(max-min inequality)* For any $f : \mathbb{R}^n \times \mathbb{R}^m$ and any $W \subseteq \mathbb{R}^n$ and $Z \subseteq \mathbb{R}^m$, the max-min inequality is given by

$$\sup_{\boldsymbol{z} \in Z} \inf_{\boldsymbol{w} \in W} f(\boldsymbol{w}, \boldsymbol{z}) \leq \inf_{\boldsymbol{w} \in W} \sup_{\boldsymbol{z} \in Z} f(\boldsymbol{w}, \boldsymbol{z}). \qquad (9.116)$$

This inequality holds with the equality if f is the Lagrangian of an optimization problem with zero duality gap. ☐

9.4.2   Certificate of suboptimality

Suppose that there is an algorithm that solves the primal and dual problems, jointly, by generating a sequence of primal feasible points $\mathbf{x}^{(1)}, \mathbf{x}^{(2)}, \ldots$, and dual feasible points $(\boldsymbol{\lambda}^{(1)}, \boldsymbol{\nu}^{(1)}), (\boldsymbol{\lambda}^{(2)}, \boldsymbol{\nu}^{(2)}), \ldots$. Then,

$$f_0(\mathbf{x}^{(k)}) - p^\star = f_0(\mathbf{x}^{(k)}) - d^\star \qquad \text{(by strong duality)}$$
$$\leq f_0(\mathbf{x}^{(k)}) - g(\boldsymbol{\lambda}, \boldsymbol{\nu}), \ \forall \boldsymbol{\lambda} \succeq \mathbf{0}, \ \boldsymbol{\nu} \in \mathbb{R}^p$$
$$\implies f_0(\mathbf{x}^{(k)}) - p^\star \leq f_0(\mathbf{x}^{(k)}) - g(\boldsymbol{\lambda}^{(k)}, \boldsymbol{\nu}^{(k)}) = \eta(\mathbf{x}^{(k)}, \boldsymbol{\lambda}^{(k)}, \boldsymbol{\nu}^{(k)}).$$

If the duality gap $\eta(\mathbf{x}^{(k)}, \boldsymbol{\lambda}^{(k)}, \boldsymbol{\nu}^{(k)}) \leq \epsilon$, then $f_0(\mathbf{x}^{(k)}) - p^\star \leq \epsilon$ (i.e., $\mathbf{x}^{(k)}$ is ϵ-suboptimal).

9.4.3   Complementary slackness

Let $\mathbf{x}^\star$ and $(\boldsymbol{\lambda}^\star, \boldsymbol{\nu}^\star)$ be primal and dual optimal points, respectively, implying that $f_i(\mathbf{x}^\star) \leq 0$, $i = 1, \ldots, m$, $h_j(\mathbf{x}^\star) = 0$, $j = 1, \ldots, p$, and $\boldsymbol{\lambda}^\star \succeq \mathbf{0}_m$. When strong duality holds, the following condition must be satisfied:

$$\lambda_i^\star f_i(\mathbf{x}^\star) = 0, \ i = 1, \ldots, m. \tag{9.117}$$

In other words,

$$\mathbf{0}_m \preceq \boldsymbol{\lambda}^\star \perp \boldsymbol{f}(\mathbf{x}^\star) \preceq \mathbf{0}_m, \tag{9.118}$$

where $\boldsymbol{f}(\mathbf{x}^\star) = [f_1(\mathbf{x}^\star), \ldots, f_m(\mathbf{x}^\star)]^T$. That is, the second term $\sum_{i=1}^m \lambda_i^\star f_i(\mathbf{x}^\star) = 0$ in $\mathcal{L}(\mathbf{x}^\star, \boldsymbol{\lambda}^\star, \boldsymbol{\nu}^\star)$ (cf. (9.3)). This condition implies that

$$\lambda_i^\star > 0 \Rightarrow f_i(\mathbf{x}^\star) = 0, \tag{9.119a}$$
$$f_i(\mathbf{x}^\star) < 0 \Rightarrow \lambda_i^\star = 0. \tag{9.119b}$$

Proof of (9.117): By the strong duality identity (9.43), we have

$$f_0(\mathbf{x}^\star) = g(\boldsymbol{\lambda}^\star, \boldsymbol{\nu}^\star) = \inf_{\mathbf{x} \in \mathcal{D}} \mathcal{L}(\mathbf{x}, \boldsymbol{\lambda}^\star, \boldsymbol{\nu}^\star) \tag{9.120}$$

$$= \inf_{\mathbf{x} \in \mathcal{D}} \left\{ f_0(\mathbf{x}) + \sum_{i=1}^m \lambda_i^\star f_i(\mathbf{x}) + \sum_{i=1}^p \nu_i^\star h_i(\mathbf{x}) \right\}$$

$$= f_0(\mathbf{x}^\star) + \sum_{i=1}^m \lambda_i^\star f_i(\mathbf{x}^\star) + \sum_{i=1}^p \nu_i^\star h_i(\mathbf{x}^\star),$$

implying that $\sum_{i=1}^m \lambda_i^\star f_i(\mathbf{x}^\star) = 0$, which holds only if $\lambda_i^\star f_i(\mathbf{x}^\star) = 0$ for all i. ∎

A necessary condition implied from (9.120) is given in the following fact:

Fact 9.3 If the strong duality holds for an optimization problem with the objective function f_0 and all the inequality constraint functions f_i and equality constraint functions h_i being differentiable (i.e., its Lagrangian $\mathcal{L}(\mathbf{x}, \boldsymbol{\lambda}^\star, \boldsymbol{\nu}^\star)$ is differentiable w.r.t. the unknown variable $\mathbf{x}$), it is necessary that $\nabla_{\mathbf{x}} \mathcal{L}(\mathbf{x}, \boldsymbol{\lambda}^\star, \boldsymbol{\nu}^\star) = \mathbf{0}$ for $\mathbf{x} = \mathbf{x}^\star$ by (9.120).

Remark 9.11 Solving the primal problem via its dual is also possible. Suppose that strong duality holds and an optimal $(\boldsymbol{\lambda}^\star, \boldsymbol{\nu}^\star)$ is known ahead of time (by solving the dual problem first). Then one can find the solution set of the unconstrained optimization problem $\min_{\mathbf{x}} \mathcal{L}(\mathbf{x}, \boldsymbol{\lambda}^\star, \boldsymbol{\nu}^\star)$, from which we can find the primal optimal $\mathbf{x}^\star$ with the minimum value of $f_0(\mathbf{x}^\star) = g(\boldsymbol{\lambda}^\star, \boldsymbol{\nu}^\star)$ if $\mathbf{x}^\star$ exists. As $\mathcal{L}(\mathbf{x}, \boldsymbol{\lambda}^\star, \boldsymbol{\nu}^\star)$ is differentiable, Fact 9.3 may also be helpful for finding the optimal $\mathbf{x}^\star$ even if the primal problem is nonconvex. □

9.5 Karush–Kuhn–Tucker (KKT) optimality conditions

Suppose that $f_0, f_1, \ldots, f_m, h_1, \ldots, h_p$ are differentiable. Then the KKT conditions for the primal optimal point $\mathbf{x}^\star$ and dual optimal $(\boldsymbol{\lambda}^\star, \boldsymbol{\nu}^\star)$ of both problem (9.1) (which is not necessarily convex) and problem (9.65) (which is convex) are as follows:

$$\nabla f_0(\mathbf{x}^\star) + \sum_{i=1}^{m} \lambda_i^\star \nabla f_i(\mathbf{x}^\star) + \sum_{i=1}^{p} \nu_i^\star \nabla h_i(\mathbf{x}^\star) = \mathbf{0}, \tag{9.121a}$$

$$f_i(\mathbf{x}^\star) \leq 0, \ i = 1, \ldots, m, \tag{9.121b}$$

$$h_i(\mathbf{x}^\star) = 0, \ i = 1, \ldots, p, \tag{9.121c}$$

$$\lambda_i^\star \geq 0, \ i = 1, \ldots, m, \tag{9.121d}$$

$$\lambda_i^\star f_i(\mathbf{x}^\star) = 0, \ i = 1, \ldots, m. \tag{9.121e}$$

In general, the KKT conditions given by (9.121b) and (9.121c) are actually the inequality and equality constraints of the primal problem, while those given by (9.121a) (equality constraints w.r.t. $(\boldsymbol{\lambda}, \boldsymbol{\nu})$) and (9.121d) (inequality constraints) are actually the constraints of the dual problem, and the complementary slackness given by (9.121e) couples the inequality constraint functions of the primal problem and the dual problem. Next, let us prove the KKT conditions and discuss their impacts on optimization problems.

- For problems with strong duality, the KKT conditions (9.121) are necessary optimality conditions as proved in the previous section. Namely, if $\mathbf{x}^\star$ and $(\boldsymbol{\lambda}^\star, \boldsymbol{\nu}^\star)$ are primal and dual optimal, then the KKT conditions must hold. Specifically, (9.121a) is due to $\nabla_{\mathbf{x}} \mathcal{L}(\mathbf{x}, \boldsymbol{\lambda}^\star, \boldsymbol{\nu}^\star) = \mathbf{0}$ for $\mathbf{x} = \mathbf{x}^\star$ (by Fact 9.3); (9.121b) and (9.121c) are due to that $\mathbf{x}^\star$ is primal feasible; (9.121d) is due to that $(\boldsymbol{\lambda}^\star, \boldsymbol{\nu}^\star)$ is dual feasible; (9.121e) is due to complementary slackness (9.117).

- For problems without strong duality, the KKT conditions are necessary for local optimality under a mild assumption [Ber99]: If $\mathbf{x}^\star$ is locally optimal and $\mathbf{x}^\star$ is regular (see Remark 9.12 below), then there exists a $(\boldsymbol{\lambda}^\star = (\lambda_1^\star, \ldots, \lambda_m^\star) \succeq \mathbf{0}, \boldsymbol{\nu}^\star)$ and $\lambda_i^\star = 0$ for $i \notin \mathcal{I}$ (cf. (9.122)) such that the KKT conditions hold. Some more discussions are given in Remark 9.15.

Remark 9.12 Let $\mathbf{x}^\star$ satisfy $f_i(\mathbf{x}^\star) \leq 0$, $i = 1, \ldots, m$, $h_j(\mathbf{x}^\star) = 0$, $j = 1, \ldots, p$, and let $\mathcal{I}(\mathbf{x}^\star)$ be the index set of active inequality constraints:

$$\mathcal{I}(\mathbf{x}^\star) \triangleq \{i \mid f_i(\mathbf{x}^\star) = 0\}. \tag{9.122}$$

Then, we say that $\mathbf{x}^\star$ is a *regular point* [Ber99] if the vectors $\nabla f_i(\mathbf{x}^\star)$, $\nabla h_j(\mathbf{x}^\star)$, for all $i \in \mathcal{I}(\mathbf{x}^\star)$, $1 \leq j \leq p$ are linearly independent.   □

- For convex problems with strong duality (e.g., when the Slater's condition is satisfied), the KKT conditions are the sufficient and necessary optimality conditions, i.e., $\mathbf{x}^\star$ and $(\boldsymbol{\lambda}^\star, \boldsymbol{\nu}^\star)$ are primal and dual optimal if and only if the KKT conditions hold.

Proof for sufficiency. For the convex problem (9.65), $\mathcal{L}(\mathbf{x}, \boldsymbol{\lambda}^\star, \boldsymbol{\nu}^\star)$ given by (9.3) is convex in $\mathbf{x} \in \mathcal{D}$ (since $\boldsymbol{\lambda}^\star \succeq \mathbf{0}$ by (9.121d)). We have

$$d^\star = g(\boldsymbol{\lambda}^\star, \boldsymbol{\nu}^\star) = \inf_{\mathbf{x} \in \mathcal{D}} \mathcal{L}(\mathbf{x}, \boldsymbol{\lambda}^\star, \boldsymbol{\nu}^\star)$$

$$= \mathcal{L}(\mathbf{x}^\star, \boldsymbol{\lambda}^\star, \boldsymbol{\nu}^\star) > -\infty \quad \text{(by (9.121a) and (4.28))},$$

which also implies that $(\boldsymbol{\lambda}^\star, \boldsymbol{\nu}^\star)$ is dual feasible. Moreover, $\mathbf{x}^\star$ is primal feasible due to (9.121b) and (9.121c). By (9.121e) and (9.121c), the above equation can be further simplified as

$$d^\star = f_0(\mathbf{x}^\star) + \sum_{i=1}^m \lambda_i^\star f_i(\mathbf{x}^\star) + \sum_{i=1}^p \nu_i^\star h_i(\mathbf{x}^\star) = f_0(\mathbf{x}^\star) = p^\star.$$

Therefore, strong duality holds, and $\mathbf{x}^\star$ and $(\boldsymbol{\lambda}^\star, \boldsymbol{\nu}^\star)$ are primal and dual optimal, respectively. ∎

The pair of the primal and dual convex problems with zero duality gap and all the objective functions and constraint functions differentiable can be solved simultaneously by solving the associated KKT conditions (9.121) that are necessary and sufficient for the optimality of the two problems. Surely closed-form solutions for the primal-dual optimal $(\mathbf{x}^\star, \boldsymbol{\lambda}^\star, \boldsymbol{\nu}^\star)$ are most desirable due to their high accuracy and minimal computational complexity in general. In practical applications, a closed-form solution for the optimal $(\mathbf{x}^\star, \boldsymbol{\lambda}^\star, \boldsymbol{\nu}^\star)$ may be hard to obtain by analytically solving the KKT conditions. Then one can use off-the-shelf convex solvers CVX and SeDuMi, which were developed using interior-point methods to be introduced in Chapter 10, yielding numerical solutions for the optimal $(\mathbf{x}^\star, \boldsymbol{\lambda}^\star, \boldsymbol{\nu}^\star)$. However, these convex solvers are general-purpose convex solvers, which usually are not very computationally efficient but very useful for performance evaluation of the problem to be solved during the research stage. A practical custom-designed fast algorithm will be developed in the ensuing implementation stage if it is needed.

Example 9.10 *(Minimum-norm solution for linear equations)* This convex problem has been defined in (9.8) (in Example 9.1 in Subsection 9.1.1) for which strong duality holds by Slater's condition. The KKT conditions are simply $\mathbf{Ax} = \mathbf{b}$ (by (9.121c)) and

$$\nabla_{\mathbf{x}} \mathcal{L}(\mathbf{x}, \boldsymbol{\nu}) = 2\mathbf{x} + \mathbf{A}^T \boldsymbol{\nu} = \mathbf{0} \quad \left(\text{or } \mathbf{x} = -\frac{1}{2} \mathbf{A}^T \boldsymbol{\nu}\right)$$

by (9.9) and (9.121a). From these two KKT conditions, we can easily obtain the optimum

$$\mathbf{x}^\star = \mathbf{A}^T (\mathbf{A}\mathbf{A}^T)^{-1} \mathbf{b} = \mathbf{A}^\dagger \mathbf{b} \tag{9.123}$$

(cf. (1.126)) which is exactly the minimum-norm solution.

Alternatively, as mentioned in Remark 9.11 in Subsection 9.4.3, one can also obtain the dual optimal $\boldsymbol{\nu}^\star$ first by solving the dual problem $\max g(\boldsymbol{\nu})$ where

$g(\nu)$ is given by (9.12) and is concave. It is easy to see the optimum

$$\nu^\star = -2(\mathbf{A}\mathbf{A}^T)^{-1}\mathbf{b}$$

by the optimality condition $\nabla g(\nu) = \mathbf{0}$. Then solve $\min_\mathbf{x} \mathcal{L}(\mathbf{x}, \nu^\star)$ to obtain the optimum

$$\mathbf{x}^\star = -\frac{1}{2}\mathbf{A}^T\nu^\star = \mathbf{A}^T(\mathbf{A}\mathbf{A}^T)^{-1}\mathbf{b}$$

(which is unique and primal feasible, and yielded by the optimality condition $\nabla_\mathbf{x}\mathcal{L}(\mathbf{x}, \nu^\star) = 2\mathbf{x} + \mathbf{A}^T\nu^\star = \mathbf{0}$ since $\mathcal{L}(\mathbf{x}, \nu^\star)$ given by (9.9) is convex in $\mathbf{x}$).   □

Example 9.11 *(Smallest eigenvalue problem)* This nonconvex problem has been defined in (9.81) for which strong duality holds, and its Lagrangian is given in (9.82) (in Example 9.8 in Section 9.3). Thus, the KKT conditions of this problem can be easily seen as follows:

$$\nabla_\mathbf{x}\mathcal{L}(\mathbf{x}, \nu) = 2\mathbf{C}\mathbf{x} + 2\nu\mathbf{x} = \mathbf{0} \ \Rightarrow \ \mathbf{C}\mathbf{x} = -\nu\mathbf{x} \quad \text{(by (9.121a))} \tag{9.124}$$

$$\|\mathbf{x}\|_2^2 = 1 \quad \text{(by (9.121c))}. \tag{9.125}$$

Note that only the orthonormal eigenvectors of $\mathbf{C}$ will satisfy (9.124) and (9.125), implying that the optimal solution $\mathbf{x}^\star$ must be the orthonormal eigenvector that minimizes $\mathbf{x}^T\mathbf{C}\mathbf{x}$ with the optimal value equal to the minimum eigenvalue of $\mathbf{C}$. Even all the other orthonormal eigenvectors also satisfy all the KKT conditions (necessary conditions), but they are not optimal to problem (9.81). Nevertheless, all the possible solution candidates inferred from the KKT conditions have been proved to lie in the set of orthonormal eigenvectors of $\mathbf{C}$, and so the optimal solution can be easily found from this set.   □

Example 9.12 *(Entropy maximization)* The entropy maximization problem has been defined in (9.34) in Example 9.4 in Subsection 9.1.3, and can be equivalently written as

$$\min_{\mathbf{x} \in \mathbb{R}_+^n} \ \sum_{i=1}^n x_i \log x_i$$
$$\text{s.t.} \ \sum_{i=1}^n x_i = 1 \tag{9.126}$$

which is a convex optimization problem and strong duality holds by Slater's condition. The Lagrangian of the above problem is given by

$$\mathcal{L}(\mathbf{x}, \boldsymbol{\lambda}, \nu) = \sum_{i=1}^n x_i \log x_i - \boldsymbol{\lambda}^T\mathbf{x} + \nu\left(\sum_{i=1}^n x_i - 1\right) \tag{9.127}$$

and the optimal solution $\mathbf{x}^\star$ can be easily obtained from the KKT conditions as follows:

$$\frac{\partial \mathcal{L}(\mathbf{x}, \boldsymbol{\lambda}, \nu)}{\partial x_i} = \log x_i + x_i \frac{1}{x_i} + \nu - \lambda_i = 0, \ i = 1, \ldots, n \quad \text{(by (9.121a))}$$

$$\Rightarrow x_i = e^{-1-\nu+\lambda_i} > 0 \ \Rightarrow \ \begin{cases} \lambda_i = 0, \ i = 1, \ldots, n \\ x_1 = \cdots = x_n = e^{-1-\nu} \end{cases} \quad \text{(by (9.121e))}$$

which together with the KKT condition (9.121c) yields the closed-form optimal solution

$$\sum_{i=1}^{n} x_i = 1 \Rightarrow n x_i = 1 \Rightarrow x_i^\star = \frac{1}{n}, \ \forall i.$$

Alternatively, one can solve the dual optimal $(\boldsymbol{\lambda}^\star, \nu^\star)$ as follows:

$$(\boldsymbol{\lambda}^\star, \nu^\star) = \arg \max_{\boldsymbol{\lambda} \succeq \mathbf{0}, \nu} g(\boldsymbol{\lambda}, \nu)$$

$$= \arg \min_{\boldsymbol{\lambda} \succeq \mathbf{0}, \nu} -g(\boldsymbol{\lambda}, \nu)$$

$$= \arg \min_{\boldsymbol{\lambda} \succeq \mathbf{0}, \nu} \left\{ \nu + \sum_{i=1}^{n} e^{\lambda_i - \nu - 1} \right\} \quad \text{(by (9.36))}$$

$$= \left(\mathbf{0}_n, (\log n) - 1 \right) \quad \text{(by (4.57))}.$$

Then substituting $(\boldsymbol{\lambda}^\star, \nu^\star)$ into $x_i = e^{-1-\nu+\lambda_i}$ (resulting from (9.121a)) yields $x_i^\star = 1/n$ (cf. Remark 9.11). $\qquad\square$

Example 9.13 *(Water-filling in channel capacity maximization)* The water-filling problem has been a widely known problem in MIMO wireless communications. Let us consider a channel capacity maximization problem defined as

$$\max \ \sum_{i=1}^{n} \log \left(1 + \frac{p_i}{\sigma_i^2} \right) \tag{9.128}$$
$$\text{s.t. } \mathbf{p} \succeq \mathbf{0}, \ \sum_{i=1}^{n} p_i = P$$

where the objective function represents the channel capacity of a system consisting of n parallel subchannels (such as in an OFDM system), p_i is the signal power of subchannel i, and σ_i^2 is the noise power in subchannel i. The above problem is equivalent to the following convex optimization problem

$$\min \ -\sum_{i=1}^{n} \log \left(1 + \frac{p_i}{\sigma_i^2} \right) \tag{9.129}$$
$$\text{s.t. } \mathbf{p} \succeq \mathbf{0}, \ \sum_{i=1}^{n} p_i = P$$

for which strong duality holds true by Slater's condition. The Lagrangian can be easily seen to be

$$\mathcal{L}(\mathbf{p}, \boldsymbol{\lambda}, \nu) = -\sum_{i=1}^{n} \log\left(1 + \frac{p_i}{\sigma_i^2}\right) - \boldsymbol{\lambda}^T \mathbf{p} + \nu\left(\sum_{i=1}^{n} p_i - P\right). \tag{9.130}$$

By the KKT conditions given by (9.121a),

$$\frac{\partial \mathcal{L}(\mathbf{p}, \boldsymbol{\lambda}, \nu)}{\partial p_i} = \frac{-1}{1 + \frac{p_i}{\sigma_i^2}} \frac{1}{\sigma_i^2} - \lambda_i + \nu = 0, \tag{9.131}$$

which leads to

$$\lambda_i = \nu - \frac{1}{p_i + \sigma_i^2}. \tag{9.132}$$

By (9.121b), (9.121d), (9.121e), and (9.132), we come up with the following two cases for solving the optimal p_i:

- Case 1: $\lambda_i > 0$ and $p_i = 0 \Rightarrow \lambda_i = \nu - \frac{1}{\sigma_i^2} > 0 \Rightarrow \frac{1}{\nu} < \sigma_i^2$.
- Case 2: $\lambda_i = 0$ and $p_i \geq 0 \Rightarrow \nu = \frac{1}{p_i + \sigma_i^2} \Rightarrow p_i = \frac{1}{\nu} - \sigma_i^2 \geq 0$.

It can be inferred from the above two cases that

$$p_i^\star = \max\left\{0, \frac{1}{\nu^\star} - \sigma_i^2\right\}, \tag{9.133}$$

where the optimal $1/\nu^\star$ can be solved from (9.121c) as follows.

$$\sum_{i=1}^{n} p_i^\star = \sum_{i=1}^{n} \max\left\{0, \frac{1}{\nu^\star} - \sigma_i^2\right\} = P. \tag{9.134}$$

Solving (9.134) may need some trials by assuming $p_i > 0$ for all i first (i.e., $\frac{1}{\nu} - \sigma_i^2 > 0$ for all i) and then finding the solution $\frac{1}{\nu^\star}$ of (9.134). If no feasible solution exists, then we obtain $p_\ell^\star = 0$ where $\ell = \text{argmax}_i\{\sigma_i^2\}$, and then solve (9.134) for $\frac{1}{\nu^\star}$ again. This procedure is repeated by obtaining one more subchannel power at each trial (which is equal to zero) associated with the largest noise variance over the remaining subchannels until $\frac{1}{\nu^\star}$ and all the remaining positive optimal $p_i^\star > 0$ are obtained. The solution obtained through the above analytical procedure is called the centralized solution, denoted as $\mathbf{p}^\star$. This solution is also optimal to problem (4.133) where $\lambda_1 = \cdots = \lambda_n$ and Pareto optimal to the convex vector optimization problem (4.132), with the associated objective value

$$\left(R_1^\star = \log\left(1 + \frac{p_1^\star}{\sigma_1^2}\right), \ldots, R_n^\star = \log\left(1 + \frac{p_n^\star}{\sigma_n^2}\right)\right)$$

on the Pareto boundary (cf. Figure 4.7) as well.

An illustration for (9.133) is shown in Figure 9.4, where $n = 8$, $p_2^\star = p_7^\star = 0$, and the rest $p_i^\star > 0$, and $1/\nu^\star$ is the optimal water level.   □

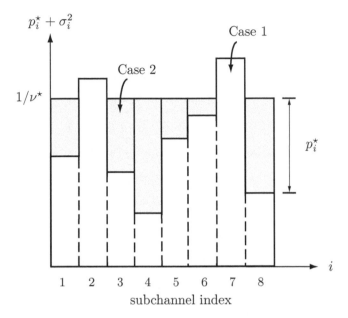

Figure 9.4 Illustration of the water-filling problem, where the signal powers for sub-channel 2 and subchannel 7 are equal to zero (no shaded regions in the two subchannels).

Remark 9.13 *(KKT conditions for complex variables)* If the objective function and all the constraint functions of (9.1) are real-valued functions but the unknown variable $\mathbf{x}$ is complex, are the KKT conditions the same as given by (9.121)? By (9.120) under strong duality of (9.1), one can see that the Lagrangian $\mathcal{L}(\mathbf{x}, \boldsymbol{\lambda}, \boldsymbol{\nu})$ defined in (9.3) must be a real-valued function in spite of the primal variable $\mathbf{x} \in \mathbb{C}^n$. Hence, we need to equivalently express $\mathcal{L}(\mathbf{x}, \boldsymbol{\lambda}, \boldsymbol{\nu})$ in a real-valued form w.r.t. $\mathbf{x} \in \mathbb{C}^n$, so that $\nabla_{\mathbf{x}}\mathcal{L} = 2\nabla_{\mathbf{x}^*}\mathcal{L}$ (cf. (1.43)) can be applied. As a result, all the KKT conditions given by (9.121) can be proven true through a procedure similar to the proof for the real-variable case above. This is illustrated by the following example. □

Example 9.14 *(Minimum variance beamforming)* This problem has been presented in Subsections 6.7.1 and 7.5.1, where we have classified it into a QP in the former or an SOCP in the latter without discussing how to find its closed-form solution. The first-order optimality condition (4.98) may not be very efficient to obtain the optimal solution though the problem is strictly convex. However, the optimal solution to this problem can be obtained by the KKT conditions as presented in this example. The problem is repeated here for convenience as follows.

$$\min_{\mathbf{x} \in \mathbb{C}^P} \ \mathbf{x}^H \mathbf{R} \mathbf{x}$$
$$\text{s.t. } \mathbf{a}^H \mathbf{x} = 1 \tag{9.135}$$

where $\mathbf{R} \in \mathbb{H}_{++}^{P}$ and $\mathbf{a} \in \mathbb{C}^{P}$. This problem is a convex optimization problem and strong duality holds by Slater's condition (cf. Remark 9.6). Note that the equality constraint $\mathbf{a}^{H}\mathbf{x} = 1$ can be decomposed as

$$\begin{cases} \operatorname{Re}\{\mathbf{a}^{H}\mathbf{x}\} = \frac{1}{2}(\mathbf{a}^{H}\mathbf{x} + \mathbf{x}^{H}\mathbf{a}) = 1 \\ \operatorname{Im}\{\mathbf{a}^{H}\mathbf{x}\} = \frac{1}{2}(-j\mathbf{a}^{H}\mathbf{x} + j\mathbf{x}^{H}\mathbf{a}) = 0 \end{cases} \quad \text{(cf. (4.101)).} \tag{9.136}$$

Then the Lagrangian of problem (9.135) is given by

$$\mathcal{L}(\mathbf{x}, \nu_1, \nu_2) = \mathbf{x}^{H}\mathbf{R}\mathbf{x} + \frac{\nu_1}{2}(\mathbf{a}^{H}\mathbf{x} + \mathbf{x}^{H}\mathbf{a} - 2)$$
$$+ \frac{\nu_2}{2}(-j\mathbf{a}^{H}\mathbf{x} + j\mathbf{x}^{H}\mathbf{a}), \quad \nu_1, \nu_2 \in \mathbb{R}, \tag{9.137}$$

which has been a real-valued function of $\mathbf{x} \in \mathbb{C}^{P}$. Then we have

$$\nabla_{\mathbf{x}}\mathcal{L}(\mathbf{x}, \nu_1, \nu_2) = 2\nabla_{\mathbf{x}^{\star}}\mathcal{L}(\mathbf{x}, \nu_1, \nu_2)$$
$$= 2\mathbf{R}\mathbf{x} + (\nu_1 + j\nu_2)\mathbf{a} = \mathbf{0} \quad \left(\text{by (9.121a) and (1.43)}\right)$$
$$\Rightarrow \mathbf{x} = -\frac{\nu_1 + j\nu_2}{2}\mathbf{R}^{-1}\mathbf{a}. \tag{9.138}$$

Substituting $\mathbf{x} = -\frac{\nu_1 + j\nu_2}{2}\mathbf{R}^{-1}\mathbf{a}$ into the constraint $\mathbf{a}^{H}\mathbf{x} = 1$ (by (9.121c)), we can get

$$-\frac{\nu_1 + j\nu_2}{2}\mathbf{a}^{H}\mathbf{R}^{-1}\mathbf{a} = 1 \Rightarrow -\frac{\nu_1 + j\nu_2}{2} = \frac{1}{\mathbf{a}^{H}\mathbf{R}^{-1}\mathbf{a}}. \tag{9.139}$$

So the optimal solution is obtained as

$$\mathbf{x}^{\star} = \frac{\mathbf{R}^{-1}\mathbf{a}}{\mathbf{a}^{H}\mathbf{R}^{-1}\mathbf{a}} \quad \text{(by (9.138) and (9.139)).}$$

Note that the optimal value $p^{\star} = 1/\mathbf{a}^{H}\mathbf{R}^{-1}\mathbf{a}$ remains the same for any feasible point $\mathbf{x}^{\star}e^{j\phi}$ where $\phi \in [0, 2\pi)$. In other words, the same optimal value can be attained under the constraint $|\mathbf{a}^{H}\mathbf{x}| = 1$ for which $\mathbf{a}^{H}\mathbf{x} = 1$ is nothing but a special case.   □

Let us conclude this section with the following two remarks:

Remark 9.14 (*Equivalence of KKT conditions and the first-order optimality condition for convex problems*) As stated in (4.23) for the convex problem given by (9.65) with the problem domain $\mathcal{D} \subseteq \mathbb{R}^{n}$, the first-order optimality condition for an optimal solution $\mathbf{x}^{\star}$ has been shown in Chapter 4 (without involving KKT conditions) to be

$$\nabla f_0(\mathbf{x}^{\star})^{T}(\mathbf{x} - \mathbf{x}^{\star}) \geq 0, \quad \forall \mathbf{x} \in \mathcal{C}, \tag{9.140}$$

where $\mathcal{C}$ denotes the feasible set of the problem, i.e.,

$$\mathcal{C} = \{\mathbf{x} \mid \mathbf{x} \in \mathcal{D}, f_i(\mathbf{x}) \leq 0, \ i = 1, \ldots, m, \ (h_1(\mathbf{x}), \ldots, h_p(\mathbf{x})) = \mathbf{A}\mathbf{x} - \mathbf{b} = \mathbf{0}\}$$

in which each $f_i(\mathbf{x})$ is a differentiable convex function, and each $h_j(\mathbf{x}) = \mathbf{a}_j^{T}\mathbf{x} - b_j$ (with $\mathbf{a}_j^{T}$ and b_j denoting the jth row of $\mathbf{A} \in \mathbb{R}^{p \times n}$ and the jth entry of $\mathbf{b} \in \mathbb{R}^{p}$, respectively). Suppose that Slater's condition holds (implying **relint** $\mathcal{C} \neq \emptyset$) and **A** is of full row rank. Then the condition (9.140) and the KKT conditions in (9.121) are equivalent.

Proof: First of all, let us prove that (9.140) can be proven true from the KKT conditions (9.121) as follows:

$$\nabla f_0(\mathbf{x}^\star)^T(\mathbf{x}-\mathbf{x}^\star) = -\left(\sum_{i=1}^{m}\lambda_i^\star\nabla f_i(\mathbf{x}^\star)^T + (\boldsymbol{\nu}^\star)^T\mathbf{A}\right)(\mathbf{x}-\mathbf{x}^\star) \quad \text{(by (9.121a))}$$

$$= -\sum_{i=1}^{m}\lambda_i^\star\nabla f_i(\mathbf{x}^\star)^T(\mathbf{x}-\mathbf{x}^\star) \quad \text{(by (9.121c) and } \mathbf{x}\in\mathcal{C})$$

$$\geq -\sum_{i=1}^{m}\lambda_i^\star(f_i(\mathbf{x})-f_i(\mathbf{x}^\star)) \quad \text{(since } f_i(\mathbf{x}) \text{ is convex)}$$

$$= -\sum_{i=1}^{m}\lambda_i^\star f_i(\mathbf{x}) \quad \text{(by (9.121e))}$$

$$\geq 0 \quad \text{(by (9.121d) and } \mathbf{x}\in\mathcal{C}).$$

Next, let us present that the KKT conditions (9.121) can be proven true from (9.140). For ease of the proof, let us define the following convex cones:

$$K_i^f \triangleq \{\kappa_i^f\cdot(\mathbf{x}-\mathbf{x}^\star) \mid \kappa_i^f \geq 0,\ f_i(\mathbf{x}) \leq 0\},\ i=1,\dots,m,$$
$$K_i^h \triangleq \{\kappa_i^h\cdot(\mathbf{x}-\mathbf{x}^\star) \mid \kappa_i^h \geq 0,\ h_i(\mathbf{x}) = 0\},\ i=1,\dots,p, \qquad (9.141)$$
$$K \triangleq \{\cap_i K_i^f\}\cap\{\cap_i K_i^h\} = \{\cap_{i\in\mathcal{I}} K_i^f\}\cap\{\cap_i K_i^h\},$$

where $\mathcal{I}$ is the index set defined in (9.122) (i.e., $f_i(\mathbf{x}^\star)=0$ for $i\in\mathcal{I}$), and we have used the fact that $K_i^f = \mathbb{R}^n$ for $i\notin\mathcal{I}$ since $\mathbf{x}^\star\in\mathbf{int}\{\mathbf{x}\in\mathbb{R}^n \mid f_i(\mathbf{x})\leq 0\}$ for $i\notin\mathcal{I}$. Furthermore, due to the convexity of the feasible set $\mathcal{C}$, it can be inferred that

$$K = \mathbf{conic}\ (\mathcal{C}-\{\mathbf{x}^\star\}) \implies \nabla f_0(\mathbf{x}^\star)\in K^\star \quad \text{(due to (9.140))}. \qquad (9.142)$$

Because each f_i is differentiable and convex, and because Slater's condition holds, $\{\mathbf{x}\in\mathbb{R}^n \mid f_i(\mathbf{x})\leq 0\}$ must be a closed convex set with nonempty interior. Hence the convex cone K_i^f for $i\in\mathcal{I}$ is either a closed halfspace with the origin on its boundary, or the union of the origin and an open halfspace with the origin on its boundary (which is neither an open set nor a closed set), implying that $\mathbf{cl}\,K_i^f$ is a closed halfspace for all $i\in\mathcal{I}$. Furthermore, by Slater's condition again, we have

$$\cap_{i\in\mathcal{I}}(\mathbf{int}\ K_i^f)\neq\emptyset. \qquad (9.143)$$

By the first-order condition of convex functions, it can be inferred that

$$0\geq f_i(\mathbf{x})\geq f_i(\mathbf{x}^\star)+\nabla f_i(\mathbf{x}^\star)^T(\mathbf{x}-\mathbf{x}^\star),\ \forall\mathbf{x}\in\{\boldsymbol{x} \mid f_i(\boldsymbol{x})\leq 0\},\ i=1,\dots,m.$$
$$\implies -\nabla f_i(\mathbf{x}^\star)^T(\mathbf{x}-\mathbf{x}^\star)\geq 0,\ \forall\mathbf{x}\in\{\boldsymbol{x} \mid f_i(\boldsymbol{x})\leq 0\},\ i\in\mathcal{I} \quad \text{(by (9.122))}.$$

That is, the dual cone

$$K_i^{f*} = \{-\alpha_i \nabla f_i(\mathbf{x}^*) \mid \alpha_i \geq 0\} \text{ (a ray)}, \ i \in \mathcal{I}. \tag{9.144}$$

On the other hand, it can be seen from (9.141) that all K_i^h are hyperplanes passing the origin (also $(n-1)$-dimensional subspaces). So their dual cones are given by

$$K_i^{h*} = \{\beta_i \nabla h_i(\mathbf{x}^*) \mid \beta_i \in \mathbb{R}\} \text{ (a line)}, \ i = 1, \ldots, p. \tag{9.145}$$

Let us only present the proof for the case that $\mathbf{x}^* \in \mathbf{relbd}\ \mathcal{C}$ is a regular point, i.e., $\mathcal{I} \neq \emptyset$ (cf. Remark 9.12) since the corresponding proof for the case of $\mathbf{x}^* \in \mathbf{relint}\ \mathcal{C}$ is a degenerate case of $\mathcal{I}$, i.e., $\mathcal{I} = \emptyset$. Since $\mathbf{x}^* \in \mathcal{C}$, conditions (9.121b) and (9.121c) automatically hold. It can be inferred that (9.121e) and (9.121a) are equivalent to the following equality

$$\nabla f_0(\mathbf{x}^*) + \sum_{i \in \mathcal{I}} \lambda_i^* \nabla f_i(\mathbf{x}^*) + \sum_{i=1}^p \nu_i^* \nabla h_i(\mathbf{x}^*) = \mathbf{0}, \tag{9.146}$$

where (9.121e) has been enforced in the second summation with all the terms involving $\lambda_i = 0$ for $i \notin \mathcal{I}$ removed regardless of $\nabla f_i(\mathbf{x}^*)$ being zero or not for $i \notin \mathcal{I}$. Hence, to show (9.121a), (9.121d), and (9.121e), it suffices to show that

$$\nabla f_0(\mathbf{x}^*) \in \mathcal{K} \triangleq \mathbf{conic}\{\{-\nabla f_i(\mathbf{x}^*)\}_{i \in \mathcal{I}}, \{\pm \nabla h_i(\mathbf{x}^*)\}_{i=1}^p\}$$

$$= \mathbf{conic}\left\{\{\cup_{i \in \mathcal{I}}\ K_i^{f*}\} \cup \{\cup_{i=1}^p K_i^{h*}\}\right\} \tag{9.147}$$

where we have applied (9.144) and (9.145) in the derivation of (9.147). Furthermore, by (9.142), one can prove (9.147) by showing

$$K^* \subseteq \mathcal{K} = \mathcal{K}^{**}, \tag{9.148}$$

where the equality is because $\mathcal{K}$ is closed and convex (by (**d5**) in Subsection 2.5.2).

To prove (9.148), let

$$\bar{K} \triangleq \left\{\cap_{i \in \mathcal{I}}(\mathbf{cl}\ K_i^f)\right\} \cap \left\{\cap_i(\mathbf{cl}\ K_i^h)\right\} \tag{9.149}$$

$$= \left\{\cap_{i \in \mathcal{I}}(\mathbf{cl}\ K_i^f)\right\} \cap \left\{\cap_i K_i^h\right\} \quad (K_i^h \text{ is a hyperplane}).$$

Next, let us show that $K^* = \bar{K}^*$. Since $K \subseteq \bar{K}$ (by (9.141) and (9.149)), we have $\bar{K}^* \subseteq K^*$ (by (**d2**) in Subsection 2.5.2). So we only need to prove $K^* \subseteq \bar{K}^*$. Suppose that there exists a $\boldsymbol{y} \in K^*$ but $\boldsymbol{y} \notin \bar{K}^*$. Then, we have

$$\boldsymbol{y}^T \boldsymbol{x} \geq 0, \ \forall \boldsymbol{x} \in K, \text{ and } \boldsymbol{y}^T \bar{\boldsymbol{x}} < 0, \text{ for some } \bar{\boldsymbol{x}} \in \bar{K}. \tag{9.150}$$

By (9.141) and (9.143), there exists an

$$\hat{\boldsymbol{x}} \in \left\{\cap_{i \in \mathcal{I}}(\mathbf{int}\ K_i^f)\right\} \cap \{\cap_i K_i^h\} \subset K,$$

and

$$\theta\widehat{\boldsymbol{x}} + (1-\theta)\bar{\boldsymbol{x}} \in \left\{\cap_{i\in\mathcal{I}}(\text{int } K_i^f)\right\} \cap \left\{\cap_i K_i^h\right\} \subset K, \ \forall\theta \in (0,1].$$

Let $\theta \to 0$, we have $\theta\widehat{\boldsymbol{x}} + (1-\theta)\bar{\boldsymbol{x}} \in K$, but $\boldsymbol{y}^T(\theta\widehat{\boldsymbol{x}} + (1-\theta)\bar{\boldsymbol{x}}) \to \boldsymbol{y}^T\bar{\boldsymbol{x}} < 0$, which contradicts (9.150). Hence, $K^* \subseteq \bar{K}^*$, and we have proven that $K^* = \bar{K}^*$.

Now we are ready to prove (9.148). Instead, we show that $\mathcal{K}^* \subseteq \bar{K}$, which implies $\bar{K}^* \subseteq \mathcal{K}^{**} = \mathcal{K}$ (by (**d2**) in Subsection 2.5.2). Given an arbitrary point $\boldsymbol{x} \in \mathcal{K}^*$, it holds true by (9.147) that

$$\boldsymbol{x}^T\boldsymbol{y}_i^f \geq 0, \ \boldsymbol{x}^T\boldsymbol{y}_j^h \geq 0, \ \forall\boldsymbol{y}_i^f \in K_i^{f*}, \ \boldsymbol{y}_j^h \in K_j^{h*}, \ i \in \mathcal{I}, \ j = 1,\dots,p,$$

implying that

$$\boldsymbol{x} \in \left\{\cap_{i\in\mathcal{I}}K_i^{f**}\right\} \cap \left\{\cap_i K_i^{h**}\right\}$$
$$= \left\{\cap_{i\in\mathcal{I}}(\text{cl } K_i^f)\right\} \cap \left\{\cap_i(\text{cl } K_i^h)\right\} = \bar{K} \quad (\text{by } (\boldsymbol{d5}) \text{ in Subsection 2.5.2})$$

and, therefore, $\mathcal{K}^* \subseteq \bar{K}$, thereby leading to $\bar{K}^* \subseteq \mathcal{K}$. Thus, we have completed the proof of $K^* = \bar{K}^* \subseteq \mathcal{K}$, i.e., (9.148). ∎ □

Remark 9.15 The proof of the equivalence of KKT conditions and the first-order optimality condition for convex problems in Remark 9.14 only requires the convexity of all the constraint functions, but never requires the objective function f_0 to be convex. Hence the equivalence also holds true even when f_0 is nonconvex, and the KKT points are also stationary-point solutions for this case provided that the Slater's condition is satisfied and the feasible set $\mathcal{C}$ is convex and closed. □

9.6 Lagrange dual optimization

Doubtlessly, the KKT conditions introduced in the previous section are the backbones for directly solving a convex optimization problem; they are also essential in the development of effective and efficient algorithms for solving convex problems, especially when the KKT conditions are too intricate to find analytical solutions of the convex problem under consideration (e.g., the problem size is very large). However, it is also possible to find the primal optimal solution and the dual optimal solution by solving problem (9.151) below. This approach updates the primal variable and the dual variable alternatively until convergence, and thus both the inner minimization for updating the primal variable and the outer maximization for updating the dual variable are easier to handle (e.g., smaller problem size or easier to solve).

For the convex problem (9.65) where the objective function and all the constraint functions are differentiable and strong duality holds,

$$(\mathbf{x}^\star, \boldsymbol{\lambda}^\star, \boldsymbol{\nu}^\star) = \arg \left\{ \max_{\boldsymbol{\lambda} \succeq 0, \boldsymbol{\nu}} \ \min_{\mathbf{x} \in \mathcal{D}} \mathcal{L}(\mathbf{x}, \boldsymbol{\lambda}, \boldsymbol{\nu}) \right\} \tag{9.151}$$

is a primal-dual optimal solution if $\mathbf{x}^\star$ is primal feasible.

Proof of (9.151): We will prove that $(\mathbf{x}^\star, \boldsymbol{\lambda}^\star, \boldsymbol{\nu}^\star)$ given by (9.151) satisfies all the KKT conditions given by (9.121). Owing to finite

$$\mathcal{L}(\mathbf{x}^\star, \boldsymbol{\lambda}^\star, \boldsymbol{\nu}^\star) = \max_{\boldsymbol{\lambda} \succeq 0, \boldsymbol{\nu}} \ \min_{\mathbf{x} \in \mathcal{D}} \mathcal{L}(\mathbf{x}, \boldsymbol{\lambda}, \boldsymbol{\nu})$$

$$= \max_{\boldsymbol{\lambda} \succeq 0, \boldsymbol{\nu}} \ g(\boldsymbol{\lambda}, \boldsymbol{\nu}) = g(\boldsymbol{\lambda}^\star, \boldsymbol{\nu}^\star) = d^\star, \tag{9.152}$$

$(\boldsymbol{\lambda}^\star, \boldsymbol{\nu}^\star)$ must be dual feasible and thus (9.121d) is true (due to $\boldsymbol{\lambda}^\star \succeq 0$). Because $\mathbf{x}^\star$ is primal feasible (by assumption), $\mathbf{x}^\star \in \mathcal{C}$ where $\mathcal{C}$ is the constraint set of (9.65) and thus (9.121b) and (9.121c) are true.

By strong duality,

$$p^\star = d^\star = \mathcal{L}(\mathbf{x}^\star, \boldsymbol{\lambda}^\star, \boldsymbol{\nu}^\star) = g(\boldsymbol{\lambda}^\star, \boldsymbol{\nu}^\star) = \min_{\mathbf{x} \in \mathcal{D}} \ \mathcal{L}(\mathbf{x}, \boldsymbol{\lambda}^\star, \boldsymbol{\nu}^\star), \tag{9.153}$$

(cf. (9.120)) which implies that $\nabla_\mathbf{x} \mathcal{L}(\mathbf{x}^\star, \boldsymbol{\lambda}^\star, \boldsymbol{\nu}^\star) = \mathbf{0}$ since $\mathcal{L}(\mathbf{x}, \boldsymbol{\lambda}^\star, \boldsymbol{\nu}^\star)$ is convex and differentiable in $\mathbf{x}$. Hence (9.121a) is true.

Let us further analyze (9.153) as follows:

$$p^\star = \min_{\mathbf{x} \in \mathcal{C}} \ \mathcal{L}(\mathbf{x}, \boldsymbol{\lambda}^\star, \boldsymbol{\nu}^\star) \quad (\text{since } \mathbf{x}^\star \in \mathcal{C} \subseteq \mathcal{D})$$

$$= \min_{\mathbf{x} \in \mathcal{C}} \ \left\{ f_0(\mathbf{x}) + \sum_{i=1}^{m} \lambda_i^\star f_i(\mathbf{x}) + \sum_{i=1}^{p} \nu_i^\star h_i(\mathbf{x}) \right\}$$

$$\leq \left\{ \min_{\mathbf{x} \in \mathcal{C}} \ f_0(\mathbf{x}) \right\} = f_0(\mathbf{x}^\star) = p^\star \quad (\text{since } \boldsymbol{\lambda}^\star \succeq 0)$$

$$\Rightarrow \sum_{i=1}^{m} \lambda_i^\star f_i(\mathbf{x}^\star) = 0$$

$$\Rightarrow \lambda_i^\star f_i(\mathbf{x}^\star) = 0 \quad \forall i, \tag{9.154}$$

which implies that (9.121e) is true. Therefore, we have proved that all the KKT conditions are satisfied by $(\mathbf{x}^\star, \boldsymbol{\lambda}^\star, \boldsymbol{\nu}^\star)$ and so it is primal-dual optimal to problem (9.65). ∎

Problem (9.151) is called *dual decomposition*, or *dual optimization method*, for solving problem (9.65). The inner minimization problem in (9.151) is convex in $\mathbf{x}$ for any fixed $(\boldsymbol{\lambda} \succeq 0, \boldsymbol{\nu})$ (with all the inequality and equality constraints relaxed), and the outer maximization problem is concave in $(\boldsymbol{\lambda}, \boldsymbol{\nu})$ for any fixed $\mathbf{x} \in \mathcal{D}$. However, the problem size may become very large, which is surely another concern in finding a primal-dual optimal solution.

All the uncoupling constraints of the problem can be relaxed by redefining the problem domain $\mathcal{D}$ as the intersection of the original problem domain and the set satisfying all the uncoupling constraints such that the resulting Lagrangian will not involve any dual variables associated with these uncoupling constraints. Doing this thereby can reduce the number of dual variables (i.e., problem size reduced) in the outer maximization on the one hand, and meanwhile will make the resulting inner minimization a constrained minimization problem with the solution in a distributed fashion on the other hand.

Rather than solving problem (9.65) that involves intricate equality and inequality constraints, solving the inner moderately constrained minimization problem by the first-order optimality condition, and then properly updating the dual variables of the problem size reduced outer maximization problem of (9.151) in an alternative fashion, has been effectively applied in resource allocation and interference management in wireless communications and networking, provided that the convergence to the primal-dual optimal solution can be guaranteed. Meanwhile, the dual decomposition may come up with a distributed implementation which is particularly needed in cooperative or coordinated systems. Let us conclude this section by the water-filling example presented in the previous section, using the above dual optimization method.

Example 9.15 *(Water-filling (revisited))* Let us solve the problem (9.129) by the dual optimization method. First of all, let us reformulate the problem as

$$\min \ -\sum_{i=1}^{n} \log\left(1 + \frac{p_i}{\sigma_i^2}\right)$$
$$\text{s.t. } \sum_{i=1}^{n} p_i \le P \tag{9.155}$$

with the problem domain redefined as $\mathcal{D} = \mathbb{R}_+^n$; namely, all the uncoupling constraints $p_i \ge 0$, $i = 1, \ldots, n$ have been incorporated in the problem domain. Then the associated Lagrangian can be seen to be

$$\mathcal{L}(\mathbf{p}, \nu) = -\sum_{i=1}^{n} \log\left(1 + \frac{p_i}{\sigma_i^2}\right) + \nu\left(\sum_{i=1}^{n} p_i - P\right), \tag{9.156}$$

where the number of the dual variables has been reduced from $n+1$ to one (cf. (9.130)). Now let us solve the problem (9.155) by dual optimization to obtain

$$(\mathbf{p}^\star, \nu^\star) = \arg\left\{\max_{\nu \ge 0} \ \min_{\mathbf{p} \succeq \mathbf{0}} \mathcal{L}(\mathbf{p}, \nu)\right\}. \tag{9.157}$$

By (9.156), the inner minimization in (9.157) can be simplified as

$$g(\nu) \triangleq \min_{\mathbf{p} \succeq \mathbf{0}} \mathcal{L}(\mathbf{p}, \nu) = \sum_{i=1}^{n} \min_{p_i \ge 0} g_i(\nu, p_i) - \nu P, \tag{9.158}$$

(which is actually the dual function of (9.155)) where

$$g_i(\nu, p_i) = \nu p_i - \log\left(1 + \frac{p_i}{\sigma_i^2}\right). \tag{9.159}$$

Note that the inner minimization defined in (9.158) has been decomposed into n convex one-dimensional subproblems for any fixed ν, which can be solved using the first-order optimization condition given by (4.23), i.e.,

$$\frac{\partial g_i(\nu, p_i)}{\partial p_i} = \nu - \frac{1}{p_i + \sigma_i^2} \begin{cases} = 0, & \text{if } p_i > 0 \\ \geq 0, & \text{if } p_i = 0. \end{cases} \tag{9.160}$$

Thus, the optimal $p_i^\star$ is given by

$$p_i^\star(\nu) = \left[\frac{1}{\nu} - \sigma_i^2\right]^+, \quad i = 1, 2, ..., n, \tag{9.161}$$

where $[a]^+ = \max\{a, 0\}$. Note that the optimal $p_i^\star$ given by (9.161) has the same form as the centralized solution given by (9.133).

Substituting (9.161) into (9.159) and then substituting the result into (9.158) yields

$$g(\nu) = \mathcal{L}(\mathbf{p}^\star(\nu), \nu) = \sum_{j \in \mathcal{J}(\nu)} (1 - \sigma_j^2 \nu) + \sum_{j \in \mathcal{J}(\nu)} \log(\sigma_j^2 \nu) - \nu P, \tag{9.162}$$

where

$$\mathcal{J}(\nu) = \{j \mid p_j^\star(\nu) > 0\} = \{j \mid 1 - \sigma_j^2 \nu > 0\},$$

implying that the number of terms in each summation in (9.162) depends on ν. Hence, although $g(\nu)$ given by (9.162) is a continuous concave function, it is not differentiable but piecewise differentiable. Thus the outer maximization problem

$$\max_{\nu \geq 0} g(\nu) \tag{9.163}$$

cannot be solved using the first-order optimality condition (4.23). Instead, we can resort to the projected subgradient method [Ber99] to iteratively update the dual variable ν with the optimal solution for ν obtained after convergence (cf. Remark 4.4).

A subgradient of $g(\nu)$ can be shown to be

$$\bar{\nabla} g(\nu) = -\sum_{j \in \mathcal{J}(\nu)} \sigma_j^2 + \frac{|\mathcal{J}(\nu)|}{\nu} - P \tag{9.164}$$

(see the proof given in Remark 9.16 below). Given the previous iterate $\nu^{(k)}$, the current iterate $\nu^{(k+1)}$ is updated as

$$\nu^{(k+1)} = \left[\nu^{(k)} + s^{(k)} \bar{\nabla} g(\nu^{(k)})\right]^+, \tag{9.165}$$

where $s^{(k)}$ is a positive step size. There are many choices for the step size $s^{(k)}$ to guarantee the convergence of $\nu^{(k)}$ to the optimal dual solution of the outer maximization problem (9.163). For a constant step size $s^{(k)} = s$ for all k, the subgradient algorithm can converge and yield an optimal dual solution provided that the step size is sufficiently small.

Therefore, we end up with an iterative algorithm given in Algorithm 9.1 for obtaining a solution of (9.157). Although this algorithm is an alternative method for solving the same problem (9.128), it exhibits a distributed fashion, namely, at each iteration, the information of $\text{sgn}(p_i^\star(\nu^{(k)}))$ is the common information needed by each local equipment (e.g., nodes in wireless networks and base stations in wireless communications) if each $p_i^\star$ must be computed by the ith local equipment to which only the local CSI σ_i^2 is known. Surely, the information exchange of $\text{sgn}(p_i^\star(\nu^{(k)}))$ among local equipments is the extra overhead, though the obtained solution by the above algorithm is theoretically the same as the centralized solution obtained by directly solving the KKT conditions. □

Algorithm 9.1 Dual optimization algorithm for solving (9.157)

1: Given an initial value for ν.
2: **repeat**
3: Each local equipment i synchronously updates $p_i^\star$ by (9.161) and then passes the information of $\text{sgn}(p_i^\star)$ to all the other local equipments.
4: Each local equipment i updates $\nu^\star$ by (9.165) and (9.164).
5: **until** a preassigned convergence criterion is met.

Remark 9.16 The subgradient given by (9.164) appears to be the derivative of $g(\nu)$ given by (9.162) as if the latter were differentiable. Let us prove that the former is indeed a subgradient of the latter in this remark.

A subgradient $\bar{\nabla}f(\tilde{\mathbf{x}})$ of a concave function $f(\mathbf{x})$ at $\mathbf{x} = \tilde{\mathbf{x}}$ is a column vector that satisfies the following first-order condition

$$f(\mathbf{x}) \leq f(\tilde{\mathbf{x}}) + \bar{\nabla}f(\tilde{\mathbf{x}})^T(\mathbf{x} - \tilde{\mathbf{x}}) \ \ (\text{cf. } (3.21)), \tag{9.166}$$

where $\bar{\nabla}f(\tilde{\mathbf{x}}) = \nabla f(\tilde{\mathbf{x}})$ when f is differentiable at $\mathbf{x} = \tilde{\mathbf{x}}$; otherwise it is nonunique. The subgradient given by (9.164) can be proved as follows.

$$g(\nu) \triangleq \min_{\mathbf{p} \succeq \mathbf{0}} \mathcal{L}(\mathbf{p}, \nu) \quad \text{(by (9.158) and (9.156))}$$

$$= -\sum_{i=1}^{n} \log\left(1 + \frac{p_i^{\star}(\nu)}{\sigma_i^2}\right) + \nu\left(\sum_{i=1}^{n} p_i^{\star}(\nu) - P\right)$$

$$\leq -\sum_{i=1}^{n} \log\left(1 + \frac{p_i^{\star}(\tilde{\nu})}{\sigma_i^2}\right) + \nu\left(\sum_{i=1}^{n} p_i^{\star}(\tilde{\nu}) - P\right)$$

$$= \sum_{j\in\mathcal{J}(\tilde{\nu})} \log(\sigma_j^2 \tilde{\nu}) + \nu\left\{\sum_{j\in\mathcal{J}(\tilde{\nu})} \left(\frac{1}{\tilde{\nu}} - \sigma_j^2\right) - P\right\} \quad \text{(by (9.161))} \qquad (9.167\text{a})$$

$$= \sum_{j\in\mathcal{J}(\tilde{\nu})} (1 - \sigma_j^2 \tilde{\nu}) + \sum_{j\in\mathcal{J}(\tilde{\nu})} \log(\sigma_j^2 \tilde{\nu}) - \tilde{\nu}P$$

$$+ \left\{\sum_{j\in\mathcal{J}(\tilde{\nu})} \left(\frac{1}{\tilde{\nu}} - \sigma_j^2\right) - P\right\}(\nu - \tilde{\nu}) \qquad (9.167\text{b})$$

$$= g(\tilde{\nu}) + \left\{\sum_{j\in\mathcal{J}(\tilde{\nu})} \left(\frac{1}{\tilde{\nu}} - \sigma_j^2\right) - P\right\}(\nu - \tilde{\nu}) \quad \text{(by (9.162))}, \qquad (9.167\text{c})$$

where (9.167b) is obtained by setting $\nu = \nu + \tilde{\nu} - \tilde{\nu}$ in (9.167a). Then, by comparing (9.167c) and (9.166), $\bar{\nabla}g(\tilde{\nu})$ given by (9.164) is a subgradient of $g(\nu)$ at $\nu = \tilde{\nu}$. $\qquad\square$

9.7   Alternating direction method of multipliers (ADMM)

As mentioned in the previous section, the dual optimization has been considered as an effective means for the design of distributed or coordinated algorithms in wireless communications and networking. For instance, the robust MCBF problem (8.77) introduced in Subsection 8.5.5 provides a centralized solution, while a distributed implementation is preferred in practical applications. In this section, let us introduce a distributed convex optimization technique known as *alternating direction method of multipliers* (ADMM) [BT89a, BPC+10], which is developed via dual optimization and has been used for distributed beamforming designs in multicell networks [SCW+12].

Consider the following convex optimization problem [BT89b]:

$$\min_{\mathbf{x}\in\mathbb{R}^n, \mathbf{z}\in\mathbb{R}^m} \quad f_1(\mathbf{x}) + f_2(\mathbf{z})$$
$$\text{s.t.} \quad \mathbf{x} \in \mathcal{S}_1, \ \mathbf{z} \in \mathcal{S}_2 \qquad\qquad (9.168)$$
$$\mathbf{z} = \mathbf{A}\mathbf{x}$$

where $f_1 : \mathbb{R}^n \mapsto \mathbb{R}$ and $f_2 : \mathbb{R}^m \mapsto \mathbb{R}$ are convex functions, $\mathbf{A}$ is an $m \times n$ matrix, and $\mathcal{S}_1 \subset \mathbb{R}^n$ and $\mathcal{S}_2 \subset \mathbb{R}^m$ are nonempty convex sets. Note that the ADMM can be extended to the case when the objective function of problem (9.168) is a sum of more than two convex functions of disjoint block variables [BPC+10]. Assume that problem (9.168) is solvable and strong duality holds.

ADMM considers the following penalty augmented problem

$$\min_{\mathbf{x}\in\mathbb{R}^n,\mathbf{z}\in\mathbb{R}^m} \ f_1(\mathbf{x}) + f_2(\mathbf{z}) + \frac{c}{2}\|\mathbf{A}\mathbf{x} - \mathbf{z}\|_2^2$$
$$\text{s.t. } \mathbf{x}\in\mathcal{S}_1,\ \mathbf{z}\in\mathcal{S}_2 \tag{9.169}$$
$$\mathbf{z} = \mathbf{A}\mathbf{x}$$

where $c > 0$ is a penalty parameter. It is easy to see that (9.169) is essentially equivalent to (9.168). The term $\frac{c}{2}\|\mathbf{A}\mathbf{x} - \mathbf{z}\|_2^2$, whose Hessian matrix is PSD w.r.t. $(\mathbf{x},\mathbf{z})$ (cf. (1.48)), guarantees strict convexity of the objective function w.r.t. $\mathbf{x}$ (as $\mathbf{A}$ is of full column rank) and $\mathbf{z}$, respectively, which is required to ensure the convergence of ADMM algorithm.

The second ingredient of ADMM is dual decomposition with the two uncoupling constraints $\mathbf{x}\in\mathcal{S}_1$, $\mathbf{z}\in\mathcal{S}_2$ incorporated in the problem domain. The considered dual problem of (9.169) is given by

$$\max_{\boldsymbol{\nu}\in\mathbb{R}^m} g(\boldsymbol{\nu}), \tag{9.170}$$

in which the dual function $g(\boldsymbol{\nu})$ is given by

$$g(\boldsymbol{\nu}) = \min_{\mathbf{x}\in\mathcal{S}_1,\mathbf{z}\in\mathcal{S}_2} \left\{ f_1(\mathbf{x}) + f_2(\mathbf{z}) + \frac{c}{2}\|\mathbf{A}\mathbf{x} - \mathbf{z}\|_2^2 + \boldsymbol{\nu}^T(\mathbf{A}\mathbf{x} - \mathbf{z}) \right\}, \tag{9.171}$$

where $\boldsymbol{\nu}$ is the dual variable associated with the equality constraint in (9.169). As the value of the dual variable $\boldsymbol{\nu}$ is fixed, the inner minimization problem of (9.170) (i.e., problem (9.171)) is convex in $(\mathbf{x},\mathbf{z})$ and can be efficiently solved. The dual variable $\boldsymbol{\nu}$ can be updated by the projected subgradient method since the dual function $g(\boldsymbol{\nu})$ may be nonsmooth. In a standard dual optimization procedure, one usually updates the dual variable $\boldsymbol{\nu}$ after the primal variable $(\mathbf{x},\mathbf{z})$ is updated by solving (9.171) at each iteration. However, for the ADMM, at the $(q+1)$th iteration, the following two convex subproblems for updating $(\mathbf{x},\mathbf{z})$ are solved instead:

$$\mathbf{z}(q+1) = \arg\min_{\mathbf{z}\in\mathcal{S}_2} \left\{ f_2(\mathbf{z}) - \boldsymbol{\nu}(q)^T\mathbf{z} + \frac{c}{2}\|\mathbf{A}\mathbf{x}(q) - \mathbf{z}\|_2^2 \right\}, \tag{9.172a}$$

$$\mathbf{x}(q+1) = \arg\min_{\mathbf{x}\in\mathcal{S}_1} \left\{ f_1(\mathbf{x}) + \boldsymbol{\nu}(q)^T\mathbf{A}\mathbf{x} + \frac{c}{2}\|\mathbf{A}\mathbf{x} - \mathbf{z}(q+1)\|_2^2 \right\}. \tag{9.172b}$$

As indicated by its name, at each iteration, ADMM alternatively updates the primal variable $(\mathbf{x},\mathbf{z})$ by solving (9.172a) and (9.172b) (inner minimization), and then updates the dual variable $\boldsymbol{\nu}$ in the outer maximization of (9.170) (equivalently maximizing $\boldsymbol{\nu}^T(\mathbf{A}\mathbf{x}(q+1) - \mathbf{z}(q+1))$) using the projected subgradient method, as summarized in Algorithm 9.2. It has been proven that ADMM can converge and yield the global optimal solution of problem (9.168) even though $\mathbf{z}(q+1)$ and $\mathbf{x}(q+1)$ are sequentially (rather than jointly) updated under relatively mild conditions as stated in the following lemma [BT89b]:

Lemma 9.2. *Assume that $\mathcal{S}_1$ is bounded or that $\mathbf{A}^T\mathbf{A}$ is invertible. A sequence $\{\mathbf{x}(q), \mathbf{z}(q), \boldsymbol{\nu}(q)\}$ generated by Algorithm 9.2 is bounded, and every limit point of $\{\mathbf{x}(q), \mathbf{z}(q)\}$ is an optimal solution of problem (9.168).*

ADMM is guaranteed to converge to an optimal solution for any constant positive value of the penalty parameter c, but the convergence rate may depend on c and how to choose a suitable value of c for fast convergence is still unknown. Moreover, the chosen value of c in Algorithm 9.2 is not required to be constant for every iteration, i.e., it can be a function of q (iteration number). However, the best choice for $c(q)$ for fast convergence, which may be problem dependent by our experience, is still unknown and worth further study.

Algorithm 9.2 ADMM

1: Set $q = 0$, choose $c > 0$.
2: Initialize $\boldsymbol{\nu}(q)$ and $\mathbf{x}(q)$.
3: **repeat**
4:    Solve the two subproblems in (9.172a) and (9.172b) for obtaining $\mathbf{z}(q+1)$ and $\mathbf{x}(q+1)$;
5:    $\boldsymbol{\nu}(q+1) = \boldsymbol{\nu}(q) + c\,(\mathbf{A}\mathbf{x}(q+1) - \mathbf{z}(q+1))$;
6:    $q := q+1$;
7: **until** the predefined stopping criterion is satisfied.

Let us revisit the robust MCBF problem (8.77) in Subsection 8.5.5, where a set of the beamforming vectors

$$\{\mathbf{w}_{ik}\} \triangleq \{\mathbf{w}_{ik} \mid i \in I(N_c), k \in I(K)\}$$

(where $I(M) = \{1, 2, \ldots, M\}$ and $\mathbf{w}_{ik}$ are associated with the kth user in the cell i) is obtained by solving the large-size SDR problem in (8.84), which relies on a powerful control center for obtaining the optimal beamforming solution using all the CSI estimates of MSs (i.e., $\{\bar{\mathbf{h}}_{ijk} \mid i, j \in I(N_c), k \in I(K)\}$), thus rendering an impractical heavy computation load to the system. Instead, it is preferable and practical to obtain the multicell beamforming solution $\{\mathbf{w}_{ik}\}$ in a decentralized fashion (namely, a distributed implementation). In other words, the ith BS among coordinated N_c BSs needs to solve for $\{\mathbf{w}_{ik} \mid k \in I(K)\}$ of a small-size SDR subproblem with only local CSI estimates used (i.e., only the CSI $\{\bar{\mathbf{h}}_{ijk} \mid j \in I(N_c), k \in I(K)\}$ used by the ith BS) and limited information provided by other BSs, and the optimal centralized beamforming solution $\{\mathbf{w}_{ik}\}$ can be obtained through a coordinated scheme after convergence. Two distributed beamforming implementations using the ADMM, each providing a realization of this scheme, have been proposed in [SCW+12], and thereby the control center is no longer needed.

9.8 Duality of problems with generalized inequalities

9.8.1 Lagrange dual and KKT conditions

Consider a primal optimization problem based on generalized inequalities (as given by (4.106))

$$
\begin{aligned}
p^* = \min\ & f_0(\mathbf{x}) \\
\text{s.t.}\ & \mathbf{f}_i(\mathbf{x}) \preceq_{K_i} \mathbf{0},\ i = 1,\dots,m \\
& h_i(\mathbf{x}) = 0,\ i = 1,\dots,p
\end{aligned}
\tag{9.173}
$$

with the problem domain

$$
\mathcal{D} = (\mathbf{dom}\ f_0) \cap \left(\bigcap_{i=1}^{m} \mathbf{dom}\ \mathbf{f}_i \right) \cap \left(\bigcap_{i=1}^{p} \mathbf{dom}\ h_i \right),
\tag{9.174}
$$

where $K_i \subset \mathbb{R}^{k_i}$, $i = 1,\dots,m$, are proper cones, $f_0(\mathbf{x})$ is the objective function, $\mathbf{f}_i(\mathbf{x})$, $i = 1,\dots,m$ are generalized inequality constraint functions defined on K_i, and $h_i(\mathbf{x})$, $i = 1,\dots,p$, are equality constraint functions. All the preceding results on duality of optimization problems with ordinary inequalities can be extended to those with generalized inequalities. Next, let us discuss the corresponding results on duality for problem (9.173).

- The Lagrangian is given by

$$
\mathcal{L}(\mathbf{x},\boldsymbol{\lambda},\boldsymbol{\nu}) = f_0(\mathbf{x}) + \sum_{i=1}^{m} \boldsymbol{\lambda}_i^T \mathbf{f}_i(\mathbf{x}) + \sum_{i=1}^{p} \nu_i h_i(\mathbf{x}),
\tag{9.175}
$$

where $\boldsymbol{\lambda}_i \in \mathbb{R}^{k_i}$, and

$$
\boldsymbol{\lambda} = [\boldsymbol{\lambda}_1^T, \dots, \boldsymbol{\lambda}_m^T]^T, \quad \boldsymbol{\nu} = [\nu_1, \dots, \nu_p]^T
$$

are the Lagrange multiplier vectors.

- Dual function:

$$
g(\boldsymbol{\lambda},\boldsymbol{\nu}) \triangleq \inf_{\mathbf{x}\in\mathcal{D}} \mathcal{L}(\mathbf{x},\boldsymbol{\lambda},\boldsymbol{\nu})
\tag{9.176}
$$

with $\mathbf{dom}\ g$ defined in (9.5). For any primal feasible point $\tilde{\mathbf{x}}$ and $\boldsymbol{\lambda}_i \succeq_{K_i^*} \mathbf{0}$ (where K_i^* denotes the dual cone of K_i),

$$
g(\boldsymbol{\lambda},\boldsymbol{\nu}) = \inf_{\mathbf{x}\in\mathcal{D}} \left(f_0(\mathbf{x}) + \sum_{i=1}^{m} \boldsymbol{\lambda}_i^T \mathbf{f}_i(\mathbf{x}) + \sum_{i=1}^{p} \nu_i h_i(\mathbf{x}) \right)
$$
$$
\leq \mathcal{L}(\tilde{\mathbf{x}},\boldsymbol{\lambda},\boldsymbol{\nu}) \leq f_0(\tilde{\mathbf{x}})
\tag{9.177}
$$

where we have used the fact $\boldsymbol{\lambda}_i^T \mathbf{f}_i(\tilde{\mathbf{x}}) \leq 0$ and $h_i(\tilde{\mathbf{x}}) = 0$ for all i.

- Dual problem:

$$d^\star = \max_{(\boldsymbol{\lambda},\boldsymbol{\nu})\in\textbf{dom}\,g} g(\boldsymbol{\lambda},\boldsymbol{\nu})$$

$$\text{s.t. } \boldsymbol{\lambda}_i \succeq_{K_i^*} \mathbf{0},\ i=1,\dots,m. \tag{9.178}$$

Then we have $d^\star \le p^\star$ by (9.177).

- Slater's condition and strong duality $(d^\star = p^\star)$:

  As for the convex problem with generalized inequalities

$$\min\ f_0(\mathbf{x}) \tag{9.179a}$$

$$\text{s.t. } \boldsymbol{f}_i(\mathbf{x}) \preceq_{K_i} \mathbf{0},\ i=1,\dots,m, \tag{9.179b}$$

$$\mathbf{A}\mathbf{x}=\mathbf{b}, \tag{9.179c}$$

where f_0 is convex and each $\boldsymbol{f}_i$ is K_i-convex, the Slater's condition is that if there exists an $\mathbf{x} \in \textbf{relint}\ \mathcal{D}$ with $\mathbf{A}\mathbf{x}=\mathbf{b}$ and $\boldsymbol{f}_i(\mathbf{x}) \prec_{K_i} \mathbf{0},\ i=1,\dots,m$, then strong duality holds.

- Complementary slackness:

  For any $\boldsymbol{\lambda}_i \in K_i^*$ and $-\boldsymbol{f}_i(\mathbf{x}) \in K_i$, it is true that $\boldsymbol{\lambda}_i^T \boldsymbol{f}_i(\mathbf{x}) \le 0$ (by (2.110)). Suppose that strong duality for problem (9.173) holds. Then,

$$\boldsymbol{\lambda}_i^{\star T} \boldsymbol{f}_i(\mathbf{x}^\star) = 0,\ i=1,\dots,m \tag{9.180}$$

(i.e., $\boldsymbol{\lambda}_i^\star \perp \boldsymbol{f}_i(\mathbf{x}^\star)$) from which we can conclude that

$$\boldsymbol{\lambda}_i^\star \succ_{K_i^*} \mathbf{0} \Rightarrow \boldsymbol{f}_i(\mathbf{x}^\star)=\mathbf{0}, \tag{9.181a}$$

$$\boldsymbol{f}_i(\mathbf{x}^\star) \prec_{K_i} \mathbf{0} \Rightarrow \boldsymbol{\lambda}_i^\star = \mathbf{0} \tag{9.181b}$$

(which can be proved by Remark 9.17 below). However, in contrast to problems with ordinary inequality it is possible that (9.180) holds with $\boldsymbol{\lambda}_i^\star \ne \mathbf{0}$, $\boldsymbol{f}_i(\mathbf{x}^\star) \ne \mathbf{0}$.

Remark 9.17 Assume that K is a proper cone. The interior of K^* can be inferred, by *(P2)* (cf. (2.121b)), to be

$$\textbf{int}\ K^* = \{\mathbf{y} \mid \mathbf{y}^T\mathbf{x} > 0\ \forall \mathbf{x} \in K,\ \mathbf{x} \ne \mathbf{0}\}.$$

Therefore, if $\mathbf{y} \succ_{K^*} \mathbf{0}$, i.e., $\mathbf{y} \in \textbf{int}\ K^*$, then $\mathbf{y}^T\mathbf{x}=0$, $\mathbf{x} \in K$ is true only when $\mathbf{x}=\mathbf{0}$. Owing to this result, it is straightforward to prove the complementary slackness (see (9.181a) and (9.181b)) of the optimization problem with strong duality and generalized inequalities. $\qquad\square$

- KKT optimality conditions:

  Suppose that $f_0, \boldsymbol{f}_1, \ldots, \boldsymbol{f}_m, h_1, \ldots, h_p$ are differentiable. Then the KKT conditions associated with (9.173) are as follows:

$$\nabla f_0(\mathbf{x}^\star) + \sum_{i=1}^m \nabla\left(\boldsymbol{f}_i(\mathbf{x}^\star)^T \boldsymbol{\lambda}_i^\star\right) + \sum_{i=1}^p \nu_i^\star \nabla h_i(\mathbf{x}^\star) = \mathbf{0}, \tag{9.182a}$$

$$\boldsymbol{f}_i(\mathbf{x}^\star) \preceq_{K_i} \mathbf{0}, \ i = 1, \ldots, m, \tag{9.182b}$$

$$h_i(\mathbf{x}^\star) = 0, \ i = 1, \ldots, p, \tag{9.182c}$$

$$\boldsymbol{\lambda}_i^\star \succeq_{K_i^*} \mathbf{0}, \ i = 1, \ldots, m, \tag{9.182d}$$

$$(\boldsymbol{\lambda}_i^\star)^T \boldsymbol{f}_i(\mathbf{x}^\star) = 0, \ i = 1, \ldots, m. \tag{9.182e}$$

The KKT conditions given by (9.182) are necessary for strong duality and the proof is similar to the proof for (9.121) associated with the optimization problem with ordinary inequalities. If the primal problem is convex and the Slater's condition holds (i.e., strong duality holds), then the KKT conditions are sufficient and necessary conditions for the primal optimal $\mathbf{x}^\star$ and dual optimal $(\boldsymbol{\lambda}^\star, \boldsymbol{\nu}^\star)$, and meanwhile

$$p^\star = d^\star = \mathcal{L}(\mathbf{x}^\star, \boldsymbol{\lambda}^\star, \boldsymbol{\nu}^\star). \tag{9.183}$$

Remark 9.18 Problem (9.1) and problem (9.65) are special cases of problem (9.173) and problem (9.179), respectively, with only one generalized inequality $\boldsymbol{f}(\mathbf{x}) = [f_1(\mathbf{x}), \ldots, f_m(\mathbf{x})]^T \preceq_K \mathbf{0}$ where $K = \mathbb{R}_+^m$. □

Remark 9.19 *(Bidual of problem* (9.173)*)* Problem (9.178) can be re-expressed as the same form as (9.46), and then its dual (or the bidual of the primal problem (9.173)) can be seen to be

$$- \max_{\boldsymbol{w}_i \succeq_{K_i} \mathbf{0}, i=1,\ldots,m} \inf\left\{ -g(\boldsymbol{\lambda}, \boldsymbol{\nu}) - \sum_{i=1}^m \boldsymbol{w}_i^T \boldsymbol{\lambda}_i \mid (\boldsymbol{\lambda}, \boldsymbol{\nu}) \in \mathbf{dom}\, g \right\}$$

$$= \min_{\boldsymbol{w}_i \succeq_{K_i} \mathbf{0}, i=1,\ldots,m} - \inf\left\{ -g(\boldsymbol{\lambda}, \boldsymbol{\nu}) - \sum_{i=1}^m \boldsymbol{w}_i^T \boldsymbol{\lambda}_i \mid (\boldsymbol{\lambda}, \boldsymbol{\nu}) \in \mathbf{dom}\, g \right\} \tag{9.184}$$

$$= \min_{\boldsymbol{w}_i \succeq_{K_i} \mathbf{0}, i=1,\ldots,m} \sup\left\{ g(\boldsymbol{\lambda}, \boldsymbol{\nu}) + \sum_{i=1}^m \boldsymbol{w}_i^T \boldsymbol{\lambda}_i \mid (\boldsymbol{\lambda}, \boldsymbol{\nu}) \in \mathbf{dom}\, g \right\}.$$

It can be proven that the bidual of the convex problem (9.179) is also the same as itself, provided that $f_0(\mathbf{x})$ is closed and $\boldsymbol{f}_i(\mathbf{x})$ is differentiable for all i (cf. Example 9.6). □

As mentioned above, the KKT conditions can be used to find the analytical solutions, but closed-form solutions for problems involving generalized inequalities are in general hard to find. Surely, one can employ the available software packages (e.g, SeDuMi or CVX) or design an interior-point algorithm to efficiently obtain optimal numerical solutions.

We have previously shown that LP in standard form (see (9.13)) is a convex problem (involving ordinary inequalities) and its dual is actually the problem in inequality form (see (9.48)). This also applies to convex problems involving generalized inequalities by Remark 9.19. In each of the following two subsections, we present a specific optimization problem involving generalized inequalities. We will illustrate how the dual function and the KKT conditions can be obtained, especially for the SDP problem where the unknown variable is a PSD matrix rather than a real vector.

9.8.2   Lagrange dual of cone program and KKT conditions

The standard form of a cone program given by (4.107) is repeated here for convenience.

$$\min \ \mathbf{c}^T\mathbf{x}$$
$$\text{s.t. } \mathbf{A}\mathbf{x} = \mathbf{b} \tag{9.185}$$
$$\mathbf{x} \succeq_K \mathbf{0}$$

where $\mathbf{A} \in \mathbb{R}^{m \times n}$, $\mathbf{b} \in \mathbb{R}^m$, $K \subseteq \mathbb{R}^n$ is a proper cone. The Lagrangian can be seen to be

$$\mathcal{L}(\mathbf{x}, \boldsymbol{\lambda}, \boldsymbol{\nu}) = \mathbf{c}^T\mathbf{x} - \boldsymbol{\lambda}^T\mathbf{x} + \boldsymbol{\nu}^T(\mathbf{A}\mathbf{x} - \mathbf{b}), \tag{9.186}$$

which is differentiable and

$$\nabla_\mathbf{x}\mathcal{L}(\mathbf{x}, \boldsymbol{\lambda}, \boldsymbol{\nu}) = \mathbf{A}^T\boldsymbol{\nu} - \boldsymbol{\lambda} + \mathbf{c};$$

so the dual function is given by

$$g(\boldsymbol{\lambda}, \boldsymbol{\nu}) = \inf_{\mathbf{x} \in \mathbb{R}^n} \mathcal{L}(\mathbf{x}, \boldsymbol{\lambda}, \boldsymbol{\nu})$$
$$= \begin{cases} -\mathbf{b}^T\boldsymbol{\nu}, & \mathbf{A}^T\boldsymbol{\nu} - \boldsymbol{\lambda} + \mathbf{c} = \mathbf{0} \\ -\infty, & \text{otherwise.} \end{cases} \tag{9.187}$$

The dual problem can also be expressed as

$$\max \ -\mathbf{b}^T\boldsymbol{\nu}$$
$$\text{s.t. } \mathbf{A}^T\boldsymbol{\nu} + \mathbf{c} = \boldsymbol{\lambda} \tag{9.188}$$
$$\boldsymbol{\lambda} \succeq_{K^*} \mathbf{0}$$

where K^* is the dual cone of the proper cone K. By eliminating $\boldsymbol{\lambda}$ and defining $\mathbf{y} = -\boldsymbol{\nu}$, this problem can be simplified to

$$\max \ \mathbf{b}^T\mathbf{y}$$
$$\text{s.t. } \mathbf{A}^T\mathbf{y} \preceq_{K^*} \mathbf{c} \tag{9.189}$$

which is a cone program in inequality form (cf. (4.108)), involving the generalized inequality defined on the dual cone K^*. Strong duality is valid if the Slater condition holds, i.e., there is an $\mathbf{x} \succ_K \mathbf{0}$ with $\mathbf{A}\mathbf{x} = \mathbf{b}$. It can be shown using

(9.184) that the dual of problem (9.189) is exactly problem (9.185), namely, the bidual of a cone program is the original cone program.

The KKT conditions given by (9.182) for the cone problem can be easily seen to consist of the two constraints of the primal problem (9.185) and the two constraints of the dual problem (9.188) and the complementary slackness $\boldsymbol{\lambda}^T\mathbf{x} = 0$, which are summarized as follows:

$$\mathbf{A}\mathbf{x}^\star = \mathbf{b}, \tag{9.190a}$$

$$\mathbf{x}^\star \succeq_K \mathbf{0}, \tag{9.190b}$$

$$\mathbf{A}^T\boldsymbol{\nu}^\star + \mathbf{c} = \boldsymbol{\lambda}^\star, \tag{9.190c}$$

$$\boldsymbol{\lambda}^\star \succeq_{K^*} \mathbf{0}, \tag{9.190d}$$

$$\boldsymbol{\lambda}^{\star T}\mathbf{x}^\star = 0. \tag{9.190e}$$

Example 9.16 Consider the following SOCP (cf. (4.109) and (7.1))

$$\begin{aligned} \min\ & \mathbf{f}^T\mathbf{x} \\ \text{s.t.}\ & \left\|\mathbf{A}_i\mathbf{x} + \mathbf{b}_i\right\|_2 \le \mathbf{c}_i^T\mathbf{x} + d_i,\ i = 1,\ldots,m \end{aligned} \tag{9.191}$$

where variables $\mathbf{x} \in \mathbb{R}^n$, $\mathbf{f} \in \mathbb{R}^n$, $\mathbf{A}_i \in \mathbb{R}^{n_i \times n}$, $\mathbf{b}_i \in \mathbb{R}^{n_i}$, $\mathbf{c}_i \in \mathbb{R}^n$, and $d_i \in \mathbb{R}$, $i = 1,\ldots,m$. Next, we find the dual of this SOCP and the KKT conditions.

Let us express the SOCP (9.191) in conic form first as follows:

$$\begin{aligned} \min\ & \mathbf{f}^T\mathbf{x} \\ \text{s.t.}\ & -\left(\mathbf{A}_i\mathbf{x} + \mathbf{b}_i, \mathbf{c}_i^T\mathbf{x} + d_i\right) \preceq_{K_i} \mathbf{0}_{n_i+1},\ i = 1,\ldots,m \end{aligned} \tag{9.192}$$

where $K_i \subset \mathbb{R}^{n_i+1}$ are second-order cones. The Lagrangian is

$$\mathcal{L}(\mathbf{x}, (\boldsymbol{\lambda}_1, \nu_1), \ldots, (\boldsymbol{\lambda}_m, \nu_m)) = \mathbf{f}^T\mathbf{x} - \left\{\sum_{i=1}^m \boldsymbol{\lambda}_i^T\left(\mathbf{A}_i\mathbf{x} + \mathbf{b}_i\right) + \nu_i\left(\mathbf{c}_i^T\mathbf{x} + d_i\right)\right\}.$$

The dual function can be seen to be

$$g(\boldsymbol{\lambda}, \boldsymbol{\nu}) = \begin{cases} -\sum_{i=1}^m \left(\boldsymbol{\lambda}_i^T\mathbf{b}_i + \nu_i d_i\right), & \text{if } \sum_{i=1}^m \mathbf{A}_i^T\boldsymbol{\lambda}_i + \nu_i\mathbf{c}_i = \mathbf{f} \\ -\infty, & \text{otherwise.} \end{cases}$$

Therefore, the dual problem is

$$\begin{aligned} \max\ & -\sum_{i=1}^m \left(\boldsymbol{\lambda}_i^T\mathbf{b}_i + \nu_i d_i\right) \\ \text{s.t.}\ & \sum_{i=1}^m \left(\mathbf{A}_i^T\boldsymbol{\lambda}_i + \nu_i\mathbf{c}_i\right) = \mathbf{f} \\ & (\boldsymbol{\lambda}_i, \nu_i) \succeq_{K_i^*} \mathbf{0}_{n_i+1},\ i = 1,\ldots,m. \end{aligned} \tag{9.193}$$

Note that $K_i^* = K_i$ since second-order cones are self-dual as proven in Example 2.14 of Subsection 2.5.2, and the dual problem (9.193) is also an SOCP. Moreover, the m inequality constraints in problem (9.192), and m inequality constraints in problem (9.193), and the equality constraint in problem (9.193), and the complementary slackness condition

$$\boldsymbol{\lambda}_i^T(\mathbf{A}_i\mathbf{x} + \mathbf{b}_i) + \nu_i(\mathbf{c}_i^T\mathbf{x} + d_i) = 0, \quad i = 1, \ldots, m,$$

constitute the KKT conditions of problem (9.192). Strong duality holds when there exists an $\mathbf{x}$ such that $(\mathbf{A}_i\mathbf{x} + \mathbf{b}_i, \mathbf{c}_i^T\mathbf{x} + d_i) \in \text{int } K_i, \ i = 1, \ldots, m.$ ☐

9.8.3 Lagrange dual of SDP and KKT conditions

The standard form of SDP defined in (4.110) and (8.3) is repeated here for convenience.

$$\min \ \text{Tr}(\mathbf{CX}) \tag{9.194a}$$

$$\text{s.t. } \text{Tr}(\mathbf{A}_i\mathbf{X}) = b_i, \ i = 1, \ldots, m, \tag{9.194b}$$

$$\mathbf{X} \succeq_K \mathbf{0}, \tag{9.194c}$$

where $\mathbf{C} \in \mathbb{S}^n$, $\mathbf{A}_i \in \mathbb{S}^n$, and $K = \mathbb{S}^n_+$. The Lagrangian is

$$\mathcal{L}(\mathbf{X}, \mathbf{Z}, \boldsymbol{\nu}) = \ \text{Tr}(\mathbf{CX}) - \text{Tr}(\mathbf{ZX}) + \sum_{i=1}^{m} \nu_i\big[\text{Tr}(\mathbf{A}_i\mathbf{X}) - b_i\big] \tag{9.195a}$$

$$= \ \text{Tr}\big[(\mathbf{C} - \mathbf{Z} + \sum_{i=1}^{m} \nu_i\mathbf{A}_i)\mathbf{X}\big] - \mathbf{b}^T\boldsymbol{\nu}, \ \mathbf{Z} \in \mathbb{S}^n, \ \boldsymbol{\nu} \in \mathbb{R}^m \tag{9.195b}$$

which is affine in $\mathbf{X}$, and bounded below only if

$$\nabla_{\mathbf{X}}\mathcal{L}(\mathbf{X}, \mathbf{Z}, \boldsymbol{\nu}) = \mathbf{C} - \mathbf{Z} + \sum_{i=1}^{m} \nu_i\mathbf{A}_i = \mathbf{0}.$$

Therefore, the dual function is given by

$$g(\mathbf{Z}, \boldsymbol{\nu}) = \inf_{\mathbf{X}} \mathcal{L}(\mathbf{X}, \mathbf{Z}, \boldsymbol{\nu}) = \begin{cases} -\mathbf{b}^T\boldsymbol{\nu}, & \mathbf{C} - \mathbf{Z} + \sum_{i=1}^{m} \nu_i\mathbf{A}_i = \mathbf{0} \\ -\infty, & \text{otherwise.} \end{cases} \tag{9.196}$$

Then, the corresponding dual problem can be expressed as

$$\max \ -\mathbf{b}^T\boldsymbol{\nu} \tag{9.197a}$$

$$\text{s.t. } \mathbf{C} - \mathbf{Z} + \sum_{i=1}^{m} \nu_i\mathbf{A}_i = \mathbf{0}, \tag{9.197b}$$

$$\mathbf{Z} \succeq_{K^*} \mathbf{0}, \tag{9.197c}$$

where $K^* = K = \mathbb{S}^n_+$. By eliminating $\mathbf{Z}$, the problem can be simplified as

$$\max \ -\mathbf{b}^T\boldsymbol{\nu}$$
$$\text{s.t. } \mathbf{C} + \sum_{i=1}^{m} \nu_i\mathbf{A}_i \succeq_{K^*} \mathbf{0} \tag{9.198}$$

which is the inequality form of SDP (cf. (8.1)). Strong duality is valid if the Slater condition holds, i.e., there is an $\mathbf{X} \succ_K \mathbf{0}$ with $\text{Tr}(\mathbf{A}_i\mathbf{X}) = b_i$, $i = 1, \ldots, m$. It can be also shown that the dual of problem (9.198) is exactly problem (9.194); namely, the bidual of the SDP program (9.194) is the original SDP program.

The KKT conditions given by (9.182) can be seen to be the two constraints (9.194b) and (9.194c) of the primal problem (9.194), and the two constraints (9.197b) and (9.197c) of the dual problem (9.197), but the complementary slackness is as follows:

$$\text{Tr}(\mathbf{Z}\mathbf{X}) = 0 \tag{9.199}$$

(i.e., $\mathbf{Z}$ and $\mathbf{X}$ are orthogonal to each other), which can be proven through the same procedure as the proof of the complementary slackness given by (9.117) for the optimization problem with strong duality and ordinary inequalities. Note that the complementary slackness of this SDP problem is exactly the equation by setting the second term of the Lagrangian in (9.195a) to zero. The strong duality also guarantees

$$p^\star = \text{Tr}(\mathbf{C}\mathbf{X}^\star) = \mathcal{L}(\mathbf{X}^\star, \mathbf{Z}^\star, \boldsymbol{\nu}^\star) = d^\star. \tag{9.200}$$

The complementary slackness for the SDP problem given by (9.199) can be further simplified as

$$\mathbf{Z}\mathbf{X} = \mathbf{0}. \tag{9.201}$$

If $\mathbf{Z} \succ \mathbf{0}$, then $\mathbf{X} = \mathbf{0}$; if $\mathbf{X} \succ \mathbf{0}$, then $\mathbf{Z} = \mathbf{0}$. It is also possible that $\mathbf{Z}\mathbf{X} = \mathbf{0}$, but $\mathbf{Z} \neq \mathbf{0}$ and $\mathbf{X} \neq \mathbf{0}$ in the meantime. This occurs as both $\mathbf{Z}$ and $\mathbf{X}$ are not of full rank.

Proof of (9.201): Let

$$\mathbf{X} = \sum_{i=1}^{m} \lambda_i \mathbf{q}_i \mathbf{q}_i^T \in K = \mathbb{S}_+^n$$

$$\mathbf{Z} = \sum_{j=1}^{\ell} \xi_i \mathbf{v}_j \mathbf{v}_j^T \in \mathbb{S}_+^n$$

(EVD of $\mathbf{X}$ and EVD of $\mathbf{Z}$) with $\lambda_i > 0$ and $m \leq n$, and $\xi_j > 0$ and $\ell \leq n$. Note that $m = n$ means $\mathbf{X} \succ \mathbf{0}$, and $m < n$ means $\mathbf{X} \succeq \mathbf{0}$ but $\mathbf{X} \not\succ \mathbf{0}$. Substituting this EVD of $\mathbf{X}$ into (9.199) gives rise to

$$\text{Tr}(\mathbf{Z}\mathbf{X}) = \text{Tr}\left(\sum_{i=1}^{m} \lambda_i \mathbf{Z}\mathbf{q}_i\mathbf{q}_i^T\right) = \sum_{i=1}^{m} \lambda_i \mathbf{q}_i^T \mathbf{Z}\mathbf{q}_i = 0,$$

which implies that $\mathbf{q}_i^T \mathbf{Z}\mathbf{q}_i = 0$ for all $i = 1, \ldots, m$, due to $\mathbf{Z} \succeq \mathbf{0}$ and $\lambda_i > 0$. Furthermore,

$$\mathbf{q}_i^T \mathbf{Z}\mathbf{q}_i = \sum_{j=1}^{\ell} \xi_j (\mathbf{q}_i^T \mathbf{v}_j)^2 = 0, \quad i = 1, \ldots, m,$$

together with the fact of $\xi_j > 0$, $j = 1, \ldots, \ell$, implies that $\mathbf{q}_i^T \mathbf{v}_j = 0$ for all $j = 1, \ldots, \ell$, namely, $\mathbf{q}_i \in \mathrm{span}([\mathbf{v}_1, \ldots, \mathbf{v}_\ell])^\perp$ (which is exactly the $(n - \ell)$-dimensional subspace spanned by all the eigenvectors of $\mathbf{Z}$ associated with the zero eigenvalue) for all $i = 1, \ldots, m$. Thus,

$$\mathbf{Z}\mathbf{q}_i = \mathbf{0}, \ \forall i = 1, \ldots, m.$$

Therefore, we have proven $\mathbf{Z}\mathbf{X} = \sum_{i=1}^m \lambda_i (\mathbf{Z}\mathbf{q}_i)\mathbf{q}_i^T = \mathbf{0}$.   ∎

The KKT conditions for the SDP (9.194) are summarized as follows:

$$\mathrm{Tr}(\mathbf{A}_i \mathbf{X}^\star) = b_i, \ i = 1, \ldots, m, \tag{9.202a}$$

$$\mathbf{X}^\star \succeq_K \mathbf{0}, \tag{9.202b}$$

$$\mathbf{C} - \mathbf{Z}^\star + \sum_{i=1}^m \nu_i^\star \mathbf{A}_i = \mathbf{0}, \tag{9.202c}$$

$$\mathbf{Z}^\star \succeq_{K^\star} \mathbf{0}, \tag{9.202d}$$

$$\mathbf{Z}^\star \mathbf{X}^\star = \mathbf{0}, \tag{9.202e}$$

where $K = K^* = \mathbb{S}_+^n$.

Example 9.17 *(Trust region problem (revisited))* In Example 9.9, we have shown that the strong duality of the trust region problem defined in (9.88) (a QCQP) holds even when $\mathbf{A}$ is indefinite. We have also derived its bidual (an SDP) as given by (9.96). In this example, let us present the KKT conditions of (9.96), from which some perspectives about optimal solutions can be deduced.

It can be seen that the Lagrangian of (9.96) is given by

$$\mathcal{L}(\mathbf{X}, \mathbf{x}, \mathbf{Z}, \lambda) = \mathrm{Tr}(\mathbf{A}\mathbf{X}) + 2\mathbf{b}^T \mathbf{x} - \mathrm{Tr}\big(\mathbf{Z}(\mathbf{X} - \mathbf{x}\mathbf{x}^T)\big) + \lambda\big(\mathrm{Tr}(\mathbf{X}) - 1\big)$$
$$= \mathrm{Tr}\big((\mathbf{A} + \lambda\mathbf{I} - \mathbf{Z})\mathbf{X}\big) + \mathbf{x}^T \mathbf{Z}\mathbf{x} + 2\mathbf{b}^T \mathbf{x} - \lambda, \tag{9.203}$$

where $\mathbf{Z} \in \mathbb{S}^n$ and λ are Lagrange multipliers. Then the dual function can be shown to be

$$g(\mathbf{Z}, \lambda) = \inf_{\mathbf{X}, \mathbf{x}} \mathcal{L}(\mathbf{X}, \mathbf{x}, \mathbf{Z}, \lambda)$$
$$= \begin{cases} -\lambda - \mathbf{b}^T \mathbf{Z}^\dagger \mathbf{b}, & \mathbf{Z} = \mathbf{A} + \lambda\mathbf{I} \succeq \mathbf{0}, \ \mathbf{b} \in \mathcal{R}(\mathbf{Z}) \\ -\infty, & \text{otherwise.} \end{cases} \tag{9.204}$$

Thus, the dual of problem (9.96) is given by (9.91). Since Slater's condition holds for (9.96), the strong duality also holds. The associated KKT conditions are then given as follows:

$$\begin{cases} \mathrm{Tr}(\mathbf{X}^\star) - 1 \leq 0 \\ \mathbf{X}^\star - \mathbf{x}^\star(\mathbf{x}^\star)^T \succeq \mathbf{0} \\ \lambda^\star \geq 0, \ \lambda^\star(\mathrm{Tr}(\mathbf{X}^\star) - 1) = 0 \\ \mathbf{Z}^\star = \mathbf{A} + \lambda^\star\mathbf{I} \succeq \mathbf{0}, \ \mathbf{b} \in \mathcal{R}(\mathbf{Z}^\star) \\ \mathbf{Z}^\star \big(\mathbf{X}^\star - \mathbf{x}^\star(\mathbf{x}^\star)^T\big) = \mathbf{0}. \end{cases} \tag{9.205}$$

Some solutions to problem (9.96) can be inferred via the above KKT conditions. When $\lambda^\star > 0$ and $\mathbf{Z}^\star \succ \mathbf{0}$, we have $\mathbf{X}^\star = \mathbf{x}^\star(\mathbf{x}^\star)^T$ and $(\mathbf{x}^\star)^T\mathbf{x}^\star = 1$. Then

$$
\begin{aligned}
p^\star &= \mathrm{Tr}(\mathbf{A}\mathbf{X}^\star) + 2\mathbf{b}^T\mathbf{x}^\star \\
&= (\mathbf{x}^\star)^T\mathbf{A}\mathbf{x}^\star + 2\mathbf{b}^T\mathbf{x}^\star \\
&= (\mathbf{x}^\star)^T\mathbf{Z}^\star\mathbf{x}^\star + 2\mathbf{b}^T\mathbf{x}^\star - \lambda^\star \\
&= d^\star = -\lambda^\star - \mathbf{b}^T(\mathbf{Z}^\star)^\dagger\mathbf{b} \quad (\text{cf. } (9.91))
\end{aligned}
$$

which holds true only if $\mathbf{x}^\star = -(\mathbf{Z}^\star)^\dagger\mathbf{b} \in \mathcal{R}(\mathbf{Z}^\star)$. Thus, $(\mathbf{X}^\star = \mathbf{x}^\star(\mathbf{x}^\star)^T, \mathbf{x}^\star)$ is also the unique solution to (9.96). However when $\lambda^\star > 0$, $\mathbf{Z}^\star \succeq \mathbf{0}$ but $\mathbf{Z}^\star \not\succ \mathbf{0}$, multiple solutions for $(\mathbf{X}^\star, \mathbf{x}^\star)$ exist. $\qquad\square$

Remark 9.20 Since SDP has been widely used in physical layer MIMO wireless communications, e.g., transmit beamforming, where the unknown variable $\mathbf{X}$ is often a Hermitian PSD matrix, an intriguing question is: What will be the associated KKT conditions? Let us reconsider the SDP problem (9.194) again, where $\mathbf{X} = \mathbf{X}^H \in \mathbb{C}^{n\times n}$, $\mathbf{C} = \mathbf{C}^H \in \mathbb{C}^{n\times n}$, $\mathbf{A}_i = \mathbf{A}_i^H \in \mathbb{C}^{n\times n}$, and $b_i \in \mathbb{R}$, and the proper cone $K = \mathbb{H}_+^n$. It can be easily shown that $K^* = K = \mathbb{H}_+^n$ due to $\mathrm{Tr}(\mathbf{A}\mathbf{B}) \geq 0$ for any Hermitian PSD matrices $\mathbf{A}$ and $\mathbf{B}$ (cf. Remark 1.20).

Note that, the objective function $\mathrm{Tr}(\mathbf{C}\mathbf{X})$ of problem (9.194) itself is a real-valued function of $\mathbf{X}$ due to $\mathbf{C}, \mathbf{X} \in \mathbb{H}^n$ (cf. Remark 1.20), and so it can be expressed as

$$
\mathrm{Tr}(\mathbf{C}\mathbf{X}) = \frac{1}{2}\mathrm{Tr}(\mathbf{C}\mathbf{X}) + \frac{1}{2}\mathrm{Tr}(\mathbf{C}^*\mathbf{X}^*), \tag{9.206}
$$

which corresponds to a real-valued function for all $\mathbf{X} \in \mathbb{C}^n$. Moreover, the constraint $\mathbf{X} \succeq \mathbf{0}$ is a complex matrix inequality since $\mathbf{X} \in \mathbb{H}^n$. What will be the corresponding term in the Lagrangian? Following the same procedure in Remark 9.5 for the real case of $\mathbf{X} \succeq \mathbf{0}$, one can come up with

$$
\begin{aligned}
\mathcal{L}(\mathbf{X}, \mathbf{Z}, \boldsymbol{\nu}) =\ & \mathrm{Tr}(\mathbf{C}\mathbf{X}) - \mathrm{Tr}(\mathbf{Z}\mathbf{X}) + \sum_{i=1}^m \nu_i\big[\mathrm{Tr}(\mathbf{A}_i\mathbf{X}) - b_i\big] \\
=\ & \mathrm{Tr}\big[(\mathbf{C} - \mathbf{Z} + \sum_{i=1}^m \nu_i\mathbf{A}_i)\mathbf{X}\big] - \mathbf{b}^T\boldsymbol{\nu},\ \mathbf{Z} \in \mathbb{H}^n,\ \boldsymbol{\nu} \in \mathbb{R}^m \\
=\ & \frac{1}{2}\big[\mathrm{Tr}(\mathbf{V}\mathbf{X}) + \mathrm{Tr}(\mathbf{V}^*\mathbf{X}^*)\big] - \mathbf{b}^T\boldsymbol{\nu} \quad (\text{cf. } (9.206)),
\end{aligned}
$$

where $\mathbf{V} = \mathbf{C} - \mathbf{Z} + \sum_{i=1}^m \nu_i\mathbf{A}_i \in \mathbb{H}^n$. Now $\mathcal{L}(\mathbf{X}, \mathbf{Z}, \boldsymbol{\nu})$ can be viewed as a real-valued function of $\mathbf{X} \in \mathbb{C}^{n\times n}$. Then $\nabla_{\mathbf{X}}\mathcal{L}(\mathbf{X}, \mathbf{Z}, \boldsymbol{\nu}) = 2\nabla_{\mathbf{X}^*}\mathcal{L}(\mathbf{X}, \mathbf{Z}, \boldsymbol{\nu}) = \mathbf{0}$ also leads to the same result $\mathbf{V} = \mathbf{C} - \mathbf{Z} + \sum_{i=1}^m \nu_i\mathbf{A}_i = \mathbf{0}$ as the real-variable case. Hence, the KKT conditions given in (9.202) also apply to the complex-variable case except that the primal variable $\mathbf{X}$ and dual variable $\mathbf{Z}$ are Hermitian PSD matrices. $\qquad\square$

♦ **Power Minimization Transmit Beamforming via SDR: Analysis for rank-1 solutions by KKT conditions**

In Example 9.8, we have proven the strong duality for the nonconvex smallest eigenvalue problem of a matrix $\mathbf{C} \in \mathbb{R}^n$ defined in (9.81). Moreover, the rank-one solution of its SDR (9.86) has also been shown unique via the associated KKT conditions when the multiplicity of minimum eigenvalue of $\mathbf{C}$ is equal to one; otherwise solutions with rank higher than one also exist.

In Chapter 8, we have presented quite some transmit beamforming designs (complex-variable optimization problems) in MIMO wireless communications using SDP and SDR, for which the rank-1 analysis for the designed complex beamformer has been discussed without showing any detailed proof. Analysis of whether or not the resulting SDP problem via SDR can yield rank-1 solutions of the original problem heavily relies on the associated KKT conditions. To illustrate this, let us revisit the transmit beamforming problem (7.40) (which was previously formulated as an SOCP given by (7.47)), which is repeated here for convenience

$$\min \ \sum_{i=1}^{n} \|\mathbf{f}_i\|_2^2$$

$$\text{s.t.} \ \frac{\left|\mathbf{h}_i^H \mathbf{f}_i\right|^2}{\sum_{j=1,j\neq i}^{n} \left|\mathbf{h}_i^H \mathbf{f}_j\right|^2 + \sigma_i^2} \geq \gamma_0, \ i = 1, \ldots, n.$$

This complex-variable problem can also be reformulated into an SDP by applying SDR, i.e., by replacing $\mathbf{f}_i \mathbf{f}_i^H$ with an $m \times m$ Hermitian PSD matrix $\mathbf{F}_i$. Then the following SDR problem results:

$$\min \ \sum_{i=1}^{n} \text{Tr}(\mathbf{F}_i) \tag{9.207a}$$

$$\text{s.t.} \ \frac{1}{\gamma_0} \mathbf{h}_i^H \mathbf{F}_i \mathbf{h}_i \geq \sum_{j=1,j\neq i}^{n} \mathbf{h}_i^H \mathbf{F}_j \mathbf{h}_i + \sigma_i^2, \ i = 1, \ldots, n, \tag{9.207b}$$

$$\mathbf{F}_i \succeq \mathbf{0}, \ i = 1, \ldots, n. \tag{9.207c}$$

Next, let us prove that the obtained optimal $\{\mathbf{F}_i^\star\}_{i=1}^n$ of problem (9.207) is of rank one, i.e., $\mathbf{F}_i^\star = \mathbf{f}_i^\star (\mathbf{f}_i^\star)^H$; thus $\{\mathbf{f}_i^\star\}_{i=1}^n$ must be an optimal solution of problem (7.40).

The Lagrangian of problem (9.207) can be easily shown to be

$$\mathcal{L}(\{\mathbf{F}_i, \lambda_i, \mathbf{Z}_i\}_{i=1}^n)$$

$$= \sum_{i=1}^n \text{Tr}\left(\left[\mathbf{I}_m - \frac{\lambda_i}{\gamma_0}\mathbf{h}_i\mathbf{h}_i^H - \mathbf{Z}_i\right]\mathbf{F}_i\right) + \sum_{i=1}^n \sum_{j=1,j\neq i}^n \text{Tr}(\lambda_i\mathbf{h}_i\mathbf{h}_i^H\mathbf{F}_j) + \sum_{i=1}^n \lambda_i\sigma_i^2$$

$$= \sum_{i=1}^n \text{Tr}\left(\left[\mathbf{I}_m - \frac{\lambda_i}{\gamma_0}\mathbf{h}_i\mathbf{h}_i^H - \mathbf{Z}_i\right]\mathbf{F}_i\right) + \sum_{j=1}^n \sum_{i=1,i\neq j}^n \text{Tr}(\lambda_j\mathbf{h}_j\mathbf{h}_j^H\mathbf{F}_i) + \sum_{i=1}^n \lambda_i\sigma_i^2$$

$$= \sum_{i=1}^n \text{Tr}\left(\left[\mathbf{I}_m - \frac{\lambda_i}{\gamma_0}\mathbf{h}_i\mathbf{h}_i^H + \sum_{j=1,j\neq i}^n \lambda_j\mathbf{h}_j\mathbf{h}_j^H - \mathbf{Z}_i\right]\mathbf{F}_i\right) + \sum_{i=1}^n \lambda_i\sigma_i^2, \qquad (9.208)$$

where $\lambda_i \in \mathbb{R}$ and $\mathbf{Z}_i \in \mathbb{H}^m$ are the dual variables associated with (9.207b) and (9.207c), respectively.

Let

$$\mathbf{A}_i = \mathbf{I}_m - \frac{\lambda_i}{\gamma_0}\mathbf{h}_i\mathbf{h}_i^H + \sum_{j=1,j\neq i}^n \lambda_j\mathbf{h}_j\mathbf{h}_j^H - \mathbf{Z}_i \in \mathbb{H}^m.$$

Then the Lagrangian given by (9.208) can be re-expressed as the following real-valued function:

$$\mathcal{L}(\{\mathbf{F}_i, \lambda_i, \mathbf{Z}_i\}_{i=1}^n) = \sum_{i=1}^n \frac{1}{2}\left[\text{Tr}(\mathbf{A}_i\mathbf{F}_i) + \text{Tr}(\mathbf{A}_i^*\mathbf{F}_i^*)\right] + \sum_{i=1}^n \lambda_i\sigma_i^2, \qquad (9.209)$$

and the dual function of problem (9.207) can be easily seen to be

$$g(\{\mathbf{Z}_i, \lambda_i\}_{i=1}^n) = \inf_{\mathbf{F}_i\in\mathbb{C}^m,\forall i} \mathcal{L}(\{\mathbf{F}_i, \lambda_i, \mathbf{Z}_i\}_{i=1}^n)$$

$$= \begin{cases} \sum_{i=1}^n \lambda_i\sigma_i^2, & \mathbf{A}_i = \mathbf{0} \; \forall i \\ -\infty, & \text{otherwise.} \end{cases} \quad \text{(by (4.99) and (1.45)).} \qquad (9.210)$$

Thus, the dual of problem (9.207) is given by

$$\max \sum_{i=1}^n \lambda_i\sigma_i^2 \qquad (9.211)$$

$$\text{s.t. } \mathbf{A}_i = \mathbf{0}, \; \mathbf{Z}_i \succeq \mathbf{0}, \; \lambda_i \geq 0 \; \forall i$$

Since problem (9.207) is convex and Slater's condition holds true, KKT conditions are the sufficient and necessary optimality conditions. Some KKT conditions needed in the proof are as follows:

$$\frac{1}{\gamma_0}\mathbf{h}_i^H\mathbf{F}_i^*\mathbf{h}_i \geq \sum_{j=1,j\neq i}^n \mathbf{h}_i^H\mathbf{F}_j^*\mathbf{h}_i + \sigma_i^2, \; i = 1,\ldots,n, \qquad (9.212a)$$

$$\mathbf{Z}_i^* = \mathbf{I}_m - \frac{\lambda_i^*}{\gamma_0}\mathbf{h}_i\mathbf{h}_i^H + \sum_{j=1,j\neq i}^n \lambda_j^*\mathbf{h}_j\mathbf{h}_j^H, \; i = 1,\ldots,n, \qquad (9.212b)$$

$$\mathbf{Z}_i^*\mathbf{F}_i^* = \mathbf{0}, \; i = 1,\ldots,n. \qquad (9.212c)$$

Note that $\lambda_i \geq 0$, $\mathbf{Z}_i^\star \succeq \mathbf{0}$, and $\mathbf{F}_i^\star \succeq \mathbf{0}$ are also KKT conditions.

Since $\mathbf{F}_i^\star \neq \mathbf{0}$ (by (9.212a)), the rank of $\mathbf{Z}_i^\star$ must be less than or equal to $m - 1$ (by (9.212c)), i.e.,

$$\operatorname{rank}(\mathbf{Z}_i^\star) \leq m - 1. \tag{9.213}$$

By defining

$$\mathbf{B} = \mathbf{I}_m + \sum_{j=1, j \neq i}^{n} \lambda_j^\star \mathbf{h}_j \mathbf{h}_j^H = \left(\mathbf{B}^{1/2}\right)^2 \succ \mathbf{0}$$

where $\mathbf{B}^{1/2} = (\mathbf{B}^{1/2})^H \succ \mathbf{0}$, the rank of $\mathbf{Z}_i^\star$ can be further inferred as follows:

$$
\begin{aligned}
\operatorname{rank}(\mathbf{Z}_i^\star) &= \operatorname{rank}\left(\mathbf{B} - \frac{\lambda_i^\star}{\gamma_0} \mathbf{h}_i \mathbf{h}_i^H\right) \quad \text{(by (9.212b))} \\
&= \operatorname{rank}\left(\mathbf{B}^{1/2}\left[\mathbf{I}_m - \frac{\lambda_i^\star}{\gamma_0}\mathbf{B}^{-1/2}\mathbf{h}_i\mathbf{h}_i^H\mathbf{B}^{-1/2}\right]\mathbf{B}^{1/2}\right) \\
&= \operatorname{rank}\left(\mathbf{I}_m - \frac{\lambda_i^\star}{\gamma_0}\mathbf{B}^{-1/2}\mathbf{h}_i\mathbf{h}_i^H\mathbf{B}^{-1/2}\right) \quad \text{(by (1.64))} \\
&\geq m - 1 \quad \text{(cf. (1.99))}. \tag{9.214}
\end{aligned}
$$

From (9.213) and (9.214), it can be inferred that $\operatorname{rank}(\mathbf{Z}_i^\star) = m - 1$. Then by (9.212c) and (1.71), we have

$$\operatorname{rank}(\mathbf{F}_i^\star) \leq \dim(\mathcal{N}(\mathbf{Z}_i^\star)) = m - \operatorname{rank}(\mathbf{Z}_i^\star) = 1,$$
$$\Rightarrow \operatorname{rank}(\mathbf{F}_i^\star) = 1 \quad \text{(since } \mathbf{F}_i^\star \neq \mathbf{0}\text{)}.$$

Therefore, the optimal solution $\{\mathbf{F}_i^\star, i = 1, \ldots, n\}$ of the SDP (9.207) must yield the optimal solution $\{\mathbf{f}_i^\star, i = 1, \ldots, n\}$ of the transmit beamforming problem (7.40) via the rank-one decomposition $\mathbf{F}_i^\star = \mathbf{f}_i^\star(\mathbf{f}_i^\star)^H$.

9.9   Theorems of alternatives

In this section, either via the weak duality or via the strong duality of advisable convex optimization problems, we will introduce the weak alternatives and strong alternatives of two systems constituted by equalities and inequalities, meaning that the feasibility of one system implies that the other system is infeasible as they are weak alternatives (i.e., sufficient but not necessary conditions); one system is feasible if and only if the other system is infeasible as they are strong alternatives. We will then prove Farkas lemma based on the theorems of alternatives. Then we conclude this section by proving the S-procedure also using the theorems of alternatives.

9.9.1   Weak alternatives

For systems with equalities and inequalities: Consider the following system of nonstrict inequalities and equalities

$$f_i(\mathbf{x}) \leq 0, \ i = 1,\ldots,m, \quad h_i(\mathbf{x}) = 0, \ i = 1,\ldots,p, \tag{9.215}$$

with nonempty domain

$$D = \Big(\bigcap_{i=1}^{m} \mathbf{dom}\, f_i\Big) \cap \Big(\bigcap_{i=1}^{p} \mathbf{dom}\, h_i\Big).$$

The system in (9.215) can be equivalently written in the form of a feasibility problem as in (4.1), with the objective function $f_0(\mathbf{x}) = 0$, i.e.,

$$p^\star = \min \ 0$$
$$\text{s.t.} \ f_i(\mathbf{x}) \leq 0, \ i = 1,\ldots,m \tag{9.216}$$
$$h_i(\mathbf{x}) = 0, \ i = 1,\ldots,p.$$

The optimal value of the above optimization problem is

$$p^\star = \begin{cases} 0, & \text{if (9.215) is feasible} \\ \infty, & \text{if (9.215) is infeasible.} \end{cases} \tag{9.217}$$

Next let us consider the dual function of (9.216)

$$g(\boldsymbol{\lambda}, \boldsymbol{\nu}) = \inf_{\mathbf{x} \in D} \left\{ \sum_{i=1}^{m} \lambda_i f_i(\mathbf{x}) + \sum_{i=1}^{p} \nu_i h_i(\mathbf{x}) \right\}. \tag{9.218}$$

This dual function is positive homogeneous in $(\boldsymbol{\lambda}, \boldsymbol{\nu})$, i.e., for any $\alpha > 0$,

$$g(\alpha\boldsymbol{\lambda}, \alpha\boldsymbol{\nu}) = \alpha g(\boldsymbol{\lambda}, \boldsymbol{\nu}).$$

The corresponding dual problem is to maximize $g(\boldsymbol{\lambda}, \boldsymbol{\nu})$ subject to the constraint that $\boldsymbol{\lambda} \succeq \mathbf{0}$. Therefore the dual optimal value is

$$d^\star = \sup_{\boldsymbol{\lambda}\succeq 0,\boldsymbol{\nu}} g(\boldsymbol{\lambda}, \boldsymbol{\nu}) = \begin{cases} \infty, & \text{if } \boldsymbol{\lambda} \succeq \mathbf{0}, \ g(\boldsymbol{\lambda}, \boldsymbol{\nu}) > 0 \text{ is feasible} \\ 0, & \text{if } \boldsymbol{\lambda} \succeq \mathbf{0}, \ g(\boldsymbol{\lambda}, \boldsymbol{\nu}) > 0 \text{ is infeasible.} \end{cases} \tag{9.219}$$

Weak duality (see (9.39)) says that the primal optimal value ($p^\star$) is no less than the dual optimal value ($d^\star$). Combining this concept of weak duality with (9.217) and (9.219) we can conclude that if the inequality system

$$\boldsymbol{\lambda} \succeq \mathbf{0}, \quad g(\boldsymbol{\lambda}, \boldsymbol{\nu}) > 0, \tag{9.220}$$

is feasible, then the system in (9.215) is infeasible as both of these cases have the same optimal value $d^\star = \infty = p^\star$. Conversely, we can also state that if the system in (9.215) is feasible, then the system in (9.220) is infeasible due to $p^\star = 0 = d^\star$. Hence, the two systems considered here, (9.215) and (9.220), are called *weak alternatives* of each other.

Certificate of infeasibility: To show that (9.215) is infeasible, it is sufficient to show a feasible point of (9.220); similarly to show the infeasibility of (9.220) it is sufficient to show a feasible point of (9.215). In other words, (9.215) and (9.220) cannot hold true (i.e., feasible) simultaneously; otherwise some contradictions must occur. However, it cannot be inferred that (9.220) is feasible or infeasible if (9.215) is infeasible and vice versa.

On the other hand, for the following system with strict inequalities

$$f_i(\mathbf{x}) < 0, \ i = 1, \ldots, m, \quad h_i(\mathbf{x}) = 0, \ i = 1, \ldots, p, \tag{9.221}$$

the alternative inequality system is given by

$$\boldsymbol{\lambda} \succeq \mathbf{0}, \ \boldsymbol{\lambda} \neq \mathbf{0}, \ g(\boldsymbol{\lambda}, \boldsymbol{\nu}) \geq 0. \tag{9.222}$$

It can be shown that (9.221) and (9.222) are weak alternatives. To see this, assume that (9.221) is feasible and let $\tilde{\mathbf{x}}$ be a feasible point. Then for any $\boldsymbol{\lambda} \succeq \mathbf{0}$, $\boldsymbol{\lambda} \neq \mathbf{0}$, and $\boldsymbol{\nu}$, we have

$$\sum_{i=1}^{m} \lambda_i f_i(\tilde{\mathbf{x}}) + \sum_{i=1}^{p} \nu_i h_i(\tilde{\mathbf{x}}) < 0. \tag{9.223}$$

Since $g(\boldsymbol{\lambda}, \boldsymbol{\nu})$ defined in (9.218) is upper bounded by $\sum_{i=1}^{m} \lambda_i f_i(\tilde{\mathbf{x}}) + \sum_{i=1}^{p} \nu_i h_i(\tilde{\mathbf{x}})$, we have $g(\boldsymbol{\lambda}, \boldsymbol{\nu}) < 0$, which contradicts with (9.222). Thus the feasibility of (9.221) implies the infeasibility of (9.222). Similarly it can be proved that the feasibility of (9.222) implies the infeasibility of (9.221).

For systems with equalities and generalized inequalities: The notion of weak alternatives can also be extended to the systems of equalities and generalized inequalities. Consider the following system of equalities and generalized inequalities

$$\boldsymbol{f}_i(\mathbf{x}) \preceq_{K_i} \mathbf{0}, \ i = 1, \ldots, m, \quad h_i(\mathbf{x}) = 0, \ i = 1, \ldots, p \tag{9.224}$$

with nonempty domain

$$\mathcal{D} = \Big(\bigcap_{i=1}^{m} \mathbf{dom} \, \boldsymbol{f}_i \Big) \cap \Big(\bigcap_{i=1}^{p} \mathbf{dom} \, h_i \Big),$$

where $K_i \subseteq \mathbb{R}^{k_i}$ are proper cones. The above system of equalities and generalized inequalities can be equivalently written as the following feasibility problem

$$p^\star = \min \ 0$$
$$\text{s.t. } \boldsymbol{f}_i(\mathbf{x}) \preceq_{K_i} \mathbf{0}, \ i = 1, \ldots, m \tag{9.225}$$
$$h_i(\mathbf{x}) = 0, \ i = 1, \ldots, p.$$

The optimal value of (9.225) is

$$p^\star = \begin{cases} 0, & \text{if (9.224) is feasible} \\ \infty, & \text{if (9.224) is infeasible.} \end{cases} \tag{9.226}$$

The dual function of (9.225) is

$$g(\boldsymbol{\lambda}, \boldsymbol{\nu}) = \inf_{\mathbf{x} \in \mathcal{D}} \left\{ \sum_{i=1}^{m} \boldsymbol{\lambda}_i^T \boldsymbol{f}_i(\mathbf{x}) + \sum_{i=1}^{p} \nu_i h_i(\mathbf{x}) \right\}, \qquad (9.227)$$

where $\boldsymbol{\lambda}_i \in \mathbb{R}^{k_i}$, and $\boldsymbol{\lambda} = [\boldsymbol{\lambda}_1^T, \dots, \boldsymbol{\lambda}_m^T]^T$ is the Lagrange multiplier vector. Note that the dual function is positive homogeneous in $(\boldsymbol{\lambda}, \boldsymbol{\nu})$, so the dual optimal value is

$$d^{\star} = \sup_{\boldsymbol{\lambda} \succeq 0, \boldsymbol{\nu}} g(\boldsymbol{\lambda}, \boldsymbol{\nu})$$
$$= \begin{cases} \infty, & \text{if } \boldsymbol{\lambda} \succeq_{K_i^*} \mathbf{0}, \ g(\boldsymbol{\lambda}, \boldsymbol{\nu}) > 0 \text{ is feasible} \\ 0, & \text{if } \boldsymbol{\lambda} \succeq_{K_i^*} \mathbf{0}, \ g(\boldsymbol{\lambda}, \boldsymbol{\nu}) > 0 \text{ is infeasible}, \end{cases} \qquad (9.228)$$

Now, based on weak duality defined in (9.39), it can be concluded that if the generalized inequality system

$$\boldsymbol{\lambda}_i \succeq_{K_i^*} \mathbf{0}, \ i = 1, \dots, m, \quad g(\boldsymbol{\lambda}, \boldsymbol{\nu}) > 0 \qquad (9.229)$$

is feasible, then the system in (9.224) is infeasible. Similarly, if the system in (9.224) is feasible, then the system in (9.229) is infeasible. Hence, the system in (9.224) and the system in (9.229) are weak alternatives of each other.

On the other hand, for the following system with strict generalized inequalities and equalities

$$\boldsymbol{f}_i(\mathbf{x}) \prec_{K_i} \mathbf{0}, \ i = 1, \dots, m, \quad h_i(\mathbf{x}) = 0, \ i = 1, \dots, p, \qquad (9.230)$$

the system

$$\boldsymbol{\lambda}_i \succeq_{K_i^*} \mathbf{0}, \ i = 1, \dots, m, \quad \boldsymbol{\lambda} \neq \mathbf{0}, \quad g(\boldsymbol{\lambda}, \boldsymbol{\nu}) \geq 0 \qquad (9.231)$$

serves as the weak alternative, and vice versa. The proof is similar to the proof of the weak alternatives for a system with equalities and strict ordinary inequalities, except that $g(\boldsymbol{\lambda}, \boldsymbol{\nu})$ is now defined by (9.227).

9.9.2   Strong alternatives

For systems with equalities and inequalities: If all the constraints in (9.215) are convex (implying $h_i(\mathbf{x})$ must be affine for all i) and if there exists an $\mathbf{x} \in \mathbf{relint}\ D$, then the inequality system in (9.215) and the inequality system given by (9.220) are called *strong alternatives*, which means that exactly one of the two alternatives holds true. In other words, (9.215) is feasible if and only if (9.220) is infeasible, and vice versa. On the other hand, if there exists an $\mathbf{x} \in \mathbf{relint}\ D$, then the following systems of inequalities and equalities

$$f_i(\mathbf{x}) < 0, \ i = 1, \dots, m, \ \mathbf{A}\mathbf{x} = \mathbf{b}, \qquad (9.232)$$

and

$$\boldsymbol{\lambda} \succeq \mathbf{0}, \ \boldsymbol{\lambda} \neq \mathbf{0}, \ g(\boldsymbol{\lambda}, \boldsymbol{\nu}) \geq 0 \qquad (9.233)$$

are also strong alternatives, where $g(\boldsymbol{\lambda}, \boldsymbol{\nu})$ is defined in (9.218). The proof is omitted here since it is similar to the case for systems with equalities and generalized inequalities (9.240) and (9.241) to be presented below.

Farkas lemma: *The system of inequalities*

$$\mathbf{Ax} \preceq \mathbf{0}, \ \mathbf{c}^T \mathbf{x} < 0, \tag{9.234}$$

and the following system of equalities and inequalities

$$\mathbf{A}^T \boldsymbol{\lambda} + \mathbf{c} = \mathbf{0}, \ \boldsymbol{\lambda} \succeq \mathbf{0}, \tag{9.235}$$

are strong alternatives of each other.

Proof. The lemma can be easily proved by the notion of LP duality. Let us consider the following LP

$$\begin{aligned} p^\star = \min \ & \mathbf{c}^T \mathbf{x} \\ \text{s.t. } & \mathbf{Ax} \preceq \mathbf{0}. \end{aligned} \tag{9.236}$$

The primal optimal value of problem (9.236) is

$$p^\star = \begin{cases} -\infty, & \text{if (9.234) is feasible,} \\ 0, & \text{if (9.234) is infeasible.} \end{cases} \tag{9.237}$$

The dual problem of (9.236) is given by

$$\begin{aligned} d^\star = \max \ & 0 \\ \text{s.t. } & \mathbf{A}^T \boldsymbol{\lambda} + \mathbf{c} = \mathbf{0} \\ & \boldsymbol{\lambda} \succeq \mathbf{0} \end{aligned} \tag{9.238}$$

and its dual optimal value is given by

$$d^\star = \begin{cases} 0, & \text{if (9.235) is feasible,} \\ -\infty, & \text{if (9.235) is infeasible.} \end{cases} \tag{9.239}$$

As strong duality holds between (9.236) and (9.238) as $\mathbf{x} = \mathbf{0}$ is feasible in (9.236), satisfying the refined Slater's condition (cf. Remark 9.6), we have $p^\star = d^\star$. Thus from (9.237) and (9.239) we can conclude that (9.234) and (9.235) are strong alternatives. ∎

For systems with equalities and generalized inequalities: The concept of strong alternatives can also be extended to the systems involving equalities and generalized inequalities. If all the constraints in (9.224) are convex, i.e., if $\boldsymbol{f}_i(\mathbf{x})$ is K_i-convex for $i = 1, \ldots, m$, and $h_i(\mathbf{x})$ is affine for $i = 1, \ldots, p$, then the following systems of equalities and strict generalized inequalities

$$\boldsymbol{f}_i(\mathbf{x}) \prec_{K_i} \mathbf{0}, \ i = 1, \ldots, m, \quad \mathbf{Ax} = \mathbf{b}, \tag{9.240}$$

and

$$\boldsymbol{\lambda}_i \succeq_{K_i^*} \mathbf{0}, \ i = 1, \ldots, m, \quad \boldsymbol{\lambda} \neq \mathbf{0}, \ g(\boldsymbol{\lambda}, \boldsymbol{\nu}) \geq 0 \tag{9.241}$$

are strong alternatives, where $g(\boldsymbol{\lambda}, \boldsymbol{\nu})$ is given by (9.227). This can be proved by considering the following problem:

$$p^\star = \min \ s$$
$$\text{s.t. } \boldsymbol{f}_i(\mathbf{x}) \preceq_{K_i} s\mathbf{a}_i, \ i = 1, \dots, m \qquad (9.242)$$
$$\mathbf{Ax} = \mathbf{b}$$

where $\mathbf{a}_i \succ_{K_i} \mathbf{0}$, and $\mathbf{x}$ and s are decision variables. The Lagrangian of (9.242) is given by

$$\mathcal{L}(s, \mathbf{x}, \boldsymbol{\lambda}, \boldsymbol{\nu}) = s + \sum_{i=1}^{m} \boldsymbol{\lambda}_i^T (\boldsymbol{f}_i(\mathbf{x}) - s\mathbf{a}_i) + \boldsymbol{\nu}^T(\mathbf{Ax} - \mathbf{b})$$
$$= s\left(1 - \sum_{i=1}^{m} \boldsymbol{\lambda}_i^T \mathbf{a}_i\right) + \sum_{i=1}^{m} \boldsymbol{\lambda}_i^T \boldsymbol{f}_i(\mathbf{x}) + \boldsymbol{\nu}^T(\mathbf{Ax} - \mathbf{b}), \qquad (9.243)$$

and the corresponding dual function is given by

$$\inf_{s, \mathbf{x} \in \mathcal{D}} \mathcal{L}(s, \mathbf{x}, \boldsymbol{\lambda}, \boldsymbol{\nu}) = \begin{cases} g(\boldsymbol{\lambda}, \boldsymbol{\nu}), & \sum_{i=1}^{m} \boldsymbol{\lambda}_i^T \mathbf{a}_i = 1, \\ -\infty, & \text{otherwise}, \end{cases} \qquad (9.244)$$

where $g(\boldsymbol{\lambda}, \boldsymbol{\nu})$ is given by (9.227) with $\sum_{i=1}^{p} \nu_i h_i(\mathbf{x})$ replaced by $\boldsymbol{\nu}^T(\mathbf{Ax} - \mathbf{b})$. Therefore, the dual problem is

$$d^\star = \max \ g(\boldsymbol{\lambda}, \boldsymbol{\nu})$$
$$\text{s.t. } \boldsymbol{\lambda}_i \succeq_{K_i^\star} \mathbf{0}, \ i = 1, \dots, m \qquad (9.245)$$
$$\sum_{i=1}^{m} \boldsymbol{\lambda}_i^T \mathbf{a}_i = 1.$$

In (9.242), since Slater's condition is satisfied for sufficiently large s, strong duality holds true (i.e., $p^\star = d^\star$). Hence, we have the following inferences:

- If (9.240) is infeasible, then $p^\star = s^\star \geq 0$. Then, as per (9.245), there exists a $(\boldsymbol{\lambda}, \boldsymbol{\nu})$ that satisfies (9.241). Note that there must exist a $\boldsymbol{\lambda} = [\boldsymbol{\lambda}_1^T, \dots, \boldsymbol{\lambda}_m^T]^T \neq \mathbf{0}$, such that $\sum_{i=1}^{m} \boldsymbol{\lambda}_i^T \mathbf{a}_i = 1$ in (9.245), which therefore can be relaxed.
- On the other hand, if (9.240) is feasible, then $p^\star = s^\star < 0$, which makes (9.241) infeasible, as $g(\boldsymbol{\lambda}, \boldsymbol{\nu})$ in (9.245) can never be nonnegative.

In the similar fashion, for $\boldsymbol{f}_i(\mathbf{x})$ being K_i-convex for $i = 1, \dots, m$, the following system of equalities and generalized inequalities

$$\boldsymbol{f}_i(\mathbf{x}) \preceq_{K_i} \mathbf{0}, \ i = 1, \dots, m, \quad \mathbf{Ax} = \mathbf{b}, \qquad (9.246)$$

and the system of generalized inequalities and one ordinary inequality

$$\boldsymbol{\lambda}_i \succeq_{K_i^\star} \mathbf{0}, \ i = 1, \dots, m, \ g(\boldsymbol{\lambda}, \boldsymbol{\nu}) > 0 \qquad (9.247)$$

can be shown to be strong alternatives, where $g(\boldsymbol{\lambda}, \boldsymbol{\nu})$ is given by (9.227).

Next, let us present an example where the two systems of equalities and generalized inequalities involving linear matrix inequalities are strong alternatives.

Example 9.18 The system

$$\mathbb{F}(\mathbf{x}) = x_1\mathbf{F}_1 + \cdots + x_n\mathbf{F}_n + \mathbf{G} \preceq \mathbf{0}, \tag{9.248}$$

where $\mathbf{F}_i \in \mathbb{S}^k$ and $\mathbf{F}_i \neq \mathbf{0}$ for all i, and $\mathbf{G} \in \mathbb{S}^k$, and the system

$$\mathbf{Z} \succeq \mathbf{0}, \ \text{Tr}(\mathbf{GZ}) > 0, \ \text{Tr}(\mathbf{F}_i\mathbf{Z}) = 0, \ i = 1, \ldots, n, \tag{9.249}$$

are strong alternatives, provided that the following implication holds

$$\sum_{i=1}^{n} v_i\mathbf{F}_i \succeq \mathbf{0} \Rightarrow \sum_{i=1}^{n} v_i\mathbf{F}_i = \mathbf{0}. \tag{9.250}$$

Note that under the implication given by (9.250), $\mathbb{F}(\mathbf{x}) - \mathbf{G} = x_1\mathbf{F}_1 + \cdots + x_n\mathbf{F}_n$ must be indefinite or a zero matrix.

Proof of strong alternatives of (9.248) and (9.249): By the strong alternatives of (9.246) and (9.247) with $\boldsymbol{f}_i(\mathbf{x}) \preceq_{K_i} \mathbf{0}$ replaced by $\mathbb{F}(\mathbf{x}) \preceq \mathbf{0}$, and $\boldsymbol{\lambda}_i \succeq_{K_i^*} \mathbf{0}$ replaced by $\mathbf{Z} \succeq \mathbf{0}$, it is straightforward to prove that (9.248) and (9.249) are strong alternatives. Instead, as in the preceding proof of strong alternatives for (9.240) and (9.241), let us consider the following problem

$$\begin{aligned} p^\star = \min \ &s \\ \text{s.t. } &\mathbb{F}(\mathbf{x}) \preceq s\mathbf{I}_k \end{aligned} \tag{9.251}$$

whose Lagrangian $\mathcal{L}(s, \mathbf{x}, \mathbf{Z})$ and dual function $g(\mathbf{Z})$ are as follows:

$$\mathcal{L}(s, \mathbf{x}, \mathbf{Z}) = s + \text{Tr}\big(\mathbf{Z}(\mathbb{F}(\mathbf{x}) - s\mathbf{I}_k)\big)$$

$$= s(1 - \text{Tr}(\mathbf{Z})) + \sum_{i=1}^{n} x_i\text{Tr}(\mathbf{ZF}_i) + \text{Tr}(\mathbf{ZG})$$

$$\Rightarrow g(\mathbf{Z}) = \inf_{s,\mathbf{x}} \mathcal{L}(s, \mathbf{x}, \mathbf{Z}) = \begin{cases} \text{Tr}(\mathbf{ZG}), & \text{Tr}(\mathbf{Z}) = 1, \ \text{Tr}(\mathbf{ZF}_i) = 0 \ \forall i, \\ -\infty, & \text{otherwise.} \end{cases}$$

Then the dual problem is

$$\begin{aligned} d^\star = \max_{\mathbf{Z} \succeq \mathbf{0}} \ &g(\mathbf{Z}) \\ = \max \ &\big\{\text{Tr}(\mathbf{ZG}) \mid \mathbf{Z} \succeq \mathbf{0}, \text{Tr}(\mathbf{Z}) = 1, \text{Tr}(\mathbf{ZF}_i) = 0 \ \forall i\big\}. \end{aligned} \tag{9.252}$$

Note that problem (9.251) is feasible and Slater's condition holds, implying that $p^\star < \infty$ and strong duality holds between (9.251) and (9.252). Let $(\mathbf{x}^\star, s^\star)$ be an optimal minimizer of (9.251), and $\boldsymbol{v}$ be an orthonormal eigenvector of $\mathbb{F}(\mathbf{x}^\star) - \mathbf{G}$ associated with $\lambda_{\max}(\mathbb{F}(\mathbf{x}^\star) - \mathbf{G}) \geq 0$ (since $\mathbb{F}(\mathbf{x}^\star) - \mathbf{G}$ is either indefinite or a zero matrix). Then the optimal solution to (9.251) satisfies

$$
\begin{aligned}
s^\star &= \lambda_{\max}(\mathbb{F}(\mathbf{x}^\star)) \quad \text{(cf. (8.8))} \\
&\geq \boldsymbol{v}^T(\mathbb{F}(\mathbf{x}^\star) - \mathbf{G})\boldsymbol{v} + \boldsymbol{v}^T\mathbf{G}\boldsymbol{v} \\
&= \lambda_{\max}(\mathbb{F}(\mathbf{x}^\star) - \mathbf{G}) + \boldsymbol{v}^T\mathbf{G}\boldsymbol{v} \\
&\geq \lambda_{\max}(\mathbb{F}(\mathbf{x}^\star) - \mathbf{G}) + \lambda_{\min}(\mathbf{G}) > -\infty.
\end{aligned}
$$

Thus we have finite $p^\star = d^\star = s^\star = \lambda_{\max}(\mathbb{F}(\mathbf{x}^\star))$, and the following inferences:

- If $s^\star = \lambda_{\max}(\mathbb{F}(\mathbf{x}^\star)) > 0$, then $\mathbb{F}(\mathbf{x}) \preceq \mathbf{0}$ is infeasible (as $\lambda_{\max}(\mathbb{F}(\mathbf{x})) \geq s^\star > 0$ for all $\mathbf{x}$ in the feasible set). Meanwhile, due to $d^\star > 0$, it can be concluded that $\text{Tr}(\mathbf{ZG}) > 0$, $\text{Tr}(\mathbf{ZF}_i) = 0$, $\mathbf{Z} \succeq \mathbf{0}$ is feasible, in spite of the constraint $\text{Tr}(\mathbf{Z}) = 1$ relaxed.

- If $s^\star = \lambda_{\max}(\mathbb{F}(\mathbf{x}^\star)) \leq 0$, then $\mathbb{F}(\mathbf{x}) \preceq \mathbf{0}$ is feasible. Again, due to $d^\star \leq 0$, it can be concluded that $\text{Tr}(\mathbf{ZG}) > 0$, $\text{Tr}(\mathbf{ZF}_i) = 0$, $\mathbf{Z} \succeq \mathbf{0}$ is infeasible.

Thus we have completed the proof that (9.248) and (9.249) are strong alternatives of each other.   ■ □

Remark 9.21 Without the premise of (9.250), $\lambda_{\max}(\mathbb{F}(\mathbf{x}))$ can be unbounded below. Thus, problem (9.251) in the previous example is feasible with $p^\star = -\infty$ (or (9.248) is always feasible), thereby leading to $p^\star = d^\star = -\infty$ (or (9.249) is always infeasible), which is a trivial case for the strong alternatives of (9.248) and (9.249).   □

9.9.3   Proof of S-procedure

Finally, let us prove the S-procedure introduced in Section 8.3, i.e., prove that (8.18) and (8.19) (for the real case) are strong alternatives of each other by using the result in Example 9.18 above.

Let us consider the following inequality system

$$
\lambda \geq 0, \quad \lambda\mathbf{A}_1 + \mathbf{A}_2 \succeq \mathbf{0}, \tag{9.253}
$$

where $\mathbf{A}_1, \mathbf{A}_2 \in \mathbb{S}^n$. This system can also be equivalently expressed as

$$
\mathbb{F}(\lambda) = \lambda\mathbf{F} + \mathbf{G} \preceq \mathbf{0}, \tag{9.254}
$$

where

$$
\mathbf{F} = \mathbf{DIAG}(-1, -\mathbf{A}_1) \in \mathbb{S}^{n+1}, \quad \mathbf{G} = \mathbf{DIAG}(0, -\mathbf{A}_2) \in \mathbb{S}^{n+1}. \tag{9.255}
$$

By (9.248) and (9.249), one can see that (9.254) and the following system

$$
\mathbf{Z} \triangleq \begin{bmatrix} z & \mathbf{a}^T \\ \mathbf{a} & \mathbf{X} \end{bmatrix} \succeq \mathbf{0}, \ \text{Tr}(\mathbf{GZ}) > 0, \ \text{Tr}(\mathbf{FZ}) = 0, \tag{9.256}
$$

are strong alternatives. Substituting (9.255) into (9.256) yields

$$\mathbf{X} \succeq \mathbf{0}, \ \mathrm{Tr}(\mathbf{X}\mathbf{A}_2) < 0, \ \mathrm{Tr}(\mathbf{X}\mathbf{A}_1) \leq 0. \tag{9.257}$$

Then by using (2.71), (9.257) can be equivalently expressed as

$$\boldsymbol{x}^T \mathbf{A}_2 \boldsymbol{x} < 0, \quad \boldsymbol{x}^T \mathbf{A}_1 \boldsymbol{x} \leq 0, \ \boldsymbol{x} \in \mathbb{R}^n. \tag{9.258}$$

Thus, we have the result that (9.253) and (9.258) must be strong alternatives, provided that $\mathbf{F}$ given in (9.255) is indefinite by (9.250). Precisely speaking, when the premise that

$$\mathbf{A}_1 \ \textit{must have a negative eigenvalue at least} \tag{9.259}$$

holds true, (9.253) and (9.258) will be strong alternatives.

Let $\mathbf{A}_1$ and $\mathbf{A}_2$ be partitioned in the following forms

$$\mathbf{A}_1 = \begin{bmatrix} \mathbf{F}_1 & \mathbf{g}_1 \\ \mathbf{g}_1^T & h_1 \end{bmatrix}, \quad \mathbf{A}_2 = -\begin{bmatrix} \mathbf{F}_2 & \mathbf{g}_2 \\ \mathbf{g}_2^T & h_2 \end{bmatrix}. \tag{9.260}$$

Substituting (9.260) into (9.253) gives rise to

$$\begin{bmatrix} \mathbf{F}_2 & \mathbf{g}_2 \\ \mathbf{g}_2^T & h_2 \end{bmatrix} \preceq \lambda \begin{bmatrix} \mathbf{F}_1 & \mathbf{g}_1 \\ \mathbf{g}_1^T & h_1 \end{bmatrix}, \quad \lambda \geq 0. \tag{9.261}$$

On the other hand, it can be shown (see Remark 9.22 below) that

$$\begin{cases} \mathbf{x}^T \mathbf{F}_1 \mathbf{x} + 2\mathbf{g}_1^T \mathbf{x} + h_1 \leq 0 \\ \mathbf{x}^T \mathbf{F}_2 \mathbf{x} + 2\mathbf{g}_2^T \mathbf{x} + h_2 > 0 \end{cases} \text{ where } \mathbf{x} \in \mathbb{R}^{n-1} \tag{9.262}$$

and (9.258) are fesiblity equivalent, provided that the premise (9.259) holds true, namely, there exists an $\widehat{\mathbf{x}}$ such that

$$\widehat{\mathbf{x}}^T \mathbf{F}_1 \widehat{\mathbf{x}} + 2\mathbf{g}_1^T \widehat{\mathbf{x}} + h_1 < 0. \tag{9.263}$$

Hence, (9.262) and (9.261) are strong alternatives under the premise (9.263). Moreover, the infeasibility of (9.262) implies that the following implication is true:

$$\mathbf{x}^T \mathbf{F}_1 \mathbf{x} + 2\mathbf{g}_1^T \mathbf{x} + h_1 \leq 0 \Longrightarrow \mathbf{x}^T \mathbf{F}_2 \mathbf{x} + 2\mathbf{g}_2^T \mathbf{x} + h_2 \leq 0. \tag{9.264}$$

In other words, the system given by (9.264) holds true if and only if there exists a λ satisfying the system given by (9.261) provided that (9.263) is true. Thus we have completed the proof of the S-procedure introduced in Section 8.3.

Remark 9.22 *(Proof of the feasibility equivalence of (9.258) and (9.262))* Substituting (9.260) and $\boldsymbol{x} = (\mathbf{v}, w) \in \mathbb{R}^n$ into (9.258) yields

$$\begin{cases} q_1(\tilde{\mathbf{v}}) \triangleq \tilde{\mathbf{v}}^T \mathbf{F}_1 \tilde{\mathbf{v}} + 2\mathbf{g}_1^T \tilde{\mathbf{v}} + h_1 \leq 0 \\ q_2(\tilde{\mathbf{v}}) \triangleq \tilde{\mathbf{v}}^T \mathbf{F}_2 \tilde{\mathbf{v}} + 2\mathbf{g}_2^T \tilde{\mathbf{v}} + h_2 > 0 \end{cases} \text{ if } w \neq 0 \text{ and } \tilde{\mathbf{v}} = \mathbf{v}/w, \tag{9.265a}$$

$$\begin{cases} \mathbf{v}^T \mathbf{F}_1 \mathbf{v} \leq 0 \\ \mathbf{v}^T \mathbf{F}_2 \mathbf{v} > 0 \end{cases} \text{ if } w = 0. \tag{9.265b}$$

Hence (9.258) is feasible, if and only if either (9.265a) (where $\tilde{\mathbf{v}} \in \mathbb{R}^{n-1}$) or (9.265b) (where $\mathbf{v} \in \mathbb{R}^{n-1}$) is feasible. Next, we prove that the feasibility of (9.265b) ($w = 0$) also leads to the feasibility of (9.265a) under the premise (9.263) (i.e., $q_1(\hat{\mathbf{x}}) < 0$).

Replacing $\tilde{\mathbf{v}}$ in (9.265a) with $\mathbf{x} = \hat{\mathbf{x}} + t\mathbf{v} \in \mathbb{R}^{n-1}$ yields

$$q_1(\mathbf{x}) = q_1(\hat{\mathbf{x}}) + t^2\mathbf{v}^T\mathbf{F}_1\mathbf{v} + 2t(\mathbf{F}_1\hat{\mathbf{x}} + \mathbf{g}_1)^T\mathbf{v}$$

$$\leq q_1(\hat{\mathbf{x}}) + 2t(\mathbf{F}_1\hat{\mathbf{x}} + \mathbf{g}_1)^T\mathbf{v} \quad \text{(by (9.265b))} \tag{9.266}$$

$$q_2(\mathbf{x}) = q_2(\hat{\mathbf{x}}) + t^2\mathbf{v}^T\mathbf{F}_2\mathbf{v} + 2t(\mathbf{F}_2\hat{\mathbf{x}} + \mathbf{g}_2)^T\mathbf{v}. \tag{9.267}$$

Suppose that $(\mathbf{F}_1\hat{\mathbf{x}} + \mathbf{g}_1)^T\mathbf{v} \neq 0$. By letting $t \to \pm\infty$ in both (9.266) and (9.267) (depending on the sign of $(\mathbf{F}_1\hat{\mathbf{x}} + \mathbf{g}_1)^T\mathbf{v}$), and meanwhile applying (9.265b) to (9.267), one can infer that

$$\begin{cases} q_1(\mathbf{x}) = \mathbf{x}^T\mathbf{F}_1\mathbf{x} + 2\mathbf{g}_1^T\mathbf{x} + h_1 \leq 0 \\ q_2(\mathbf{x}) = \mathbf{x}^T\mathbf{F}_2\mathbf{x} + 2\mathbf{g}_2^T\mathbf{x} + h_2 > 0 \end{cases} \tag{9.268}$$

(which is exactly (9.265a) and (9.262)) must be feasible if (9.265b) is feasible under the premise (9.263). Moreover, it can be easily seen that, when $(\mathbf{F}_1\hat{\mathbf{x}} + \mathbf{g}_1)^T\mathbf{v} = 0$, (9.266) reduces to $q_1(\mathbf{x}) \leq q_1(\hat{\mathbf{x}}) < 0$. So (9.265a) is feasible for this case as well. Therefore, we have completed the proof that (9.262) is feasible if and only if (9.258) is feasible, provided that (9.263) is true. □

9.10   Summary and discussion

In this chapter, for a general optimization problem (called the primal problem) with equality constraints and ordinary inequality constraints, we have introduced the concept of duality along with the associated Slater's condition and KKT conditions. We have illustrated the concept of duality by using suitable examples, wherein we have derived dual and bidual of convex and nonconvex optimization problems. Like the Fourier transform pair of a signal in the time domain and the frequency domain, the dual problem provides another perspective to the primal problem with the (strong or weak) duality as an essential linkage. The duality concept along with the associated Slater's condition and KKT conditions was also extended to the optimization problem involving generalized inequalities. Since the KKT conditions are necessary and sufficient conditions for optimal primal-dual solutions of convex problems under strong duality, they are the foundation both in finding the optimal solutions of a convex problem analytically and in efficient algorithm development (e.g., interior-point methods to be presented in Chapter 10) for finding optimal numerical solutions. The theorems of alternatives based on the duality were also introduced, relating the feasibility of a set of equalities and inequalities to the feasibility of another set of equalities and inequalities when they are strong alternatives. The widely used S-procedure was also proven by theorems of alternatives. An alternative and simpler proof for the

S-procedure has also been presented by the S-lemma, via the strong duality of a specific convex problem.

The KKT conditions also play a central role in the analysis of the optimal solutions of an optimization problem. For instance, solving an SDP which arises as an approximation of the original problem via SDR has been prevalent in the transmit beamforming algorithm design in MIMO wireless communications, but one may face a challenge of theoretically or analytically showing whether the SDP solution $\mathbf{X}^\star$ is of rank one or not, even if it is strongly supported by extensive simulation tests. This is very important because the rank-one $\mathbf{X}^\star$ confirms that it is also the optimal solution of the original problem. However, to prove or disprove if the solutions of SDP are of rank one, or to identify some scenarios under which $\mathbf{X}^\star$ is of rank one, is often challenging and intricate, whereas the KKT conditions are always the backbone of the involved proof and analysis.

References for Chapter 9

[Apo07]   T. M. Apostol, *Mathematical Analysis*, 2nd ed. Pearson Edu. Taiwan Ltd., 2007.

[Ber99]   D. P. Bertsekas, *Nonlinear Programming*, 2nd ed. Belmont, MA: Athena Scientific, 1999.

[Ber03]   ——, *Convex Analysis and Optimization*. Athena Scientific, 2003.

[BPC⁺10]   S. Boyd, N. Parikh, E. Chu, B. Peleato, and J. Eckstein, "Distributed optimization and statistical learning via the alternating direction method of multipliers," *Foundations and Trends in Machine Learning*, vol. 3, no. 1, pp. 1–122, 2010.

[BSS06]   M. S. Bazaraa, H. D. Sherali, and C. M. Shetty, *Nonlinear Programming: Theory and Algorithms*, 3rd ed. John Wiley & Sons, Inc., 2006.

[BT89a]   D. P. Bertsekas and J. N. Tsitsiklis, *Parallel and Distributed Computation: Numerical Methods*. Upper Saddle River, NJ: Prentice-Hall, 1989.

[BT89b]   ——, *Parallel and Distributed Computation: Numerical Methods*. Upper Saddle River, NJ: Prentice-Hall, 1989.

[Fal69]   J. Falk, "Lagrange multipliers and nonconvex programs," *SIAM J. Control*, vol. 7, no. 4, pp. 534–545, Nov. 1969.

[SCW⁺12]   C. Shen, T.-H. Chang, K.-Y. Wang, Z. Qiu, and C.-Y. Chi, "Distributed robust multi-cell coordinated beamforming with imperfect CSI: An ADMM approach," *IEEE Trans. Signal Process.*, vol. 60, no. 6, pp. 2988–3003, June 2012.

10 Interior-point Methods

This chapter introduces the interior-point method (IPM) that has been widely used to solve various convex optimization problems. First of all, we focus on an IPM, called the *barrier method*. This IPM solves a constrained convex optimization problem (with equality and inequality constraints) by reducing it to a sequence of linear equality constrained problems, which can be effectively solved using Newton's method. With the foundation of the introduced barrier method, we then introduce the *primal-dual interior-point method*. Both methods heavily involve the strong duality and the KKT conditions of the convex optimization problem under consideration. Some general-purpose convex optimization solvers are available on-line, such as SeDuMi (http://sedumi.ie.lehigh.edu/) and CVX (http://www.stanford.edu/~boyd/cvxbook/), that employ IPMs to solve convex optimization problems. However, a tailored interior-point algorithm is often much faster than these general-purpose solvers, and thus developing a customized algorithm using IPMs is essential to efficient hardware/software implementation. For notational simplicity, the gradient $\nabla_{\mathbf{x}} f(\mathbf{x})$ may sometimes simply be denoted as $\nabla f(\mathbf{x})$ without confusion in this chapter.

10.1 Inequality and equality constrained convex problems

Consider the following convex optimization problem (with ordinary inequality constraints):

$$
\begin{aligned}
p^{\star} = \min \ & f_0(\mathbf{x}) \\
\text{s.t. } & f_i(\mathbf{x}) \leq 0, \ i = 1, \dots, m \\
& \mathbf{A}\mathbf{x} = \mathbf{b}
\end{aligned}
\tag{10.1}
$$

where $f_0, \dots, f_m : \mathbb{R}^n \to \mathbb{R}$ are convex and twice continuously differentiable, $\mathbf{b} \in \mathbb{R}^p$, and $\mathbf{A} \in \mathbb{R}^{p \times n}$ with $\text{rank}(\mathbf{A}) = p < n$. The Lagrangian is known to be

$$
\mathcal{L}(\mathbf{x}, \boldsymbol{\lambda}, \boldsymbol{\nu}) = f_0(\mathbf{x}) + \sum_{i=1}^{m} \lambda_i f_i(\mathbf{x}) + \boldsymbol{\nu}^T (\mathbf{A}\mathbf{x} - \mathbf{b}),
\tag{10.2}
$$

where $\boldsymbol{\lambda} = (\lambda_1, \ldots, \lambda_m) \in \mathbb{R}^m$ and $\boldsymbol{\nu} = (\nu_1, \ldots, \nu_p) \in \mathbb{R}^p$, and the Lagrange dual function is given by

$$g(\boldsymbol{\lambda}, \boldsymbol{\nu}) = \inf_{\mathbf{x} \in \mathcal{D}} \mathcal{L}(\mathbf{x}, \boldsymbol{\lambda}, \boldsymbol{\nu}), \tag{10.3}$$

where

$$\mathcal{D} = \left(\bigcap_{i=0}^{m} \text{dom } f_i \right) \bigcap \text{dom } h$$

(in which $h(\mathbf{x}) = \mathbf{A}\mathbf{x} - \mathbf{b}$) denotes the problem domain of (10.1).

Let $\mathcal{C}$ denote the feasible set of the problem (10.1), i.e.,

$$\mathcal{C} = \bigcap_{i=1}^{m} \{\mathbf{x} \mid f_i(\mathbf{x}) \leq 0\} \bigcap \{\mathbf{x} \mid \mathbf{A}\mathbf{x} = \mathbf{b}\}.$$

Assume that problem (10.1) is solvable and let $\mathbf{x}^\star \in \mathcal{C}$ be the optimal solution and $p^\star = f_0(\mathbf{x}^\star)$ be the optimal value. Further, assume that there exists a strictly feasible point in the relative interior of $\mathcal{D}$, which means that the Slater's constraint qualification holds, i.e., the strong duality holds. Thus,

$$p^\star = f_0(\mathbf{x}^\star) = d^\star = \sup_{\boldsymbol{\lambda} \succeq \mathbf{0}, \boldsymbol{\nu} \in \mathbb{R}^p} g(\boldsymbol{\lambda}, \boldsymbol{\nu}) = g(\boldsymbol{\lambda}^\star, \boldsymbol{\nu}^\star) = \mathcal{L}(\mathbf{x}^\star, \boldsymbol{\lambda}^\star, \boldsymbol{\nu}^\star), \tag{10.4}$$

where dual optimal $\boldsymbol{\lambda}^\star \in \mathbb{R}^m$ and $\boldsymbol{\nu}^\star \in \mathbb{R}^p$ exist, which together with $\mathbf{x}^\star$ satisfy the KKT conditions as given below:

$$\mathbf{A}\mathbf{x}^\star = \mathbf{b}, \tag{10.5a}$$

$$f_i(\mathbf{x}^\star) \leq 0, \; i = 1, \ldots, m, \tag{10.5b}$$

$$\boldsymbol{\lambda}^\star \succeq \mathbf{0}, \tag{10.5c}$$

$$\nabla_{\mathbf{x}} f_0(\mathbf{x}^\star) + \sum_{i=1}^{m} \lambda_i^\star \nabla_{\mathbf{x}} f_i(\mathbf{x}^\star) + \mathbf{A}^T \boldsymbol{\nu}^\star = \mathbf{0}, \tag{10.5d}$$

$$\lambda_i^\star f_i(\mathbf{x}^\star) = 0, \; i = 1, \ldots, m. \tag{10.5e}$$

Note that among the KKT conditions in (10.5), the two conditions (10.5a) and (10.5b) guarantee the feasibility of $\mathbf{x}^\star$; the two conditions (10.5c) and (10.5d) guarantee the feasibility of $(\boldsymbol{\lambda}^\star, \boldsymbol{\nu}^\star)$; the last condition (10.5e) guarantees the zero duality gap. Moreover, the condition (10.5d) is identical to

$$\nabla_{\mathbf{x}} \mathcal{L}(\mathbf{x}^\star, \boldsymbol{\lambda}^\star, \boldsymbol{\nu}^\star) = \mathbf{0} \quad \text{(by (10.2))}.$$

IPMs solve the problem (10.1) (or the associated KKT conditions (10.5)) by applying Newton's method to a sequence of equality constrained problems, or to a sequence of modified versions of the KKT conditions.

10.2 Newton's method and barrier function

The barrier method to be introduced in the next section is an iterative method that, at each iteration, solves an updated convex optimization problem with only the equality constraints using the Newton method. The convex optimization problem at each iteration is redefined from the original primal problem (10.1) with all the inequality constraints removed, and a barrier function (determined by all the inequality constraint functions) added in the objective function in the meantime. Next, let us present the Newton's method and the barrier function, respectively.

10.2.1 Newton's method for equality constrained problems

Consider the following equality constrained convex problem:

$$\min\ f(\mathbf{x})$$
$$\text{s.t. } \mathbf{A}\mathbf{x} = \mathbf{b}. \tag{10.6}$$

Suppose that $\bar{\mathbf{x}}$ is a feasible point of this problem, obtained at the previous iteration. How the Newton's method updates $\bar{\mathbf{x}}$ at the current iteration is presented next.

Let $\hat{f}(\mathbf{x})$ denote the second-order Taylor series approximation of $f(\mathbf{x})$ at the feasible point $\mathbf{x} = \bar{\mathbf{x}}$, i.e.,

$$\hat{f}(\mathbf{x}) = f(\bar{\mathbf{x}}) + \nabla f(\bar{\mathbf{x}})^T(\mathbf{x} - \bar{\mathbf{x}}) + \frac{1}{2}(\mathbf{x} - \bar{\mathbf{x}})^T\nabla^2 f(\bar{\mathbf{x}})(\mathbf{x} - \bar{\mathbf{x}}). \tag{10.7}$$

Define the following QP

$$\min\ \hat{f}(\mathbf{x})$$
$$\text{s.t. } \mathbf{A}\mathbf{x} = \mathbf{b} \tag{10.8}$$

which is convex since $\nabla^2 f(\bar{\mathbf{x}}) \succeq \mathbf{0}$. Problem (10.8) satisfies the Slater's condition, and hence can be solved by the KKT conditions as follows:

$$\nabla f(\bar{\mathbf{x}}) + \nabla^2 f(\bar{\mathbf{x}})(\mathbf{x} - \bar{\mathbf{x}}) + \mathbf{A}^T\boldsymbol{\nu} = \mathbf{0}, \tag{10.9a}$$
$$\mathbf{A}\mathbf{x} = \mathbf{b}, \tag{10.9b}$$

where $\boldsymbol{\nu}$ is the dual variable associated with the equality constraint in (10.8). The KKT conditions given by (10.9) can be expressed as the following linear equation system:

$$\begin{bmatrix} \nabla^2 f(\bar{\mathbf{x}}) & \mathbf{A}^T \\ \mathbf{A} & \mathbf{0} \end{bmatrix} \begin{bmatrix} \mathbf{x} \\ \boldsymbol{\nu} \end{bmatrix} = \begin{bmatrix} \nabla^2 f(\bar{\mathbf{x}})\bar{\mathbf{x}} - \nabla f(\bar{\mathbf{x}}) \\ \mathbf{b} \end{bmatrix}. \tag{10.10}$$

Suppose that this linear system is solvable. Then the primal-dual optimal solution of (10.8) is given by

$$\begin{bmatrix} \mathbf{x}^\star \\ \nu^\star \end{bmatrix} = \begin{bmatrix} \nabla^2 f(\bar{\mathbf{x}}) & \mathbf{A}^T \\ \mathbf{A} & 0 \end{bmatrix}^\dagger \begin{bmatrix} \nabla^2 f(\bar{\mathbf{x}})\bar{\mathbf{x}} - \nabla f(\bar{\mathbf{x}}) \\ \mathbf{b} \end{bmatrix} + \mathbf{y}, \; \mathbf{y} \in \mathcal{N}\left(\begin{bmatrix} \nabla^2 f(\bar{\mathbf{x}}) & \mathbf{A}^T \\ \mathbf{A} & 0 \end{bmatrix} \right).$$
(10.11)

For the case of no equality constraint in problem (10.6), i.e., $\mathbf{A} = \mathbf{0}$ and $\mathbf{b} = \mathbf{0}$, the resulting primal solution of (10.8) reduces to

$$\mathbf{x}^\star = \left(\nabla^2 f(\bar{\mathbf{x}}) \right)^\dagger \left(\nabla^2 f(\bar{\mathbf{x}})\bar{\mathbf{x}} - \nabla f(\bar{\mathbf{x}}) \right) + \tilde{\mathbf{y}}$$
$$= \bar{\mathbf{x}} - \left(\nabla^2 f(\bar{\mathbf{x}}) \right)^\dagger \nabla f(\bar{\mathbf{x}}) + \tilde{\mathbf{y}}, \; \tilde{\mathbf{y}} \in \mathcal{N}\left(\nabla^2 f(\bar{\mathbf{x}}) \right). \quad (10.12)$$

The iterate $\bar{\mathbf{x}}$ will be updated along a chosen direction, denoted as $\mathbf{d}_{\bar{\mathbf{x}}}$, such that the objective function f decreases. Supposing that $\mathbf{d}_{\bar{\mathbf{x}}}$ satisfies $\nabla f(\bar{\mathbf{x}})^T \mathbf{d}_{\bar{\mathbf{x}}} < 0$, the iterate $\bar{\mathbf{x}}$ can be updated by

$$\bar{\mathbf{x}} := \bar{\mathbf{x}} + t \cdot \mathbf{d}_{\bar{\mathbf{x}}}, \quad (10.13)$$

where t is the step size parameter determined by the backtracking line search. The backtracking line search is summarized in Algorithm 10.1 that yields the value of t with two preassigned parameters, $\alpha \in (0, 0.5)$, $\beta \in (0, 1)$. The parameter α describes to what extent the function f is decreased as predicted by linear approximation. The other parameter β determines the accuracy of the backtracking line search. In particular, larger β implies higher accuracy, and hence corresponds to higher computational complexity.

Algorithm 10.1 Backtracking line search

1: Given a descending direction $\mathbf{d}_{\bar{\mathbf{x}}}$ of f at $\bar{\mathbf{x}} \in \mathbf{dom}\, f$, $\alpha \in (0, 0.5)$, $\beta \in (0, 1)$, $t = 1$;
2: **if** $f(\bar{\mathbf{x}} + t\mathbf{d}_{\bar{\mathbf{x}}}) > f(\bar{\mathbf{x}}) + \alpha t \nabla f(\bar{\mathbf{x}})^T \mathbf{d}_{\bar{\mathbf{x}}}$ **then**
3:     **repeat**
4:        $t := \beta t$;
5:     **until** $f(\bar{\mathbf{x}} + t\mathbf{d}_{\bar{\mathbf{x}}}) \leq f(\bar{\mathbf{x}}) + \alpha t \nabla f(\bar{\mathbf{x}})^T \mathbf{d}_{\bar{\mathbf{x}}}$
6: **end if**
7: Output t as the step size for decreasing f in $\mathbf{d}_{\bar{\mathbf{x}}}$.

To illustrate how Algorithm 10.1 operates, let us consider a one-dimensional unconstrained convex problem with the objective function and the associated linear approximation given by

$$f(x) = e^{-2x} + x - 0.5 \quad (10.14)$$
$$\text{Linear Approximation}: f(\bar{x}) + \alpha f'(\bar{x})(x - \bar{x}) \quad (10.15)$$

that are shown in Figure 10.1 by a blue solid line for the former, and a black dashed line ($\alpha = 0.25$) for the latter. A suitable point $x = \bar{x} + t d_{\bar{x}}$ will be found by iteratively updating t by βt such that $f(\bar{x} + td_{\bar{x}})$ is below the linear approximation for the preassigned $\alpha \in (0, 0.5)$. Another dashed line denotes the linear approximation for $\alpha = 1$ (though not admissible for Algorithm 10.1) is actually the first-order tight lower bound to both f and $\hat{f}$ (cf. (10.7)), also implying that no step size t satisfying $f(\bar{x} + td_{\bar{x}}) < f(\bar{x}) + \alpha f'(\bar{x})td_{\bar{x}}$ exists for $\alpha = 1$.

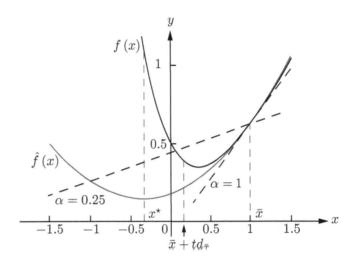

Figure 10.1 Illustration of backtracking line search by Algorithm 10.1, where $f(x)$ given by (10.14) (blue solid line), the associated quadratic convex function $\hat{f}$ given by (10.7) for $\bar{x} = 1$ (red solid line), and the associated linear approximations (dashed lines, one for $\alpha = 0.25$ and one for $\alpha = 1$) given by (10.15) are depicted. For this case, $x^{\star} = -0.35$ (the optimal solution of $\hat{f}$) and $d_{\bar{x}} = x^{\star} - \bar{x} = -1.35$; an admissible point $\bar{x} + td_{\bar{x}}$ for $t = 0.64$ is also indicated by an arrow.

The Newton's method for solving the equality constrained convex problem (10.6) is summarized in Algorithm 10.2, where the search direction vector

$$\mathbf{d}_{\bar{x}} = \mathbf{x}^{\star} - \bar{x}, \tag{10.16}$$

called the Newton step, is used provided that (10.11) is solvable with $\mathbf{x}^{\star} \neq \bar{x}$. By the first-order optimality condition of the convex QP (10.8) and the fact of $\nabla^2 f(\bar{x}) \succeq \mathbf{0}$, we have

$$\nabla \hat{f}(\mathbf{x}^{\star})^T(\mathbf{x}^{\star} - \bar{x}) = (\nabla f(\bar{x}) + \nabla^2 f(\bar{x})(\mathbf{x}^{\star} - \bar{x}))^T(\mathbf{x}^{\star} - \bar{x}) \leq 0$$
$$\Rightarrow \nabla f(\bar{x})^T(\mathbf{x}^{\star} - \bar{x}) \leq -(\mathbf{x}^{\star} - \bar{x})^T \nabla^2 f(\bar{x})(\mathbf{x}^{\star} - \bar{x}) \leq 0$$
$$\Rightarrow \nabla f(\bar{x})^T \mathbf{d}_{\bar{x}} < 0$$

provided that $\mathbf{d}_{\bar{x}} = \mathbf{x}^{\star} - \bar{x} \notin \mathcal{N}(\nabla^2 f(\bar{x}))$ (cf. (1.90)). Moreover, since $\mathbf{Ad}_{\bar{x}} = \mathbf{0}$ (due to $\mathbf{Ax}^{\star} = \mathbf{A}\bar{x} = \mathbf{b}$), the updated $\bar{x}$ using (10.13) is also feasible to problem (10.6). Consequently, Algorithm 10.2 will yield a new iterate $\bar{x} + t\mathbf{d}_{\bar{x}}$ between $\bar{x}$ (corresponding to $t = 0$) and $\mathbf{x}^{\star}$ (corresponding to $t = 1$) at every iteration until convergence (cf. Figure 10.1, where $f(x^{\star}) > f(\bar{x})$). On the other hand, when

$\mathbf{x}^\star = \bar{\mathbf{x}}$, the KKT conditions (10.9) of problem (10.8) can be easily shown to be equivalent to $\nabla f(\bar{\mathbf{x}})^T(\mathbf{x} - \bar{\mathbf{x}}) = 0$ for all $\mathbf{x} \in \{\boldsymbol{x} \mid \mathbf{A}\boldsymbol{x} - \mathbf{b} = \mathbf{0}\}$, implying that $\nabla f(\bar{\mathbf{x}}) = \mathbf{0}$; namely, $\bar{\mathbf{x}}$ is also the optimal solution to problem (10.6) and the convergence of Algorithm 10.2 has been achieved in the meantime.

Algorithm 10.2 Newton's method for convex problems with equality constraints

---

1: Given a feasible point $\bar{\mathbf{x}}$ of problem (10.6);
2: **repeat**
3:    Obtain the optimal solution $\mathbf{x}^\star$ of problem (10.8) by (10.11) or (10.12);
4:    Choose step size t by the backtracking line search (Algorithm 10.1) in the descending direction of f being $\mathbf{d}_{\bar{\mathbf{x}}} = \mathbf{x}^\star - \bar{\mathbf{x}}$;
5:    Update $\bar{\mathbf{x}} := \bar{\mathbf{x}} + t\mathbf{d}_{\bar{\mathbf{x}}}$;
6: **until** the predefined stopping criterion is met.
7: Output $\bar{\mathbf{x}}$ as the solution of problem (10.6).

10.2.2   Barrier function

In this subsection, we aim to approximately reformulate the original problem (10.1) as an equality constrained problem to which Newton's method can be applied. For this purpose we consider to remove the inequality constraints in (10.1), and add a compensated term in the objective function due to this removal. Then we come up with an equivalent problem defined as

$$\min \left\{ \tilde{f}_0(\mathbf{x}) \triangleq f_0(\mathbf{x}) + \sum_{i=1}^{m} I_-(f_i(\mathbf{x})) \right\} \tag{10.17}$$

$$\text{s.t. } \mathbf{A}\mathbf{x} = \mathbf{b}$$

where $I_- = I_{\mathcal{B}=-\mathbb{R}_+}$ is an indicator function defined in (9.19), i.e.,

$$I_-(u) = \begin{cases} 0, & u \le 0 \\ \infty, & u > 0. \end{cases} \tag{10.18}$$

Note that this problem is convex but the objective function $\tilde{f}_0(\mathbf{x})$ is not differentiable. If this problem can be solved, the optimal solutions will be exactly the same as those of problem (10.1). However, we will not try to solve this problem, while we intend to solve an approximate problem to problem (10.17) as discussed below.

Although problem (10.17) has no inequality constraints, the objective function $\tilde{f}_0(\mathbf{x})$ actually remains identical to $f_0(\mathbf{x})$ for any strictly feasible point $\mathbf{x} \in \mathcal{C}$ of problem (10.1), otherwise ∞, implying that this compensated term leads to a binary decision of either $\tilde{f}_0(\mathbf{x}) = \infty$ or $\tilde{f}_0(\mathbf{x}) = f_0(\mathbf{x})$, i.e., without alert for the closeness of any feasible point to the barrier, defined as

$$\mathcal{B} \triangleq \bigcup_{i=1}^{m} \{\mathbf{x} \mid f_i(\mathbf{x}) = 0\}, \tag{10.19}$$

which corresponds to a high barrier that prevents the objective value from increasing towards infinity in search of a minimizer of problem (10.17). This can be likened to a hard decision in decoding the information-bearing signal at the receiving end of a communication system, while soft decision (according to a smoothly increasing likelihood function) turns out to be more decoding performance reliable and thus widely used in communication systems. Similarly, we make use of the so-called *logarithmic barrier* method, which approximates the indicator function by the following:

$$\hat{I}_-(u) = \begin{cases} -\dfrac{1}{t}\log(-u), & u < 0 \\ \infty, & u \geq 0 \end{cases}$$

where $t > 0$ is a parameter that determines the accuracy of the approximation to $I_-(u)$ (see Figure 10.2). The approximate indicator function $\hat{I}_-(u)$ is convex and strictly increasing, and becomes ∞ for $u \geq 0$. In addition, $\hat{I}_-(u)$ is twice differentiable for $u < 0$ and closed (cf. Remark 3.23).

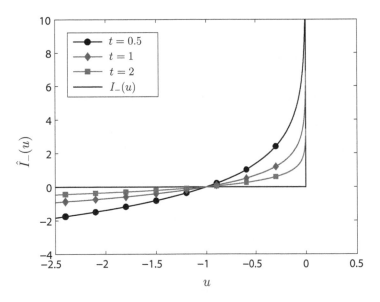

Figure 10.2 The black line shows the indicator function $I_-(u)$, and the other curves show $\hat{I}_-(u) = -(1/t)\log(-u)$ for $t = 0.5, 1, 2$. The curve for $t = 2$ gives the best approximation to $I_-(u)$.

Substituting $\hat{I}_-(u)$ for $I_-(u)$ into problem (10.17) yields

$$\min \; f_0(\mathbf{x}) + \frac{1}{t} \cdot \phi(\mathbf{x})$$
$$\text{s.t. } \mathbf{Ax} = \mathbf{b}$$

(10.20)

where the function

$$\phi(\mathbf{x}) \triangleq t \sum_{i=1}^{m} \hat{I}_-(f_i(\mathbf{x})) = - \sum_{i=1}^{m} \log(-f_i(\mathbf{x}))$$

(10.21)

with

$$\mathbf{dom} \; \phi = \{\mathbf{x} \in \mathbb{R}^n \mid f_i(\mathbf{x}) < 0, \; i = 1, \ldots, m\}$$

is called the *logarithmic barrier* or *log barrier*. Note that the objective function $f_0(\mathbf{x}) + \phi(\mathbf{x})/t$ of problem (10.20) takes the same value as $f_0(\mathbf{x})$ for all $\mathbf{x} \in \{\mathbf{x} \mid \phi(\mathbf{x}) = 0\}$ regardless of the value of t.

Let us consider a simple example to illustrate problem (10.20) as follows:

$$\min \; \left\{ f_0(x) \triangleq e^{-x} + \frac{x^3}{3} \right\}$$
$$\text{s.t. } |x| \le 2.$$

(10.22)

Then the corresponding log barrier is given by

$$\phi(x) = -\log(-x+2) - \log(x+2), \text{ and } \{x \mid \phi(x) = 0\} = \{ \pm \sqrt{3} \}. \quad (10.23)$$

An illustration for the objective function and the solutions of the corresponding problem (10.20) is shown in Figure 10.3 for different values of t. Some perspective insights to problem (10.20) can also be observed from this figure. The objective value of this problem increases as the decision variable x is closer to a barrier, and the closer the x to a barrier, the faster the increasing rate of the objective value for smaller t. Conceptually, the point $x = \sqrt{3}$, at which all the objective functions $f_0(x) + \phi(x)/t$ (for any $t > 0$) intersects, can be thought of as a roadsign beyond which a barrier (i.e., $f_0(x) + \phi(x)/t \to \infty$) is ahead at $x = 2$, and meanwhile all the objective values increase fast as x approaches the barrer. The same situation applies to the point $x = -\sqrt{3}$. Next, let us present why this problem will be the core problem in the iterative barrier method to be introduced later.

Obviously, the solutions of problem (10.20) are just approximate solutions to problem (10.1). Nevertheless, the second term $\phi(\mathbf{x})/t$ of the objective function in (10.20) now plays the role of a regularization term (determined by the inequality constraint functions f_i due to their removal from the original optimization problem). Furthermore, because $\phi(\mathbf{x})$ is larger for $\mathbf{x}$ closer to $\mathcal{B}$, the minimizer of (10.20) tends to be farther from the barrier $\mathcal{B}$ (cf. (10.19)) for smaller t. The regularization term will approach zero for any strictly feasible point as t goes to infinity. Moreover, the optimal minimizer of (10.20) with a finite optimal value, denoted as $\mathbf{x}^\star(t)$, is strictly feasible to the original optimization problem (10.1),

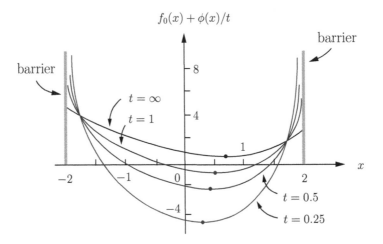

$$f_0(x) + \phi(x)/t$$

Figure 10.3 Illustration of problem (10.20) for different values of t, where $f_0(x)$ is given in (10.22) and the log barrier $\phi(x)$ is given in (10.23). The optimal solutions to the corresponding unconstrained convex problem (10.20) are $x = 0.310$ for $t = 0.25$, $x = 0.435$ for $t = 0.5$, $x = 0.539$ for $t = 1$, and $x = 0.7035$ for $t = \infty$, together with the associated optimal values of $f_0(x) + \phi(x)/t$ denoted by bullets. Note that the four associated curves intersect at $x = \pm\sqrt{3}$ due to $\phi(\pm\sqrt{3}) = 0$.

i.e., $\mathbf{x}^\star(t) \in \mathbf{relint}\ \mathcal{C}$. Hence, with a starting point value of t and a feasible initial point $\mathbf{x}(t)$, one can iteratively solve problem (10.20) (with the value of t increased at each iteration) to obtain a strictly feasible ϵ-optimal solution to the original optimization problem (10.1), though a nonstrictly feasible solution is admissible theoretically.

The objective function in problem (10.20) is convex, because $\phi(\mathbf{x})$ is convex (by composition of the increasing convex function $\hat{I}_-$ and the convex function f_i, cf. (3.76a)) and twice differentiable in $\mathbf{x}$. Problem (10.20) now can be solved using the Newton's method. The gradient and Hessian of the log barrier function ϕ are, respectively, given by

$$\nabla\phi(\mathbf{x}) = \sum_{i=1}^{m} \frac{1}{-f_i(\mathbf{x})} \nabla f_i(\mathbf{x}), \tag{10.24}$$

$$\nabla^2\phi(\mathbf{x}) = \sum_{i=1}^{m} \frac{1}{f_i^2(\mathbf{x})} \nabla f_i(\mathbf{x}) \nabla f_i(\mathbf{x})^T + \sum_{i=1}^{m} \frac{1}{-f_i(\mathbf{x})} \nabla^2 f_i(\mathbf{x}) \succeq \mathbf{0}. \tag{10.25}$$

10.3   Central path

Consider the equivalent form of the convex problem (10.20) (that is obtained after multiplying the objective function of (10.20) by t) as follows:

$$\min \ t f_0(\mathbf{x}) + \phi(\mathbf{x})$$
$$\text{s.t. } \mathbf{Ax} = \mathbf{b} \tag{10.26}$$

which has the same minimizers as problem (10.20). Assume that the convex problem (10.26) can be solved by using Newton's method and it has a unique solution $\mathbf{x}^\star(t)$ for each $t > 0$. Therefore, the associated Lagrangian only involves the p equality constrained functions as follows

$$L(\mathbf{x}, \boldsymbol{\nu}) = t f_0(\mathbf{x}) + \phi(\mathbf{x}) + \boldsymbol{\nu}^T (\mathbf{Ax} - \mathbf{b}), \tag{10.27}$$

where $\boldsymbol{\nu} \in \mathbb{R}^p$.

The *central path* associated with the primal convex problem (10.1) is defined as the set of optimal points $\mathbf{x}^\star(t) \in \mathcal{C}$ for $t > 0$, called *central points*, of problem (10.26). Therefore, central points, each associated with a finite optimal value of problem (10.26), must satisfy the following conditions

$$\mathbf{Ax}^\star(t) = \mathbf{b}, \ f_i(\mathbf{x}^\star(t)) < 0, \ i = 1, \dots, m, \tag{10.28}$$

i.e., $\mathbf{x}^\star(t)$ is strictly feasible to problem (10.1), and there exists a $\widehat{\boldsymbol{\nu}} \in \mathbb{R}^p$ such that

$$\begin{aligned}
\mathbf{0} &= \nabla_{\mathbf{x}} L \left(\mathbf{x}^\star(t), \widehat{\boldsymbol{\nu}} \right) = t \nabla_{\mathbf{x}} f_0(\mathbf{x}^\star(t)) + \nabla_{\mathbf{x}} \phi(\mathbf{x}^\star(t)) + \mathbf{A}^T \widehat{\boldsymbol{\nu}} \\
&= t \nabla_{\mathbf{x}} f_0(\mathbf{x}^\star(t)) + \sum_{i=1}^{m} \frac{1}{-f_i(\mathbf{x}^\star(t))} \nabla_{\mathbf{x}} f_i(\mathbf{x}^\star(t)) + \mathbf{A}^T \widehat{\boldsymbol{\nu}}
\end{aligned} \tag{10.29}$$

which actually has the same form as (10.5d) after some suitable variable changes.

For a given t, $\mathbf{x}^\star(t)$ and $\widehat{\boldsymbol{\nu}}$ are a primal-dual optimal solution of (10.26) since problem (10.26) is convex (for any given $t > 0$) with the primal optimal value equal to the dual optimal value (due to strong duality). The optimal $\mathbf{x}^\star(t)$ and $\widehat{\boldsymbol{\nu}}$ to problem (10.26) also provide a primal-dual feasible point to problem (10.1) as stated in the following fact.

Fact 10.1 Every point $\mathbf{x}^\star(t)$ in the central path of problem (10.1) is a strictly feasible point which yields a dual feasible point $(\boldsymbol{\lambda}^\star(t), \boldsymbol{\nu}^\star(t))$ given by

$$\lambda_i^\star(t) = -\frac{1}{t f_i(\mathbf{x}^\star(t))} \geq 0, \ i = 1, \dots, m, \tag{10.30a}$$

$$\boldsymbol{\nu}^\star(t) = \widehat{\boldsymbol{\nu}}/t, \tag{10.30b}$$

owing to

$$\nabla_{\mathbf{x}} L \left(\mathbf{x}^\star(t), \widehat{\boldsymbol{\nu}} \right)/t = \nabla_{\mathbf{x}} \mathcal{L} \left(\mathbf{x}^\star(t), \boldsymbol{\lambda}^\star(t), \boldsymbol{\nu}^\star(t) \right) = \mathbf{0} \ \text{(by (10.29))}. \tag{10.31}$$

Furthermore, the duality gap associated with $\mathbf{x}^\star(t)$ and the dual feasible point $(\boldsymbol{\lambda}^\star(t), \boldsymbol{\nu}^\star(t))$ is simply

$$\eta(t) = f_0(\mathbf{x}^\star(t)) - g(\boldsymbol{\lambda}^\star(t), \boldsymbol{\nu}^\star(t))$$
$$= f_0(\mathbf{x}^\star(t)) - \mathcal{L}(\mathbf{x}^\star(t), \boldsymbol{\lambda}^\star(t), \boldsymbol{\nu}^\star(t)) = m/t \qquad (10.32)$$

(where m is the total number of inequality constraints) and therefore $\mathbf{x}^\star(t)$ is m/t-suboptimal.

Proof of (10.32): For a given $\mathbf{x}^\star(t)$, which satisfies (10.28) and (10.29), it follows that $\mathbf{x}^\star(t)$ also satisfies (10.5a) and (10.5b) and $(\boldsymbol{\lambda}^\star(t), \boldsymbol{\nu}^\star(t))$ (defined in (10.30a) and (10.30b)) satisfies (10.5c) and (10.5d) by (10.31), implying that $(\boldsymbol{\lambda}^\star(t), \boldsymbol{\nu}^\star(t))$ is dual feasible to problem (10.1). Besides, one can infer that the dual function associated with problem (10.1) is given by

$$g(\boldsymbol{\lambda}^\star(t), \boldsymbol{\nu}^\star(t)) = \inf_{\mathbf{x}\in\mathcal{D}} \mathcal{L}(\mathbf{x}, \boldsymbol{\lambda}^\star(t), \boldsymbol{\nu}^\star(t)) = \mathcal{L}(\mathbf{x}^\star(t), \boldsymbol{\lambda}^\star(t), \boldsymbol{\nu}^\star(t)) \qquad (10.33)$$

$$= f_0(\mathbf{x}^\star(t)) + \sum_{i=1}^{m} \lambda_i^\star(t) f_i(\mathbf{x}^\star(t)) + \boldsymbol{\nu}^\star(t)^T(\mathbf{A}\mathbf{x}^\star(t) - \mathbf{b})$$

$$= f_0(\mathbf{x}^\star(t)) - m/t \quad \text{(by (10.30))}.$$

Hence we can conclude that $\mathbf{x}^\star(t)$ is a primal feasible point and $(\boldsymbol{\lambda}^\star(t), \boldsymbol{\nu}^\star(t))$ is a dual feasible point of (10.1) with duality gap $\eta(t) = m/t$. ∎

Next, the central path condition given by (10.28) and (10.29) can be interpreted as a continuous deformation of the KKT optimality conditions (10.5). Recall from Fact 10.1 and (10.28) that a point $\mathbf{x}$ is a central point $\mathbf{x}^\star(t)$ if and only if there exists a pair $(\boldsymbol{\lambda}, \boldsymbol{\nu})$ along with the point $\mathbf{x}$ such that

$$\mathbf{A}\mathbf{x} = \mathbf{b}, \qquad (10.34a)$$

$$f_i(\mathbf{x}) \leq 0, \ i = 1, \ldots, m, \qquad (10.34b)$$

$$\boldsymbol{\lambda} \succeq \mathbf{0}, \qquad (10.34c)$$

$$\nabla f_0(\mathbf{x}) + \sum_{i=1}^{m} \lambda_i \nabla f_i(\mathbf{x}) + \mathbf{A}^T \boldsymbol{\nu} = \mathbf{0}, \qquad (10.34d)$$

$$-\lambda_i f_i(\mathbf{x}) = 1/t, \ i = 1, \ldots, m. \qquad (10.34e)$$

The only difference between the above centrality conditions and the original KKT conditions (10.5) is that the complementary slackness condition $\lambda_i f_i(\mathbf{x}) = 0$ is replaced by the condition $-\lambda_i f_i(\mathbf{x}) = 1/t$. This implies that, as t is large, $\mathbf{x}^\star(t)$ and the associated dual feasible point $(\boldsymbol{\lambda}^\star(t), \boldsymbol{\nu}^\star(t))$ "almost" satisfy the KKT optimality conditions associated with the problem (10.1).

10.4   Barrier method

As mentioned before, $\mathbf{x}^\star(t)$ is m/t-suboptimal (cf. Fact 10.1) and the certificate of this accuracy is provided by the dual feasible pair $(\boldsymbol{\lambda}^\star(t), \boldsymbol{\nu}^\star(t))$. This implies that problem (10.1) can be solved with a guaranteed accuracy $\epsilon = m/t$. One can set $t = m/\epsilon$ and solve the following equality constrained problem

$$\min \quad \frac{m}{\epsilon} f_0(\mathbf{x}) + \phi(\mathbf{x})$$
$$\text{s.t. } \mathbf{A}\mathbf{x} = \mathbf{b} \tag{10.35}$$

using Newton's method. The above procedure yields moderately accurate solutions for small-size problems with good starting points, but does not work well in other cases. The reason is that a small ϵ may make (10.35) "ill-conditioned" such that Newton's method hardly converges to the global optimal point. Hence we go for a simple extension of the above procedure which is called *the barrier method*.

This method is also called the *sequential unconstrained minimization technique* (SUMT) or *path-following method*. In this method, a sequence of unconstrained (or linearly constrained) minimization problems is solved (since linear constraints can be relaxed through an equivalent representation as presented in Chapter 4), using the previously found optimal point as the starting point for the next unconstrained minimization problem. That is to say, $\mathbf{x}^\star(t)$ is computed for a sequence of increasing values of t, until $t > m/\epsilon$, which guarantees that an ϵ-suboptimal solution of the original problem can be obtained. The steps of the algorithm are summarized in Algorithm 10.3.

Algorithm 10.3 Barrier method for solving problem (10.1)

1: Given strictly feasible $\boldsymbol{x}$, $t^{(0)} > 0$, $\mu > 1$, and tolerance $\epsilon > 0$.
2: Set $t := t^{(0)}/\mu$.
3: **repeat**
4:    Set $t := \mu t$.
5:    Centering step. Compute $\mathbf{x}^\star(t)$ by minimizing $tf_0 + \phi$, subject to $\mathbf{A}\mathbf{x} = \mathbf{b}$, starting at $\boldsymbol{x}$.
6:    Update. $\boldsymbol{x} := \mathbf{x}^\star(t)$.
7: **until** Duality gap $m/t < \epsilon$.

We have the following remarks regarding the barrier method:

- At each iteration (except the first one) the central point $\mathbf{x}^\star(t)$ is calculated (starting from the previously computed central point) and then the parameter t is increased by the factor $\mu > 1$.

- The algorithm can also return a dual feasible point $(\boldsymbol{\lambda}^\star(t), \boldsymbol{\nu}^\star(t))$ given by (10.30) and (10.29), in addition to the m/t-suboptimal point $\mathbf{x}^\star(t)$ in Step 5.

- Each execution of Step 5 (for different t) is referred to as a *centering step* (since a central point is computed) or an *outer iteration*.
- In Step 5, Algorithm 10.2 (Newton's method) can be used (although any method to solve a linearly constrained minimization problem can be used). The Newton iterations or steps executed within the centering step are called *inner iterations*.
- The strictly feasible $\boldsymbol{x}$ for initializing Algorithm 10.3 can be obtained by solving

$$\min_{s,\boldsymbol{x}} \ s \tag{10.36a}$$

$$\text{s.t.} \ f_i(\boldsymbol{x}) \le s, \ i=1,\dots,m \tag{10.36b}$$

$$\mathbf{A}\boldsymbol{x} = \mathbf{b}. \tag{10.36c}$$

A strictly feasible point $(\boldsymbol{x},s)$ to problem (10.36) can be easily obtained by setting s to a sufficiently large value so that the optimal solution of (10.36) can also be obtained using Algorithm 10.3. Note that, as our goal is just to find a strictly feasible point $\boldsymbol{x}$ of (10.1), it suffices to solve (10.36) for a feasible point $(\boldsymbol{x},s<0)$, so the stopping criterion can be modified into $s<0$ for reducing the computational complexity.

For a more detailed discussion on parameter settings, please refer to Section 10.3.1 of [BV04]. Next, let us give an example to illustrate the barrier method.

Example 10.1 Consider the following convex optimization problem:

$$\min_{\mathbf{x}\in\mathbb{R}^n} \ \left\{f_0(\mathbf{x}) \triangleq e^{\mathbf{a}^T\mathbf{x}} + e^{\mathbf{b}^T\mathbf{x}}\right\}$$
$$\text{s.t.} \ \mathbf{c}^T\mathbf{x} \ge 1, \ \mathbf{x} \succeq \mathbf{0} \tag{10.37}$$

where $\mathbf{a}\in\mathbb{R}^n$, $\mathbf{b}\in\mathbb{R}^n$, and $\mathbf{c}\in\mathbb{R}^n_+$.

In the centering step (i.e., Step 5), the Newton's method is used to solve the following unconstrained problem (cf. (10.26)):

$$\min_{\mathbf{x}\in\mathbb{R}^n} \ \Big\{g_t(\mathbf{x}) \triangleq tf_0(\mathbf{x}) + \phi(\mathbf{x})$$
$$= t\left(e^{\mathbf{a}^T\mathbf{x}} + e^{\mathbf{b}^T\mathbf{x}}\right) - \log\left(\mathbf{c}^T\mathbf{x}-1\right) - \sum_{i=1}^n \log x_i\Big\}, \tag{10.38}$$

and the gradient and Hessian matrix of $g_t(\mathbf{x})$ are, respectively, given by

$$\nabla g_t(\mathbf{x}) = t\left(e^{\mathbf{a}^T\mathbf{x}}\cdot\mathbf{a} + e^{\mathbf{b}^T\mathbf{x}}\cdot\mathbf{b}\right) - \frac{\mathbf{c}}{\mathbf{c}^T\mathbf{x}-1} - \left[x_1^{-1}, x_2^{-1},\dots,x_n^{-1}\right]^T$$

$$\nabla^2 g_t(\mathbf{x}) = t\left(e^{\mathbf{a}^T\mathbf{x}}\cdot\mathbf{a}\mathbf{a}^T + e^{\mathbf{b}^T\mathbf{x}}\cdot\mathbf{b}\mathbf{b}^T\right) + \frac{\mathbf{c}\mathbf{c}^T}{(\mathbf{c}^T\mathbf{x}-1)^2} + \mathbf{Diag}\left(x_1^{-2},\dots,x_n^{-2}\right).$$

Thus, in each inner iteration in Step 5 of Algorithm 10.2, we compute

$$\boldsymbol{x}^\star = \bar{\mathbf{x}} - s\cdot\left(\nabla^2 g_t(\bar{\mathbf{x}})\right)^\dagger \nabla g_t(\bar{\mathbf{x}}) \ \text{(by (10.12))},$$

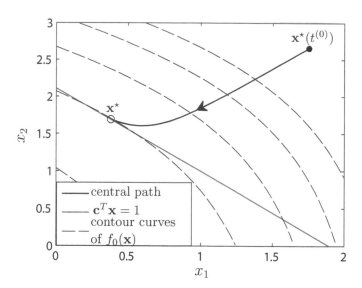

Figure 10.4 The central path of problem (10.37) with $n = 2$, $\mathbf{a} = (1.006, -0.006)$, $\mathbf{b} = (0.329, 0.671)$, and $\mathbf{c} = (0.527, 0.473)$, where the starting point is $\mathbf{x}^\star(t^{(0)})$ for $t^{(0)} = 0.1$, and the parameters for backtracking line search are $\alpha = 0.1$, $\beta = 0.5$.

where $s > 0$ is the step size determined by backtracking line search (Algorithm 10.1), and update $\bar{\mathbf{x}} := \boldsymbol{x}^\star$ until convergence (cf. Newton's method in Algorithm 10.2). Then the centering step at the current outer iteration ends with $\mathbf{x}^\star(t) = \boldsymbol{x}^\star$ (i.e., the solution of problem (10.38)). The central path $\mathbf{x}^\star(t)$ of a problem instance of (10.37) for $n = 2$, $\mathbf{a} = (1.006, -0.006)$, $\mathbf{b} = (0.329, 0.671)$, and $\mathbf{c} = (0.527, 0.473)$ is illustrated in Figure 10.4. $\square$

Finally, we conclude this section with a discussion on the complexity of the barrier method. The complexity analysis of the barrier method has been reported in Section 11.5 of [BV04]. Here, we only briefly summarize this complexity result. Suppose that the convex problem (10.1) satisfies the following conditions:

(C1) The function $t f_0(\mathbf{x}) + \phi(\mathbf{x})$ is closed (i.e., **epi** $t f_0 + \phi$ is a closed set), and self-concordant for all $t \geq t^{(0)}$ as defined next.

- A convex function $f : \mathbb{R}^n \to \mathbb{R}$ is self-concordant if, for all $\mathbf{x} \in \mathbf{dom}\ f$ and $\mathbf{v} \in \mathbb{R}^n$,

$$\tilde{f}(t) \triangleq f(\mathbf{x} + t\mathbf{v}), \quad \mathbf{dom}\ \tilde{f} = \{t \mid \mathbf{x} + t\mathbf{v} \in \mathbf{dom}\ f\}$$

satisfies

$$\left|\tilde{f}'''(t)\right| \leq 2\tilde{f}''(t)^{3/2}, \ \forall t \in \mathbf{dom}\ \tilde{f}.$$

(C2) The constraint set associated with all the inequality constraints of (10.1) is bounded, i.e., $\{\mathbf{x} \in \mathbb{R}^n \mid f_i(\mathbf{x}) \leq 0, \ i = 1, \ldots, m\} \subseteq B(\mathbf{0}, R)$ (a 2-norm ball) for some $R > 0$.

Then, the total number of Newton iterations needed for solving problem (10.1) by the barrier method is bounded by $N_1 + N_2$ (cf. (11.36) and (11.37) in [BV04]), where

$$N_1 = \left\lceil \sqrt{m+2} \log_2 \left(\frac{(m+1)(m+2)GR}{|s^\star|} \right) \right\rceil \left(\frac{1}{2\gamma} + c \right), \qquad (10.39)$$

$$N_2 = \left\lceil \sqrt{m+1} \log_2 \left(\frac{(m+1)(\bar{p} - p^\star)}{\epsilon} \right) \right\rceil \left(\frac{1}{2\gamma} + c \right). \qquad (10.40)$$

For the former, N_1 denotes the number of Newton iterations needed for initializing the barrier method, i.e., obtaining the initial $\mathbf{x}$ and t (cf. Algorithm 10.3), where

$$G \triangleq \max_{i=1,\ldots,m} \|\nabla f_i(\mathbf{0})\|_2,$$

$$s^\star \triangleq \inf_{\mathbf{Ax}=\mathbf{b}} \max_{i=1,\ldots,m} f_i(\mathbf{x}) \quad \text{(optimal value of problem (10.36))},$$

and γ, c are constants depending on the backtracking parameters α, β (cf. Algorithm 10.1) and the solution accuracy of the Newton's method. On the other hand, N_2 is the total number of Newton iterations spent in performing Step 4 to Step 6 until convergence of Algorithm 10.3, where $\bar{p}$ is an upper bound, e.g., $f_0(\mathbf{x}^\star(t^{(0)}))$, to the optimal value of problem (10.1).

Considering either of addition and multiplication of real scalars as an elementary operation, the total number of elementary operations in a Newton iteration, i.e., computing (10.11), is bounded by some polynomial functions of the problem dimensions, i.e., dimension of unknown variable n, number of inequality constraints m and the number of equality constraints p. Therefore, under a mild assumption that G, R, and $s^\star$ do not grow faster than the exponential of polynomial function of n, m, and p, the total number of elementary operations needed (in the worst case) for solving the convex problem (10.1) (that satisfies the above two conditions (C1) and (C2)) using the barrier method is bounded by some polynomial functions of the problem dimensions, and hence the convex problem (10.1) is polynomial-time solvable.

10.5   Primal-dual interior-point method

First of all, let us delineate the conceptual and operational distinctions between the primal-dual IPM and the preceding barrier method though both of them aim to seek an ϵ-optimal solution of the convex problem (10.1).

- In the barrier method, the modified KKT equations (10.34) are solved by using Newton's method after eliminating the dual variable $\boldsymbol{\lambda}$ from those equations, whereas, in the primal-dual IPM, the modified KKT equations are solved directly using Newton's method.

- Unlike the barrier method, here there are no distinctions between inner and outer iterations (i.e., there is only one loop). At each iteration, both the primal and dual variables are updated simultaneously.

- In contrast to the barrier method (for which at each iteration the iterate $\mathbf{x}^\star(t)$ obtained by solving the convex problem (10.26) is a central point to the original convex problem (10.1)), at each iteration for the primal-dual IPM, $\mathbf{x}^\star(t)$ is no longer a central point, and thus the duality gap $\eta(t) = m/t$ for the former does not apply to the latter.

10.5.1   Primal-dual search direction

We start with the modified KKT conditions given by (10.34), expressed as $\mathbf{r}_t(\mathbf{x}, \boldsymbol{\lambda}, \boldsymbol{\nu}) = \mathbf{0}_{n+m+p}$ (only those involving equality constraints). Here we define

$$\mathbf{r}_t(\mathbf{x}, \boldsymbol{\lambda}, \boldsymbol{\nu}) = \begin{bmatrix} \mathbf{r}_{\text{dual}} \\ \mathbf{r}_{\text{cent}} \\ \mathbf{r}_{\text{pri}} \end{bmatrix} = \begin{bmatrix} \nabla f_0(\mathbf{x}) + \left(D\boldsymbol{f}(\mathbf{x})\right)^T \boldsymbol{\lambda} + \mathbf{A}^T \boldsymbol{\nu} \\ -\mathbf{Diag}(\boldsymbol{\lambda})\boldsymbol{f}(\mathbf{x}) - (1/t)\mathbf{1}_m \\ \mathbf{Ax} - \mathbf{b} \end{bmatrix}, \tag{10.41}$$

and $t > 0$. In (10.41), $\boldsymbol{f} : \mathbb{R}^n \to \mathbb{R}^m$ and its derivative matrix $D\boldsymbol{f}$ are, respectively, defined and given by

$$\boldsymbol{f}(\mathbf{x}) \triangleq \begin{bmatrix} f_1(\mathbf{x}) \\ \vdots \\ f_m(\mathbf{x}) \end{bmatrix}, \quad D\boldsymbol{f}(\mathbf{x}) = \begin{bmatrix} Df_1(\mathbf{x}) \\ \vdots \\ Df_m(\mathbf{x}) \end{bmatrix} = \begin{bmatrix} \nabla f_1(\mathbf{x})^T \\ \vdots \\ \nabla f_m(\mathbf{x})^T \end{bmatrix}.$$

If $\mathbf{x}^\star(t), \boldsymbol{\lambda}^\star(t), \boldsymbol{\nu}^\star(t)$ satisfies $\mathbf{r}_t(\mathbf{x}, \boldsymbol{\lambda}, \boldsymbol{\nu}) = \mathbf{0}_{n+m+p}$ and $f_i(\mathbf{x}^\star(t)) < 0$ for all i (implying $\lambda_i^\star(t) > 0$, for all i), then $\mathbf{x}^\star(t)$ must be a central point and a primal feasible point, and $(\boldsymbol{\lambda}^\star(t), \boldsymbol{\nu}^\star(t))$ must be a dual feasible point of problem (10.1), with duality gap $\eta(t) = m/t$.

The first block component of $\mathbf{r}_t$,

$$\mathbf{r}_{\text{dual}} = \nabla f_0(\mathbf{x}) + \left(D\boldsymbol{f}(\mathbf{x})\right)^T \boldsymbol{\lambda} + \mathbf{A}^T \boldsymbol{\nu},$$

is called the *dual residual*, and the last block component,

$$\mathbf{r}_{\text{pri}} = \mathbf{Ax} - \mathbf{b},$$

is called the *primal residual*. The middle block component,

$$\mathbf{r}_{\text{cent}} = -\mathbf{Diag}(\boldsymbol{\lambda})\boldsymbol{f}(\mathbf{x}) - (1/t)\mathbf{1}_m,$$

is called the *centrality residual*, i.e., the residual for the modified complementary slackness condition.

Now consider the Newton step for solving the nonlinear equations $\mathbf{r}_t(\mathbf{x}, \boldsymbol{\lambda}, \boldsymbol{\nu}) = \mathbf{0}$, for fixed t (without elimination of $\boldsymbol{\lambda}$ beforehand), and meanwhile the solution of which should also satisfy $\boldsymbol{f}(\mathbf{x}) \prec \mathbf{0}$ and $\boldsymbol{\lambda} \succ \mathbf{0}$. The current point $\mathbf{y}$ and Newton step $\Delta \mathbf{y}$ are, respectively, denoted as

$$\mathbf{y} = (\mathbf{x}, \boldsymbol{\lambda}, \boldsymbol{\nu}), \quad \Delta \mathbf{y} = (\Delta\mathbf{x}, \Delta\boldsymbol{\lambda}, \Delta\boldsymbol{\nu}).$$

The Newton step is characterized by the first-order approximated linear equations of the Taylor series expansion of $\mathbf{r}_t$:

$$\mathbf{r}_t(\mathbf{y} + \Delta\mathbf{y}) \approx \mathbf{r}_t(\mathbf{y}) + D\mathbf{r}_t(\mathbf{y})\Delta\mathbf{y} = \mathbf{0}.$$

In terms of $\mathbf{x}, \boldsymbol{\lambda}$, and $\boldsymbol{\nu}$, we have

$$\begin{bmatrix} \nabla^2 f_0(\mathbf{x}) + \sum_{i=1}^{m} \lambda_i \nabla^2 f_i(\mathbf{x}) & (D\boldsymbol{f}(\mathbf{x}))^T & \mathbf{A}^T \\ -\mathbf{Diag}(\boldsymbol{\lambda})D\boldsymbol{f}(\mathbf{x}) & -\mathbf{Diag}(\boldsymbol{f}(\mathbf{x})) \, \mathbf{0}_{m\times p} \\ \mathbf{A} & \mathbf{0}_{p\times m} & \mathbf{0}_{p\times p} \end{bmatrix} \begin{bmatrix} \Delta\mathbf{x} \\ \Delta\boldsymbol{\lambda} \\ \Delta\boldsymbol{\nu} \end{bmatrix} = -\mathbf{r}_t(\mathbf{y}). \quad (10.42)$$

The primal-dual search direction $\Delta\mathbf{y}_{pd} = (\Delta\mathbf{x}, \Delta\boldsymbol{\lambda}, \Delta\boldsymbol{\nu})$ is defined as the solution of (10.42). Note that the primal and dual search directions are coupled, both through the coefficient matrix in (10.42) and the residuals $\mathbf{r}_{\text{dual}}, \mathbf{r}_{\text{cent}}$, and $\mathbf{r}_{\text{pri}}$.

10.5.2 Surrogate duality gap

In the primal-dual IPM, the iterates $\mathbf{x}^{(k)}, \boldsymbol{\lambda}^{(k)}$, and $\boldsymbol{\nu}^{(k)}$ are not necessarily feasible, except in the limit as the algorithm converges. This means that a duality gap $\eta^{(k)}$ associated with iteration k of the algorithm cannot be easily evaluated as in (the outer steps of) the barrier method. Instead we define the surrogate duality gap, for any $\mathbf{x}$ that satisfies $\boldsymbol{f}(\mathbf{x}) \prec \mathbf{0}$ and $\boldsymbol{\lambda} \succeq \mathbf{0}$, as (from the modified KKT conditions)

$$\hat{\eta}(\mathbf{x}, \boldsymbol{\lambda}) = -\boldsymbol{f}(\mathbf{x})^T \boldsymbol{\lambda} = -\sum_{i=1}^{m} \lambda_i f_i(\mathbf{x}). \quad (10.43)$$

The surrogate duality gap $\hat{\eta}$ would be the duality gap, if $\mathbf{x}$ were primal feasible and $\boldsymbol{\lambda}$ and $\boldsymbol{\nu}$ were dual feasible, i.e., if $\mathbf{r}_{\text{pri}} = \mathbf{0}$ and $\mathbf{r}_{\text{dual}} = \mathbf{0}$. Note that the value of the parameter t that corresponds to the surrogate duality gap $\hat{\eta}$ is $m/\hat{\eta}$ (cf. (10.32)).

10.5.3 Primal-dual interior-point algorithm

The primal-dual IPM is summarized in Algorithm 10.4. In Step 3, the parameter t is set to a factor μ times $m/\hat{\eta}$, where $m/\hat{\eta}$ is the value of t associated with the current surrogate duality gap $\hat{\eta}$. If $\mathbf{x}, \boldsymbol{\lambda}$, and $\boldsymbol{\nu}$ were central (and therefore with duality gap m/t), then in Step 3 we would increase t by the factor μ, which is exactly the same update used in the barrier method. The primal-dual interior-point algorithm terminates when $\mathbf{x}$ is primal feasible and $(\boldsymbol{\lambda}, \boldsymbol{\nu})$ is dual feasible and the surrogate duality gap $\hat{\eta}$ is smaller than the tolerance ϵ.

Algorithm 10.4 Primal-dual interior-point algorithm for solving problem (10.1)

1: Given $\mathbf{x}$ that satisfies $f_1(\mathbf{x}) < 0, \ldots, f_m(\mathbf{x}) < 0$, $\boldsymbol{\lambda} \succ \mathbf{0}$, $\mu > 1$, $\epsilon > 0$.
2: **repeat**
3:    Determine t. Set $t := \mu m / \hat{\eta}$.
4:    Compute primal-dual search direction $\Delta \mathbf{y}_{pd}$ where $\Delta \mathbf{y}_{pd}$ is the solution of (10.42).
5:    Line search and update. Determine step size $s > 0$ and update

$$\mathbf{y} := \mathbf{y} + s\Delta \mathbf{y}_{pd},$$

   and then compute the surrogate duality gap $\hat{\eta}$ given by (10.43).
6: **until** $\hat{\eta} \le \epsilon$.

Since the primal-dual IPM often converges in a faster than linear convergence manner, it is common to choose small ϵ. Linear convergence in the context of iterative numerical methods means that the error

$$\rho_k \triangleq f(\mathbf{x}_k) - f(\mathbf{x}^\star)$$
$$\propto r^k \text{ for large } k,$$

lies below a line on a log-linear plot of ρ_k versus k, where f is the objective function to be minimized, $\mathbf{x}^\star$ denotes the optimum minimizer, $\mathbf{x}_k$ denotes the iterate obtained at the kth iteration, and $0 < r < 1$ is a constant.

◆ **Line search in Step 5**

The line search used in the above algorithm is the backtracking line search (cf. Algorithm 10.1), based on the residual, and suitably modified to ensure that $\boldsymbol{\lambda} \succ \mathbf{0}$ and $\boldsymbol{f}(\mathbf{x}) \prec \mathbf{0}$. Let the current iterates be denoted as $\mathbf{x}$, $\boldsymbol{\lambda}$, $\boldsymbol{\nu}$ and let the next iterates be denoted as $\mathbf{x}^+$, $\boldsymbol{\lambda}^+$, $\boldsymbol{\nu}^+$. That is,

$$\mathbf{x}^+ = \mathbf{x} + s\Delta\mathbf{x}, \ \boldsymbol{\lambda}^+ = \boldsymbol{\lambda} + s\Delta\boldsymbol{\lambda}, \ \boldsymbol{\nu}^+ = \boldsymbol{\nu} + s\Delta\boldsymbol{\nu}. \qquad (10.44)$$

The first step is to compute the largest positive step size, not exceeding one, that gives $\boldsymbol{\lambda}^+ \succeq \mathbf{0}$. That is,

$$s_{\max} = \sup \left\{ s \in [0, 1] \mid \boldsymbol{\lambda} + s\Delta\boldsymbol{\lambda} \succeq \mathbf{0} \right\}$$
$$= \min \left\{ 1, \ \min\{-\lambda_i/\Delta\lambda_i \mid \Delta\lambda_i < 0\} \right\}. \qquad (10.45)$$

Now, let $s = 0.99 s_{\max}$, and multiply s by $\beta \in (0, 1)$ until $\boldsymbol{f}(\mathbf{x}^+) \prec \mathbf{0}$. Then, the multiplication of s by β is continued until

$$\|\mathbf{r}_t(\mathbf{x}^+, \boldsymbol{\lambda}^+, \boldsymbol{\nu}^+)\|_2 \le (1 - \alpha s)\|\mathbf{r}_t(\mathbf{x}, \boldsymbol{\lambda}, \boldsymbol{\nu})\|_2. \qquad (10.46)$$

Common choices for the backtracking parameters α and β are the same as those for Newton's method: α is typically chosen in the range 0.01 to 0.1, and β

is typically chosen in the range 0.3 to 0.8. The effectiveness and efficiency of the primal-dual interior-point algorithm introduced above may also need suitable modifications depending on the specific convex optimization problem under consideration, so efficient reliable algorithm implementations (in terms of complexity and running speed) may also rely on practical algorithm design experience. Complexity analysis for Algorithm 10.4 is discussed in the following remark.

Remark 10.1 The complexity of Algorithm 10.4, a generic IPM, consists of two parts:

1. *Iteration Complexity*: Given an $\epsilon > 0$, the number of iterations required to reach an ϵ-optimal solution to the convex problem (10.1) is on the order of $\sqrt{\beta} \cdot \log(1/\epsilon)$, where β depends on the problem size parameters (e.g., n, m, and p).
2. *Per-Iteration Computation Cost*: In each iteration, a search direction is found by solving the system of $k = n + m + p$ linear equations in k unknowns given by (10.42). The computation cost is dominated by (i) the formation of the $k \times k$ coefficient matrix of the linear system, yielding computation cost C_{form}, and (ii) the numerical approach used for solving (10.42), yielding computation cost C_{sol}. Hence, the total computation cost per iteration is on the order of $C_{\text{form}} + C_{\text{sol}}$.

By combining the above two parts, it follows that the complexity of Algorithm 10.4 for solving (10.1) is on the order of $\sqrt{\beta} \cdot (C_{\text{form}} + C_{\text{sol}}) \cdot \log(1/\epsilon)$. The detailed complexity order of an IPM in terms of the problem size parameters depends on the problem under consideration (e.g., LP, QP, SOCP, SDP). The author is referred to [BTN01] for the worst-case complexity analysis of solving various convex problems using IPM. $\square$

Example 10.2 *(Primal-dual IPM for solving an LP [CSA+11])* Consider solving the following LP:

$$\min_{\beta \in \mathbb{R}^N} \quad -\mathbf{b}^T \beta$$
$$\text{s.t.} \quad -\mathbf{C}\beta - \mathbf{d} \preceq \mathbf{0}, \tag{10.47}$$

where $\beta \in \mathbb{R}^N$ is the variable vector, and $\mathbf{b} \in \mathbb{R}^N$, $\mathbf{C} \in \mathbb{R}^{M \times N}$, and $\mathbf{d} \in \mathbb{R}^M$ are given. The primal-dual IPM iteratively updates the primal-dual variable (β, λ) by $(\beta + s\Delta\beta, \lambda + s\Delta\lambda)$ where $(\Delta\beta, \Delta\lambda)$ and s denote the search direction and the step size, respectively. By solving the modified KKT conditions with the first-order approximation given by (10.42), the search direction $\Delta\beta$ and $\Delta\lambda$ can be obtained as follows:

$$\Delta\beta = (\mathbf{C}^T \mathbf{D} \mathbf{C})^{-1} (\mathbf{C}^T \mathbf{D} \mathbf{r}_2 - \mathbf{r}_1), \tag{10.48a}$$
$$\Delta\lambda = \mathbf{D}(\mathbf{r}_2 - \mathbf{C}\Delta\beta), \tag{10.48b}$$

where

$$\mathbf{D} = \mathbf{Diag}(\boldsymbol{\lambda})\big(\mathbf{Diag}(\mathbf{C}\boldsymbol{\beta} + \mathbf{d})\big)^{-1}, \tag{10.49}$$

$$\mathbf{r}_1 = -\mathbf{b} - \mathbf{C}^T\boldsymbol{\lambda}, \tag{10.50}$$

$$\mathbf{r}_2 = -(\mathbf{C}\boldsymbol{\beta} + \mathbf{d}) + \frac{1}{t}[1/\lambda_1, \ldots, 1/\lambda_M]^T, \quad t > 0. \tag{10.51}$$

The step size $s \in (0,1]$ can be chosen as any value such that $\boldsymbol{\lambda} + s\Delta\boldsymbol{\lambda} \succ \mathbf{0}$ and $\mathbf{C}(\boldsymbol{\beta} + s\Delta\boldsymbol{\beta}) + \mathbf{d} \succ \mathbf{0}$ are satisfied simultaneously. Instead of finding the step size s sequentially as presented above, we first obtain the corresponding largest step size $s_{\max}$ as follows

$$s_{\max} = \sup\left\{s \in (0,1] \mid \boldsymbol{\lambda} + s\Delta\boldsymbol{\lambda} \succeq \mathbf{0}, \mathbf{C}(\boldsymbol{\beta} + s\Delta\boldsymbol{\beta}) + \mathbf{d} \succeq \mathbf{0}\right\}$$

$$= \min\left\{1, \min\left\{-\frac{[\boldsymbol{\lambda}]_i}{[\Delta\boldsymbol{\lambda}]_i} \,\Big|\, [\Delta\boldsymbol{\lambda}]_i < 0\right\},\right.$$

$$\left.\min\left\{-\frac{[\mathbf{C}\boldsymbol{\beta} + \mathbf{d}]_i}{[\mathbf{C}\Delta\boldsymbol{\beta}]_i} \,\Big|\, [\mathbf{C}\Delta\boldsymbol{\beta}]_i < 0\right\}\right\}. \tag{10.52}$$

Then, a step size can be determined as $s = 0.99s_{\max}$ to ensure $\boldsymbol{\lambda} + s\Delta\boldsymbol{\lambda} \succ \mathbf{0}$ and $\mathbf{C}(\boldsymbol{\beta} + s\Delta\boldsymbol{\beta}) + \mathbf{d} \succ \mathbf{0}$. With the surrogate duality gap given in (10.43), the customized primal-dual interior-point algorithm for (10.47) is summarized in Algorithm 10.5. □

Algorithm 10.5 Customized primal-dual interior-point algorithm for (10.47).

1: Given a primal-dual strictly feasible initial point $(\boldsymbol{\beta}, \boldsymbol{\lambda})$, $\mu = 10$, and a solution accuracy $\epsilon > 0$.
2: **repeat**
3:    Calculate the surrogate duality gap $\hat{\eta}(\boldsymbol{\beta}, \boldsymbol{\lambda}) = (\mathbf{C}\boldsymbol{\beta} + \mathbf{d})^T\boldsymbol{\lambda}$ and determine $t := \mu M/\hat{\eta}(\boldsymbol{\beta}, \boldsymbol{\lambda})$.
4:    Compute $(\Delta\boldsymbol{\beta}, \Delta\boldsymbol{\lambda})$ given by (10.48).
5:    Compute $s_{\max}$ by (10.52) and the step size $s = 0.99s_{\max}$.
6:    Update $\boldsymbol{\beta} := \boldsymbol{\beta} + s\Delta\boldsymbol{\beta}$ and $\boldsymbol{\lambda} := \boldsymbol{\lambda} + s\Delta\boldsymbol{\lambda}$.
7: **until** $\hat{\eta}(\boldsymbol{\beta}, \boldsymbol{\lambda}) \leq \epsilon$.

10.5.4   Primal-dual interior-point method for solving SDP

Consider the standard form of an SDP (see (9.194)):

$$\begin{aligned} \min \ &\mathrm{Tr}(\mathbf{C}\mathbf{X}) \\ \text{s.t. } &\mathbf{X} \succeq \mathbf{0}, \ \mathrm{Tr}(\mathbf{A}_i\mathbf{X}) = b_i, \ i = 1, \ldots, p \end{aligned} \tag{10.53}$$

where $\mathbf{C} \in \mathbb{S}^m$, $\mathbf{A}_i \in \mathbb{S}^m$, and $b_i \in \mathbb{R}$. Note that the problem domain is $\mathcal{D} = \mathbb{S}^m$ and the generalized inequality here is associated with the PSD cone (a proper cone). The Lagrangian of problem (10.53) is given by

$$\mathcal{L}(\mathbf{X}, \mathbf{Z}, \boldsymbol{\nu}) = \mathrm{Tr}(\mathbf{CX}) - \mathrm{Tr}(\mathbf{ZX}) + \sum_{i=1}^{p} \nu_i(\mathrm{Tr}(\mathbf{A}_i\mathbf{X}) - b_i). \quad \text{(by (9.195))}$$

(10.54)

The corresponding dual problem is given by (cf. (9.197))

$$\max \ -\mathbf{b}^T\boldsymbol{\nu}$$

$$\text{s.t. } \mathbf{Z} \succeq \mathbf{0}, \ \mathbf{C} - \mathbf{Z} + \sum_{i=1}^{p} \nu_i\mathbf{A}_i = \mathbf{0}.$$

(10.55)

Moreover, the KKT conditions for the SDP problem (10.53) are summarized as follows:

$$\mathrm{Tr}(\mathbf{A}_i\mathbf{X}) - b_i = 0, \ i = 1, \ldots, p, \tag{10.56a}$$

$$\mathbf{X} \succeq \mathbf{0}, \tag{10.56b}$$

$$\mathbf{Z} \succeq \mathbf{0}, \tag{10.56c}$$

$$\nabla_{\mathbf{X}}\mathcal{L}(\mathbf{X}, \mathbf{Z}, \boldsymbol{\nu}) = \mathbf{C} - \mathbf{Z} + \sum_{i=1}^{p} \nu_i\mathbf{A}_i = \mathbf{0}, \tag{10.56d}$$

$$\mathbf{ZX} = \mathbf{0} \ \text{(complementary slackness)}. \tag{10.56e}$$

However, rather than solving the KKT conditions given by (10.56), the primal-dual IPM is to solve the modified or deformed KKT conditions using Newton's method, which are derived next.

Similar to the preceding formulation of (10.26), by using the following log barrier function

$$\phi(\mathbf{X}) = -\log\det(\mathbf{X}) \tag{10.57}$$

for $\mathbf{X} \succ \mathbf{0}$, the SDP problem (10.53) can be approximated as a linear equality constrained convex problem for a given $t > 0$ (cf. (10.26))

$$\min \ t \cdot \mathrm{Tr}(\mathbf{CX}) + \left(-\log\det(\mathbf{X})\right)$$

$$\text{s.t. } \mathrm{Tr}(\mathbf{A}_i\mathbf{X}) = b_i, \ i = 1, \ldots, p. \tag{10.58}$$

The Lagrangian of problem (10.58) can be seen to be

$$L(\mathbf{X}, \boldsymbol{\nu}) = t \cdot \mathrm{Tr}(\mathbf{CX}) - \log\det(\mathbf{X}) + \sum_{i=1}^{p} \nu_i(\mathrm{Tr}(\mathbf{A}_i\mathbf{X}) - b_i). \tag{10.59}$$

Also note that

$$\nabla_{\mathbf{X}}L(\mathbf{X}, \boldsymbol{\nu}) = t\mathbf{C} - \mathbf{X}^{-1} + \sum_{i=1}^{p} \nu_i\mathbf{A}_i, \tag{10.60}$$

where we have used the results $\nabla_{\mathbf{X}}\mathrm{Tr}(\mathbf{CX}) = \mathbf{C}$ (by (3.40)) and $\nabla_{\mathbf{X}}\log\det(\mathbf{X}) = \mathbf{X}^{-1}$ (by (3.46)).

Let $\mathbf{X}^\star(t)$ be an optimal solution of problem (10.58), or a central point of problem (10.53). Then it must be primal feasible, i.e., $\mathrm{Tr}(\mathbf{A}_i\mathbf{X}^\star(t)) = b_i$, $i = 1, \ldots, p$, and $\mathbf{X}^\star(t) \succ \mathbf{0}$. Moreover, $\mathbf{X}^\star(t)$ must satisfy the KKT condition $\nabla_{\mathbf{X}} L(\mathbf{X}^\star(t), \boldsymbol{\nu}) = \mathbf{0}$, which means that there exists a $\widehat{\boldsymbol{\nu}} \in \mathbb{R}^p$, such that

$$\nabla_{\mathbf{X}} L(\mathbf{X}^\star(t), \widehat{\boldsymbol{\nu}}) = t\mathbf{C} - \mathbf{X}^\star(t)^{-1} + \sum_{i=1}^{p} \widehat{\nu}_i \mathbf{A}_i = \mathbf{0} \qquad \text{(by (10.60))}$$

$$\Rightarrow \mathbf{C} - \frac{\mathbf{X}^\star(t)^{-1}}{t} + \sum_{i=1}^{p} \frac{\widehat{\nu}_i}{t} \mathbf{A}_i = \mathbf{0}. \tag{10.61}$$

Analogous to the ones defined in (10.30), let us define

$$\mathbf{Z}^\star(t) = \frac{1}{t}\mathbf{X}^\star(t)^{-1} \succ \mathbf{0} \tag{10.62}$$

$$\nu_i^\star(t) = \widehat{\nu}_i/t. \tag{10.63}$$

Then (10.61) becomes

$$\nabla_{\mathbf{X}} \mathcal{L}(\mathbf{X}^\star(t), \mathbf{Z}^\star(t), \boldsymbol{\nu}^\star(t)) = \mathbf{C} - \mathbf{Z}^\star(t) + \sum_{i=1}^{p} \nu_i^\star(t)\mathbf{A}_i = \mathbf{0}. \tag{10.64}$$

In other words, $\mathbf{X}^\star(t)$ is primal feasible, and $(\mathbf{Z}^\star(t), \boldsymbol{\nu}^\star(t))$ is dual feasible to problem (10.53), since they actually satisfy almost all the KKT conditions in (10.56), except for the complementary slackness. It can be easily inferred that the duality gap $\eta(t)$ is given by

$$\begin{aligned} \eta(t) &= \mathrm{Tr}(\mathbf{C}\mathbf{X}^\star(t)) + \mathbf{b}^T \boldsymbol{\nu}^\star(t) \quad \text{(by (10.53) and (10.55))} \\ &= \mathrm{Tr}(\mathbf{C}\mathbf{X}^\star(t)) - \mathcal{L}(\mathbf{X}^\star(t), \mathbf{Z}^\star(t), \boldsymbol{\nu}^\star(t)) \quad \text{(by (10.54) and (10.64))} \\ &= \mathrm{Tr}(\mathbf{Z}^\star(t)\mathbf{X}^\star(t)) = \mathrm{Tr}(\mathbf{I}_m/t) = m/t \quad \text{(by (10.62)).} \end{aligned} \tag{10.65}$$

Consequently, we come up with the following deformed KKT conditions for $\mathbf{X}$ to be a central point of the SDP problem (10.53) if and only if there exists a pair $(\mathbf{Z}, \boldsymbol{\nu})$ together with the $\mathbf{X}$ such that

$$\mathrm{Tr}(\mathbf{A}_i\mathbf{X}) - b_i = 0, \ i = 1, \ldots, p, \tag{10.66a}$$

$$\mathbf{X} \succeq \mathbf{0}, \tag{10.66b}$$

$$\mathbf{Z} \succeq \mathbf{0}, \tag{10.66c}$$

$$\mathbf{C} - \mathbf{Z} + \sum_{i=1}^{p} \nu_i \mathbf{A}_i = \mathbf{0}, \tag{10.66d}$$

$$\mathbf{Z}\mathbf{X} = \frac{1}{t}\mathbf{I}_m. \tag{10.66e}$$

Note that the deformed KKT conditions (10.66) (analogous to (10.34) for non-SDP problems) differ from the original KKT conditions (10.56) only in the complementary slackness condition.

The surrogate duality gap $\hat{\eta}$ is defined as

$$\hat{\eta} = \mathrm{Tr}(\mathbf{ZX}), \tag{10.67}$$

which would be the duality gap being m/t, if $\mathbf{X}$ were primal feasible (i.e., (10.66a) and (10.66b) satisfied) and $\mathbf{Z}$ and $\boldsymbol{\nu}$ were dual feasible (i.e., (10.66c) and (10.66d) satisfied). Note that the value of the parameter t that corresponds to the surrogate duality gap $\hat{\eta}$ is $m/\hat{\eta}$.

Having defined the barrier function for the SDP problem, the reformulated SDP problem (10.58), and its associated deformed KKT conditions (10.66), we now present the primal-dual IPM for solving the SDP problem (10.53), which, again, is to iteratively solve the linear equations (10.66a), (10.66d), and (10.66e), meanwhile maintaining that the inequalities (10.66b) and (10.66c) are satisfied. This method, which therefore is quite similar to the one for solving problem (10.1) (cf. Algorithm 10.4) due to the same philosophies, is given in Algorithm 10.6.

Algorithm 10.6 Primal-dual IPM for solving SDP problem (10.53)

1: Given an initial strictly feasible point $(\mathbf{X}, \mathbf{Z}, \boldsymbol{\nu}) = (\mathbf{X}^{(0)}, \mathbf{Z}^{(0)}, \boldsymbol{\nu}^{(0)})$, $\epsilon > 0$, $\mu > 1$.

2: **repeat**

3:     Compute the current $t := \mu m/\hat{\eta} = \mu m/\mathrm{Tr}(\mathbf{ZX})$.

4:     Compute primal-dual search direction (refer to Subsection 10.5.1, that is, compute $(\Delta\mathbf{X}, \Delta\mathbf{Z}, \Delta\boldsymbol{\nu})$ by solving the following deformed KKT equations (given by (10.66)):

$$(\mathbf{Z} + \Delta\mathbf{Z})(\mathbf{X} + \Delta\mathbf{X}) - \mathbf{I}_m/t = \mathbf{0}, \tag{10.68a}$$

$$\mathbf{C} - (\mathbf{Z} + \Delta\mathbf{Z}) + \sum_{i=1}^{p}(\nu_i + \Delta\nu_i)\mathbf{A}_i = \mathbf{0}, \tag{10.68b}$$

$$\mathrm{Tr}(\mathbf{A}_i(\mathbf{X} + \Delta\mathbf{X})) - b_i = 0, \ i = 1, \ldots, p \tag{10.68c}$$

(only those involving equality constraints) through the first-order Taylor series approximation (to be discussed below).

5:     Line search. Find the step size s such that

$$\mathbf{X} + s\Delta\mathbf{X} \succ \mathbf{0} \text{ and } \mathbf{Z} + s\Delta\mathbf{Z} \succ \mathbf{0}.$$

6:     Update $\mathbf{X} := \mathbf{X} + s\Delta\mathbf{X}$, $\mathbf{Z} := \mathbf{Z} + s\Delta\mathbf{Z}$, $\boldsymbol{\nu} := \boldsymbol{\nu} + s\Delta\boldsymbol{\nu}$.

7: **until** the surrogate duality gap $\hat{\eta} = \mathrm{Tr}(\mathbf{XZ}) < \epsilon$.

In Step 4, we need to solve $\Delta\mathbf{X}$, $\Delta\mathbf{Z}$, and $\Delta\boldsymbol{\nu}$. It can be observed that equations (10.68b) and (10.68c) in Step 4 are linear and are easy to handle. The difficulty mainly lies in (10.68a). So we can apply the first-order Taylor series approximation to (10.68a) as follows

$$(\mathbf{Z} + \Delta\mathbf{Z})(\mathbf{X} + \Delta\mathbf{X}) - \mathbf{I}_m/t = \mathbf{0}$$
$$\Rightarrow \mathbf{Z}\mathbf{X} + \mathbf{Z}\Delta\mathbf{X} + \Delta\mathbf{Z}\mathbf{X} + \Delta\mathbf{Z}\Delta\mathbf{X} - \mathbf{I}_m/t = \mathbf{0}$$
$$\Rightarrow \mathbf{Z}\mathbf{X} + \mathbf{Z}\Delta\mathbf{X} + \Delta\mathbf{Z}\mathbf{X} - \mathbf{I}_m/t = \mathbf{0} \quad \text{(first-order approximation)}. \quad (10.69)$$

Now, (10.68b), (10.68c), and (10.69) are all linear in $(\Delta\mathbf{X}, \Delta\mathbf{Z}, \Delta\boldsymbol{\nu})$ and can be solved straightforwardly. We then use the line search to find the step size s (in Step 5), and thereby update $(\mathbf{X}, \mathbf{Z}, \boldsymbol{\nu})$ (in Step 6). The procedure is repeated until the stopping criterion (in Step 7) is achieved.

The above interior-point algorithm design for solving the SDP given by (10.53) is merely an illustrative example. In practical applications, a convex problem under consideration could involve various ordinary inequality constraints and generalized inequality constraints, and equality constraints simultaneously. One can follow the same design philosophies for the desired interior-point algorithm. Let us conclude this section with a practical design example below.

In Subsection 8.4.5, a LFSDR approach to noncohenernt ML detection of higher-order QAM OSTBC [CHMC10] is reformulated into an SDP (i.e., problem (8.49)), which is repeated here for convenience:

$$
\begin{aligned}
\max_{\mathbf{Z} \in \mathbb{R}^{n \times n}} \quad & \text{Tr}(\mathbf{G}\mathbf{Z}) \\
\text{s.t.} \quad & \text{Tr}(\mathbf{D}\mathbf{Z}) = 1 \\
& [\mathbf{Z}]_{n,n} \le [\mathbf{Z}]_{k,k} \le (2^q - 1)^2 [\mathbf{Z}]_{n,n}, \ k = 1, \dots, n-1 \\
& \mathbf{Z} \succeq \mathbf{0}
\end{aligned}
\quad (10.70)
$$

for which strong duality holds true by Slater's condition. The matrices $\mathbf{G}$ and $\mathbf{D}$ in (10.70) were defined in (8.44). The optimal solution can surely be obtained by using CVX or SeDuMi.

Essentially we consider solving (10.70) by solving its dual which can be shown (cf. (9.47)) to be

$$
\begin{aligned}
\min \ & \nu \\
\text{s.t.} \ & \nu \in \mathbb{R}, \ \mathbf{t} \in \mathbb{R}^{2(n-1)}, \ \mathbf{Y} \in \mathbb{R}^{n \times n} \\
& \mathbf{Y} \succeq \mathbf{0}, \ \mathbf{t} \succeq \mathbf{0} \\
& \mathbf{Y} = \text{Diag}\left(\begin{bmatrix} \nu \mathbf{1}_{n-1} - \mathbf{t}_1 + \mathbf{t}_2 \\ s_1^2 \nu + \mathbf{1}_{n-1}^T(\mathbf{t}_1 - (2^q - 1)^2 \mathbf{t}_2) \end{bmatrix} \right) - \mathbf{G}
\end{aligned}
\quad (10.71)
$$

where $(\mathbf{Y}, \mathbf{t}, \nu)$ are the dual variables of (10.70), and $\mathbf{t}_1, \mathbf{t}_2 \in \mathbb{R}^{n-1}$, respectively, represent the upper and lower part of $\mathbf{t}$; i.e., $\mathbf{t} = [\ \mathbf{t}_1^T \ \mathbf{t}_2^T \ ]^T$. The idea is to apply a logarithmetic barrier approximation to (10.71) to implicitly handle the constraints $\mathbf{Y} \succeq \mathbf{0}$ and $\mathbf{t} \succeq \mathbf{0}$:

$$\min \; \nu - \mu \left(\log \det(\mathbf{Y}) + \sum_{i=1}^{2(n-1)} \log t_i \right)$$

$$\text{s.t. } \nu \in \mathbb{R}, \quad \mathbf{t} \in \mathbb{R}^{2(n-1)}, \quad \mathbf{Y} \in \mathbb{R}^{n \times n}, \qquad (10.72)$$

$$\mathbf{Y} = \mathbf{Diag}\left(\begin{bmatrix} \nu \mathbf{1}_{n-1} - \mathbf{t}_1 + \mathbf{t}_2 \\ s_1^2 \nu + \mathbf{1}_{n-1}^T (\mathbf{t}_1 - (2^q - 1)^2 \mathbf{t}_2) \end{bmatrix} \right) - \mathbf{G}$$

where the parameter $\mu > 0$ corresponds to $1/t$ in the barrier method. It is known that (10.72) approaches (10.71) as $\mu \to 0$. Following the preceding design procedure, one should be able to design a specialized interior-point algorithm for simultaneously solving (10.70) and (10.71). One such algorithm is also given in [CHMC10], which turns out to be around ten times faster than SeDuMi by some simulation results, showing its advantages in practical implementations.

10.6   Summary and discussion

In this chapter we have introduced the barrier method and the primal-dual IPM and the associated theories. These methods are basically derived from the KKT conditions of the convex optimization problem under consideration. Efficient utilization of IPMs for solving convex optimization problems will make the designed algorithm significantly faster than using off-the-shelf general-purpose solvers (e.g., CVX and SeDuMi), and thus IPMs are considered very useful for a specific algorithm design with stringent requirements on hardware and/or software implementation, in spite of the same performance result obtained no matter if CVX or SeDuMi is used. We finally introduced the primal-dual IPMs for solving an SDP problem.

References for Chapter 10

[BTN01]   A. Ben-Tal and A. Nemirovsk, *Lectures on Modern Convex Optimization: Analysis, Algorithms, and Engineering Applications*. Philadelphia, PA: MPSSIAM Series on Optimization, 2001.

[BV04]   S. Boyd and L. Vandenberghe, *Convex Optimization*. Cambridge, UK: Cambridge University Press, 2004.

[CHMC10] T.-H. Chang, C.-W. Hsin, W.-K. Ma, and C.-Y. Chi, "A linear fractional semidefinite relaxation approach to maximum-likelihood detection of higher order QAM OSTBC in unknown channels," *IEEE Trans. Signal Process.*, vol. 58, no. 4, pp. 2315–2326, Apr. 2010.

[CSA+11] T.-H. Chan, C.-J. Song, A. Ambikapathi, C.-Y. Chi, and W.-K. Ma, "Fast alternating volume maximization algorithm for blind separation of non-negative sources," in

Proc. 2011 IEEE International Workshop on Machine Learning for Signal Processing (MLSP), Beijing, China, Sept. 18–21, 2011.

A   Appendix: Convex Optimization Solvers

In this Appendix, we will briefly introduce two MATLAB-based softwares, namely, SeDuMi and CVX, to solve convex optimization problems. SeDuMi is based on interior-point methods (introduced in Chapter 10) and CVX utilizes either SeDuMi solver or SDPT3 solver, which clearly indicates that both softwares are based on interior-point methods, but in a different way. A finite impulse response (FIR) filter design example is used to illustrate the programming methodology of these software packages.

A.1   SeDuMi

SeDuMi stands for *self-dual minimization* and it is a powerful software for solving a variety of convex optimization problems, namely the LP, SOCP, SDP, and their combinations. It runs under the MATLAB environment, and is very easy to use. It can be downloaded from `http://sedumi.ie.lehigh.edu/` (for free). To use SeDuMi, it is necessary to formulate your problem into either the primal standard form of the cone program (see (9.185)):

$$\begin{aligned} \min \ & \mathbf{c}^T \mathbf{x} \\ \text{s.t. } & \mathbf{A}\mathbf{x} = \mathbf{b} \\ & \mathbf{x} \in K \end{aligned} \tag{A.1}$$

or the dual standard form (see (9.189)):

$$\begin{aligned} \max \ & \mathbf{b}^T \mathbf{y} \\ \text{s.t. } & \mathbf{c} - \mathbf{A}^T \mathbf{y} \in K^* \end{aligned} \tag{A.2}$$

where K is a proper cone defined by the user, and K^* is the associated dual cone. The calling sequence for solving the primal-dual pair is

$$[\mathbf{x}, \mathbf{y}] = \text{sedumi}(\mathbf{A}, \mathbf{b}, \mathbf{c}, K).$$

If K is not specified, then by default $K = \mathbb{R}_+^n$; that is, the LP. A powerful feature of SeDuMi is that it allows us to define K so that it can be a concatenation of nonnegative orthants, SOCs, and PSD cones; see the SeDuMi user guide for the programming details. Next, we will show two examples to illustrate this feature.

Example A.1 Consider the problem

$$\min\ \mathbf{c}^T\mathbf{x}$$
$$\text{s.t. } \mathbf{Ax} = \mathbf{b},\ \mathbf{x} = [\mathbf{x}_1^T, \mathbf{x}_2^T]^T,\ \mathbf{x}_1 \in \mathbb{R}^{n_1},\ \mathbf{x}_2 \in \mathbb{R}^{n_2} \qquad (A.3)$$
$$\mathbf{x}_1 \succeq \mathbf{0},\ \mathbf{x}_2 \in K_2$$

where $K_2 = \{(z_1, \mathbf{z}_2) \in \mathbb{R} \times \mathbb{R}^{n_2-1} \mid z_1 \geq \|\mathbf{z}_2\|_2\}$ is a SOC. We can implement this problem using SeDuMi by setting $K.l = n_1$ and $K.q = n_2$, where $K.l$ means the dimension of the nonnegative orthant cone, and $K.q$ means the dimension of the SOC. Then $[\mathbf{x}, \mathbf{y}] = \text{sedumi}(\mathbf{A}, \mathbf{b}, \mathbf{c}, K)$ provides a primal-dual solution. □

Example A.2 Consider the problem

$$\min\ \mathbf{c}^T\mathbf{x}$$
$$\text{s.t. } \mathbf{Ax} = \mathbf{b},\ \mathbf{x} = [\mathbf{x}_1^T, \mathbf{x}_2^T]^T,\ \mathbf{x}_1 \in \mathbb{R}^{n_1},\ \mathbf{x}_2 \in \mathbb{R}^{n_2} \qquad (A.4)$$
$$\mathbf{x}_1 \in K_1,\ \mathbf{x}_2 \in K_2$$

where K_1 and K_2 both are SOCs. We can implement this problem using SeDuMi by setting $K.q = [n_1, n_2]$. This means that there are two SOCs, one having a dimension of n_1, another having a dimension of n_2. Then $[\mathbf{x}, \mathbf{y}] = \text{sedumi}(\mathbf{A}, \mathbf{b}, \mathbf{c}, K)$ provides a primal-dual solution. □

In the above two examples, the solver SeDuMi is used to solve real convex problems. It can also be directly applied to solve convex problems as the unknown variables are complex, by some extra complex variable declarations, e.g., $K.xcomplex = [n_1 + 1 : n_1 + n_2]$ if $\mathbf{x}_2$ is complex in Example A.1; $K.xcomplex = [1 : n_1 + n_2]$ if both $\mathbf{x}_1$ and $\mathbf{x}_2$ are complex in Example A.2.

A.2   CVX

In contrast to SeDuMi, CVX provides a much more straightforward input interface of the problem formulation, and supports an even wider range of convex conic problems. Like SeDuMi, CVX also works under a MATLAB platform. It can be downloaded from `http://www.stanford.edu/~boyd/cvx/` (for free). CVX employs a methodology called disciplined convex programming, which can rapidly and automatically convert the verified convex problems into a solvable form, suitable for its implementation. The core solvers embedded in CVX include SeDuMi and SDPT3 (the default one). Apart from standard conic problems (e.g., LP, SOCP, SDP) which can be exactly handled by the embedded solvers, CVX also supports some special convex functions, e.g., log-sum-exp, entropy, etc. (solved by powerful successive approximation methods).

CVX solver is more user friendly than SeDuMi because the latter requires the inputs (variables and constraints) expressed in a certain (real or complex) problem form, while the former directly describes the problem in terms of variables, the objective function, and constraints, as long as the problem is (real or complex) convex. However, the running time for CVX will be longer than for SeDuMi mainly due to an extra process of problem reformulation into a certain SeDuMi recognized convex problem form.

Example A.3 In Chapter 4, we have shown some solutions to an SDP problem (4.127), which is repeated here for convenience:

$$\min \mathrm{Tr}(\mathbf{WX})$$
$$\text{s.t. } \mathbf{X} \succeq \mathbf{A}_i, \ i = 1, \ldots, m,$$

where $\mathbf{W} \in \mathbb{R}^{n \times n}$ and $\mathbf{A}_i \in \mathbb{S}^n$, $i = 1, \ldots, m$, are given. This SDP problem can be readily implemented on the MATLAB platform with the following CVX codes:

```
cvx_begin
    variable X(n, n) symmetric;
    minimize( trace(W * X) );
    subject to
        for i = 1 : m
            X - Ai == semidefinite(n);
        end
cvx_end
```

Two solutions for $n = 2$ and $m = 3$ are shown in Figure 4.6. As illustrated in this example, variables, the objective function, and constraints can be directly and easily described in the CVX codes, without requiring any specific forms as in the SeDuMi recognized convex problem form. □

A.3   Finite impulse response (FIR) filter design

In this section, we will present a Type-I linear-phase, noncausal FIR filter design problem and implement the problem using SeDuMi, and then using CVX, in the following subsections.

Consider a Type-I linear-phase, noncausal FIR filter with an input-output relation

$$y_k = \sum_{i=-n}^{n} h_i x_{k-i}, \tag{A.5}$$

where x_k is the input sequence, y_k is the output sequence, and $\{h_i\}_{i=-n}^{n}$ are the filter coefficients. The filter frequency response is given by

$$H(\omega) = \sum_{i=-n}^{n} h_i e^{-j\omega i}, \ \omega \in [0, \pi]. \tag{A.6}$$

A linear-phase filter is symmetric w.r.t. its midpoint, i.e.,

$$h_i = h_{-i} \;\forall i = 1, \ldots, n, \tag{A.7}$$

implying that $H(\omega)$ is real and is given by

$$H(\omega) = h_0 + 2 \sum_{i=1}^{n} h_i \cos(\omega i). \tag{A.8}$$

A.3.1   Problem formulation

Let $\mathbf{h} = [h_0, h_1, \ldots, h_n]^T \in \mathbb{R}^{n+1}$ and consider the Chebychev (minimax) filter design problem: Given a desired filter response $H_{\text{des}}(\omega)$, we need to solve the following optimization problem

$$\min_{\mathbf{h}} \; \max_{\omega \in [0,\pi]} |H(\omega) - H_{\text{des}}(\omega)|, \tag{A.9}$$

where $H_{\text{des}}(\omega)$ is the given desired frequency response. This filter design problem can be approximated by discretization

$$\min_{\mathbf{h}} \; \max_{p=1,\ldots,P} |H(\omega_p) - H_{\text{des}}(\omega_p)|, \tag{A.10}$$

where $\{\omega_1, \ldots, \omega_P\} \in [0, \pi]$ is a set of frequency sample points (usually uniformly distributed), and $P \gg n$ is the total number of sample points. The approximated filter design problem (A.9) can be rewritten as (epigraph form)

$$\begin{aligned} \min_{\mathbf{h},t} \; & t \\ \text{s.t. } & \left|H(\omega_p) - H_{\text{des}}(\omega_p)\right| \leq t, \; p = 1, \ldots, P \end{aligned} \tag{A.11}$$

which can be further reformulated as an LP:

$$\begin{aligned} \min_{\mathbf{h},t} \; & t \\ \text{s.t. } & H(\omega_p) - H_{\text{des}}(\omega_p) \leq t, \; p = 1, \ldots, P \\ & - H(\omega_p) + H_{\text{des}}(\omega_p) \leq t, \; p = 1, \ldots, P. \end{aligned} \tag{A.12}$$

With the substitution of the frequency response $H(\omega_p)$ given by (A.8) in the inequality constraints of the LP problem (A.12), an equivalent form of (A.12) can be obtained as follows:

$$\begin{aligned} \max \; & -t \\ \text{s.t. } & H_{\text{des}}(\omega_p) - \left(h_0 + 2 \sum_{i=1}^{n} h_i \cos(\omega_p i)\right) + t \geq 0, \; p = 1, \ldots, P \\ & - H_{\text{des}}(\omega_p) + \left(h_0 + 2 \sum_{i=1}^{n} h_i \cos(\omega_p i)\right) + t \geq 0, \; p = 1, \ldots, P. \end{aligned} \tag{A.13}$$

Next, let us present how to solve this problem using SeDuMi and CVX, respectively.

A.3.2   Problem implementation using SeDuMi

To use SeDuMi to perform the above filter design, one can reformulate the problem into either a primal standard form or a dual standard form. In this example, we will consider employing the dual form implementation (i.e., (A.2)), since the problem is already quite similar to the dual form. Let

$$\mathbf{y} = [\mathbf{h}^T, t]^T \in \mathbb{R}^{n+2}. \tag{A.14}$$

Set

$$\mathbf{b} = [\mathbf{0}_{n+1}^T, -1]^T \in \mathbb{R}^{n+2}. \tag{A.15}$$

Define, for $p = 1, \ldots, P$,

$$\mathbf{a}_p = \begin{bmatrix} 1 \\ 2\cos(\omega_p) \\ \vdots \\ 2\cos(\omega_p n) \\ -1 \end{bmatrix}, \quad \mathbf{a}_{p+P} = \begin{bmatrix} -1 \\ -2\cos(\omega_p) \\ \vdots \\ -2\cos(\omega_p n) \\ -1 \end{bmatrix} \tag{A.16}$$

$$c_p = H_{\text{des}}(\omega_p), \quad c_{p+P} = -H_{\text{des}}(\omega_p). \tag{A.17}$$

The dual standard form corresponding to this problem is

$$\begin{aligned} \max \ & \mathbf{b}^T \mathbf{y} \\ \text{s.t. } & \mathbf{c} - \mathbf{A}^T \mathbf{y} \in K^* \end{aligned} \tag{A.18}$$

with $K^* = K = \mathbb{R}_+^{2P}$, where

$$\mathbf{A} = [\mathbf{a}_1, \mathbf{a}_2, ..., \mathbf{a}_P, \mathbf{a}_{P+1}, ..., \mathbf{a}_{2P}], \tag{A.19}$$

and

$$\mathbf{c} = [c_1, c_2, ..., c_P, c_{P+1}, ..., c_{2P}]^T. \tag{A.20}$$

By executing

$$[\mathbf{x}, \mathbf{y}] = \text{sedumi}(\mathbf{A}, \mathbf{b}, \mathbf{c}, K)$$

in MATLAB, one can optimally solve the FIR design problem.

A.3.3   Problem implementation using CVX

CVX is relatively very simple to use. By CVX for solving (A.11), we simply type the following MATLAB codes:

```
cvx_begin
    variables h(n + 1) t;
    minimize(t);
    subject to
        for p = 1 : P
            abs(H_des(p) − h(1) − 2 ∑_{i=2}^{n+1} h(i)cos(ω_p(i − 1))) ≤ t;
        end
cvx_end
```

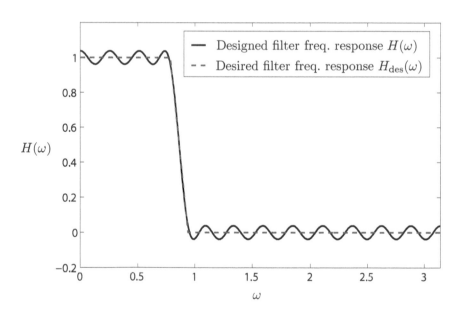

Figure A.1 The frequency response of the designed filter with $n = 10$ and $P = 100$.

Note that the modification in the constraint set of the above CVX implementation is due to the fact that MATLAB cannot support zero indexing. Figure A.1 shows the frequency response of the designed filter. A better matching (i.e., smaller minimax error) between the designed filter frequency response and the desired filter frequency response could be obtained by further increasing n and P. Let us conclude this section with the following remark:

Remark A.1 As the problem to be solved is a complex problem, one only needs to add an instruction declaring those complex variables, while all the rest of the program will be basically the same as the corresponding real problem. For instance, if variables $h(n+1)$ in the above MATLAB program were complex, we need to replace "variables $h(n+1)$ t;" with "variable $h(n+1)$ complex;" and "variable t;" and the rest of the program remains the same.   □

A.4   Conclusion

In this Appendix we have briefly introduced two convex optimization solvers, namely, SeDuMi and CVX. Though CVX may appear more user friendly than SeDuMi, the former consumes more running time than the latter using a personal computer, as CVX itself is based on either SeDuMi or SDPT3 solvers. However, one should not forget the fact that the customized interior-point methods are yet a faster way of implementing convex optimization problems. For detailed explanations regarding these softwares, interested readers can refer to the tutorial papers that come along with the software packages [GBY08, Stu99].

References for Appendix A

[GBY08]   M. Grant, S. Boyd, and Y. Ye, "CVX: MATLAB software for disciplined convex programming," 2008. [Online]. Available: http://cvxr.com/cvx.

[Stu99]   J. F. Sturm, "Using SeDuMi 1.02, a MATLAB toolbox for optimization over symmetric cones," *Optimization Methods and Software*, vol. 11, no. 4, pp. 625–653, 1999.

Index

Milton Keynes UK
Ingram Content Group UK Ltd.
UKHW050456071024
449327UK00015B/400

9 780367 573928